COLLECTED PAPERS

COLLECTED PAPERS
Volume II

Robert J. Aumann

QA
269
A882
2000
v.2

The MIT Press
Cambridge, Massachusetts
London, England

© 2000 Massachusetts Institute of Technology

All rights reserved. No part of this book may be reproduced in any form by any electronic or mechanical means (including photocopying, recording, or information storage and retrieval) without permission in writing from the publisher.

This book was set in Times New Roman by Asco Typesetters, Hong Kong.

Printed and bound in the United States of America.

Library of Congress Cataloging-in-Publication Data

Aumann, Robert J.
 [Selections. 2000]
 Collected papers / Robert J. Aumann.
   p. cm.
 Includes bibliographical references and index.
 ISBN 0-262-01154-9 (v. 1 : alk. paper).—ISBN 0-262-01155-7 (v. 2 : alk. paper)
 1. Game theory. I. Title.
QA269.A882   2000
519.3—dc20   96-6332
                                                                            CIP

Contents of Volume II

	Preface	xi
VII	**COALITIONAL GAMES: THE NTU FORM**	1
38	Von Neumann–Morgenstern Solutions to Cooperative Games without Side Payments *with B. Peleg*	5
39	The Core of a Cooperative Game without Side Payments	13
40	Introduction to "Some Thoughts on the Theory of Cooperative Games"	27
41	A Survey of Cooperative Games without Side Payments	31
VIII	**COALITIONAL GAMES: BARGAINING SET, KERNEL, NUCLEOLUS**	57
42	The Bargaining Set for Cooperative Games *with M. Maschler*	63
43	A Method of Computing the Kernel of n-Person Games *with B. Peleg and P. Rabinowitz*	91
44	Cooperative Games with Coalition Structures *with J. Dreze*	113
45	Game-Theoretic Analysis of a Bankruptcy Problem from the Talmud *with M. Maschler*	135
IX	**COALITIONAL GAMES: CORE AND EQUILIBRIA OF MARKETS**	155
46	Markets with a Continuum of Traders	159
47	Existence of Competitive Equilibria in Markets with a Continuum of Traders	171
48	Disadvantageous Monopolies	189
49	A Note on Gale's Example *with B. Peleg*	201
50	On the Rate of Convergence of the Core	205

X	**COALITIONAL GAMES: ECONOMIC AND POLITICAL APPLICATIONS OF THE SHAPLEY VALUE**	215
51	Values of Markets with a Continuum of Traders	221
52	Power and Taxes *with M. Kurz*	257
53	Power and Taxes in a Multi-Commodity Economy *with M. Kurz*	285
54	Core and Value for a Public Goods Economy: An Example *with R. J. Gardner and R. W. Rosenthal*	335
55	Power and Public Goods *with M. Kurz and A. Neyman*	339
56	Voting for Public Goods *with M. Kurz and A. Neyman*	359
57	Values of Markets with Satiation or Fixed Prices *with J. Dreze*	383
58	Economic Applications of the Shapley Value	431
59	Endogenous Formation of Links between Players and of Coalitions: An Application of the Shapley Value *with R. Myerson*	447
XI	**COALITIONAL GAMES: FOUNDATIONS OF THE NTU SHAPLEY VALUE**	465
60	An Axiomatization of the Non-Transferable Utility Value	469
61a	Values for Games without Side Payments: Some Difficulties with Current Concepts, by Alvin Roth	485
61b	On the Existence and Interpretation of Value Allocation, by Wayne Shafer	495
61c	On the Non-Transferable Utility Value: A Comment on the Roth–Shafer Examples	507
61d	On the Non-Transferable Utility Value: A Reply to Aumann, by Alvin Roth	523
61e	Rejoinder	529
62a	Non-Symmetric Cardinal Value Allocations, by Allen Scafuri and Nicholas Yannelis	537

62b	Value, Symmetry, and Equal Treatment: A Comment on Scafuri and Yannelis	541
XII	**COALITIONAL GAMES: SURVEYS OF VALUE THEORY**	547
63	Recent Developments in the Theory of the Shapley Value	551
64	The Shapley Value	563
XIII	**MATHEMATICAL METHODS**	571
65	Spaces of Measurable Transformations	577
66	Borel Structures for Function Spaces	581
67	On Choosing a Function at Random	599
68	Integrals of Set-Valued Functions	607
69	An Elementary Proof that Integration Preserves Uppersemicontinuity	619
70	A Variational Problem Arising in Economics *with M. Perles*	623
71	Random Measure Preserving Transformations	639
72	Orderable Set Functions and Continuity III: Orderability and Absolute Continuity *with U. Rothblum*	645
73	Bi-Convexity and Bi-Martingales *with S. Hart*	653
	Author Index	675
	Journal Index	677
	Citation Index	679
	Name Index	759
	Subject Index	767

Contents of Volume I

Preface
I General
1 What Is Game Theory Trying to Accomplish?
2 Game Theory
3 The Game of Politics: A Review of Rapoport's *Fights, Games, and Debates*
4 CORE as a Macrocosm of Game-Theoretic Research, 1967–1987
5 Arrow—the Breadth, Depth, and Conscience of the Scholar: An Interview
6 Economic Theory and Mathematical Method: An Interview
7 Report of the Committee on Election Procedures for Fellows *with M. Bruno, F. Hahn, and A. Sen*
8 Foreword to *A General Theory of Equilibrium Selection in Games*
9 Foreword to *Two-Sided Matching: A Study in Game-Theoretic Modeling and Analysis*
II Knot Theory
10 Asphericity of Alternating Knots
III Decision Theory: Utility and Subjective Probability
11 The Coefficients in an Allocation Problem *with J. B. Kruskal*
12 Assigning Quantitative Values to Qualitative Factors in the Naval Electronics Problem *with J. B. Kruskal*
13 Subjective Programming
14a Utility Theory without the Completeness Axiom
14b Utility Theory without the Completeness Axiom: A Correction
15 Measurable Utility and the Measurable Choice Theorem
16 Linearity of Unrestrictedly Transferable Utilities
17 A Definition of Subjective Probability *with F. J. Anscombe*
18 Letter from Robert Aumann to Leonard Savage and Letter from Leonard Savage to Robert Aumann
19 The St. Petersburg Paradox: A Discussion of Some Recent Comments
IV Strategic Games: Repeated
20 Acceptable Points in General Cooperative n-Person Games
21 Acceptable Points in Games of Perfect Information
22 Long-Term Competition—A Game-Theoretic Analysis *with L. S. Shapley*
23 Survey of Repeated Games
24 Cooperation and Bounded Recall *with S. Sorin*
25 Rationality and Bounded Rationality
V Strategic Games: Extensive
26 A Characterization of Game Structures of Perfect Information
27 Almost Strictly Competitive Games
28 Mixed and Behavior Strategies in Infinite Extensive Games

29 Some Thoughts on the Minimax Principle *with M. Maschler*

30 Approximate Purification of Mixed Strategies *with Y. Katznelson, R. Radner, R. Rosenthal, and B. Weiss*

VI Strategic Equilibrium and the Theory of Knowledge

31 Subjectivity and Correlation in Randomized Strategies

32 Agreeing to Disagree

33 Correlated Equilibrium as an Expression of Bayesian Rationality

34 Nash Equilibria Are Not Self-Enforcing

35 Irrationality in Game Theory

36 Backward Induction and Common Knowledge of Rationality

37 Epistemic Conditions for Nash Equilibrium *with A. Brandenburger*

A duplicate Author Index, Journal Index, Citation Index, Name Index, and Subject Index appear in each volume.

Preface

These two volumes contain all my papers through January 1995.[1] They are grouped into several categories: "general," knot theory, decision theory,[2] strategic games, coalitional games, and mathematical methods.[3] Though designed to fit the natural contours of my work, the grouping remains to some extent arbitrary: there are many cross-relationships between the papers in the various categories, and the arrangement could easily have been different. Within the groups, the ordering is roughly chronological; but there are frequent departures from this rule to enable related articles to appear together, to underscore the conceptual development, or for other pertinent reasons. Each group is preceded by an introduction that briefly describes the content and background of each paper in it—provides motivation, discusses the research process, and relates it to other papers in the collection and to work by others.

Except for chapter 10, which is about knots, all the papers in the collection concern game theory, its applications and its tools. Beyond the subject matter, they also share a common methodological theme: they deal with *relationships*. Science is often characterized as a quest for truth, where truth is something absolute, which exists outside of the observer. But I view science more as a quest for *understanding*, where the understanding is that of the observer, the scientist. Such understanding is best gained by studying relations—relations between different ideas, relations between different phenomena, relations between ideas and phenomena.[4] Rather than asking "How does this phenomenon work?" we ask "How does this phenomenon resemble others with which we are familiar?" Rather than asking "Does this idea make sense?" we ask, "How does this idea resemble other ideas?"

My early work on decision theory (group III) stresses the relation between various notions of utility—the common structure that they share (see,[5] e.g., chapter 13). The definition of subjective probability in chapter 17 is based on the relation between subjective and objective probabilities, and on von Neumann–Morgenstern utility theory; thus it differs from Savage's approach, which starts from "scratch." My first papers in game

---

1. No substantive changes have been made in the papers. Minor errors and misprints have been corrected, and bibliographic references brought up to date; i.e., where a paper originally cited as a preprint was subsequently published, we here cite the published version. In references to my own papers, chapter numbers in this collection have been added (in square brackets).

2. Utility and subjective probability.

3. The categories of strategic and coalitional games are subdivided further.

4. Compare chapter 1, sections 2 and 3.

5. The reader is invited to refer also to the discussions, in the introductions to the various groups, of the papers cited in this preface.

theory proper (chapters 20, 21) concerned the relation between strategic equilibrium and the core, or more generally, between strategic ("noncooperative") and coalitional ("cooperative") game theory. There followed papers on the "equivalence principle" (chapters 46, 51), which are about the relation between (indeed the coincidence of) game-theoretic concepts like core and value, and the economic concept of competitive equilibrium, in large markets. The work on epistemic models (group VI) explores the relation between decision-theoretic notions like knowledge and rationality,[6] and game-theoretic notions like correlated or Nash or subgame perfect equilibrium. Underlying much of my work on the Bargaining Set and its relatives (group VIII) is the notion of consistency, a notion that turns up again and again in the study of widely diverse solution notions (see chapter 44). Chapter 45 concerns the relationship between game-theoretic ideas—nucleolus, kernel, consistency—and ancient Talmudic law. Chapter 57 explores the relation of the value to previously defined notions of equilibrium in fixed price markets. The idea of relationship pervades my work.

Indeed, the idea of relationship is fundamental to game theory. Disciplines like economics or political science use disparate models to analyze monopoly, oligopoly, perfect competition, public goods, elections, coalition formation, and so on. In contrast, game theory uses the *same* tools in all these applications. The nucleolus yields the competitive solution in large markets (chapter 46), the homogeneous weights in parliaments,[7] and the Talmudic solution in bankruptcy games (chapter 45). Perhaps the most exciting advance in game theory in recent years has been the connection with evolution: The realization that when properly interpreted, the fundamental notion of Nash equilibrium, which a priori reflects the behavior of consciously maximizing agents, is the *same* as an equilibrium of populations that reproduce blindly without regard to maximizing anything.[8]

The great American naturalist and explorer John Muir said, "When you look closely at anything in the universe, you find it hitched to everything else." Though Muir was talking about the natural universe, this applies also to scientific ideas—how we *understand* our universe. When you look closely at one scientific idea, you find it hitched to all others. It is these hitches that I have tried to study.

---

6. In the sense of utility maximization.
7. B. Peleg, "On weights of constant-sum majority games," *SIAM J. App. Math.* 16 (1968), 527–532.
8. See chapter 2, 1970–1986, (i).

My doctoral thesis (chapter 10) is about knots; about strands of rope—"components of the knot"—that won't come apart, are inextricably tied together. It symbolizes my subsequent work, which has studied how ideas—strands of thought—are inextricably tied together.

How does one write acknowledgments for a collection like this, which encompasses the work of a lifetime? To start with, these papers directly represent not only my work, but also that of twenty-three others—my coauthors, with whom it has been a great privilege and a great pleasure to work for close to forty years. Next, I would like to acknowledge the encouragement and help of three wonderful individuals at the MIT Press—Terry Vaughn, whose idea it was to publish this collection; Ann Sochi, who with her gentle persistence made it a reality; and Deborah Cantor-Adams, who saw it through to publication. And without my assistant Mike Borns, who did all the technical work on the Jerusalem end, this collection would never have appeared.

But that is only the beginning. I must go on to thank all the many individuals who helped me and influenced me throughout these many years —parents, brother, wife, children, grandchildren, other family, teachers, students, coworkers, colleagues, others whose work I have read or heard, referees, editors, friends—the list is enormous. For what it may be worth, this collection represents their work as well as mine; to paraphrase Muir, my work is hitched to that of all these others.

Jerusalem, Israel

VII COALITIONAL GAMES: THE NTU FORM

Chapters 20 and 21 (Volume I) established a relationship between equilibria of a repeated game and the core of the corresponding one-shot game. The proper representation of these one-shot games in coalitional form is as NTU (nontransferable utility) games, as distinguished from the TU (transferable utility, or "side payment") games that had theretofore been the main concern of cooperative game theory. In 1959, B. Peleg and I undertook a study of NTU games "for their own sake"—that is, not just for their relation with repeated games. The results of this study were reported in chapters 38 and 39, and in a paper by B. Peleg entitled "Solutions to Cooperative Games without Side Payments."[1] Chapter 38, which is joint with Peleg, provides a general definition of NTU games and of the core and the von Neumann–Morgenstern solutions (or "stable sets") for such games, and announces some basic results of the study. Chapter 39 is devoted to the core of NTU games. Inter alia, it is in this paper that the connection between the equilibria of a repeated game and the core of the corresponding one-shot game is made explicit; in chapter 20 it is only implicit.

Chapter 40 is an introduction to a paper by Gerd Jentzsch that appeared in *Advances in Game Theory*[2] in 1964. The author died in 1959 while still a young man. The dittoed manuscript came to my attention while I was on sabbatical at Princeton in 1960–61, and I immediately recognized that it was a contribution of outstanding originality and importance. So when papers for the "Advances" were being gathered, I suggested that Jentzsch's paper be translated into English and included, and offered to see it through the process. The editors also asked me to provide an introduction to put the paper in context, and this is chapter 40.

NTU games "caught on," and by October 1961, there was sufficient material to justify a survey lecture on the subject at the conference on "Recent Advances in Game Theory" held in Princeton that month. The text of the lecture appeared in the proceedings of the conference, which was privately distributed to the attendees. By six years later, much more material had accumulated, and a considerably augmented version of the text appeared in the Morgenstern Festschrift in 1967. This augmented version is chapter 41.

1. *Trans. Amer. Math. Soc.* 106 (1963), 280–292.
2. Edited by M. Dresher, L. S. Shapley, and A. W. Tucker (Princeton University Press).

38 Von Neumann–Morgenstern Solutions to Cooperative Games without Side Payments
with B. Peleg

The use of side payments in the classical[1] theory of n-person games involves three restrictive assumptions. First, there must be a common medium of exchange (such as money) in which the side payments may be effected; next, the side payments must be physically and legally feasible; and finally, it is assumed that utility is "unrestrictedly transferable," i.e. that each player's utility for money[2] is a linear function of the amount of money.[3] These assumptions severely limit the applicability of the classical theory; in particular, the last assumption has been characterized by Luce and Raiffa [2, p. 233] as being "exceedingly restrictive—for many purposes it renders n-person theory next to useless." It is the purpose of this paper to present the outline of a theory that parallels the classical theory, but *makes no use of side payments*.[4] Our definitions are related to those given in [2, p. 234] and in [3], but whereas the previous work went no further than proposing definitions, the theory outlined here contains results which generalize a considerable portion of the classical theory. It thus demonstrates that the restrictive side payment assumption is not necessary for the development of a theory based on the ideas of von Neumann and Morgenstern. Only a general description of the theory and statements of the more important theorems will be included here; details and proofs will be published elsewhere.

---

[1] We will use the word "classical" to denote the von Neumann-Morgenstern theory as described in [1] and in [4].

[2] Or any other medium of exchange.

[3] See [2, p. 168]. It can be proved that when $n \geq 3$, linearity of the utilities in money is necessary and sufficient for the existence of an unrestrictedly transferable utility.

[4] In particular, our theory is of course also applicable to the case in which side payments are permitted. When, in addition, utility is unrestrictedly transferable, then our theory reduces to the classical theory.

This chapter originally appeared in *Bulletin of the American Mathematical Society* 66 (1960): 173–179. Reprinted with permission.

1. Effectiveness. Let us fix attention on a given finite n-person game, and let N denote the set of players. Let E^N denote an n-dimensional euclidean space, and let us index the coordinates of points in E^N by the members of N. The points of E^N will be called *payoff vectors*; if $x \in E^N$ and $i \in N$, then x_i will denote the coordinate of x corresponding to player i, and will be called the *payoff* to i.

Intuitively, a coalition B is *effective* for a payoff vector x if the members of B, by joining forces, can play so that each player i in B receives at least x_i. This intuitive definition is open to a number of interpretations. The rather conservative one adopted by von Neumann and Morgenstern assumes that the most the members of B can count on is what they can get if the players of $N-B$ form a coalition whose purpose it is to minimize the payoff to B. There are at least two generalizations of this notion of effectiveness to the case in which there are no side payments:

(i) *A coalition B is said to be α-effective for the payoff vector x if there is a strategy[5] for B, such that for each strategy used by $N-B$, each member i of B receives at least x_i.*

(ii) *A coalition B is said to be β-effective for the payoff vector x, if for each strategy used by $N-B$, there is a strategy for B such that each member i of B receives at least x_i.*

Roughly, α-effectiveness means that B can assure itself of its portion of x independently of the actions of $N-B$, whereas β-effectiveness means that $N-B$ cannot prevent B from obtaining its (B's) portion of x. In the classical theory the two notions are equivalent, but this is not the case when side payments are forbidden. Which of the two definitions is preferable is a matter of taste; both have appeared, in more or less disguised form, in the previous literature [2, p. 175; 3; 5]. There seems to be a tendency to consider α-effectiveness as intuitively more appealing; on the other hand, there is evidence that β-effectiveness may eventually turn out to be the more significant concept.[6] The present theory applies equally well to both notions.

2. Axiomatic treatment. It is possible to define many of the basic notions of n-person theory—domination, solution, core, etc.—in terms of effectiveness. Of course the objects we will get will usually depend on what kind of effectiveness we started with. Thus we will define the α-core and the β-core, but for a given game they usually

---

[5] The word "strategy" as used in this paper means what has been variously called "correlated mixed strategy" [2, p. 116], "joint randomized strategy" [2, p. 116], "cooperative strategy" [2, p. 175], and "correlated strategy B-vector" [5].

[6] See §6.

differ; similarly with α-solutions and β-solutions, etc. Nevertheless, it is possible to prove a considerable number of general theorems which hold for either kind of effectiveness; the proofs of these theorems make use only of certain basic properties common to both kinds. The situation invites axiomatic treatment.

An n-person "characteristic function" is a set N with n members, together with a function v that carries each subset B of N into a subset $v(B)$ of E^N so that
  (1) *$v(B)$ is convex;*
  (2) *$v(B)$ is closed;*
  (3) *$v(\emptyset) = E^N$;*[7]
  (4) *if $x \in v(B)$, $y \in E^N$, and for all $i \in B$, $y_i \leq x_i$, then $y \in v(B)$; and*
  (5) *if B_1 and B_2 are disjoint, then $v(B_1 \cup B_2) \supset v(B_1) \cap v(B_2)$.*

An n-person "game" is an n-person characteristic function (N, v) together with a convex compact polyhedral subset H of $v(N)$.

An n-person game as just defined actually represents more than a game in the usual sense; it is a game together with a concept of effectiveness. The set H is the set of all "feasible" or "attainable" payoff vectors, i.e. the set of all payoff vectors which can be attained by a joint strategy of N. $v(B)$ represents the set of all payoff vectors for which B is effective. Conditions (1), (2), and (3) are self-explanatory. Condition (4) says that if a coalition B is effective for a payoff vector x, then it is also effective for any payoff vector with smaller (or equal) payoffs to its members. Condition (5) is the natural generalization of super-additivity of the characteristic function in the classical theory.[8]

In order to justify these definitions, it is necessary to show that an arbitrary finite game, when combined with the concept either of α-effectiveness or of β-effectiveness, satisfies our definition of a game.[9] For the most part this is straightforward; the only deep part occurs in verifying condition (5) in the case of β-effectiveness, where use is made of Kakutani's fixed point theorem.

One of the chief advantages of the axiomatic approach is its flexibility: it can be used not only with the notions of effectiveness described in §1, which are based on the conservative approach that characterizes the classical theory, but also with many other notions of effectiveness. For example, we may prefer an effectiveness notion based on the ideas of ψ-stability [2, pp. 163–168, 174–176, 220–236].

---

[7] $\emptyset$ denotes the empty set.

[8] Condition (5) is not needed for any of the results stated in this paper. It was included in order to underscore the parallelism with the classical theory, and with the hope that stronger axioms will eventually yield a richer theory.

[9] Where $v(B)$ and H have the meanings described in the previous paragraph.

Such notions can be constructed in a number of ways; the general idea would be that in order for a coalition B to be effective for a pair consisting of a payoff vector x and a coalition structure τ, the coalition B must be attainable from the given coalition structure τ, and no attainable combination of coalitions in $N-B$ should be able to prevent B from obtaining its portion of the payoff vector x. Specifically, we would say that B is effective for (x, τ) if $B \in \psi(\tau)$ and there is a strategy for B such that for each partition $(B_1, \cdots, B_k)$ of $N-B$ into members of $\psi(\tau)$, and each k-tuple of strategies used by the B_j, each member i of B receives at least x_i. A related but different notion can be obtained if we reverse the quantifiers. For fixed τ these two notions of effectiveness satisfy all the axioms except (5), and therefore the results stated in this paper hold for them as well (cf. footnote 8).

3. **Domination and solution.** Fix an n-person game $G=(N, v, H)$. A payoff vector x is said to *dominate* a payoff vector y *via* B if $x \in v(B)$ and $x_i > y_i$ for all $i \in B$; x is said to *dominate* y if there is a B such that x dominates y via B. If K is an arbitrary set of payoff vectors, we define dom K to be the set of all payoff vectors dominated by at least one member of K. If P is an arbitrary set of payoff vectors, then a subset K of P is said to be *P-stable* if $K \cap \text{dom } K$ is empty and $K \cup \text{dom } K \supset P$. The set $P - \text{dom } P$ is called the *P-core*. A payoff vector x is said to *majorize* a payoff vector y if x dominates y and all z that dominate x also dominate y. All the lemmas and theorems of §1 of [4] concerning domination, P-stability, the P-core and majorization remain true in this context; the proofs go through essentially unchanged.

It is easy to show that for each $i \in N$, there is an extended real number[10] v_i such that $v(\{i\}) = \{x \in E^N : x_i \leq v_i\}$. A payoff vector x is called *individually rational* if $x_i \geq v_i$ for each $i \in N$. x is called *group rational* if there is no $y \in H$ such that $y_i > x_i$ for each $i \in N$. Let us denote by $\bar{A}$ the set of individually rational members of H, and by A the set of members of $\bar{A}$ that are also group rational. Then it can be proved that a subset of H is A-stable if and only if it is $\bar{A}$-stable. This justifies us in defining a *solution* of G to be an A-stable set.[11]

The next step is to investigate which games are solvable and what their solutions are. First of all, it is easy to show that all 2-person games have a unique solution,[12] namely all of A. We next investigate

---

[10] A real number or $+\infty$ or $-\infty$.

[11] If K is a solution of G, we will also say that K *solves* G, and that G is *solvable*.

[12] This solution is closely related to the *negotiation set* of a 2-person cooperative game [2, p. 118], but is *not* the same thing.

3-person zero-sum[13] games. Unlike the situation in the classical theory, we now find a large number of essentially different games, whose solutions exhibit the greatest variety and complexity. The basic theorem is

THEOREM 1. *Every 3-person zero-sum game is solvable.*

The proof, which is rather involved, proceeds by dividing A into regions, solving each region separately, and then combining the regional solutions into a solution for all of A. The shapes of the regions depend on the $v(B)$ and on their interrelationships. The question of the solvability of 3-person general-sum or 4-person zero-sum games remains open.

Incidentally, Theorem 1 is the only one of our results for which the assumption that the $v(B)$ be convex (condition (1)) is required.[14]

4. Composition. Let $G_1 = (N_1, v_1, H_1)$ and $G_2 = (N_2, v_2, H_2)$ be games whose player sets N_1 and N_2 are disjoint. Intuitively, the *composition* G of G_1 and G_2 is the game each play of which consists of a play of G_1 and a play of G_2, played without any interconnection. Formally, we define[15] $G = (N, v, H)$, where $N = N_1 \cup N_2$, $H = H_1 \times H_2$, and for each $B \subset N$, $v(B) = v_1(B \cap N_1) \times v_2(B \cap N_2)$.

THEOREM 2. *A necessary and sufficient condition that a subset K of H solve G is that it be of the form $K_1 \times K_2$, where K_1 solves G_1 and K_2 solves G_2.*

The simplicity of this result is somewhat surprising, in view of the fact that the corresponding result in the classical theory is much more complicated. The complexity of the classical result is explained by the fact that it permits side payments between members of N_1 and members of N_2, whereas no such intercourse can be possible in our framework; thus although the classical theory is a special case of our theory, the composition of two games in the classical sense yields a game which is in general not the composition of the games in our sense. All of our solutions appear in the classical theory, but the converse is not true. Our solutions are precisely those in which no "tribute" is paid by either group of players (cf. [1, §46.11.2, p. 401]).

5. The core.

THEOREM 3. *The A-core and the $\overline{A}$-core coincide.*

---

[13] A game is said to be *zero-sum* if H is contained in the plane $\sum_{i \in N} x_i = 0$.

[14] The question as to whether the theorem holds without this assumption remains open; certainly the proof does not go through.

[15] $\times$ denotes cartesian product.

This theorem justifies us in defining the *core* of a game to be its A-core. The proof makes essential use of the fact that H is polyhedral (this is the only theorem for which this assumption is needed); indeed, if this assumption is dropped, the theorem becomes false.

Shapley [6] has conjectured that in the classical theory, the intersection of all solutions is the core. This is not true in our theory; indeed, there is a 3-person zero-sum game with a unique solution, which strictly includes the core.[16]

6. The β-core and the supergame. The *supergame* of a game[17] Γ is the game each play of which consists of an infinite sequence of plays of Γ. A *strong equilibrium point* in an n-person game [5] is, roughly speaking, an n-tuple $\{\xi_i\}_{i \in N}$ of strategies with the property that if the members j of any coalition B use strategies different from the ξ_j, while the players not in B keep using the ξ_i, then at least one player in B will not profit from the change, i.e. will get no more than he would have gotten had all the players used the ξ_i. Strong equilibrium points are strengthened forms of the Nash equilibrium points [7]; at a Nash equilibrium point there is no direct incentive for any *individual* to change his strategy, whereas at a strong equilibrium point there is no direct incentive for any *coalition* to change its strategy.

In [5] the concept of a c-acceptable payoff vector is defined, and it is shown that a payoff vector is c-acceptable in a given finite game if and only if it is the vector of payoffs to a strong equilibrium point in the corresponding supergame. It can be shown that the set of c-acceptable payoff vectors coincides with the β-core.[18] Hence *the β-core of a finite game is precisely the set of payoff vectors to strong equilibrium points in the corresponding supergame.*

7. The "extended" theory. The definition of an *extended game* is similar to that of a game, with the single exception that H is not assumed to be a subset of $v(N)$ but merely of E^N. In the classical theory extended games are important as a theoretical tool in composition theory; they were first considered by von Neumann and Morgenstern [1, Chapter X].

Theorems 1 and 2 remain true as they stand for extended games. Theorem 3 must be adjusted to read "The A-core is the intersection

---

[16] This solution is disconnected, so that it also provides a counter-example to another conjecture of Shapley, namely that the union of all solutions is connected. Of course this example does not yet settle these questions for the classical theory.

[17] We are now referring to "game" in the ordinary sense of the word, not that defined in §2.

[18] I.e. the core if we use β-effectiveness as our definition of effectiveness.

of the $\overline{A}$-core with A." §1 of [4] generalizes as before; however, we have not succeeded in obtaining a relation between the A-stable sets and the $\overline{A}$-stable sets in extended games.

References

1. J. von Neumann and O. Morgenstern, *Theory of games and economic behavior*, Princeton University Press, 1943, 2nd ed., 1947.

2. R. D. Luce and H. Raiffa, *Games and decisions*, John Wiley, 1957.

3. L. S. Shapley and M. Shubik, *Solutions of n-person games with ordinal utilities* (abstract), Econometrica vol. 21 (1953) p. 348.

4. D. B. Gillies, *Solutions to general non-zero-sum games*, Contributions to the Theory of Games IV, Princeton University Press, 1959, pp. 47–85.

5. R. J. Aumann, *Acceptable points in general cooperative n-person games*, ibid., pp. 287–324 [Chapter 20].

6. L. S. Shapley, *Open questions* (dittoed), Report of an Informal Conference on the Theory of n-Person Games held at Princeton University, March 20–21, 1953, p. 15.

7. J. F. Nash, *Non-cooperative games*, Ann. of Math. vol. 54 (1951) pp. 286–295.

The Hebrew University,
    Jerusalem, Israel

39 The Core of a Cooperative Game without Side Payments

The *core* of an *n*-person game[1], though used already by von Neumann and Morgenstern [15], was first explicitly defined by Gillies [5]. Gillies's definition is restricted to cooperative games with side payments and unrestrictedly transferable utilities[2], but the basic idea is very simple and natural, and appears in many approaches to game theory. We consider a certain set of "outcomes" to a game, and define a relation of "dominance" (usually not transitive) on this set. The core is then defined to be the subset of outcomes maximal with respect to the dominance relation; in other words, the subset of outcomes from which there is no tendency to move away—the equilibrium states.

To turn this intuitive description of the core notion into a mathematical definition, we need precise characterizations of

(a) the kind of game-theoretic situation to which we are referring (cooperative game, noncooperative game, etc.);

(b) what we mean by "outcome"; and

(c) what we mean by "dominance."

Different ways of interpreting these three elements yield different applications of the generalized "core" notion, many of them well-known in game theory. Gillies's core, Luce's ψ-stability [10], Nash's equilibrium points [12], Nash's solution to the bargaining problem [13][3], and the idea of Pareto optimality—to mention only some of the applications—can all be obtained in this way.

Here we shall be concerned exclusively with cooperative games without side payments[4]. Our procedure will be to generalize von Neumann's fundamental notion of characteristic function to this case, and on the basis of this generalization to define the core in a way that generalizes and parallels the core in the classical theory—i.e., Gillies's core. The generalization of the characteristic function is of interest for its own sake also; for example, a theory of "solutions" has been developed that generalizes and parallels the classical theory of solutions and is based on the characteristic function [3; 16].

---

[1] Most of the results proved here were announced in [3], to which the reader may refer for additional introductory and background material. The basic ideas of this paper were conceived jointly with B. Peleg, to whom the author is greatly indebted.

[2] Such games will be called *classical* games in the sequel, and the theory described in [5; 15] will be called the *classical* theory.

[3] Cf. [6].

[4] Classical games are known to be special cases of these games.

As in the classical theory, our "outcomes" will all be payoff vectors. We leave aside for the moment the question as to which particular set of payoff vectors we wish formally to consider as our set of outcomes. This brings us to the question of how to characterize the notion of "dominance."

Although formally it is simpler to define the characteristic function first and then to base on it the definition of dominance, the more intuitive procedure is the reverse: We must first state what we require from the dominance relation, and this will enable us to motivate our definition of characteristic function. Following the classical theory, then, we will say that a payoff vector x dominates another one y if

(i) there is a coalition S that prefers x to y, and
(ii) this preference is "not idle," i.e. S can actually achieve at least its portion of x.

What is meant by condition (i) is clear; each member of S must get more in x than in y. As for condition (ii), its exact meaning depends on how we wish to interpret the words "can actually achieve"; or to say the same thing in more technical language, it depends on when we wish to consider the coalition S "effective" for the payoff vector x.

In the sequel we will give a number of different definitions of effectiveness, each one leading to a different notion of dominance and hence to a different core. An alternative procedure is to assume that we already know for each coalition S which are the payoff vectors x for which S is effective; on the basis of this information we can then determine the core, without having to know the normal form of the game or the definition of effectiveness. A game presented in this form is said to be in *characteristic function* form. The characteristic function form of a game can always be calculated from its normal form and a particular definition of effectiveness. Note the similarity with the characteristic function of the classical theory; there there is associated with each coalition S a number $v(S)$, and the vectors x for which S is effective are precisely those for which[5] $\sum_{i \in S} x^i \leq v(S)$. Here the set of x for which S is effective need not have such a simple form, and cannot be characterized by a single number; we therefore define $v(S)$ to be the set itself, rather than a number that characterizes the set. A considerable part of the theory can be developed on the basis of the characteristic function, without referring to the original game or to the particular notion of effectiveness we are using. As in the classical theory, some assumptions must be made about $v(S)$ to justify this development; these assumptions are natural ones, and we will establish that they hold for the particular definitions of effectiveness that we will wish to use.

We now return to the question of which payoff vectors we wish to consider as "outcomes." One possibility is the set H of all those payoff vectors that can be obtained by means of some correlated mixed strategy of the set

---

[5] The coordinates of the payoff vectors are indexed with superscripts.

N of all players. It is also possible to impose various more or less natural restrictions on the set of outcomes. There are two such restrictions that have received special attention in the literature, namely "individual rationality" and "group rationality." The former restricts the outcomes to payoff vectors in which each individual player gets at least what he can guarantee himself without any aid from the other players; under the latter restriction, a payoff vector is not called an "outcome" if there is another payoff vector in H which yields more to each player. These two restrictions can be imposed on the "outcome" concept in various combinations, so that we obtain four possibilities for this concept. In the classical theory it is easily established that all four lead to the same core; in the present theory this is also true, but the proof is no longer trivial. An interesting sidelight on this theorem is that its proof depends essentially on the assumption that H is a polyhedron (this assumption always holds if we start out with a finite game). If we replace H by a non-polyhedral convex set, the theorem becomes false; such a situation can actually be realized in the case of games with infinite strategy sets.

The paper is divided into two parts: the first part (§§1–7) deals with the theory of games in characteristic function form; the second part (§§8–10) deals with applications to games in normal form. §1 is devoted to a review of notation. In §2 we give the formal definition of a game in characteristic function form. §3 is devoted to the definition of various basic concepts such as domination, individual and group rationality, and core. §§4 and 5 are devoted to the statement and proof of the theorem that all the sets of outcomes discussed above lead to the same core. In §6 we give the counter-example to this theorem when H is not polyhedral. In §7 we discuss the composition of two games, and remark that the core of the composition is the cartesian product of the cores of the components. In §8 we pass to the normal form. We define two kinds of effectiveness, both generalizations of the classical definition, and show that they are different. In §9 we show that both these definitions lead to characteristic functions that satisfy the conditions of §2. In §10 we discuss the connection between the supergame([6]) of a game and its various cores; in particular we shall show that the set of acceptable payoff vectors of a game [1; 2] coincides with the core for one of the two definitions alluded to above.

1. **Notation.** N will denote a fixed finite set with n members, who will be called *players*. E^N will denote euclidean space of n dimensions, the coordinates of the points being indexed by the members of N; formally, E^N may be considered the set of functions from N to the reals. The points of E^N will be called payoff vectors. If $x \in E^N$, the coordinates of x will be denoted by x^i, where $i \in N$. For fixed $x \in E^N$ and $S \subset N$, we will call the S-tuple $\{x^i\}_{i \in S}$ an *S-vector* and denote it by x^S.([7]) Note that $x = x^N$. If x^S and y^S are S-vectors, then any

---

([6]) The game each play of which consists of an infinite sequence of the plays of the original game.

([7]) x^S is the projection of x on E^S; if x is considered a function, then x^S is x restricted to S.

relation between x^S and y^S is to be understood coordinate-wise; e.g., $x^S \geq y^S$ means $x^i \geq y^i$ for all $i \in S$. If $S \subset N$ then (x^S, y^{N-S}) denotes the payoff vector z such that $z^S = x^S$ and $z^{N-S} = y^{N-S}$.

Subsets of N will be called coalitions, and will be denoted by S and T. Lower case latin letters towards the end of the alphabet will denote payoff vectors. $\emptyset$ denotes the empty set. In addition to its usual meaning, 0 will sometimes denote a vector all of whose components are 0; no confusion will result. The letter i always denotes a player. Unless the contrary is specifically indicated, summation, the taking of maxima or minima, etc., will be over i; for instance, $\sum_S$ means $\sum_{i \in S}$. The symbol $\times$ denotes the cartesian product.

We shall need a norm on E^N. Any norm with reasonable properties would serve our purposes; we shall use the maximum, defined by $\|x\| = \max_N |x^i|$. In addition to the usual norm properties, we note

(1)   if $x > 0$ and $y > 0$, then $\|x + y\| > \max(\|x\|, \|y\|)$.

Similar to the definition of norm on E^N, we define a norm on E^S by $\|x^S\| = \max_S |x^i|$.

The numbering of formulas, theorems, etc., starts from the beginning in each section; references from one section to another specify the section number as well as the formula number.

2. The definition of a game in characteristic function form.

DEFINITION. A *characteristic function*([8]) is a pair (N, v), where N is a finite set and v is a function that carries each subset S of N into a subset $v(S)$ of E^N so that

(1) *$v(S)$ is convex*;
(2) *$v(S)$ is closed*;
(3) *$v(\emptyset) = E^N$*;
(4) *if $x \in v(S)$ and $y^S \leq x^S$, then $y \in v(S)$*;
(5) *if $S \cap T = \emptyset$, then $v(S \cup T) \supset v(S) \cap v(T)$*.

A *game in characteristic function form*, or simply a *game*, is a triple (N, v, H), where (N, v) is a characteristic function and

(6) *H is a convex compact polyhedral subset of E^N*.

Condition (5) is the natural generalization of the classical notion of superadditivity: it says that if a certain outcome can be achieved by the disjoint coalitions S and T when acting separately, then it can also be achieved by them when acting in concert.

We shall say that (N, v, H) is an *ordinary* game if

(7) *$x \in v(N)$ if and only if there is a $y \in H$ such that $x \leq y$*.

This condition is easily justified intuitively, if we consider the interpretations of H and $v(N)$: H is the set of all payoff vectors that can be achieved by a joint strategy of all of N, whereas $v(N)$ is the set of payoff vectors x such that

---

([8]) Note the similarity with the "end games" used by Isbell [8] in a somewhat different context. (This work is independent of Isbell's.)

N can jointly achieve at least x. The notion of game as originally defined (without (7)) provides a generalization of von Neumann and Morgenstern's "extended" game [15][9]; this is why (7) was not included in the original definition[10].

3. **Domination, core, rationality.** Fix a game (N, v, H). A payoff vector x is said to *dominate* a payoff vector y via S (notation: $x \succ_S y$) if $x \in v(S)$ and $x^S > y^S$; x is said to *dominate* y (notation: $x \succ y$) if there is an S such that $x \succ_S y$. If R is an arbitrary set of payoff vectors, we define the *R-core* $\mathcal{C}(R)$ to be the set of all members of R not dominated by any other member of R.

It is easy to show that for each $i \in N$, there is an extended real number[11] v^i such that $v(\{i\}) = \{x : x^i < v^i\}$. A payoff vector x is called *individually rational* if $x \geq v^N$. x is called *group rational* if there is no $y \in H$ such that $y > x$. We will denote by E the set of group rational payoff vectors in H, and by $\overline{A}$ the set of individually rational payoff vectors in H; also, we set[12] $A = E \cap \overline{A}$ and $\overline{E} = H$.

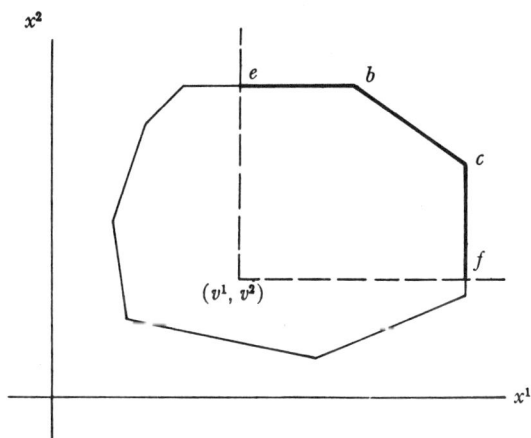

FIG. 1

We will consider the R-cores for $R = E$, $\overline{E}$, A, and $\overline{A}$. For two-person games, all these cores turn out to be equal to A. This is a set which is related to what has been called the "negotiation set" [11, p. 118], but is *not* always the same thing. (See Figure 1, in which the negotiation set is the line bc, whereas the set A is the broken line $ebcf$. Note that eb is horizontal and cf is vertical.)

---

[9] See [3, §7].

[10] Half of (7)—the "only if" half—was included in the definition of "game" as given in [3, §2]. What we call "game" here is called "extended game" in [3], and what we call "ordinary game" here is slightly stronger than either of the definitions in [3].

[11] A real number or $+\infty$ or $-\infty$.

[12] Following the notation of [5, p. 58].

4. A lemma on polyhedral sets. If $B \subset E^N$, denote by $I(B)$ the set[13] of all members x of B for which there is no y such that $y > x$. For example, $A = I(\bar{A})$ and $E = I(\bar{E})$. The lemma we shall establish in this section may be described as follows:

If B is a bounded polyhedron in E^N, then for each x in B but not in $I(B)$, there is a payoff vector x'' in $I(B)$ such that if we move along the ray connecting x to x'' at a constant speed (which is the same for all $x \in B$), then the rate of increase of each of the coordinates is uniformly bounded away from zero (for all $x \in B$).

If B is not a polyhedron this need not be true, as we shall see in an example.

For $x > 0$, define $f(x) = \max_{i,j \in N}(x^i/x^j)$. We have

(1) $$f(x + y) \leq \max(f(x), f(y)).$$

LEMMA 2. *For every closed polyhedron B in E^N, there is a positive number K such that for all $x \in B - I(B)$, there is an $x' \in B$ such that*

(3) $$x' > x \quad \text{and} \quad f(x' - x) \leq K.$$

Proof. Suppose B to be defined by the set of linear inequalities $L_1(x) \geq b_1, \cdots, L_m(x) \geq b_m$; we denote this set by M. Each subset Q of M defines a subset B_Q of B, namely the set of those elements of B which satisfy the inequalities in Q strictly, and the inequalities in $M - Q$ as equalities. Some of the B_Q may be empty; but those that are not are distinct, and we have $B = \bigcup_{Q \subset M} B_Q$. (Geometrically, the B_Q are the interiors of the faces of B.)

For each Q such that $B_Q - I(B) \neq \emptyset$, choose a payoff vector x_Q in $B_Q - I(B)$. Then there is a payoff vector $y_Q \in B$ such that $y_Q > x_Q$. Now let x be an arbitrary element of B_Q; define $y_\delta = x + \delta(y_Q - x_Q)$. For sufficiently small positive δ, y_δ satisfies the inequalities in Q strictly; the inequalities in $M - Q$ are satisfied by y_δ for *all* positive δ. Hence for positive δ sufficiently small, $y_\delta \in B$ and $y_\delta > x$; we define $x' = y_\delta$ for this δ. Then $x' - x = \delta(y_Q - x_Q)$, so that $f(x' - x) = f(y_Q - x_Q)$. Setting $K = \max_{Q \subset M} f(y_Q - x_Q)$, we obtain $f(x' - x) \leq K$ for all those x that are located in some B_Q for which $B_Q - I(B) \neq \emptyset$. But since every $x \in B - I(B)$ is located in some such B_Q, our proof is complete.

LEMMA 4. *For every compact polyhedron B in E^N, there is a positive number K such that for all $x \in B - I(B)$, there is an $x'' \in I(B)$ such that $x'' > x$ and $f(x'' - x) \leq K$.*

Proof. For each $x \in B - I(B)$, let F_x be the union of the single point x with the set of all $x' \in B$ satisfying (3). F_x is compact, and therefore the function $\|y - x\|$, considered as a function of y, attains its maximum in F_x, say

---

[13] If B is closed and convex, then $I(B)$ is the weak top of B over its base as defined in [7].

at the point x''. Suppose $x'' \notin I(B)$. Then by Lemma 2 there is a point $y \in B$ such that $y > x''$ and $f(y-x'') \leq K$. From (1) it then follows that $f(y-x) \leq K$ and from (1.1) that $\|y-x\| = \|(y-x'') + (x''-x)\| > \|x''-x\|$; hence x'' does not have the maximum property by which it was defined, which is a contradiction. This completes the proof.

COROLLARY 5. *If B is a compact polyhedron in E^N, then there is a positive number K, such that for each $x \in B - I(B)$, there is an $x'' \in I(B)$ such that $x'' > x$ and for each $i \in N$, $x''^i - x^i \geq \|x'' - x\|/K$.*

5. **Relations between the R-cores for $R = E$, $\overline{E}$, A, and $\overline{A}$.** Fix a game (N, v, H). The crux of this section is the following theorem:

THEOREM 1. *Let B be a compact polyhedron in E^N, and let $y \in I(B)$. If there is a $z \in B$ which dominates y, then there is also a $w \in I(B)$ which dominates y.*

Proof. We may assume without loss of generality that $y = 0$. Let V denote the closed positive orthant $\{x : x \geq 0\}$. Since $0 = y \in I(B)$, B cannot intersect the interior of V, and there is therefore a hyperplane $g(x) = \sum_N c^i x^i = 0$ which separates B from V. W. l. o. g. $g(x) \leq 0$ for all $x \in B$, and $g(x) \geq 0$ for all $x \in V$; from the latter fact it follows that $c^N \geq 0$. Note that if $x \in B$ and $g(x) \geq 0$, then $x \in I(B)$; otherwise we would have an $x_1 \in B$ such that $x_1 > x$, and since not all the c^i vanish, it would follow that $g(x) < g(x_1) \leq 0$.

Let the effective set for the domination of z over 0 be S; set $h(x) = \sum_{N-S} c^i x^i$. Suppose there is an $x \in B$ such that $x \succ_S 0$ and $h(x) \geq 0$. Then since $x^S > 0$ and $c^S \geq 0$ it follows that $\sum_S c^i x^i \geq 0$. Hence $g(x) \geq 0$, and therefore $x \in I(B)$; but then we are finished (set $w = x$). Therefore we may assume without loss of generality that

(2) $\qquad$ if $x \in B$ and $x \succ_S 0$, then $h(x) < 0$.

Let $k = (\min_S z^i)/2$; note that $k > 0$. Let C be the set of those x in B for which $x^S \geq 0$ and $\|x^S\| = k$. C is compact, and therefore $h(x)$ attains its maximum in C at a point x_1 in C. If $x_1 \notin I(B)$, then there is an $x_2 \in B$ for which $x_2 > x_1$; hence $\|x_2^S\| > \|x_1^S\| = k$, and

(3) $\qquad\qquad\qquad h(x_2) \geq h(x_1).$

Set $x_3 = (k/\|x_2^S\|) x_2$. Then $\|x_3^S\| = k$, $x_3^S > 0$, and since B is convex, $x_3 \in B$. Hence $x_3 \in C$, and therefore

(4) $\qquad\qquad\qquad h(x_1) \geq h(x_3).$

Since $\|x_3^S\| = k$, it follows that $x_3^S < z^S$; but since $z \in v(S)$, it follows from (2.4) that $x_3 \in v(S)$. Hence from $x_3^S > 0$, it follows that $x_3 \succ_S 0$. Hence by (2), $h(x_3) < 0$. But $h(x_3) = (k/\|x_2^S\|) h(x_2)$, and $k/\|x_2^S\| < 1$; hence $h(x_2) < h(x_3)$, which contradicts (3) and (4). We conclude that $x_1 \in I(B)$.

Now since $\|x_1^S\| = k$, it follows that $x_1^S < z^S$; therefore $x_1 \in v(S)$. If $x_1^S > 0$ it then follows that $x_1 >_S 0$, and since $x_1 \in I(B)$, we are finished ($w = x_1$). It therefore remains only to deal with the case in which one of the coordinates of x_1^S vanishes. In this case, set $\delta = k/(K+1)$, the K being that of Corollary 4.5. Let $x_4 \in B$ be such that $x_4^S > 0$ and $\|x_4 - x_1\| \leq \delta$; such an x_4 can be constructed by choosing a point sufficiently close to x_1 on the line segment joining z to x_1. Define x_4'' in accordance with Corollary 4.5. If $\|x_4'' - x_4\| > K\delta$, then $x_4''^i - x_4^i > \delta$ for each $i \in N$; therefore for $i \in N$ we have

$$x_4''^i - x_1^i = (x_4''^i - x_4^i) + (x_4^i - x_1^i) > \delta - |x_4^i - x_1^i| \geq \delta - \|x_4 - x_1\| \geq 0.$$

Hence $x_4'' > x_1$ and $x_4'' \in I(B)$, contradicting $x_1 \in I(B)$. Hence $\|x_4'' - x_4\| \leq K\delta$. But then

$$\|x_4'' - x_1\| \leq \|x_4'' - x_4\| + \|x_4 - x_1\| \leq K\delta + \delta = k;$$

hence

$$\|x_4''^S\| \leq \|x_4'' - x_1\| + \|x_1^S\| \leq 2k = \min_S z^i.$$

Hence $x_4''^S \leq z^S$ and from (2.4) we deduce $x_4'' \in v(S)$. Since $x_4''^S > x_1^S \geq 0$, we obtain $x_4'' >_S 0$ and $x_4'' \in I(B)$. The proof of Theorem 1 is now complete ($w = x_4''$).

COROLLARY 5. *If B is a compact polyhedron, then $\mathcal{C}(I(B)) = \mathcal{C}(B) \cap I(B)$.*

Proof. If $y \in \mathcal{C}(I(B))$, then surely $y \in I(B)$. If y were not in $\mathcal{C}(B)$, then it would be dominated by a member z of B, and hence by Theorem 1, by a member w of $I(B)$; but then it would not be in $\mathcal{C}(I(B))$. Hence $y \in \mathcal{C}(B)$ also.

Conversely, if $y \in I(B)$ and is not dominated by any member of B, then a fortiori it is not dominated by any member of $I(B)$. Hence $y \in \mathcal{C}(I(B))$.

COROLLARY 6. $\mathcal{C}(E) = \mathcal{C}(\overline{E}) \cap E$; $\mathcal{C}(A) = \mathcal{C}(\overline{A}) \cap A$.

COROLLARY 7. *If (N, v, H) is an ordinary game, then $\mathcal{C}(E) = \mathcal{C}(\overline{E})$ and $\mathcal{C}(A) = \mathcal{C}(\overline{A})$.*

Proof. We need only remark that if $B \subset v(N)$ then $\mathcal{C}(B) \subset I(B)$; for any payoff vector not in $I(B)$ is dominated via N by some other payoff vector. Our result now follows by applying Corollary 5 with $B = \overline{E}$ or $\overline{A}$.

THEOREM 8. *If (N, v, H) is an ordinary game, then $\mathcal{C}(E) = \mathcal{C}(\overline{E}) = \mathcal{C}(A) = \mathcal{C}(\overline{A})$.*

Proof. It is sufficient to prove that $\mathcal{C}(\overline{E}) = \mathcal{C}(\overline{A})$. Clearly $v^N \in \bigcap_N v(\{i\})$; hence by (2.5), $v^N \in v(N)$. Hence by (2.7) there is a $y \in H$ such that $y \geq v^N$. Now let $x \in \mathcal{C}(\overline{E})$. If $x \notin \overline{A}$, then for some i, $x^i < v^i \leq y^i$. Let y_1 be on the line segment connecting y to x, but so close to x so that $v^i \geq y_1^i > x^i$. Since $\overline{E}$ is convex and both x and y are in $\overline{E}$ (which is the same as H), so is y_1; but then

$y_1 >_{\{i\}} x$, and therefore $x \notin \mathcal{C}(\overline{E})$, a contradiction. Therefore $x \in \overline{A}$. Therefore if $x \notin \mathcal{C}(\overline{A})$ then there is a $z \in \overline{A}$ such that $z > x$; but since $\overline{A} \subset \overline{E}$, it follows that $z \in \overline{E}$, and therefore $x \notin \mathcal{C}(\overline{E})$, again a contradiction. We have shown that $\mathcal{C}(\overline{E}) \subset \mathcal{C}(\overline{A})$.

Conversely if $x \in \mathcal{C}(\overline{A})$, then surely $x \in \overline{E}$, since $\overline{E} \supset \overline{A} \supset \mathcal{C}(\overline{A})$. Hence if $x \notin \mathcal{C}(\overline{E})$, there must be a $y \in \overline{E}$ and an $S \subset N$ such that $y >_S x$. In particular, $y^S > x^S$ and $y \in v(S)$. Set $z = (y^S, v^{N-S})$; then by (2.4), $z \in v(S)$, and therefore $z \in v(S) \cap \bigcap_{N-S} v(\{i\})$. Hence by (2.5), $z \in v(N)$. Hence by (2.7) there is a $w \in H$ such that $w \geq z$. In particular, $w^S \geq z^S = y^S > x^S \geq v^S$ (since $x \in \overline{A}$), and $w^{N-S} \geq z^{N-S} = v^{N-S}$; hence $w \in \overline{A}$. Now let w_1 be on the line segment connecting w to x, but so close to x so that $y^S > w_1^S > x^S$. Since $\overline{A}$ is convex and both x and w are in $\overline{A}$, so is w_1; but since $y \in v(S)$, it follows from (2.4) that $w_1 \in v(S)$. Since $w_1^S > x^S$ and $w_1 \in v(S)$, it follows that $w_1 >_S x$; therefore since $w_1 \in \overline{A}$, it follows that $x \notin \mathcal{C}(\overline{A})$, a contradiction. Hence $x \in \mathcal{C}(\overline{E})$, and the proof is complete.

If G is an ordinary game, we shall call the common value of $\mathcal{C}(\overline{E})$, $\mathcal{C}(E)$, $\mathcal{C}(\overline{A})$, and $\mathcal{C}(A)$ the *core* of G.

6. A counter-example. The results of §§4 and 5 may fail if B (or H) is not polyhedral. In the case of Lemma 4.2 a circle in two dimensions is a counter-example. In the case of Theorem 5.1, let $N = \{1, 2, 3\}$ and let B be the convex hull of the sets C and D, where

$$C = \{x: x^1 \geq 0, x^2 \geq 0, x^3 = 0, (x^1)^2 + (x^2)^2 \leq 1\},$$
$$D = \{x: x^1 \geq 0, x^2 \geq 0, x^3 = 1, (x^1)^2 + (x^2 + 1)^2 \leq 4\}.$$

Then

$$I(B) = D \cup \{x: x^1 = 0, x^2 = 1, 0 \leq x^3 \leq 1\}.$$

Define the characteristic function v by

$$v^N = 0, \qquad v(\{ij\}) = \{x: x^i \leq 1/2 \text{ and } x^j \leq 1/2\},$$
$$v(N) = \{x: \text{there is a } y \in H \text{ such that } y \geq x\}.$$

We have $(1/2, 1/2, 1/2) \in B$ and $(1/2, 1/2, 1/2) >_{\{13\}} (0, 1, 0)$, but there is no member of $I(B)$ that dominates $(0, 1, 0)$. If we set $H = B$ we obtain counter-examples to the other results of §5.

7. Composition. Let $G_1 = (N_1, v_1, H_1)$ and $G_2 = (N_2, v_2, H_2)$ be games whose player sets N_1 and N_2 are disjoint. Intuitively, the *composition* G of G_1 and G_2 is the game each play of which consists of a play of G_1 and a play of G_2, played without any interconnection. Formally, we define $G = (N, v, H)$, where $N = N_1 \cup N_2$, $H = H_1 \times H_2$, and for each $S \subset N$, $v(S) = v_1(S \cap N_1) \times v_2(S \cap N_2)$.

Let $R_1 \subset E^{N_1}$ and $R_2 \subset E^{N_2}$, and set $R = R_1 \times R_2$. Then it is easily seen that $\mathcal{C}(R) = \mathcal{C}_1(R_1) \times \mathcal{C}_2(R_2)$. Furthermore, if G_1 and G_2 are ordinary, then so is G.

It follows that in this case, *the core of G is the cartesian product of the cores of G_1 and G_2.*

8. α-effectiveness and β-effectiveness. Up to now we have been treating games in characteristic function form only; we now turn to games given in normal form, and ask how we may obtain the characteristic function form from the normal form. As we remarked in the introduction, this may be done in a number of ways, depending on our definition of effectiveness. Here we shall give two such definitions.

A (finite) game Γ in normal form consists of a finite set N, called the set of *players*, a finite set P^i for each $i \in N$, called the set of *pure strategies* for player i, and a function F from the cartesian product P of all the P^i to E^N; F is called the *payoff* function[14], and its ith coordinate F^i is the *payoff to i*. If $S \subset N$, we write $P^S = \prod_S P^i$, the cartesian product being meant. A probability measure on P^S will be called a *c-strategy S-vector* (c for correlated); the set of all c-strategy S-vectors will be denoted C^S. Note that a c-strategy $\{i\}$-vector is the same as a mixed strategy for player i. If $c^N \in C^N$, then $F(c^N)$ will denote the expected payoff if the c-strategy N-vector c^N is played. If $S, T \subset N$, $S \cap T = \emptyset$, then $(c^S \times c^T)$ denotes the product measure[15] on $P^{S \cup T} = P^S \times P^T$ induced by c^S and c^T. Occasionally we shall have cause to consider a topological and a convex structure on C^S; in this case C^S will be considered a subset of E^{P^S}.

DEFINITION. (1) A coalition S is said to be *α-effective* for the payoff vector x if there is a $c^S \in C^S$ such that for each $c^{N-S} \in C^{N-S}$, we have $F^S(c^S \times c^{N-S}) \geq x^S$.

(2) S is said to be *β-effective* for x if for each $c^{N-S} \in C^{N-S}$ there is a $c^S \in C^S$ such that $F^S(c^S \times c^{N-S}) \geq x^S$.

Intuitively, α-effectiveness means that S can assure itself, independently of the actions of $N-S$, that each of its members i will receive at least his coordinate x^i of x. β-effectiveness means that S can always act so that each of its members i receives at least x^i, but the strategy that it must use to achieve this end may depend on the strategy used by $N-S$; in other words, $N-S$ cannot effectively prevent S from obtaining at least[16] x^S. Although α-effectiveness seems at first to be the intuitively more straightforward concept, technically speaking β-effectiveness possesses certain interesting properties not shared by α-effectiveness (see §10) which lead one to think that it may eventually turn out to be the more significant concept.

To construct a game in which α-effectiveness and β-effectiveness are not the same, let $N = \{1, 2, 3\}$ and $S = \{1, 2\}$. Let P^S have two members p_1^S and p_2^S, and P^3 two members p_1^3 and p_2^3. Define F^S by the matrix

---

[14] Denoted by H in [1; 2].

[15] Denoted (c^S, c^T) in [1].

[16] The difference between the two kinds of effectiveness may be formulated as the difference between a maxmin and a minmax; for 2-person games it follows from the von Neumann theorem that the two concepts coincide, but this does not generalize to more players.

	p_1^3		p_2^3	
p_1^S	1,	−1	0,	0
p_2^S	0,	0	−1,	1

;

the values of F^3 need not concern us. Then for $(0, 0, 0)$, S is β-effective but not α-effective.

9. Passage from the normal form to the characteristic function form. Fix a game Γ in normal form, and for each $S \subset N$, let $v_\alpha(S)$ be the set of payoff vectors for which S is α-effective. Define $v_\beta(S)$ similarly, using β-effectiveness instead of α-effectiveness. Define $H = F(C^N)$; H is the convex hull of all the payoff vectors of the form $F(p)$, where $p \in P$. Both (N, v_α, H) and (N, v_β, H) are ordinary games (though they may be different, as we saw in the previous section); except for Condition 2.5 in the case of β-effectiveness, all the conditions of §2 are easily verified for both these games. To establish Condition 2.5 for (N, v_β, H), let $x \in v_\beta(S) \cap v_\beta(T)$ and $c^{N-S-T} \in C^{N-S-T}$. Define subsets U and V of $C^S \times C^T$ as follows:

$$U = \{(c^S, c^T): F^T(c^S \times c^T \times c^{N-S-T}) \geq x^T\},$$
$$V = \{(c^S, c^T): F^S(c^S \times c^T \times c^{N-S-T}) \geq x^S\}.$$

Applying the von Neumann-Kakutani fixed point theorem[17], we obtain the existence of a point (c_0^S, c_0^T) in $U \cap V$. Setting $c_0^{S \cup T} = c_0^S \times c_0^T$, we obtain $F^{S \cup T}(c_0^{S \cup T} \times c^{N-S-T}) \geq x^{S \cup T}$, and it follows that $x \in v_\beta(S \cup T)$.

From (2.6) and (2.7) it follows that in an ordinary game, $v(N)$ must be polyhedral. The reader may suspect that in the characteristic function form of a finite game in normal form, $v(S)$ must be polyhedral for all S. This is true for $v_\alpha(S)$, but not for $v_\beta(S)$. The example is the same as in the previous section, except that $F^1(p_1^S, p_1^3) = F^2(p_2^S, p_2^3) = 0$ rather than 1. (See Figure 2; $v_\beta(S)$ is the cylinder whose cross-section is the shaded area.)

Note that we always have $v_\alpha(N) = v_\beta(N)$ and $v_\alpha^N = v_\beta^N$ (the former is trivial, the latter follows from the minimax theorem for 2-person zero-sum games [15]). In particular, α- and β-effectiveness are equivalent for all 2-person games.

If Γ_1 and Γ_2 are games in normal form with disjoint player sets, we may define their *composition* Γ by $N = N_1 \cup N_2$, $P = P_1 \times P_2$, $F(p_1, p_2) = (F_1(p_1), F_2(p_2))$. It is easily established that either definition of effectiveness yields a characteristic function form for Γ that is the composition of the corresponding characteristic function forms of Γ_1 and Γ_2 in the sense of §7.

10. The supergame. Nash's notion of equilibrium point for noncooperative games [12] is an example of the core notion as described in the introduc-

---

[17] [9, Theorem 2]; see also [14].

tion. The "outcomes" are strategy n-tuples; one strategy n-tuple f "dominates" another one g if they coincide for all but one of the players, and that one player prefers([18]) f to g. Thus an equilibrium point is a strategy n-tuple f with the property that if all the players have reason to believe that f will be played, then no player will be tempted to deviate from f. In the context of cooperative games, it is natural to broaden the definition of dominance so that f dominates g whenever they coincide for all players not in a certain coalition S, and the players in S each prefer f to g. When dominance between strategy n-tuples is defined in this way, members of the core are called *strong equilibrium points*([19]).

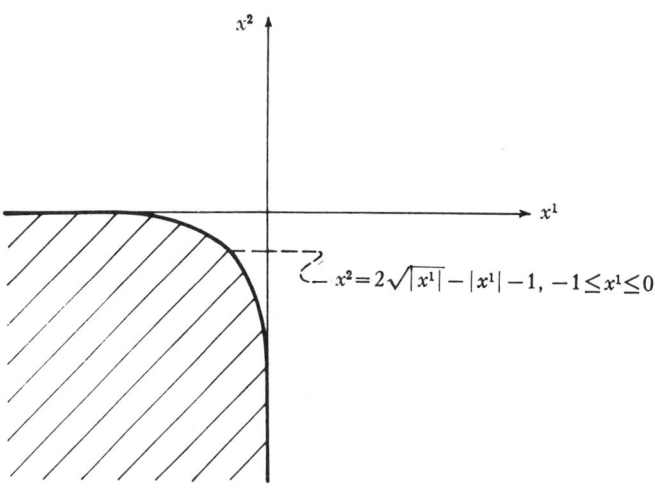

$x^2 = 2\sqrt{|x^1|} - |x^1| - 1,\ -1 \leq x^1 \leq 0$

FIG. 2

This definition of dominance seems well justified for a single play of a game which is not to be repeated. If the game is to be played repeatedly, though, then a player or group of players may be unwilling to deviate even if the deviation *will* yield a temporary advantage, for fear of future retaliation. If future retaliation is to be ruled out, then S must be able to *maintain* its payoff at the level of f; that is, S must be effective for f. We are thus led to the conclusion that a strategy n-tuple for one of a long sequence of plays of a game Γ should be considered in equilibrium if its payoff is in the core of the characteristic function form of Γ.

The question now arises, is it the α-core or the β-core that is appropriate for use in this context, or possibly we should use an altogether different notion of effectiveness? To answer this question, we consider a long sequence of plays of Γ as a single play of a game Γ^*, which we call the *supergame* of Γ [1; 11]. It stands to reason that equilibrium behavior for Γ, knowing that

---
([18]) I.e., receives a higher payoff when g is played.
([19]) A related definition is given in [4].

there will be more plays of Γ in the future, should correspond to equilibrium behavior in Γ^*, provided that Γ^* is not repeated. But for games that are not repeated, we have a perfectly well-defined equilibrium notion, namely that of strong equilibrium point. And it turns out that it is precisely the β-core of the characteristic function form of Γ that corresponds to the strong equilibrium points in Γ^*.

Formally[20], the supergame Γ^* is the game each play of which is an infinite sequence of plays of Γ. As in finite games, an n-tuple f of supergame strategies is said to *dominate* another n-tuple g if they coincide for all players not in a certain coalition S, and the players in S each prefer f to g. To define the word "prefer" in this context, we consider a sequence $\Gamma_1, \Gamma_2, \cdots$ of plays of Γ, and look at the average payoff for all the plays up to the kth. A number of definitions of preference are now possible, of which the following are the two "extreme" possibilities:

(a) S prefers f to g if the probability is positive that infinitely often the average payoff to each member of S will be uniformly[21] larger if f is used than if g is used.

(b) S prefers f to g if it is certain[22] that from a certain play Γ_k onwards, the average payoff to each member of S will always be uniformly larger if f is used than if g is used.

An n-tuple f of supergame strategies is said to have the payoff x if with probability 1 the average payoffs[23] tend to x. f is said to *correspond*[24] to a c-strategy vector c in Γ if the payoff to f exists and is the same as the payoff to c in Γ. An n-tuple of supergame strategies is said to be a *strong equilibrium point* if it is undominated and if it possesses a payoff. Actually we get two sets of strong equilibrium points, one for each of the two (inequivalent) notions of preference defined above. However, it turns out that both these sets correspond to the same set of c-strategy vectors, called *acceptable points* [1, §4]. By making use of Lemma 9.1 of [1], it is not difficult to show that the set of payoff vectors to acceptable points—the set of acceptable payoff vectors—coincides with the β-core[25]. Hence the β-core of a finite

---

[20] For a more detailed treatment of the supergame, see [1].

[21] The difference must be larger than a fixed (independent of k) positive S-vector.

[22] The probability is 1.

[23] Care should be taken to differentiate between the average payoff for the first k plays, and the expected payoff (for the latter, probability statements would of course be meaningless). Because of the law of large numbers, the existence of a payoff to f is quite plausible. For example, if a "steady state" in strategy choices on the individual Γ_k is ever reached, then f certainly has a payoff. As we have shown in [1, §12], considerations of expected payoff are inappropriate for Γ^*.

[24] The correspondence must be defined via the payoffs because there is no direct method for comparing strategies for individual games with supergame strategies.

[25] Definitions "between" (a) and (b) yield the same set of payoff vectors. For example, in either of the definitions we could substitute "with probability at least 1/2" for the respective probability statements.

game is the set of payoff vectors to strong equilibrium points in its supergame.

REFERENCES

1. R. J. Aumann, *Acceptable points in general cooperative n-person games*, in [17, pp. 287–324] [Chapter 20].

2. ———, *Acceptable points in games of perfect information*, Pacific J. Math. vol. 10 (1960) pp. 381–417 [Chapter 21].

3. R. J. Aumann and B. Peleg, *Von Neumann—Morgenstern solutions to cooperative games without side payments*, Bull. Amer. Math. Soc. vol. 66 (1960) pp. 173–179 [Chapter 38].

4. R. Farquharson, *Sur une généralisation de la notion d'equilibrium*, C. R. Acad. Sci. Paris vol. 240 (1955) pp. 46–48.

5. D. B. Gillies, *Solutions to general non-zero-sum games*, in [17], pp. 47–83.

6. J. C. Harsanyi, *Approaches to the bargaining problem before and after the theory of games: a critical discussion of Zeuthen's, Hicks', and Nash's theories*, Econometrica vol. 24 (1956) pp. 144–157.

7. J. R. Isbell, *Absolute games*, in [17], pp. 357–396.

8. ———, *A modification of Harsanyi's bargaining model*, Bull. Amer. Math. Soc. vol. 66 (1960) pp. 70–73.

9. S. Kakutani, *A generalization of Brouwer's fixed point theorem*, Duke Math. J. vol. 8 (1941) pp. 457–459.

10. R. D. Luce, *A definition of stability for n-person games*, Ann. of Math. vol. 59 (1954) pp. 357–366.

11. R. D. Luce and H. Raiffa, *Games and decisions*, New York, John Wiley, 1957.

12. J. F. Nash, *Non-cooperative games*, Ann. of Math. vol. 54 (1951) pp. 286–295.

13. ———, *The bargaining problem*, Econometrica vol. 18 (1950) pp. 155–162.

14. J. von Neumann, *Über ein ökonomisches Gleichungssystem und eine Verallgemeinerung des Brouwerschen Fixpunktsatzes*, Ergebnisse eines Math. Kolloquiums vol. 8 (1937) pp. 73–83.

15. J. von Neumann and O. Morgenstern, *Theory of games and economic behavior*, Princeton, N. J., Princeton University Press, 1944, 2nd ed., 1947.

16. B. Peleg, *Solutions to cooperative games without side payments*, Trans. Amer. Math. Soc. vol. 106 (1963) pp. 280–292.

17. A. W. Tucker and R. D. Luce, editors, *Contributions to the theory of games*. IV, Annals of Mathematics Studies, no. 40, Princeton, N. J., Princeton University Press, 1959.

40. Introduction to "Some Thoughts on the Theory of Cooperative Games"

The author of this paper, Gerd Jentzsch, died while still a young man on March 26, 1959. This is apparently his only publication. Judging from its originality and all-around brilliance, his death was a loss of the first magnitude to game theory.

The von Neumann-Morgenstern (N-M) theory of n-person games [5] is concerned with cooperative games in which side payments are permitted and utility is "unrestrictedly transferable"—that is, each player's utility for money is linear in money.[1] Jentzsch's investigations grew out of an attempt to generalize the N-M theory either by eliminating the requirement that the utility functions[2] be linear, or more generally, by eliminating side payments altogether. He notices at the outset that the notion of "effectiveness"—which is crucial in the N-M theory—does not generalize in a straightforward manner. In the classical theory, a coalition K is effective for a payoff vector f if, roughly speaking, the coalition can assure itself of getting at least f. An equivalent definition of effectiveness is that the *opposition*—the complement of K—cannot prevent K from obtaining at least f. But when utilities are nonlinear in money or side payments are forbidden, these two definitions of effectiveness are in general no longer equivalent—in Jentzsch's terminology, the game need not be "clear" (Example 4). Jentzsch addresses himself to the task of broadening the class of games considered by von Neumann and Morgenstern, while still retaining the clearness property.

The chief result is Theorem 21. Rather than stating it here in its most general form, we will describe its application to games with side payments ("money games" for short) in which the utility functions need not be linear. The problem that Jentzsch considers is, what kinds of utility functions of the players will always lead to clear games (as linear utility functions do)? More precisely, what conditions, when placed on the utility functions of the players, will ensure that all money games in which these players participate are clear? The answer is that each coalition must have a kind of "social utility function" for money. For example, this involves the demand that $50 be indifferent—from the point of view of the coalition as a whole—to some probability combination of 0 dollars and $100 (though not necessarily the 1/2-1/2 combination). The sums of money involved ($50, $0, $100) are not given to the individual players,

---

This chapter is an introduction to [4]. It originally appeared in *Advances in Game Theory*, Annals of Mathematics Studies 52, edited by M. Dresher, L. S. Shapley, and A. W. Tucker, pp. 407–409, Princeton University Press, Princeton, 1964. Reprinted with permission.

1. See R. D. Luce and H. Raiffa, *Games and Decisions*, p. 168.
2. By the phrase "utility function" we shall henceforth *always* mean "utility of money as a function of money."

but to the coalition as a whole for distribution among its members. "Indifferent" has a very precise meaning here: The two sets of (utility) payoff vectors that can result from the two possibilities must coincide.

The existence of such a "social utility function" is a considerable restriction. Jentzsch remarks without proof that Bernoullian (i.e., logarithmic) individual utility functions lead to a social utility function (Examples 11, 16) and that other individual utility functions that lead to a social utility function can be obtained as solutions of a third-order differential equation with one parameter (which he does not specify). These questions must certainly be investigated further. But on the whole, Jentzsch's result shows that clearness is the exception rather than the rule—that games with nonlinear utility functions or without side payments cannot be "expected" to be clear.

The difference between the two kinds of effectiveness was appreciated by others, working independently of Jentzsch, as far back as 1957—which is probably the approximate date of Jentzsch's investigations.[3] It was explicitly mentioned by Aumann and Peleg [2], who used the names α- and β-effectiveness for the two kinds. A survey of the whole field of cooperative games without side payments is given in [1], which has a bibliography of 51 items; but of this work, Jentzsch knew only of the pioneering investigation of Shapley and Shubik [6]. This is another example of the known phenomenon of the intrinsic "ripeness" of a scientific concept—leading to simultaneous independent discovery by widely separated investigators. It should be emphasized, though, that it is only in the basic recognition of the difference between α- and β-effectiveness that Jentzsch's work overlaps that of others; the main result of this paper has not been found by anybody else, and appears here for the first time. Indeed other workers have approached the subject from a somewhat different viewpoint—they have tried to "live with" the difference, whereas Jentzsch characterized the conditions under which it could be eliminated (see [1]).

An attempt has been made to keep editorial comment separate from Jentzsch's original text. All the footnotes are the editor's, as are the two "Editor's Notes." The long formal proofs given by Jentzsch for Theorems 10 and 13 have been replaced by short intuitive sketches. There has been some rearranging of the material, and the more straightforward proofs have been left out. Those are all the changes.

---

3. In fact, the idea is related to Blackwell's approachability-excludability theory [3] which appeared already in 1956.

Since Jentzsch is interested only in the question of effectiveness, he fixes once and for all a coalition K, and considers only the joint strategies of the coalition, the joint strategies of the opposition, and the payoff to the coalition. The resulting formal object is called a "K-game," and this is the object of investigation throughout.[4] The individual strategies of members of the coalition and of the opposition, and the payoff to the opposition, are of no interest in this context, and are therefore suppressed in the formal model.

References

[1] Aumann, R. J., "Cooperative Games without Side Payments," in *Essays in Mathematical Economics in Honor of Oskar Morgenstern*, M. Shubik (ed.), Princeton University Press, Princeton, 1967, pp. 3–27 [Chapter 41].

[2] Aumann, R. J. and B. Peleg, "Von Neumann–Morgenstern Solutions to Cooperative Games without Side Payments," *Bull. Amer. Math. Soc.* 66, 1960, pp. 173–179 [Chapter 38].

[3] Blackwell, D., "An Analog of the Minimax Theorem for Vector Payoffs," *Pac. J. Math.* 6, 1956, pp. 1–8.

[4] Jentzsch, G., "Some Thoughts on the Theory of Cooperative Games," in *Advances in Game Theory*, Annals of Mathematics Studies 52, M. Dresher, L. S. Shapley, and A. W. Tucker (eds.), Princeton University Press, Princeton, 1964, pp. 407–442.

[5] Von Neumann, J. and O. Morgenstern, *Theory of Games and Economic Behavior*, Princeton University Press, Princeton, 1944.

[6] Shapley, L. S. and M. Shubik, "Solutions of n-Person Games with Ordinal Utilities" (abstract), *Econometrica* 21, 1953, p. 348.

---

4. It is formally identical to Blackwell's "game with vector payoffs" [3].

41 A Survey of Cooperative Games without Side Payments

1. SCOPE OF THE PAPER

We begin by giving intuitive definitions of our terms. Start out with a game, either in extensive or in normal form. The game becomes *cooperative* if we allow the players to communicate before each play and to make binding agreements about the strategies they will use (either mixed or pure). We say that *side payments* are allowed if there is a common medium of exchange, such as money, which can be transferred between the players before or after the play. We say that *utility is transferable* if the increment to the payoff of a player caused by a transfer of money is proportional to the amount of money transferred [33]. The classical theory of *n*-person games as first conceived by von Neumann and Morgenstern [45], and later elaborated upon by many other writers, is concerned exclusively with cooperative games in which side payments are allowed and utility is transferable. It is commonly assumed that this involves an interpersonal comparison of utility, but this is false; it is only necessary that each individual's utility be an increasing linear function of money, and nothing need be said about the constant of proportionality (indeed any statement about the constant of proportionality is meaningless within the framework of N-M[1] utility theory). However, it *is* true that mathematically, N-M games can be treated as if the payoff were in money rather than in utility.

It is also often assumed that the N-M theory and its subsequent elaborations depend in an essential way on side payments and transferable utility; this is also false, as is shown by the small but growing body of recent work which parallels the N-M theory but deals with cooperative games in which side payments are either altogether forbidden, or are allowed but utility is not transferable. It is this body of work that I wish to survey here. Incidentally, recall that noncooperative games include cooperative games as a special case, cooperative games without side payments include cooperative games with side payments, and the case of transferable utility is the most special of all. Cooperative games without side payments and cooperative games with side payments but without transferable utility present many of the same problems, and since the former are more general we restrict much of our discussion to them.

Revised version of a lecture delivered at the Princeton Conference on Recent Advances in Game Theory, October 1961.

[1] von Neumann—Morgenstern.

This chapter originally appeared in *Essays in Mathematical Economics in Honor of Oskar Morgenstern*, edited by M. Shubik, pp. 3–27, Princeton University Press, Princeton, 1967. Reprinted with permission.

2. MOTIVATION

Cooperative games without side payments are of considerable importance in the applications. In some situations side payments are impossible because there is no common medium of exchange, or such a medium, if it exists, is irrelevant; think of the international situation. In other cases side payments are called bribes and are ruled out for ethical or legal reasons, while cooperation is considered perfectly all right. Finally, even when side payments are legal, utility usually is nonlinear in money, and this may result in a situation which is not covered by the N-M theory. It is this last fact that caused Luce and Raiffa to state that the N-M theory is "for many purposes next to useless" [19, p. 233]. We do not share this view, because if money is substituted for utility the N-M theory still applies to any situation in which probabilistic considerations are considered irrelevant;[2] but we do feel that an extension of the N-M theory to the no-side-payment case is useful.

3. THE CHARACTERISTIC FUNCTION

Let us now begin with a description of some of the work that has been done in this field. There are three widely used models for studying n-person games: the extensive form, the normal form, and the characteristic function.

The *extensive* form is essentially a mathematical representation of the rules of the game. The *normal* form is the "payoff matrix"—a list of strategies for each player, together with a payoff vector for each n-tuple of strategies. The *characteristic function* gives for each coalition the set of payoff vectors that that coalition can "assure" its players. There are, of course, connections between the various forms; the normal form can be calculated from the extensive form, and the characteristic function from the normal form. However, each form is suited for different kinds of investigations. We will begin our study of cooperative games without side payments with the characteristic function.

Let us represent the payoff to each player by a coordinate of Euclidean space; thus we will be working in Euclidean space of dimension equal to the number of players, and in its subspaces. We denote by N the set of players, by E^N the Euclidean space in which we are working, and by E^S the subspace of E^N spanned by the axes belonging to the players in a subset S of N. Points of E^N are called *payoff vectors*, of E^S payoff *S-vectors*. The *characteristic function* associates with each $S \subset N$ a subset $v(S)$ of E^S. Intuitively, $v(S)$ represents the set of payoffs that S can assure itself. When side payments are allowed and utility is transferable (this will henceforth be called

---

[2] Even when they are relevant the N-M theory applies in a much wider range of situations than has often been supposed; see §8.

the N-M case), $v(S)$ is the closed half-space

$$\left\{x \in E^S : \sum_{i \in S} x^i \leq f(S)\right\},$$

where $f(S)$ is the N-M characteristic function (i.e., the total amount of money that S can assure itself). A typical such half-space is illustrated in Figure 1 for the 2-player case; $v(S)$ is the whole area to the "southwest" of the line $x^1 + x^2 = f(S)$.

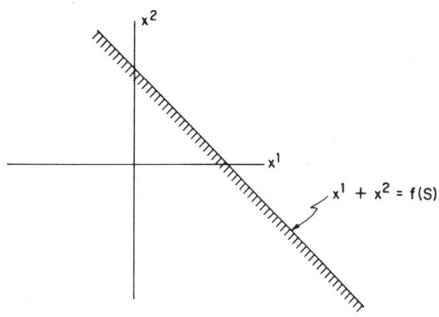

FIGURE 1

Returning to the no-side-payment case, we assume the following conditions for our characteristic function:

$$v(S) \text{ is convex, closed, and non-empty.} \tag{1}$$

$$x \in v(S), y \in E^S, x \geq y \Rightarrow y \in v(S). \tag{2}$$

$$v(S \cup T) \supset v(S) \times v(T) \text{ for } S \text{ and } T \text{ disjoint.} \tag{3}$$

The vector inequality in (2), like all subsequent vector inequalities, is meant to hold for each coordinate.

Intuitively, convexity follows from the fact that players can mix and correlate their strategies. Closedness is mainly a question of mathematical convenience and is satisfied in all applications that I can think of. Condition (2) says that if a coalition can assure itself of a payoff vector x, it can also assure itself of anything coordinate-wise less. The last condition is superadditivity; any vector whose components can be obtained by each of two disjoint coalitions acting separately can also be obtained by them when acting together. In Figure 2 we show a typical set of the form $v(S)$ in two dimensions.

We have defined a characteristic function; in order to define a *game in characteristic function form*, we need an additional concept that is not needed in the N-M theory. This is the set H of outcomes that "can actually occur." H has a very close connection with $v(N)$: its "top" coincides with

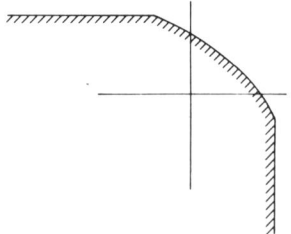

FIGURE 2

the "top" of $v(N)$ (see Figure 3), or in precise terms

$$v(N) = \{x \in E^N : \exists\ y \in H \text{ such that } y \geq x\}. \tag{4}$$

The difference between $v(N)$ and H is that $v(N)$ consists of those vectors x that N can "guarantee," in the sense that it can get *at least* x; whereas H is the set of vectors such that N can get *exactly* x.

Summing up, a game in *characteristic function form* is a pair (v, H), where v is a characteristic function obeying (1), (2), and (3), and H is a convex proper subset of E^N satisfying (4).

Sometimes it will be assumed that H is a convex compact polyhedron; this is justified if one thinks of the game in characteristic function form as being generated from a finite game in normal form. On other occasions, it is more convenient to assume that $H = v(N)$; this is justified, for example, if one makes an assumption of "free disposal." The latter assumption is the one more suited to the N-M case.

The set of conditions ((1) through (4)) that we have assumed for v and H is by no means the only possible one, and in fact almost every paper on the subject uses a different variant of the set of assumptions. In particular, for many purposes super-additivity (3) is unnecessary, sometimes convexity is not needed either, and for other purposes condition (4) is unnecessary. The conditions given here have been chosen for convenience in exposition,

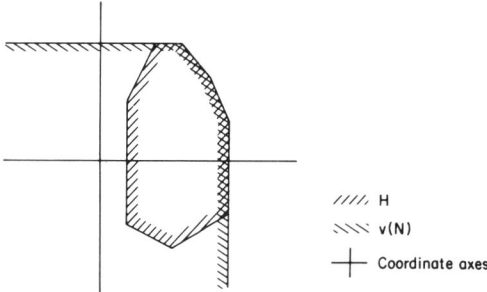

FIGURE 3

and so for many of the theorems they are stronger than necessary. The reader is referred to the original papers for statements of alternative conditions under which the various theorems hold.

In the N-M theory, various kinds of payoff vectors are distinguished according to their "rationality" attributes. A payoff vector is said to be *individually rational* if each player gets at least what he can guarantee himself, and *group rational* if the whole group cannot play in such a way that each player gets more. The same notions can be defined in our context; precisely, x is individually rational if for each player i, we have $x^i \geq \max v(\{i\})$; and x is group rational if there is no payoff vector $y \in v(N)$ such that $y > x$. Let us denote by H_{ig}, H_i, and H_g the subsets of H obtained by imposing individual and group rationality in various combinations; these sets, together with H, correspond to the sets of payoff vectors that have been studied in the N-M theory [39, 50]. In particular H_{ig} corresponds to what is usually called the set of "imputations."

Note that the characteristic function $v(S)$ does not necessarily have to be interpreted as the set of payoff vectors that S can *assure* itself; if preferred, it may be interpreted in any other way, such as what a coalition "thinks it can get." It is also possible that a game is given *a priori* in characteristic function form, like the following voting game:

Let the number of players be odd, and let C be a convex compact subset of E^N. The game consists of the players "voting" for a member of C by majority rule. If a majority agrees on a point x of C, then x is the payoff vector (to all players); otherwise each player i gets only his personal minimum m^i in C, i.e. $\min \{x^i : x \in C\}$. It is easy to see how this can be generalized to weighted majority games and to simple games in general.

To describe the characteristic function, let C^S denote the projection of C on E^S, and let m^S denote the S-vector $\{m^i\}_{i \in S}$. Then

$$v(S) = \begin{cases} \text{(if } S \text{ is winning) the set of all } S\text{-vectors that} \\ \quad \text{are } \leq \text{ a member of } C^S; \\ \text{(if } S \text{ is losing) the set of all } S\text{-vectors that} \\ \quad \text{are } \leq m^S. \end{cases}$$

4. THE VON NEUMANN-MORGENSTERN SOLUTION

We can now develop a theory of games parallel to the N-M theory. The two most important elements of that theory are the solution and the core. First, we define domination:

Let $x, y \in E^N$, and let x^S denote the projection of x on E^S. Then

$$x \succ_S y \Leftrightarrow_{df} x^S \in v(S), x^S > y^S$$
$$x \succ y \Leftrightarrow_{df} x \succ_S y \text{ for some } S.$$

Let $K \subset E^N$. Just as in the N-M theory, a *solution* of K is a subset V of K such that no two members of K dominate each other, and every member of K not in V is dominated by some member of V. The *core* of K (denoted by $C(K)$) is the set of members of K not dominated by other members of K.

THEOREM 1. *A solution of H_i is a solution of H_{ig}, and conversely.*

This corresponds to a theorem in the N-M theory first proved by Shapley [50]. The proof, which is not difficult, is given in [23]. Henceforth a *solution of a game* is a solution of H_{ig} for that game.

THEOREM 2. *Every 2-person game has a unique solution, namely all of H_{ig}. This is also the core of H_{ig}.*

This too is easy to prove. The first difficult theorem is:

THEOREM 3. *Every 3-person 0-sum game has a solution.*

A 3-person 0-sum game is one in which H is contained in the hyperplane $\sum_{i=1}^{3} x^i = 0$. Theorem 3 is proved in Peleg [23]. The 0-sum restriction may seem somewhat strange in a non-side-payment context; however, it makes sense if one assumes that the payoff to a game is in money, that no money enters or leaves the game from outside, that chance and mixed strategies are irrelevant, and that side payments, though obviously possible, are illegal. In addition, the proof was a considerable technical achievement, and pointed the way to the subsequent:

THEOREM 4. *Every 3-person game for which H is a polyhedron has a solution.*

This is proved in Stearns [30]. In the same place Stearns classifies all solutions to 3-person games.

The biggest problem left open by von Neumann and Morgenstern in their book [45] was that of the existence of a solution for an arbitrary *n*-person cooperative game with side payments and transferable utility. The problem remains unsolved to this day. One of the methods they used to attack this problem [45, pp. 266-271 and pp. 587-603] was to define the notion of solution for an abstract relation defined on an abstract set (abstracting from the game situation, where it is defined for the domination relation on the set of imputations). They then studied the solution notion in this abstract framework, seeking conditions of a general nature that would ensure the existence of a solution and that would be satisfied in the game context. This work was carried on by Richardson and others (see for example [40, 47]), but though many interesting sufficient conditions for existence were found, none could be proved to apply to the game context. In 1959, Kalisch and Nering [41] constructed a game with a countable infinity of players and showed that it has no solution, thus showing that the completely "abstract" approach to proving the existence theorem could not work. However, the imputation space in the Kalisch-Nering example is not compact. Thus there remained the hope that a "modified abstract" approach could be made to work, in which account would be taken of topological properties of the imputation space and the domination relation. This hope has recently been shattered by Stearns (unpublished[3]), who proved:

THEOREM 5. *There is a 7-person game with no solution.*

The original problem proposed by von Neumann and Morgenstern—for games with side payments and transferable utility—remains open.

We mention that it is possible to construct a theory of composition of games that parallels the N-M theory, but that yields simpler and more intuitive results [3, 5, 7].

Isbell [16] has constructed a theory of cooperative games without side payments in which he makes use of the notion of N-M solution. However, his work is not based on the characteristic function model presented in §3.

5. THE CORE

Let $K \subset E^N$. The *Core* of K (denoted by $C(K)$) is the set of members of K not dominated by other members of K.

THEOREM 6. *Assume either that H is a convex compact polyhedron, or that $H = v(N)$. Then $C(H) = C(H_i) = C(H_g) = C(H_{ig})$.*

In other words, all the "interesting" cores are equal, so we are justified in referring to the "core of a game." This is trivial in the N-M theory, but no longer so in the current theory. Under the first of the two assumptions, the proof was first published in [3]; subsequently it was considerably simplified by Stearns (unpublished). We sketch the simplified proof here.

The difficult part is to prove that imposing group rationality, either on H or on H_i, does not change the core. For example, take H; we must prove that $C(H) = C(H_g)$. $C(H) \subset C(H_g)$ is easily established. The crux of the proof is the opposite inclusion. For $x \in E^N$, denote $\max_i |x^i|$ by $\|x\|$. We need the following

LEMMA. *There is a positive number M (depending on H only) such that for all $z \in H - H_g$, there is a $\hat{z} \in H_g$ such that $\hat{z} > z$ and for each $i \in N$, $\hat{z}^i - z^i > \|\hat{z} - z\|/M$.*

In words, the lemma states that for each z in H that is not already in the top of H, we can find a ray that leads to the top of H, and that is increasing in all coordinates at a rate that is uniformly (i.e., independently of z) bounded away from 0. The lemma is true only because H is a polyhedron; for example, in Figure 4, as the points z approach the x^1-axis, the rate of increase of x^1 along the dotted lines tends to 0. Indeed, there are counter-examples to Theorem 5 if it is not assumed that H is a

---

[3] A previous published version [31] has the disadvantage that some of the $v(S)$ are empty.

polyhedron [3]. The proof of our lemma is given in [3], and will not be repeated here.

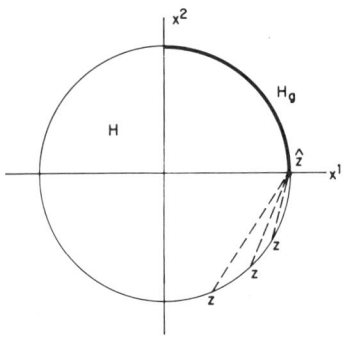

FIGURE 4

Let $x \in C(H_g)$. We will suppose that $x \notin C(H)$, i.e., x is dominated via some S by an element y of H, and will then construct an element $\hat{z}$ of H_g which also dominates x; this will be a contradiction. Roughly, this is done by taking an element z very close to x on the line segment $\overline{xy}$ which joins x and y, and constructing the corresponding $\hat{z}$. Now either

i) $\hat{z}$ is far from z, or

ii) $\hat{z}$ is close to z.

In the first case, it follows from the lemma that all the coordinates of $\hat{z}$ must be considerably greater than those of z; since z is close to x, it follows that $\hat{z} > x$, contradicting $x \in H_g$. In the second case, it follows that $\hat{z}$ is close to x. Hence from $y \succ_S x$ we deduce that $y \succ_S \hat{z}$, and hence it follows that $\hat{z}^S \in v(S)$ (from property (2) of the characteristic function). But $\hat{z}^S > z^S > x^S$, and therefore $\hat{z} \succ_S x$, which gives us the desired contradiction.

More precisely, we suppose without loss of generality that $x = 0$. Let $\sigma = \min_{i \in S} y^i$, where $y \succ_S x$. Let $z \in \overline{yx}$ be such that $|z| < \delta/(M+1)$. Then $\hat{z}^i - z^i > \|\hat{z} - z\|/M$ for all i. Hence if $\|\hat{z} - z\| \geq \delta M/(M+1)$, then $\hat{z}^i - z^i > \delta/(M+1)$. Then

$$\hat{z}^i = \hat{z}^i - z^i + z^i \geq \hat{z}^i - z^i - \|z\| > \delta/(M+1) - \frac{\delta}{(M+1)} = 0,$$

contradicting $0 \in H_g$. Hence $\|\hat{z} - z\| < \delta M/(M+1)$. Hence for all i, $\hat{z}^i = \hat{z}^i - z^i + z^i \leq \|\hat{z} - z\| + \|z\| < \delta M/(M+1) + \delta/(M+1) = \delta$. Hence for $i \in S$, $\hat{z}^i - y^i \leq \max_i \hat{z}^i - \min_i y^i < \delta - \delta = 0$. Hence $\hat{z}^S < y^S$. Hence $\hat{z}^S \in v(S)$, and $\hat{z} \succ_S x$, contradicting $x \in C(H_g)$.

Under the second of the two assumptions, Theorem 6 was proved by Burger [7]; the proof is simpler than under the first assumption. Burger's paper is the first to make systematic use of the assumption $H = v(N)$; this makes for a considerably simpler theory.

When is the core of a game non-empty? In the N-M case, a necessary and sufficient condition for the non-emptiness of the core has been given by Shapley [51], in terms of "balanced" collections of coalitions. A similar condition has been given (independently) by Bondareva [35]. Using this notion of balanced collections, Scarf [27] recently obtained a sufficient condition for the non-emptiness of the core in the no-side-payment case as well.

For each $S \subset N$, define a vector e_S in E^N by

$$e_S^i = \begin{cases} 1 & \text{if } i \in S; \\ 0 & \text{if } i \notin S. \end{cases}$$

A collection $\mathscr{S}$ of subsets S of N is called *balanced* if it is possible to assign to each S in $\mathscr{S}$ a non-negative number δ_S, such that

$$\sum_{S \in \mathscr{S}} \delta_S e_S = e_N.$$

For example, if $N = \{1, 2, 3\}$, then $\mathscr{S} = \{\{1, 2\}, \{2, 3\}, \{1, 3\}\}$ is balanced, where the δ_S are given by $\delta_{\{1,2\}} = \delta_{\{2,3\}} = \delta_{\{1,3\}} = \frac{1}{2}$.

Scarf's theorem may now be stated as follows:

THEOREM 7. *Let $H = v(N)$. Assume that for every balanced collection $\mathscr{S}$ of subsets of N, we have*

$$\bigcap_{S \in \mathscr{S}} (v(S) \times E^{N-S}) \subset v(N).$$

Then the core is non-empty.

The importance of this theorem may be illustrated by the fact that it implies the existence of a competitive equilibrium in a market (cf. §8); since the proof of Theorem 7 is "elementary" in the sense that it does not involve fixed point theorems, it follows that the existence of competitive equilibria may also be given an "elementary" proof.

6. VALUE

To motivate the notion of "value" as used in game theory, we can do no better than quote Shapley [49]: "At the foundation of the theory of games is the assumption that the players of a game can evaluate, in their utility scales, every 'prospect' that might arise as a result of a play.... One would normally expect to be permitted to include, in the class of 'prospects,' the prospect of having to play a game. The possibility of evaluating games is therefore of critical importance."

The value problem has been treated for games with side payments and transferable utility (the N-M case) by Shapley [49] and Selten [48], and in the no-side-payment case by Nash [21, 22], Harsanyi [9, 11, 12], Isbell [17], Miyasawa [20], and Shapley [28]. In all treatments the value assigns to each game one (or sometimes more than one) payoff vector. Often the treatment proceeds from the normal rather than the characteristic function form, and in at least one case [48] it proceeds direct from the extensive form. Here we will confine ourselves to discussing the Shapley value, for games in characteristic function form.

The Shapley value was defined for the N-M case in [49], as an imputation satisfying a certain set of axioms. It was proved to be unique, and was shown to have the following probabilistic interpretation: The Shapley value of player i is the expected value of the random variable $f(S \cup \{i\}) - f(S)$, where S is the set of players "before" i in a random ordering of all the players, and f is the N-M characteristic function. This definition clearly depends on a numerical value for f, and it is not at all clear how it can be generalized to a no-side-payment characteristic function as defined in §3.

Very recently Shapley [28] succeeded in giving an elegant definition of his values in the no-side-payment case by means of a reduction to the N-M case. His procedure is as follows: Let us be given a no-side-payment game with characteristic function v, and let us *imagine* what would happen if we were to allow side payments. We would then obtain an N-M game, and this would have a Shapley value, say $w = (w^1, \ldots, w^n)$. If it happens that $w \in v(N)$, i.e., that the players can attain w without actually making side payments, then w would be an excellent candidate for the Shapley value of the original game. Suppose that we now re-scale the original game, i.e., multiply the payoff of each player i by some non-negative constant p^i, and *then* allow side payments. We would then obtain another N-M game (generally different from the one discussed above), and this too would have a Shapley value. Shapley proved that for an appropriate choice of the scaling factors p^i, the Shapley value of the resulting N-M game is attainable by the players in the original (but re-scaled) no-side-payment game without actually making side payments. The scaling factors can then be eliminated and a Shapley value for the original (unscaled and no-side-payment) game results. We use the indefinite article advisedly; the value is no longer unique, because a number of different sets of scaling factors may yield attainable outcomes.

To simplify the formal description, we adopt the convention that if x and y are vectors, then xy denotes the vector whose ith coordinate is $x^i y^i$. For each vector $p \geq 0$, define a characteristic function v_p from the given one v by

$$v_p(S) = \{y \in E^S : \text{there is an } x \in v(S) \text{ such that } y \leq p^S x\}.$$

It may be verified that v_p satisfies the axioms for a characteristic function.

Now define an N-M characteristic function f_p by

$$f_p(S) = \max \left\{ \sum_{i \in S} y^i : y \in v_p(S) \right\}; \tag{5}$$

it is not difficult to show, by using (4), that the maximum is attained. Let $w(p)$ be the Shapley value for f_p. Then a pair (p, w) is called a *valuation* of the original characteristic function v if $p \neq 0$, $w(p) = pw$, and $w \in v(N)$. Shapley's theorem is:

THEOREM 8. *Every game has a valuation.*

The first investigation of what amounts to a cooperative game without side payments is due to Nash [21]; Nash's "bargaining problem" is the same thing as a 2-person game in characteristic function form, in the sense of §3. Each such game has a unique valuation (p, w), in which w is the Nash solution. Like the Nash solution in the 2-person case, the valuation in the general case is derivable from a small number of abstract axioms [28].

7. THE BARGAINING SET M_1^i

In the context of the bargaining set, the object of interest is not a payoff vector, but a *payoff configuration*, i.e., a pair consisting of a payoff vector and a coalition structure (partition of the players into disjoint coalitions). Furthermore the possibility that certain coalitions are "forbidden" (for example because of legal restrictions or communication difficulties) is admitted. For the N-M case, M_1^i is defined elsewhere in this volume [37] as a set of payoff configurations enjoying certain stability properties. Peleg [46] has proved that in the N-M case it is non-empty for each choice of a coalition structure, i.e., for each coalition structure there is a payoff vector such that the resulting pair is stable in the required sense. Unlike the situation for Shapley values, it is here quite easy to generalize the definition of M_1^i to the no-side-payment case, and in fact this can be done in two ways; the more appropriate of the two generalizations is denoted $\tilde{M}_1^i$. However, the existence theorem does not generalize.

THEOREM 9. *There is a 4-person game for which $\tilde{M}_1^i$ is empty (for appropriate choice of coalition structure).*

The example is due to Peleg [24]. A positive result obtained by Peleg in the same paper is:

THEOREM 10. *In a game in which only 2-player coalitions are permitted, $\tilde{M}_1^i$ is non-empty for each coalition structure.*

The proof makes use of the Eilenberg-Montgomery fixed-point theorem [38].

8. GAMES WITH SIDE PAYMENTS BUT WITHOUT TRANSFERABLE UTILITIES

Consider a game given by an N-M characteristic function f, i.e., a numerical function defined on the set of all subsets of N satisfying the super-additivity condition

$$f(S \cup T) \geq f(S) + f(T) \quad \text{for} \quad S \cap T = \phi.$$

Give this game the following interpretation: Each coalition S may go to a "referee" and receive exactly $f(S)$ dollars, on condition that it has agreed beforehand on how this money should be divided.

For each player i, let $u_i(b)$ be[4] the utility of player i for b dollars, in the sense of N-M ([45, pp. 617–632]; also [19, pp. 12–38]). We will assume that u_i is bounded,[5] continuous, and strictly increasing in b. Define a function u from E^N to itself by

$$u^i(x) = u_i(x^i) \tag{6}$$

for all $x \in E^N$ and $i \in N$. Let v' be the function defined on the subsets S of N by

$$v'(S) = \left\{ y \in E^S : \text{There is an } x \text{ in } E^N \text{ such that} \right.$$
$$\left. \sum_{i \in S} x^i = f(S) \text{ and } y \leq u^S(x) \right\}. \tag{7}$$

Intuitively, $v'(S)$ is the set of payoff vectors, *expressed in terms of utilities*, that are attainable by the coalition S. However, v' is not a characteristic function in the sense of §3, because $v'(S)$ may fail to be convex. We therefore replace $v'(S)$ by its convex hull; intuitively, this means that the coalition S will in general agree on a *lottery* that will determine the division of the $f(S)$ dollars, rather than agreeing on a specific division. We thus define a function v by

$$v(S) = \text{convex hull } v'(S). \tag{8}$$

Then v satisfies conditions (1)–(4) (where for convenience we take $H = v(N)$; it is of course neither compact nor polyhedral).

Suppose now that in the original game f, we exclude the possibility that the players will use lotteries to divide the payoffs. In that case the utility functions of the players become irrelevant, because their purpose is to represent preferences between *lotteries*. To represent preferences between actual sums of money (as distinguished from lotteries over such sums), utilities are not needed, as the dollar amount is a perfectly good measure for this purpose. And in fact, the reasoning leading to the N-M solution is then valid when the payoffs are expressed in money. Therefore we may calculate the N-M solutions (or the core, bargaining set, ψ-stable payoff configurations, and so on) of the characteristic function f as given, expressing the result in dollar terms, and the intuitive validity of the result is not

---

[4] u_i is determined only up to an additive and a positive multiplicative constant. These constants may be chosen independently for the various players (indeed there is no meaningful way of correlating them).

[5] The boundedness assumption is not strictly necessary but simplifies the discussion considerably; moreover it is intuitively very acceptable (cf. Isbell [16], p. 360).

based on any consideration of "linear utility," "transferable utility," "comparable utility," or indeed any utility whatsoever.

All this is based on the assumption that probabilistic considerations are for some reason excluded. If they are admitted, then utilities become relevant and indeed crucial; we must therefore replace the function f by the function v defined in (7) and (8), and use the corresponding definition of solution (§4). The question then arises: What is the relation, if any, between the solutions to f and the solutions to v?

THEOREM 11. *If the utility functions u_i are concave, then a subset F of E^N is a solution to f if and only if $u(F)$ is a solution to v.*

Theorem 11 asserts that if the utility functions are concave, then the same utility distributions—and so also the same money distributions—result when the characteristic function of §3 is used rather than the original N-M characteristic function. It follows that for the validity of solution theory as described in [45] it is not necessary to assume that utilities are *linear* in money, as is usually supposed,[6] but only that they are concave. The concavity assumption is an eminently reasonable one, and is often made in the literature.

The theorem is intuitively not surprising, because concave utilities mean that the players never prefer a gamble to its expectation, and hence the function v does not offer them different possibilities than the function f. The proof is very simple. From the concavity of the u_i it follows that $v' = v$, and hence u is a *domination-preserving* 1-1 correspondence from the space I_{ig} of imputations in the game f onto the corresponding space H_{ig} for v. Theorem 11 follows from this property of u.

It is rather curious that for simple games f (i.e., f taking the values 0 and 1 only), a result superficially similar to Theorem 11 holds in the diametrically opposed case, when the utility functions are all convex (i.e., the players always like a gamble at least as well as its expectation). Assume the utility functions are normalized so that $u_i(0) = 0$ and $u_i(1) = 1$. Then (in general) $u(I_{ig}) \neq H_{ig}$, and hence u does not provide a correspondence between I_{ig} and H_{ig}; but I_{ig} and H_{ig} are formally equal, and the *identity* is a domination preserving 1-1 correspondence between them. Here again, the result is easy to understand intuitively: a coalition of these gamblers will never split the money, always preferring a lottery in which one member gets all with a certain probability; the probabilities in the solutions to v then correspond to sums of money in the solutions to f.

Is there *always* a domination-preserving 1-1 correspondence between I_{ig} and H_{ig}? The answer is no. Consider the 3-person simple majority game ($f(S) = 0$ or 1 according as S has one or more members). Let the utility functions be the piecewise linear functions graphed in Figure 5.

---

[6] cf. the quotation from Luce and Raiffa in §2.

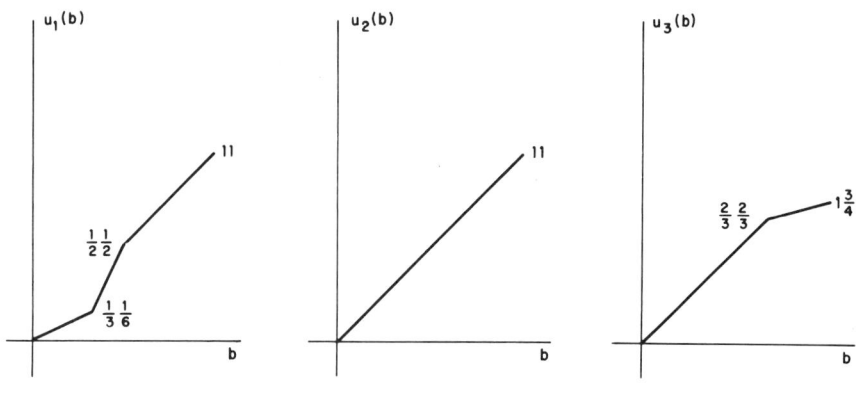

FIGURE 5

H_{ig} is pictured in Figure 6. A 1-1 domination-preserving mapping from H_{ig} onto I_{ig} would have to take both the points $(\frac{1}{6}, \frac{1}{3}, \frac{1}{2})$ and $(\frac{1}{4}, \frac{1}{4}, \frac{1}{2})$ of H_{ig} onto the point $(\frac{1}{4}, \frac{1}{4}, \frac{1}{2})$ of I_{ig}, an absurdity (parentheses and commas are omitted in the figures and henceforth in the text).

We close this section with an example of what happens when a pessimist (concave utility) plays a simple majority game with two optimists (convex

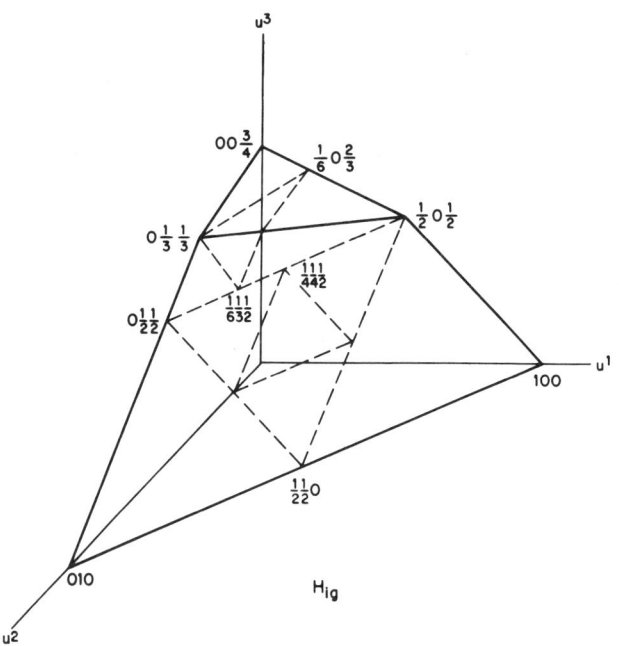

FIGURE 6, PART I

A Survey of Cooperative Games without Side Payments

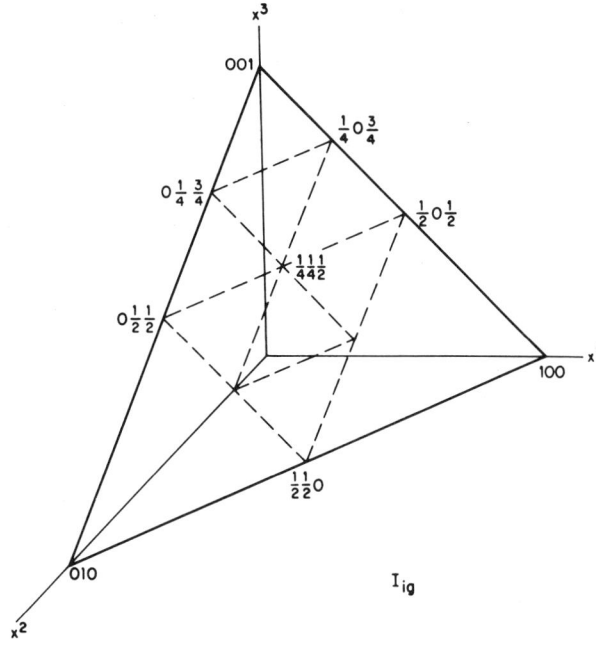

FIGURE 6, PART II

utility). Let f be as in the previous paragraph, and let the utility functions be as graphed in Figure 7. The 3-point solution of f is $\frac{1}{2}\frac{1}{2}0, \frac{1}{2}0\frac{1}{2}, 0\frac{1}{2}\frac{1}{2}$, it being understood that the coordinates of the vectors in the solution are expressed in dollars. This solution applies when lotteries are excluded. When lotteries are admitted, we must pass from f to v. $u(I_{ig})$ is pictured in Figure 8; it may be seen that H_{ig} is formally equal to I_{ig}. Hence v also has the 3-point solution $\frac{1}{2}\frac{1}{2}0, \frac{1}{2}0\frac{1}{2}, 0\frac{1}{2}\frac{1}{2}$, but this time the coordinates of the

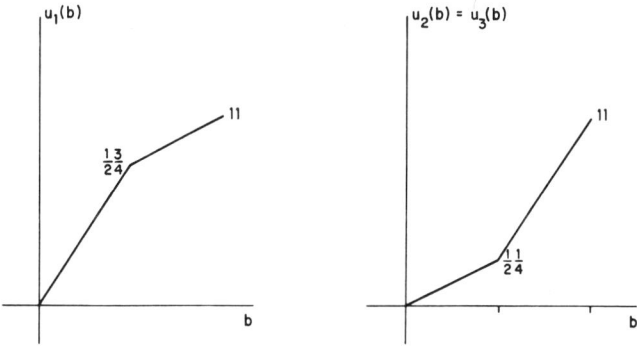

FIGURE 7

vectors are expressed in utility units rather than in dollars. If we translate back into dollars, we find that the solution is $1\frac{1}{4}\frac{3}{4}0$, $\frac{1}{4}0\frac{3}{4}$, $0\frac{3}{4}\frac{3}{4}$. The point $0\frac{3}{4}\frac{3}{4}$ cannot be attained by a distribution of money, but is attained by a $\frac{1}{2}$-$\frac{1}{2}$ lottery for $f(23)$ between players 2 and 3. However, $1\frac{1}{4}\frac{3}{4}0$ *is* attainable by a distribution of the sum $f(12)$, without recourse to lotteries. It follows that if the coalition 12—consisting of the pessimist and an optimist—forms,

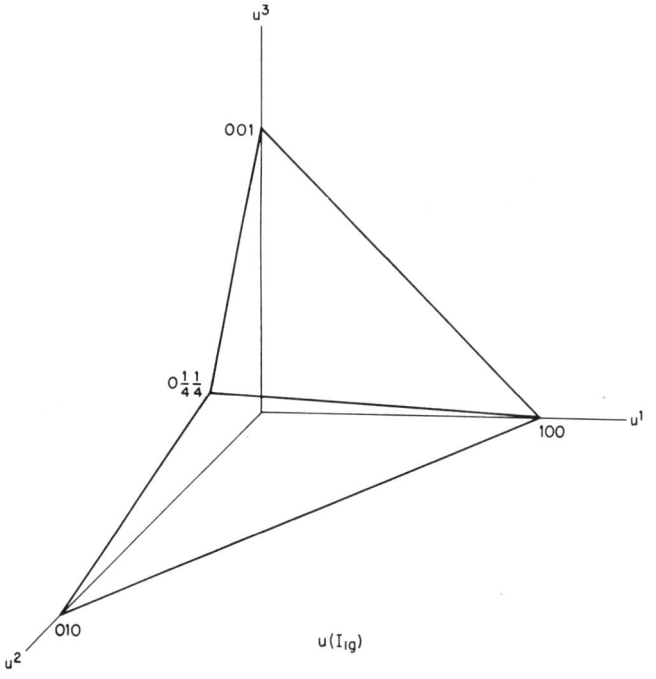

FIGURE 8

then the willingness of the optimist to take risks puts the pessimist at a very considerable material disadvantage, even though in the end no risks are taken by either player.

If *all* the points of the solution of v had been attainable without recourse to lotteries, then this phenomenon would not have occurred. This follows from the fact that in an arbitrary n-person game, *if a solution V of v is included in $u(I_{ig})$, then $u^{-1}(V)$ solves f.* The proof is an easy consequence of Theorem 1.

The results of this section, hitherto unpublished, are the outcome of conversations between M. Maschler and the author. Though they are not deep, they shed light on the relation between utilities and n-person games.

9. MARKET GAMES

Recently a good deal of attention has been paid to market games, which are in fact cooperative games without side payments, essentially in characteristic function form. The game-theoretic tool most significant in this connection is the core; it has been shown that for markets with "many" traders (this notion has been formalized in a number of different ways) the notion of core is essentially equivalent to that of Walrasian "competitive equilibrium." Under certain conditions it appears that Shapley's valuation (see §6) is closely connected with the competitive equilibrium as well. Finally, Scarf's core-theorem (Theorem 7) yields the non-emptiness of the core of a market game under wide conditions, and by using this an "elementary" existence proof for the competitive equilibrium can be obtained. For details, the reader is referred to the original papers (Debreu and Scarf [8], Aumann [4, 34], Vind [32], Shapley [28], Scarf [27]).

10. THE NORMAL FORM

The passage from the normal to the characteristic function form is not without its pitfalls. Even in the case of games with side payments and transferable utility (the N-M case), it is not generally agreed that the characteristic function as derived from the normal form by von Neumann and Morgenstern adequately represents the game; this is chiefly because for games that are not constant-sum, it does not always take adequate account of threats. Nevertheless the N-M definition is useful for some purposes, and we now examine how it can be generalized to the no-side-payment game.

In the N-M case, if f is the characteristic function and S is a coalition, then $f(S)$ is the maximum amount that S can guarantee itself; by the minimax theorem $N - S$ can prevent S from getting more. Here these two approaches—what S can guarantee itself, and what $N - S$ can prevent—are no longer equivalent. We write down definitions corresponding to both approaches.[7]

$v_\alpha(S) =$ the set of all payoff S-vectors x such that S can guarantee that it will get at least x.

$v_\beta(S) =$ the set of all payoff S-vectors x such that $N - S$ cannot prevent S from getting at least x.

By "getting at least x" we mean getting an amount that is at least x^i for each player i; and by "can guarantee" or "can prevent" we mean the existence of a single (correlated) strategy that guarantees or prevents, independent of the actions of the other players.

[7] These definitions were first explicitly given in [5].

When side payments are allowed,
$$v_\alpha(S) = v_\beta(S) = \left\{ x \in E^S : \sum_i x^i \leq f(S) \right\},$$
where $f(S)$ is the N-M value of S. That v_α and v_β are in general different is seen by means of the 3-person game,[8]

1, −1	0, 0
0, 0	−1, 1

where the rows represent strategies of the coalition $\{1, 2\}$, the columns represent strategies of player 3, and the entries are payoff $\{1, 2\}$-vectors (the payoffs to player 3 are irrelevant). A picture of $v_\alpha(\{1, 2\})$ and $v_\beta(\{1, 2\})$ is given in Figure 9.

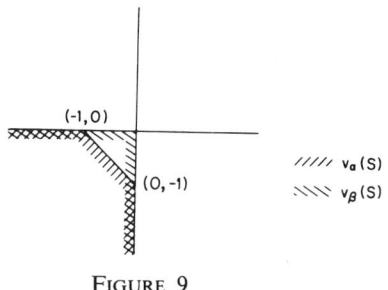

FIGURE 9

Whether the α-notion or the β-notion is preferable is a matter of taste. Both satisfy the axioms for characteristic functions [3]; that is an advantage of the axiomatic treatment. The α-notion seems to be intuitively more appealing, but as we shall see the β-notion has a certain technical advantage.

Some authors have considered the discrepancy between $v_\alpha(S)$ and $v_\beta(S)$ to be a disturbing phenomenon. We noted above that $v_\alpha(S) = v_\beta(S)$ for games with transferable utility side payments. Jentzsch [18] has investigated the possibility of obtaining a wider class of games with the same property. The general tenor of his result is negative, i.e., v_α and v_β cannot be expected to coincide unless one is talking about games that to start with are very similar to games with transferable utility side payments. To give a more precise description of his result, let us for the moment fix attention on a single coalition S. If we are interested in this coalition only, then in the (transferable utility) side-payment case we can substitute the following "adjusted normal form" for the usual, normal form: The "adjusted normal form" is a matrix whose rows are pure strategies of S,

---

[8] See Jentzsch [18] and Aumann [3].

whose columns are pure strategies of $N - S$, and whose entries are the total payoff to S for the appropriate strategy n-tuples. What this really means is that after S and $N - S$ have chosen strategies, S can pick any vector whose sum does not exceed the entry (the redistribution is made possible by the side payments). Use of mixed (i.e., correlated) strategies on the parts of S and $N - S$ will yield a numerical payoff for S which is the appropriate mixture of the pure payoffs, and which S can also allocate between its members as it sees fit. It is exactly the "value" of this adjusted normal form, when considered as a 2-person 0-sum game, that gives the N-M characteristic function $f(S)$.

The adjusted normal form can be generalized to cover games with nontransferable utility side payments. Suppose that a pair (p, q) of strategies (of S and $N - S$) yields a payoff S-vector x. By the use of side payments, S can redistribute the income from x among its members, but because utility is not assumed to be transferable, total utility will not be conserved in the redistribution. The set of all payoff S-vectors that can be obtained from x by means of redistributions of this kind will be a subset of E^S that satisfies conditions (1) and (2) for characteristic functions (closedness, convexity, unboundedness towards the southwest); such subsets of E^S will be called S-catalogues, or simply *catalogues*.[9] The adjusted normal form for games with nontransferable utility side payments is thus a matrix whose entries are catalogues rather than numbers. As in the previous case, it is possible to consider mixed outcomes, corresponding to mixed strategies: a mixture of two catalogues (with specified probabilities) is simply the set of mixtures of its members (with the same probabilities), and is itself a catalogue.

If we now restrict the catalogues in the entries to be half-spaces of the form $\Sigma x^i \leq k$, then we are back in the transferable utility case, and it follows that $v_\alpha(S) = v_\beta(S)$. Jentzsch asked whether we could not still assure $v_\alpha(S) = v_\beta(S)$ by imposing a weaker restriction on the catalogues that appear in the adjusted normal form. More precisely, consider a family $\mathscr{F}$ of S-catalogues; let us call a game *adjusted to* $\mathscr{F}$ if in its adjusted normal form, all payoffs—including the mixed ones—are in $\mathscr{F}$. Then what conditions must be placed on $\mathscr{F}$ in order to ensure that for every game that is adjusted to $\mathscr{F}$, we have $v_\alpha(S) = v_\beta(S)$?

Let us call an $\mathscr{F}$ for which this holds *determinate*. Jentzsch showed that the condition of being determinate is a very strong one:

THEOREM 12. *Let us call a catalogue F regular if it has a supporting hyperplane in each positive direction,[10] and assume that every member of*

---

[9] Presumably because you can pick from them whatever you want.

[10] I.e., for each vector x with positive coordinates, there is a hyperplane that supports $\mathscr{F}$ and is perpendicular to x. Figure 2 illustrates a regular catalogue, and Figure 1 illustrates one that is not regular.

$\mathscr{F}$ is regular. *If $\mathscr{F}$ is determinate, then of every three members of $\mathscr{F}$, one is a probability mixture of the other two (and in particular lies between the other two).*

The theorem does not apply directly to the N-M case, because half-spaces are not regular. But in all practical cases side payments are limited, and this makes the catalogues regular and the theorem applicable.

Since the motivation for this work derives from an analysis of games with side payments but without transferable utility, it is natural to ask for conditions on the utility functions u_i of the players i in S that will lead to games for which[11] $v_\alpha(S) = v_\beta(S)$. Suppose, for example, that the utility function of each player for a (positive) amount of money b is that suggested by Bernoulli, namely $\log b$. Then if S has a total amount of money d to divide between its members (d positive), the resulting catalogue is

$$\left\{ y \in E^S \colon \sum_{i \in S} \exp(y^i) \leq d \right\},$$

where exp is the exponential function ($\exp(b) = e^b$). Jentzsch remarked that it can be shown from his results that the family of all such catalogues, as d varies, is determinate. This means that in side-payment games played by players all of whom have the utility function $\log b$, we have $v_\alpha(S) = v_\beta(S)$.

What other S-tuples of utility functions have this property? This question was answered by B. Peleg in [25]. It turns out that there are very few of them. His chief result is as follows: Suppose that the utility functions of the players i in S are concave and have the property that $v_\alpha(S) = v_\beta(S)$ in any game in which these players participate, providing that the players all have personal minima of 0 (i.e., $v_\alpha(\{i\}) = v_\beta(\{i\}) = (-\infty, 0]$). Then *either $u_i(b) = \log b$ for all i in S, or there is a λ obeying $0 < \lambda \leq 1$ such that for all i in S, $u_i(b) = b^\lambda$, or there is a λ obeying $\lambda < 0$ such that for all i in S, $u_i(b) = -b^\lambda$.* This underscores the fact that equivalence between the α- and the β-notions is a very rare event.

The concepts we have described may be applied to supergames, i.e., long sequences of plays of a cooperative game without side payments. We look for stable behavior in such games. There are two ways of approaching this problem. One is to treat the entire supergame as a single game, and use stability criteria appropriate for a single game. The other way is to speculate as to what kind of behavior in the individual plays constituting the supergame would lead to stability in the long run. Now if we are going to follow the first method, then one of the concepts we could use would be Nash's equilibrium point [44]. Recall that this is a strategy n-tuple, or

---

[11] This question is not connected with that investigated in §6, where the characteristic function was given and there was no question of mixed strategies and of the difference between v_α and v_β.

"point," such that no individual can gain by deviating from it, while the others retain the same strategies they were previously using. Since we are discussing cooperative games, it would be more appropriate to consider a point such that no *coalition* can gain by deviating from it while the others retain the same strategies. Let us call such a point a *strong* equilibrium point.

Let us now try the other approach. If a coalition is expecting more plays, then the question of whether it can improve its lot for fixed strategies of the other players becomes irrelevant, because the other players will not keep their strategies fixed, but will take counteraction on subsequent plays. The question is: when can a coalition be *sure* of a higher payoff? Obviously some kind of core notion is involved here; the surprising fact is that it is not the core according to the α-notion but rather according to the β-notion.

THEOREM[12] 13. *The β-core of a game coincides with the set of payoff vectors to strong equilibrium points in its supergame.*

This is the "technical advantage" of the β-notion to which we previously referred.

We outline the proof of this theorem. For this purpose we should first define the payoff in the supergame. But the precise definition is complicated and not important at the moment; the general idea is that the payoff to a superplay is some kind of average of the payoffs to the individual plays, and this is all that we shall need.

First suppose that x is in the β-core of the game (because of Theorem 5 we do not have to specify which one of the β-cores). We will build a strong equilibrium point whose payoff is x. Now it is possible to prove[13] from the definition of v_β and with a little fussing that to say $x \in \beta$-core is equivalent to saying that *each coalition can prevent its complement from getting more than it (the complement) does at x*.

This being the case, let us construct a strong equilibrium point as follows: First find a correlated strategy n-tuple whose payoff is x; call this c^N. Next, for each coalition S, let c^S *be a correlated strategy for S that prevents the complement of S from obtaining more than it does at x*. Now each player adopts the following strategy in the supergame: He starts out by proposing c^N for the first play, and continues to propose this, play after play, as long as the other players agree. If, however, there is a set of players that disagree—let us call them "disloyal" players—then our player will propose c^S, where S is the set of loyal players. Once a player has become disloyal, he will no longer be accepted in the set of loyal players. The result is that if everybody plays along with this equilibrium point, then x is the result, but if some set of players does not, then eventually

---

[12] First proved in [1]. For a description that is more precise than the present one and more readable than that of [1], see the end of [3]; but no proof is sketched there.
[13] Cf. Lemma 9.1, p. 304 of [1].

each of its members will be found in the disloyal set; from then on all the loyal players S will be playing c^S against the disloyal players, and by the definition of c^S at least one disloyal player i will get no more than x^i. This shows (modulo some glossed-over difficulties) that the β-core is a subset of the strong equilibrium payoffs.

Conversely, suppose that x is *not* in the β-core; then it is β-dominated. This means that there is a coalition S and a $y \in E^S$ such that $N - S$ cannot prevent S from getting y, and $y > x^S$. Suppose there were a strong equilibrium point f in the supergame whose payoff is x. Let f^{N-S} denote the part that $N - S$ has in f, i.e., an $(N - S)$-tuple of strategies in the supergame, one for each member of $N - S$. On the first play f^{N-S} dictates a certain set of strategies in $N - S$. For this set of strategies, there exists a strategy for S which yields at least y (this is what is meant by saying that $N - S$ cannot prevent S from getting at least y). On the second play, f^{N-S} again dictates a strategy set for $N - S$, based on the history of the previous play. For this strategy, there again exists a strategy for S that yields at least y; it may be different from the strategy of S on the first play. We can continue in this way; no matter what f^{N-S} dictates, there exists a strategy for S that yields at least y on every play. Since $y > x^S$, this shows that by deviating from f, S can gain, so f cannot be a strong equilibrium point.

This last part of the proof has a curious flavor, because of course S cannot know which supergame strategy $N - S$ is using. However, it does definitely prove that f cannot be in equilibrium, which really involves nobody wanting to deviate even if he knows what the others are playing. Theorem 13 is related to Blackwell's work on games with vector payoffs [6].

The Zermelo-von Neumann-Kuhn theorem about pure-strategy equilibrium points in games of perfect information has the following analogue for supergames of cooperative games with side payments.

THEOREM 14. *If the supergame of a game of perfect information has any strong equilibrium points at all, then it also has strong equilibrium points which involve only pure strategies.*[14]

In §6, we discussed the close connection between no-side-payment characteristic functions and Nash's bargaining problem [21], and pointed out that Shapley's valuation generalizes Nash's solution to the bargaining problem. Nash followed up the work in [21] by a paper on 2-person games in normal form [22], for which he proposed a "value" taking threat possibilities into account. This work was generalized by Raiffa in [26], but he too treated only 2-person games. The problem of defining a value for n-person games (both with and without side payments) that will take threats into account has been treated by several authors (cf. §6). Isbell [16] has constructed a theory of games without side payments that parallels the N-M solution theory but takes threat possibilities into account.

[14] For the precise statement and proof see [2].

11. HISTORICAL REMARKS

Shapley and Shubik [29] were the first to suggest that N-M solutions could be defined even in the absence of a transferable utility. Their definition of dominance is very similar to ours; but rather than explicitly using a characteristic function, it depends directly on the α-notion. Shapley and Shubik imply that they must have (nontransferable utility) side payments to make their definition work, but actually it is perfectly general. They proved no theorems, confining themselves to general definitions.

Luce and Raiffa [19, p. 234] also gave a definition of dominance and solution for cooperative games without side payments. Their definition has some restrictive and complicating features, which in the light of later work turn out to have been unnecessary. They, too, proved no theorems, mentioning that "next to nothing is known about these definitions."

Functions that are very similar in form to the characteristic functions of §3 were first described by Isbell [16, 17]. He called them *end-games*, and used them to characterize, for each given payoff vector, the redistributions of utility that are made possible by means of nontransferable utility side payments. This use is related to Jentzsch's catalogues rather than to the development of §3; the latter is due to Aumann and Peleg [5]. The form of v in this survey differs slightly from that in [5]; the $v(S)$ of [5] would be $v(S) \times E^{N-S}$ in the notation of this paper.

Other historical references may be found in the body of the paper.

12. A Bibliography of Cooperative Games Without Side Payments

[1] R. J. Aumann, "Acceptable points in general cooperative n-person games," in [43], pp. 287–324 [Chapter 20].

[2] ——, "Acceptable points in games of perfect information," *Pac. J. Math.* 10 (1960), pp. 381–417 [Chapter 21].

[3] ——, "The core of a cooperative game without side payments," *Trans. Amer. Math. Soc.* 98 (1961), pp. 539–552 [Chapter 39].

[4] ——, "Markets with a continuum of traders," *Econometrica* 32 (1964), pp. 39–50 [Chapter 46].

[5] R. J. Aumann and B. Peleg, "Von Neumann-Morgenstern solutions to cooperative games without side payments," *Bull. Amer. Math. Soc.* 66 (1960), pp. 173–179 [Chapter 38].

[6] D. Blackwell, "An analog of the minimax theorem for vector payoffs," *Pac. J. Math.* 6 (1956), pp. 1–8.

[7] E. Burger, "Bemerkungen zum Aumannschen Core-Theorem," *Zeitschrift für Wahrscheinlichkeitstheorie* 3 (1964), pp. 148–153.

[8] G. Debreu and H. Scarf, "A limit theorem on the core of an economy," *Int. Econ. Rev.* 4 (1963), pp. 235–246.

[9] J. C. Harsanyi, "A bargaining model for the cooperative n-person game," in [43], pp. 325–356.

[10] ——, "Measurement of social power in n-person reciprocal power situations," *Behavioral Science* 7 (1962), pp. 81–91.
[11] ——, "A simplified bargaining model for the n-person cooperative game," *Int. Econ. Rev.* 4 (1963), pp. 194–220.
[12] ——, "Approaches to the bargaining problem before and after the theory of games: a critical discussion of Zeuthen's, Hicks' and Nash's theories," *Econometrica* 24 (1956), pp. 144–157.
[13] ——, "On the rationality postulates underlying the theory of cooperative games," *J. Conflict Resolution* 5 (1961), pp. 179–196.
[14] ——, "Bargaining in ignorance of the opponent's utility function," *J. Conflict Resolution* 6 (1962), pp. 29–38.
[15] ——, "A bargaining model for social status in informal groups and formal organizations," *Behavioral Science* 11 (1966), pp. 357–369.
[16] J. R. Isbell, "Absolute games," in [43], pp. 357–396.
[17] ——, "A modification of Harsanyi's bargaining model," *Bull. Amer. Math. Soc.* 66 (1960), pp. 70–73.
[18] G. Jentzsch, "Some thoughts on the theory of cooperative games," in [36], pp. 407–442.
[19] R. D. Luce and H. Raiffa, *Games and Decisions*, New York, John Wiley, 1957.
[20] K. Miyasawa, "The n-person bargaining game," in [36], pp. 547–575.
[21] J. F. Nash, "The bargaining problem," *Econometrica* 18 (1950), pp. 155–162.
[22] ——, "Two-person cooperative games," *Econometrica* 21 (1953), pp. 128–140.
[23] B. Peleg, "Solutions to cooperative games without side payments," *Trans. Amer. Math. Soc.* 106 (1963), pp. 280–292.
[24] ——, "Bargaining sets of cooperative games without side payments," *Israel J. Math.* 1 (1963), pp. 197–200.
[25] ——, "Utility functions of money for clear games," *Nav. Res. Log. Quart.* 12 (1965), pp. 57–63.
[26] H. Raiffa, "Arbitration schemes for generalized two-person games," in [42], pp. 361–387.
[27] H. Scarf, "The core of an n-person game," *Econometrica* 35 (1967), pp. 50–69.
[28] L. S. Shapley, "Values of large market games: Status of the problem," The RAND Corporation RM3957PR, February 1964.
[29] L. S. Shapley and M. Shubik, "Solutions of n-person games with ordinal utilities" (abstract), *Econometrica* 21 (1953), p. 348.
[30] R. Stearns, "Three-person cooperative games without side payments," in [36], pp. 377–406.
[31] ——, "On the axioms for a cooperative game without side payments," *Proc. Amer. Math. Soc.* 15 (1964), pp. 82–86.
[32] K. Vind, "Edgeworth-allocations in an exchange economy with many traders," *Int. Econ. Rev.* 5 (1964), pp. 165–177.

Other References

[33] R. J. Aumann, "Linearity of unrestrictedly transferable utilities," *Nav. Res. Log. Quart.* 7 (1960), pp. 281–284 [Chapter 16].
[34] ——, "Existence of competitive equilibria in markets with a continuum of traders," *Econometrica* 34 (1966), pp. 1–17 [Chapter 47].

[35] O. Bondareva, "The core of an n-person game," *Vestnik Leningrad Univ.* 17 (1962), pp. 141–142.
[36] M. Dresher, L. S. Shapley, and A. W. Tucker, editors, *Advances in Game Theory*, Ann. of Math. Study 52, Princeton, New Jersey, Princeton University Press, 1964.
[37] M. Davis and M. Maschler, "Existence of stable payoff configurations for cooperative games," in M. Shubik, editor, *Essays in Mathematical Economics in Honor of Oskar Morgenstern*, Princeton, New Jersey, Princeton University Press, 1967, pp. 39–52.
[38] S. Eilenberg and D. Montgomery, "Fixed point theorems for multi-valued transformations," *Amer. J. Math.* 68 (1946), pp. 214–222.
[39] D. B. Gillies, "Solutions to general non-zero-sum games," in [43], pp. 47–85.
[40] F. Harary and M. Richardson, "A matrix algorithm for solutions and r-bases of a finite irreflexive relation," *Nav. Res. Log. Quart.* 6 (1959), pp. 307–314.
[41] G. K. Kalisch and E. D. Nering, "Countably infinitely many person games," in [43], pp. 43–45.
[42] H. W. Kuhn and A. W. Tucker, editors, *Contributions to the Theory of Games II*, Ann. of Math. Study 28, Princeton, New Jersey, Princeton University Press, 1953.
[43] R. D. Luce and A. W. Tucker, editors, *Contributions to the Theory of Games IV*, Ann. of Math. Study 40, Princeton, New Jersey, Princeton University Press, 1959.
[44] J. F. Nash, "Non-cooperative games," *Ann. of Math.* 54 (1951), pp. 286–295.
[45] J. von Neumann and O. Morgenstern, *Theory of Games and Economic Behavior*, Princeton, New Jersey, Princeton University Press, 1944, Third Edition rev., 1953.
[46] B. Peleg, "Existence theorem for the bargaining set $M_1^{(i)}$," in M. Shubik, editor, *Essays in Mathematical Economics in Honor of Oskar Morgenstern*, Princeton, New Jersey, Princeton University Press, 1967, pp. 53–56; also, *Bull. Amer. Math. Soc.* 69 (1963), pp. 109–110.
[47] M. Richardson, "Solutions of irreflexive relations," *Ann. of Math.* 58 (1953), pp. 573–590.
[48] R. Selten, "Valuation of n-person games," in [36], pp. 577–626.
[49] L. S. Shapley, "A value for n-person games," in [42], pp. 307–317.
[50] ———, "Notes on the n-person game III: Some variants of the von Neumann-Morgenstern definition of solution," The RAND Corporation, RM817, April 1952.
[51] ———, "On balanced sets and cores," *Nav. Res. Log. Quart.* 14 (1967), pp. 453–460.

VIII COALITIONAL GAMES: BARGAINING SET, KERNEL, NUCLEOLUS

At some time in 1959 (probably in the spring), I gave a lecture at the mathematics colloquium of the Hebrew University in Jerusalem, at which I discussed the von Neumann–Morgenstern solution (or "stable set") for TU games. Michael Maschler, who at that time was a pure mathematician specializing in functions of a complex variable, was in the audience. At the end of the lecture, he raised his hand to criticize the definition. I responded in some way, but Mike was not satisfied; after the lecture, the discussion continued for a long time. Exasperated by his persistence, I finally asked whether he could propose a definition that is more satisfactory than von Neumann's. Unfazed, he proposed something; I don't remember what, but after some discussion, we agreed that it was no good. This continued for about a year; Mike would propose a definition, and I would shoot it down.[1] By the end of the year, a definition had emerged that he liked and I didn't. He wrote a paper, and asked me to be a coauthor. I refused; but Mike can be very persistent, and eventually I caved in. The result is chapter 42.

As it turned out, we were both right. The Bargaining Set proposed here—the set $\mathcal{M}$—is an ugly, unwieldy creature. Justifiably, it itself attracted little attention; it never "caught on." But the set $\mathcal{M}_1^{(i)}$, a variant of $\mathcal{M}$ proposed by Davis and Maschler[2] shortly after chapter 42 was published, *did* catch on in a big way; a beautiful and very extensive theory developed from it, including the theory of the kernel and nucleolus. The ugly duckling turned into a beautiful swan; chapter 42 became one of my most frequently cited publications, not for its own sake, but as the granddaddy of this rich and deep theory.

Chapter 43, joint with B. Peleg and P. Rabinowitz, deals with the computation of the kernel. At the time this was done, in the early sixties, computers had little power, and considerable ingenuity was required to devise a computational method that would be practical even for moderately sized games. Using this method, we tabulated the kernels of all weighted majority games up to five players, all extreme constant sum games with five players, and several other categories of games. One result was that the conjecture that kernels must be convex was settled in the negative. The tabulation gave impetus to subsequent deep theoretical

---

1. Of course, definitions are not "right" or "wrong." But it often happens that a definition that at first looks attractive fails to live up to its promise in one way or another; it may yield absurd results in some cases, or fail to express what it was meant to express, or fail to exist when one would have liked it to, and so on.

2. "Existence of stable payoff configurations for cooperative games," *Bull. Amer. Math. Soc.* 69 (1963), 106–108.

studies by Maschler and Peleg on the structure of the kernel,[3] and these, in turn, led to the discovery of the nucleolus by Schmeidler.[4]

Chapter 44, written jointly with J. Dreze, is not exclusively about the Bargaining Set, Kernel, and Nucleolus; it deals also with the Shapley value, core, and von Neumann–Morgenstern solution (or "stable set"). The paper studies what each of these six solution concepts predicts when applied to a "coalition structure" $\mathcal{B}$—a partition of the all-player set N into subcoalitions. In all six cases, one can characterize the solution for $\mathcal{B}$ by means of auxiliary coalitional worth functions defined on each of the coalitions S that constitute $\mathcal{B}$. The underlying idea of this paper foreshadows the notion of "consistency" that was to become the cornerstone of Sobolev's axiomatization of the nucleolus, of Peleg's axiomatizations of the core, kernel, and related concepts, and indeed of "consistency" in general.[5] It is this consistency notion that plays the central role in chapter 45—one of my own favorites—to which we now turn.

While on sabbatical at Stanford during 1980–81, I ran across a preprint by Barry O'Neill entitled "A Problem of Rights Arbitration from the Talmud."[6] It aroused my interest, and I sent a copy to my son Shlomo (subsequently killed in action in Operation Peace for Galilee), who was at that time studying at a Talmudical Academy in Jerusalem. In his response, Shlomo directed me to a fascinating Talmudic passage (Ketubot 93a) related to that studied by O'Neill, but far more subtle. The passage proposes a division of the assets of a bankrupt debtor who owes a total of 600 zuz to three creditors: 100 to one, 200 to another, and 300 to a third. Three separate cases are considered: when the assets total 100 zuz, 200 zuz, or 300 zuz. On the face of it, the proposed divisions seem wildly inconsistent, and indeed their logic is hard to fathom even when considered individually. For close to two millennia, the best Talmudic scholars did not succeed in coming up with a convincing explanation of this passage.

When I returned to Jerusalem, Mike Maschler and I decided to see whether we couldn't bring the modern theory of games to bear on the problem. We tried many different solution concepts,[7] but nothing worked.

---

3. "A characterization, existence proof and dimension bounds for the kernel of a game," *Pac. J. Math.* 18 (1966), 289–328, and "The structure of the kernel of a cooperative game," *SIAM J. Appl. Math.* 15 (1967), 569–604.

4. "The nucleolus of a cooperative game," *SIAM J. Appl. Math.* 17 (1969), 1163–1170.

5. See subsection (vi) of the 1970–1986 section of chapter 2.

6. Published in 1982 in *Mathematical Social Sciences* 2, 345–371.

7. A bankruptcy problem is not a game; before applying solution concepts, one must formulate an appropriate coalitional game. However, this was relatively easy to do in a natural way.

One day, when we were sitting together in my office, one of us said, "Let's try the nucleolus; it can't hurt." The calculation (by hand) took about twenty minutes, and the result was startling: the nucleolus yielded *precisely* the nine strange numbers appearing in the Talmud.

Of course, this was only the beginning. The nucleolus is one of the more sophisticated game-theoretic solution concepts. Astute as they were, it is highly unlikely that the sages of the Talmud had anticipated Schmeidler's discovery of the nucleolus; they simply did not have the necessary technical equipment. What *did* seem possible was that there is some underlying principle common to both the nucleolus and to the sages' way of thinking that determines the division of assets in these bankruptcy problems. Mike and I set ourselves the task of finding such a principle.

A good place to look for "principles" is an axiomatization. A few years earlier, A. I. Sobolev[8] had discovered an axiomatization of the nucleolus using the principle of "consistency." This was the key to the puzzle, though it took many more months to work out. In the end, we were able to explain the Talmudic division on the basis of reasoning which, while relatively sophisticated, was well within the reach of the sages of the Talmud, and fits nicely with other Talmudic principles.

For thousands of years, the commentators on the Talmud—who were much more astute than we—had been trying to make sense of this passage. Why do we think that we succeeded where they failed? The reason is that modern game theory gave us a tool that they didn't have. Though the solution can be stated without game theory, it would not have been *reached* without it.

There is a profound lesson here. Up to now, most of the implications of game theory and economic theory have been not quantitative but qualitative. For such an implication to be convincing, it is important that it be supportable in common sense terms. Game Theory is most satisfying when the formal analysis suggests new insights—insights that, while not obvious, do eventually make sense on the common sense, verbal plane. Chapter 45 is a prime example of that kind.

---

8. "Characterization of the principle of optimality for cooperative games through functional equations," in *Matematicheskie Metody v Socialnix Naukax*, N. N. Vorobiev, ed. (Vilnius: Vipusk 6, 1975), 94–151 (in Russian).

42 The Bargaining Set for Cooperative Games
with M. Maschler

1 Introduction

This paper grew out of an attempt to translate into mathematical formulas what people may argue when faced with a cooperative *n*-person game described by a characteristic function.

The basic difficulty in *n*-person game theory is due to the lack of a clear meaning as to what is the purpose of the game. Certainly, the purpose is not just to get the maximum amount of profits, because if every player will demand the maximum he can get in a coalition, no agreement will be reached. Thus, one decides that the purpose of the game is to reach some kind of stability, to which the players would or should agree if they want any agreement to be enforced. This stability should reflect in some sense the power of each player, which results from the rules of the game.

In this paper, we assume that all the players can "bargain" together, with perfect communication, and settle at a "stable" outcome which is based on the "threats" and "counter threats" that they possess. The set of all the stable outcomes, called the *bargaining set*,[1] is defined in Section 2 and some of its properties are discussed. In particular, this set can always be determined by solving systems of algebraic linear inequalities.

The bargaining sets for the 2- and 3-person games are fully described (Sections 3, 4, 5) and some cases of 4-person games are treated, in which not all the coalitions are permissible (Sections 6, 8).

Some counterexamples for various conjectures, as well as existence theorems, are treated in Section 7, and possible modifications are suggested in Sections 9 and 10. In Section 11 we discuss some similarities and deviations between our theory and other solution concepts, such as W. Vickrey's concept of self-policing patterns [8] and D. Luce's concept of ψ-stability [3]. We also outline how our theory can be modified if the game is given by Thrall's characteristic function [7] or by the Aumann-Peleg characteristic function for cooperative games without side payments.

We conclude by pointing out that "central parts" of the von Neumann-Morgenstern solution for some games appear also in the bargaining set. The reason for this phenomenon is obscure to us.

This chapter originally appeared in *Advances in Game Theory*, Annals of Mathematics Studies 52, edited by M. Dresher, L. S. Shapley, and A. W. Tucker, pp. 443–476, Princeton University Press, Princeton, 1964. Reprinted with permission.

This research was sponsored in part by the Office of Naval Research under Nonr-1858(16), and in part by the Carnegie Corporation of New York.

1. The definition of the bargaining set appeared in [1].

2 The Bargaining Set

We shall consider an n-person cooperative game Γ, described by its characteristic function. More precisely, a set $N = \{1, 2, \ldots, n\}$ of n players is given, together with a collection $\{B\}$ of non-empty subsets B of N, called *permissible coalitions*. For each B, $B \in \{B\}$, a number $v(B)$ is given and it is called *the value of the coalition B*.

For the sake of simplicity, we shall assume throughout this paper that all 1-person coalitions are in $\{B\}$ and have a zero value, i.e.,

$$i \in \{B\}, \quad v(i) = 0. \tag{2.1.a}$$

In addition, we shall also assume that

$$v(B) \geqslant 0, \quad B \in \{B\}. \tag{2.1.b}$$

It will turn out later that no essential change will occur if we add to $\{B\}$ all other, non-permissible coalitions, assigning them the value zero.

A *payoff configuration* (p.c.) will now be defined as an expression of the form

$$(x; \mathscr{B}) \equiv (x_1, x_2, \ldots, x_n; B_1, B_2, \ldots, B_m), \tag{2.2}$$

where $B_1, B_2, \ldots, B_m$ are mutually disjoint sets of $\{B\}$ whose union is N, i.e.,

$$B_j \cap B_k = \phi, \quad j \neq k; \quad \bigcup_{j=1}^{m} B_j = N; \tag{2.3}$$

and the x_i are real numbers which satisfy

$$\sum_{i \in B_j} x_i = v(B_j); \quad j = 1, 2, \ldots, m. \tag{2.4}$$

A p.c. is therefore a representation of a possible outcome of the game, in which the players divide themselves into the coalitions $B_1, B_2, \ldots, B_m$, each coalition shares its value among its members, and each player receives the amount x_i, $i = 1, 2, \ldots, n$.

When people are faced with such a game, each one trying to get as high an amount as he thinks he can get, it is reasonable to expect that some of the p.c.'s will never form. E.g., one does not expect that a p.c. will occur with $x_{i_0} < 0$, since the player i_0 alone can secure more by playing as a 1-person coalition. We are willing to make a strong assumption, namely, that the outcome (2.2) will be a *coalitionally rational p.c.* (c.r.p.c.),[2] i.e., for each B, $B \in \{B\}$, $B \subset B_j, j = 1, 2, \ldots, m$,

---

2. In R. J. Aumann and M. Maschler [1], a c.r.p.c. is called a p.c.

$$\sum_{i \in B} x_i \geq v(B). \tag{2.5}$$

Thus, we assume that a coalition will not form if some of its members can obtain more by themselves forming a permissible coalition.

The assumption of coalitional rationality differs from the assumption of belonging to the core by the restricting condition $B \subset B_j$. This restriction avoids some of the difficulties which arise when dealing with games whose core is empty. (See R. D. Luce and H. Raiffa [3].)

In itself, the coalitional rationality assumption is a very strong one, as it forces the game to be essentially superadditive within those coalitions which are actually formed. Indeed, a coalition whose value is less than the sum of the values of disjoint subcoalitions cannot occur in any c.r.p.c., and can as well be declared non-permissible or have its value be replaced by zero. Moreover, this assumption is open to the same theoretical objections which are discussed at length in R. D. Luce and H. Raiffa [3]. As a matter of fact, our theory can be developed without the coalitional rationality assumption, as indicated in Section 10. Nevertheless, as we are only interested in "stable" outcomes, we feel it instructive to make this assumption.

Several phenomena can be observed when watching people who are confronted with a game such as described above. Usually, negotiations start, each one tries to get at least as much as he expects, and at the same time there is an attempt to enter into a "safe" coalition. This latter factor applies, in particular, to those coalitions which are planned to operate for a long period. The search for "safety" gives rise to feelings of sympathy and antipathy which play an important role in the final decisions. Guarantees of all kinds are demanded, contracts are signed, etc. *If people do not feel safe enough, they often do not enter a coalition even if they can win more in it.*

The demand for safety is usually considered legitimate and a sound way to convince the partners to get a smaller amount of profit *in order that no one in the coalition will feel deprived.* There is a desire for "fair play," which can be achieved in various ways. Often it is accepted that "if all things are equal" it is "fair" to divide the profits equally. Sometimes people share the profits according to some fixed ratio established by other precedents, etc.

If "all things" are *not* equal, people will still be happy with their coalition if they agree that the "stronger" partners will get more. Thus, during the negotiations prior to the coalition formation, each player will try to convince his partners that in some sense he is strong. This he can try in various ways, among which an important factor is his ability to show that he has other, perhaps better, alternatives. His partners, besides pointing out their own alternatives, may argue in return that even without his help they can perhaps keep their proposed shares. Thus, a negotiation quite often takes the form of a sequence of "threats" and "counter threats," or "objections" against "counter objections." It is this principle that we shall try to formulate mathematically. It seems that a certain kind of stability is reached if all objections can be answered by counter objections.

Perhaps it is not enough that any objection by one person could be met. It is possible that a subset of the players of a coalition unite, during the negotiation period, and threaten another subset. If we insist on a strong stability, we have to take care also of such threats. This, in fact, will be done.

To be sure, there are other means used during the bargaining period, such as threats based on the so-called "interpersonal comparison of utilities," sanctions in other games, propaganda, etc. These will be ignored in this paper.

The following example will illustrate our purpose. Let $n = 3$, $v(1) = v(2) = v(3) = v(123) = 0$, $v(12) = 100$, $v(13) = 100$, $v(23) \doteq 50$. Consider the p.c.

$$(80, 20, 0; 12, 3). \tag{2.6}$$

Now, player 2 can object by pointing out that in the p.c.

$$(0, 21, 29; 1, 23) \tag{2.7}$$

he and player 3 get more. Player 1 has no counter objection because he cannot keep his 80 while offering player 3 at least 29. Thus, (2.6) is unstable. On the other hand,

$$(75, 25, 0; 12, 3) \tag{2.8}$$

is stable. An objection of player 2, e.g.,

$$(0, 26, 24; 1, 23), \tag{2.9}$$

can be met by a counter objection

$$(75, 0, 25; 13, 2); \tag{2.10}$$

or an objection of player 1, e.g.,

$$(76, 0, 24; 13, 2) \tag{2.11}$$

can be met by the counter objection

$$(0, 25, 25; 1, 23). \tag{2.12}$$

In these counter objections, the threatened player can keep his share and offer his partners at least what the player who objects offered. It will turn out that the only stable p.c.'s in this game are

$$(0,0,0;1,2,3); \quad (75,25,0;12,3); \quad (75,0,25;13,2); \quad (0,25,25;1,23). \tag{2.13}$$

Let it be said at once that our paper was largely motivated by the fact that most of our friends, to whom this game was presented, started their considerations from the p.c.'s (2.8) and (2.10). We tried to find what characterizes these p.c.'s and how they can be generalized to more complicated cases.

Let Γ be a game, as described above. Let K be a non-empty subset of the set of players N. A player i will be called a *partner of K in a p.c.* $(x; \mathscr{B})$, if he is a member of a coalition in $\mathscr{B}$ which intersects K. The set $P[K; (x; \mathscr{B})]$ of all the partners of K in $(x, \mathscr{B})$ is, therefore,

$$P[K; (x; \mathscr{B})] \equiv \{i \mid i \in B_j, B_j \cap K \neq \phi\}. \tag{2.14}$$

Note that $K \subset P[K; (x; \mathscr{B})]$; i.e., each member of K is also a partner of K, contrary to everyday usage. K needs only the consent of its partners in order to get its part of x.

DEFINITION 2.1 Let $(x; \mathscr{B})$ be a coalitionally rational payoff configuration (2.2), (2.5) for a game Γ. Let K and L be non-empty disjoint subsets of a coalition B_j which appears in $(x; \mathscr{B})$. An *objection* of K against L in $(x, \mathscr{B})$ will be a c.r.p.c.

$$(y; \mathscr{C}) \equiv (y_1, y_2, \ldots, y_n; C_1, C_2, \ldots, C_\ell) \tag{2.15}$$

for which

$$P[K; (y; \mathscr{C})] \cap L = \phi, \tag{2.16}$$

$$y_i > x_i \quad \text{for all } i, i \in K, \tag{2.17}$$

$$y_i \geq x_i \quad \text{for all } i, i \in P[K; (y; \mathscr{C})]. \tag{2.18}$$

Verbally, in their objection, players K claim that, without the aid of players L ((2.16)), they can get more in another c.r.p.c. ((2.17)), and the new situation is reasonable because their new partners do not get less than what they got in the previous p.c. ((2.18)).

DEFINITION 2.2 Let $(x, \mathscr{B})$ be a coalitionally rational payoff configuration (2.2), (2.5) in a game Γ, and let $(y; \mathscr{C})$ be an objection of a set K against a set L in $(x; \mathscr{B})$, $K, L \subset B_j$. A *counter objection* of L against K is a c.r.p.c.

$$(z; \mathcal{D}) \equiv (z_1, z_2, \ldots, z_n; D_1, D_2, \ldots, D_k) \qquad (2.19)$$

for which

$$P[L; (z; \mathcal{D})] \not\supset K, \qquad (2.20)$$

$$z_i \geq x_i \quad \text{for all } i, i \in P[L; (z; \mathcal{D})], \qquad (2.21)$$

$$z_i \geq y_i \quad \text{for all } i, i \in P[L; (z; \mathcal{D})] \cap P[K; (\mathcal{Y}; \mathcal{C})]. \qquad (2.22)$$

Verbally, in their counter objection, players L claim that they can hold their original properties ((2.21)), promise their partners at least their original share ((2.21)), and if they need partners of K in his objection, they can give them not less than what they were offered in the objection ((2.22)). Sometimes, the members of L have to use the tactics of "divide and rule" by using members of K as partners, but they may not use *all* members of K ((2.20)).

DEFINITION 2.3   A c.r.p.c $(x, \mathcal{B})$ is called stable if for each objection of a K against an L in $(x, \mathcal{B})$ there is a counter objection of L against K. The *bargaining set* $\mathfrak{M}$ of a game Γ is the set of all stable c.r.p.c.'s.

The feeling of "safety" suggested by this definition lies in the assurance that all threats *within a coalition* can be met. It may be felt, perhaps, that there is a lack of symmetry when comparing (2.16) and (2.20), but the situation is not symmetric in the first place. An objection (2.15) can serve, in general, as an objection of K (or another "K") against various groups "L," each one of which has to have a counter objection.

To be sure, even if there is a desire for a stability as demanded in the p.c.'s of the bargaining set, this does not mean that the outcome will belong to $\mathfrak{M}$. A player, e.g., may agree to sacrifice some of his profits in order to make sure that he enters a coalition. Other factors, mentioned above, may also cause deviations from $\mathfrak{M}$. However, if the demand for stability is strong enough, we hope that the outcome will not be too far from $\mathfrak{M}$; in this sense the theory has a normative aspect. Moreover, as the number of the players increases, there arise many possible threats, and, using the concepts involved in the definition, one may compute and show the players where they are "safe" and what threats they do possess. This is another normative aspect.[3]

The bargaining set is never empty. Indeed, $(0, 0, \ldots, 0; 1, 2, \ldots, n)$ always belongs to $\mathfrak{M}$.

---

3. Results close to the bargaining set have been observed in an experimental study [4].

In a coalition of zero value, any objection (if there is one) can be countered by the other players playing as 1-person coalitions.

A dummy always gets zero in a c.r.p.c., therefore he cannot belong to any objecting K. On the other hand, he can always keep his zero by playing alone. He can be of no use for any objection or counter objection, since the same can be effected without his help. Thus, a dummy has no essential effect on $\mathfrak{M}$.

The definition of $\mathfrak{M}$ does not use "interpersonal comparisons of utilities" and it is independent of the names of the players.

THEOREM 2.1 The bargaining set $\mathfrak{M}$ of a game Γ can be represented as the set of solutions of a conjunctive–disjunctive[4] system of linear inequalities involving the x_i as unknowns. It is, therefore, a union of a finite number of polyhedral convex sets in the n-space with coordinates $(x_1, x_2, \ldots, x_n)$.

Proof.[5] In any finite expression with coordinates which has the form of quantifiers followed by linear inequalities connected by the words "or" and "and," the free variables—if such exist—which satisfy the expression are those and only those which satisfy a certain disjunctive–conjunctive system of linear inequalities. This is a known theorem in logic, but for the sake of completeness we sketch the proof. It is sufficient to prove the theorem when there is only one quantifier. Moreover, we may assume that this quantifier is $\exists$, because $\forall = \sim \exists \sim$. The theorem now follows from the fact that the projection of a polyhedron is a polyhedron.

3 The Two-Person Game

The bargaining set $\mathfrak{M}$ for the game:

$$v(1) = v(2) = 0 \quad v(12) = a \geq 0, \tag{3.1}$$

consists of all possible c.r.p.c.'s; i.e.,

$$\begin{cases} (0, 0; 1, 2) \\ (x_1, x_2; 12) \quad x_1 + x_2 = a, \quad x_1 \geq 0, \quad x_2 \geq 0. \end{cases} \tag{3.2}$$

Indeed, there are no possible objections.

---

4. I.e., a system of linear inequalities connected by the words "or" and "and".
5. We are indebted to Professor M. Rabin and Professor A. Robinson for pointing this out.

4 The 3-Person Game. Permissible Coalitions of Less Than Three Players

In this section we shall study the game:

$$v(1) = v(2) = v(3) = 0; \quad v(12) = a; \quad v(23) = b; \quad v(13) = c; \quad a, b, c \geqslant 0. \tag{4.1}$$

THEOREM 4.1 In the game (4.1), essentially two cases arise.

Case A. If a, b, c satisfy the "triangle inequality"

$$a \leqslant b + c, \quad b \leqslant a + c, \quad c \leqslant a + b, \tag{4.2}$$

then the bargaining set $\mathfrak{M}$ is:

$$\begin{cases} (\quad 0 \quad , \quad 0 \quad , \quad 0 \quad ; \; 1, 2, 3) \\ (\dfrac{a+c-b}{2}, \; \dfrac{a+b-c}{2}, \quad 0 \quad ; \; 12, 3 \quad) \\ (\dfrac{a+c-b}{2}, \quad 0 \quad , \; \dfrac{c+b-a}{2}; \; 13, 2 \quad) \\ (\quad 0 \quad , \; \dfrac{a+b-c}{2}, \; \dfrac{c+b-a}{2}; \; 1, 23 \quad) \end{cases} \tag{4.3}$$

Case B. If, e.g., $a > b + c$, then the bargaining set $\mathfrak{M}$ is:

$$\begin{cases} (\quad 0 \quad , \quad 0 \quad , \; 0; \; 1, 2, 3) \\ (c \leqslant x_1 \leqslant a - b, \; a - x_1, \; 0; \; 12, 3 \quad) \\ (\quad c \quad , \quad 0 \quad , \; 0; \; 13, 2 \quad) \\ (\quad 0 \quad , \quad b \quad , \; 0; \; 1, 23 \quad) \end{cases} \tag{4.4}$$

Before proving this theorem, we shall give some illustrations which will throw some light on the nature of the bargaining sets.

Example 1 Let $a = 100$, $b = 100$, $c = 50$. The triangle inequality is satisfied, and therefore $\mathfrak{M}$ is discrete: $\{(0, 0, 0; 1, 2, 3), (25, 75, 0; 12, 3) (25, 0, 25; 13, 2) (0, 75, 25; 1, 23)\}$.

One can approach this solution also by the following intuitive argument: Suppose that player 1 receives α, then player 2 gets $100 - \alpha$ and he is thus willing to pay player 3 at most $100 - (100 - \alpha) = \alpha$. Thus, player 3 will be willing to pay player 1 at most $50 - \alpha$. If $50 - \alpha > \alpha$, then player 1 will prefer to join player 3. This will cause player 2 to agree to get less. If $50 - \alpha < \alpha$, player 2 will demand more as he will get more from player

3 if player 1 insists on getting α. Thus an equilibrium will be reached only if $\alpha = 50 - \alpha$, in which case $\alpha = 25$.

Example 2 The above argument fails in the case $a = 20$, $b = 30$, $c = 100$. Here one obtains $\alpha = 45$ in which case player 2 will lose money. This he can avoid by playing alone. Our bargaining set is no longer discrete: $\{(0, 0, 0; 1, 2, 3)\ (20 \leqslant x_1 \leqslant 70, 0, 100 - x_1; 13, 2)\ (20, 0, 0; 12, 3)\ (0, 0, 30; 1, 23)\}$. One can reason as follows: Player 1, being in the coalition 13, will not be satisfied in getting less than 20, since otherwise he will do better by joining player 2. Similarly, player 3 will demand at least 30. Fortunately, both demands can be satisfied, and player 2 cannot cause any harm since he is a weak player.

Example 3 Let $a = 100$, $b = 100$, $c = 0$. We observe that the bargaining set is again discrete: $\{(0, 0, 0; 1, 2, 3)\ (0, 100, 0; 12, 3)\ (0, 0, 0; 13, 2)\ (0, 100, 0; 1, 23)\}$. This solution reflects the character of an "unrestricted competition" in our bargaining set. Indeed, player 2 can practically receive the amount 100 because whatever the positive demand of player 1 will be, player 3 will be "satisfied" in getting less, and vice versa. One observes that our theory does not take into account the psychological threat that player 2 may also "lose" his profit 100, and probably will therefore be willing to pay some amount in order to be in a coalition with player 1 or with player 3. In practical situations several side conditions may come into consideration such as: (1) It may be *customary* not to enter a coalition unless a certain minimum amount or percentage of profit is guaranteed in advance. (2) A "cartel" agreement is decided between player 1 and player 3, in which both of them declare not to enter a coalition with player 2 without getting at least a certain amount of profit. (3) There is enforced a "cartel" or an "anti-cartel" law in the country. (4) It is known that in order to ensure a certain profit, one is willing to give up a certain amount or percentage in order to "push" an equilibrium situation to one's side.

Proof of Theorem 4.1 Certainly, $(0, 0, 0; 1, 2, 3) \in \mathfrak{M}$.

Next, let us examine under what circumstances a payoff configuration $(x_1, x_2, 0; 12, 3)$ can belong to the bargaining set. It should be coalitionally rational, and therefore it must satisfy

$$x_1 \geqslant 0, \quad x_2 \geqslant 0; \quad x_1 + x_2 = v(12). \tag{4.5}$$

LEMMA 1 A necessary and sufficient condition that player 1 has no objection is:

$$x_1 \geqslant v(13). \tag{4.6}$$

Proof Indeed, if $x_1 \geq v(13)$, then player 1 has no objection either by playing alone (see (4.5)) or by participating in the coalition 13. If $x_1 < v(13)$, then player 1 can suggest the objection

$$\left(\frac{v(13)+x_1}{2}, 0, \frac{v(13)-x_1}{2}; 13, 2\right). \tag{4.7}$$

This is a coalitionally rational payoff configuration.

LEMMA 2   A necessary and sufficient condition that player 1 has an objection and to each such objection player 2 has a counter objection, is:

$$x_1 < v(13), \tag{4.8}$$

$$x_1 - x_2 \geq v(13) - v(23) \quad \text{or } x_2 = 0. \tag{4.9}$$

Proof Indeed, if (4.8) and (4.9) hold, then, by Lemma 1, player 1 has an objection. This can only be (see (4.5)) of the form

$$(x_1 + \varepsilon, 0, v(13) - x_1 - \varepsilon; 13, 2), \tag{4.10}$$

where ε is a sufficiently small positive number. If $x_2 = 0$, then (4.10) is itself also a counter objection; otherwise,

$$(0, v(23) - v(13) + x_1 + \varepsilon, v(13) - x_1 - \varepsilon; 1, 23) \tag{4.11}$$

is a possible counter objection. By (4.8), player 2 will now receive even more than x_2. If (4.8) does not hold, then there is no objection for player 1, by Lemma 1. If (4.8) holds, but

$$x_2 > 0 \quad \text{and} \quad x_1 - x_2 < v(13) - v(23), \tag{4.12}$$

then player 1 can object by (4.10), choosing ε so small that $v(23) - v(13) + x_1 + \varepsilon < x_2$. Now, player 2 does not have any counter objection, either by playing alone or by forming a coalition with player 3.

Summing up, and making the necessary permutations, we obtain:

LEMMA 3   A necessary and sufficient condition that a payoff configuration $(x_1, x_2, 0; 12, 3)$ will belong to the bargaining set $\mathfrak{M}$, is that x_1 and x_2 will satisfy (4.5) as well as at least one of the following columns:

$$\begin{array}{|c|c|c|} x_1 \geq v(13) & x_1 < v(13) & x_1 < v(13) \\ & x_2 = 0 & x_1 - x_2 \geq v(13) - v(23) \end{array} \tag{4.13}$$

and also at least one of the following columns:

$$\begin{array}{|c|c|c|} x_2 \geq v(23) & x_2 < v(23) & x_2 < v(23) \\ & x_1 = 0 & x_2 - x_1 \geq v(23) - v(13) \end{array} \tag{4.14}$$

Taking into account that $x_1 + x_2 = a$, these inequalities reduce to

$$
\begin{array}{c|c|c}
0 \leqslant x_1 \leqslant a & x_1 < c & 0 \leqslant x_1 \leqslant a \\
c \leqslant x_1 & x_1 = a & \dfrac{a+c-b}{2} \leqslant x_1 < c
\end{array}
\tag{4.15}
$$

$$
\begin{array}{c|c|c}
x_1 \leqslant a-b & a-b < x_1 & a-b < x_1 \leqslant \dfrac{a+c-b}{2} \\
 & x_1 = 0 &
\end{array}
\tag{4.16}
$$

We now use the assumption $a, b, c \geqslant 0$, and the inequalities (4.15), (4.16). A detailed calculation yields the following results:

Case A. If a, b, c satisfy the "triangle inequalities" (4.2), then

$$x_1 = \frac{a+c-b}{2} \tag{4.17}$$

is the only solution.

Case B. If $a > b + c$, then each x_1 satisfying

$$c \leqslant x_1 \leqslant a - b, \tag{4.18}$$

is a solution, and there are no other solutions.

Case C. If $b > a + c$, then $x_1 = 0$ is the only solution.

Case D. If $c > a + b$, then $x_1 = a$ is the only solution.

These are the only possible cases, they exclude each other, and therefore the proof of Theorem 4.1 has been completed.

5 The General 3-Person Game

Let us add the coalition 123 with its value $v(123) = d \geqslant 0$ to the game treated in Section 4. This coalition will have no effect on the previous p.c.'s of the bargaining set. Indeed, this coalition cannot be used for objections and counter objections, because it contains all the players L and K. Thus, it only remains to find out under what condition does a p.c. $(x_1, x_2, x_3; 123)$ belong to the new bargaining set.

As it should be coalitionally rational, it is necessary that x_1, x_2, x_3 satisfy:

$$x_1, x_2, x_3 \geqslant 0; \quad x_1 + x_2 \geqslant a, \quad x_2 + x_3 \geqslant b, \quad x_1 + x_3 \geqslant c;$$

$$x_1 + x_2 + x_3 = d. \tag{5.1}$$

On the other hand, if (5.1) is satisfied, there can be no objection and hence this pair belongs to $\mathfrak{M}$.

In order that the inequalities (5.1) have at least one solution, it is necessary and sufficient that

$$d \geq a, b, c, \quad d \geq \frac{a+b+c}{2}. \tag{5.2}$$

We have thus proved:

THEOREM 5.1 In the 3-person game for which

$$v(1) = v(2) = v(3) = 0, \quad v(12) = a, \quad v(23) = b,$$

$$v(13) = c, \quad v(123) = d, \quad a, b, c, d \geq 0,$$

the bargaining set $\mathfrak{M}$ consists of the p.c.'s given by Theorem 4.1, and also of the p.c.'s $(x_1, x_2, x_3; 123)$ which satisfy (5.1). The latter p.c.'s exist if and only if (5.2) is satisfied.

6 The 4-Person Game. Coalitions of 1 Person and 3 Persons

Consider the four-person game, in which the permissible coalitions are all the single-person and the three-person coalitions. Let their values be

$$\begin{cases} v(1) = v(2) = v(3) = v(4) = 0, \quad v(123) = a, \quad v(124) = b, \\ v(134) = c, \quad v(234) = d, \quad a, b, c, d \geq 0. \end{cases} \tag{6.1}$$

Evidently $(0, 0, 0, 0; 1, 2, 3, 4)$ belongs to the bargaining set $\mathfrak{M}$. Similar considerations to those which were used in Section 4 lead to the inequalities which are listed in Appendix 1. These inequalities express a necessary and sufficient condition in order that the payoff configuration $(x_1, x_2, x_3; 123, 4)$ belongs to the bargaining set.

We omit the calculations, which are somewhat lengthy but easy, and state the results. There are essentially four different cases:

Case A. If

$$\begin{cases} 2a \leq b + c + d, \\ 2b \leq a + c + d, \\ 2c \leq a + b + d, \\ 2d \leq a + b + c, \end{cases} \tag{6.2}$$

then the bargaining set is

$$\begin{cases} (& 0 & 0 & 0 & 0 & ; & 1,2,3,4) \\ (\dfrac{a+b+c-2d}{3} & \dfrac{a+b+d-2c}{3} & \dfrac{a+c+d-2b}{3} & 0 & ; & 123,4) \\ (\dfrac{a+b+c-2d}{3} & \dfrac{a+b+d-2c}{3} & 0 & \dfrac{b+c+d-2a}{3} & ; & 124,3) \\ (\dfrac{a+b+c-2d}{3} & 0 & \dfrac{a+c+d-2b}{3} & \dfrac{b+c+d-2a}{3} & ; & 134,2) \\ (& 0 & \dfrac{a+b+d-2c}{3} & \dfrac{a+c+d-2b}{3} & \dfrac{b+c+d-2a}{3} & ; & 234,1) \end{cases}$$

(6.3)

Case B. If

$$\begin{cases} 2a > b+c+d, \\ 2b \leqslant a+c+d, & b \leqslant c+d, \\ 2c \leqslant a+b+d, & c \leqslant b+d, \\ 2d \leqslant a+b+c, & d \leqslant b+c, \end{cases}$$

(6.4)

then the bargaining set is:

$$\begin{cases} (& 0 & 0 & 0 & 0; & 1,2,3,4) \\ (& x_1 & a-x_1-x_3 & x_3 & 0; & 123,4) \\ (\dfrac{b+c-d}{2} & \dfrac{b+d-c}{2} & 0 & 0; & 124,3) \\ (\dfrac{b+c-d}{2} & 0 & \dfrac{c+d-b}{2} & 0; & 134,2) \\ (& 0 & \dfrac{b+d-c}{2} & \dfrac{c+d-b}{2} & 0; & 234,1) \end{cases}$$

(6.5)

Here, x_1 and x_3 satisfy the inequalities

$$0 \leqslant x_1 \leqslant a-d, \quad 0 \leqslant x_3 \leqslant a-b, \quad c \leqslant x_1 + x_3 \leqslant a. \tag{6.6}$$

Case C. If

$$\begin{cases} 2a > b+c+d, \\ 2b \leqslant a+c+d, & b > c+d, \\ 2c \leqslant a+b+d, \\ 2d \leqslant a+b+c, \end{cases}$$

(6.7)

then the bargaining set is:

$$\begin{cases} (& 0 & 0 & 0 & 0; & 1,2,3,4) \\ (& x_1 & a-x_1-x_3 & x_3 & 0; & 123,4) \\ (c \leqslant \xi_1 \leqslant b-d & b-\xi_1 & 0 & 0; & 124,3) \\ (& c & 0 & 0 & 0; & 134,2) \\ (& 0 & d & 0 & 0; & 234,1) \end{cases} \quad (6.8)$$

Here, x_1 and x_3 satisfy the inequalities (6.6).

Case D. If

$$\begin{cases} 2a > b+c+d, \\ 2b > a+c+d, \\ a \geqslant b, \end{cases} \quad (6.9)$$

the bargaining set is the same as in Case C.

Only Case A is completely discrete; all other cases contain the continuum (6.6). Equations (6.2) can be considered as a generalization of the triangle inequalities. In fact, it follows from (6.2) that any three of the numbers a, b, c, d satisfy the triangle inequalities. Moreover, an equality $a = b + c$, for example, can occur only if $a = d$. The converse does not hold. (E.g., $a = 8$, $b = c = d = 5$.)

It is possible to approach the bargaining set in Case A as follows: If players 1 and 2 get α and β, respectively, then player 3 gets $a - \alpha - \beta$ in the coalition 123. With these values, player 4 will get $b - \alpha - \beta$ in the coalition 124, $c - a + \beta$ in the coalition 134, and $d - a + \alpha$ in the coalition 234. In order that no coalition can exert threats on others, it is necessary and sufficient that

$$b - \alpha - \beta = c - a + \beta = d - a + \alpha. \quad (6.10)$$

Hence

$$\alpha = \frac{a+b+c-2d}{3}, \quad \beta = \frac{a+b+d-2c}{3}. \quad (6.11)$$

If

$$2a \geqslant b+c+d, \quad a \geqslant b,c,d, \quad (6.12)$$

then the coalition 123 is strong and player 4 cannot get more than zero. If we decide to omit him from the game, and look at each of the remaining two persons in a coalition which contained him as a new 2-person coalition which has the same value as before, we get a 3-person game in which $v(123) = a$, $v(12) = b$, $v(23) = d$, $v(13) = c$, $v(1) = v(2) = v(3) = 0$.

A comparison with the previous two sections shows that the bargaining set of the new game is essentially the same as the game treated in Cases B, C, D, and Case A, if equality holds in the first relation of (6.12).

(Note that each of the systems (6.4) and (6.7) implies $a > b, c, d$, (6.9) implies $a \geqslant b$, $a > c, d$, and (6.7) as well as (6.9) implies $b > c + d$.)

We can therefore conclude:

THEOREM 6.1 A 4-person game, in which the permissible coalitions are all the 1-person and 3-person coalitions always has a bargaining set in which all possible partitions into coalitions appear. The set is discrete if and only if (6.2) is satisfied. If (6.12) holds, the bargaining set is essentially the same as the one of a full 3-person game obtained from the original one by deleting a player who belongs to the complement of a maximal valued coalition. This player always gets 0.

Remark 1 The same situation occurs in a 3-person game in which the only non-permissible coalition is the 3-person coalition. If the triangle inequality does not hold, then one coalition is strong enough to reduce the game to a 2-person game with essentially the same bargaining set. The weak player gets 0.

Remark 2 The conditions (6.12) are necessary and sufficient for the existence of a c.r.p.c. $(x_1, x_2, x_3, 0; 123, 4)$ such that

$$x_1, x_2, x_3 \geqslant 0, \quad x_1 + x_2 \geqslant b, \quad x_1 + x_3 \geqslant c, \quad x_2 + x_3 \geqslant d. \tag{6.13}$$

Obviously, such p.c.'s cannot be objected against. However, in any c.r.p.c, in which player 4 is in a 3-person coalition and receives more than 0, there exists an objection against player 4 which cannot be countered. Thus, the coalition 123 "dictates" everything; this is why player 4 cannot claim more than 0.

Remark 3 The following *ad hoc* rule serves for the discrete case: The value of each coalition is equally divided among its members. If a person enters a coalition he gets the sum of "his shares" minus the sum of the "shares" which his partners get from coalitions which do not include him. For example: The first player's shares are $a/3, b/3, c/3$. If he is entering the coalition 123, his partners have their shares $d/3, d/3$ from the coalition 234, which is that coalition that does not contain player 1; therefore, this player gets

$$\frac{a}{3} + \frac{b}{3} + \frac{c}{3} - \frac{d}{3} - \frac{d}{3} \tag{6.14}$$

if he enters the coalition 123.

The same rule applies also to the 3-person game, with 1- and 2-person coalitions, in the discrete case.

Remark 4 The discrete case exhibits a game with a nonnegative "3-quota." Each player always receives his "quota" in the bargaining set if he succeeds in becoming a member of a 3-person coalition. Quota games are treated in [5], and the results of this section are further generalized in [6].

7 Existence Theorems. Counterexamples

DEFINITION 7.1 A permissible coalition in a game Γ will be called *effective* if it is possible to divide its value among its members in such a way that no permissible sub-coalition can alone make more.

Condition (5.2), for example, is a necessary and sufficient condition that the coalition 123 will be effective in the game treated in Section 5.

Clearly, we can assume that all subsets of N are permissible coalitions and that those having a positive value are effective, since we are dealing only with c.r.p.c.'s. The zero-valued coalitions will be called *trivial coalitions*.

The first question which may arise is whether each partition of the players, in which the only trivial coalitions are 1-person coalitions, is represented in $\mathfrak{M}$. The answer is *no*.

Example 7.1 $n = 5$, the non-trivial coalitions are 12, 35, 134, 2345, with values:

$$v(12) = 10, \quad v(35) = 85, \quad v(134) = 148, \quad v(2345) = 160. \tag{7.1}$$

Consider the coalitionally rational payoff configuration

$$(\alpha, \beta, 0, 0, 0; 12, 3, 4, 5), \tag{7.2}$$

where, of course, $0 \leq \alpha \leq 10$, $\alpha + \beta = 10$. Now, player 1 can object by

$$(11, 0, 29, 108, 0; 134, 2, 5). \tag{7.3}$$

This objection is *justified*—i.e., player 2 has no counter objection—if $\alpha < 10$. Indeed, any attempt of player 2 to keep his positive share β will end with a coalitionally non-rational p.c. Thus, (7.2) can belong to $\mathfrak{M}$ only if $\alpha = 10$, $\beta = 0$. But this case is also ruled out, since now player 2 has a justified objection: $(0, 1, 100, 44, 15; 1, 2345)$.

Let Γ be a game, some of the values of the coalitions of which are positive. Is it possible that no p.c. belongs to $\mathfrak{M}$ unless all the players get zero? In other words—is it possible that, in spite of some coalitions having a positive value, it would be worthless to enter into such coalitions, if one insists on the stability demanded by the definition of $\mathfrak{M}$? This, in fact, may happen as the following example shows:

Example 7.2[6]

$$\begin{cases} v(12b) = 1, & b = 3, 4, 5, 6. \\ v(1ab) = 1, & a = 3, 4; \quad b = 5, 6. \\ v(2pq) = 1, & p = 3, \quad q = 4 \quad \text{or} \quad p = 5, \quad q = 6. \\ v(3456) = 1, \\ v(B) = 1, & B \text{ contains at least one of the} \\ & \text{above-mentioned coalitions.} \\ v(B) = 0, & \text{otherwise.} \end{cases} \quad (7.4)$$

It is a long but easy computation to verify that for this game $(x; \mathcal{B}) \in \mathfrak{M}$ implies $x_i = 0$, $i = 1, 2, \ldots, 6$.

The following theorem might be helpful in gaining some more insight into the nature of the bargaining set $\mathfrak{M}$.

THEOREM 7.1 Let Γ be an n-person game, in which 12 is a permissible coalition. Let $\mathcal{B}^0 \equiv 12, B_2, \ldots, B_m$ be a fixed partition. Let $(x; \mathcal{B}^0) \equiv (x_1, x_2, \ldots, x_n; \mathcal{B}^0)$ be a c.r.p.c. and let J be the set of all the numbers σ_1, $0 \leqslant \sigma_1 \leqslant v(12)$, such that player 1 has a justified objection[7] against player 2, in $(\sigma_1, v(12) - \sigma_1, x_3, x_4, \ldots, x_n; \mathcal{B}^0)$; then J is an open set relative to the closed interval $[0, v(12)]$.

Proof If

$$(x_1, x_2, \ldots, x_n; 12, B_2, \ldots, B_m) \quad (7.5)$$

is a coalitionally rational payoff configuration, then so is also

$$(x_1 + \varepsilon, x_2 - \varepsilon, x_3, \ldots, x_n; 12, B_2, \ldots, B_m), \quad (7.6)$$

provided that $-x_1 \leqslant \varepsilon \leqslant v(12) - x_1$.

If $x_1 \in J$, then $\delta \equiv v(12) - x_1 > 0$, since otherwise player 2 can counter object by playing alone.

Let $(y; \mathcal{C})$ be an objection of player 1 against player 2; then $y_1 > x_1$. Let z_2 be the maximum that player 2 can get by joining a coalition such

---

6. This game was given by J. von Neumann and O. Morgenstern [9], pp. 467–469, as an example of a simple game which is not a weighted majority game and for which no main simple solution exists.

7. By the term "a justified objection" we mean an objection which has no counter objection.

that his partners (if such exist) get what they are supposed to get in a counter objection. Obviously such a maximum exists, and $z_2 < x_2$, because $x_1 \in J$. Choose ε such that

$$-x_1 \le \varepsilon < \min(\delta, y_1 - x_1, x_2 - z_2); \tag{7.7}$$

then $x_1 + \varepsilon$ will also belong to J. Indeed, (7.6) will be coalitionally rational; $(\mathscr{y}; \mathscr{C})$ will remain an objection which is justified.

Thus, if $x_1 \in J$, then so are all the points on the interval $[0, x_1 + \varepsilon]$.

THEOREM 7.2 Let Γ be an n-person game, in which 12 is a permissible coalition and all the permissible coalitions are 1, 2 and 3-person coalitions; then, if $(x, \mathscr{B}^0)$ is a c.r.p.c., there exists a c.r.p.c.

$$(\xi_1, \xi_2, x_3, x_4, \ldots, x_n; 12, B_2, \ldots, B_m) \tag{7.8}$$

such that neither player 1 nor player 2 has any justified objection. Here $\mathscr{B}^0 \equiv 12, B_2, \ldots, B_m$.

Proof We proved in Theorem 7.1 that the numbers x_1, for which player 1 has a justified objection, form an open set T_1 with respect to $[0, v(12)]$. Similarly, the numbers x_1 for which player 2 has a justified objection form an open set T_2 with respect to the same interval ($x_3, \ldots, x_n$ remain fixed). We shall show that T_1 and T_2 are disjoint, from which it will follow that there is a point ξ_1 in $[0, v(12)]$, which is neither in T_1, nor in T_2, and therefore (7.8) will satisfy the requirements. (None of the sets is the closed interval because $v(12) \notin T_1$, $0 \notin T_2$.)

Indeed, suppose that

$$(\sigma_1, \sigma_2, x_3, \ldots, x_n; 12, B_2, \ldots, B_m) \tag{7.9}$$

is a c.r.p.c. in which both players have justified objections. Player 1, in his objection, must join a coalition C which contains more than one person and does not contain player 2. Similarly, player 2 must join, in his objection, a coalition D which consists of more than one person and does not contain player 1. If $C \cap D = \phi$, then player 2's objection can serve as a counter objection for player 1's objection, the latter being therefore not justified. If $C \cap D \equiv E = \phi$, then E contains one or two members. Without loss of generality, we can assume that the total amount that the players in E got in player 2's objection was not less than what they got from player 1's objection. If E contains one member, player 2's objection is a counter objection to player 1's objection. If E contains two members, this is not always true, because in order to counter object, player 2 has to modify, perhaps, his payments to the members of E. By doing so, a payoff configuration may result, which is not coalitionally rational; i.e., one player in E and player 2 can now make more by together forming a coalition. But if this is the case, then this coalition can serve in a counter objection; e.g., by player 2 taking σ_2 for himself and letting the other player get the rest.

Remark The theorem fails to hold if we remove the restriction on the the number of the players in the permissible coalitions. A counterexample is provided in Example 7.1.

Application Suppose that each one of two men got a license to build a gasoline station. Each one considers the possibility of taking at most two partners. They expect various profits from the corresponding possible coalitions. The other partners do not have licenses. Of course, the two men consider also their joint coalition. Under these assumptions, Theorem 7.2 says that the coalition of the two licensees is represented in the bargaining set.

8 The 4-Person Game in Which Only 1- and 2-Person Coalitions Are Permissible

The inequalities that determine under what condition is $(x_1, x_2, x_3, x_4; 12, 34)$ in $\mathfrak{M}$, for the game

$$\begin{cases} v(1) = v(2) = v(3) = v(4) = 0, \quad v(12) = a, \quad v(23) = b, \quad v(34) = c, \\ v(13) = d, \quad v(24) = e, \quad v(14) = f, \quad a, b, c, d, e, f \geq 0 \end{cases}$$
(8.1)

are given in Appendix 2.

Theorem 7.2 ensures that any partition which contains only one 2-person coalition is represented in $\mathfrak{M}$. We shall now study the case of partitions into two couples. Our object is to prove that some such partitions appear in $\mathfrak{M}$. It turns out that this can be proved even if we limit ourselves to *maximal* partitions, i.e., to those partitions in which the sum of the values of the coalitions is maximal. This restriction helps us by reducing the number of inequalities which need to be examined.

THEOREM 8.1 Let Γ be the game (8.1), where

$$a + c \geq d + e, \quad a + c \geq b + f;$$
(8.2)

then there always exists a p.c. $(x_1, x_2, x_3, x_4; 12, 34)$ in the bargaining set $\mathfrak{M}$.

Proof We omit the calculations, but state the various cases.

Case A. If
$$a \leqslant b+d, \quad b \leqslant a+d, \quad d \leqslant a+b, \quad 2c \geqslant b+d-a, \tag{8.3}$$
then
$$\left(\frac{a+d-b}{2}, \frac{a+b-d}{2}, \frac{b+d-a}{2}, \frac{2c+a-b-d}{2}; 12, 34\right) \in \mathfrak{M}. \tag{8.4}$$
If
$$a \leqslant b+d, \quad b \leqslant a+d, \quad d \leqslant a+b, \quad 2c < b+d-a, \tag{8.5}$$
then
$$\left(\frac{a+d-b}{2}, \frac{a+b-d}{2}, c, 0; 12, 34\right) \in \mathfrak{M}. \tag{8.6}$$

Case B. If
$$a > b+d, \quad c > d+f, \tag{8.7}$$
then
$$(d, a-d, 0, c; 12, 34) \in \mathfrak{M}. \tag{8.8}$$

Case C. If
$$a > b+d, \quad f > c+d, \quad b+c \geqslant e, \tag{8.9}$$
then
$$(f-c, a+c-f, 0, c; 12, 34) \in \mathfrak{M}. \tag{8.10}$$
(Indeed, (8.9) and (8.2) imply $c \geqslant e-b \geqslant e-(a+c-f)$, hence $a+2c \geqslant e+f$. The rest follows directly.) If
$$a > b+d, \quad f > c+d, \quad e > b+c, \tag{8.11}$$
then
$$\left(\frac{a+f-e}{2}, \frac{a+e-f}{2}, 0, c; 12, 34\right) \in \mathfrak{M}. \tag{8.12}$$
(Indeed, (8.2) and (8.11) imply $2d+e \leqslant d+a+c < a+f$. Also $2b+f \leqslant b+a+c < a+e$.)

Case D. If
$$a > b+d, \quad d > f+c, \quad b+c \geqslant e, \tag{8.13}$$
then
$$(a-b, b, 0, c; 12, 34) \in \mathfrak{M}. \tag{8.14}$$

If
$$a > b+d, \quad d > f+c, \quad e > b+c, \tag{8.15}$$
then
$$(a+c-e, e-c, 0, c; 12, 34) \in \mathfrak{M}. \tag{8.16}$$

Case E. If
$$d > a+b, \quad d > c+f, \tag{8.17}$$
then
$$(a, 0, d-a, c+a-d; 12, 34) \in \mathfrak{M}. \tag{8.18}$$

All other cases are either not maximal partitions, or they can be reduced to these cases by permuting the players:
$$1 \leftrightarrow 3, 2 \leftrightarrow 4. \tag{8.19}$$

9 The Restricted Bargaining Set

In a given game there are in general many stable p.c.'s. Though we do not possess a criterion for choosing between them, there are cases in which it is clear that some p.c.'s in $\mathfrak{M}$ are "better" than others. We therefore suggest that the latter should be deleted from $\mathfrak{M}$, thus giving rise to the restricted bargaining set $\mathfrak{M}^*$.

A p.c. $(x; \mathscr{B})$ in $\mathfrak{M}$ should be deleted if one of the following cases occurs:

(i) There exists in $\mathfrak{M}$ a p.c. $(x^*; \mathscr{B}^*)$ with
$$x_i^* > x_i; \quad i = 1, 2, \ldots, n. \tag{9.1}$$

(ii) There exists in $\mathfrak{M}$ a p.c. $(x^{**}; \mathscr{B}^{**})$, where the coalitions in $\mathscr{B}^{**}$ are unions of coalitions in $\mathscr{B}$, such that
$$x_i^{**} > x_i \tag{9.2}$$
for all the players i which belong to a union of more than one coalition of $\mathscr{B}$ and
$$x_i^{**} \geq x_i \tag{9.3}$$
for all the other players.

One sees that in the examples given in the previous sections, only those coalitions which have relatively big values (if such exist), will appear in $\mathfrak{M}^*$.

10 Possible Modifications

Inasmuch as our theory tries to cope with "reality," it is flexible enough to allow for some modifications.

For instance, if players are faced with the game treated in Example 7.2, they may claim that the demand for stability is too strong. They would rather relax this demand and still gain something from the game.

One can then offer them the following definition of a bargaining set $\mathfrak{M}_1$:

DEFINITION 10.1 A c.r.p.c. $(x; \mathscr{B})$ belongs to the bargaining set $\mathfrak{M}_1$, if for any objection K against L, there is somebody in L who can counterobject.

According to this definition, each player in a coalition B_j which contains K, who does not belong to the partners of K, is required to be able to counterobject; but several such players may perhaps be unable to protect their shares simultaneously. Clearly, the resulting bargaining set $\mathfrak{M}_1$ *includes* $\mathfrak{M}$, since the number of sets which is required to counterobject is reduced. In this case, e.g., the players of the game treated in Example 7.2 may agree to

$$(\tfrac{1}{3}, \tfrac{1}{3}, \tfrac{1}{3}, 0, 0, 0; 123, 4, 5, 6), \tag{10.1}$$

which belongs to $\mathfrak{M}_1$.

In some other real-life cases, one can estimate and tell in advance which coalitions may object and which coalitions may counterobject. This leads to various bargaining sets and brings us to the circle of ideas surrounding ψ-stability. (See Luce and Raiffa [3], pp. 163–168, 174–176, 220–236.)

One may limit K to be always one-person and L to be the remaining members of the coalition, except for K's partners. This type of stability of one against the rest, which generates a bargaining set $\mathfrak{M}_2$, is still different from the stability demanded in $\mathfrak{M}$, as the following example shows:

Example 10.1 Consider the game:

$$n = 5, \quad v(i) = 0, \quad v(123) = 30, \quad v(24) = v(35) = 50,$$
$$v(1245) = v(1345) = 60. \tag{10.2}$$

Let

$$(x; \mathscr{B}) \equiv (10, 10, 10, 0, 0; 123, 4, 5). \tag{10.3}$$

If $K = 1, 2,$ or 3, then the remaining players which belong to the same coalition and are not among his partners can always counterobject; but the objection for $K = 23$,

$$(0, 11, 11, 39, 39; 1, 24, 35), \tag{10.4}$$

has no counter objection because player 1 cannot keep his profits. Thus $(x; \mathscr{B}) \in \mathfrak{M}_2$ but $\notin \mathfrak{M}$.

It is easy to show that $\mathfrak{M} \subset \mathfrak{M}_2$.

Sometimes people would like to feel safe not only within their coalitions but also from "outside" threats. It may happen, e.g., that several players from *various coalitions* will threaten together other people from these coalitions. A reasonable way to cope with this strong demand for stability would be to allow K and L to belong to several coalitions, provided that K and L are required to intersect the same coalitions. This will bring us to a bargaining set $\mathfrak{M}_0$ which is *included* in $\mathfrak{M}$. Let us remark that $\mathfrak{M}_0 = \mathfrak{M}$ for the 2- and 3-person games, as well as for the 4-person game with only 1-, 3-, and 4-person coalitions permissible. If $n = 4$, where 1- and 2-person coalitions are permissible, one has to replace the inequalities of Appendix 2 by those listed in Appendix 3. Fortunately, these inequalities are satisfied in all the examples given in Section 8, and therefore Theorem 8.1 is valid if one replaces $\mathfrak{M}$ by $\mathfrak{M}_0$.

Finally, we would like to question the assumption of coalitional rationality. If we drop this condition, we may arrive at negative values in the bargaining set, but this does not have to bother us, since $(0, 0, \ldots, 0; 1, 2, \ldots, n)$ will certainly remain in the bargaining set, and therefore we can *demand* that the *restricted* bargaining set will contain only individually rational p.c.'s. However, we shall show in Example 10.2 that the resulting restricted bargaining set may still contain non-coalitionally rational p.c.'s.

Example 10.2 Let Γ be the game

$$v(i) = 0, \quad v(12) = v(45) = v(46) = v(56) = v(123) = 30, \quad v(34) = 10. \tag{10.5}$$

In this game, the non-coalitionally rational p.c.

$$(10, 10, 10, 0, 15, 15; 123, 4, 56) \tag{10.6}$$

is stable if one drops the condition of coalitional rationality. In fact, it then belongs to the restricted bargaining set, since otherwise there exists a p.c. $(x; \mathscr{B})$ in the bargaining set with

$$\sum_{i=1}^{6} x_i > 60. \tag{10.7}$$

This can only occur if the coalition 34 is formed. Since, in addition, $x_3 \geq 10$, $x_4 \geq 0$, player 3 gets 10. This is impossible because in this case, player 4 has a justified objection, due to the fact that $x_1 + x_2 = 30$.

We have thus shown that the restricted bargaining set may contain a non-coalitionally rational p.c., if this condition is dropped from the definition of the bargaining set.

11 Concluding Remarks

Perhaps the nearest to our theory is W. Vickrey's concept of self-policing patterns [8]. His objections—called "heretical imputations"—are similar to ours; however, his counter objections—named "penalizing policing imputations"—are quite different.

Both the heretical and the penalizing policing imputations are in Vickrey's case *imputations*, whereas this is not the case in our theory. His penalizing policing imputation insists that at least one member of the "heretical coalition" is punished, whereas we only demand that the set L will be able to hold on to its property. However, the main difference lies, perhaps, in the fact that Vickrey is looking for a *set* of imputations—"self-policing patterns"—which are stable as a whole,[8] while our bargaining set consists of payoff configurations, each one of which is stable in itself.

It has been pointed out in Section 10 that if a lack of communication is known to exist, one can incorporate ideas from ψ-stability theory into ours. Both theories stress the dependence of an outcome on the coalition structure which actually forms. However, ψ-stability theory requires the payoffs to be imputations,[9] whereas we require that the outcome satisfies (2.4). (See, e.g., R. D. Luce and H. Raiffa [3], p. 222.) The coalitional rationality requirement (2.5) is a special case of a ψ-stability requirement, if one requires that all the subsets of the coalitions in the coalition structure τ are values of $\psi(\tau)$. A similar requirement appears also in Milnor's class L of reasonable outcomes ([3], pp. 240–242).

In many practical situations, the characteristic function is not the best way to describe a game. It would rather be better to apply the "Thrall characteristic function" (see R. M. Thrall [7]), which associates with each *coalition structure* a value for each coalition appearing in that structure. One can try to define the concepts of objections and counter objections, for such cases, and it is possible to do so in various ways.

---

8. In particular, he looks for "strong solutions."
9. Or at least feasible individually rational n-tuples ([3], p. 226).

It is also possible to apply the notions described in this paper to the Aumann-Peleg characteristic function for cooperative games without side payments [2], essentially without change.

Finally, we should like to point out that our theory gives in many cases answers similar to those appearing in classical theories. Thus, e.g., the bargaining set in the discrete case of the 3-person non-zero-sum game[10] consists essentially of the "central" three points of the non-discriminatory von Neumann–Morgenstern solution (see [9], pp. 550–554), but does not contain the additional "wiggles" that occur in their solution. The bargaining set for the non-discrete case is essentially the core. This suggests a pattern in which the bargaining set forms the "central" or "intuitive" part of a von Neumann–Morgenstern solution, whereas the "complications" disappear.

Appendix 1

Let Γ be a 4-person game, the coalitions of which, and their values, are given by (6.1). In order that the pair $(x_1, x_2, x_3, 0; 123, 4)$ belong to the bargaining set $\mathfrak{M}$, it is necessary and sufficient that

$$0 \leqslant x_1, \quad 0 \leqslant x_3, \quad x_1 + x_3 \leqslant a, \quad x_2 = a - x_1 - x_3, \tag{A1.1}$$

and that at least one inequality (or equality) in each of the following rows should be satisfied.

$x_1 + x_3 \geqslant c$	$2x_1 + x_3 \geqslant a + c - d$	$x_1 + x_3 = a$
$x_3 \leqslant a - b$	$x_3 - x_1 \leqslant d - b$	$x_3 = 0$
$x_1 \leqslant a - d$	$2x_1 + x_3 \leqslant a + c - d$	$x_1 = 0$
$x_3 \leqslant a - b$	$x_1 + 2x_3 \leqslant a + c - b$	$x_3 = 0$
$x_1 + x_3 \geqslant c$	$x_1 + 2x_3 \geqslant a + c - b$	$x_1 + x_3 = a$
$x_1 \leqslant a - d$	$x_3 - x_1 \geqslant d - b$	$x_1 = 0$

Appendix 2

Let Γ be a 4-person game, the coalitions of which, and their values, are given by (8.1). In order that the pair $(x_1, x_2, x_3, x_4; 12, 34)$ belong to the

---

10. Clearly, one has to modify the characteristic function in the obvious way so as to get superadditivity.

bargaining set $\mathfrak{M}$, it is necessary and sufficient that

$$0 \leqslant x_1 \leqslant a, \quad 0 \leqslant x_3 \leqslant c, \quad x_1 + x_2 = a, \quad x_3 + x_4 = c, \qquad \text{(A2.1)}$$

and that at least one inequality (or equality) in each row be satisfied.

If the partition 12, 34 is maximal (in the sense of (8.2)), the last column can be omitted.

$x_1 = a$	$x_1 + x_3 \geqslant d$	$2x_1 \geqslant a + d - b$	$x_1 + x_3 \geqslant a + c - e$
$x_1 = a$	$x_1 - x_3 \geqslant f - c$	$2x_1 \geqslant a + f - e$	$x_1 - x_3 \geqslant a - b$
$x_1 = 0$	$x_1 - x_3 \leqslant a - b$	$2x_1 \leqslant a + d - b$	$x_1 - x_3 \leqslant f - c$
$x_1 = 0$	$x_1 + x_3 \leqslant a + c - e$	$2x_1 \leqslant a + f - e$	$x_1 + x_3 \leqslant d$
$x_3 = c$	$x_1 + x_3 \geqslant d$	$2x_3 \geqslant c + d - f$	$x_1 + x_3 \geqslant a + c - e$
$x_3 = c$	$x_1 - x_3 \leqslant a - b$	$2x_3 \geqslant c + b - e$	$x_1 - x_3 \leqslant f - c$
$x_3 = 0$	$x_1 - x_3 \geqslant f - c$	$2x_3 \leqslant c + d - f$	$x_1 - x_3 \geqslant a - b$
$x_3 = 0$	$x_1 + x_3 \leqslant a + c - e$	$2x_3 \leqslant b + c - e$	$x_1 + x_3 \leqslant d$

Appendix 3

The following inequalities replace those given in Appendix 2, if one desires that $(x_1, x_2, x_3, x_4; 12, 34)$ shall belong to $\mathfrak{M}_0$. (See Section 10.) Again, at least one inequality in each row should be satisfied, as well as those given in (A2.1).

$x_1 + x_3 \geqslant d$	$x_1 = a$ $2x_3 \geqslant c + d - f$	$x_3 = c$ $2x_1 \geqslant a + d - b$	$x_1 + x_3 \geqslant a + c - e$
$x_1 - x_3 \geqslant f - c$	$x_1 = a$ $2x_3 \leqslant c + d - f$	$x_3 = 0$ $2x_1 \geqslant a + f - e$	$x_3 - x_1 \leqslant b - a$
$x_1 - x_3 \leqslant a - b$	$x_1 = 0$ $2x_3 \geqslant b + c - e$	$x_3 = c$ $2x_1 \leqslant a + d - b$	$x_1 - x_3 \leqslant f - c$
$x_1 + x_3 \leqslant a + c - e$	$x_1 = 0$ $2x_3 \leqslant b + c - e$	$x_3 = 0$ $2x_1 \leqslant a + f - e$	$x_1 + x_3 \leqslant d$

Bibliography

[1] Aumann, R. J. and Maschler, M., "An equilibrium theory for *n*-person cooperative games," *Notices American Math. Soc.*, Vol. 8 (1961), p. 261.

[2] Aumann, R. J. and Peleg, B., "von Neumann-Morgenstern solutions to cooperative games without side payments," *Bull. Amer. Math. Soc.* 66 (1960), pp. 173–179 [Chapter 38].

[3] Luce, R. D. and Raiffa, H., *Games and Decisions*, John Wiley and Sons, New York (1957).

[4] Maschler, M., "Playing an n-person game: an experiment," in *Contributions to Experimental Economics 8, Coalition Forming Behavior*, H. Sauermann (ed.), J. C. B. Mohr (Paul Siebeck), Tübingen, 1978, pp. 231–328.

[5] Maschler, M., "Stable payoff configurations for quota games," *Advances in Game Theory*, Annals of Mathematics Studies No. 52, Princeton University Press, Princeton (1964), pp. 477–499.

[6] Maschler, M., "n-Person games with only 1, $n-1$, and n-person permissible coalitions," *J. Math. Analysis and Applications*, Vol. 6 (1963), pp. 230–256.

[7] Thrall, R. M., "Generalized characteristic functions for n-person games," *Recent Advances in Game Theory, Proceedings of a Princeton University Conference, October 4–6, 1961*, ed. by M. Maschler. Philadelphia: Ivy Curtis Press. (1962), pp. 157–160.

[8] Vickrey, W., "Self-policing properties of certain imputation sets," *Contributions to the Theory of Games*, Vol. IV, Annals of Mathematics Studies, No. 40, Princeton University Press, Princeton (1959), pp. 213–246.

[9] Von Neumann, J. and Morgenstern, O., *Theory of Games and Economic Behavior*, 2nd ed., Princeton University Press, Princeton (1947).

43 A Method of Computing the Kernel of *n*-Person Games
with B. Peleg and P. Rabinowitz

1. Introduction. During the last three years we have been witnessing the growth of a new theory for n-person characteristic-function games—the theory of bargaining sets. This theory tries to answer the following basic question: Given a partition of the players into coalitions, how should the payoff to each of these coalitions be divided among its members?

Various answers are given in [1], [4] and [3], where several bargaining sets are defined. The bargaining sets are sets of "stable" payoff distributions; roughly speaking, a payoff distribution is "stable" if the players in each coalition can effectively counter any threats by their coalition partners to obtain higher payoffs by leaving the old coalition in order to combine with outside players. This intuitive notion can be made precise in a number of different ways, leading to the various bargaining sets.

For several bargaining sets there are general existence theorems ([4], [11] and [3]); for other bargaining sets, general existence theorems cannot be proved [1, Section 7]. After the existence question had been settled, the problem of computing the bargaining sets arose. Since every bargaining set is given by a system of linear inequalities in the space of the payoffs [1, Theorem 2.1] there is no difficulty in principle, and the only problem is to find a practical method of computation. In this paper we give such a method for the kernel,[1] a bargaining set that was defined and investigated by Davis and Maschler in [3], and has been the subject of considerable subsequent research [9, 12, 13]. The kernel is the easiest of the bargaining sets to compute.

The computational method described here is practical for games with 5 or fewer players, whose characteristic functions take small integer values. For such games, a computer program was written for the CDC-1604A at the Weizmann Institute of Science; all computations reported here were performed on that machine. The program was used to compute the kernels of all weighted majority games with 5 or fewer players, and all extreme zero-sum games with 5 players; the results of these computations are tabulated in this paper.

The kernel is defined in §2. In §3 we discuss the problems connected with computing the kernel, and describe our method. §4 contains several estimates needed for the programming of the method. In §§5 and 6 a detailed description of the program is given. The tables mentioned above are in §§7 and 8, together with descriptive material; in particular, in §7 it is shown how the table in that section may be applied to certain games with more than 5 players.

It should be emphasized that the method used here is applicable for a class of

---

[1] The research described in this paper was partially supported by the U. S. Office of Naval Research, Logistics and Mathematical Statistics Branch, under Contract Number N62558-3586.

This chapter originally appeared in *Mathematics of Computation* 19 (1965): 531–551. Reprinted with permission.

computational problems of which the bargaining set is typical, but that need have no relation with game theory. Specifically, if a set in euclidean space is defined by a (possibly large) number of linear inequalities connected (possibly in a very complicated way) by "or" and "and", and if it is desired to express the set as a union of convex polyhedra, then the method of this paper may be applicable if the dimension is sufficiently small and the coefficients of the inequalities are small integers.

2. Definitions. In this section we give the basic definitions that are used in the rest of the paper.

An *n person game* is a set N with n members, together with a real function v, defined on the subsets of N. v is the *characteristic function* of the game. The members of N are called *players*, and will be denoted by the numbers $1, \cdots, n$. Subsets of N are called *coalitions*. We assume that v satisfies the following conditions: $v(\{i\}) = 0$ for $i = 1, \cdots, n$, $v(\varnothing) = 0$, and $v(S) \geq 0$ for all $S \subset N$.

Let (N, v) be an n-person game. A *coalition structure* is a partition of N. Intuitively, when (N, v) is played, the players are partitioned according to a certain coalition structure, and each coalition S in the partition divides its share $v(S)$ among its members. A *payoff vector* is an n-tuple $\{x_i\}_{i \in N}$ of real numbers,[2] one number for each $i \in N$. If each player refuses to receive less than what he can get alone, namely zero, then a possible outcome of the game, which will be called an *individually rational payoff configuration*, is a pair $(x, \mathcal{B})$, where $\mathcal{B}$ is a coalition structure, and x, the distribution of the payoffs, is a payoff vector that satisfies: $x_i \geq 0$ for $i = 1, \cdots, n$, and $\sum_B x_i = v(B)$ for all $B \in \mathcal{B}$.

Let $(x, \mathcal{B})$ be an individually rational payoff configuration. For each coalition S, set

$$e(S, x) = v(S) - \sum_S x_k.$$

Now let $i, j \in B \in \mathcal{B}$ and $i \neq j$; set

$$T_{ij} = \{S : S \subset N, i \in S \text{ and } j \notin S\},$$

and

$$s_{ij}(x) = \max \{e(S, x) : S \in T_{ij}\}.$$

We say that i *outweighs* j if $x_j > 0$ and $s_{ij}(x) > s_{ji}(x)$. The payoff vector x is *balanced* if there exists no pair of players h and k in the same B in $\mathcal{B}$ such that h outweighs k. The *kernel* is the set of all balanced payoff vectors. It depends, of course, on the coalition structure $\mathcal{B}$.

Intuitively, when i and j are in the same coalition of $\mathcal{B}$ and $s_{ij}(x)$ is non-negative, then it is the maximum amount with which i can "manouevre" (i.e. take for himself or offer to his partners) in case he wants to set up a new coalition that will exclude j. When $s_{ij}(x)$ is negative it no longer has the above meaning, but we think it is still a measure of i's power to threaten j. Of course player j is immune from threats if $x_j = 0$, for he can get 0 without help from any other player. But if $x_j > 0$, then i outweighs j if i's power to threaten j exceeds j's power to threaten i.

---

[2] Formally, it is a function from N to the real numbers.

The payoff vector is balanced if no player outweighs another player in the same coalition.

The reader who wishes a more extensive intuitive and theoretical (rather than computational) discussion of the kernel concept, as well as a detailed discussion of the kernel of a particular game, is referred to [3]. Also, in [3] it is proved that the kernel is never void, i.e. if (N, v) is an n-person game and $\mathcal{B}$ is a coalition structure, then there always exists a payoff vector x that is balanced [3, Theorem 5.4]. However, most of the methods of this paper do not depend on [3], and a reader not familiar with [3] should have no difficulty in understanding this paper.

3. Method of Computation. Let (N, v) be an n-person game and let $\mathcal{B}$ be a coalition structure. Denote by $X(\mathcal{B})$ the set of all payoff vectors x such that $(x, \mathcal{B})$ is an individually rational payoff configuration. Explicitly,

$$X(\mathcal{B}) = \{x : x_i \geq 0 \text{ for } i = 1, \cdots, n, \text{ and } \sum_B x_i = v(B) \text{ for } B \in \mathcal{B}\}.$$

Our problem is to find those payoff vectors in $X(\mathcal{B})$ that belong to the kernel.

The computational method that we use is based on the fact that for each $\mathcal{B}$, the kernel is a polyhedron that is not necessarily convex; that is, it is a finite union of convex polyhedra. To see this, note that a vector x is in the kernel if and only if

(3.1) $\quad x \in X(\mathcal{B})$ and for $i = 1, \cdots, n$, $x_i = 0$ or $s_{ij}(x) \geq s_{ji}(x)$

for all $j \in B - \{i\}$, where B is that coalition of $\mathcal{B}$ that contains i.

Now $s_{ij}(x)$ (and therefore also $s_{ji}(x)$) is the maximum of a finite number of linear functions of x, i.e. functions of the form $a \cdot x + c$, where a is a vector and c a scalar. Therefore (3.1) may be stated as a sentence $\mathcal{S}$ built from weak linear inequalities, by means of iterated conjunctions and disjunctions (i.e. by using the connectives "and" and "or"). Now any sentence built in such a way from a number of "primitive statements" (in our case linear inequalities), no matter how complex its structure, is equivalent to an appropriately chosen disjunction of conjunctions of the primitive statements. Therefore there are linear functions $g_{pq}(x)$, where p and q run over finite index sets (say from 1 to P and 1 to Q respectively), such that (3.1) is equivalent to:

(3.2) $\quad$ There is a p such that for all q, $g_{pq}(x) \geq 0$.

Each of the $g_{pq}(x)$ appears in the sentence $\mathcal{S}$, and it may be verified that each one is therefore either of the form $\sum_S x_i - v(S)$ for some S, or of the form $\sum_S x_i - \sum_T x_i - (v(S) - v(T))$ for some S and T; in particular, the g_{pq} have rational coefficients only, except possibly for the constant term. From the form (3.2) it follows immediately that the kernel is the union of P convex polyhedra, each determined by Q linear inequalities.

The process of reducing a complex sentence like (3.1) to the form (3.2) will be familiar to the reader from the Propositional Calculus; it consists essentially of repeated applications of the distributive law

$$a \text{ and } (b \text{ or } c) \Leftrightarrow (a \text{ and } b) \text{ or } (a \text{ and } c).$$

For fixed n and $\mathcal{B}$ it is of course possible explicitly to carry out this procedure, thereby obtaining numerical values for P and Q and listing all the g_{pq} (with the values of v entering as parameters). But our purpose in the foregoing was merely to show that the kernel is the union of finitely many convex polyhedra; this we have done, and for this purpose it is not necessary to exhibit P, Q, and the g_{pq} explicitly.

We wish to "compute" the kernel; what does this mean? If the kernel consists of one point, we would like to know the coordinates of that point; if it is an interval we would like the end points of the interval; if it is the union of two intervals, we would like the end points of each of the intervals. As we have seen, the kernel is a finite union of convex polyhedra; the end result of a "computation" of the kernel should be a listing of these convex polyhedra, where each one of them is described by listing its extreme points (i.e. vertices). Theoretically, we know how to achieve a breakdown into convex polyhedra: by using the rules of the propositional calculus to change (3.1) to (3.2). Possibly this approach could be used on a practical, computational level as well. Essentially, (3.2) presents the kernel as the union of P **convex polyhedra, each one defined by a set of linear inequalities.** A practical method for finding the vertices of a convex polyhedron defined by a set of inequalities is given by Balinski in [2]. Couldn't we find (3.2) explicitly, then use Balinski's method to calculate the vertices of each of the P resulting convex polyhedra?

The answer is no, and the reason is that P is very large—for $n = 5$, probably on the order of 10^4 or 10^5. To apply Balinski's method to so many polyhedra would be prohibitive. The very large value for P might lead the reader to think that it is really necessary to use many convex polyhedra to describe a kernel, but nothing could be further from the truth. The kernel is usually very simply described; indeed, it often consists of a single point. In the tables in §§7 and 8, there are only three instances in which the kernel is not itself convex, and in those cases it is the union of 2 convex polyhedra. The representation (3.2) is inefficient, because it holds for all games simultaneously; the characteristic function enters into the individual g_{pq} only as a parameter, and affects neither P nor Q. For a particular game, most of the P convex polyhedra are *empty*, or coincide with each other. What is needed is some way of computing the kernel without first writing it explicitly in the form (3.2). We shall now outline our solution to this problem.

From now on it will be assumed that (N, v) is *integral*, i.e. that the characteristic function v takes integer values only. In theory, everything in the sequel applies to arbitrary integral games, and so by S-equivalence [8, p. 197] to arbitrary rational games. In practice, though, the method is useable only when the values of v are *small* integers, probably 5 or 6 at most.

For the given game (N, v) and coalition structure $\mathcal{B}$, denote by $C_1, \cdots, C_P$ the convex polyhedra defined by (3.2); that is,

$$C_p = \{x \in E^n : g_{pq}(x) \geqq 0 \text{ for } q = 1, \cdots, Q\}.$$

A given C_p may be empty, or it may coincide with others having different indices. The purpose of the computation is to provide a list of the nonempty C_p's, each one exactly once, and each one described by listing its extreme points. The straightforward way of doing this would seem to be first somehow to characterize the C_p's in question, then to determine the extreme points of each one. For the reasons explained above, this straightforward way is impractical. The approach we use is apparently "upside-down": first *all* the extreme points of *all* the C_p are listed, and only afterwards are they sorted out.

Let x be one of the extreme points in question. x is the unique solution of a system of n linear equations, which have one of the following two forms: $\sum_S x_i = v(S)$, or $\sum_S x_i - \sum_T x_i = v(S) - v(T)$. Since the game is integral, x must be a rational point, and it has a denominator which does not exceed a bound K_n, which is not greater than the maximal value attained by a determinant of order n whose entries are 1, 0, or -1. Now, if for each rational point in $X(\mathcal{B})$ with denominator[3] $\leqq K_n$, we determine whether or not that point is in the kernel (i.e. satisfies the inequalities (3.1)), we obtain a list of points that contains all the extreme points in question. Moreover, if with each point x that we have found in this way we list the values $e(S, x)$ for all $S \subset N$, then we shall be able to group the points according to the various C_p to which they belong. It remains only to determine whether a given x is actually extreme in at least one of the C_p; how this is done is described below.

Strictly speaking, the method just outlined can be exploited only to solve 4-person games; when $n > 4$, there are too many rational points with denominator $\leqq K_n$. To overcome this difficulty, we use the following procedure: first, for a certain natural K we examine all the rational points with denominator K, to find which of them are *near* the kernel, i.e. satisfy the inequalities (3.1) approximately (within ϵ, say). This part of the procedure is called *step* A. We then note that the kernel itself must be contained in a neighborhood of the set of points found in step A, the size of the neighborhood being determined by ϵ. Now the extreme points that we are seeking must all be in this neighborhood, must be rational with denominator not exceeding K_n, and must satisfy the inequalities (3.1)—or equivalently, (3.2)—precisely. Moreover, for each such extreme point x, the rank of those of the linear inequalities in (3.2) that are actually satisfied as equalities by x must be n; indeed this is a necessary and sufficient condition for extremality. We therefore check the inequalities (3.1) for all points in $X(\mathcal{B})$ with denominators $\leqq K_n$, that are in the neighborhood determined by step A. Those points that are found to satisfy the inequalities are then examined for extremality by the rank method. This second (and last) part of the procedure is called *step* B.

To avoid round-off, integer arithmetic was used throughout the program.

4. Various Estimates. In this section we shall compute certain bounds needed for the programming of steps A and B for five person games.

We denote by Q_n the maximal value attained by a determinant of order n whose entries are 1, 0, or -1. Hadamard's inequality yields $Q_n \leqq n^{n/2}$, so $Q_5 \leqq 5^{5/2} \approx 55.9$. In fact, $Q_5 = 48$. Actually we are interested not in Q_5 but in K_5.

LEMMA 4.1. $K_5 \leqq 36$.

Proof. Let A be a 5×5 regular matrix whose entries are 0, 1 or -1, b an integral vector and x the solution of the system of equations $Ax = b$. If all the zeros of A are contained in at most two rows or two columns, then $|A| \equiv 0 \pmod{4}$, and each minor of order 4 is a multiple of 2. So in this case the denominator of x does not exceed $48/2 = 24$. Next, suppose that the zeros of A are not in two rows or two columns. Clearly, it is sufficient to show that when there are exactly three zeros,[4]

---

[3] I.e., least common denominator of all the coordinates.
[4] If there are four zeros, not contained in two rows or two columns, then Hadamard's inequality gives $|A| \leqq 36$.

$|A| \leq 36$. W.l.o.g. we may assume that $|A|$ has the following form:

$$\begin{vmatrix} 0 & a_{12} & a_{13} & a_{14} & 1 \\ a_{21} & 0 & a_{23} & a_{24} & 1 \\ a_{31} & a_{32} & 0 & a_{34} & 1 \\ a_{41} & a_{42} & a_{43} & a_{44} & 1 \\ 1 & 1 & 1 & 1 & 1 \end{vmatrix}.$$

We now develop $|A|$ with respect to the last two rows, using Laplace's rule. It is not difficult to see that at most six of the 2×2 determinants that appear are non-zero, and that of the three 3×3 determinants with only one zero, only one can equal 4, and the other two are not greater than 2; the other determinants are not greater than 3. Using these remarks we obtain $|K_5| \leq 36$.

The exact value of K_5 is not known.

Let (N, v) be an integral 5-person game, and let $\mathcal{B}$ be a coalition structure. The number of rational points in $X(\mathcal{B})$ with denominator K depends on the values $v(B)$ for $B \in \mathcal{B}$. We shall now compute it in the following simple case: $\mathcal{B} = \{N\}$ and $v(N) = 1$. Denote by $Z(n, K)$ the number of integral points satisfying $x_i \geq 0$, $i = 1, \cdots, n$ and $\sum_{i=1}^n x_i = K$. Then

(4.2) $$Z(n, K) = \sum_{j=0}^{K} Z(n-1, j).$$

Using this relation we obtain

(4.3) $$Z(5, K) = (K^4 + 10K^3 + 35K^2 + 50K + 24)/24.$$

The number of rational points in $X(\mathcal{B})$ with denominator not exceeding K_5 is bounded by $\sum_{k=1}^{36} Z(5, K)$, which is of the same order of magnitude as $\sum_{k=1}^{36} K^4/4! \approx 36^5/5!$.

The following lemma gives an estimate of the density of the rational points in $X(\mathcal{B})$ with given denominator.

LEMMA 4.4. *Let K be a natural number. If $x \in X(\mathcal{B})$, then there is a rational point $y \in X(\mathcal{B})$ with denominator K, such that*:

(4.4.1) $\qquad\qquad$ *if* $x_i = 0$ *then* $y_i = 0$, *and*

(4.4.2) $\qquad |x_i - y_i| \leq 4/(5K) = \delta(K)$, *for* $i = 1, \cdots, 5$.

The proof, which is elementary, is omitted. We remark that the bound $4/(5K)$ cannot be improved.

We now want to estimate the rate of variation of the functions $s_{ij}(x)$. A convenient norm is $\|x\| = \max\{|x_1|, \cdots, |x_5|\}$.

LEMMA 4.5. *Let $i, j \in B \in \mathcal{B}$, $i \neq j$, and let $h_{ij}(x) = s_{ij}(x) - s_{ji}(x)$, for $x \in X(\mathcal{B})$. If x and x' are in $X(B)$, then $|h(x) - h(x')| \leq 4\|x - x'\|$.*

Proof. Let $s_{ij}(x) = e(S, x)$, $s_{ij}(x') = e(S', x')$, $s_{ji}(x) = e(T, x)$ and $s_{ji}(x') = e(T', x')$. We may assume that $h(x) \geq h(x')$; then we have $|h(x) - h(x')| = h(x) - h(x') = e(S, x) - e(T, x) - \{e(S', x') - e(T', x')\} \leq e(S, x) - e(S, x') + e(T', x') - e(T', x) = \sum_S (x_k' - x_k) + \sum_{T'} (x_k - x_k')$. If $S \cap T' \neq \emptyset$ or $S \cup T' \neq N$, then the number of the differences that are not cancelled does not exceed 4. So, in this case $|h(x) - h(x')| \leq 4\|x - x'\|$. If $S \cap T' = \emptyset$ and

$S \cup T' = N$, we have $\sum_S(x_k' - x_k) = \sum_{T'}(x_k - x_k')$. One of the sets, S or T', does not contain more than two players; so also in this case $|h(x) - h(x')| \leq 4 \|x - x'\|$. This completes the proof of the lemma.

The following definitions and lemma are the core of step A.

A subset A of $X(\mathfrak{B})$ is a δ-*kernel*, if for each x in the kernel there is a $y \in A$ such that $\|y - x\| \leq \delta$. A point $x \in X(\mathfrak{B})$ is ϵ-*balanced* if it satisfies the following systems of inequalities:

(4.6)
$$\text{for } i = 1, \cdots, n \quad x_i = 0, \quad \text{or} \quad s_{ij}(x) \geq s_{ji}(x) - \epsilon$$
$$\text{for all } j \in B - \{i\}, \text{ where } B \text{ is that coalition of } \mathfrak{B} \text{ that contains } i.$$

LEMMA 4.7. *Let K be a natural number. The set of rational points in $X(\mathfrak{B})$ with denominator K that are $3/K$-balanced, is a $4/5$-kernel.*

Proof. Let $x \in X(\mathfrak{B})$ be balanced. Using Lemma 4.4 we obtain a rational y with denominator K, with the same zero coordinates as x and such that $|x_j - y_j| \leq 4/(5K)$, $j = 1, \cdots, 5$. We shall complete the proof by showing that y is $3/K$-balanced. If y is not $3/K$-balanced then there exist players $i, j \in B \in \mathfrak{B}$, $i \neq j$, such that $y_i > 0$ and $h_{ij}(y) = s_{ij}(y) - s_{ji}(y) < -3/K$. Since $h_{ij}(y)$ is a rational number with denominator K, we must have $h_{ij}(y) \leq -4/K$. On the other hand, using Lemma 4.5, we have

$$h_{ij}(y) = h_{ij}(x) + (h_{ij}(y) - h_{ij}(x))$$
$$\geq h_{ij}(x) - |h_{ij}(y) - h_{ij}(x)| \geq - |h_{ij}(y) - h_{ij}(x)|$$
$$\geq -4 \|x - y\| \geq -4 \cdot 4/(5K) > -4/K$$

a contradiction which shows that the assumption that y is not $3/K$-balanced is false.

5. Step A. Let (N, v) be an integral five person game and $\mathfrak{B}$ a coalition structure. The problem that confronts us in step A is how to choose from all the rational points in $X(\mathfrak{B})$, a set of points, with the same (not too large) denominator K, that will be a δ-kernel with volume as small as possible. Given K, Lemma 4.4 and the subsequent remark tell us that the best possible δ is $\delta(K) = 4/(5K)$. Having determined δ we have to decide upon the way of choosing the points, so as to obtain a $\delta(K)$-kernel with a minimum number of points. We do this by choosing all ϵ-balanced points, where ϵ is determined by Lemma 4.7, i.e., $\epsilon = \epsilon(K) = 3/K$.

We can now describe step A formally. Let K be a natural number and $\epsilon = 3/K$. Step A consists of finding all rational points in $X(\mathfrak{B})$ with denominator K that are ϵ-balanced.

There remains only the problem of determining K. If K is determined then so is ϵ, and therefore the set of points that will pass step A is determined. In step B we shall have to check all the rational points with denominator not exceeding K_5, that are in a $\delta(K)$-neighborhood of this set. Increasing K increases the number of points that must be checked in step A, but, on the other hand, decreases ϵ, and therefore decreases the proportion that will pass step A; in addition $\delta(K)$ also decreases, so that we can expect that the "volume" we shall have to inspect in step B decreases. This conjecture was verified in the experiments that we performed.

We carried out step A for various games and coalition structures, with denominators K in the range 12–24, and observed that the number of points that passed step A was independent of K in this range; we concluded that increasing K decreases the time needed for step B. Clearly, increasing K increases the time needed for step A itself. Now, what we want, is to minimize the total time needed for steps A and B, so we have to find a compromise. A description of how this worked out for simple games may be found toward the end of §7.

The number of points that passed step A varied greatly in the various experiments that we performed. We checked only characteristic functions v and coalition structures $\mathcal{B}$ where the values $v(B)$ for $B \in \mathcal{B}$ were either 1 or 0. For the denominator 24 we received from 4 to 240 points. The calculations in §4 (see (4.3)) show that when $\mathcal{B} = \{N\}$ and $v(N) = 1$, the number of points that must be checked in step A is about 22000, so that the proportion that passed step A did not exceed approximately one percent.

We shall now proceed to describe the computations. The data of step A is the characteristic function v, the coalition structure $\mathcal{B}$, the denominator K and $\epsilon = 3/K$. The computer multiplies v and ϵ by K, so that in all the computations *only integral numbers appear*. The set $X(\mathcal{B})$ is thus transformed into $X_K(\mathcal{B}) = \{x : x_i \geq 0$ for $i = 1, \cdots, 5,$ and $\sum_B x_i = Kv(B)$ for $B \in \mathcal{B}\}$. Then the computer examines all the integral points in $X_K(\mathcal{B})$, taken in lexicographic order; those points that satisfy equations (4.6) with $\epsilon = 3$ are stored in the memory for use in step B.

The flow chart in Figure 1 describes the order of the computations for step A. All the expressions $e(S, x)$ are computed immediately after a point x is generated. The functions $s_{ij}(x)$ are computed only when needed in the course of the check.

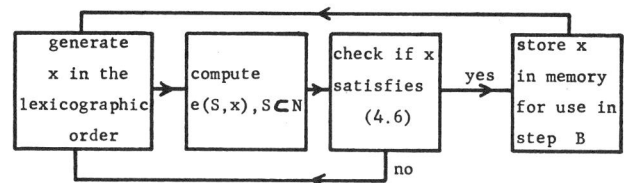

FIGURE 1. Order of computations for step A.

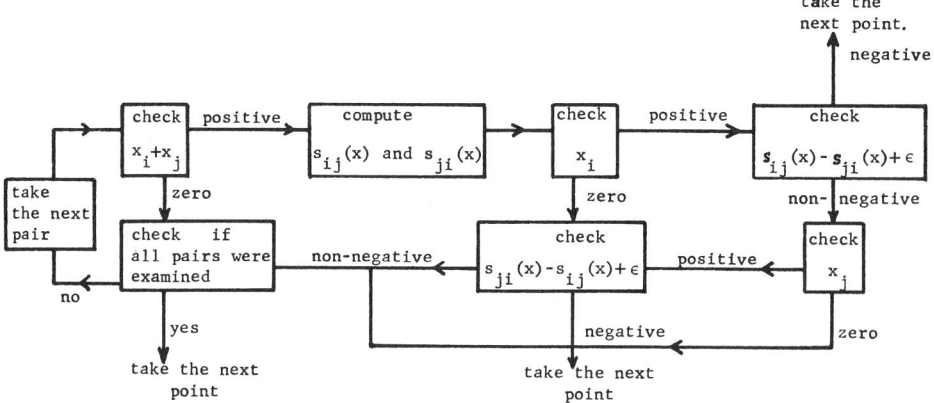

FIGURE 2. Check if x satisfies (4.6).

The check deserves special attention: it consists of comparing the functions $s_{ij}(x)$ and $s_{ji}(x)$, for all the pairs $\{i,j\}$ of different players that belong to the same coalition $B \in \mathcal{B}$. The details are given in the flow chart in Figure 2.

When the point stored in the memory differs from its predecessor only in the last two coordinates, which is the case most frequently encountered, the expressions $e(S, x)$ are computed directly from those of the predecessor, in an easy way.

Another device to save time is the following: when a point is rejected, the check of the next point begins by examining the last pair that was inspected before, i.e. the pair that caused the rejection.

6. Step B. The data of step B consists of the data of Step A plus the list of the points that were found $3/K$-balanced in step A, where K is the denominator of step A. These points form a set which is a $\delta(K)$-kernel. Step B consists of finding all the rational points with denominator not exceeding K_5 that are in the intersection of the kernel and the $\delta(K)$-neighborhood of the kernel determined by step A, and determining which of these points are extreme in the polyhedra defined by (3.2).

The new problems in step B are how to generate the points to be examined, and how to check extremality. The procedure of actually checking whether the point is in the kernel is the same as that of step A, except that now we examine if the points satisfy (3.1) and not (4.6).

The order of the computations is as follows. We take the denominators we have to examine in their natural order. When we reach a denominator K_1 we multiply v by K_1, so that only integral numbers will appear in the computations. After the multiplication, the points that passed step A are taken from the list, and for each point we examine all the integral points of $X_{K_1}(\mathcal{B})$ that are in a $\delta(K)K_1$-neighborhood of it, where K is the denominator of step A, and

$$X_{K_1}(\mathcal{B}) = \{x_i : x_i \geq 0 \text{ for } i = 1, \cdots, 5 \text{ and } \sum_B x_i = K_1 v(B) \text{ for } B \in \mathcal{B}\}.$$

For a point taken from the list of step A, we compute first the bounds that define its $\delta(K)K_1$-neighborhood and afterwards we check all the integral points in the neighborhood, taken in the lexicographic order, for membership in the kernel.

Each point x that emerges from this procedure is in the kernel. The next task is to check for extremality, i.e. to check whether x is an extreme point of one of the polyhedra C_p defined by (3.2). Now it is possible that x is in more than one of these polyhedra. Let E_x be the intersection of all the C_p to which x belongs. Certainly, if x is extreme in one of the C_p, then it is extreme in E_x; the converse is also true.[5] What we must therefore check is whether x is extreme in E_x.

Among the linear inequalities $a \cdot y \geq b$ defining E_x, some are satisfied as equalities by x; denote these by

$$a_1 \cdot y \geq c_1, \quad a_2 \cdot y \geq c_2, \cdots, a_m \cdot y \geq c_m,$$

where the a_i are vectors depending on x and the c_i are scalars depending on x. Thus we have

(6.1) $$a_1 \cdot x = c_1, \quad a_2 \cdot x = c_2, \cdots, a_m \cdot x = c_m.$$

---

[5] This follows from the fact that the intersection of any two C_p is a face of both.

The point x is extreme in E_x if and only if there are 5 linearly independent a_i, i.e. the rank of the matrix A_x formed by the a_i is 5. To find the rank of A_x, we form a matrix $D(x)$ of zeros and ones as follows: to each pair of distinct players i and j in the same coalition $B \in \mathfrak{B}$, there is a row $D_{(ij)}$ in the matrix. Let the coalitions[6] in T_{ij} be $S_{ij}^1, \cdots, S_{ij}^8$; then $D_{(ij)} = (d_{(ij)}^1, \cdots, d_{(ij)}^8)$, where $d_{(ij)}^k = 0$ or 1, according as $e(S_{ij}^k) < s_{ij}(x)$ or $e(S_{ij}^k) = s_{ij}(x)$. Consider the system of linear equations

$$e(S_{ij}^k) = e(S_{ji}^{k'}),$$

where k, k', i, j, range over values such that

(6.2) $\qquad d_{(ij)}^k = d_{(ji)}^{k'} = 1$ and $s_{ij}(x) = s_{ji}(x)$.

This system is closely related to the system (6.1). More precisely, let f_{ij}^k be the incidence vector[7] of the coalition S_{ij}^k, and let F_x be the matrix whose rows are the unit vectors e^i for i obeying $x_i = 0$, the incidence vectors of the coalitions B in the coalition structure $\mathfrak{B}$, and the vectors $f_{ij}^k - f_{ji}^{k'}$, where k, k', i, j, range over values such that (6.2) holds; then F_x has the same rank as A_x. The test of extremality that we use consists of finding the rank of F_x for each x. Once a point x has been found to be extreme (i.e. F_x has been found to have rank 5), it is stored in the memory together with the matrix $D(x)$, until all the extreme points have been found.

There remains only the problem of grouping the extreme points according to the various polyhedra C_p to which they belong. Now it can be proved that the kernel of a five person game is one-dimensional, i.e. consists of a finite union of points and line segments. So the only question we have to settle is which pairs of extreme points can be joined by a line segment in the kernel. Let x and y be extreme points, and let $D(x)$ and $D(y)$ be their matrices. Then there is a convex polyhedron C_p that contains both x and y, if and only if for each two players i and j belonging to the same coalition $B \in \mathfrak{B}$, either $x_i = y_i = 0$ or $\sum_{k=1}^{8} d_{(ij)}^k(x) d_{(ij)}^k(y) > 0$. So the question of "pairing" can be solved by computing the scalar products of the corresponding rows of $D(x)$ and $D(y)$.

The output is a list of extreme points, together with a list of all pairs such that the line segment joining them is in the kernel; it gives a complete picture of the kernel.

7. Weighted Majority Games. We recall a number of definitions and elementary facts. A game $G = (N, v)$ is called *superadditive* if for every pair of disjoint coalitions S and T, we have $v(S \cup T) \geq v(S) + v(T)$; *monotonic* if for every pair of coalitions S and T such that $S \supset T$, we have $v(S) \geq v(T)$; *constant-sum* if $v(S) + v(N - S) = v(N)$ for all S. A *simple* game is one whose characteristic function takes the values 0 and 1 only; then S is called *winning* if $v(S) = 1$, and *losing* if $v(S) = 0$. A *weighted majority game* is a simple game for which there exist n non-negative numbers $w_1, \cdots, w_n$, and a real number q (usually non-negative) such that S is winning if and only if $\sum_S w_i \geq q$. Then $w_1, \cdots, w_n$ are called *weights* for the players, and q is called a *quota*; $[q; w_1, \cdots, w_n]$ is called a *representation* of the game, though often we will use it to refer to the game itself. The representation of

---

[6] Because $n = 5$ there are exactly 8 coalitions in each T_{ij}.

[7] The vector whose rth member is 0 or 1 according as r is or is not in S_{ij}^k.

a given weighted majority game is in no sense unique. Two representations of the same game are called *equivalent*. A simple game is superadditive if and only if there is no pair of disjoint winning coalitions. A weighted majority game is always monotonic, but it need be neither superadditive nor constant-sum; for example, [3; 1, 1, 2, 2] is neither. Of course, both superadditivity and constant-sumness are *intrinsic* properties, i.e. they depend on the game (characteristic function) and not on its representation.

A player i in a game (N, v) is called *at least as desirable* as another player j (in symbols $i \gtrsim j$) if $v(S \cup \{i\}) \geq v(S \cup \{j\})$ for all S containing neither i nor j. If $i \gtrsim j$ and $j \gtrsim i$ then i and j are called *symmetric* ($i \sim j$). Every weighted majority game has a representation $[q; w_1, \cdots, w_n]$ that completely reflects the desirability relation, i.e. such that $i \gtrsim j$ if and only if $w_i \geq w_j$; but not every representation need do this. Two games (N, v) and (M, u) are said to have the same *desirability pattern* if there is a one-one mapping φ of N onto M that preserves desirability, i.e. so that $i \gtrsim j$ if and only if $\varphi(i) \gtrsim \varphi(j)$. The desirability pattern is an intrinsic property of the game.

A *veto player* i in a simple game (N, v) is a player such that any coalition S not containing i is losing. There may be more than one veto player in a game. A *dummy* is a player i such that $v(S \cup \{i\}) = v(S)$. If i is a dummy in a weighted majority game then the game has a representation in which $w_i = 0$.

In this section we tabulate the kernels of all weighted majority games with at most 5 players, for the coalition structure consisting of the set N of all players only. From this table it is possible to compute the kernel of any superadditive weighted majority game (regardless of the number of players), for coalition structures in which the winning coalition[8] has at most 5 players. We first describe the arrangement of the table, then show how to compute the kernels of any weighted majority game (with the above restriction) from it.

In the table, the games are identified by representations. Since the kernel is an intrinsic property of the game, each game appears in only one representation. However, a user of the table may wish to find the kernel of a game for which he has a representation that does not appear in the table. In order to simplify the task of finding an equivalent representation in the table, it is desirable, insofar as possible, to classify the games in the table by intrinsic properties. This is done in two ways. *First*, the table is divided into three parts: Games with veto players; superadditive games without veto players; and nonsuperadditive games without veto players (all three properties are intrinsic). *Second*, the games in the last two parts are classified by desirability pattern. Though the desirability pattern cannot always be read off at once from a given representation of a game, for a 5-person game it can usually be determined rather quickly.

We first describe the second and third parts, which are the more interesting. The first column of the table is a serial number (to simplify references to the table). The second column signifies the desirability pattern; the notation used is self-explanatory. The symbol for the desirability pattern is printed only next to the first game with this pattern, to make the table easier to read. The third column gives a representation of the game. The kernel of the game, with coalition structure

---

[8] Because of superadditivity there is at most one.

N, is given in the fourth column. The kernel usually consists of a point or an interval; there is one exception (game 5-71) in which it is V-shaped. When the kernel consists of a single point, the point is given; when it is an interval, the two endpoints are given; in the V-shaped case, all three vertices are given, with the one common to the two intervals between the other two. A common denominator is signified by /; thus 12234/12 signifies the point $(1/12, 1/6, 1/6, 1/4, 1/3)$.

In the first part, the games with veto players are listed and given serial numbers, but the kernels are not given nor are the games classified by desirability pattern. It can be shown that the kernel of a game with veto players (for coalition structure $\{N\}$) consists of a single point, in which the veto players share the amount 1 equally, and the remaining players get 0. (Example: The kernel of [5; 1, 1, 2, 2] is 0011/2.) Thus the kernel can be determined immediately as soon as the game is given, and there is no need for tabulation. The list is included for the secondary purpose of completing the list of weighted majority games with at most 5 players.

Games with dummies are excluded. A weighted majority game in which player 1 is a dummy has a representation of the form $[q; 0, w_2, \cdots, w_n]$, and its kernel is the set of all payoff vectors of the form $(0, x)$, where x is in the kernel of $[q; w_2, \cdots, w_n]$.

To show how the table can be used in computing the kernels of superadditive weighted majority games for coalition structures other than $\{N\}$, we now discuss the notions of "pseudo-kernel" and "reduced game" due to Davis and Maschler [3, §7]. Suppose (N, v) is a game that does not necessarily satisfy the normalization assumptions $v(\{i\}) = 0$; as we will soon see, such games arise naturally in the process of computing the kernel of normalized games for certain coalition structures other than $\{N\}$. The kernel of a non-normalized game (N, v) is defined by renormalizing. More precisely, define a game (N, u) by $u(S) = v(S) - \sum_S v(\{i\})$, and let K be the kernel of (N, u) for coalition structure $\mathcal{B}$; then the kernel of (N, v) for $\mathcal{B}$ is defined to be $K + (v(\{1\}), \cdots, v(\{n\}))$. Equivalently, the kernel of (N, v) may be defined by replacing the condition $x_j > 0$ in the definition by $x_j > v(\{j\})$. As it happens, though, this is not the definition that is appropriate for our current purpose. What is needed here is the *pseudo-kernel*, which is defined in literally the same way as in Section 2, retaining the condition $x_j > 0$ even though $v(\{j\})$ may differ from 0.

To explain the notion of reduced game, we need the following notational device: If x is a payoff vector to the player set N, and if $B \subset N$, denote the vector $\{x_i\}_{i \in B}$ of payoffs to B by x_B. If x is viewed as a function from N to the reals, then x_B is the restriction of x to B.

Now let (N, v) be a monotonic (normalized) game, $\mathcal{B} = \{B_1, \cdots, B_k\}$ a coalition structure such that $v(B_2) = \cdots = v(B_k) = 0$. Define the *reduced game* (B_1, v^*) by $v^*(B_1) = v(B)$ and $v^*(B) = v(B \cup B_2 \cup \cdots \cup B_k)$ for all B that are strictly included in B_1; the game (B_1, v^*) will in general be neither monotonic nor normalized. It is now easily seen that x is in the kernel of (N, v) for coalition structure $\mathcal{B}$ if and only if $x_i = 0$ for $i \notin B_1$ and x_{B_1} is in the pseudo-kernel of (B_1, v^*) for coalition structure $\{B_1\}$. Of course, if (B_1, v^*) happens to be normalized, then its pseudo-kernel and its true kernel coincide.

Suppose now that $[q; w_1, \cdots, w_n]$ is a superadditive weighted majority game, $\mathcal{B}$ a coalition structure. Because of superadditivity there can be at most one winning coalition in $\mathcal{B}$. If there are none the kernel is trivial; assume therefore that B is the unique winning coalition. Then the reduced game corresponding to the structure $\mathfrak{B}$ and the coalition B is the game $[q - \sum_{i \notin B} w_i \,;\, w_B]$. The reduced game may or may not have some players whose quotas w_i exceed the reduced quota $q - \sum_{i \notin B} w_i$. If it does, then those players will win by themselves, so that the reduced game will not be normalized. In either case, the kernel of the original game for the coalition structure $\mathfrak{B}$ can be deduced from the pseudo-kernel of the reduced game for coalition structure $\{B\}$, with the understanding that the pseudo-kernel coincides with the kernel when the reduced game is normalized.

For example, if we wish to find the kernel of [4; 1, 1, 2, 2] for the coalition structure (123, 4), we are led to the reduced game [2; 1, 1, 2], in which player 3 wins by himself. The pseudo-kernel of the latter game for the coalition structure $\{123\}$ is 112/4, so that the kernel of the original game for $\{123, 4\}$ is 1120/4. The true kernel of [2; 1, 1, 2] for $\{123\}$ is 001, but this is irrelevant.

The pseudo-kernels of weighted majority games with at most 5 players and at least 1 winning player are listed in the fourth part of the table. By using the entire table and the technique described above, one can obtain the kernel of any superadditive weighted majority game, with arbitrarily many players, for coalition structures in which the winning coalition has at most 5 players.

The remarks made above about kernels of games with dummies apply also to pseudo-kernels.

Example 7.3. Calculate the kernel of [7; 1, 2, 2, 3, 3] (game no. 5-41) for all coalition structures.

Solution. It is necessary to consider only coalition structures of the form $\{B, N - B\}$, where B is winning. The winning coalitions are 12345, 2345, 1345, 1234, 145, 234, 345, and coalitions obtained from these by replacing players by symmetric players. For $B = N$ the kernel can be found by looking at the table. For $B = \{2345\}$ we obtain $[7 - w_1\,;\, 2, 2, 3, 3] = [6; 2, 2, 3, 3]$. This representation does not appear in the table. To find an equivalent representation in the table, we note that it is superadditive and that its desirability pattern is *aabb*. The only game with these specifications is [4; 1, 1, 2, 2] (no. 4-8), and its kernel is 1122/6. Hence the kernel of our game for coalition structure $\{1, 2345\}$ is 01122/6. For the coalition structure $\{1234, 5\}$, the reduced game is [4; 1, 2, 2, 3]. This representation appears in the table itself (no. 4-12), and so we obtain 12230/8 for the kernel. For $\{145, 23\}$, the reduced game is [3; 1, 3, 3]. This is a non-normalized game in which the first player is a dummy. Removing the dummy, we obtain [3; 3, 3], which is equivalent to [1; 1, 1] (game no. 2-2). Hence the pseudo-kernel of [3; 1, 3, 3] is 011/2, and that of the original game for the coalition structure $\{145, 23\}$ is 00011/2. The kernels for the other coalition structures are similarly obtained. The final result is as follows:

Coalition structure	Kernel
12345	01122/6
2345, 1	*1122/6
1345, 2	0*111/3
1234, 5	1223*/8
145, 23	0**11/2
234, 15	**112*/4
345, 12	**111/3

TABLE 1. *Weighted Majority Games With at Most 5 Players and Their Kernels*

Serial number	Game	Serial number	Game	Serial number	Game
1. GAMES WITH VETO PLAYERS					
2-1	[2; 1, 1]	5-1	[5; 1, 1, 1, 1, 1]	5-10	[7; 1, 1, 2, 2, 2]
3-1	[3; 1, 1, 1]	5-2	[5; 1, 1, 1, 1, 2]	5-11	[7; 1, 1, 2, 2, 3]
3-2	[3; 1, 1, 2]	5-3	[5; 1, 1, 1, 1, 3]	5-12	[7; 1, 1, 2, 2, 4]
4-1	[4; 1, 1, 1, 1]	5-4	[5; 1, 1, 1, 1, 4]	5-13	[7; 1, 1, 2, 2, 5]
4-2	[4; 1, 1, 1, 2]	5-5	[6; 1, 1, 1, 2, 2]	5-14	[8; 1, 1, 2, 3, 3]
4-3	[4; 1, 1, 1, 3]	5-6	[7; 1, 1, 1, 3, 3]	5-15	[8; 1, 1, 2, 3, 5]
4-4	[5; 1, 1, 2, 2]	5-7	[6; 1, 1, 1, 2, 3]	5-16	[9; 1, 2, 2, 3, 4]
4-5	[5; 1, 1, 2, 3]	5-8	[6; 1, 1, 1, 2, 4]	5-17	[9; 1, 2, 2, 3, 5]
		5-9	[7; 1, 1, 1, 3, 4]		

Serial number	Desirability pattern	Game	Kernel
2. SUPERADDITIVE GAMES WITHOUT VETO PLAYERS			
3-3	aaa	[2; 1, 1, 1]	111/3
4-6	aaaa	[3; 1, 1, 1, 1]	1111/4
4-7	aaab	[3; 1, 1, 1, 2]	1112/5
4-8	aabb	[4; 1, 1, 2, 2]	1122/6
4-9	abbc	[5; 1, 2, 2, 3]	0111/3
5-18	aaaaa	[3; 1, 1, 1, 1, 1]	11111/5
5-19		[4; 1, 1, 1, 1, 1]	11111/5
5-20	aaaab	[4; 1, 1, 1, 1, 2]	11112/6
5-21		[4; 1, 1, 1, 1, 3]	11113/7
5-22	aaabb	[4; 1, 1, 1, 2, 2]	22244/14
			00077/14
5-23		[5; 1, 1, 1, 2, 2]	11122/7
5-24		[6; 1, 1, 1, 3, 3]	11133/9
5-25		[7; 2, 2, 2, 3, 3]	22222/10
			00055/10
5-26	aaabc	[5; 1, 1, 1, 2, 3]	11123/8
5-27	aabbb	[5; 1, 1, 2, 2, 2]	33333/15
			00555/15
5-28		[6; 1, 1, 2, 2, 2]	11222/8
5-29	aabbc	[5; 1, 1, 2, 2, 3]	11223/9
5-30		[6; 1, 1, 2, 2, 3]	11112/6
			00222/6
5-31		[6; 1, 1, 2, 2, 4]	11224/10
5-32		[7; 1, 1, 3, 3, 4]	00111/3
5-33		[8; 1, 1, 3, 3, 5]	00111/3
5-34		[8; 2, 2, 3, 3, 4]	11111/5
5-35	aabcc	[6; 1, 1, 2, 3, 3]	00111/3
5-36	aabcd	[7; 1, 1, 2, 3, 4]	00111/3
5-37		[9; 2, 2, 3, 4, 5]	11122/7
5-38	abbbc	[6; 1, 2, 2, 2, 3]	12223/10
5-39		[7; 1, 2, 2, 2, 3]	01111/4
5-40		[7; 1, 2, 2, 2, 5]	01112/5

TABLE 1—*Continued*

Serial number	Desirability pattern	Game	Kernel
2. SUPERADDITIVE GAMES WITHOUT VETO PLAYERS—*Continued*			
5-41	abbcc	[7; 1, 2, 2, 3, 3]	01122/6
5-42		[8; 1, 2, 2, 3, 3]	01111/4
5-43	abbcd	[7; 1, 2, 2, 3, 4]	01112/5
5-44		[8; 1, 2, 2, 3, 4]	12234/12
5-45		[9; 1, 2, 2, 4, 5]	01122/6
5-46	abccd	[8; 1, 2, 3, 3, 4]	00111/3
5-47		[8; 1, 2, 3, 3, 5]	01112/5
5-48	abcde	[9; 1, 2, 3, 4, 5]	00111/3
3. NONSUPERADDITIVE GAMES			
4-10	aaaa	[2; 1, 1, 1, 1]	1111/4
4-11	aabb	[3; 1, 1, 2, 2]	1111/4
			0022/4
4-12	abbc	[4; 1, 2, 2, 3]	1223/8
5-49	aaaaa	[2; 1, 1, 1, 1, 1]	11111/5
5-50	aaaab	[3; 1, 1, 1, 1, 2]	11112/6
5-51	aaabb	[3; 1, 1, 1, 2, 2]	11122/7
			11111/5
5-52		[4; 1, 1, 1, 3, 3]	22222/10
			00055/10
5-53		[6; 2, 2, 2, 3, 3]	22233/12
5-54	aaabc	[4; 1, 1, 1, 2, 3]	11112/6
			00033/6
5-55	aabbb	[3; 1, 1, 2, 2, 2]	33333/15
			00555/15
5-56		[4; 1, 1, 2, 2, 2]	11222/8
5-57	aabbc	[4; 1, 1, 2, 2, 3]	11223/9
5-58		[5; 1, 1, 2, 2, 4]	00336/12
			22224/12
5-59		[6; 1, 1, 3, 3, 4]	11334/12
5-60		[6; 1, 1, 3, 3, 5]	00112/4
5-61		[7; 2, 2, 3, 3, 4]	44448/24
			33666/24
5-62	aabcc	[5; 1, 1, 2, 3, 3]	00044/8
			11222/8
5-63	aabcd	[5; 1, 1, 2, 3, 4]	00011/2
5-64		[8; 2, 2, 3, 4, 5]	00088/16
			22345/16
5-65	abbbc	[4; 1, 2, 2, 2, 3]	12223/10
5-66		[5; 1, 2, 2, 2, 3]	03333/12
			22224/12
5-67		[6; 1, 2, 2, 2, 5]	12225/12
5-68	abbcc	[4; 1, 2, 2, 3, 3]	11111/5
5-69		[5; 1, 2, 2, 3, 3]	01111/4
5-70	abbcd	[5; 1, 2, 2, 3, 4]	11122/7
5-71		[6; 1, 2, 2, 3, 4]	03333/12
			12234/12
			00066/12

TABLE 1—*Continued*

Serial number	Desirability pattern	Game	Kernel
3. NONSUPERADDITIVE GAMES—*Continued*			
5-72		[6; 1, 2, 2, 4, 5]	00011/2
5-73	abccd	[6; 1, 2, 3, 3, 4]	01111/4
5-74		[7; 1, 2, 3, 3, 5]	12335/14
5-75	abcde	[7; 1, 2, 3, 4, 5]	00022/4
			01111/4

Serial number	Desirability pattern	Game	Pseudo-kernel
4. GAMES WITH WINNING PLAYERS AND THEIR PSEUDO-KERNELS			
1-1	a	[1; 1]	1/1
2-2	aa	[1; 1, 1]	11/2
3-4	aaa	[1; 1, 1, 1]	111/3
3-5	aab	[2; 1, 1, 2]	112/4
4-13	aaaa	[1; 1, 1, 1, 1]	1111/4
4-14	aaab	[2; 1, 1, 1, 2]	1112/5
4-15		[3; 1, 1, 1, 3]	1113/6
4-16	aabb	[2; 1, 1, 2, 2]	1122/6
4-17	aabc	[3; 1, 1, 2, 3]	0011/2
5-76	aaaaa	[1; 1, 1, 1, 1, 1]	11111/5
5-77	aaaab	[2; 1, 1, 1, 1, 2]	11112/6
5-78		[3; 1, 1, 1, 1, 3]	11113/7
5-79		[4; 1, 1, 1, 1, 4]	11114/8
5-80	aaabb	[2; 1, 1, 1, 2, 2]	11122/7
5-81		[3; 1, 1, 1, 3, 3]	11133/9
5-82	aaabc	[3; 1, 1, 1, 2, 3]	11123/8
5-83		[4; 1, 1, 1, 2, 4]	00011/2
5-84		[4; 1, 1, 1, 3, 4]	00011/2
5-85	aabbb	[2; 1, 1, 2, 2, 2]	11222/8
5-86	aabbc	[3; 1, 1, 2, 2, 3]	00222/6
			11112/6
5-87		[4; 1, 1, 2, 2, 4]	11224/10
5-88		[5; 1, 1, 2, 2, 5]	00112/4
5-89	aabcc	[3; 1, 1, 2, 3, 3]	00111/3
5-90	aabcd	[5; 1, 1, 2, 3, 5]	00011/2
5-91	abbcd	[4; 1, 2, 2, 3, 4]	12234/12
5-92		[5; 1, 2, 2, 3, 5]	01112/5

The stars indicate players not in the winning coalition of the coalition structure, who therefore automatically get 0 (as distinguished from players who *are* in the winning coalition, but get 0 anyway).

The authors have tabulated the kernels of all superadditive 5-person games for all coalition structures, by the above procedure. The table is available from them upon request.

We remark that all kernels and pseudo-kernels appearing in the table appear as the kernel of some 6-person superadditive game, with an appropriate coalition structure.

The computations of the kernels in Table 1 were carried out on the CDC-1604-A. Trial and error led to the conclusion that the denominator 24 is fairly efficient for step A. With this denominator, the time needed for the entire calculation of the kernel of a simple game (both steps) is approximately 55 seconds (the addition time of the CDC-1604-A is 7.2 μ sec.). Of this, step A takes 18–27 seconds, and step B the remainder.

To minimize the human factor in the computation of the table, and consequent mistakes, a computer program was written to compute the characteristic function of a weighted majority game from its representation. It was therefore possible to use the representations as input to the computer.

A word is in order about the compilation of the games that appear in the table. For games with up to four players we used Shapley's list [14]. For games with 5 players, we used Isbell's list [7] of 6- and 7-person superadditive constant-sum weighted majority games. The work was divided into two steps:

(i) Deriving all the 6-person superadditive weighted majority games from Isbell's list.

(ii) Deriving all the 5-person weighted majority games from the 6-person superadditive ones.

To accomplish step (i), we made use of the "zero-sum $(n+1)$-person extension" of an n-person game due to von Neumann and Morgenstern[9] [10, p. 506]. The zero-sum extension is always unique. For weighted majority games, if $[q; w_1, \cdots, w_{n+1}]$ is zero-sum, then it is the zero-sum extension of $[q; w_1, \cdots, w_{j-1}, w_{j+1}, \cdots, w_{n+1}]$ (which is always superadditive) for every j; and conversely, if $[q; w_1, \cdots, w_n]$ is superadditive, then there is a w (not necessarily $\geq w_n$) such that $[q; w_1, \cdots, w_n, w]$ is its zero-sum extension. Step (i) consisted of compiling all the 6-person games $[q; w_1, \cdots, w_{j-1}, w_{j+1}, \cdots, w_7]$ for each $[q; w_1, \cdots, w_7]$ in Isbell's list and each j between 1 and 7, and adding to them the games in Isbell's list of 6-person games. It can be deduced from the fact that the "zero-sum extension" is defined in an intrinsic way that this yields *all* the 6-person weighted majority superadditive games.

To accomplish step (ii), we made use of the Maschler-Davis "reduced game" (defined above); this too is an intrinsic process. Our procedure was to compile all the 5-person games $[q - w_j; w_1, \cdots, w_{j-1}, w_{j+1}, \cdots, w_6]$ for all $[q; w_1, \cdots, w_6]$ resulting from step (i) and all j from 1 to 6. This yields all weighted majority 5-person games.

Steps (i) and (ii) were programmed on the CDC-1604-A. Of course the final list contained many duplications. Duplications due to multiple appearances of the same representation were removed first; then the characteristic function of each representation was computed and duplications due to a characteristic function having different representations were removed.

8. Extreme Games. For a fixed player set N with n members, the set of all superadditive games $G = (N, v)$ may be considered a cone in euclidean space of

---

[9] The "zero-sum extension" of the superadditive game (N, v) is the game $(N \cup \{n + 1\}, w)$, where $w(S) = v(S)$ for $n + 1 \notin S$ and $w(S) = v(N) - v(N - S)$ for $n + 1 \in S$.

dimension $2^n - n - 1$ (2^n is the number of coalitions, and v is fixed on the empty coalition and on the 1-member coalitions). A game (N, v) is called *extreme* if it is on an extreme ray of this cone; that is, if v is not the sum of two unproportional superadditive characteristic functions. All superadditive simple games are extreme.

Constant-sum extreme games have received particular attention, in papers by Griesmer [5] and Gurk [6]. When $n \leq 4$ all constant-sum extreme games are simple; however, when $n \geq 5$ there are non-simple constant-sum extreme games as well. Gurk [6] found a method of determining all constant-sum extreme games with five players; it turns out that in addition to the simple ones, there are 8 essentially different such games.[10] In all these games the characteristic function v takes exactly three values—in the normalization we adopt here, they are 0, 1/2, and 1.

We have constructed these 8 games by Gurk's method, and computed their kernels, for all coalition structures, with the program described in the previous sections; the results are tabulated in Table 2. For general integral games, the computer time needed for the computation of the kernel corresponding to a coalition structure $\mathcal{B}$ depends monotonically on the values of v, and in particular on the $v(B)$ for B in $\mathcal{B}$. In the integral games corresponding to the 8 extreme games under consideration, v takes the values 0, 1, and 2; after simple games, therefore, these may be expected to be the games whose kernels may be most quickly computed. As it turned out, the time needed for the coalition structure containing the all-player coalition only was usually about 4 or 5 minutes (compared with about 1 minute for simple games). For other coalition structures, the computation time was a matter of seconds.

From the theoretical viewpoint, the kernels tabulated here are interesting because there are coalition structures in which more than one coalition gets positive payoff; for superadditive simple games, this is impossible.

We now describe the construction of the table and its use. Each of the eight games is identified by the set of 3-person coalitions whose value is 1; the value of each of the remaining 3-person coalitions is 1/2. These conditions determine a unique constant-sum 5-person game such that $v(N) = 1$.

For each game there are two columns; in the left column there is a list of coalition structures, and in the right column the kernels corresponding to these coalition structures are tabulated. In no case do we list all the coalition structures; in the next paragraph we will discuss how the kernels corresponding to the unlisted coalition structures may be obtained. When the kernel consists of a single point, the point is given; when it is an interval, the two end-points are given; when it is V-shaped, all three vertices are given, with the one common to the two intervals between the other two. In this table we do not use the "common denominator" notation adopted in §7. When a coalition consists of a single player, that player necessarily gets 0 in the kernel; in that case his payoff in the kernel is indicated by a star rather than a 0. When there is more than one non-flat[11] coalition in a structure, then one of these is picked arbitrarily and its members (in the left column) and payoffs (in the right column) are italicized. Thus the coalition structure in

---

[10] I.e. every 5-person non-simple constant-sum extreme game is obtained from one of these eight by permuting the players and multiplying the characteristic function by a non-negative constant.

[11] A coalition S is *flat* if $v(S) = 0$.

TABLE 2
Kernels of Five Person Constant-Sum Non-Simple Extreme Games

Coalition structure	Kernel				
Game No. 1 $v(S) = 1/2$ for all S with 3 members					
12345	1/5	1/5	1/5	1/5	1/5
1234, 5	1/4	1/4	1/4	1/4	*
123, 45	1/6	1/6	1/6	1/4	1/4
12, 34, 5	1/4	1/4	1/4	1/4	*
Game No. 2 $v(\{2, 3, 4\}) = 1$					
12345	1/8	1/4	1/4	1/4	1/8
1, 2345	*	1/4	1/4	1/4	1/4
2, 1345	1/6	*	1/3	1/3	1/6
123, 45	0	1/4	1/4	1/4	1/4
125, 34	1/8	1/4	1/4	1/4	1/8
1, 23, 45	*	1/4	1/4	1/4	1/4
3, 12, 45	1/6	1/3	*	1/3	1/6
Game No. 3 $v(\{2, 3, 4\}) = v(\{3, 4, 5\}) = 1$					
12345	0	1/4	1/4	1/4	1/4
1, 2345	*	1/4	1/4	1/4	1/4
2, 1345	1/12	*	1/3	1/3	1/4
3, 1245	1/8	1/4	*	3/8	1/4
123, 45	0	1/4	1/4	1/4	1/4
134, 25	0	1/4	1/4	1/4	1/4
235, 14	1/5	1/10	3/10	3/10	1/10
125, 34	0	1/4	1/4	1/4	1/4
1, 23, 45	*	1/4	1/4	1/4	1/4
1, 25, 34	*	1/4	1/4	1/4	1/4
2, 13, 45	1/6	*	1/3	1/3	1/6
3, 14, 25	1/8	1/4	*	3/8	1/4
Game No. 4 $v(\{1, 2, 3\}) = v(\{3, 4, 5\}) = 1$					
12345	1/6	1/6	1/3	1/6	1/6
1, 2345	*	1/3	1/3	1/6	1/6
3, 1245	1/4	1/4	*	1/4	1/4
134, 25	0	1/3	1/3	1/6	1/6
	1/6	1/6	1/3	0	1/3
125, 34	1/6	1/6	1/3	1/6	1/6
3, 14, 25	0	1/2	*	1/2	0
	1/2	0	*	0	1/2
1, 34, 25	*	1/3	1/3	1/6	1/6
Game No. 5 $v(\{2, 3, 4\}) = v(\{3, 4, 5\}) = v(\{1, 4, 5\}) = 1$					
12345	1/9	1/9	2/9	1/3	2/9
1, 2345	*	1/12	1/4	1/3	1/3
3, 1245	1/8	1/4	*	3/8	1/4
4, 1235	1/6	1/6	1/3	*	1/3
123, 45	1/6	1/12	1/4	1/3	1/6
135, 24	1/12	1/6	1/6	1/3	1/4
124, 35	0	1/6	1/6	1/3	1/3
	1/6	0	1/3	1/3	1/6
134, 25	0	1/6	1/6	1/3	1/3
	1/12	1/4	1/12	1/3	1/4
1, 25, 34	*	1/6	1/6	1/3	1/3
1, 24, 35	*	1/6	1/6	1/3	1/3
3, 14, 25	1/8	1/4	*	3/8	1/4
4, 13, 25	0	0	1/2	*	1/2
	1/4	1/4	1/4	*	1/4
Game No. 6 $v(\{1, 2, 3\}) = v(\{3, 4, 5\}) = v(\{1, 2, 5\}) = 1$					
12345	1/6	1/6	1/3	0	1/3
1, 2345	*	3/8	1/4	1/8	1/4
3, 1245	1/4	1/4	*	1/4	1/4
4, 1235	1/6	1/6	1/3	*	1/3
124, 35	3/16	3/16	1/4	1/8	1/4
134, 25	1/6	1/6	1/3	0	1/3
135, 24	1/2	0	0	1/2	0
	0	3/8	1/4	1/8	1/4
1, 24, 35	*	3/8	1/4	1/8	1/4
3, 15, 24	0	1/2	*	0	1/2
	1/2	0	*	1/2	0
4, 13, 25	1/6	1/6	1/3	*	1/3
Game No. 7 $v(\{1, 2, 3\}) = v(\{2, 3, 4\}) = v(\{3, 4, 5\}) = v(\{1, 4, 5\}) = 1$					
12345	2/7	1/14	2/7	2/7	1/14
1, 2345	*	1/4	1/4	1/4	1/4
2, 1345	1/4	*	3/8	1/4	1/8
3, 1245	1/6	1/3	*	1/3	1/6
134, 25	0	1/2	0	1/2	0
	1/6	1/4	1/6	1/6	1/4
	0	0	1/2	0	1/2
135, 24	0	1/2	0	0	1/2
	1/4	1/8	1/4	3/8	0
235, 14	1/4	1/16	5/16	1/4	1/8
125, 34	1/4	1/8	1/4	1/4	1/8
1, 24, 35	*	0	1/2	1/2	0
	*	1/2	0	0	1/2
2, 14, 35	1/4	*	3/8	1/4	1/8
3, 14, 25	0	1/2	*	1/2	0
	1/4	1/4	*	1/4	1/4
1, 25, 34	*	0	1/2	0	1/2
	*	1/2	0	1/2	0
Game No. 8 $v(\{1, 2, 3\}) = v(\{2, 3, 4\}) = v(\{3, 4, 5\}) = v(\{1, 4, 5\}) = v(\{1, 2, 5\}) = 1$					
12345	1/5	1/5	1/5	1/5	1/5
1, 2345	*	1/3	1/6	1/6	1/3
124, 35	0	1/2	0	0	1/2
	1/6	1/6	1/4	1/6	1/4
	1/2	0	1/2	0	0
1, 24, 35	*	1/2	0	0	1/2
	*	1/4	1/4	1/4	1/4

TABLE 3
Pseudo-Kernels of Certain 3-Person Games
$v(\{12\}) = v(\{13\}) = v(\{23\}) = 1$

$v(\{123\})$	$v(\{1\})$	$v(\{2\})$	$v(\{3\})$	Pseudo-Kernel		
1/2	1/2	1/2	1/2	1/6	1/6	1/6
1/2	1/2	1/2	1	1/8	1/8	1/4
1/2	1/2	1	1	0	1/4	1/4
1/2	1	1	1	1/6	1/6	1/6
1	1/2	1/2	1/2	1/3	1/3	1/3
1	1/2	1/2	1	1/4	1/4	1/2
1	1/2	1	1	0	1/2	1/2
1	1	1	1	1/3	1/3	1/3

the right column can be determined at a glance from the type font, without having to refer to the left column; the left column is retained chiefly to facilitate reference to the table.

The coalition structures that are listed for each game are of the types $\{abcde\}$ $\{a, bcde\}$, $\{abc, de\}$, and $\{a, bc, de\}$. The kernels corresponding to some of the omitted ones can be obtained by applying the symmetries of the game to the tabulated kernels. In any other omitted structure, there is at most one non-flat coalition, and it has at most 3 players. Therefore by the Davis-Maschler "reduced game" technique explained in the previous section, the calculation of the corresponding kernel is equivalent to the calculation of the pseudo-kernel of a game with at most 3 players. If the game has exactly 3 players, then all 2-player coalitions have value 1. The pseudo-kernel may then be read off from Table 3, after applying the appropriate permutation. If the game has 2 players, then all 1-player coalitions have value 1, and the pseudo-kernel is $(1/4, 1/4)$ or $(0, 0)$ according as the 2-player coalition has value 1/2 or 0.

The Hebrew University of Jerusalem
Weizmann Institute of Science,
Rehovoth, Israel

1. R. J. AUMANN & M. MASCHLER, "The bargaining set for cooperative games," *Advances in Game Theory*, M. Dresher, L. S. Shapley and A. W. Tucker, (Eds.), Annals of Mathematics Studies, No. 52, Princeton Univ. Press, Princeton, N. J., 1964, pp. 443–476 [Chapter 42].
2. M. L. BALINSKI, "An algorithm for finding all vertices of convex polyhedral sets," *J. Soc. Indust. Appl. Math.*, v. 9, 1961, pp. 72–88. MR 25 #5451.
3. M. DAVIS & M. MASCHLER, "The kernel of a cooperative game," *Nav. Res. Log. Quart.*, v. 12, 1965, pp. 223–259.
4. M. DAVIS & M. MASCHLER, "Existence of stable payoff configurations for cooperative games," *Bull. Amer. Math. Soc.*, v. 69, 1963, pp. 106–108; also in *Essays in Mathematical Economics in Honor of Oskar Morgenstern*, M. Shubik, (Ed.), Princeton Univ. Press, Princeton, N. J., 1967, pp. 39–52.
5. J. H. GRIESMER, "Extreme games with three values," *Contributions to the Theory of Games, Vol. IV*, A. W. Tucker and R. D. Luce, (Eds.), Annals of Mathematics Studies, No. 40, Princeton Univ. Press, Princeton, N. J., 1959, pp. 189–212. MR 21 #2539.
6. H. M. GURK, "Five-person constant-sum extreme games," *Contributions to the Theory of Games, Vol. IV*, A. W. Tucker and R. D. Luce, (Eds.), Annals of Mathematics Studies, No. 40, Princeton University Press, Princeton, N. J., 1959, pp. 179–189. MR 21 #1909.
7. J. R. ISBELL, "On the enumeration of majority games," *MTAC*, v. 13, 1959, pp. 21–28. MR 21 #1912.
8. R. D. LUCE & H. RAIFFA, *Games and Decisions: Introduction and Critical Survey*, Wiley, New York, 1957. MR 19, 373.

9. M. Maschler & B. Peleg, "A characterization, existence proof, and dimension bounds for the kernel of a game," *Pacific J. Math.*, v. 18, 1966, pp. 289–328.

10. J. von Neumann & O. Morgenstern, *Theory of Games and Economic Behavior*, Princeton Univ. Press, Princeton, N. J., 1944; 3rd ed., 1953. MR, 6 235; MR **9**, 50.

11. B. Peleg, "Existence theorem for the bargaining set $M_1^{(i)}$,"[12] *Bull. Amer. Math. Soc.*, v. 69, 1963, pp. 109–110. MR 26 #2333; also in *Essays in Mathematical Economics in Honor of Oskar Morgenstern*, M. Shubik, (Ed.), Princeton Univ. Press, Princeton, N. J., 1967, pp. 53–56.

12. B. Peleg, "On the kernel of constant-sum games with homogeneous weights," *Illinois J. Math.*, v. 10, 1966, pp. 39–48.

13. B. Peleg, "The kernel of m-quota games," *Canad. J. Math.* v. 17, 1965, pp. 239–244.

14. L. S. Shapley, "Simple games: an outline of the descriptive theory," *Behavioral Sci.*, v. 7, 1962, pp. 59–66. MR 24 #B2490.

44 Cooperative Games with Coalition Structures
with J. Dreze

1. Introduction

A coalition structure in an n-person game is a partition of the set of players. Coalition structures have been used in defining the various solution notions that constitute the bargaining set family, i.e. the various bargaining sets [AUMANN and MASCHLER, 1964; DAVIS and MASCHLER, 1967], the kernel [DAVIS and MASCHLER, 1965] and the nucleolus [SCHMEIDLER, 1969]; in effect, these notions are defined separately for each coalition structure. By contrast, the value [SHAPLEY, 1953], core [GILLIES, 1959] and VON NEUMANN-MORGENSTERN solutions [1944] are not a priori defined with reference to a coalition structure[1]). Moreover, much of the theory that has been developed for the bargaining set family refers to the coalition structure containing the all-player set only.

This contrast between the bargaining set family and the other solution notions is, however, merely a historical accident; it is easy to define the value, core and VON NEUMANN-MORGENSTERN solutions with respect to a given coalition structure. In this paper, we will establish theorems that connect a given solution notion defined for a coalition structure $\mathscr{B}$ with the same solution notion — applied to appropriately defined games on each of the coalitions in the coalition structure. In the case of the kernel, such a theorem has already been proved by MASCHLER and PELEG [1967].

Perhaps the most remarkable aspect of our results is that there is a single function — the function v_x^* defined in (2.4) — that plays the central role in the theorems dealing with 5 out of the 6 solution notions in question (all except the value), though each of these 5 notions is entirely different. Moreover, this function

---

[1]) The games in "partition function form" of THRALL and LUCAS [1963] are not analogous to games with coalition structures as used in the bargaining set family.

This chapter originally appeared in *International Journal of Game Theory* 4 (1975): 217–237. Reprinted with permission.

enters into the theorems in a completely natural way, which is essentially the same in all 5 cases. This is an extraordinary — and unusual — instance of a game theoretic phenomenon that does not depend on a particular solution notion, but holds "across" a wide class of such notions.

Section 2 collects some basic definitions. In Section 3, we define the value for a game with an arbitrary, given coalition structure $\mathscr{B}$ and relate it to the values defined separately on each element of $\mathscr{B}$. In sections 4 to 8, we present a similar analysis for the nucleolus, the core, the von NEUMANN-MORGENSTERN solutions, the bargaining set and the kernel. The order in which these solution concepts are reviewed is motivated by convenience of exposition. In section 9, we present a condition under which a payoff vector in the core entails equal treatment for players who are substitutes but belong to different elements of the partition $\mathscr{B}$. In section 10, we show that the core of a game with a coalition structure, when not empty, is equal to the core of the superadditive cover of the game. Section 11 is devoted to two examples, of economic (and academic) interest, in which some of the results of the previous sections are applied. Section 12 is devoted to general discussion. The rationale for studying games with a coalition structure is reviewed there in some detail.

It should be made clear that we have not attempted to be absolutely comprehensive; there are important solution concepts not covered by our analysis (see for example SELTEN [1972]).

The numbering system in this paper is keyed to the numbering of the sections. Thus the theorem in Section 4 is called Theorem 4, and the corollary in Section 5 is called Corollary 5; and there is no Theorem 1 or Theorem 2.

The authors are grateful to MOSHE JUSTMAN, MICHAEL MASCHLER, BEZALEL PELEG, and DIETER SONDERMANN for critical and constructive comments on the material of this paper, including several substantial contributions.

2. Definitions

A game in characteristic function form, or simply a *game*, is a pair (N, v), where N is a finite set (the set of *players*), and v is a real-valued function on the family of subsets of N with $v(\emptyset) = 0$. The function v itself will also be called a game, or *a game on N*. The set of all games on N is denoted G^N; G^N is a EUCLIDean space of dimension $2^{|N|} - 1$, where $|N|$ is the cardinality of N.

A *payoff vector* for N is a real-valued function x on N; it may be thought of as a vector whose coordinates are indexed by the players. If $S \subset N$, write $x(S) = \sum_{i \in S} x(i)$. The set of all payoff vectors for N is denoted E^N. It is sometimes useful to constrain the set of payoff vectors under consideration to a subset X of E^N; we therefore define a *constrained game*[2]) to be a triple (N, v, X), where (N, v) is

---

[2]) This is by no means a new idea. The core and N-M solutions were first defined in terms of an arbitrary X by GILLIES [1959]; the nucleolus was first defined in this way by SCHMEIDLER [1969].

a game and $X \subset E^N$. When there is no constraint, then $X = E^N$; thus (N,v) may be identified with (N,v,E^N). We will use the term "game" for a constrained game as well; no confusion should result.

A *coalition structure* $\mathscr{B}$ on N is a partition of N, the generic element of which will be denoted B_k. A *game with coalition structure* $\mathscr{B}$ is a triple $(N,v,\mathscr{B})$. The analysis of $(N,v,\mathscr{B})$ differs from that of (N,v) in two respects:

(a) Payoff vectors associated with $(N,v,\mathscr{B})$ usually satisfy the conditions $x(B_k) = v(B_k)$ for all k (no side-payments between coalitions); in particular, these conditions are imposed by all the solution concepts considered below.

(b) In addition, the partition $\mathscr{B}$ enters directly into the definition of certain of the solution concepts (namely, the value, the bargaining set and the kernel).

The conditions stated in (a) may easily be replaced by constraints on the set of payoff vectors. Given a game (N,v), define:

$$X_\mathscr{B} = \{x \in E^N : x(B_k) = v(B_k) \text{ for all } k \text{ and } x_i \geq v(\{i\}) \text{ for all } i\}. \quad (2.1)$$

As will be seen below, the games $(N,v,\mathscr{B})$ and $(N,v,X_\mathscr{B})$ are equivalent from the point of view of some, but not all, solution concepts.

We also find it convenient to define

$$X_k = \{x \in E^{B_k} : x(B_k) = v(B_k) \text{ and } x_i \geq 0 \text{ for all } i \text{ in } B_k\}. \quad (2.2)$$

A *0-normalized* game is a game for which $v(\{i\}) = 0$ for all i. If (N,v) is a 0-normalized game, then[3]) $X_\mathscr{B} = \times_k X_k$. In general, however, there is a distinction between the definition of $X_\mathscr{B}$, which includes the conditions $x_i \geq v(\{i\})$, and the definition of X_k, which includes the conditions $x_i \geq 0$. (See the remark in section 8.)

In section 3, we use the following definitions.

A *permutation* π of N is a one-one function from N onto itself. For $S \subset N$, write $\pi S = \{\pi i : i \in S\}$. If v is a game on N, define a game $\pi_* v$ on N by

$$(\pi_* v)(S) = v(\pi S).$$

Call a coalition structure $\mathscr{B} = (B_1, \ldots, B_p)$ *invariant* under π if $\pi B_j = B_j$ for all j.

Player i is *null* if $v(S \cup \{i\}) = v(S)$ for all $S \subset N$.

In sections 4 through 8, we use the following definitions. Given a vector x in E^N, the *excess* $e(x,S)$ of the coalition S is defined by

$$e(x,S) = v(S) - x(S). \quad (2.3)$$

Three solution concepts are defined in terms of excesses, namely the core, the kernel, and the nucleolus.

Given a game $(N,v,\mathscr{B})$, a payoff vector x, and a coalition B_k in $\mathscr{B}$, define a game (B_k, v_x^*) by

---

[3]) Thus, $X_\mathscr{B} \neq \emptyset$ implies $X_k \neq \emptyset$ for all k, a property that does not hold in general.

$$v_x^*(S) = \begin{cases} \max_{T \subset N \setminus B_k} (v(S \cup T) - x(T)), & \text{for } S \subset B_k, S \neq \emptyset, S \neq B_k \\ v(S), & \text{for } S = \emptyset \text{ or } S = B_k. \end{cases} \quad (2.4)$$

Obviously, $v_x^*(S) \geq v(S)$ for every x. Note that v_x^* need *not* be 0-normalized, even when v is.

Let $\mathscr{B} = (B_1, \ldots, B_p)$ be a partition of N. The game (N, v) is called *decomposable with partition* $\mathscr{B}$ if for all S,

$$v(S) = \sum_{k=1}^{p} v(S \cap B_k).$$

Finally, let Z be a subset of E^N and B a subset of N. For every y in the projection of Z on $E^{N \setminus B}$, we define the *section of Z at y* as $\{w \in E^B : (w, y) \in Z\}$. (See Figure 1.)

3. The Shapley Value

Fix N and $\mathscr{B}$. A $\mathscr{B}$-*value* is a function $\varphi_{\mathscr{B}}$ from G^N to E^N — i.e. a function that associates with each game a payoff vector — obeying the following conditions:

Relative efficiency: For all k, $(\varphi_{\mathscr{B}} v)(B_k) = v(B_k)$. (3.1)

Symmetry: For all permutations π of N under which $\mathscr{B}$ is invariant, (3.2)
$(\varphi_{\mathscr{B}}(\pi_* v))(S) = (\varphi_{\mathscr{B}} v)(\pi S)$.

Additivity: $\varphi_{\mathscr{B}}(v + w) = \varphi_{\mathscr{B}} v + \varphi_{\mathscr{B}} w$. (3.3)

Null-Player condition: If i is a null-player, then $(\varphi_{\mathscr{B}} v)(i) = 0$. (3.4)

When $\mathscr{B} = \{N\}$, it is known that there is a unique function $\varphi_{\mathscr{B}}$ satisfying (3.1) through (3.4), namely the usual SHAPLEY value of the game [SHAPLEY, 1953]; it will be denoted by φ. This notation will be maintained even for games whose player set differs from N; thus if v is a game with player set M, φv is defined to be $\varphi_{\mathscr{B}} v$, where $\mathscr{B} = \{M\}$.

For each $S \subset N$, denote by $v|S$ the game on S defined for all $T \subset S$ by $(v|S)(T) = v(T)$.

Theorem 3:

Fix N and $\mathscr{B} = (B_1, \ldots, B_p)$. Then there is a unique $\mathscr{B}$-value, and it is given for all $k = 1, \ldots, p$, and all $i \in B_k$, by

$$(\varphi_{\mathscr{B}} v)(i) = (\varphi(v|B_k))(i). \quad (3.5)$$

Remark:

(3.5) asserts that the restriction to B_k of the value $\varphi_{\mathscr{B}}$ for the game (N, v) is the value φ for the game $(B_k, v|B_k)$. In other words, the value of a game with coalition structure $\mathscr{B}$ has the "restriction property": The restriction of the value is the value of the restriction of the game. An important implication of this property is that $\varphi_{\mathscr{B}}$ can be computed by computing separately $\varphi(v|B_k)$ for each k.

Proof:

The operator defined by (3.5) satisfies (3.1) through (3.4), so there is at least one $\mathscr{B}$-value. We must prove that there is only one. For each non-empty $T \subset N$, define the *T-unanimity-game* v_T by

$$v_T(S) = \begin{cases} 1 & \text{if } S \supset T \\ 0 & \text{otherwise}. \end{cases}$$

We first show that the games v_T are linearly independent. Indeed, suppose there is a linear relation among them; let T_0 be a set of minimal cardinality such that v_{T_0} appears with non-vanishing coefficient in this linear relation. We then have $v_{T_0} = \sum \alpha_T v_T$, where all the T appearing on the right side are different from T_0 and have cardinality at least that of T_0. Therefore T_0 does not contain any of these T, and hence $1 = v_{T_0}(T_0) = \sum \alpha_T v_T(T_0) = 0$.

This shows that the v_T are linearly independent; since there are $2^{|N|} - 1$ different v_T, and $2^{|N|} - 1$ is the cardinality of G^N, it follows that they form a basis for G^N; therefore every game on N is a linear combination of the games v_T. By the additivity axiom, it then follows that if the $\mathscr{B}$-value is unique on all games of the form αv_T, where α is a constant, then it is unique.

Consider therefore a game of the form αv_T. By (3.4), $(\varphi_{\mathscr{B}}(\alpha v_T))(i) = 0$ whenever $i \notin T$. From (3.2) it follows that if i and j are in T and in the same member B_k of $\mathscr{B}$, then

$$(\varphi_{\mathscr{B}}(\alpha v_T))(i) = (\varphi_{\mathscr{B}}(\alpha v_T))(j).$$

Hence from (3.2) it follows that if $i \in B_k$, then

$$(\varphi_{\mathscr{B}}(\alpha v_T))(i) = \begin{cases} \alpha/|T| & \text{if } T \subset B_k \\ 0 & \text{otherwise}. \end{cases}$$

This determines $\varphi_{\mathscr{B}}(\alpha v_T)$, and so completes the proof.

4. The Nucleolus

Let (N, v, X) be a constrained game. For each $x \in X$, let $\theta(x)$ be a vector in $E^{2^{|N|}}$, the elements of which are the excesses $e(x, S)$ for $S \subset N$, arranged in order of non-increasing magnitude; i.e. $\theta_s(x) \geq \theta_t(x)$ whenever $t > s$. Write $\theta(y) \geq \cdot \theta(x)$ (or $\theta(y) > \cdot \theta(x)$) if and only if $\theta(x)$ is not greater (or is smaller) than $\theta(y)$ in the lexicographic order on $E^{2^{|N|}}$. The *nucleolus*, w.r.t. the set X, is then defined by

$$Nu(N, v, X) = \{x \in X : \theta(y) \geq \cdot \theta(x) \text{ for all } y \in X\}.$$

For a coalition structure $\mathscr{B}$, we define $\text{Nu}(N, v, \mathscr{B}) = \text{Nu}(N, v, X_{\mathscr{B}})$. In particular, when $\mathscr{B} = \{N\}$, we write $\text{Nu}(N, v) = \text{Nu}(N, v, \{N\})$.

When $X \neq \emptyset$, the nucleolus consists of a single element [SCHMEIDLER, 1969; KOHLBERG, 1971]; this element, as well as the set of which it is the only member, will also be called the nucleolus. Thus, like the value and unlike other solution concepts, the nucleolus assigns to each game precisely one payoff vector.

We saw in Section 3 that the restriction to B_k of the value for $(N, v, \mathscr{B})$ is the value for $(B_k, v|B_k)$. Does a similar property hold for the nucleolus? The answer is no, as the following example shows.

Example 4:

Consider the weighted majority game with $|N| = 4$, $w_1 = w_2 = w_3 = 1$, $w_4 = 2$ and

$$v(S) = \begin{cases} 1, & \sum_{i \in S} w_i \geq 3 \\ 0, & \text{otherwise}. \end{cases}$$

Let $\mathscr{B} = \{(1), (2,3,4)\}$. Then $\text{Nu}(N, v, \mathscr{B}) = (0, \frac{1}{4}, \frac{1}{4}, \frac{1}{2})$, whereas $\text{Nu}(B_2, v|B_2) = (0, 0, 1)$.

The reason for this negative answer is easily understood: excesses of coalitions S not included in B_k (e.g. $S = \{1, 2, 3\}$ in example 4) may play a crucial role in determining the payoff vector $x|B_k \in X_k$, when x is the nucleolus[4]). The characteristic function v_x^* was defined in (2.4) in a way which captures the influence on $x|B_k$ of coalitions not included in B_k, when x is the nucleolus. (Reminder: knowledge of x is required to compute v_x^*).

Theorem 4:

Let (N, v) be a 0-normalized game, and let $x = \text{Nu}(N, v, \mathscr{B})$. Then $\text{Nu}(N, v, \mathscr{B})|B_k = \text{Nu}(B_k, v_x^*, X_k)$.

Proof[5]):

For $S \subset B_k$ and $y \in B_k$, let

$$e^*(y, S) = v_x^*(S) - y(S)$$

and let $\theta^*(y)$ be the vector of these $2^{|B_k|}$ excesses arranged in non-increasing order. Let $x^* = x|B_k$, and let y^* in X be different from x^*. We show that $\theta^*(y^*) >_\cdot \theta^*(x^*)$, from which it follows that $x^* = \text{Nu}(B_k, v_x^*, X_k)$.

Define $y \in X_\mathscr{B}$ by

$$y_i = \begin{cases} y_i^* & \text{for } i \in B_k \\ x_i & \text{for } i \in N \setminus B_k. \end{cases}$$

Set

$$\alpha = \max \{e(x, R): R \subset N, e(x, R) \neq e(y, R)\}. \tag{4.1}$$

---

[4]) In this example, $v(1, 2, 3) = 1$, $(v|B_2)(2, 3) = 0$
$\phantom{^4)\text{ In this example, }}v(1, 4) = 1$ $(v|B_2)(4) = 0$.
The nucleolus of $(N, v, \mathscr{B})$ is determined by the excesses of the coalitions $(1, 2, 3)$ and $(1, 4)$; the nucleolus of $(B_2, v|B_2)$ is determined by the excesses of the coalitions $(2), (3), (2, 4)$ and $(3, 4)$.

[5]) This proof is due to M. JUSTMAN. We are thankful for his permission to use it here.

Let there be q coordinates in $\theta(x)$ larger than α, and r coordinates equal to α. For $\varepsilon > 0$ sufficiently small, the first q coordinates of $\theta((1 - \varepsilon)x + \varepsilon y)$ equal the corresponding coordinates of $\theta(x)$. Suppose that for all R with $e(x,R) = \alpha$ we have

$$e(y, R) \leq e(x, R). \tag{4.2}$$

By the definition of α there is then at least one such R with $e(y,R) < e(x,R)$. Hence for $\varepsilon > 0$ sufficiently small, the $(q + 1)$-st through $(q + r)$-th coordinates of $\theta((1 - \varepsilon)x + \varepsilon y)$ are all $\leq \alpha$, and at least one of them is $< \alpha$. Hence $\theta(x) > \cdot \theta((1 - \varepsilon)x + \varepsilon y)$, contradicting $x = \text{Nu}(N,v,\mathscr{B})$. Hence (4.2) is false, i.e. there is at least one coalition — call it U — with

$$e(x, U) = \alpha \quad \text{and} \quad e(y, U) > e(x, U). \tag{4.3}$$

From (4.3) and $x|N\backslash B_k = y|N\backslash B_k$ it follows that

$$(x^* - y^*)(U \cap B_k) = (x - y)(U) = e(y, U) - e(x, U) > 0. \tag{4.4}$$

Now let $(S_1, \ldots, S_l)$ be an ordering of the subsets of B_k so that

$$\theta^*(x^*) = (e^*(x^*, S_1), \ldots, e^*(x^*, S_l))$$

and if $U \cap B_k = S_p$, then

$$i < p \Rightarrow e^*(x^*, S_i) > e^*(x^*, S_p). \tag{4.5}$$

It is easy to see that if a vector z' is obtained from a vector z by arranging the coordinates of z in non-increasing order, then $z' \geq \cdot z$. Hence

$$\theta^*(y^*) - \theta^*(x^*) \geq \cdot (e^*(y^*, S_1), \ldots, e^*(y^*, S_l)) - \theta^*(x^*) \\ = ((x^* - y^*)(S_1), \ldots, (x^* - y^*)(S_p), \ldots, (x^* - y^*)(S_l)). \tag{4.6}$$

Now if $i < p$ and $\emptyset \neq S_i \neq B_k$, then from (4.5) and $S_p = U \cap B_k$ it follows that for some $T_i \subset N\backslash B_k$, we have

$$e(x, S_i \cup T_i) = (v(S_i \cup T_i) - x(T_i)) - x(S_i) \\ > \max\{v(S_p \cup T) - x(T) : T \subset N\backslash B_k\} - x(S_p) \\ \geq v(U) - x(U\backslash B_k) - x(U \cap B_k) = e(x, U).$$

Hence by (4.3) and (4.1), $e(x, S_i \cup T_i) = e(y, S_i \cup T_i)$. From this and $x|N\backslash B_k = y|N\backslash B_k$ it follows that $x(S_i) = y(S_i)$, i.e.

$$x^*(S_i) = y^*(S_i) \tag{4.7}$$

in this case. If $S_i = \emptyset$ or $S_i = B_k$, then (4.7) holds trivially. Hence (4.7) holds for all $i < p$. But by (4.4) and $S_p = U \cap B_k$, we have

$$x^*(S_p) - y^*(S_p) > 0.$$

Hence by (4.6),
$$\theta^*(y^*) - \theta^*(x^*) > \cdot 0,$$
so that $\theta^*(y^*) > \cdot \theta^*(x^*)$, as was to be proved.

Corollary 4:

Let (N,v) be a 0-normalized game, decomposable with partition $\mathscr{B} = (B_1, \ldots, B_p)$. Then $\mathrm{Nu}(N,v,\mathscr{B}) = \underset{k=1}{\overset{p}{\times}} \mathrm{Nu}(B_k, v|B_k, X_k)$.

Proof:

By theorem 4, $\mathrm{Nu}(N,v,\mathscr{B}) = \underset{k=1}{\overset{p}{\times}} \mathrm{Nu}(B_k, v_x^*, X_k)$. Because (N,v) is decomposable, we have
$$v_x^*(S) = \max_{T \subset N \setminus B_k} \{v(S) + v(T) - x(T)\} = v(S) + \max_{T \subset N \setminus B_k} \{v(T) - x(T)\}.$$
It follows that $\mathrm{Nu}(B_k, v_x^*, X_k) = \mathrm{Nu}(B_k, v|B_k, X_k)$.

Remark:

A similar result holds for the SHAPLEY value; but in that case, it holds for all games, not only decomposable games (see Theorem 3).

5. The Core

The *core* of the game (N, v, X) is defined by
$$\mathrm{Co}(N,v,X) = \{x \in X : e(x,S) \leqq 0 \text{ for all } S \subset N\}.$$
For a coalition structure $\mathscr{B}$, we define $\mathrm{Co}(N,v,\mathscr{B}) = \mathrm{Co}(N,v,X_{\mathscr{B}})$. In particular, when $\mathscr{B} = \{N\}$, we write $\mathrm{Co}(N,v) = \mathrm{Co}(N,v,\{N\})$.

The core does not have the uniqueness property of the nucleolus. Accordingly, it could not have the "restriction property" of the value. But one could raise questions such as the following:
(i) Does $x \in \mathrm{Co}(N,v,\mathscr{B})$ imply $x|B_k \in \mathrm{Co}(B_k, v|B_k)$?
(ii) Does $y \in \mathrm{Co}(B_k, v|B_k)$ imply $y = x|B_k$ for some $x \in \mathrm{Co}(N,v,\mathscr{B})$?
The answer to question (i) is positive, but the answer to question (ii) is negative. Indeed, in example 4, $\mathrm{Co}(N,v,\mathscr{B}) = \emptyset$, whereas $\mathrm{Co}(B_2, v|B_2) = (0,0,1)$.

The definition of the characteristic function v_x^* in (2.4) is again relevant, in relating $\mathrm{Co}(N,v,\mathscr{B})$ to the cores of appropriately defined games on the B_k's.

Theorem 5:

Let (N,v) be a 0-normalized game, and let $x \in \mathrm{Co}(N,v,\mathscr{B})$. Then the section of $\mathrm{Co}(N,v,\mathscr{B})$ at $x|N \setminus B_k$ is $\mathrm{Co}(B_k, v_x^*, X_k)$.

Remarks:

1. The conclusion is perhaps most easily understood with the help of Figure 1.
   Thus, for all $x \in \mathrm{Co}(N,v,\mathscr{B})$, the section of the core at $x|N \setminus B_k$ defines the core

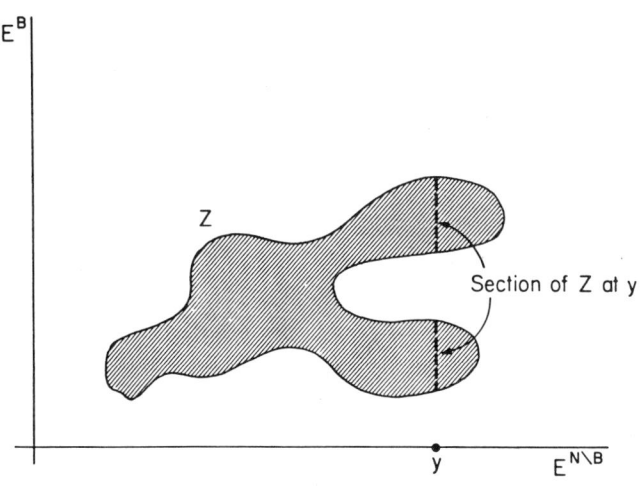

Fig. 1

of a game on B_k; however, the relevant characteristic function is not $v|B_k$, but v_x^*, which depends upon x.

2. When X consists of a single element, then the section of X at $x|N\backslash B_k$ is $x|B_k$. Our result for the nucleolus is thus of the same form as our result for the core, i.e., the conclusion in Theorem 4 could be stated in the following form: if $x \in \text{Nu}(N,v,\mathscr{B})$, then the section of $\text{Nu}(N,v,\mathscr{B})$ at $x|N\backslash B_k$ is $\text{Nu}(B_k,v_x^*,X_k)$.

3. The conclusion does *not* include the property: $x|B_k \in \text{Co}(B_k,v_x^*,X_k)$ implies $x \in \text{Co}(N,v,\mathscr{B})$. Indeed $\text{Co}(B_k,v_x^*,X_k)$ may well be non-empty, when $x|N\backslash B_k$ does not belong to the projection of $\text{Co}(N,v,\mathscr{B})$ on $E^{N\backslash B_k}$.

Proof:

An alternative statement of the conclusion, in terms of which the proof will be presented, is:

$$x \in \text{Co}(N,v,\mathscr{B}) \text{ implies:} \tag{5.1}$$
(a) $x|B_k \in \text{Co}(B_k,v_x^*,X_k)$;
(b) $z \in \text{Co}(N,v,\mathscr{B})$ for all $z \in E^N$ for which
$z|N\backslash B_k = x|N\backslash B_k$ and $z|B_k \in \text{Co}(B_k,v_x^*,X_k)$.

(a) From $x \in \text{Co}(N,v,\mathscr{B})$, it follows that $v(S \cup T) - x(S) - x(T) \leq 0$ for all $S \subset B_k$, $T \subset N\backslash B_k$. Hence $v_x^*(S) - x(S) \leq 0$ for all $S \subset B_k$, and so $x|B_k \in \text{Co}(B_k,v_x^*,X_k)$.

(b) From $z|B_k \in \text{Co}(B_k,v_x^*,X_k)$, it follows that $v_x^*(S) - z(S) \leq 0$ for all $S \subset B_k$. Hence

$$v(S \cup T) - z(S) - x(T) \leq 0 \quad \text{for all} \quad S \subset B_k \text{ with } S \neq \emptyset, \quad \text{and for all} \quad T \subset N\backslash B_k. \tag{5.2}$$

For the case $S = \emptyset$, we have

$$v(T) - x(T) \leq 0 \quad \text{for all} \quad T \subset N \backslash B_k, \tag{5.3}$$

since $x \in \text{Co}(N, v, \mathscr{B})$. From (5.2), (5.3), and $z \mid N \backslash B_k = x \mid N \backslash B_k$, it then follows that $v(S \cup T) - z(S \cup T) \leq 0$ for all $S \subset B_k$ and $T \subset N \backslash B_k$, and so $z \in \text{Co}(N, v, \mathscr{B})$.

Corollary 5:
Let (N, v) be a 0-normalized game, decomposable with partition $\mathscr{B} = (B_1, \ldots, B_p)$. Then $\text{Co}(N, v, \mathscr{B}) = \underset{k=1}{\overset{p}{\times}} \text{Co}(B_k, X_k)$.

This result was proved by MASCHLER, PELEG, and SHAPLEY [1972], as Lemma 2.9. It also follows from Theorem 5 and decomposability upon noting that, when $x \in \text{Co}(N, v, \mathscr{B})$,

$$v_x^*(S) = v(S) + \max_{T \subset N \backslash B_k} \{v(T) - x(T)\} = v(S).$$

6. VON NEUMANN-MORGENSTERN Solutions

The VON NEUMANN-MORGENSTERN solutions rely on the notion of "domination". As before, let X be a compact convex subset of E^N, and let (N, v) be a game. Let x and y be in X. Then *x dominates y with respect to* the coalition T if and only if $x_i > y_i$ for all $i \in T$, and $v(T) \geq x(T)$. One then writes $x \succ_T y$. Next, *x dominates y* if there exists a $T \subset N$ for which $x \succ_T y$; one then writes $x \succ y$. A VON NEUMANN-MORGENSTERN *solution of* (N, v, X) is a set $Q \subset X$ such that
(i) there do not exist $x, x' \in Q$ with $x \succ x'$ (internal consistency);
(ii) for every $x' \in X \backslash Q$, there exists an $x \in Q$ such that $x \succ x'$ (external domination).

If $\mathscr{B}$ is a coalition structure, then a VON NEUMANN-MORGENSTERN *solution of* $(N, v, \mathscr{B})$ is a VON NEUMANN-MORGENSTERN solution of $(N, v, X_{\mathscr{B}})$.

Theorem 6.
Let (N, v) be a 0-normalized game, let Q be a VON NEUMANN-MORGENSTERN solution of $(N, v, \mathscr{B})$, and let $x \in Q$. Then the section of Q at $x \mid N \backslash B_k$ is a VON NEUMANN-MORGENSTERN *solution of* (B_k, v_x^*, X_k).

Proof:
Internal consistency and external domination in (B_k, v_x^*, X_k) follow immediately from the properties of Q and the definition of sections upon remembering that $x(B_k) = v(B_k)$ for all $x \in X$.

Thus the relationship of VON NEUMANN-MORGENSTERN solutions for $(N, v, \mathscr{B})$ to VON NEUMANN-MORGENSTERN solutions for (B_k, v_x^*, X_k) is identical to that for the core; the only difference is that a game may have many VON NEUMANN-MORGENSTERN solutions, whereas it has one core.

7. The Bargaining Set

We turn next to the bargaining set, which is defined in terms of the notion of "objection". Let (N,v,X) be a game, $\mathscr{B}$ a coalition structure. If i and j are elements of B_k, and $x \in X$, an *objection* of i against j at x consists of a payoff vector x' in E^N and a coalition $S \subset N$ such that $i \in S$, $j \notin S$, $x'_i > x_i$, $x'_l \geq x_l$ for all $l \in S$ and $x'(S) \leq v(S)$. A *counterobjection* of j to such an objection consists of a payoff vector x'' in E^N and a coalition $T \subset N$ such that $i \notin T$, $j \in T$, $x''_l \geq x_l$ for all $l \in T\setminus S$, $x''_l \geq x'_l$ for all $l \in T \cap S$ and $x''(T) \leq v(T)$. The *bargaining set* $M(N,v,X,\mathscr{B})$ is the set of payoff vectors $x \in X$ such that for all k and all $i,j \in B_k$, j has a counterobjection to every objection of i against j at x. Define[6]

$$M(N,v,\mathscr{B}) = M(N,v,X_\mathscr{B},\mathscr{B}) \quad \text{and} \quad M(N,v,X) = M(N,v,X,\{N\}).$$

The following theorem is due to B. PELEG; we gratefully acknowledge his permission for publishing it here.

Theorem 7:
Let (N,v) be a 0-normalized game, and let $x \in M(N,v,\mathscr{B})$. Then the section of $M(N,v,\mathscr{B})$ at $x|N\setminus B_k$ is included in $M(B_k,v_x^*,X_k)$.

Proof:
For $x \in M(N,v,\mathscr{B})$ let (y,S) (with $y \in E^{B_k}$ and $S \subset B_k$) be an objection of i against j at $x|B_k$ in the game (B_k,v_x^*,X_k). Remember that $v_x^*(S) = v(S \cup T) - x(T)$ for an appropriate $T \subset N\setminus B_k$, so that $j \notin T$. Then (y,S) may be extended into an objection $(z, S \cup T)$ of i against j at x in the game $(N,v,\mathscr{B})$, where $z \in E^N$ is given by $z|B_k = y$, $z|N\setminus B_k = x|N\setminus B_k$. Because $x \in M(N,v,\mathscr{B})$, there exists a counterobjection (z',T') of j to i's objection $(z, S \cup T)$ in the game $(N,v,\mathscr{B})$. It is readily verified that $(z'|B_k, T' \cap B_k)$ defines a counterobjection of j to i's objection (y,S) in the game (B_k,v_x^*,X_k).

Indeed, $i \notin T' \cap B_k$, $j \in T' \cap B_k$, $z'_l \geq x_l$ for all $l \in T'\setminus(S \cup T) \supset (T' \cap B_k)\setminus S$, $z'_l \geq y_l$ for all $l \in (T' \cap B_k) \cap S$ and $z'(T' \cap B_k) \leq v_x^*(T' \cap B_k)$. The last inequality follows from

$$v_x^*(T' \cap B_k) \geq v(T') - x(T'\setminus B_k) \geq z'(T') - x(T'\setminus B_k)$$
$$\geq z'(T') - z'(T'\setminus B_k) = z'(T' \cap B_k).$$

Remark:
The conclusion in Theorem 7 corresponds to the conclusion in Theorem 5, except for the fact that the equality in theorem 5 becomes an *inclusion* here. The reverse inclusion is false, as can be verified by means of Example 4. Indeed, in that example, the bargaining set for $(N,v,\mathscr{B})$ is defined by the condition $1/4 \geq x_2 = x_3 \geq 1/5$; whereas the bargaining set for (B_2,v_x^*,X_2) is defined by the condition $1/4 \geq x_2 = x_3 \geq 0$.

---

[6] $M(N,v,\mathscr{B})$ is denoted $M_1^{(i)}(N,v,\mathscr{B})$ by AUMANN and MASCHLER [1964], and by DAVIS and MASCHLER [1967].

8. The Kernel

Finally, for the sake of completeness, we quote an earlier result of MASCHLER and PELEG [1967] regarding the kernel. It is of the same form as our result for the core.

Let $i, j \in B_k$, $x \in E^N$. Define $\delta_{ij}(x)$ to be the maximal excess, with respect to x, of a coalition containing i but not j; i.e.,

$$\delta_{ij}(x) = \max \{e(x, S): S \subset N, i \in S, j \notin S\}.$$

Define

$$K(N, v, X, \mathscr{B}) = \{x \in X : (\forall K)(\forall i, j \in B_k)(\delta_{ij}(x) \geq \delta_{ji}(x) \text{ or } x_i = v(\{i\}))\},$$

$$K(N, v, \mathscr{B}) = K(N, v, X_\mathscr{B}, \mathscr{B}), \quad \text{and} \quad K(N, v, X) = K(N, v, X, \{N\}).$$

$K(N, v, \mathscr{B})$ is called the *kernel* of $(N, v, \mathscr{B})$.

Theorem 8 (MASCHLER and PELEG [1967, Theorem 2.9]):

Let (N, v) be a 0-normalized game, and let $x \in K(N, v, \mathscr{B})$. Then the section of $K(N, v, \mathscr{B})$ at $x | N \setminus B_k$ is $K(B_k, v_x^*, X_k)$.

Remark:

The definition of v_x^* is due to MASCHLER and PELEG [1967, p. 599], who call $K(B_k, v_x^*, X_k)$ the *pseudo-kernel* of (B_k, v_x^*). The kernel of (B_k, v_x^*) would require, in addition, that $x_i \geq v_x^*(\{i\})$. One could define similarly the "pseudo-core" and "pseudo-nucleolus" of (B_k, v_x^*), to be, respectively, the core and the nucleolus of (B_k, v_x^*, X_k).

Corollary 8:

Let (N, v) be a 0-normalized game, decomposable with partition $\mathscr{B} = (B_1, \ldots B_p)$. Then $K(N, v, \mathscr{B}) = \underset{k=1}{\overset{p}{\bigtimes}} K(B_k, v | B_k, X_k)$.

Proof:

By decomposability, $v_x^*(S) = v(S) + \underset{T \subset N \setminus B_k}{\max} \{v(T) - x(T)\}$. Hence

$$\underset{\substack{S \subset B_k \\ i \in S \\ j \notin S}}{\max} \{v_x^*(S) - x(S)\} - \underset{\substack{S' \subset B_k \\ i \notin S' \\ j \in S'}}{\max} \{v_x^*(S') - x(S')\}$$

$$= \underset{\substack{S \subset B_k \\ i \in S \\ j \notin S}}{\max} \{v(S) - x(S)\} - \underset{\substack{S' \subset B_k \\ i \notin S' \\ j \in S'}}{\max} \{v(S') - x(S')\}$$

and so $K(B_k, v_x^*, X_k) = K(B_k, v | B_k, X_k)$. The corollary then follows from Theorem 8.

9. Equal Treatment

Two players i and j are called *substitutes* if $v(S \cup i) = v(S \cup j)$ for all $S \subset N$ such that $i \notin S, j \notin S$. Whenever a solution concept imposes that substitutes receive the same payoff, that solution concept is said to have the *equal treatment property*. The SHAPLEY value, kernel and nucleolus of (N,v) have that property. Theorems 4 and 8 imply that the kernel and nucleolus of $(N,v,\mathscr{B})$ impose equal treatment for substitutes belonging to the same element B_k of the partition $\mathscr{B}$. Example 4 shows that the kernel and the nucleolus do *not* impose equal treatment for substitutes who belong to different elements of $\mathscr{B}$. On the other hand, it is well known that the core does not impose equal treatment for substitutes belonging to the same element[7]) of $\mathscr{B}$.

The following theorem shows that the core of $(N,v,\mathscr{B})$ imposes equal treatment for substitutes who belong to *different* elements of the partition $\mathscr{B}$. Because $\text{Nu} \in \text{Co}$ and $K \cap \text{Co} \neq \emptyset$ whenever $\text{Co} \neq \emptyset$, the condition has obvious implications for the nucleolus and the kernel as well.

Theorem 9:
Let $x \in \text{Co}(N,v,\mathscr{B})$. If i and j are substitutes in (N,v), $i \in B_k$, and $j \notin B_k$, then $x_i = x_j$.

Proof:
We have
$$0 \geq v(\{i\} \cup N\setminus B_k\setminus\{j\}) - x(\{i\} \cup N\setminus B_k\setminus\{j\}) =$$
$$= v(N\setminus B_k) - \{x(N\setminus B_k) - x_j + x_i\} = x_j - x_i > 0.$$
Similarly $x_i - x_j \leq 0$, and hence $x_i - x_j = 0$.

10. The Superadditive Cover

A game v on N is called *superadditive* if $S \cap T = \emptyset$ implies $v(S \cup T) \geq v(S) + v(T)$. The *superadditive cover* of a game v is the game $\hat{v}$ defined by

$$\hat{v}(S) = \max \left\{ \sum_{i=1}^{p} v(S_i) : (S_1,\ldots,S_p) \text{ is a partition of } S \right\}.$$

Note that the superadditive cover is itself superadditive. In fact, if one defines a relation $\geq$ between games on N by $v \geq w$ if: $v(S) \geq w(S)$ for all S, then $\hat{v}$ is the minimal superadditive game that is $\geq v$.

Theorem 10:
If $\text{Co}(N,v,\mathscr{B}) \neq \emptyset$, then $\text{Co}(N,v,\mathscr{B}) = \text{Co}(N,\hat{v})$.

Proof:
1. First we prove $\text{Co}(N,v,\mathscr{B}) = \text{Co}(N,\hat{v},\mathscr{B})$. Indeed, let $x \in \text{Co}(N,\hat{v},\mathscr{B})$; then $x(S) \geq \hat{v}(S) \geq v(S)$ and $x \in X_\mathscr{B}$, so $x \in \text{Co}(N,v,\mathscr{B})$. Conversely let $x \in \text{Co}(N,v,\mathscr{B})$; so that $x \in X_\mathscr{B}$. For any S, let $(S_1,\ldots,S_l)$ be a partition of S such that $\hat{v}(S) = \sum_{i=1}^{l} v(S_i)$. Because $x(S_i) \geq v(S_i)$, $i = 1,\ldots,l$, it follows that $x(S) = \sum_{i=1}^{l} x(S_i) \geq \hat{v}(S)$, and so $x \geq \text{Co}(N,\hat{v},\mathscr{B})$.

---
[7]) Example: $N = \{1,2\}$, $v(N) = 1$, $v(\{1\}) = v(\{2\}) = 0$, $\mathscr{B} = \{N\}$.

2. Next, we show that if $\text{Co}(N,v,\mathcal{B}) \neq \emptyset$, then $\sum_k v(B_k) \geq v(N)$ and $\sum_k \hat{v}(B_k) = \hat{v}(N)$. Indeed, $\hat{v}(N) \geq \sum_k \hat{v}(B_k)$ by superadditivity. If $\hat{v}(N) > \sum_k \hat{v}(B_k)$, then there exists a partition $(S_1,\ldots,S_l)$ such that for all x in $\text{Co}(N,v,\mathcal{B})$, $\sum_{i=1}^l v(S_i) > \sum_k \hat{v}(B_k) \geq \sum_k v(B_k) = \sum_k x(B_k)$. Hence, for some i, $v(S_i) > x(S_i)$, contradicting $x \in \text{Co}(N,v,\mathcal{B})$.

3. Finally, we show that if $\text{Co}(N,\hat{v},\mathcal{B}) \neq \emptyset$, then $\text{Co}(N,\hat{v},\mathcal{B}) = \text{Co}(N,\hat{v})$. Indeed, let $x \in \text{Co}(N,\hat{v},\mathcal{B}) \neq \emptyset$. Then $x(N) = \hat{v}(N)$ by 2., and $x(S) \geq \hat{v}(S)$ for all S, so that $x \in \text{Co}(N,\hat{v})$. Conversely, let $x \in \text{Co}(N,\hat{v})$, and suppose that $x \notin X_\mathcal{B}$. Then, there exists a B_k such that $x(B_k) < \hat{v}(B_k)$, contradicting $x \in \text{Co}(N,\hat{v})$.

11. Examples

In this section we consider two games that illustrate some of the methods and results of this paper.

1. Consider a system of p universities with a total of n professors, including m game theorists. Model this by a game with n players and a coalition structure $\mathcal{B}$, in which B_k corresponds to a university. Assume that $m < p$; that no university employs more than one game theorist; that $\mathcal{B}$ is *efficient*, i.e. $\sum_k v(B_k) = v(N)$; and that the game theorists are substitutes for each other. Number the professors and the universities so that the game theorists are numbered $1,\ldots,m$, and $i \in B_i$ for $i = 1,\ldots,m$.

There is no reason to expect that

$$v(B_i) - v(B_i\setminus\{i\}) = v(B_j) - v(B_j\setminus\{j\}), \ i,j = 1,\ldots,m\ ;$$

but from efficiency it follows that

$$v(B_i) - v(B_i\setminus\{i\}) \geq v(B_k \cup \{i\}) - v(B_k), \ i = 1,\ldots,n, k = n+1,\ldots,p.$$

Theorem 9 implies that for all x in $\text{Co}(N,v,\mathcal{B})$,

$$\min_{i=1,\ldots,n} \{v(B_i) - v(B_i\setminus\{i\})\} \geq x_1 = x_2 = \cdots = x_n \geq \max_{k=n+1,\ldots,p} \{v(B_k \cup \{i\}) - v(B_k)\}.$$

In economic terms, one would say that, in the core, salaries of game theorists are equal in all universities, with an upper limit given by the productivity of the marginal game theorist employed as such and a lower limit given by the "alternative productivity" of a game theorist in a university that does not currently employ one.

2. In this example the players are again professors, but the coalitions in the coalition structure $\mathcal{B}$ are countries. We assume that v is superadditive, so that

$$v(N) - v(N\setminus B_l) \geq v(B_l)$$

for all l; and that $\mathcal{B}$ is *inefficient*, so that strict inequality holds for at least one l, say $l = k$. For simplicity of exposition, give k a name, say "Israel" (see the discussion in Subsection 5 of Section 12). Intuitively, the strict inequality says that the total value added by Israeli professors to their country is less than the total value that they could add to the rest of the world. Or, in terms of salaries, the total salary that Israeli professors command abroad is larger than the total they command at home. For all x in $X_\mathscr{B}$, we have

$$v_x^*(B_k) \geq v(B_k \cup (N\setminus B_k)) - x(N\setminus B_k) = v(N) - v(N\setminus B_k) > v(B_k) = x(B_k).$$

Let $x\setminus B_k$ be the nucleolus of (B_k, v_x^*, X_k). Assume further that, for all $S \subset B_k$, $v_x^*(S) = \sum_{i \in S} v_x^*(\{i\})$: the salaries that Israeli professors could earn abroad are unaffected by the presence there of other Israeli professors. It is readily verified that the nucleolus of (B_k, v_x^*, X_k) is such that $x_i = v_x^*(\{i\}) - c$ for all $i \in B_k$, where

$$c = \left(\sum_{i \in B_k} v_x^*(\{i\}) - v(B_k) \right) / |B_k| \,;$$

this property holds whenever $v_x^*(\{i\}) \geq c$ for all $i \in B_k$. That is: each Israeli professor would, in the nucleolus, receive a salary equal to his "opportunity cost" (the salary which he could earn abroad), minus a flat deduction which is the same for all Israeli professors.

12. Discussion

1. For a given characteristic function v, the major novel element introduced by the coalition structure $\mathscr{B}$ lies in the conditions $x(B_k) = v(B_k)$, which constrain the solution to allocate exactly among the members of each coalition the total payoff of that coalition. As a consequence of this, the bargaining over the payoff inside coalition B_k will involve a mixture of considerations which are endogenous to B_k and of considerations which are exogenous to B_k and reflect the "outside opportunities" of the members of B_k.

In so far as the value is concerned, the solution is determined entirely by considerations which are endogenous to each coalition B_k (Theorem 3). In so far as the other solution concepts reviewed here are concerned, considerations exogenous to B_k are relevant, but they are fully described (conditionally on the outside imputation $x|N\setminus B_k$) by the characteristic function v_x^* of the auxiliary game (B_k, v_x^*, X_k).

2. Whereas the *implications* of a coalition structure are quite clear, the *idea* of a coalition structure needs some clarification. On the one hand, the players are constrained to "form" the coalitions $B_1, \ldots, B_p$ that make up the structure $\mathscr{B}$. On the other hand, considerations of other coalitions, including those that "cut across" the B_k, is by no means excluded. Such coalitions are used to dominate as in the definition of core and VON NEUMANN-MORGENSTERN solution, and to object as in the bargaining set and its relatives; the excesses of these coalitions enter into the definition of nucleolus and kernel. This raises the question: what, precisely, does the "constraint" to the structure $\mathscr{B}$ mean?

The scenario usually associated with the coalition structure idea is as follows: the players consider forming the coalitions $B_1, \ldots, B_p$; one may think of them as going to business lunches in p different groups, each B_k forming a group. At these lunches they negotiate the division of the payoff, on the assumption that the coalitions $B_1, \ldots, B_p$ will be formed. In such negotiations, it is perfectly reasonable for each coalition B_k to base the division of the payoff on the opportunities that its members have outside of B_k. The negotiations at the lunch may of course break down, and at no time is it asserted that they will or even should succeed. What is being asserted is only that *if* the structure $\mathscr{B}$ forms, then the B_k should divide the payoff in whatever way the particular solution concept under consideration dictates.

What this scenario does not explain is why the groups B_k would form, or why the process of formation of the groups B_k should be separated from the bargaining for the payoff.

In attempting to answer that question, we will consider first some of the traditional explanations for the formation of coalition structures, and show that they do not survive close examination (Subsection 3). We will then advance three different arguments that show why coalition structures might arise. The most transparent one of these (Subsection 4) is valid only for games that are *not* superadditive. The other two arguments (Subsections 5 and 6) are more subtle; but they apply to all games, superadditive or not.

3. The arguments traditionally advanced for the formation of coalition structures include a) difficulties of communication; b) legal barriers such as anti-trust laws; c) personal, family, patriotic, geographical or professional relationships. Unfortunately, they do not survive close examination. The arguments advanced under a) and b) apply to potential coalitions S — those used in dominating, objecting, threatening, and so on — to exactly the same extent as they apply to the coalitions B_k constituting the structure. If a coalition is difficult or impossible to form because of communication difficulties, then these difficulties should also be taken into account when the players are comparing their opportunities. If antitrust laws forbid the formation of a coalition, and this coalition is thereby excluded from participating in a coalition structure, it should by the same token be excluded from considerations affecting the payoff.

There remain the groupings formed for a variety of "personal" reasons — item c) in the above list. It might be argued that one uses such considerations to choose the people one deals with, but there would be no hesitation about breaking up such a grouping if it would lead to material gain. When examined carefully, this argument is seen to imply a kind of lexicographic utility, with personal relationships coming out on the short end, i.e. counting for next to nothing. Most people value their personal relationships more highly than this indicates. But even if one accepts such a lexicographic utility, it would seem better to allow the entire lexicographic utility to determine the payoff, and not arbitrarily to use one component for the coalition structure and another component for bargaining.

4. In certain non-superadditive games, one can give a fairly straightforward explanation for the formation of coalition structures. But before we do that,

let us try to understand the phenomenon of non-superadditivity — to see how it is that a game arising in applications might actually fail to be superadditive. After all, superadditivity is intuitively rather compelling; why shouldn't disjoint coalitions, when acting together, get at least as much as they can when acting separately?

The answer is that the very act of "acting together" may be difficult, costly, or illegal, or the players may, for various "personal" reasons, not wish to do so. In other words, we return to points a), b), and c) of the previous subsection — but as explanations for non-superadditivity of the game, rather than for the formation of coalition structures. In fact, they are simply restrictions on the formations of coalitions. As such, it is perfectly natural to embody them in the definition of the characteristic function v; and in general, a non-superadditive v will result. For instance, if coalitions of n players or more are forbidden by the law, then "$v(S) = -\infty$ whenever $|S| \geq n$" describes the situation correctly.

Another point is that there is more involved than just "acting together"; there is also the matter of side payments. If $v(S \cup T) \geq v(S) + v(T)$, then S can transfer some of its own payoff to T. Even when communication is unrestricted, such transfers may be illegal, restricted, or subject to transaction costs. Moreover, "acting together" and sharing the proceeds may change the nature of the game. For example, if two independent farmers were to merge their activities and share the proceeds, both of them might work with less care and energy; the resulting output might be less than under independent operations, in spite of a possibly more efficient division of labour.

Suppose then, that v is a game on N, and let $\hat{v}$ be its superadditive cover (see Section 10). Call a coalition structure $\mathscr{B}$ efficient if $\sum_k v(B_k) = \hat{v}(N)$. Obviously $\hat{v}(N) \geq v(N)$. If $\hat{v}(N) = v(N)$ — as is the case in all superadditive games — then one could consider it reasonable for the coalition structure $\{N\}$ to form. If, however, $\hat{v}(N) > v(N)$, then the coalition structure $\{N\}$ is inefficient, and so a major incentive for its formation is absent. In that case it is possible that an efficient coalition structure $\mathscr{B}$ will form; but since a major cause of non-superadditivity is lack of communication for one reason or another, it might well happen, too, that $\mathscr{B}$ will be inefficient. In any case, when $\hat{v}(N) > v(N)$, one may expect the formation of a $\mathscr{B}$ other than $\{N\}$. Thus one sees that points a), b), and c) in the previous subsection, though not directly valid as an explanation for coalition structures, nevertheless are involved in an indirect manner; they explain non-superadditivity, and this in turn leads to coalition structures[8].

If the reader wishes, he may view the analysis in this subsection as part of a broader analysis, which would consider simultaneously the process of coalition

---

[8] In some cases, one might wish to model the situation by means of a game in "partition function form" [THRALL and LUCAS, 1963], where the characteristic function is a vector-valued function on the family of partitions of N. For instance, if the law requires the existence of at least m coalitions, then only coalition structures with m elements or more should be considered. The theory of games in partition function form raises substantial technical difficulties, and is not considered in this paper.

formation and the bargaining for the payoff. Let a given coalition structure $\mathscr{B}$, and a given payoff x consistent with it, provide a "solution" to the game (N,v); then certainly the payoff x must provide a "solution" to the game $(N,v,\mathscr{B})$. Our analysis has been concerned with this last topic, and should thus be understood as a contribution to partial equilibrium analysis.

5. If the game (N,v) is superadditive, the arguments for the formation of the structure $\{N\}$ sound rather compelling. Must we then abandon altogether the concept of coalition structure, in the context of superadditive games?

We think not. *Coalition structures represent groupings formed for reasons that are important and weighty, but whose impact is*
a) *difficult to measure, and/or*
b) *difficult to communicate believably, and/or*
c) *consciously excluded by the players from bargaining considerations.*

As an example that illustrates all three points, consider the world academic community, which we may think of as having partitioned itself into countries. From the point of view of material payoff, this partition is inefficient. Thus the United States could probably absorb the entire academic community of a small country like Israel, and pay each of its members a salary considerably higher than the one he is now getting. The professors do not move because for personal reasons they prefer to live in their own country. In competing for payoff within their own country, though, they may very well cite the opportunities they have abroad (see Example 2 in Section 11).

Now the question arises, why do not the professors incorporate their preferences for living in their own country into their utility functions explicitly, and bargain accordingly? One answer is that it is difficult to measure such preferences on the same terms as salary, another is that because of their subjectivity it is difficult to communicate such preferences to one's colleagues, or at least to communicate them in a convincing and believable manner.

But the most important consideration, perhaps, is of the third kind. We can at least imagine a situation in which one's love for one's country can be assessed in monetary terms; it is simply a question of deciding what increase in salary abroad would make a person want to move. We could even imagine situations in which these assessments could be believably communicated to one's colleagues. Consider now the consequences of using such assessments in the bargaining: The result would be that people who value their home country highly would be penalized. Indeed, the worth $v(S)$ of coalitions including such people would tend to be lower than the worth of coalitions containing their colleagues who are relatively indifferent, and the excesses would of course behave in the same way. This is quite understandable, as the indifferent professor is in a better position to exert pressure by means of threats to leave than the professor who values highly the place in which he lives and is unwilling to consider a move. Nevertheless it is quite conceivable that all concerned would agree that this kind of consideration should not enter into salary determinations. Given such an understanding, it is clear that the situation would best be analyzed by a game with a coalition structure determined by the countries, in which the worths $v(S)$ of the coalitions are expressed in more or less "objective" monetary terms.

A coalition structure may thus reflect considerations that are excluded from the formal description of the game by necessity (impossibility to measure or communicate) or by choice. This situation may arise in the non-superadditive as well as in the superadditive case — but in the former case it is not the only possible explanation of the coalition structure. Finally, the difficulty to measure or communicate may exist at the level of the scientist and not only at the level of the players. This illustrates a basic principle of modelling; following SAVAGE [1954], one might call it the "small worlds" principle. A model cannot always be expected to take into account in a systematic and consistent manner all the complexities of a complex situation. It is often necessary — or if not necessary at least convenient — to treat basically similar phenomena in a methodologically different fashion. Thus in SAVAGE's theory of subjective probabilities, a very clear distinction is made between the concepts of "act", "consequence", and "state of the world", and this distinction is basic to his theory. But in the real world these distinctions blur, and it is sometimes difficult to distinguish between these concepts. Our situation here is similar; we cannot say exactly which considerations go into determining a coalition structure, and which go into bargaining; we are not even sure that these two elements can be clearly separated from each other; but we feel that there are situations in which the two elements are present, and are better treated separately.

6. In the previous subsection we adduced an *exogenous* argument for the formation of a coalition structure $\mathscr{B}$, i.e. we took the coalition structure to be based on factors *not* taken into account in the characteristic function v. On the other hand, in Subsection 4 we used an *endogenous* argument, i.e. we explained the formation of $\mathscr{B}$ in terms of v itself. In this subsection, we will adduce another endogenous argument; but the current argument, unlike that in Subsection 4, is valid for superadditive as well as non-superadditive games[9]).

Briefly, the point is that the coalition structure $\mathscr{B}$ might arise because for the players in some of the B_k, it might be more worthwhile to bargain in the framework of $\mathscr{B}$ than in the framework of the all-player coalition N.

For example, consider the game v on $\{1,2,3\}$ defined by

$$v(S) = \begin{cases} 0 & \text{if } |S| = 1 \\ 8 & \text{if } |S| = 2 \\ 9 & \text{if } |S| = 3. \end{cases}$$

---

[9]) The argument in this subsection is due to MICHAEL MASCHLER, and we are grateful to him for permission to publish it here.

Assume that there are no significant asymmetries between the players (i. e. asymmetries not included in the description of the situation by v). Suppose that, for some reason, players 1 and 2 find themselves at "lunch" (see Subsection 2) without player 3. There is little doubt that they would quickly seize the opportunity to form the coalition $(1,2)$ and collect a payoff of 4 each. Thus the outcome would be $(4,4,0)$, with $\mathscr{B} = \{(1,2),(3)\}$. This would happen in spite of its inefficiency. The reason is that if players 1 and 2 were to invite player 3 to the lunch, the outcome would presumably be $(3,3,3)$. Neither would they want to risk inviting him and offering him, say, $\frac{1}{2}$ (and dividing the remaining $8\frac{1}{2}$ among themselves); because each of the two players 1 and 2 would realize that once player 3 is invited to participate in the negotiations, the situation turns "wide open" – anything can happen.

All this if players 1 and 2 happen "to find themselves at lunch". But even if this does not happen by chance, it is now fairly clear that the players would seek to form pairs for the purpose of negotiation, and not negotiate in the all-player framework.

Our example is particularly convincing because of its symmetry. Even in unsymmetric cases, though, it is clear that the framework of negotiations plays an important role in the outcome, so that individual players and groups of players will seek frameworks that are advantageous to them; and this may well lead to inefficient coalition structures. The phenomenon of seeking an advantageous framework for negotiating is also well-known in the real world at many levels – from decision making within an organization like a corporation or a university, to international negotiations.

The remarks at the end of Subsection 4 about the partial equilibrium nature of the analysis apply here as well. Incidentally, considerations of coalition formation are implicit in the VON NEUMANN-MORGENSTERN solution, even when applied only to the coalition structure $\{N\}$. But we know that the VON NEUMANN-MORGENSTERN approach to game theory is only one of a number of possible approaches, and it is worthwhile to study coalition structures explicitly and in the context of notions other than just the VON NEUMANN-MORGENSTERN solution.

References

AUMANN, R. J., and M. MASCHLER: The Bargaining Set for Cooperative Games, in: Advances in Game Theory, M. DRESHER, L. S. SHAPLEY, and A. W. TUCKER, eds., Annals of Mathematics Studies, No. 52, pp. 443–476, Princeton 1964 [Chapter 42].

DAVIS, M., and M. MASCHLER: Existence of Stable Payoff Configurations for Cooperative Games, in: Essays in Mathematical Economics in Honor of Oskar Morgenstern, M. Shubik, ed., pp. 39–52, Princeton 1967.

–: The Kernel of a Cooperative Game, Naval Research Logistics Quarterly **12**, 223–259, 1965.

GILLIES, D. B.: Solutions to General Non-zero-sum Games, in: Contributions to the Theory of Games IV, R. D. LUCE, and A. W. TUCKER, eds., Annals of Mathematics Studies, No. 40, pp. 47–85, Princeton 1959.

KOHLBERG, E.: On the Nucleolus of a Characteristic Function Game, SIAM Journal of Applied Mathematics, **20**, 62–66, 1971.

MASCHLER, M., and B. PELEG: The Structure of the Kernel of a Cooperative Game, SIAM Journal of Applied Mathematics, **15**, 569–604, 1967.

MASCHLER, M., B. PELEG, and L. SHAPLEY: The Kernel and Bargaining Set for Convex Games, International Journal of Game Theory **1**, 73–93, 1972.

SAVAGE, L. J.: Foundations of Statistics; Wiley, New York 1954.

SCHMEIDLER, D.: The Nucleolus of a Characteristic Function Game, SIAM Journal of Applied Mathematics **17**, 1163–1170, 1969.

SELTEN, R.: Equal Share Analysis of Characteristic Function Experiments, in: Contributions to Experimental Economics III, H. SAUERMANN and J. C. B. MOHR, eds., pp. 130–165, Tübingen 1972.

SHAPLEY, L. S.: A Value for n-person Games, in: Contributions to the Theory of Games II, H. W. KUHN, and A. W. TUCKER, eds., Annals of Mathematics Studies no. 52, pp. 307–317, Princeton 1953.

THRALL, R., and W. F. LUCAS: Games in Partition Function Form, Naval Research Logistics Quarterly **10**, 281–298, 1963.

VON NEUMANN, J., and O. MORGENSTERN: Theory of Games and Economic Behaviour, Princeton 1944.

Note added in proof: Theorem 6 (p. 128) is wrong. See Chang, C.: A Note on the von Neumann–Morgenstern Solution, International Journal of Game Theory **17**, 311–314, 1988.

45 Game-Theoretic Analysis of a Bankruptcy Problem from the Talmud
with M. Maschler

DEDICATED TO THE MEMORY OF SHLOMO AUMANN, TALMUDIC SCHOLAR AND MAN OF THE WORLD, KILLED IN ACTION NEAR KHUSH-E-DNEIBA, LEBANON, ON THE EVE OF THE NINETEENTH OF SIVAN, 5742 (JUNE 9, 1982).

For three different bankruptcy problems, the 2000-year-old Babylonian Talmud prescribes solutions that equal precisely the nucleoli of the corresponding coalitional games. A rationale for these solutions that is independent of game theory is given in terms of the Talmudic principle of equal division of the contested amount; this rationale leads to a unique solution for all bankruptcy problems, which always coincides with the nucleolus. Two other rationales for the same rule are suggested, in terms of other Talmudic principles. (Needless to say, the rule in question is not proportional division.) *Journal of Economic Literature Classification Numbers*: 022, 026, 031, 043, 213. © 1985 Academic Press, Inc.

1. INTRODUCTION

A man dies, leaving debts $d_1,...,d_n$ totalling more than his estate E. How should the estate be divided among the creditors?

A frequent solution in modern law is proportional division. The rationale is that each dollar of debt should be treated in the same way; one looks at dollars rather than people. Yet it is by no means obvious that this is the only equitable or reasonable system. For example, if the estate does not exceed the smallest debt, equal division among the creditors makes good sense. Any amount of debt to one person that goes beyond the entire estate might well be considered irrelevant; you cannot get more than there is.

A fascinating discussion of bankruptcy occurs in the Babylonian

---

This work was supported by National Science Foundation Grant SES 83-20453 at the Institute for Mathematical Studies in the Social Sciences, Stanford University, and by the Institute for Mathematics and its Applications at the University of Minnesota.

This chapter originally appeared in *Journal of Economic Theory* 36 (1985): 195–213. © Academic Press. Reprinted with permission.

TABLE I

		Debt		
		100	200	300
	100	$33\frac{1}{3}$	$33\frac{1}{3}$	$33\frac{1}{3}$
Estate	200	50	75	75
	300	50	100	150

Talmud[1] (Kethubot 93a). There are three creditors; the debts are 100, 200, and 300. Three cases are considered, corresponding to estates of 100, 200, and 300. The *Mishna*[2] stipulates the divisions shown in Table I.

The reader is invited to study Table I. When $E = 100$, the estate equals the smallest debt; as pointed out above, equal division then makes good sense. The case $E = 300$ appears based on the different—and inconsistent—principle of proportional division. The figures for $E = 200$ look mysterious; but whatever they may mean, they do not fit any obvious extension of either equal *or* proportional division. A common rationale for all three cases is not apparent.

Over two millenia, this Mishna[3] has spawned a large literature. Many authorities disagree with it outright. Others attribute the figures to special circumstances, not made explicit in the Mishna. A few have attempted direct rationalizations of the figures as such, mostly with little success. One modern scholar, exasperated by his inability to make sense of the text, suggested errors in transcription.[4] In brief, the passage is notoriously difficult.

This paper presents a game-theoretic analysis of the general bankruptcy problem, for arbitrary $d_1, ..., d_n$ and E. We obtain (Sect. 6) an explicit characterization of the nucleolus of the coalitional game that is naturally associated with this problem. For the three cases considered in the Talmud,

---

[1] A 2,000-year-old document that forms the basis for Jewish civil, criminal, and religious law.

[2] The basic text that forms the starting point for the discussions recorded in the Talmud.

[3] The word "Mishna" is used both for the entire text on which the Talmud is based, and for specific portions of it dealing with particular issues. Similar ambiguities occur in many languages. One may say "My son studied law" as well as "Yesterday Congress passed a law."

[4] Lewy [8, p. 106], near the end of the long footnote.

the nucleolus prescribes precisely the numbers of Table I—those in the Mishna.

Of course, it is unlikely that the sages of the Mishna were familiar with the general notion of a coalitional game, to say nothing of the nucleolus. In Sections 3, 4, and 5, we present three different justifications of the solution to the bankruptcy problem that the nucleolus prescribes, in terms that are independent of each other and of game theory, and that were well within the reach of the sages of the Mishna. The justifications also fit in well with other Talmudic principles, a consideration that is no less significant than innate reasonableness in explaining the text.

To emphasize the independence from game theory, we start with the non-game theoretic analysis. In the research process, however, the order was reversed. Only after realizing that the numbers in the Mishna correspond to the nucleolus did we find independent rationales. Without the game theory, it is unlikely that we would have hit on the analysis presented in Sections 2 through 5.

This paper addresses two related but distinct questions, "what" and "why." The Mishna explicitly gives only three numerical examples; the first, most basic question is, *what* general rule did it have in mind, what awards would it actually assign to the creditors in an arbitrary bankruptcy problem? The second question is *why* did it choose the rule that it chose, what reasoning guided Rabbi Nathan (the author of this particular Mishna)? We have three different answers to the "why" question, all of them leading to the same answer to the "what" question. And while we cannot be sure that any particular one of our answers to the "why" question really represents Rabbi Nathan's thinking, all of them together leave little doubt that our answer to the "what" question is the correct one.

It is hoped that the research reported here will be of interest in two spheres—in the study of the Talmud, and in game theory. In this paper we concentrate on the mathematical, game-theoretic side. For motivation we will, when appropriate, present and explain underlying Talmudic principles; but there will be no careful textual analysis of Talmudic passages, no lengthy citation of authorities.[5] An analysis of the latter kind—concentrating on the validity of our explanation from the source viewpoint, but skimping on the mathematics—will be published elsewhere.[5a]

---

[5] In particular, this article should on no account be used as a source for Talmudic law. In citing Talmudic dicta, we mention only those of their aspects that are directly relevant to the matter at hand; additional conditions and circumstances are omitted. For more information the reader should refer to the sources, which we have been at pains to cite explicitly.

[5a] Aumann, Y., "On the Man with Three Wives," *Moriah* 22, 3-4, Tevet 5759, 1999, pp. 98-107; in Hebrew.

2. The Contested Garment

A famous Mishna (Baba Metzia 2a) states: "Two hold a garment; one claims it all, the other claims half. Then the one is awarded $\frac{3}{4}$, the other $\frac{1}{4}$."

The principle is clear. The lesser claimant concedes half the garment to the greater one. It is only the remaining half that is at issue; this remaining half is therefore divided equally.[6] Note that this is quite different from proportional division.

Let us transpose this principle to the 2-creditor bankruptcy problem with estate E and claims d_1, d_2. The amount that each claimant i concedes to the other claimant j is $(E - d_i)_+$, where

$$\theta_+ = \max(\theta, 0).$$

The amount at issue is therefore

$$E - (E - d_1)_+ - (E - d_2)_+;$$

it is shared equally between the two claimants, and, in addition, each claimant receives the amount conceded to her by the other one. Thus the total amount awarded to i is

$$x_i = \frac{E - (E - d_1)_+ - (E - d_2)_+}{2} + (E - d_j)_+. \tag{2.1}$$

We will say that this division (of E for claims d_1, d_2) is prescribed by the CG (contested garment) principle.[7,8]

If one views the solution as a function of E, one obtains the following process. Let $d_1 \leq d_2$. When E is small, it is divided equally. This continues until each claimant has received $d_1/2$. Each additional dollar goes to the greater claimant, until each claimant has received all but $d_1/2$ of her claim. Beyond that, each additional dollar is again divided equally. Note that the principle is *monotonic*, in the sense that for fixed claims d_1 and d_2, each of the two awards is a monotonic function of the estate E.

---

[6] This explanation is explicit in the eleventh-century commentary of Rabbi Shlomo Yitzhaki (Rashi). Alternatively, one could say that the claims total $1\frac{1}{2}$, whereas the worth of the garment is only 1; the loss is shared equally.

[7] An additional instance of this rule may be found in the Tosefta to the first chapter of Baba Metzia, where $E = d_1 = 1$, $d_2 = \frac{1}{3}$. (The Tosefta is a secondary source that is contemporaneous with the Mishna.)

[8] Alternatively, one may argue that neither claimant i can ask for more than $\min(E, d_i)$. If each claimant is awarded this amount, the total payment may exceed the estate; the excess is deducted in equal shares from the claimants' awards. This procedure leads to the same payoff as (2.1).

The legal circumstances of the contested garment are somewhat different from those of bankruptcy. In the garment case, there is uncertainty about the validity of the claims; they cannot both be justified. In the bankruptcy problem, all claims are definitely valid; there simply is not enough money to go around. Therefore, some authorities have held that the Mishna in Baba Metzia (about the garment) is not relevant to the bankruptcy problem.

While this certainly constitutes an important difference between the cases, it is not clear why it would make the principle of equal division of the contested amount inapplicable. Indeed, the early medieval authority Rabbi Hai Gaon (10th century) did express the opinion[9] that the Mishna in Kethubot (about bankruptcy) should be explained on the basis of that in Baba Metzia. He did not, however, make an explicit connection, and in subsequent years, this line of attack was abandoned.

3. Consistency

A *bankruptcy problem* is defined as a pair $(E; d)$, where $d = (d_1,..., d_n)$, $0 \leq d_1 \leq \cdots \leq d_n$ and $0 \leq E \leq d_1 + \cdots + d_n$. A *solution* to such a problem is an n-tuple $x = (x_1,..., x_n)$ of real numbers with

$$x_1 + \cdots + x_n = E$$

(x_i is the amount assigned to claimant i). A solution is called *CG-consistent*, or simply *consistent*, if for all $i \neq j$, the division of $x_i + x_j$ prescribed by the contested garment principle for claims d_i, d_j is (x_i, x_j).

Intuitively, a solution is consistent if any two claimants i, j use the contested garment principle to divide between them the total amount $x_i + x_j$ awarded to them by the solution. It may be verified that the solutions in Table I are consistent.

THEOREM A. *Each bankruptcy problem has a unique consistent solution.*

Proof. First we prove that there is at most one consistent solution. If there were more, we could find consistent solutions x and y, and creditors i and j, with $y_i > x_i$, $y_j < x_j$, and $y_i + y_j \geq x_i + x_j$. Consistency implies that if just i and j are involved, the CG principle awards y_j to j when the total estate is $y_i + y_j$, and x_j when it is $x_i + x_j$. Since $y_i + y_j \geq x_i + x_j$, the

---

[9] Quoted by Rabbi Isaac Alfasi (1013–1103) in his commentary on our Mishna in Kethubot.

monotonicity of the CG principle then implies $y_j \geq x_j$, contradicting[10] $y_j < x_j$.

To show that there is at least one consistent solution, we exhibit it as a function of the estate E (for fixed debts $d_1,..., d_n$). Let us think of the estate as gradually growing. When it is small, all n claimants divide it equally. This continues until 1 has received $d_1/2$; for the time being she[11] then stops receiving payments, and each additional dollar is divided equally between the remaining $n-1$ claimants. This, in turn, continues until 2 has received $d_2/2$, at which point *she* stops receiving payments for the time being, and each additional dollar is divided equally between the remaining $n-2$ claimants. The process continues until each claimant has received half her claim. This happens when $E = D/2$, where

$$D := d_1 + \cdots + d_n = \text{the total debt}.$$

When $E \geq D/2$, the process is the mirror image of the above. Instead of thinking in terms of i's award x_i, one thinks in terms of her loss $d_i - x_i$, the amount by which her award falls short of her claim. When the total loss $D - E$ is small, it is shared equally between all creditors, so that creditor i receives her claim d_i less $(D-E)/n$. The creditors continue sharing each additional dollar of total loss equally, until 1 has lost $d_1/2$ (which is the same as receiving $d_1/2$). For the time being she then stops losing, and each additional dollar of total loss is divided equally between the remaining $n-1$ claimants. This, in turn, continues until 2 has lost $d_2/2$ (= received $d_2/2$), at which point *she* stops losing for the time being, and each additional dollar is divided equally between the remaining $n-2$ claimants. The process continues until each claimant has lost half her claim, which happens when $E = D/2$. This is precisely to where we got in describing the first part of the procedure; we have dug the tunnel from both its ends, and have met in the middle.

It will be useful to give an alternative description of the procedure, in terms of increasing award. Recall that when E was slightly less than $D/2$, claimant n was receiving all of each additional dollar of the estate. She continues to do so as E passes $D/2$, until she has received a total of $d_n - (d_{n-1}/2)$, i.e., all but $d_{n-1}/2$ of her claim. At this point, $n-1$ reenters the picture, and each additional dollar is shared equally between n and

---

[10] This part of the proof was generated at a seminar presentation at the IMA in Minneapolis; it replaces a more devious proof that we previously had. Mainly responsible are Y. Kannai and D. Kleitman.

[11] While we are sympathetic with the feminist movement, the reader should not conclude from our use of "she" that we write half our papers in the feminine gender. Much of the Talmud is couched in terms of case law; and while the passage under discussion does form the basis of bankruptcy law in general, the creditors in this particular case were women.

$n-1$. This continues until n and $n-1$ have each received all but $d_{n-2}/2$ of their claims (which happens at the same instant). At this point $n-2$ reenters the picture, and so on. Creditor 1 reenters the picture when *all* creditors have received all but $d_1/2$ of their claims; each additional dollar of estate is shared equally between all creditors.

Consider now two claimants i and j, where $d_i \leqslant d_j$. When E is small, they receive equal amounts. This continues until i has received $d_i/2$. Beyond that, i leaves the picture for the time being, and only j may receive any part of each additional dollar; that is, each additional dollar received by both together goes to j. This continues until j has received all but $d_i/2$ of her claim; beyond that, i and j again receive equal shares of each additional dollar. But this is precisely the verbal description of the CG solution given in the previous section. This shows that the solution we have exhibited is indeed consistent, and completes the proof of Theorem A.

Define a *rule* as a function that assigns a solution to each bankruptcy problem. The *CG-consistent* (or simply *consistent*) rule is the one that assigns the CG-consistent solution to each bankruptcy problem. A rule f is called *self-consistent* if

$$f(E;d) = x \quad \text{implies} \quad f(x(S); d|S) = x|S$$

for each set S of creditors, where $x|S$ means "x restricted to S," and $x(S)$ is short for $\sum_{i \in S} x_i$. In words, *any* subset S of the set of all creditors (not only a 2-person subset) uses the rule f to divide among its members the total amount $x(S)$ that it gets when the rule f is applied to the original bankruptcy problem.

COROLLARY 3.1. *The CG-consistent rule is self-consistent.*

Proof. Let $(E;d)$ be a bankruptcy problem, x its CG-consistent solution, g the function that assigns to each 2-person bankruptcy problem its CG solution, and S a set of creditors. For *any* i, j, the CG-consistency of x yields $(x_i, x_j) = g(x_i + x_j; d_i, d_j)$; in particular, this is so for $i, j \in S$. But that means that $x|S$ is the CG-consistent solution of $(x(S); d|S)$.   Q.E.D.

Self-consistency and CG-consistency are totally different kinds of concepts. Self-consistency applies to *rules*, CG-consistency to *individual solutions*. If three creditors come to a judge and ask him to divide an estate between them, and he does so in some specific way, then they cannot complain that he is not self-consistent; to do that they would have to know what he would have decided in other situations. But CG-consistency can be checked directly for each proposed solution of each case separately. If one thinks of the principle of CG-consistency as "just," then one can complain about the injustice of one particular decision; the corresponding statement cannot be made for self-consistency.

The CG-consistent rule is by no means the only self-consistent one. Others include division in proportion to the claims, and the constrained equal division solutions in the next section; these rules also play an important role in the Talmudic discussion of bankruptcy.

Various consistency conditions that are similar in spirit to those discussed here play an important role in game theory and bargaining theory. For examples, see Sections 6 and 7 below.

4. Self Duality and Constrained Equal Division

Define the *dual* f^* of a rule f by

$$f^*(E; d) := d - f(D - E; d);$$

f^* assigns awards in the same way that f assigns losses. A *self-dual* rule is one with $f^* = f$; such a rule treats losses and awards in the same way.

Two prominent features of the consistent rule, both of which follow from the explicit characterization in the proof of Theorem A, are its self-duality, and the qualitative change in the rule that occurs at $E = D/2$. In this section we discuss these features and show that they are strongly rooted in the Talmudic literature. We end the section with an alternative characterization (Theorem B) of the consistent rule in terms of these features, a characterization not directly related to the CG principle.

The basic idea behind duality is that there are certain types of division problem in which it is natural to think in terms of dividing the award (amount received, gain), and other problems in which it is more natural to think in terms of dividing the loss. All other things being equal, it seems appropriate to apply dual solution rules to these problems. In still other problems, it is equally natural to think in terms of losses or of gains; in such cases a self-dual rule is called for.

We start by illustrating the notion of duality with some rules other than the consistent one. A *constrained equal award* (CEA) solution of a bankruptcy problem $(E; d)$ is one of the form $(\alpha \wedge d_1, ..., \alpha \wedge d_n)$, where $a \wedge b := \min(a, b)$. In words, this means that all claimants get the same award α, except that those who claim less than α get their claims. Note that

each bankruptcy problem has a unique CEA solution. (4.1)

Indeed, $\sum_{i=1}^{n} \alpha \wedge d_i$ is a continuous strictly increasing function of α on the interval $[0, d_n]$, and maps this interval onto $[0, D]$; hence every point in $[0, D]$ is attained precisely once, proving (4.1).

In our Mishna, the CEA rule prescribes equal awards to all creditors up to $E = 300$; it prescribes $(100, 150, 150)$ for $E = 400$, and $(100, 200, 200)$ for

$E = 500$. The rule, which divides each additional dollar equally between those claimants who still have an outstanding claim, seems natural enough; it has been adopted as law by most major codifiers, including Maimonides[12] (1135–1204).

Maimonides's great intellectual adversary, Rabad,[13] adopted a different rule. This provides equal awards for all creditors when $E \leq d_1$. When E passes d_1, 1 leaves the picture, and each additional dollar is divided equally among the remaining $n-1$ creditors. When E passes d_2, also 2 leaves the picture, and each additional dollar is divided equally among the remaining $n-2$ creditors; and so on. The rule is not defined beyond $E = d_n$. This, too, seems quite natural; when $E \leq d_n$ one might say that all creditors have a claim on the first d_1 dollars, only $2,\ldots,n$ on the next $d_2 - d_1$ dollars, and so on.[14]

We would not discuss these rules in such detail if it were not for the remarkable fact that their duals also make an explicit appearance in the Talmud (Erakhin 27b).[15] At an auction there are n bidders, who bid $b_1 < b_2 < \cdots < b_n$. If n reneges, the object is acquired by $n-1$, and the seller sustains a loss of $b_n - b_{n-1}$; this loss must be paid by n, as the price of being allowed out of his contract. Suppose now that all n bidders renege, and that for one reason or another the object cannot be sold to anyone else; then the loss to the seller is b_n, and this must somehow be shared among the bidders. How?

In this case Maimonides says that the loss is divided equally among all bidders, subject, of course, to no bidder paying more than his bid.[16] Rabad,[17] on the other hand, divides the loss into the n successive increments $b_1, b_2 - b_1, b_3 - b_2, \ldots, b_n - b_{n-1}$. The first increment is paid by all bidders in equal shares, the second in equal shares by all except 1, and

---

[12] *The Laws of Lending and Borrowing*, Chapter 20, Section 4.

[13] Acronym for Rabbi Abraham ben David (1125–1198).

[14] This rule is implicit already in the Babylonian Talmud's discussion of our Mishna, and so goes back at least to the third or fourth century. It first appears explicitly in Alfasi (op. cit.), and is also mentioned by Rashi and many other medieval commentators. Only Rabad, though, seems to have adopted it as law; see his gloss on Alfasi (op. cit.). The rule is also mentioned by the mathematician Abraham Ibn Ezra in connection with certain types of inheritance problem [3, p. 60ff.]; cf. Rabinovitch [17, p. 162] and O'Neill [14]. In modern times it has surfaced again as the solution to the airport landing problem [9]; it is closely connected with the Shapley value [19], a game-theoretic solution concept that is conceptually quite different from the nucleolus.

[15] We are grateful to Y. Aumann for bringing this reference to our attention.

[16] *The Laws of Appraisal*, Chapter 8, Section 4. Unlike for bankruptcy, Maimonides here gives no clear general rule; for the specific numerical example treated in the Talmud (and by Maimonides), his division is equal. But it is difficult to imagine that Maimonides would ever require a reneging bidder to pay more than his bid.

[17] See his gloss on Maimonides, op. cit.

so on. These rules are the exact duals of the rules adopted by these same authorities for bankruptcy.

It is apparent that these authorities had precise ideas on how to deal with division problems, and that they applied them to the award or the loss according to whether the funds were to be received (as in bankruptcy) or paid (as in the auction). But one can also approach the problem not so much from the technical viewpoint of the direction in which money flows at the specific time of the court decision, but from the more substantive viewpoint of whether the protagonists themselves consider the transaction an award or a loss. In bankruptcy, for example, the creditors will in the end receive checks as a result of the court proceedings. Nevertheless, they are worse off than before making the loan, and they may well conceive of the transaction as a loss rather than an award.

This suggests a rule in which (i) awards and losses are treated dually, and (ii) it makes no difference whether we think of the outcome as an award or a loss.[18] Together, (i) and (ii) call for a self-dual rule.

Self-duality was just one of the two prominent features of the consistent rule mentioned at the beginning of this section. The other was the qualitative change in the rule that occurs at the "halfway point," $E = D/2$. This, too, is strongly rooted in the Talmud; there are dozens—perhaps hundreds—of discussions hinging on the principle that "more than half[19] is like the whole" (Hulin 27a). For example, kosher slaughter of an animal calls for cutting through the windpipe and the foodpipe; but as long as more than half of each pipe is cut, the meat is still kosher (op. cit.).

Another example is based on the Talmudic principle that in general, a lender automatically has a lien on the borrower's real property. But when his entire property is worth less than half the loan, the borrower may in certain cases dispose of it "free and clear" (Erakhin 23b). Rashi explains that when the property is worth more than half the loan, the lien is of considerable importance, and the lender relies on it as a guarantee. But when it is worth less than half the loan, the lien will not help very much anyway; the loan was presumably made "on trust," and we are not justified in repossessing the property from the bona fide recipient.

Again, the principle involved here is "more than half is like the whole;" property amounting to more than half the loan is conceptually close to covering it all, and cannot be ignored. Less than half is like nothing; property covering less than half the loan is inconsiderable, need not be taken into account. This is not merely a legal convention, but is explicitly based on a psychological presumption. In the "less-than-half" case, the len-

---

[18] Specifically, whether we think of the outcome to Creditor i as an award of x_i or a loss of $d_i - x_i$.

[19] "Rov."

der is presumed not to rely on the lien, to lend "on trust." Psychological presumptions of this kind often play an important role in Talmudic law.[20]

In the bankruptcy problem, too, the half-way point is a psychological watershed. If you get more than half your claim, your mind focusses on the full debt, and your concern is with the size of your loss. If you get less than half, your mind writes off the debt entirely, and is "happy" with whatever it can get; your concern is with your award. Moreover, it is socially unjust for different creditors to be on opposite sides of this watershed; for one creditor to get most of his claim, while another one loses most of his. Subject to this constraint, therefore, the losses are divided equally when $E \geqslant D/2$, the awards when $E \leqslant D/2$. In brief, we have

THEOREM B. *The consistent rule is the unique self-dual rule that, when $E \leqslant D/2$, assigns to $(E; d)$ the constrained equal award solution of $(E; d/2)$.*

Mathematically, this theorem is simply a concise expression of the explicit construction in the proof of Theorem A. Conceptually, though, it provides an additional, independent characterization of the consistent rule.

We end this section by mentioning two properties of the consistent rule that will be useful in the sequel. Call a rule f *monotonic* if $f_i(E; d)$ is a non-decreasing function of E when i and d are held fixed; that is, no claimant loses from an increase in the estate. Call a solution x to a bankruptcy problem $(E; d)$ *order-preserving* if

$$0 \leqslant x_1 \leqslant x_2 \leqslant \cdots \leqslant x_n \quad \text{and} \quad 0 \leqslant d_1 - x_1 \leqslant \cdots \leqslant d_n - x_n.$$

That is, a person with a higher claim than another gets an award that is no smaller and suffers a loss that is no smaller;[21] ordinarily, one might say, both the award and the loss are scaled to the claim. As we have said, the consistent rule is monotonic and yields order-preserving solutions. So are the other rules we have considered here: Division in proportion to the claims, the CEA rule, its dual, and the solution adopted by Rabad (insofar as it is defined).

---

[20] E.g., the finder's right to found property depends on whether the loser can be presumed to have "despaired" of regaining it (Baba Metzia 21a ff.).

[21] Also, awards are non-negative and do not exceed the claims; but this statement follows from the other if we include outsiders, whose claim is 0.

5. COALITION FORMATION AND THE JERUSALEM TALMUD

In discussing our Mishna, the Jerusalem Talmud[22] says as follows: "Samuel says, the Mishna takes it that the creditors empower each other; specifically, that the third empowers the second to deal with the first. She may say to her, 'Your claim is 100, right? Take 50 and go.'"

Samuel is obviously referring to the cases $E = 200$ and $E = 300$. The second and third creditors (whose claims are 200 and 300, respectively) form a coalition "against" the first (whose claim is 100). This leaves two effective protagonists, with claims of 500 and 100, respectively; applying the contested garment rule yields the first creditor 50, with the remainder going to the coalition. If the coalition again applies the contested garment rule to divide its award among its members, the numbers in the Mishna result.

If applied to the case $E = 100$, this procedure would lead to the payoff vector (50, 25, 25), which is not order-preserving: 1's award is larger than that of 2 or of 3. It begins yielding order-preserving results at $E = 150$; as E rises, it continues to do so until $E = 450$. Beyond that, order preservation again fails, this time on the loss side. At $E = 500$, for example, we obtain (50, 175, 275), so that 1's loss is larger than that of 2 or of 3.

We may proceed in the same way in any problem with three creditors. First 2 and 3 pool their claims and act as a single agent vis-a-vis 1. The CG solution of the resulting problem yields awards to 1, and to the coalition of 2 and 3; to divide its award among its members, the coalition again applies the CG principle. The result is order preserving if and only if $3d_1/2 \leqslant E \leqslant D - (3d_1/2)$. If one divides the awards equally when $E \leqslant 3d_1/2$, and the losses equally when $E \geqslant D - (3d_1/2)$, one obtains precisely the consistent solution over the entire range $0 \leqslant E \leqslant D$.

By using induction, one may generalize this in a natural way to arbitrary n. Suppose we already know the solution for $(n-1)$-person problems. Depending on the values of E and d, we treat a given n-person problem in one of the following three ways:

(i) Divide E between $\{1\}$ and $\{2,...,n\}$ in accordance with the CG solution of the 2-person problem $(E; d_1, d_2 + \cdots + d_n)$, and then use the $(n-1)$-person rule, which we know by induction, to divide the amount assigned to the coalition $\{2,...,n\}$ between its members.

(ii) Assign equal awards to all creditors.

(iii) Assign equal losses to all creditors.

---

[22] The Jerusalem Talmud is based on the same source (the Mishna) as the Babylonian Talmud, and is contemporaneous with it. Although considered less authoritative, it is valued as an independent parallel source, which often sheds light on obscure passages in the Babylonian Talmud. We are very grateful to Yehonatan Aumann for calling to our attention the remarkable passage that forms the basis for this section.

Specifically, (i) is applied whenever it yields an order-preserving result, which is precisely when $nd_1/2 \leqslant E \leqslant D - (nd_1/2)$. We apply (ii) when $E \leqslant nd_1/2$, and (iii) when $E \geqslant D - (nd_1/2)$. This is called the *coalitional procedure*.

For example, let $n = 5$, $d_i = 100\,i$, $E = 510$. At the first step of the induction, the coalition $\{2, 3, 4, 5\}$ forms; its joint claim is 1400, while 1's claim is 100. Applying the CG rule yields 50 to 1, and 460 to the coalition. To divide the 460 between 2, 3, 4, and 5, one splits $\{2, 3, 4, 5\}$ into 2 and $\{3, 4, 5\}$, and again applies the CG rule. This yields 100 to 2, and 360 to $\{3, 4, 5\}$. If one were again to split $\{3, 4, 5\}$ into 3 and $\{4, 5\}$, then 3 would be awarded 150, leaving 4 and 5 only 210 to divide between them. At least one of them would therefore receive $\leqslant 105$, so that the result would not be order-preserving. At this point, therefore, the 360 are split equally between 3, 4, and 5. The final result is (50, 100, 120, 120, 120), which *is* order-preserving.

Note that this is the consistent solution of the above problem. More generally, it may be verified that

THEOREM C. *The coalitional procedure yields the consistent solution for all bankruptcy problems.*

Theorem A and C both use the CG principle to characterize the same rule; but the two characterizations are conceptually totally different. Theorem A applies the CG principle to pairs of *individuals* only, and it is applied to all $\binom{n}{2}$ such pairs. The $\binom{n}{2}$ resulting conditions are desiderata of a solution, but they do not tell us directly how we should arrive at one; it is not a priori clear that they have any simultaneous solution at all, or that they do not have more than one. Theorem C applies the CG principle to pairs of *coalitions*; and it describes an orderly step-by-step process, which by its very definition must lead to a unique result. But it uses only certain carefully selected pairs of coalitions, not all such pairs.[23]

6. THE NUCLEOLUS AND THE KERNEL

It will be recalled that a (*coalitional*) *game* is a function v that associates a real number $v(S)$ with each subset S of a finite set N. The members of $N

---

[23] As described—with the coalitions $\{1\}$ and $\{2,...,n\}$—the coalitional procedure yields a monotonic rule and order-preserving solutions. Moreover, it appears that they are the *only* such coalitions, though we have no satisfactory formulation and proof of such a result. Also, it appears that if the creditors may form coalitions as they wish, then for $E \leqslant D/2$, the incentives lead to the coalitions suggested by the coalitional procedure. These matters call for further study.

are called *players*, the sets S *coalitions*. Intuitively, $v(S)$ represents the total amount of payoff that the coalition S can get by itself, without the help of other players; it is called the *worth* of S. By convention, $v(\emptyset) = 0$. A *payoff vector* is a vector x with components indexed by the players; x_i represents the payoff to i.

Solution concepts associate payoff vectors with games; each such concept represents a specific notion of stability, expected outcome, or the like. In many cases a solution concept associates several payoff vectors with a game, or none at all. Only two of the better known solution concepts associate a unique payoff vector with each game; they are the value [19] and the nucleolus [18].

As it stands, the bankruptcy problem considered here is not a game; coalitions do not appear explicitly in its formulation. A natural way to associate a game with a bankruptcy problem $(E; d)$ is to take the worth of a coalition S to be what it can get without going to court; i.e., by accepting either nothing, or what is left of the estate E after each member i of the complementary coalition $N \setminus S$ is paid his complete claim d_i. Thus we define the *(bankruptcy) game* $v_{E;d}$ corresponding to the bankruptcy problem $(E; d)$ by

$$v_{E;d}(S) := (E - d(N \setminus S))_+ \qquad (6.1)$$

THEOREM D. *The consistent solution of a bankruptcy problem is the nucleolus of the corresponding game.*

The proof of Theorem D makes use of several concepts and results of cooperative game theory. Let v be a game, S a coalition, x a payoff vector. The *reduced game* $v^{S,x}$ is defined [1, 20, 10, 15] on the player space S as follows:

$$v^{S,x}(T) = x(T) \qquad \text{if } T = S \text{ or } T = \emptyset,$$
$$= \max\{v(Q \cup T) - x(Q) : Q \subset N \setminus S\} \qquad \text{if } \emptyset \subsetneq T \subsetneq S.$$

In the reduced game, the players of S consider how to divide the total amount assigned to them by x under the assumption that players i outside S get exactly x_i. Together, all the players of S get $x(S)$; as always, the empty set gets nothing. If a non-empty proper subcoalition T of S chooses a set Q of "partners" outside S, it will have total worth $v(Q \cup T)$; but to keep the partners satisfied, it will have to pay them the total $x(Q)$ assigned to them by x. Thus T will choose its partners Q to maximize the amount $v(Q \cup T) - x(Q)$ left for it after paying off the partners.

LEMMA 6.2. *Let x be a solution of the bankruptcy problem $(E; d)$, such that $0 \leq x_i \leq d_i$ for all i. Then for any coalition S,*

$$v_{E;d}^{S,x} = v_{x(S);d|S}. \qquad (6.2)$$

A Bankruptcy Problem from the Talmud

(In words, the reduced bankruptcy game is the game corresponding to the "reduced bankruptcy problem.")

Proof. Set $v := v_{E;d}$ and $v^S := v_{E;d}^{S;x}$. First let $\emptyset \subsetneq T \subsetneq S$, and let the maximum in the definition of $v^S(T)$ be attained at Q. Since $x_i \geq 0$ and $a_+ - b_+ \leq (a-b)_+$ for all a and b, we have

$$v^S(T) = v(T \cup Q) - x(Q) = (E - d(N\setminus(Q \cup T)))_+ - (x(Q))_+$$
$$\leq (x(N) - d(N\setminus(Q \cup T)) - x(Q))_+$$
$$= [x(S) - d(S\setminus T) - (d-x)(N\setminus(S \cup Q))]_+$$
$$\leq (x(S) - d(S\setminus T))_+, \tag{6.3}$$

where the last inequality follows from $x_i \leq d_i$. On the other hand, setting $Q = N\setminus S$ yields

$$v^S(T) \geq v(T \cup (N\setminus S)) - x(N\setminus S)$$
$$= (E - d(N\setminus(T \cup (N\setminus S))))_+ - (x(N) - x(S))$$
$$\geq (E - d(S\setminus T)) - (E - x(S)) = x(S) - d(S\setminus T); \tag{6.4}$$

and setting $Q = \emptyset$ yields

$$v^S(T) \geq v(T \cup \emptyset) - x(\emptyset) = v(T) = (E - d(N\setminus T))_+ \geq 0. \tag{6.5}$$

Formulas (6.4) and (6.5) together yield

$$v^S(T) \geq (x(S) - d(S\setminus T))_+;$$

together with (6.3), this yields

$$v^S(T) = (x(S) - d(S\setminus T))_+ = v_{x(S);d|S}(T). \tag{6.6}$$

When $T = \emptyset$ or $T = S$, formula (6.6) is immediate, so the proof of the lemma is complete.

We also make use of the solution concepts called kernel [1] and pre-kernel [10]. Let v be a game. For each payoff vector x and players i, j, define

$$s_{ij}(x) = \max\{v(S) - x(S): S \text{ contains } i \text{ but not } j\}.$$

The *pre-kernel* of v is the set of all payoff vectors x with $x(N) = v(N)$ and $s_{ij}(x) = s_{ji}(x)$ for all i and j. The *kernel* of v is the set of all payoff vectors x with $x(N) = v(N)$, $x_i \geq v(i)$ for[24] all i, and for all i and j,

$$s_{ij}(x) > s_{ji}(x) \quad \text{implies} \quad x_j = v(j).$$

---

[24] We do not distinguish between i and $\{i\}$.

One more definition is required. The *standard solution* of a 2-person game v with player set $\{1, 2\}$ is given by

$$x_i = \frac{v(12) - v(1) - v(2)}{2} + v(i). \tag{6.7}$$

Note that this is equivalent to $x_1 + x_2 = v(12)$, $x_1 - x_2 = v(1) - v(2)$. In words, the standard solution gives each player i the amount $v(i)$ that he can assure himself, and divides the remainder equally between the two players. The nucleolus and kernel,[25] the pre-kernel and the Shapley value of a 2-person game all coincide with its standard solution; so do most of the better-known bargaining solutions [13, 4, 12]. Indeed, the standard solution constitutes the only symmetric and efficient point-valued solution concept for 2-person games that is covariant under strategic equivalence.[26]

LEMMA 6.8. *Let x be in the pre-kernel of a game v, and let S be a coalition with exactly two players. Then $x|S$ is the standard solution of $v^{S,x}$.*

Proof. Let $S = \{i, j\}$. Then

$$s_{ij}(x) = \max_{Q \subset N \setminus S} (v(Q \cup i) - x(Q \cup i))$$

$$= \max_{Q \subset N \setminus S} (v(Q \cup i) - x(Q)) - x_i = v^{S,x}(i) - x_i;$$

similarly $s_{ji}(x) = v^{S,x}(j) - x_j$. Since by the definition of pre-kernel, $s_{ij}(x) = s_{ji}(x)$, it follows that $v^{S,x}(i) - x_i = v^{S,x}(j) - x_j$. Hence

$$x_i - x_j = v^{S,x}(i) - v^{S,x}(j)$$

and

$$x_i + x_j = x(i,j) = v^{S,x}(i, j),$$

which proves the lemma.[27]

Remark. The converse of this lemma is also true; i.e., if $x(N) = v(N)$ and $x|S$ is the standard solution of $v^{S,x}$ for all 2-person coalitions S, then x is in the pre-kernel of v. From this it follows that if $|N| \geq 3$, then x is in the pre-kernel of v if and only if $x(N) = v(N)$ and $x|S$ is in the pre-kernel of $v^{S,x}$ for all coalitions $S \subsetneq N$ [14].

---

[25] When they are non-empty, which is the case whenever there is at least one payoff vector x that is both individually rational ($x_i \geq v(i)$ for all i) and efficient ($x(N) = v(N)$).

[26] A similar remark is made in [15]. Rules such as proportional division or the CEA rule are not covariant in terms of the game v.

[27] This lemma is a special case of Lemma 7.1 in [1].

LEMMA 6.9. *The contested garment solution of 2-person bankruptcy problem is the standard solution of the corresponding game.*

Proof. Follows from (2.1), (6.1), and (6.7).

PROPOSITION 6.10. *The kernel of a bankruptcy game $v_{E;d}$ consists of a single point, namely the consistent solution of the problem $(E; d)$.*

Proof. Set $v = v_{E;d}$, and let x be in its kernel. By its definition (6.1), v is superadditive $(S \cap T = \varnothing \Rightarrow v(S) + v(T) \leqslant v(S \cup T))$ and hence 0-monotonic $(S \subset T \Rightarrow v(S) + \sum_{i \in T \setminus S} v(i) \leqslant v(T))$. In 0-monotonic games, the kernel coincides with the pre-kernel [11]. Hence x is in the pre-kernel of v.

Now let S be an arbitrary 2-person coalition. By Lemma 6.8, $x|S$ is the standard solution of $v^{S,x}$, and hence by Lemma 6.2, of $v_{x(S);d|S}$. Hence by Lemma 6.9, $x|S$ is the CG-solution of $(x(S); d|S)$; but that means that x is the consistent solution of $(E; d)$. Q.E.D.

Theorem D follows from Proposition 6.10, since the nucleolus is always in the kernel [18].[28]

We stated Theorem D in terms of the nucleolus because it is better known than the kernel and conceptually simpler (it is point valued). In fact, though, the idea of CG-consistency is more closely related to the kernel than it is to the nucleolus.[29] We have already noted that the contested garment solution is in fact simply the standard solution of 2-person games. Thus an appropriate generalization of the notion of a consistent solution to an arbitrary game v is a payoff vector x such that whenever S is a 2-person coalition, $x|S$ is the standard solution of the reduced game $v^{S,x}$. Lemma 6.8, together with the theorem of Peleg cited in the succeeding remark, show that with this definition, the set of all consistent solutions of an arbitrary game is precisely its pre-kernel; and as we have noted, this coincides with the kernel for 0-monotonic games.

7. HISTORICAL NOTES

In the Talmudic bankruptcy literature, the rule that is perhaps closest to ours was proposed by Piniles [16, p. 64]; it coincides with ours for $E \leqslant D/2$, but beyond that they differ. Evidently, Piniles was unaware of the connection with the contested garment, since even for the CG Mishna itself

---

[28] Since this is the only property of the nucleolus that we require, we will not cite its definition here. The interested reader may consult the original article [18], or any one of several equivalent characterizations [5, 6, 20].

[29] The nucleolus has more to do with self-consistent rules (Sect. 3) than with CG-consistent solutions. See [20].

(where $E = 1 > \frac{3}{4} = D/2$), his rule gives $(\frac{5}{8}, \frac{3}{8})$ rather than $(\frac{3}{4}, \frac{1}{4})$. We owe the reference to Erakhin 23b (see Sect. 4) to Piniles.

In bargaining theory, the idea of consistency first appeared in [2, p. 328], where Harsanyi characterized the product maximization solution to an n-person bargaining problem as the unique solution x at which each two players i, j use the Nash solution [13] to divide between them what remains if every other player k gets x_k.

In the context of apportionment, self-consistency is discussed in Balinski and Young's "Fair Representation" (Yale University Press, 1982, 43–45 and 141–149). Alexander Hamilton's apportionment method, vetoed by George Washington but nevertheless used in the U.S. from 1852 until 1901 (and in many countries until today), is not self-consistent. Just before Oklahoma became a state in 1907, the House of Representatives had 386 seats. Oklahoma was allocated 5, bringing the total to 391. Nothing else changed, so presumably the apportionment among the old states should have remained the same. Yet under Hamilton's method, Oklahoma's joining meant New York losing a seat to Maine!

Others who have recently used ideas related to consistency include H. Moulin and W. Thomson. A particularly striking result is by T. Lensberg [7], who showed that Nash's bargaining solution [13] can be characterized by a set of axioms in which the Independence of Irrelevant Alternatives is replaced by a self-consistency axiom.

8. Acknowledgments

Particular thanks are due to Y. Aumann, Y. Kannai, D. Kleitman, and B. Peleg, both for the specific contributions acknowledged in the text and footnotes, and for lengthy discussions on various other aspects of this research. We have also benefitted from discussions with many other individuals. What started us on this research was reading B. O'Neill's beautiful paper [14], and then being led to the Mishna in Kethubot 93a as the result of a correspondence on the subject of O'Neill's paper with the late S. Aumann.

References

Note. Items from the ancient and medieval Talmudic literature are not listed here; citations of this literature in the text and footnotes use the commonly accepted style.

1. M. Davis and M. Maschler, The kernel of a cooperative game, *Naval Res. Logist. Quart.* **12** (1965), 223–259.
2. J. Harsanyi, A bargaining model for the cooperative n-person game, *in* "Contributions to the Theory of Games IV," pp. 325–355, Annals of Mathematics Studies Vol. 40, Princeton Univ. Press, Princeton, N.J., 1959.

3. A. IBN EZRA, "Sefer Hamispar" ("The Book of the Number"), Verona, 1146; German translation by M. Silberberg, Kauffmann, Frankfurt a. M., 1895. [Hebrew]
4. E. KALAI AND M. SMORODINSKI, Other solutions to Nash's bargaining problem, *Econometrica* **43** (1975), 513–518.
5. E. KOHLBERG, On the nucleolus of a characteristic function game, *SIAM J. Appl. Math.* **20** (1971), 62–66.
6. E. KOHLBERG, The nucleolus as a solution to a minimization problem, *SIAM J. Appl. Math.* **23** (1972), 34–39.
7. T. Lensberg, Stability and the Nash Solution, *J. Econ. Theory* **45** (1988), 330–341.
8. I. LEWY, Interpretation des IV. Abschnittes des palast. Talmud-Traktats Nesikin (Commentary on the Fourth Chapter of the Tractate Nezikin of the Jerusalem Talmud), *Jahr. Judischen Theologischen Sem. Breslau* (1908), 101–131. [Hebrew]
9. S. C. LITTLECHILD AND G. OWEN, A simple expression for the Shapley value in a special case, *Manage. Sci.* **20** (1973), 370–372.
10. M. MASCHLER, B. PELEG, AND L. S. SHAPLEY, The kernel and bargaining set for convex games, *Int. J. Game Theory* **1** (1972), 73–93.
11. M. MASCHLER, B. PELEG, AND L. S. SHAPLEY, Geometric properties of the kernel, nucleolus, and related solution concepts, *Math. Oper. Res.* **4** (1979), 303–338.
12. M. MASCHLER AND M. PERLES, The super-additive solution for the Nash bargaining game, *Int. J. Game Theory* **10** (1981), 163–193.
13. J. F. NASH, The bargaining problem, *Econometrica* **18** (1950), 155–162.
14. B. O'NEILL, A problem of rights arbitration from the Talmud, *Math. Soc. Sci.* **2** (1982), 345–371.
15. B. PELEG, On the reduced game property and its converse, *Int. J. Game Theory* **15** (1986), 187–200.
16. H. M. PINILES, "Darkah shel Torah" ("The Way of the Law"), Forster, Vienna, 1861. [Hebrew]
17. N. L. RABINOVITCH, "Probability and Statistical Inference in Ancient and Medieval Jewish Literature," Univ. Toronto Press, Toronto, Buffalo, 1973.
18. D. SCHMEIDLER, The nucleolus of a characteristic function game, *SIAM J. Appl. Math.* **17** (1969), 1163–1170.
19. L. S. SHAPLEY, A value for n-person games, *in* "Contributions to the Theory of Games II," pp. 307–312, Annals of Mathematics Studies Vol. 28, Princeton Univ. Press, Princeton, 1953.
20. A. I. SOBOLEV, Xaraketrizatziya Trintixitof Optimalinosti v Kooprativnix Itrax Posredstvom Funktzionalnix Uqavneniyi (Characterization of the Principle of Optimality for Cooperative Games through Functional Equations), *in* "Matematicheskie Metody v Socialnix Naukax" (N. N. Vorobiev, Ed.), pp. 94–151, Vipusk 6, Vilnius, 1975. [Russian]

IX COALITIONAL GAMES: CORE AND EQUILIBRIA OF MARKETS

During 1960–61, when I was on sabbatical from the Hebrew University at Oskar Morgenstern's Econometric Research Program in Princeton, a Rand Corporation memorandum by John Milnor and Lloyd Shapley caught my attention. Entitled "Oceanic Games,"[1] it dealt with the Shapley values of voting games with a small number of relatively "large" players, who swim in an "ocean"—or continuum—of tiny, individually insignificant players. An example is a corporation with a few large stockholders and an "ocean" of small ones. In October of 1961, toward the end of my sabbatical, a conference on "Recent Advances in Game Theory" was held at Princeton. At that conference, Herb Scarf presented an equivalence theorem between the core and the competitive equilibria of markets with many small traders. Scarf's model had a denumerable infinity of traders, divided into a finite number of "types;" because the total amount of goods is infinite, the model had some rough spots, for example, in the definition of an allocation. When I heard Scarf, I remembered Milnor and Shapley's oceanic games, and said to myself, "surely, the continuum just *has* to be the right way to do this."

That was the genesis of chapter 46, which formulates a market model with a nonatomic continuum of small traders, and shows that in such markets, the core coincides with the set of competitive equilibrium allocations. The conditions are quite general—there may be infinitely many (indeed a continuum of) types, and the preferences need not be convex nor even transitive or complete.

Though chapter 46 establishes *equivalence* of the core with the competitive allocations in markets with a continuum of traders, it leaves open the question of *existence* of competitive equilibrium—that is, non-emptiness of the core—in such markets. This is settled in chapter 47, which shows that all markets with a continuum of traders possess competitive equilibria. The conditions are significantly weaker than those required for finite markets, in that the preferences need not be convex.[2]

Chapter 48 is about "mixed" markets; like Milnor and Shapley's oceanic games, they have some large, individually significant players, and an "ocean" of small players who are significant only in aggregate. Here a "large" player—or "atom"—is one with significant market share. Intuitively, it would seem that there is an advantage to largeness, at least when there is only one large player. Chapter 48 shows that this is not necessarily so; it exhibits a market with a continuum of small traders and

---

1. And published only much later, in *Mathematics of Operations Research* 3 (1978), pp. 1–9.

2. Schmeidler later showed that when there is a continuum of traders one may also dispense with complete preferences in establishing existence ("Competitive equilibria in markets with a continuum of traders and incomplete preferences," *Econometrica* 37 (1969), 578–586).

one large one, in which the competitive allocation is, from the point of view of the large trader, *best* in the core. Thus by the equivalence theorem (chapter 46), the large trader in this example stands only to gain by breaking himself up into a continuum of small traders.

Chapter 49, joint with B. Peleg, constructs an example of "disadvantageous endowments"—where larger initial holdings may, in equilibrium, work to the disadvantage of an agent. The result, which is related to work on "immiserating growth" in international trade theory, answers a question raised by D. Gale.[3]

Chapter 50 concerns the speed of convergence of the core to the competitive allocations when the number of traders in a finite market tends to infinity. In 1975, Debreu[4] showed that when utilities are sufficiently smooth, this is generically $O(1/k)$. Shortly thereafter, Shapley[5] showed that without the smoothness, convergence can be much slower. Chapter 50 shows that this can happen also with arbitrarily smooth preferences; by Debreu's result, such a phenomenon is exceptional.

3. "Exchange equilibrium and coalitions: an example," *J. Math. Econ.* 1 (1974), 63–66.
4. "The rate of convergence of the core of an economy," *J. Math. Econ.* 2, 1–7.
5. "An example of a slow-converging core," *Int. Econ. Rev.* 20 (1979), 345–351.

46 Markets with a Continuum of Traders

·1. INTRODUCTION

THE NOTION of *perfect competition* is fundamental in the treatment of economic equilibrium. The essential idea of this notion is that the economy under consideration has a "very large" number of participants, and that the influence of each individual participant is "negligible." Of course, in real life no competition is perfect; but, in economics, as in the physical sciences,[1] the study of the ideal state has proved very fruitful, though in practice it is, at best, only approximately achieved.

Though writers on economic equilibrium have traditionally assumed perfect competition,[2] they have, paradoxically, adopted a mathematical model that does not fit this assumption. Indeed, the influence of an individual participant on the economy cannot be mathematically negligible, as long as there are only finitely many participants. Thus *a mathematical model appropriate to the intuitive notion of perfect competition must contain infinitely many participants*. We submit that the most natural model for this purpose contains a *continuum* of participants, similar to the continuum of points on a line or the continuum of particles in a fluid. Very succinctly, the reason for this is that one can integrate over a continuum, and changing the integrand at a single point does not affect the value of the integral, that is, the actions of a single individual are negligible.

This paper is confined to "pure exchange economies," i.e., markets. Presumably, the results could be extended to economies with production, but we have not done this. We investigate and compare two known concepts of market equilibrium—the *competitive equilibrium* and the *core*. By means of the continuous model, it is possible to express the relation between these two concepts in a particularly simple and transparent manner. In fact, in the continuous model—but in no finite model—the two concepts are essentially equivalent.

The market model that we consider consists of a set of traders, each of whom starts out with an initial commodity bundle to be used for trading, and each of whom has a well-defined preference order on the set of all commodity bundles. A *trade* (or *allocation*) is a redistribution of the commodities in the initial bundles

---

[1] Think of a "freely falling body" (no air resistance), an "ideal gas" (the molecules do not collide), an "ideal fluid" (incompressible and nonviscous), and so on.

[2] A discussion of the literature will be found in Section 5.

This chapter originally appeared in *Econometrica* 32 (1964): 39–50. Reprinted with permission.

among the traders. This is a perfectly standard model in economic theory with the sole exception that the set of traders has heretofore been assumed finite.

A *competitive equilibrium* is a state of the market arrived at via "the law of supply and demand"; it consists of a price structure p (one price for each commodity) at which the total supply of each good exactly balances the total demand, and the allocation x that results from trading at these prices. More precisely, x is an allocation with the property that at the price structure p, no trader can, with the value of his initial bundle, buy a bundle that he prefers to his part of x. If (p,x) is a competitive equilibrium, then x is called an *equilibrium allocation*.

An allocation x is said to be in the *core* of the market if no coalition of traders can force an outcome that is better for them than x. More precisely, x is in the core if there is no group of traders that, by its own efforts alone, without help from traders not in the group, can assure each of its members of a final commodity bundle preferred to that obtained under x. What we mean by "its own efforts" is that the desired result can be obtained if the traders in the group exchange the commodities in their initial bundles among themselves only, as if the other traders were not present. The core is a generalization of Edgeworth's "contract curve" [8].

It is widely recognized that the notion of competitive equilibrium makes economic sense only if perfect competition is assumed. Otherwise, a change in an individual's offer to buy or sell can easily upset prevailing prices, so that the restriction to these prices is meaningless. The notion of core, on the other hand, does not depend on perfect competition; it is perfectly valid even for markets containing only two or three traders.

The definition of competitive equilibrium assumes that the traders allow market pressures to determine prices and that they then trade in accordance with these prices, whereas that of core ignores the price mechanism and involves only direct trading between the participants. Intuitively, one feels that money and prices are no more than a device to simplify trading, and therefore the two concepts should lead to the same allocations. It is to be expected that this will not happen in finite markets, where the notion of competitive equilibrium is not really applicable; and, indeed, though every equilibrium allocation is always in the core, the core of a finite market usually contains points that are not equilibrium allocations. But, when the notion of perfect competition *is built into the model*, that is, in a continuous market, one may expect that the core equals the set of equilibrium allocations. That this is indeed the case is the main result of this paper.

It has long been conjectured that some such theorem holds; the basic idea dates back at least to Edgeworth [8]. The usual rough statement is that "the core approaches the set of equilibrium allocations as the number of traders tends to infinity." Unfortunately, it is extremely difficult to lend precise meaning to this kind of statement, to say nothing of proving it. Very recently, Debreu and Scarf did succeed in stating and proving a theorem of this kind in a brilliant and elegant fashion [7]; even so, their theorem holds only under comparatively restrictive conditions. Their work will be discussed and compared with ours in Section 6. There we will also discuss an earlier model of Scarf [18] and Debreu [6], containing a denumerable infinity of traders.

Continuous models are nothing new in economics or game theory, but it is usually parameters such as price or strategy that are allowed to vary continuously. Models with a continuum of *players* (traders in this instance) are a relative novelty, and the references can still be counted on the fingers of one hand. Milnor and Shapley [15] and Shapley alone [19] pioneered the idea in two papers dealing with power indices (Shapley values); [19] is set in a context of considerable economic interest. The only other references of which we know are Davis [4] and Peleg [17], who treat such games from the point of view of their von Neumann-Morgenstern solutions and their "bargaining sets," respectively.

The idea of a continuum of traders may seem outlandish to the reader. Actually, it is no stranger than a continuum of prices or of strategies or a continuum of "particles" in fluid mechanics. In all these cases, the continuum can be considered an approximation to the "true" situation in which there is a large but finite number of particles (or traders or strategies or possible prices). The purpose of adopting the continuous approximation is to make available the powerful and elegant methods of the branch of mathematics called "analysis," in a situation where treatment by finite methods would be much more difficult or even hopeless (think of trying to do fluid mechanics by solving n-body problems for large n).

There is perhaps a certain psychological difference between a fluid with a continuum of particles and a market with a continuum of traders. Though we are intellectually convinced that a fluid contains only finitely many particles, to the naked eye it still looks quite continuous. The economic structure of a shopping center, on the other hand, does not look continuous at all. But, for the economic policy maker in Washington, or for any professional macroeconomist, there is no such difference. He works with figures that are summarized for geographic regions, different industries, and so on; the individual consumer (or merchant) is as anonymous to him as the individual molecule is to the physicist.

Of course, to the extent that individual consumers or merchants are in fact *not* anonymous (think of General Motors), the continuous model is inappropriate, and our results do not apply to such a situation. But, in that case, perfect competition does not obtain either. In many real markets the competition is indeed far from perfect; such markets are probably best represented by a mixed model, in which some of the traders are points in a continuum, and others are individually significant (compare [19]). The purpose of this paper is to study the extreme case in which perfect competition does obtain, i.e., there are no individually significant traders.

It should be emphasized that our consideration of a continuum of traders is not merely a mathematical exercise; it is the expression of an economic idea. This is underscored by the fact that the chief result holds *only* for a continuum of traders—it is false for any finite number. It would presumably also be possible to consider a continuum of different commodities, but, in contrast to the continuum of traders, this would serve no useful purpose (in this context). A continuum of commodities is the appropriate mathematical idealization of a real situation in which there are "many" commodities, but our result holds for any number of commodities, many or few, so there is nothing to be gained by considering only the case of "many" commodities. The continuum of traders, on the other hand, captures an idea and enables the proof of a theorem that could *not* be proved in the finite set-up.

Our chief mathematical tools are Lebesgue measure and integration, but only their most elementary properties are needed. Riemann integration can *not* be substituted (see the beginning of the proof of Lemma 4.1). Much of the proof is adapted from the work of Debreu and Scarf [6, 7], but becomes simpler and more natural in this context. Indeed, over and above the specific result obtained here, what we would like to stress is the power and simplicity of the continuum-of-players method in describing mass phenomena in economics and game theory. The present work should be considered primarily as an illustration of this method as applied to an area where no other treatment seemed completely satisfactory, and we hope that it may stimulate more extensive development and use of models with a continuum of players.

The mathematical model and the main theorem are presented in the following section, and the assumptions are briefly discussed in Section 3. Section 4 is devoted to the proof of the main theorem. Section 5 reviews the literature and compares the present work with that of other authors.

2. THE MATHEMATICAL MODEL AND THE MAIN THEOREM

We will be working in a Euclidean space R^n; the dimensionality n of the space represents the number of different commodities being traded in the market.

Superscripts will be used exclusively to denote coordinates. Following standard practice, for x and y in R^n we write $x > y$ to mean $x^i > y^i$ for all i; $x \geq y$ to mean $x^i \geq y^i$ for all i; and $x \geqq y$ to mean $x \geq y$ but not $x = y$. The integral of a vector function is to be taken as the vector of integrals of the components. The scalar product $\sum_{i=1}^{n} x^i y^i$ of two members x and y of R^n is denoted $x \cdot y$. The symbol 0 denotes the origin in R^n as well as the real number zero; no confusion will result.

A *commodity bundle* x is a point in the nonnegative orthant Ω of R^n. The set of *traders* is the closed unit interval $[0, 1]$; it will be denoted T. An *assignment* (of commodity bundles to traders) is a function x from T to Ω, each coordinate of which is Lebesgue integrable over T. As all integrals are with respect to t, we shall omit writting t, as well as the symbol dt, under the integral sign; thus $\int_T x$ means $\int_T x(t) \, dt$.

There is a fixed *initial assignment* i. Intuitively, $i(t)$ is the bundle with which trader t starts out. We assume

(2.1) $\quad \int_T i > 0$.

An *allocation* (or "final assignment" or "trade") is an assignment x for which $\int_T x = \int_T i$.

For each trader t there is defined a relation $\succ_t$ on Ω, which is called the *preference relation* of t and satisfies the following conditions:

(2.2) *Desirability (of the commodities)*: $x \geq y$ implies $x \succ_t y$.
(2.3) *Continuity (in the commodities)*: For each $y \in \Omega$, the sets $\{x: x \succ_t y\}$ and $\{x: y \succ_t x\}$ are open (relative to Ω).
(2.4) *Measurability*: If x and y are assignments, then the set $\{t: x(t) \succ_t y(t)\}$ is Lebesgue measurable in T. Note specifically that $\succ_t$ is not assumed to be complete, nor even transitive.

A *coalition* of traders is a Lebesgue measurable subset of T; if it is of measure 0, it is called *null*.[3] An allocation y *dominates* an allocation x *via* a coalition S if $y(t) \succ_t x(t)$ for each $t \in S$, and S is *effective* for y, i.e.,

$$\int_S y = \int_S i.$$

The *core* is the set of all allocations that are not dominated via any nonnull coalition.

A *price vector* p is an n-tuple of nonnegative real numbers, not all of which vanish. A *competitive equilibrium* is a pair consisting of a price vector p and an allocation x, such that for almost every trader t, $x(t)$ is maximal with respect to $\succ_t$ in t's *budget set* $\{x: p \cdot x \leq p \cdot i(t)\}$. An *equilibrium allocation* is an allocation x for which there exists a price vector p such that (p, x) is a competitive equilibrium.

THEOREM: *The core coincides with the set of equilibrium allocations.*

3. DISCUSSION OF THE MODEL

The definitions of allocation, core, and competitive equilibrium are the same as the usual ones, except that where the usual definitions sum over a set of traders, we integrate. It is useful to think of the commodity bundle $x(t)$ held by an individual trader t under an assignment x as an "infinitesimal." Only when a "significant" (i.e., nonnull) coalition pools its resources can a non-infinitesimal bundle result, namely $\int_S x$. Thus an individual trader t—and, in general, any null coalition—is without influence in the market, and can and will always be ignored. Intuitively, this is what justifies the exclusion of null coalitions in the definitions of core and competitive equilibrium.

The intuitive meaning of "S is effective for y" is that the coalition S can assure to each of its members t the bundle $y(t)$, without any help from traders not in S.

---

[3] Not to be confused with the *empty* coalition, which has no members at all. The term "almost every trader" will mean "every trader except possibly for a null set."

This it can do by an appropriate redistribution of the initial bundles $i(t)$ because of the condition $\int_S y = \int_S i$ that defines effectiveness.

Assumption (2.1) asserts that each of the commodities is actually present in the market. In other words, each commodity is initially held by some traders, though different comodities might be held by different traders, and it could well happen that no single trader initially holds a positive amount of all commodities. The finite analogue of this assumption has been used by McKenzie [14, p. 58, Assumption 5]. Though very weak, the assumption is not merely a normalization; it can be shown that the main theorem is false without it. On the other hand, for a given real-life market, it is always possible and, in fact, natural to build a model satisfying (2.1), namely, by counting as commodities in the model only those commodities actually present in the real-life market.

The desirability assumption (2.2) says that each trader always wants more of every commodity. One consequence is that satiation is never reached.[4] The continuity assumption (2.3) is the usual one; it asserts that if $x \succ_t y$, and x' and y' are sufficiently close to x and y, respectively, then also $x' \succ_t y'$. The measurability assumption (2.4) is of technical significance only and constitutes no real economic restriction. Nonmeasurable sets are extremely "pathological"; it is unlikely that they would occur in the context of an economic model.

We remark that our assumptions are far weaker than those usually assumed in market models.

The choice of the unit interval as a model for the set of traders is of no particular significance. A planar or spatial region would have done just as well. In technical terms, T can be any measure space *without atoms*.[5] The condition that T have no atoms is precisely what is needed to ensure that each individual trader have no influence.

4. PROOF OF THE MAIN THEOREM

First, we show that every equilibrium allocation is in the core. Let (p, x) be a competitive equilibrium. Suppose, contrary to the theorem, that x is dominated via a coalition S by an allocation y. Then, by the definition of competitive equilibrium, we have $p \cdot y(t) > p \cdot i(t)$ for almost all $t \in S$. Hence

$$p \cdot \int_S y = \int_S p \cdot y > \int_S p \cdot i = p \cdot \int_S i,$$

and this contradicts

$$\int_S y = \int_S i.$$

---

[4] The assumption can be dispensed with for $y > i(t)$; this would permit satiation for such y.

[5] An *atom* of a measure space is a nonnull set in the space which includes no nonnull subset of smaller measure. The best example is a "mass point," i.e., a single point that carries positive measure.

Conversely, we show that every allocation in the core is an equilibrium allocation. Let x be in the core. Define

$$F(t) = \{x : x \succ_t x(t)\},$$
$$G(t) = F(t) - i(t) = \{x - i(t) : x \in F(t)\}.$$

For each set U of traders, let $\Delta(U)$ denote the convex hull of the union $\cup_{t \in U} G(t)$. Define U to be *full*, if its complement is null.

LEMMA 4.1: *There is a full set U of traders, such that 0 is not an interior point of $\Delta(U)$.*

PROOF: For each x in R^n, let $G^{-1}(x)$ be the set of all traders t for whom $G(t)$ contains x. The notation G^{-1} is suggested by the fact that $t \in G^{-1}(x)$ if and only if $x \in G(t)$. From $G^{-1}(x) = \{t : x + i(t) \succ_t x(t)\}$ and measurability (2.4), it follows that $G^{-1}(x)$ is measurable for each x.

Let N be the set of all those rational points r in R^n (i.e., points with rational coordinates) for which $G^{-1}(r)$ is null. Obviously N is denumerable. Define $U = T \setminus \cup_{r \in N} G^{-1}(r)$, where $\setminus$ denotes set-theoretic subtraction. Then U is full.

Suppose 0 is in the interior of $\Delta(U)$. Then there is a point $x > 0$ such that $-x \in \Delta(U)$; by definition of $\Delta(U)$, $-x$ is a convex combination of finitely many points in $\cup_{t \in U} G(t)$. That is, there are traders $t_1, \ldots, t_k \in U$ (not necessarily distinct), points $x_i \in G(t_i)$, and positive numbers $\beta_1, \ldots, \beta_k$ summing to 1, such that $\Sigma_{i=1}^k \beta_i x_i = -x < 0$. By the continuity assumption (2.3), we may find rational points $r_i \in G(t_i)$ sufficiently close to the x_i, and positive rational numbers γ_i sufficiently close to the β_i, so that we still have $\Sigma_{i=1}^k \gamma_i r_i < 0$. Let $-r = \Sigma_{i=1}^k \gamma_i r_i$, and pick an arbitrary trader t_0 in U. Since $r > 0$, we have $\alpha r + i(t_0) > x(t_0)$ for sufficiently large positive rational α. Hence by the desirability assumption (2.2), $\alpha r + i(t_0) \succ_{t_0} x(t_0)$, i.e., $\alpha r \in G(t_0)$. Now set $r_0 = \alpha r$, $\alpha_0 = 1/(\alpha+1)$, $\alpha_i = \alpha \gamma_i/(\alpha+1)$ for $i = 1, \ldots, k$. Then $\alpha_i > 0$ for all i, and $\Sigma_{i=0}^k \alpha_i = 1$; furthermore

$$\Sigma_{i=0}^k \alpha_i r_i = \frac{\alpha}{\alpha+1} r + \frac{\alpha}{\alpha+1} \Sigma_{i=1}^k \gamma_i r_i = 0,$$

and $r_i \in G(t_i)$ for all i. Then $t_i \in G^{-1}(r_i)$, and, since $t_i \in U$, it follows that $r_i \notin N$. Therefore $G^{-1}(r_i)$ is of positive measure for each i. Therefore, for a sufficiently small positive number δ, we can find disjoint subsets S_i of $G^{-1}(r_i)$ such that $\mu(S_i) = \delta \alpha_i$, where μ is Lebesgue measure. Define a coalition S by $S = \cup_{i=0}^k S_i$, and an assignment y by

$$y(t) = \begin{cases} r_i + i(t) & \text{for } t \in S_i, \\ i(t) & \text{for } t \notin S. \end{cases}$$

That $y(t) \in \Omega$ is trivial for $t \notin S$, and for $t \in S_i$ it follows from $r_i \in G(t)$ (which in turn follows from $S_i \subset G^{-1}(r_i)$). Next,

$$\int_S y = \Sigma_{i=0}^n \delta \alpha_i r_i + \int_S i = 0 + \int_S i,$$

and therefore S is effective for y; since $y(t)=i(t)$ for $t\notin S$, it follows that y is an allocation. Finally, from $S_i \subset G^{-1}(r_i)$ it follows that $r_i + i(t) \succ_t x(t)$ for $t \in S_i$; in other words, $y(t) \succ_t x(t)$ for $t \in S$. Since S is of positive measure, we have shown that x is not in the core, contrary to assumption. This proves the lemma.

Let U be as in the lemma. To avoid annoying repetitions, let us agree that in the remainder of the proof, statements about traders will refer to $t \in U$. This is sufficient because U is full.

From the lemma and the supporting hyperplane theorem we obtain a hyperplane $p \cdot x = 0$ that supports $\Delta(U)$. Therefore it also supports each of the $G(t)$; thus $p \cdot x \geq 0$ for $x \in G(t)$, or

(4.2) $\qquad p \cdot x \geq p \cdot i(t) \qquad$ for $\quad x \in F(t)$.

By desirability (2.2), each $F(t)$ contains a translate of the positive orthant; therefore $p \geq 0$. We shall show that (p, x) is a competitive equilibrium.

We first show that for almost all t, $x(t)$ is in t's budget set, i.e.,

(4.3) $\qquad p \cdot x(t) \leq p \cdot i(t)$.

Indeed, because of desirability (2.2), there are bundles arbitrarily close to $x(t)$ that t prefers to $x(t)$. Therefore $x(t)$ is in the closure of $F(t)$, and so, by (4.2), $p \cdot x(t) \geq p \cdot i(t)$. If, for a nonnull t-set, we would have $p \cdot x(t) > p \cdot i(t)$, then it would follow that $\int_T p \cdot x > \int_T p \cdot i$, contrary to the assumption that x is in the core and hence is an allocation. This demonstrates (4.3) for almost all t; we may and will assume that all traders satisfy it.

To complete the proof, we must show that $x(t)$ is maximal in t's budget set, i.e., that (4.2) can be sharpened to

(4.4) $\qquad p \cdot x > p \cdot i(t) \qquad$ for $\quad x \in F(t)$.

To this end, we first establish $p > 0$. Suppose not; let $p^1 = 0$, say. Since $p \neq 0$, some coordinate of p does not vanish; let $p^2 > 0$, say. By (2.1), $\int_T i^2 > 0$. Since x is an allocation, it follows that $\int_T x^2 > 0$, so there must be a nonnull set of traders t for whom $x^2(t) > 0$. Now for any trader t, it follows from desirability (2.2) that

$$x(t) + (1, 0, \ldots, 0) \succ_t x(t).$$

Hence choosing t so that $x^2(t) > 0$, we deduce from continuity (2.3) that for sufficiently small $\delta > 0$,

$$x(t) + (1, -\delta, 0, \ldots, 0) \succ_t x(t).$$

Then by (4.2),

$$\begin{aligned} p \cdot i(t) &\leq p \cdot [x(t) + (1, -\delta, 0, \ldots, 0)] \\ &= p \cdot x(t) + p^1 - \delta p^2 \\ &= p \cdot x(t) - \delta p^2 \\ &< p \cdot x(t), \end{aligned}$$

contradicting (4.3). This proves $p > 0$.

To demonstrate (4.4), let $x \in F(t)$. Suppose first that $i(t) \geq 0$; then $p \cdot i(t) > 0$, because $p > 0$. So by (4.2), $p \cdot x > 0$; hence, there is j such that $x^j > 0$; let $j = 1$, say. From continuity (2.3), it then follows that $x - (\delta, 0, \ldots, 0) \in F(t)$ for sufficiently small $\delta > 0$. Then by (4.2),

$$p \cdot i(t) \leq p \cdot [x - (\delta, 0, \ldots, 0)] = p \cdot x - \delta p^1 < p \cdot x,$$

proving (4.4). If $i(t) = 0$ and $x \geq 0$, then clearly $p \cdot x > 0 = p \cdot i(t)$, proving (4.4). Finally, suppose $i(t) = 0$ and $x = 0$. Since $x \in F(t)$, this means that $i(t) = 0 \succ_t x(t)$. If the set S of traders t for whom this happens is null, then it can be ignored; if, on the other hand, it has positive measure, then i dominates x via S, contradicting the membership of x in the core. This completes the proof of the main theorem.

5. DISCUSSION OF THE LITERATURE

The concept of competitive equilibrium is so well known in economics that we content ourselves with citing the papers of Walras [23], Wald [22], Arrow-Debreu [1], and McKenzie [14]; which are but a selection from an enormous literature. In discussing competitive equilibrium, Wald [22] acknowledged that the economic validity of his considerations is based on the assumption that "each of the participants is of the opinion that his own transactions do not influence the prevailing prices." But, like most workers in this area, he did not consider it necessary to modify either the finiteness of the model or the definition of competitive equilibrium. Von Neumann and Morgenstern saw more deeply into the problem: "The fact that every participant is influenced by the anticipated reactions of the others to his own measures . . . is most strikingly the crux of the matter in the classical problems of duopoly, oligopoly, etc. When the number of participants becomes really great, some hope emerges that the influence of every particular participant will become negligible . . . These are, of course, the classical conditions of 'free competition' . . . . The current assertions concerning free competition appear to be very valuable surmises and inspiring anticipations of results. But they are not results, and it is scientifically unsound to treat them as such" [16, pp. 13–14]. The results of Scarf-Debreu [6, 7, 18], Shubik [21], and this paper are presumably precisely the kind of "results" to which von Neumann and Morgenstern were referring.

The concept of core is well known in game theory. Though first named and intensively studied by Gillies and Shapley,[6] it had been used repeatedly already by von Neumann and Morgenstern [16]. This work was confined to games with side payments. An extension of the core notion to games without side payments—of which the market under consideration is an example—was made by Aumann and Peleg [2, 3]. The notion of Pareto optimality is related to that of core, but only

---

[6] The term "core" was introduced by Gillies and Shapley during a study of the properties of the von Neumann-Morgenstern solutions (see [9, 10]). The core as an independent solution concept was developed by Shapley in lectures at Princeton University in the fall of 1953.

distantly: Whereas the core consists of all those outcomes of a game with the property that no coalition of players can do better through its own efforts alone, Pareto optimality demands this only for the all-player coalition.

In economics, Edgeworth [8] was the first to investigate the core, albeit under a different name ("contract curve"). He, and subsequently Shubik [21], considered markets with two commodities and two "types" of traders, where traders of the same "type" have the same initial bundles and the same preferences. Under appropriate conditions,[7] they showed that, as the number of traders of each type tends to infinity, the core of the market in a certain sense "shrinks to a limiting point"; this point can be identified with the equilibrium allocation (which is unique in the markets they were considering). Recently, Scarf and Debreu [7, 18] generalized this work considerably by allowing an arbitrary but fixed finite number k of "types" of traders, rather than just two, and an arbitrary number of commodities. As the number of traders of each type tends to ∞, the core "shrinks to a limiting set," which can be identified with the set of equilibrium allocations. In this process, the number k of types is held fixed.

These results are special cases of the rough conjecture stated in the introduction that "the core approaches the set of equilibrium allocations as the number of traders tends to infinity." Supplying a precise statement of this is a nontrivial problem. The trouble is that the core and the set of equilibrium allocations are subsets of a space whose dimension varies with the number of traders. Thus as we add traders, the space under consideration changes, and in such a context it is not clear what "approach" means. It was because of this difficulty that Debreu and Scarf [7] postulated a fixed finite number of types of traders; together with some other assumptions, this enabled them to work in a fixed finite dimensional space. The other assumptions are that the preferences are quasi-orders[8] and that the indifference surfaces are strictly convex (i.e., convex and contain no straight line segment).

The notion of finitely many types might not at first sight seem objectionable. But it involves the further assumption that there are "many" traders of each type; in fact the number of traders of each type must be very large compared to the number of types in order for their model to be applicable. This seems far from economic reality, where, in general, different traders cannot be expected to have the same initial bundles or the same preferences. The continuous model allows *all* traders to have different initial bundles and different preferences.[9] There is no problem of working in spaces of varying dimension, because we start with a space

---

[7] Which differ in the two investigations.

[8] A "quasi-order" is a transitive, reflexive, and complete relation. It is usually denoted $\gtrsim$, and is to be thought of intuitively as "preference-or-indifference."

[9] It does not, of course, insist on this; a coalition of positive measure—or even all the traders—may have the same initial bundles and preferences. The concept of "type" is simply irrelevant in our approach.

of *infinite* dimension and remain in this same space throughout the investigation. For this reason also, it is not necessary to assume that the preferences are quasi-orders or that they are convex. Incidentally, Debreu and Scarf assume that each trader starts with a positive amount of each commodity, whereas we dispense with this assumption.

As we stated in the introduction, much of our proof is adapted from that of Debreu-Scarf. The key to the proof is Lemma 4.1. Debreu-Scarf prove the corresponding lemma by using the finiteness of the number of types. Our proof uses the same basic idea, after noting that the set of traders can be partitioned into a large number of nonnull coalitions whose members, though not identical, are fairly similar. The idea involved here is the same as in the approximation of an arbitrary function by a simple function, i.e., a function taking finitely many values.

Scarf [18] and Debreu [6] have also investigated a model with a *denumerable* infinity of traders. The assumptions are similar to those of [7], including finitely many types, positivity of initial bundles, and preferences are quasi-orders and are convex. Other references for games with denumerably many players are Kalisch and Nering [11] and Shapley [20], but these are in a non-economic context.

There are two aspects of the continuous model that we have not discussed. The first is the question of the existence of a competitive equilibrium, or equivalently, the nonemptiness of the core. We have proved only that the core equals the set of equilibrium allocations, but it may well happen that both are empty. It is possible to prove an existence theorem if it is assumed that the preferences are quasi-orders[8] (but it is not necessary to assume convexity). We plan to publish details in a subsequent paper.

The other aspect we have not discussed is the economic significance of the Lebesgue measure of a coalition. This, too, we plan to discuss in a subsequent paper.

It is a pleasure to acknowledge inspiration received from the work of Debreu, Scarf, and Shapley, and from a stimulating correspondence with Professor Debreu. Also, we thank Mr. A. Shlein for an enlightening conversation.

The Hebrew University of Jerusalem

REFERENCES

[1] ARROW, K. J., AND G. DEBREU: "Existence of an Equilibrium for a Competitive Economy," *Econometrica*, Vol. 22 (1954), pp. 265–90.

[2] AUMANN, R. J., AND B. PELEG: "Von Neumann-Morgenstern Solutions to Cooperative Games without Side Payments," *Bulletin of the American Mathematical Society*, Vol. 66 (1960), pp. 173–79 [Chapter 38].

[3] AUMANN, R. J.: "The Core of a Cooperative Game without Side Payments," *Transactions of the American Mathematical Society*, Vol. 98 (1961), pp. 539–52 [Chapter 39].

[4] DAVIS, M.: "Symmetric Solutions to Symmetric Games with a Continuum of Players," in *Recent Advances in Game Theory* [13], pp. 119–26.

[5] DEBREU, G.: *Theory of Value*, John Wiley, 1959.

[6] ———: "On a Theorem of Scarf," *Review of Economic Studies*, June, 1963.
[7] DEBREU, G., AND H. SCARF: "A Limit Theorem on the Core of an Economy," *International Economic Review*, September, 1963.
[8] EDGEWORTH, F. Y.: *Mathematical Psychics*, London, 1881.
[9] GILLIES, D. B.: *Some Theorems on n-Person Games*, Ph.D. Thesis, Department of Mathematics, Princeton University, 1953.
[10] ———: "Solutions to General Non-Zero Sum Games," in *Contributions to the Theory of Games IV* [12], pp. 47–85.
[11] KALISCH, G. K., AND E. D. NERING: "Countably Infinitely Many Person Games," in *Contributions to the Theory of Games IV* [12], pp. 43–45.
[12] LUCE, R. D., AND A. W. TUCKER (editors): *Contributions to the Theory of Games IV*, Annals of Mathematics Studies No. 40, Princeton University Press, 1959.
[13] MASCHLER, M. (editor): *Recent Advances in Game Theory* (Proceedings of a Princeton University Conference, October 4–6, 1961), Philadelphia: Ivy Curtis Press, pp. 113–118.
[14] MCKENZIE, L. W.: "On the Existence of General Equilibrium for a Competitive Market," *Econometrica*, Vol. 27 (1959), pp. 54–71.
[15] MILNOR, J. W., AND L. S. SHAPLEY: "Values of Large Games II: Oceanic Games," *Mathematics of Operations Research*, Vol. 3 (1978), pp. 290–307.
[16] NEUMANN, J. VON, AND O. MORGENSTERN: *Theory of Games and Economic Behavior*, Princeton University Press, 1944 (Third edition, 1953).
[17] PELEG, B.: "Quota Games with a Continuum of Players," *Israel Journal of Mathematics*, Vol. 1 (1963), pp. 48–53.
[18] SCARF, H.: "An Analysis of Markets with a Large Number of Participants", in *Recent Advances in Game Theory* [13], pp. 127–55.
[19] SHAPLEY, L. S.: "Values of Large Games III: A Corporation with Two Large Stockholders," *The RAND Corporation*, RM 2650 PR, December, 1961 (multilithed).
[20] ———: "Values of Games with Infinitely Many Players," in *Recent Advances in Game Theory* [13], pp. 113–118.
[21] SHUBIK, M.: "Edgeworth Market Games," in *Contributions to the Theory of Games IV* [12], pp. 267–78.
[22] WALD, A.: "Ueber einige Gleichungssysteme der mathematischen Ökonomie," *Zeitschrift für Nationalökonomie*, Vol. 7 (1936), pp. 637–70 (translated as "On Some Systems of Equations of Mathematical Economics," *Econometrica*, Vol. 19 (1951) pp. 368–403).
[23] WALRAS, L.: *Mathematische Theorie der Preisbestimmung der wirtschaftlichen Güter*, Stuttgart, Ferdinand Enke, 1881.

47 Existence of Competitive Equilibria in Markets with a Continuum of Traders[1]

1. INTRODUCTION

THE PROBLEM of rigorously establishing the existence of a competitive equilibrium in a market was first brought to the attention of economists by Wald [11]. Since the appearance of his pioneering paper, other authors[2] have established the existence of competitive equilibria under various sets of assumptions. In all this work, it was invariably assumed that the traders have convex preferences.[3] Indeed, if this assumption is abandoned it is easy to give examples of markets that do not possess any competitive equilibria.

Attention has recently been called[4] to the possibility of dispensing with the convexity assumption if the market in question has a large number of traders, no individual one of whom can significantly affect the outcome of trading. In a heuristic, imprecise way it was argued that the preferences of a large number of individually insignificant traders would have a convex effect in the aggregate, even if none of the individual preferences were convex. A rigorous treatment of this theme was given very recently by Shapley and Shubik [10], though not directly in connection with the competitive equilibrium. Their work will be discussed in Section 8.

In a previous paper [2], we suggested that the most appropriate model for a market with many individually insignificant traders is one with a continuum of traders. Analogous models are used in physics, for example, when the large number of particles in a fluid are replaced for mathematical convenience by a con-

---

[1] Research partially supported by the Office of Naval Research, Logistics and Mathematical Statistics Branch, under contract No. N62558-3586. Previous research connected with this paper was supported by the Carnegie Corporation of New York through the Econometric Research Program of Princeton University and by U.S. Air Force Project RAND.

[2] Such as Arrow-Debreu [1], Gale [7], and McKenzie [9].

[3] I.e., that the set of commodity bundles preferred or indifferent to a given bundle is convex.

[4] See the articles by Bator, Farrell, Koopmans, and Rothenberg in the *Journal of Political Economy*: (Vol. 67, 1959, pp. 377–391; Vol. 68, 1960, pp. 435–468; Vol. 69, 1961, pp. 478–493).

This chapter originally appeared in *Econometrica* 34 (1966): 1–17. Reprinted with permission.

tinuum of particles. This raises the question of whether it would be possible to establish the existence of competitive equilibria in markets with a continuum of traders, even when the preferences need not be convex. The purpose of this paper is to give an affirmative answer to that question, and thus to underscore the power and scope of the continuum-of-traders approach to market theory.

We remark that the concept of competitive equilibrium is generally agreed to be significant only in a market with "perfect competition," i.e., one with a large number of individually insignificant traders. The concept makes no sense for a small number of traders. Thus, we show here that when competitive equilibria are at all relevant, convex preferences are not needed to establish their existence.

The proof is based on McKenzie's beautiful existence proof [9] for competitive equilibria in finite markets. Major modifications are required, however, because of the presence of a continuum of traders (which necessitates the use of Banach-space methods) and the nonavailability of convex preferences.

In Section 2 we give a precise statement of the model and the main theorem. Section 3 is devoted to the statement of an auxiliary theorem. In Section 4 the proof of the auxiliary theorem is outlined, and in Section 5 it is completed. In Section 6 the main theorem is deduced from the auxiliary theorem.

Section 7 is devoted to a detailed comparison of our proof with McKenzie's, and Section 8 to a discussion of the relation of our current result to that of our previous paper [2] and to the Shapley-Shubik results [10].

Our result concerns true markets only, i.e., pure exchange economies. Presumably it can be extended to economies with production (at least if one assumes constant returns to scale), but we have not done this.

2. MATHEMATICAL MODEL AND STATEMENT OF MAIN THEOREM

We shall be working in a Euclidean space E^n; the dimensionality n of the space represents the number of different commodities being traded in the market. Superscripts will be used exclusively to denote coordinates. Following standard practice, for x and y in E^n we take $x > y$ to mean $x^i > y^i$ for all i; $x \geq y$ to mean $x^i \geq y^i$ for all i; and $x \geqq y$ to mean $x \geq y$ but not $x = y$. The integral of a vector function is to be taken as the vector of integrals of the components. Superscripts will be used exclusively to denote coordinates. The scalar product $\Sigma_{i=1}^{n} x^i y^i$ of two members x and y of E^n is denoted $x \cdot y$. The symbol 0 denotes the origin in E^n as well as the real number zero; no confusion will result. The symbol \ will be used for set-theoretic subtraction, whereas − will be reserved for ordinary algebraic subtraction.

A *commodity bundle* x is a point in the nonnegative orthant Ω of E^n. The set of *traders* is the closed unit interval [0, 1]; it will be denoted T. The words "measure," "measurable," "integral," and "integrable" are to be understood in the sense of Lebesgue. All integrals are with respect to the variable t (which stands for trader), and in most cases the range of integration is all of T. In an integral we will therefore omit the symbol dt and the indication of dependence of the integrand on t, and will specifically indicate the range of integration only when it differs from all of T. Thus $\int x$ means $\int_T x(t) dt$. A *null set* is a set of measure 0. *Null sets of traders are systematically ignored throughout the paper.* Thus a statement asserted for "all" traders, or "each" trader, or "each" trader in a certain set, is to be understood to hold for all such traders except possibly for a null set of traders.

An *assignment* (of commodity bundles to traders) is an integrable function on T to Ω. There is a fixed *initial assignment* $\mathbf{i}$; intuitively, $\mathbf{i}(t)$ is the bundle with which trader t comes to market. We assume

(2.1) $\int \mathbf{i} > 0$.

Intuitively, this asserts that no commodity is totally absent from the market.

For each trader t there is defined on Ω a relation $\succsim_t$ called *preference-or-indifference*. This relation is assumed to be a *quasi-order*, i.e., transitive, reflexive, and complete.[5] From $\succsim_t$ we define relations $\succ_t$ and $\sim_t$ called *preference* and *indifference*, respectively, as follows:

$x \succ_t y$ if $x \succsim_t y$ but not $y \succsim_t x$;

$x \sim_t y$ if $x \succsim_t y$ and $y \succsim_t x$.

The following assumptions are made:

(2.2) *Desirability (of the commodities)*: $x \geq y$ implies $x \succ_t y$.

(2.3) *Continuity (in the commodities)*: For each $y \in \Omega$, the sets $\{x: x \succ_t y\}$ and $\{x: y \succ_t x\}$ are open (relative to Ω).

(2.4) *Measurability*: If x and y are assignments, then the set $\{t: x(t) \succ_t y(t)\}$ is measurable.

The intuitive content of these assumptions should be fairly clear from their names. Note that together with the assumption that $\succsim_t$ is a quasi-order, the continuity assumption (2.3) yields the existence of a continuous utility function $v_t(x)$ on Ω for each fixed trader t [4]. Then the measurability assumption[6] (2.4) says that the v_t can be chosen so that $v_t(x)$ is simultaneously measurable in t and x.

An *allocation* is an assignment x such that $\int x = \int \mathbf{i}$. A *price vector* is a member p of R^n such that $p \geq 0$; though it is in Ω, it should not be thought of as a commodity bundle. A *competitive equilibrium* is a pair consisting of a price vector p and an allocation x, such that for all traders t, $x(t)$ is maximal with respect to $\succ_t$ in the "budget set" $B_p(t) = \{x \in \Omega : p \cdot x \leq p \cdot \mathbf{i}(t)\}$.

[5] A relation $\mathscr{R}$ is called *transitive* if $x \mathscr{R} y$ and $y \mathscr{R} z$ imply $x \mathscr{R} z$; *reflexive* if $x \mathscr{R} x$ for all x; and *complete* if for all x and y, either $x \mathscr{R} y$ or $y \mathscr{R} x$.

[6] In this context (but not in [2]), the measurability assumption is equivalent to the assumption that $\{t: x \succ_t y\}$ is measurable for all x and y in Ω.

MAIN THEOREM: *Under the conditions of this section, there is a competitive equilibrium.*

3. STATEMENT OF AUXILIARY THEOREM

To prove the main theorem, we first establish an auxiliary theorem, which has some interest in its own right. Let us define a *market* $\mathscr{M}$ to consist of a positive integer n (the number of commodities), an initial assignment i, and preference-or-indifference relations $\succsim_t$ on Ω for each of the traders t. The markets that we consider here differ from those described in the previous section in a number of ways. First, condition (2.1) on the initial assignments is strengthened to read

(3.1) $\quad i(t) > 0 \quad$ for all t.

This means that a positive amount of each commodity is initially held by each trader.

Second, a bundle x is said to *saturate*, or more explicitly, to *saturate trader t's desire*, if $x \succsim_t y$ for all $y \in \Omega$. Assumption 2.2 is weakened to read as follows:

(3.2) *Weak Desirability*: Unless y saturates, $x > y$ implies $x \succ_t y$.

Notice that this is a double weakening of (2.2); the hypothesis $x \geq y$ is replaced by $x > y$, and allowance is made for saturation (saturation is impossible under (2.2)).

Third, under the auxiliary theorem we do not only permit saturation, we specifically require it. Let v be an assignment. We say that trader t's desire is *commodity-wise-saturated* at $v(t)$ if for all bundles x and commodities i such that $x^i \geq v^i(t)$, we have

$$x \sim_t (x^1, \ldots, x^{i-1}, v^i(t), x^{i+1}, \ldots, x^n).$$

In other words, changing the value of the ith coordinate above $v^i(t)$ does not change the indifference level. Intuitively, this means that desire for the ith commodity is saturated when the quantity of that commodity is $v^i(t)$, although trader t may still

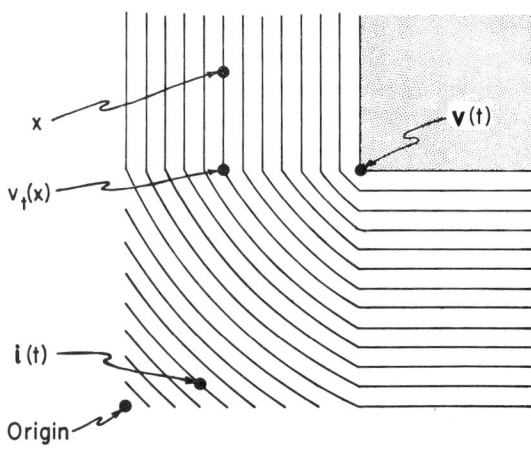

FIGURE 1

want more of other commodities j, of which he holds less than $v^j(t)$. To rephrase the condition, let $V(t) = \{x \in \Omega: x \leq v(t)\}$ be the "hyper-rectangle" of bundles that are $\leq v(t)$, and define a mapping v_t from Ω into $V(t)$ as follows: $v_t(x)$ is the bundle formed from x by replacing by $v^i(t)$ all coordinates x^i of x that exceed $v^i(t)$. Then commodity-wise saturation at $v(t)$ asserts that $v_t(x) \sim_t x$. It follows that the entire preference order is determined by its behavior in the hypercube $V(t,)$ since $x \succsim_t y$ if and only if $v_t(x) \succsim_t v_t(y)$. A preference order with commodity-wise saturation is illustrated in Figure 1.

The existence of a $v(t)$ that commodity-wise saturates desire is intuitively very acceptable; it simply means that there is an upper bound on the amount of a commodity that can be profitably used by an individual, no matter what other commodities are or are not available. The demand that v be an assignment, i.e., integrable, means that "the market as a whole can be commodity-wise saturated"; more precisely, it means that there is a bundle (namely $\int v$) that can be distributed among the traders in such a way as to commodity-wise saturate each trader's desire. We now assume

(3.3)   There is an assignment v such that each trader t's desire is commodity-wise saturated at $v(t)$.

Finally, we need the following assumption:

(3.4)   *Saturation restriction*: x cannot saturate unless $x > i(t)$.

AUXILIARY THEOREM: *Let $\mathcal{M}$ be a market satisfying the assumptions of this section as well as (2.3) and (2.4). Then $\mathcal{M}$ has a competitive equilibrium.*

4. OUTLINE OF THE PROOF OF THE AUXILIARY THEOREM

The starting point of the proof is the *preferred set* $C_p(t)$, defined for each trader t and each price vector p to be the set of commodity bundles preferred or indifferent to all elements of the budget set $B_p(t)$; formally,

$$C_p(t) = \{x \in \Omega: \text{ for all } y \in B_p(t), \; x \succsim_t y\}$$

(see Figure 2). Next, define

$$\int C_p = \{\int x: x \text{ is an assignment such that } x(t) \in C_p(t) \text{ for all } t\};$$

this is called the *aggregate preferred set*. $\int C_p$ is the set of all aggregate bundles that can be distributed among the traders in such a way that each trader is at least as satisfied as he is when he sells his initial bundle and buys the best (by his standards) that he can with the proceeds, at prices p.

Since we have made no convexity assumption on the preferences, the individual preferred sets $C_p(t)$ need not be convex. The *aggregate* preferred set $\int C_p$, on the other hand, *is* convex; as we shall see, that fact holds only because there is a continuum of traders, and it constitutes the nub of the proof. By using the convexity of the

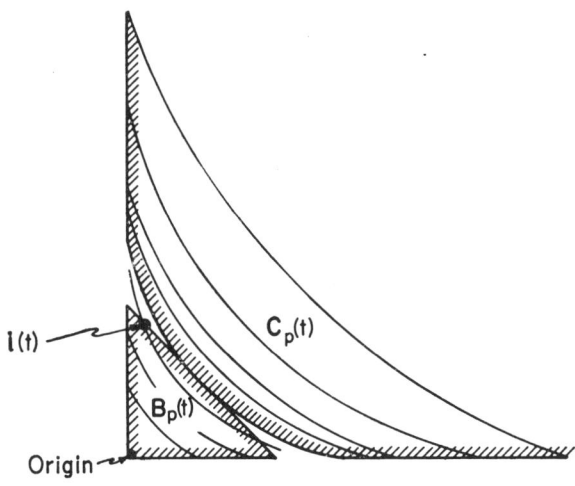

FIGURE 2

aggregate preferred set $\int C_p$, we shall be able to show that there is a unique point $c(p)$ in $\int C_p$ that is nearest to $\int i$; set $h(p) = c(p) - \int i$.

Let P be the simplex of price vectors normalized so that their sum is 1, i.e., $P = \{p \in \Omega : \Sigma_{i=1}^{n} p^i = 1\}$. The central idea of the proof is to use h to construct a continuous function f from P to itself, and then to apply Brouwer's fixed point theorem;[7] the resulting fixed point—denoted q—turns out to be an equilibrium price vector. The function f is defined by

$$f(p) = \frac{p + h(p)}{1 + \sum_{i=1}^{n} h^i(p)}.$$

We shall show later that $h(p) \geq 0$. Therefore, the denominator in the definition of f does not vanish, and so $f(p) \in P$ for all $p \in P$. Suppose q is a fixed point of f. Then

$$q\left(1 + \sum_{i=1}^{n} h^i(q)\right) = q + h(q),$$

i.e.,

(4.1) $\quad h(q) = \alpha q,$

where, because $h(p) \geq 0$,

$$\alpha = \sum_{i=1}^{n} h^i(q) \geq 0.$$

---

[7] Brouwer's theorem asserts that every continuous single-valued function f from P to itself has a fixed point, i.e., a point p such that $f(p) = p$. For a proof, see Dunford and Schwartz [5, Sec. V. 12, p. 468].

We wish to show that

(4.2) $\quad h(q) = 0$.

Indeed, suppose (4.2) is false. From the definition of h and the convexity of $\int C_p$ it follows that for all p, the hyperplane through $h(p) + \int i$ perpendicular to $h(p)$ supports[8] $\int C_p$. Applying this for $p = q$, we obtain

$$(y - \int i) \cdot h(q) \geq h(q) \cdot h(q)$$

for all $y \in \int C_q$. Because (4.2) is false, $\alpha > 0$; so by (4.1), we obtain

$$(y - \int i) \cdot \alpha q \geq \alpha^2 q \cdot q,$$

and hence

(4.3) $\quad (y - \int i) \cdot q \geq \alpha(q \cdot q) > 0 \quad \text{for all} \quad y \in \int C_q$.

Now if for each t we let $x(t)$ be a point in the budget set $B_q(t)$ that is maximal with respect to t's preference order, then on the one hand we have $(x(t) - i(t)) \cdot q \leq 0$, and on the other hand $x(t) \in C_q(t)$. Hence, by integrating we obtain $(\int x - \int i) \cdot q \leq 0$, and $\int x \in \int C_q$; this contradicts (4.3), and establishes (4.2).

Equation (4.2) says that $\int i \in \int C_q$, i.e., there is an assignment x such that $\int x = \int i$ and $x(t) \in C_q(t)$ for all t. Thus, x is an allocation, and $x(t)$ is preferred or indifferent to all elements of $B_q(t)$. To complete the proof that (q, x) is a competitive equilibrium, it is only necessary to show that $x(t)$ is in $B_q(t)$ for all t. Suppose now that $q \cdot x(t) < q \cdot i(t)$ for some t. Then x does not saturate (because of the saturation restriction (3.4)), and so from the desirability assumption (3.2), it follows that $x(t) + (\delta, \ldots, \delta) \succ_t x(t)$ for $\delta > 0$. But for δ sufficiently small, we shall still have

$$q \cdot (x(t) + (\delta, \ldots, \delta)) = q \cdot x(t) + \delta < q \cdot i(t),$$

so $x(t) + (\delta, \ldots, \delta) \in B_q(t)$; this contradicts $x(t) \in C_q(t)$. So $q \cdot x(t) < q \cdot i(t)$ is impossible, and we conclude that $q \cdot x(t) \geq q \cdot i(t)$ for all t. If the $>$ sign would hold for some t, we could deduce $\int q \cdot x > \int q \cdot i$, contradicting $\int x = \int i$. So $q \cdot x(t) = q \cdot i(t)$ for all t, and it follows that $x(t) \in B_q(t)$ for all t. So (q, x) is a competitive equilibrium.

The foregoing proof, which follows McKenzie's ideas [9] rather closely, is incomplete in two respects: The required properties of $h(p)$—existence, uniqueness, continuity, and nonnegativity—have not been established; and it has not been shown that the x whose integral $\int x$ contradicts (4.3) may be chosen to be measurable. These items will be taken up in the next section.

5. COMPLETION OF THE PROOF OF THE AUXILIARY THEOREM

In this section we make considerable use of the theory of integrals of set-valued functions, as developed in [3]. Before stating the results from [3] that are used in the sequel, we recall the necessary definitions.

---

[8] This is a standard method of constructing a supporting hyperplane. An explicit proof is given by McKenzie [9, Lemma 7 (1), p. 61].

Let F be a function defined on T whose values are subsets of Ω. Define

$$\int F = \{\int f : f \text{ is integrable and } f(t) \in F(t) \text{ for all } t\}.$$

F is called *Borel-measurable* if its *graph* $\{(x, t): x \in E^n, x \in F(t)\}$ is a Borel subset of $\Omega \times T$. F is called *integrably bounded* if there is an integrable point-valued function b such that for all t, $x \in F(t)$ implies $x \leq b(t)$. For each t, $F^*(t)$ denotes the convex hull of $F(t)$.

For each p in P, let F_p be a subset of Ω. F is said to be *upper-semicontinuous in p* if for each convergent sequence $p_1, p_2, \ldots$ in P and each convergent sequence $x_1, x_2, \ldots$ in Ω such that $x_1 \in F_{p_1}, x_2 \in F_{p_2}, \ldots$, we have $\lim x_k \in F_{\lim p_k}$. It is *lower-semicontinuous in p* if for each convergent sequence $p_1, p_2, \ldots$ in P, every point in $F_{\lim p_k}$ is the limit of a sequence $x_1, x_2, \ldots$ in Ω such that $x_1 \in F_{p_1}, x_2 \in F_{p_2}, \ldots$. It is *continuous* if it is both upper- and lower-semicontinuous.

If $F_1, F_2, \ldots$ are subsets of E^n, then $\limsup F_k$ is defined to be the set of all x in E^n such that every neighborhood of x intersects infinitely many F_k.

The following lemmas are proved in [3]:

LEMMA 5.1: *$\int F$ is convex.*

LEMMA 5.2: *If F is Borel-measurable, and $F(t)$ is non-empty for each t, then there is a measurable function f such that $f(t) \in F(t)$ for all t.*

LEMMA 5.3: *If $F_1, F_2, \ldots$ is a sequence of set-valued functions that are all bounded by the same integrable point-valued function, then $\int \limsup F_k \supset \limsup \int F_k$.*

LEMMA 5.4: *If $F_p(t)$ is continuous in p for each fixed t and Borel-measurable in t for each fixed p in P, and if all the F_p are bounded by the same integrable point-valued function, then $\int F_p$ is continuous in p.*

We now wish to establish the existence, uniqueness, continuity, and non-negativity of the function h. In principle, the first three of these properties will follow from the closedness and nonemptiness, convexity, and continuity (in p) of $\int C_p$ respectively; nonnegativity will follow from weak desirability. In carrying out the proofs, however, the unboundedness of the $C_p(t)$ causes difficulties. To circumvent these difficulties, we shall find a bounded "substitute" for C_p.

Let v be a commodity-wise saturating assignment (i.e., an assignment satisfying (3.3)), and recall the notation $V(t) = \{x: x \leq v(t)\}$. We shall work with the sets $V(t) \cap C_p(t)$, which we shall denote $D_p(t)$, passing back to the consideration of C_p itself only at the very end of the section. Note that the $D_p(t)$ are integrably bounded, uniformly in p, by the function v.

LEMMA 5.5: *For each t, $D_p(t)$ is continuous in p.*

PROOF: A similar lemma was proved by McKenzie [9, Lemma 4, pp. 57, 68]; we repeat the proof for the sake of completeness. Let $p_1, p_2, \ldots \in P$ have limit p. Suppose first that $x_k \in D_{p_k}(t)$ are such that $\lim x_k = x$. Certainly $x \in V(t)$; so if $x \notin D_p(t)$, then there is $y \in B_p(t)$ such that $y \succ_t x$. Then $p \cdot y \leq p \cdot i(t)$, and since assumption (3.1) asserts that $i(t) > 0$, it follows that $p \cdot i(t) > 0$. So we can find a z that is sufficiently close to y so that we still have $z \succ_t x$ (by continuity (2.3)), but $p \cdot z < p \cdot i(t)$. Then for k sufficiently large, we shall still have $p_k \cdot z < p_k \cdot i(t)$. Again applying continuity (2.3), we deduce from $z \succ_t x$ that for k sufficiently large $z \succ_t x_k$; but this contradicts $x_k \in D_{p_k}(t)$. Hence $x \in D_p(t)$, and upper-semicontinuity is proved.

Next, let $x \in D_p(t)$. If x saturates, then it is a member of all $D_{p_k}(t)$, so we can set $x_k = x$ in the definition of lower-semicontinuity. Assume therefore that x does not saturate. Let x_k be a point in $D_{p_k}(t)$ closest to x; the existence of x_k follows from the closedness of $D_{p_k}(t)$, which in turn follows from upper-semicontinuity. For arbitrary $\delta > 0$, set $y_\delta = v_t(x + (\delta, \ldots, \delta))$; then $y_\delta \sim_t x + (\delta, \ldots, \delta) \succ_t x$, by (3.2). Either for all δ, $y_\delta \in D_{p_k}(t)$ for all sufficiently large k, or else for some δ, there are infinitely many k such that $y_\delta \notin D_{p_k}(t)$. In the first case we have for all δ, by the definition of x_k, that $\|x_k - x\| \leq \|y_\delta - x\| \leq \delta\sqrt{n}$, where $\|\ \|$ represents the euclidean norm (i.e., the distance from the origin). Since δ can be chosen arbitrarily small, this shows that $x_k \to x$, and establishes lower-semicontinuity. In the second case, we can assume without loss of generality that $y_\delta \notin D_{p_k}(t)$ for all k. Then for each k there is a $z_k \in B_{p_k}(t) \cap V(t)$ such that $z_k \succ_t y_\delta$. Since the z_k are all in $V(t)$, they have a limit point z; again without loss of generality, we can let it be the limit. Since $p_k \to p$, and $p_k \cdot z_k \leq p_k \cdot i(t)$, it follows that $p \cdot z \leq p \cdot i(t)$, i.e., $z \in B_p(t)$. On the other hand, from continuity (2.3) it follows that $z \succsim_t y$; since $y_\delta \succ_t x$, it follows that $z \succ_t x$. But this contradicts $x \in D_p(t) \subset C_p(t)$, and completes the proof of the lemma.

The proof of upper-semicontinuity in this lemma is the only place where use is made of $i(t) > 0$ (3.1), rather than the far weaker $\int i > 0$ (2.1).

LEMMA 5.6: C_p and D_p are Borel-measurable for each fixed p.

PROOF: Since every measurable function differs on at most a null set from a Borel-measurable function, we may assume that v and i are Borel-measurable. The statement "$x \in D_p(t)$" is equivalent to "$x \leq v(t)$ and $x \in C_p(t)$." The statement "$x \in C_p(t)$" is equivalent to "for all $y \in B_p(t)$, $x \succsim_t y$"; because of continuity (2.3), this is equivalent to "for each rational point[9] $r \in B_p(t)$, $x \succsim_t r$." For fixed r, "$x \succsim_t r$" is equivalent to "not $r \succ_t x$." Because of continuity, "$r \succ_t x$" is equivalent to "there is a rational point s in Ω such that $s \geq x$ and $r \succ_t s$." Hence $\{(x, t): x \succsim_t r\}$, which equals

$$\Omega \times T \setminus \bigcup_{\text{rational } s \text{ in } \Omega} [\{x: s \geq x\} \times \{t: r \succ_t s\}],$$

is a Borel set. Hence $\{(x, t): x \in C_p(t)\}$, which equals $\bigcap_{\text{rational } r \text{ in } \Omega} [(\Omega \times \{t: p \cdot r > p \cdot i(t)\}) \cup \{(x, t): x \succsim_t r\}]$, is a Borel set, and this proves that C_p is Borel-measurable.

---

[9] I.e., point with rational coordinates.

Hence $\{(x, t): x \in D_p(t)\}$, which equals

$$\{(x, t): x \leq v(t)\} \cap \{(x, t): x \in C_p(t)\},$$

is a Borel set, and the proof is complete.

COROLLARY 5.7: *$\int D_p$ is closed, non-empty, convex, and continuous in p.*

PROOF: $\int v \in \int D_p$, so non-emptiness is proved. Convexity follows from Lemma 5.1. Since D_p is uniformly integrably bounded by v, continuity follows from Lemmas 5.4, 5.5, and 5.6. Since the values of a continuous set-valued function are always closed, the corollary is proved.

For each p in P, let $d(p)$ be the point in $\int D_p$ that is closest to $\int i$. Such a point exists because $\int D_p$ is non-empty and closed; it is unique because $\int D_p$ is convex.

LEMMA 5.8: *$d(p)$ is a continuous (point-valued) function of p.*

PROOF: A similar lemma was proved by McKenzie [9, Lemma 10, p. 62]; we repeat the proof for the sake of completeness. Let $p_k \to p$, and let x be a limit point of $d(p_k)$. From the upper-semicontinuity of $\int D_p$ it follows that $x \in \int D_p$. Suppose that there is a point $y \in \int D_p$ such that $\|y - \int i\| < \|x - \int i\|$. By the lower-semicontinuity of $\int D_p$, there is a sequence of points $y_k \in \int D_{p_k}$ converging to y. Let $\{d(p_{k_j})\}$ be a subsequence of $\{d(p_k)\}$ converging to x. Since the norm is continuous, it follows that for j sufficiently large,

$$\|y_{k_j} - \int i\| < \|d(p_{k_j}) - \int i\|,$$

contradicting the definition of $d(p_{k_j})$. Hence $x = d(p)$. So the only limit point of $\{d(p_k)\}$ is $d(p)$, and the lemma is proved.

LEMMA 5.9: *For each p in P, $d(p) \geq \int i$.*

PROOF: If not, then $d(p)$ has a coordinate—without loss of generality, we can let it be the first—such that $d^1(p) < \int i^1$. Now $d(p) = \int x$, where $x(t) \in D_p(t)$ for all t. Let $y(t) = (v^1(t), x^2(t), \ldots, x^n(t))$. Then $y(t) \geq x(t)$ and $y(t) \leq v(t)$; therefore $y(t) \in D_p(t)$ for all t. Therefore

$$(\int v^1, d^2(p), \ldots, d^n(p)) = \int y \in \int D_p.$$

Now $d^1(p) < \int i^1$, and by the saturation restriction (3.4), $\int i^1 < \int v^1$; so there is an α with $0 < \alpha < 1$ such that $\alpha \int v^1 + (1-\alpha)d^1(p) = \int i^1$. Setting $z = \alpha y + (1-\alpha)x$ and $z = \int z$, we obtain $z \in \int D_p$ (by the convexity of $\int D_p$), and $z = (\int i^1, d^2(p), \ldots, d^n(p))$. Then from $d^1(p) < \int i^1$, we deduce $\|z - \int i\|^2 = \sum_{i=2}^{n}(d^i(p) - \int i^i)^2$

$$< \sum_{i=1}^{n} (d^i(p) - \int i^i)^2 = \|d(p) - \int i\|^2.$$

Thus z is closer than $d(p)$ to $\int i$, a contradiction. This proves the lemma.

Let $g(p) = d(p) - \int i$. We have established for $g(p)$ all the properties that we set out to establish for $h(p)$: existence, uniqueness, continuity, and nonnegativity (the last by Lemma 5.9). So with the following lemma we achieve our aim:

LEMMA 5.10: $g(p) = h(p)$.

PROOF: Fix p, and write $g = g(p)$, $h = h(p)$, $c = c(p)$, $d = d(p)$. If $g = 0$ there is nothing to prove. Otherwise, by the definition of g, the hyperplane through d perpendicular to g supports $\int D_p$ (see footnote 8). This means that

(i) $\quad x \cdot g \geq \|g\|^2 \quad$ for all $\quad x \in \int D_p - \int i$.

Suppose there is a point in $\int C_p$ that is nearer to $\int i$ than d is. This means that there is a point y in $\int C_p - \int i$ that is nearer to 0 than g is. Then

(ii) $\quad \|y\|^2 < \|g\|^2$.

Furthermore $\|y\|^2 - 2y \cdot g + \|g\|^2 = \|y - g\|^2 > 0$. Hence $\|y\|^2 > y \cdot g + [y \cdot g - \|g\|^2]$. If $y \cdot g - \|g\|^2 \geq 0$, then it follows that $\|y\|^2 > y \cdot g \geq \|g\|^2$, contradicting (ii). Hence

(iii) $\quad y \cdot g < \|g\|^2$.

Formula (iii) expresses the geometrically obvious fact that any point nearer than d to $\int i$ must be on the near side of the hyperplane through d perpendicular to g.

Now $y = \int x - \int i$, where $x(t) \in C_p(t)$ for all t. Then by commodity-wise saturation, $v_t(x(t)) \in D_p(t)$ for all t. Furthermore $v_t(x(t)) \leq x(t)$, and $v_t(x(t)) \leq v(t)$. Setting $z(t) = v_t(x(t))$, we obtain $\int z \in \int D_p$ and $\int z - \int i \leq y$. Since $g \geq 0$ (Lemma 5.9), it follows that $(\int z - \int i) \cdot g \leq y \cdot g$. Hence by (iii), $(\int z - \int i) \cdot g < \|g\|^2$. But since $\int z - \int i \in \int D_p - \int i$, it follows from (i) that $(\int z - \int i) \cdot g \geq \|g\|^2$, and this is the contradiction that proves our lemma.

It remains to show that a measurable x may be chosen whose integral will contradict (4.3). According to Section 4, it is sufficient to show that there is a measurable x such that for all t, $x(t)$ is maximal in $B_q(t)$ with respect to t's preference order. Let $X(t)$ be the set of all maximal points in $B_q(t)$. As in the proof of Lemma 5.6, we may assume that i is Borel-measurable. Then

$$\{(x, t): x \in B_q(t)\} = \{(x, t): q \cdot x \leq q \cdot i(t)\}$$
$$= \Omega \times T \setminus \bigcup_\theta [\{x: q \cdot x > \theta\} \times \{t: \theta > q \cdot i(t)\}],$$

where θ runs over the rational numbers. Hence the left side is a Borel set. Applying Lemma 5.6 we deduce that $\{(x, t): x \in X(t)\}$, which equals

$$\{(x, t): x \in B_q(t)\} \cap \{(x, t): x \in C_q(t)\},$$

is a Borel set. Hence X is Borel-measurable.

Next, we show that $X(t)$ is non-empty for each t. From the compactness of $V(t) \cap B_q(t)$ and the continuity condition (2.3) for preferences, it follows that $V(t) \cap B_q(t)$

has a maximal element y. Then because of commodity-wise saturation, y is also maximal in $B_q(t)$. Indeed, suppose $z \in B_q(t)$ is such that $z \succ_t y$. Now $z \in B_q(t)$ means $q \cdot z \leq q \cdot i(t)$; therefore $q \cdot v_t(z) \leq q \cdot z \leq q \cdot i(t)$, and therefore also $v_t(z) \in B_q(t)$. But by definition, $v_t(z) \in V(t) \cap B_q(t)$. Finally, $v_t(z) \sim_t z \succ . v$. Thus $v_t(z)$ contradicts the maximality of z in $V(t) \cap B_q(t)$, proving the existence a maximal element in $B_q(t)$.

From Lemma 5.2 we may now deduce the existence of an appropriate x. This completes the proof of the auxiliary theorem.

6. PROOF OF THE MAIN THEOREM

The general idea is to approximate a given market $\mathcal{M}$ satisfying the conditions of the main theorem by a sequence of markets $\mathcal{M}_k$ satisfying the conditions of the auxiliary theorem. Then by the auxiliary theorem, the $\mathcal{M}_k$ have competitive equilibria (q_k, y_k); from these competitive equilibria we shall construct a pair (q, y) that is a competitive equilibrium in the original market $\mathcal{M}$.

To define the markets $\mathcal{M}_k$, we must specify their initial assignments i_k and their preference orders $\precsim_t^k$; the number of commodities is taken to be n in all the $\mathcal{M}_k$. Let δ_k be a monotone sequence of numbers tending to 0, and define

$$i_k(t) = i(t) + (\delta_k, \ldots, \delta_k).$$

To define the preference orders, let γ_k be a monotone sequence of numbers tending to ∞ such that $\gamma_1 > \delta_1$, let

$$v_k(t) = i(t) + (\gamma_k, \ldots, \gamma_k),$$

and let "hyper-rectangles" $V_k(t)$ and functions $v_{k,t}$ from Ω onto $V_k(t)$ be defined as in Section 3, with v_k in place of v. Now define the preference orders by

$$x \succsim_t^k y \quad \text{if and only if} \quad v_{k,t}(x) \succsim_t v_{k,t}(y).$$

It may be verified that the $\mathcal{M}_k$ satisfy the conditions of the auxiliary theorem, with v_k as the commodity-wise saturating assignment. Furthermore, note that the preference orders in $\mathcal{M}_k$ coincide with those in $\mathcal{M}$ for all x and y such that x and y are $\leq i(t) + (\gamma_k, \ldots, \gamma_k)$.

Let (q_k, y_k) be a competitive equilibrium of $\mathcal{M}_k$. Because of the compactness of P, the sequence $\{q_k\}$ has a convergent subsequence, and we may suppose without loss of generality that this subsequence is the original sequence. Let $q = \lim_k q_k$. The following is the crucial lemma of this section:

LEMMA 6.1: $q > 0$.

PROOF: Suppose, on the contrary, that some coordinate of q vanishes, say $q^1 = 0$. First we establish

(i)   if $q \cdot i(t) > 0$, then $\{y_k(t)\}$ has no limit point as $k \to \infty$.

Indeed, suppose y were such a limit point; without loss of generality, assume that it is actually the limit. Now because (q_k, y_k) is a competitive equilibrium in $\mathcal{M}_k$, we have $q_k \cdot y_k(t) \leq q_k \cdot i_k(t)$. Using this and the saturation restriction (3.4) in $\mathcal{M}_k$, we deduce that $y_k(t)$ does not saturate. Hence if $q_k \cdot y_k(t) < q_k \cdot i_k(t)$, then by weak desirability (3.2) in $\mathcal{M}_k$, it would be possible to find a member of $B_{q_k}(t)$ preferred to $y_k(t)$, contradicting the definition of competitive equilibrium. Thus $q_k \cdot y_k(t) = q_k \cdot i(t)$, and so from the hypothesis of (i) we obtain

(ii) $\quad q \cdot y = \lim_k q_k \cdot y_k(t) = \lim_k q_k \cdot i_k(t) = q \cdot i(t) > 0.$

Hence there is a coordinate j such that $y^j > 0$ and $q^j > 0$; assume without loss of generality that $j = 2$. Now by desirability (2.2), $y + \{1, 0, \ldots, 0\} \succ_t y$. If for sufficiently small $\delta > 0$ we define $z = y + \{1, -\delta, 0, \ldots, 0\}$, then $z \in \Omega$, and by continuity we deduce $z \succ_t y$. Again using continuity, we obtain $z \succ_t y_k(t)$ for k sufficiently large. Since (q_k, y_k) is a competitive equilibrium in $\mathcal{M}_k$, we obtain $q_k \cdot z > q_k \cdot i_k(t)$. Letting $k \to \infty$ and applying (ii), we get

(iii) $\quad q \cdot z = \lim_k q_k \cdot z \geq \lim q_k \cdot i_k(t) = q \cdot y.$

But since $q^1 = 0$ and $q^2 > 0$, we have

$$q \cdot z = q \cdot y + q^1 - \delta q^2 = q \cdot y - \delta q^2 < q \cdot y,$$

contradicting (iii). This proves (i).

Since $q \in P$ and $\int i > 0$ (2.1), it follows that $\int q \cdot i = q \cdot \int i > 0$. Let $S = \{t: q \cdot i(t) > 0\}$; then S is non-null, and we denote its measure by $\mu(S)$. Define

$$A = \left\{ x \in \Omega: \sum_{i=1}^n x^i \leq 2 \int \sum_{j=1}^n i^j / \mu(S) \right\}.$$

For $t \in S$, it follows from (i) and the compactness of A that $y_k(t) \in A$ for at most finitely many k; that is, for each $t \in S$ there is an integer $k(t)$ such that $\Sigma_i y_k^i(t) > 2 \int \Sigma_j i^j / \mu(S)$ for $k \geq k(t)$. Hence for $t \in S$,

(iv) $\quad \liminf_k \sum_i y_k^i(t) \geq 2 \int \sum_j i^j / \mu(S).$

Because y_k is an allocation in $\mathcal{M}_k$, we have

(v) $\quad \lim_k \int \sum_i y_k^i = \lim_k \int \sum_i i_k^i = \lim_k \left[\int \sum_i i^i + n\delta_k \right] = \int \sum_i i^i.$

But by Fatou's Lemma[10] and (iv),

$$\lim_k \int \sum_i y_k^i \geq \int \liminf_k \sum_i y_k^i \geq \int_S \liminf_k \sum_i y_k^i$$

$$\geq \int_S \left[2 \int \sum_j i^j / \mu(S) \right] = \left[2 \int \sum_j i^j \right] \int_S 1/\mu(S) = 2 \int \sum_j i^j > \int \sum_j i^j,$$

---

[10] Fatou's Lemma states that if φ_k are nonnegative measurable real functions, then $\liminf_k \int \varphi_k \geq \int \liminf_k \varphi_k$. See [5, III. 6.19, p. 152].

where the last inequality follows from $\int i > 0$ (2.1). This contradicts (v) and proves Lemma 6.1.

Since $q_k \to q > 0$, there is a $\delta > 0$ such that $q_k^i \geq \delta$ for k sufficiently large and all i. Without loss of generality, we may assume that $q_k^i \geq \delta$ for all i and k, and that $i_k^i(t) \leq i^i(t) + \delta$ for all i, k, and t. Hence for all i, k, and t,

$$\delta y_k^i(t) \leq q_k \cdot y_k(t) \leq q_k \cdot i_k(t) \leq q_k \cdot i(t) + \delta \leq \sum_{j=1}^n i^j(t) + \delta.$$

Thus we obtain

(6.2) $\quad y_k^i(t) \leq 1 + \sum_{j=1}^n \dfrac{i^j(t)}{\delta}.$

For each t, let $Y(t)$ be the set of limit points of $y_k(t)$ as $k \to \infty$. Let $Y_k(t)$ be the set consisting of the single point $y_k(t)$; then $Y(t) = \limsup Y_k(t)$. By (6.2), all the Y_k are bounded by the same integrable function. Hence by Lemma 5.3,

$$\int i = \lim \int i_k = \lim \int y_k \in \limsup \int Y_k \subset \int \limsup Y_k = \int Y.$$

Let y be such that $y(t) \in Y(t)$ for all t, and

(6.3) $\quad \int y = \int i.$

We shall show that (q, y) is a competitive equilibrium in $\mathcal{M}$.

To this end we must demonstrate that y is an allocation, that $y(t)$ belongs to $B_q(t)$ for all t, and that $y(t)$ is maximal in $B_q(t)$ for all t, i.e., that no member of $B_q(t)$ is preferred to $y(t)$. We have already shown that y is an allocation (6.3). Next, since $y(t) \in Y(t)$, it follows that $y(t)$ is a limit point of $\{y_k(t)\}$, say $y(t) = \lim_{m \to \infty} y_{k_m}(t)$. Since

$$q_{k_m} \cdot y_{k_m}(t) \leq q_{k_m} \cdot i_{k_m}(t),$$

we deduce by letting $m \to \infty$ that $q \cdot y(t) \leq q \cdot i(t)$, and so for all t,

(6.4) $\quad y(t) \in B_q(t).$

Finally, suppose that for t in a set of positive measure, there is a $z \in B_q(t)$ such that $z \succ_t y(t)$. Clearly $z \neq 0$; suppose without loss of generality that $z^1 > 0$. If for $\delta > 0$ sufficiently small we define $z_\delta = z - (\delta, 0, \ldots, 0)$, then we still have

(6.5) $\quad z_\delta \succ y(t).$

Moreover, since

$$\lim_k q_k \cdot z_\delta = q \cdot z - q^1 \delta < q \cdot z \leq q \cdot i(t) = \lim q_k \cdot i_k(t),$$

it follows that

$$q_k \cdot z_\delta < q_k \cdot i_k(t)$$

for all sufficiently large k, say for $k > k_0$. Now since $y(t)$ is a limit point of $\{y_k(t)\}$, there is a subsequence $\{y_{k_m}(t)\}$ converging to $y(t)$; hence for m sufficiently large,

$$z_\delta \succ_t y_{k_m}(t),$$

by (6.5). If we also pick m so large so that $k_m \geq k_0$, then z_δ contradicts the maximality of $y_{k_m}(t)$ in the budget set $\{x: q_{k_m} \cdot x \leq q_{k_m} \cdot i_{k_m}(t)\}$. Thus the supposition $z \succ_t y(t)$ has led to a contradiction, and we conclude that $y(t)$ is maximal in $B_q(t)$ for all t. Together with (6.3) and (6.4), this completes the proof that (q, y) is a competitive equilibrium, and with it the proof of the main theorem.

7. COMPARISON WITH MCKENZIE'S PROOF

The differences between this proof and McKenzie's are caused by the different initial equipment: we have no convexity assumption to work with, and we have a continuum of traders rather than a finite number.

McKenzie needs the convexity assumption in only one place, to show that the aggregate preferred set (in his case the sum of the individual preferred sets) is convex. This is needed to define $h(p)$ uniquely, and follows from the convexity of the individual preferred sets. In a finite model there is no getting around this: no intuitive assumption other than individual convexity would lead to the convexity of the aggregate preferred set.

In a continuous model, however, this is superfluous, because of Lemma 5.1; this says that the integral of any set-valued function over a non-atomic measure space (in our case the unit interval) is convex, even if the individual values of the function are not convex. In particular, the aggregate preferred set, as the integral of the (possibly nonconvex) individual preferred sets, is convex.

Because of the presence of a continuum of traders, the space of assignments is no longer a subset of a finite-dimensional euclidean space, but of an infinite-dimensional function space. This necessitates the use of completely new methods to justify the passage from properties proved for individual traders to the corresponding properties for the aggregate of all traders. Consider, for example, the continuity of the aggregate preferred set as a function of the price vector. In the finite case, this follows trivially from the continuity of the individual preferred sets. Here, on the other hand, it involves Lemma 5.4, which is comparatively deep. In fact, Lemmas 5.1-5.4, which have been separately published, were originally proved for the purposes of this paper, and they embody the chief mathematical difficulties. The proofs of these lemmas involve Lyapunov's theorem on the range of a vector measure [8], and the methods of functional analysis (Banach spaces) and topology.

Another significant difference between this proof and McKenzie's is in the matter of boundedness. In the proof of the auxiliary theorem, the set of bundles under consideration must be in some sense bounded in order to establish the continuity —and indeed the existence—of the individual preferred sets. McKenzie does this by noting that no individual trader can have more goods than the whole market. This is not available here, because no matter how large an individual trader's bundle is, it is still infinitesimal compared with the whole market. We therefore used the notion of commodity-wise saturation, which does the job of bounding for us. In the passage from the auxiliary to the main theorem we do not have commodity-wise saturation, but need boundedness so that the sequence of competitive equilibria of the auxiliary markets $\mathscr{M}_k$ should converge. Here we first deduce from the desirability assumption (2.2) that all prices must be nonvanishing, and this bounds the bundles under consideration to a finite simplex.

8. THE CORE

Intimately connected with the concept of competitive equilibrium is that of *core*. This is the set of all allocations with the property that no "coalition" of traders can assure each of its members of a more desirable bundle by trading within itself only, without recourse to traders not in the coalition. Formally (in our model), an allocation x is in the core if there is no measurable nonnull set S of traders, for whom there is an allocation y such that $y(t) \succ_t x(t)$ for all $t \in S$ and $\int_S y = \int_S i$. In a finite market, the integral should be replaced by a sum.

In a finite market with convex preferences, the core is never empty; but when the preferences are not convex the core may be empty. As with the competitive equilibrium, it might be conjectured that this "pathology" would "tend to disappear" as the number of traders increases. Investigating this possibility, Shapley and Shubik [10] showed that though the core itself may remain empty for any (finite) number of traders, it is possible to define a kind of approximation to the core called an ε-core; and that for any positive ε, if the number n of traders is allowed to increase in a certain way, the ε-core will become non-empty for sufficiently large n. They concluded that, heuristically speaking, the true core lies "just below the surface" for sufficiently large n. The assumptions on which their theorem is based are comparatively strong: They assumed transferable utilities, that all traders have the same utility function, and that there is a fixed finite number of distinct types of traders (where two traders are of the same "type" if they have the same initial bundles).

We shall now describe how the concepts of core and competitive equilibrium are related. Let us define an *equilibrium allocation* to be an allocation that forms a competitive equilibrium when paired with an appropriate price vector. For finite markets, the core always contains the set of equilibrium allocations, but the two sets do not usually coincide. A long-standing conjecture states, however, that as the number of players in a market increases, the core of the market "tends," in some sense, to the set of equilibrium allocations. Recently this conjecture has been formalized and proved in a number of different ways.[11] In [2] we showed that for a market with a *continuum* of traders, the core actually *equals* the set of equilibrium allocations. This was shown under conditions that are even weaker than those of

---

[11] See [2] for a brief survey of these developments.

this paper.[12] A question that was left open was the *existence* of a competitive equilibrium, or equivalently, the non-emptiness of the core; though it had been shown that the two sets coincide, the possibility that both vanish was left open. From the theorem of this paper, it now follows that the core is non-empty as well. This agrees well with the Shapley-Shubik result (which was, however, obtained under considerably stronger assumptions): Since the ε-core is non-empty for large n, it is to be expected that the true core is non-empty for "infinite n."

The Hebrew University of Jerusalem

REFERENCES

[1] ARROW, K. J., AND G. DEBREU: "Existence of an Equilibrium for a Competitive Economy," *Econometrica*, Vol. 22, 1954, pp. 265-290.
[2] AUMANN, R. J.: "Markets with a Continuum of Traders," *Econometrica*, Vol. 32, 1964, pp. 39-50 [Chapter 46].
[3] ———: "Integrals of Set-Valued Functions," *Journal of Mathematical Analysis and Applications*, Vol. 12, 1965, pp. 1-12 [Chapter 68].
[4] DEBREU, G.: *Theory of Value*, John Wiley and Sons, Inc., New York, 1959.
[5] DUNFORD, N., AND J. T. SCHWARTZ, *Linear Operators, Part I*, Interscience Publishers, Inc., New York, 1958.
[6] EGGLESTON, H. G.: *Convexity*, Cambridge University Press, 1958.
[7] GALE, D.: "The Law of Supply and Demand," *Mathematica Scandinavica*, Vol. 3, 1955, pp. 155-169.
[8] LYAPUNOV, A., "Sur les Fonctions-vecteurs complètements additives," *Bull. Acad. Sci. URSS sér. Math*, Vol. 4, 1940, pp. 465-478.
[9] MCKENZIE, L. W.: "On the Existence of General Equilibrium for a Competitive Market," *Econometrica*, Vol. 27, 1959, pp. 54-71.
[10] SHAPLEY, L. S., AND M. SHUBIK: "Quasi-Cores in a Monetary Economy with Nonconvex Preferences," *Econometrica*, Vol. 34, 1966, pp. 805-827.
[11] WALD, A.: "Über einige Gleichungssysteme der mathematischen Ökonomie," *Zeitschrift für Nationalökonomie*, Vol. 7, 1936, pp. 637-670. Translated as "On Some Systems of Equations of Mathematical Economics," *Econometrica*, Vol. 19, 1951, pp. 368-403.

---

[12] The model of [2] differs from that of this paper in that there we started out directly with preference relations $\succ_t$ rather than deriving them from preference-or-indifference relations $\succsim_t$; furthermore, unlike here, we there made no assumptions of total or even partial order for the preference relations (for example, transitivity was not assumed). Otherwise, the two models are identical.

48 Disadvantageous Monopolies

1. Introduction

It seems intuitively obvious that in a monopolistic market, the monopolist has an advantage because he can avoid competition, i.e., force an outcome that is better for himself than one that would be a result of competition. A formulation of this principle in terms of the core is the following:

CONJECTURE. In a monopolistic market, for each core allocation **x** there is a competitive allocation **y** whose utility to the monopolist[1] is $\leq$ that of **x**.

This conjecture appears[2] in [3], where it is stated in terms of a market in which the traders form a measure space with a single atom (the "monopolist") and a nonatomic part. It is proved in [3] that the conjecture is correct in the case of homogeneous markets, i.e., markets in which all traders have the same utility function, and this utility function is homogeneous. In that case the core is generally quite large, whereas there is a unique competitive allocation; the latter is, from the point of view of the monopolist, the worst allocation in the core.

In this note the above conjecture is settled, in the negative. We bring three examples. In Example A, the core is quite large, there is a unique competitive allocation, and from the monopolist's viewpoint, the competitive allocation is approximately in the middle of the core. Example B is a variant of Example A. In it, the core is again quite large, there is a unique competitive allocation, and from the monopolist's viewpoint,

---

Research partially supported by National Science Foundation Grant GS-3269 at the Institute for Mathematical Studies in the Social Sciences at Stanford University.

[1] That is, the utility of the bundle allocated to the monopolist under the allocation in question.

[2] Albeit not as a conjecture, but as an open problem.

This chapter originally appeared in *Journal of Economics Theory* 6 (1973): 1–11. © Academic Press. Reprinted with permission.

the competitive allocation is the *best* in the core. Thus the monopoly is a disadvantage; the monopolist would do well to "go competitive", i.e., split himself into many competing small traders.[3] In Example C, the core consists of exactly two points, one competitive and one not; and the competitive point is the better of the two from the monopolist's viewpoint. Thus again, the monopolist would do well to "go competitive."

The examples are formulated in terms of "mixed markets," i.e., the traders form a measure space consisting of a single atom $\{a\}$ and a non-atomic part, which will be called the "ocean." All examples are 2-commodity markets, with all of one commodity initially concentrated in the hands of the atom and all of the other initially held by the ocean. Thus a is a "monopolist" both in the sense of being an atom and in the sense that he initially holds a "corner" on one of the two commodities.

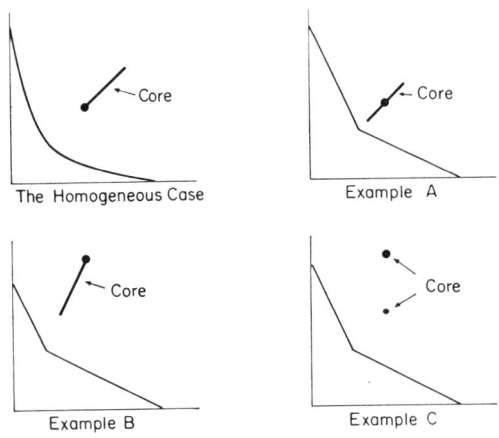

FIGURE 1

The chief features of the examples are illustrated in Fig. 1. This figure is drawn from the viewpoint of the atom only. In each case, we indicate the core (or more precisely, the monopolist's bundles in the core); the position of the competitive allocation is indicated by a heavy dot. The indifference level corresponding to the "personal minimum" of the monopolist (i.e. the utility of his initial bundle) is also shown.

The examples are presented in Section 2. Section 3 is devoted to a discussion of the implications of these examples, and the relation of the core analysis with that by means of other tools.

[3] In which case the core would consist of the competitive allocation only; see [1].

2. THE EXAMPLES

It is assumed that the reader is familiar with the terminology and notation of Core theory, in particular as applied to mixed markets; see [3], for example. The atom and the ocean are denoted $\{a\}$ and T_0 respectively; each has measure 1. In all the examples, the initial bundle density of trader t is given by

$$\begin{aligned}\mathbf{i}(t) &= (0, 1) \quad \text{if} \quad t \in T_0\,; \\ &= (1, 0) \quad \text{if} \quad t = a.\end{aligned} \tag{2.1}$$

To define the preference relations, we express the ocean as the union of two disjoint sets U and V of measure $\tfrac{1}{2}$ each; traders in U will have different preference relations from those in V.

To simplify the notation, we write $x = (\alpha, \beta)$ rather than $x = (x^1, x^2)$ for the bundle densities.

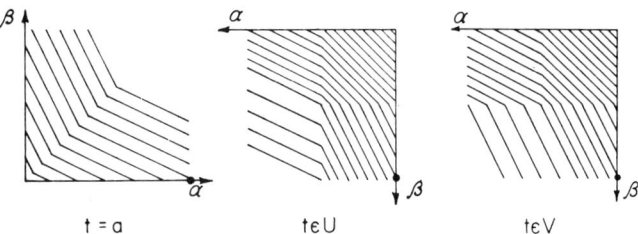

FIG. 2. Indifference maps in Example A.

EXAMPLE A. Let

$$h(\alpha, \beta) = \min(2\alpha + \beta, 2(\alpha + \beta) - 1/2, \alpha + 2\beta, (\alpha + 2\beta)/2 + 3/4).$$

Define the utility function u_t of trader t by

$$u_t(x) = u_t(\alpha, \beta)$$
$$= \begin{cases} \alpha + \beta + \min(\alpha, \beta) = \min(\beta + 2\alpha, \alpha + 2\beta), & \text{for } t = a; \\ h(\alpha, \beta), & \text{for } t \in U; \\ h(\beta, \alpha), & \text{for } t \in V. \end{cases}$$

The indifference maps and initial bundle densities are illustrated in Fig. 2; the maps for $t \in U$ and $t \in V$ have been inverted, so as to prepare the way for the use of the Edgeworth box below.

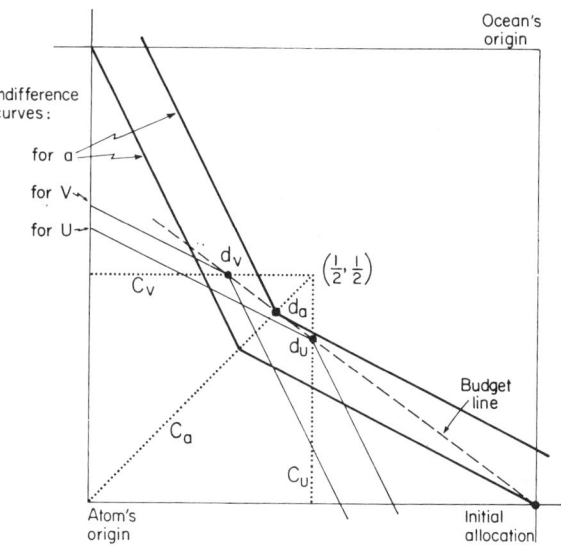

FIGURE 3

There is only one competitive allocation in this market, given by $\mathbf{x}(t) = (\tfrac{1}{2}, \tfrac{1}{2})$ for all t; the price vector is $p = (1, 1)$. It is easily verified that this is indeed a competitive equilibrium. To see that there is no other, let $p = (p^1, p^2)$ be a price vector with $1 > p^1/p^2 > \tfrac{1}{2}$, and refer to the Edgeworth Box in Fig. 3. The dotted lines C_a, C_U, and C_V are called *critical lines* for $t = a$, $t \in U$, and $t \in V$ respectively; on these lines the respective utility functions are not differentiable. The demand of each trader t is determined to be the intersection of the budget line with his critical line. Since $\mu(V) = \mu(U) = \tfrac{1}{2}$, the ocean's total demand is the midpoint of the segment connecting the demands d_U and d_V of traders in U and V respectively; since the demand d_a of the atom does not coincide[4] with this midpoint, total demand does not match total supply.

Next, we turn to the core. The atom's bundles in the core consists of the line segment connecting $(3/8, 3/8)$ to $(5/8, 5/8)$ (Fig. 1). At the top endpoint the oceanic players get their personal minimum, and so the atom cannot hope to get more. The bottom endpoint is, however, more than the atom's personal minimum $(1/3, 1/3)$. The unique competitive allocation sits squarely in the middle of the core. Since our interest lies chiefly in the points of the core below the competitive allocation, we shall concen-

---

[4] One must recall that in the Edgeworth box, the atom and the oceanic traders have different origins. Thus if in the box, d_a coincides with the point half-way between d_U and d_V, then algebraically, $d_a + (d_U + d_V)/2 = (1, 1)$ (*not* $d_a = (d_U + d_V)/2$).

trate on them; the reader may verify for himself those of our other assertions that he wishes.

We make use of Theorem A* (Theorem 5.2) of [3]; applied to a mixed market such as ours[5], this yields that an individually rational allocation $\mathbf{x}$ is in the core if and only if there is a price vector p such that

$$\text{for almost all } t, \mathbf{x}(t) \text{ is maximal in} \\ \{x \in \Omega : p \cdot x \leqslant p \cdot \mathbf{x}(t)\}, \text{ and} \tag{2.2}$$

$$\text{for almost all } t \text{ in the ocean, } p \cdot \mathbf{i}(t) \geqslant p \cdot \mathbf{x}(t). \tag{2.3}$$

If p satisfies (2.2), then it is called a vector of *efficiency prices*; the line $\{x : p \cdot x = p \cdot \mathbf{x}(t)\}$ is called the *efficiency budget line*.

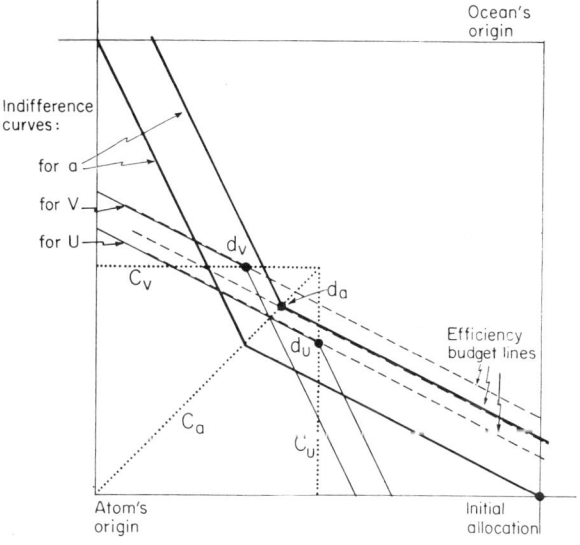

FIGURE 4

Refer to Fig. 4. For a given bundle d_a of the atom between $(3/8, 3/8)$ and $(1/2, 1/2)$, choose d_U and d_V to lie on C_U and C_V respectively, so that d_a is exactly half-way between them. The efficiency budget lines with slope $-\frac{1}{2}$ through d_a, d_U, and d_V respectively (dashed in the figure) will support the indifference curves and pass above the initial allocation; this means that (2.2) and (2.3) are satisfied, and so we get a core allocation. When d_a is under $(3/8, 3/8)$, any efficiency budget line for d_U will pass under the initial allocation, so that (2.3) is violated.

[5] That is, a market with precisely one atom, in which the ocean cannot gain by trading within itself only.

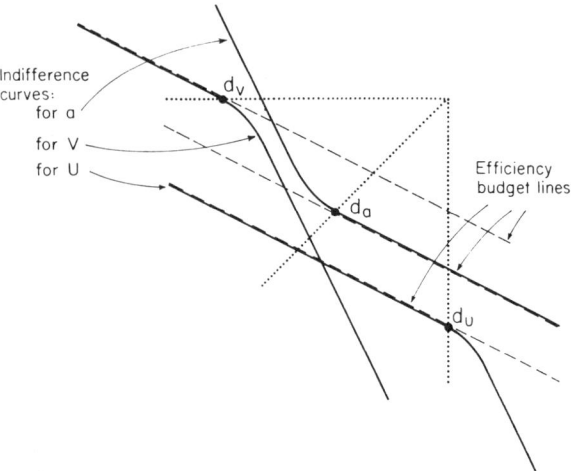

FIG. 5. Smoothing in Example A.

A differentiable version of this example is constructed by appropriate smoothing in the neighborhood of the "corners." The smoothing must be carried out so that the efficiency budget lines continue to support the indifference curves at the points d_a, d_U, and d_V respectively (Fig. 5). The above argument, proving that this yields a core allocation, then remains unchanged. Moreover, if the new indifference curves are sufficiently close to the old ones, we cannot get new competitive allocations at any appreciable distance from $(1/2, 1/2)$. Thus although it is conceivable that some competitive allocations are added, we will certainly still have a considerable part of the core yielding the atom less than the "worst" competitive allocation.

The example is also easily modified so that the utility functions are *strictly* quasiconcave (i.e., the preferences are strictly convex) as well as differentiable.

EXAMPLE B. The utility functions in this example are as follows:

$$u_t(x) = u_t(\alpha, \beta)$$
$$= \begin{cases} 4\alpha + 5\beta + 3\min(2\alpha, \beta) = \min(5(\beta + 2\alpha), 4(\alpha + 2\beta)), \\ \qquad \text{for} \quad t = a; \\ \alpha + 2\beta + 3\min(\alpha, 1/2) = \min(2(\beta + 2\alpha), \alpha + 2\beta + 3/2), \\ \qquad \text{for} \quad t \in U; \\ \beta + 2\alpha, \quad \text{for} \quad t \in V. \end{cases}$$

The initial allocation is given by (2.1).

In this example the indifference curves are similar to those in the bottom half of the box in the previous example; the critical lines are, of course, different. For $t \in V$ there is no critical line; its place is taken by the α axis (see Fig. 6).

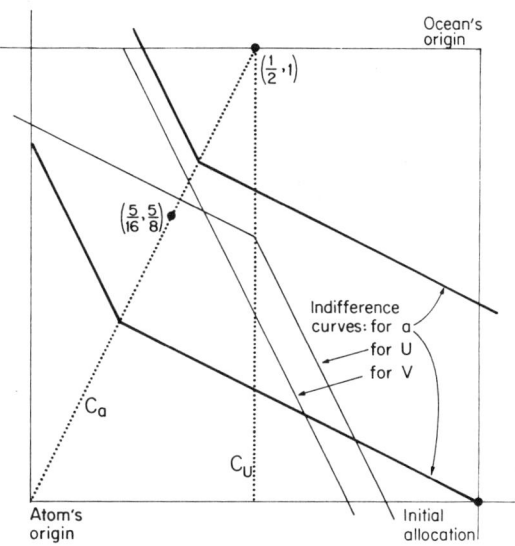

FIG. 6. Example B.

Arguments similar to those used in connection with Example A show that there is a unique competitive equilibrium, which yields the atom (1/2, 1); whereas in the core, the atom gets anything on the line connecting (5/16, 5/8) to (1/2, 1) (Fig. 6).

Like Example A, this example can also be "smoothed out," i.e., the utility functions can be made differentiable (and strictly quasiconcave). As in Example A, the smoothing may introduce new competitive allocations close to the original one. In any case, all but an arbitrarily small portion of the core will be worse for the monopolist than the worst competitive allocation.

EXAMPLE C. This example is perhaps the strangest of all. The utility functions for the oceanic players are defined as in Example B. The atom's utility function is given by

$$u_a(x) = u_a(\alpha, \beta) = \min(14(\beta + 2\alpha), 13(\alpha + 2\beta), 4(\alpha + 2\beta) + 18).$$

The initial allocation is given by (2.1). The critical lines as well as some indifference curves for the atom are given in Fig. 7.

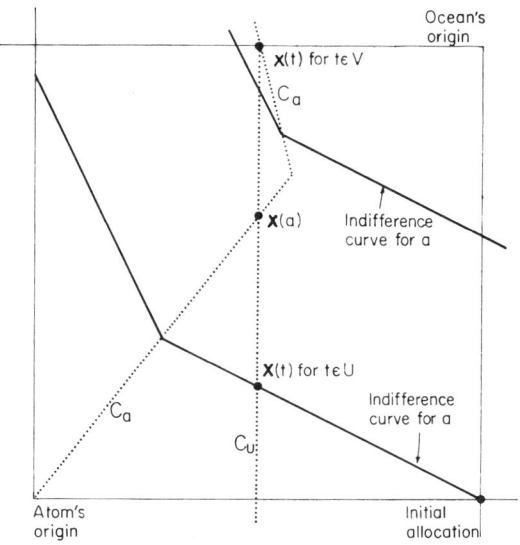

Fig. 7. Example C.

As before, there is a unique competitive allocation, which yields the atom $(1/2, 1)$. There is exactly one other allocation $\mathbf{x}$ in the core; it is given by

$$\begin{aligned}\mathbf{x}(t) &= (1/2, 5/8) &&\text{for}\quad t = a\\&= (1/2, 3/4) &&\text{for}\quad t \in U\\&= (1/2, 0) &&\text{for}\quad t \in V.\end{aligned}$$

Like the previous examples, this example is easily smoothed, i.e. made differentiable (and strictly quasiconcave). Unlike in the previous examples, the smoothing leaves both the competitive allocation and the core entirely unchanged.

3. Discussion

Our first remark concerns the relation between Examples A and B. The reader might ask why we bring Example A at all, when Example B is both simpler (in the definition of the utilities) and more striking (in that the competitive allocation is *best* in the core for the monopolist). The answer is that Example B is in a certain sense atypical, because at the unique competitive allocation, the oceanic traders get their personal minima. Thus the atom cannot hope to get more than at the competitive allocation; if the core is to extend beyond the competitive allocation at all, it can only extend "below" it. Of course, it is still very surprising that the core does in fact extend below the competitive allocation, i.e., to get the best point in the core, the atom should dissolve. Nevertheless, had we brought this example alone, some readers might have dismissed it as atypical. Example A shows that the core may extend below the competitive allocation[6] even in a case that has no discernible atypical or extreme features.

Indeed, perhaps the most disturbing aspect of these examples is their utter lack of pathology. It would be difficult to ask for more sedate, well-behaved utility functions. They can be assumed differentiable to any degree and strictly quasiconcave, and the slopes of the indifference curves are bounded away from zero and bounded; in particular, therefore, any prices associated with these markets must be bounded away from zero and bounded. What makes the examples possible is that not all the small traders have the same preferences; this again, far from being a pathology, is the "usual" situation. One is almost forced to the conclusion that monopolies that are not particularly advantageous (like Example A) are probably the rule rather than the exception. This conclusion runs counter to common sense as well as to economic theory. Perhaps what is needed at this stage is a careful reappraisal of the ideas underlying the use of the core in economic analysis.

While we are not prepared to undertake such a reappraisal within the framework of this note, we can at least hint at one aspect of it. But first, let us see how these examples appear when looked at from the viewpoint of classical economic theory. According to the classical theory, the oceanic traders in a monopoly will act like price takers, i.e., they will maximize their utility given the prices set by the monopolist. The monopolist will set the prices so that the result of price-taking on the part of the ocean will maximize *his* utility. The kind of phenomenon illustrated for the core in this note is of course impossible in the classical theory. If the monopolist sets prices, he cannot end up worse off than at the competitive equilibrium, since he always has the option of setting the prices equal to competitive prices.[7] In our examples, the monopolist can choose p^1/p^2

---

[6] In cases of this kind, it must necessarily extend above as well. Indeed, the allocation at which the atom maximizes his utility subject to the oceanic traders getting their personal minima is in the core; this is because the oceanic traders initially hold only one good, and so cannot gain by trading among themselves only.

[7] Strictly speaking, this conclusion is justified only when the preferences of the oceanic traders are *strictly* convex, so that their demands are well defined. Such an assumption is in any case necessary to make the classical theory meaningful, since well-defined demands are needed to define the result of price-taking on the part of the ocean.

close[8] to 2. In Example A this leads to a bundle close to (3/4, 1/2) for the monopolist, which he prefers to his competitive bundle. In Examples B and C it leads to a bundle close to the competitive one.

The objection against this kind of approach is that it treats the ocean and the monopolist in a fundamentally asymmetric fashion. Specifically, at what point does one start to apply this theory? When the "ocean" consists of 1,000 traders, 10 traders, 2 traders, or one trader? What one would like is a theory that is applicable in any market, and when applied to a monopoly, yields the price-taking mechanism. To put the argument differently, one feels on an intuitive, common sense level that the monopolist has a distinct advantage; but economic theory, rather than explaining this phenomenon, simply states it in a specific form. For an *explanation*, one looks to game theory; but evidently, the game-theoretic notion of core is not the proper vehicle for such an explanation.

Let us try to investigate *why* the core fails to display the monopolist's power. Most of the usual objections against the core concept are not relevant here, because they argue that the core is too small, whereas in our case it is clearly too large. For example, it is sometimes argued that the definition of the core does not take coalition-forming costs into account. But such costs make it more difficult for a coalition to block, i.e., they enlarge the core; whereas we wish to find a reason to make it smaller.

The concept of core is based on what a coalition can guarantee for itself. Monopoly power is probably not based on this at all, but rather on what the monopolist can *prevent* other coalitions from getting. His strength lies in his threat possibilities, in the bargaining power engendered by the harm he can cause by refusing to trade. Put differently, the monopolist's power—and for that matter, that of any other trader—is measured by the difference between what others can get with him and what they can get without him. This line of reasoning is entirely different from that used in the definition of core. But it is not foreign to game theory; indeed, it is closely related to the ideas underlying the Shapley value. It is known that the Shapley value is significant in some economic contexts[9], and it may well turn out to be significant in accounting for monopoly power as well.

ACKNOWLEDGMENTS

The author wishes to thank B. Shitovitz for several very helpful conversations, and, in particular, for the suggestion to use Theorem A* of [3] in the construction of the

[8] If $p^1/p^2 = 2$, the ocean's demand is not defined; see the previous footnote. As we have already noted, it is easy to modify our examples so that the preferences are strictly convex, in which case this problem will not arise.

[9] See, for example, Ref. [2].

examples. Thanks are also due to P. Champsaur, G. Laroque, and a referee for pointing out two errors in a previous version.

REFERENCES

1. R. J. AUMANN, Markets with a continuum of traders, *Econometrica* **32** (1964), 39–50 [Chapter 46].
2. L. S. SHAPLEY AND M. SHUBIK, Pure competition, coalitional power, and fair division, *Int. Econ. Rev.* **10** (1969), 337–362.
3. B. SHITOVITZ, Oligopoly in markets with a continuum of traders, *Econometrica* **41** (1973), 467–501.

49 A Note on Gale's Example
with B. Peleg

David Gale (1974) constructs an example of a pure exchange economy with 3 traders in which two of the traders, by exchanging goods among themselves only, can affect prices in the entire economy in such a way so that *both* will benefit (and so necessarily, the third trader will lose). Gale's example involves indifference curves with sharp corners, and he raises the question as to whether an example of this kind can be found with smooth preferences.

In this note we discuss a related phenomenon which at first glance is even more striking but which is even simpler to prove. In a 2-trader, 2-commodity market, it is possible for a trader simply to discard some of his initial bundle and to gain from this act – at the expense of the other trader, of course. This example may be modified to yield Gale's phenomenon – instead of discarding, he donates to a third trader. Moreover, the preferences in our example are smooth, thus settling Gale's question as to smoothness.

The idea is quite simple, and is similar to that of Gale's example. Each trader initially holds a 'corner' on one of the two commodities, i.e., the initial bundles are of the form $(\alpha, 0)$ and $(0, \beta)$. If trader 1 throws away some of his initial bundle, the price of commodity 1 goes up; as in Gale's example, this rise in the *price* of the commodity he holds is more than enough to compensate for the drop in the amount.

It seems to us that this is what acreage restrictions and similar tactics are all about.

In the example, there are 2 goods; traders 1 and 2 initially hold $(2, 0)$ and $(0, 1)$, respectively. Both traders have the same preferences, which are homothetic; at $(1, 1)$ the indifference curve has slope -1, and at $(2, 1)$ it has slope $-\frac{1}{8}$ (see fig. 1). There is of course no difficulty at all in constructing arbitrarily smooth preferences that obey these conditions (in addition to quasi-concavity, strict monotonicity, and practically whatever one wishes). A numerical example is given below.

Since the preferences are homothetic, the budget line is tangent to the indifference curve at the total initial bundle; i.e., it has slope $-\frac{1}{8}$ (see fig. 1).

This chapter originally appeared in *Journal of Mathematical Economics* 1 (1974): 209–211. Reprinted with permission.

Again because of the homotheticity, the competitive bundle of trader 1 lies on the line $x = 2y$, and hence the competitive bundle is $(\frac{2}{5}, \frac{1}{5})$. If trader 1 discards one unit of good 1 before trading starts, the total initial bundle is (1, 1), and hence the prices stand in the ratio 1 : 1 (see fig. 1). The competitive bundle of trader 1 then lies on the line $x = y$, and hence it is $(\frac{1}{2}, \frac{1}{2})$ – yielding more of *each* commodity than he got before he discarded anything.

To get an example of Gale's kind from this, add another trader (trader 3), and give him the initial bundle (1, 0) and the utility $2x+y$. If trader 1 gives trader 3 one unit of good 1, then trader 1 gains, and so does trader 3. Of course trader 2 loses.

To obtain appropriate preferences (for traders 1 and 2), consider the utility function defined to be 0 at the origin and $[(x+\alpha y)^{-3}+(\alpha x+y)^{-3}]^{-1}$ elsewhere

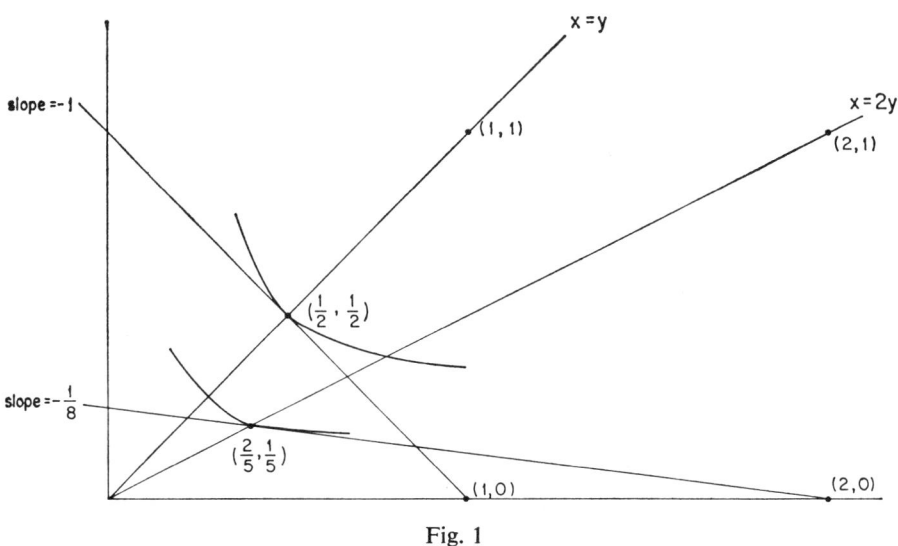

Fig. 1

in the non-negative orthant Ω, where α is a parameter that may take values in the half-open interval (0, 1]. This is continuous, quasi-concave, strictly monotonic, and induces homothetic preferences. The indifference curve has slope -1 at (1, 1) for all values of α. When α is near 0 the indifference curve at (2, 1) has slope near $-\frac{1}{16}$, and when $\alpha = 1$ it has slope -1. From continuity considerations it then follows that for an appropriate α, the indifference curve has slope $-\frac{1}{8}$ at (2, 1). The utility function is of class C^∞ in the entire non-negative orthant. The gradient is strictly positive everywhere in Ω, except at the origin, where it vanishes. There is perhaps a certain lack of smoothness at the origin, since the normalized gradient cannot be continuously extended to the origin. But this is unavoidable when the preferences are homothetic (unless the utility function is linear). In our case, if one wants to avoid this, one can re-define the preferences in an essentially arbitrary way in a neighborhood of the origin. The homotheticity will be destroyed, but if the neighborhood is sufficiently small, the example will not be affected in any essential way.

Reference

Gale, D., Exchange equilibrium and coalitions: An example, Journal of Mathematical Economics 1, 63–66.

50 On the Rate of Convergence of the Core

1. INTRODUCTION

The classical theorem of Debreu and Scarf [1963] asserts that the core of a replicated pure exchange economy tends to the set of its competitive allocations as the index k of the replication (the number of traders of each type) tends to infinity. In separate contributions that complement each other nicely, Shapley [1975] and Debreu [1975] have investigated the rate of this convergence. Shapley showed that with C^1 utility functions, the convergence can be as slow as one wishes; whereas Debreu showed that when the utility functions are C^2 and the indifference surfaces have positive Gaussian curvature, the rate of convergence is generically $O(1/k)$. Here "generically" means that for fixed utilities, the rate $O(1/k)$ will obtain except possibly when the initial allocation lies in a closed set of Lebesgue measure 0.

These developments, though they went far toward settling the question of the rate of convergence, still left a gap. Shapley's example depends critically on an essential discontinuity in the second derivative: it becomes unbounded in the neighborhood of the unique competitive allocation. In the introduction to his paper, he speculates that "a condition that bounds the second derivatives... might suffice" to assure $O(1/K)$. On the other hand, the generic nature of Debreu's theorem leaves open the possibility of slow convergence even when the second derivative is bounded and in fact continuous.

Here we will show that the condition of genericity in Debreu's theorem cannot be removed; i.e., there are C^2 (in fact C^∞) utility functions, with indifference surfaces that have positive Gaussian curvature, where the core converges as slowly as we like.[1]

2. THE EXAMPLE

First, some definitions. A *null sequence* is a sequence $\{\delta_k\}$ of positive numbers tending to 0. A *market* is a pure exchange economy with a finite number n of goods, a finite number m of agents, and consumption sets all equal to R_+^n; it is *smooth* if the utility functions are[2] C^2 and strictly quasiconcave, and have positive first derivatives and indifference surfaces with positive Gaussian curvature.

Manuscript received October 27, 1977; revised January 30, 1978.

[1] This work was supported by National Science Foundation Grant SOC 74-11446 at the Institute for Mathematical Studies in the Social Sciences, Stanford University.

[2] A function is called C^2 on R_+^n if it is C^2 on Int R_+^n, continuous on R_+^n, and the second partial derivatives can be extended to continuous functions on R_+^n.

This chapter originally appeared in *International Economic Review* 20 (1979): 349–357. Reprinted with permission.

Given a smooth market M, its k-replication M_k is the smooth market with km agents and n goods, in which there are k agents corresponding to each agent of M and having the same utility and endowment as that agent; these k agents are said to be of the same *type*. By the "equal treatment principle" (Debreu and Scarf [1963]), each allocation in the core of M_k assigns the same commodity vector to agents of the same type; therefore each core allocation in M_k corresponds to an allocation in M. The set of all allocations in M corresponding to core-allocations in M_k will be denoted C_k. Let W be the set of Walras (i.e., competitive) allocations of M; the Debreu-Scarf Theorem says that $\max\{d(x, W): x \in C_k\} \to 0$ as $k \to \infty$, where $d(x, W)$ denotes the Euclidean distance from the point x to the compact set W.

THEOREM. *Let $\{\delta_k\}$ be a null sequence. Then there is a smooth market M such that*

(1) $$\max\{d(x, W): x \in C_k\} > \delta_k$$

for all k.

PROOF. It is sufficient to prove (1) for all sufficiently large k, since by increasing the scale of the market we can multiply the left side of (1) by as large a constant as we want (simultaneously for all k); this will overcome any deficiency in (1) for any finite number of k's.

Before writing down formulas, let us describe the construction geometrically. The markets we consider will have two goods and two agents (before replication). We start out by recalling the condition for an allocation x in such a market to be in C_k. Letting e be the initial allocation, set

(2) $$h^k(x) = x + \frac{1}{k-1}(x - e),$$

$$h_k(x) = x - \frac{1}{k}(x - e).$$

Then

(3) $x \in C_k$ *if and only if $x \in C_1$ and both agents weakly prefer x to both $h^k(x)$ and $h_k(x)$.*

Essentially, this is the condition of Edgeworth [1881, p. 37]. The proof is also indicated in Shapley [1975], and moreover it is a straightforward matter to verify it directly. The condition is illustrated in the Edgeworth box of Figure 1. The point x is in C_k because both $h^k(x)$ and $h_k(x)$ are "beneath" the indifference curve for x from the point of view of both agents; whereas y is *not* in C_k, as $h_k(y)$ is above the indifference curve of y for agent 1; i.e., agent 1 prefers $h_k(y)$ to y.

Rather than proceeding immediately with the construction of M, we construct first an auxiliary market M^*, illustrated in the Edgeworth box of Figure 2. Note the contract curve, which for definiteness we can think of as the diagonal of the

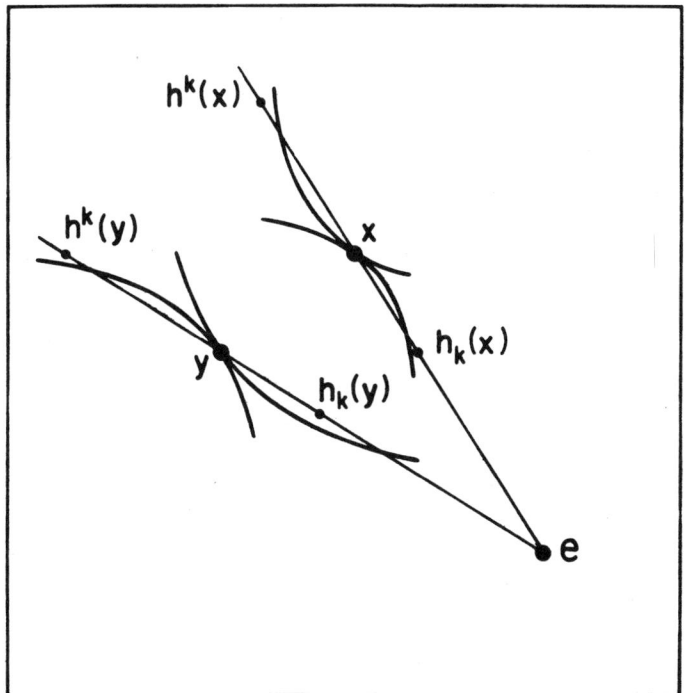

FIGURE 1

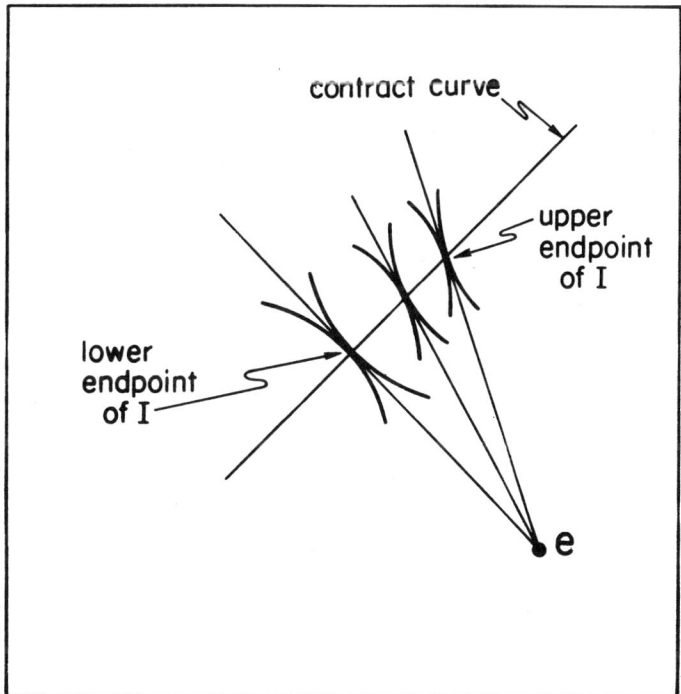

FIGURE 2

box, and the initial allocation e, which is assumed *not* to be on the contract curve. The only special feature we require of M^* is that *every* point in some non-degenerate interval I along the contract curve be a Walras allocation. Obviously these conditions can be satisfied in a smooth market.

Let us denote the lower (from agent 1's standpoint) endpoint of I by w (see Figure 3). To construct M from M^* we choose a sequence of points a_k on I whose distance from w is, for sufficiently large k, precisely δ_k. Since a_k is a Walras allocation in M^*, the point a_k is *strictly preferred* in M^* by both agents to both $h^k(a_k)$ and $h_k(a_k)$. Let us now modify the utilities so that the slopes of the two indifference curves at a_k are very slightly changed, but remain equal to each other. Then a_k remains in the contract curve but is no longer a Walras allocation; moreover, if the change in the utilities is sufficiently small, a_k will still be preferred to both $h^k(a_k)$ and $h_k(a_k)$, and so will be in C_k (see Figure 3). If we make sure that the slopes of the indifference curves at *all* points of I except w are changed (but remain equal to each other), then I remains part of the contract curve C_1, but there will be no Walras allocations in I other than w. Thus for sufficiently large k,

$$\max \{d(x, W): \ x \in C_k\} \geq \|a_k - w\| = \delta_k,$$

and so our construction is complete.

The only question that remains is whether the desired modification of the

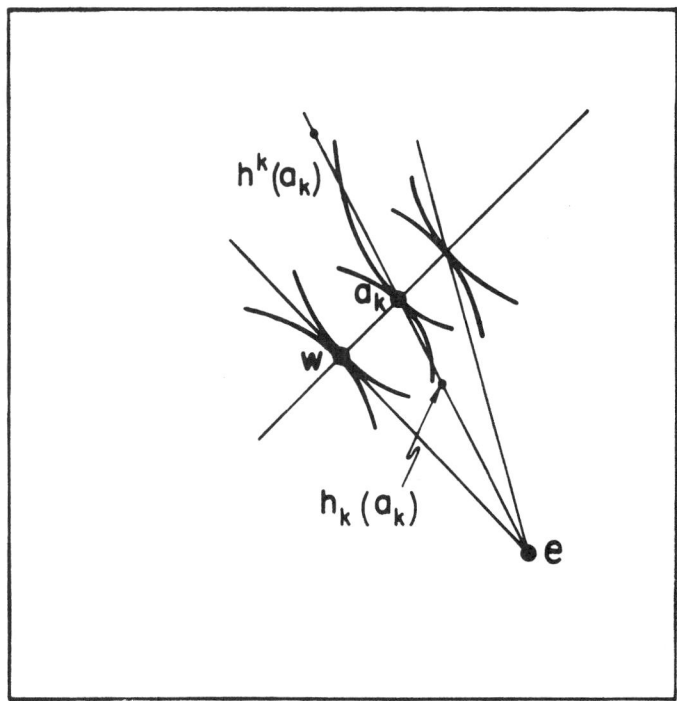

FIGURE 3

utility functions can really be carried out, simultaneously for all k, so that the resulting market is smooth. This we will now show by carrying out the construction more or less explicitly.

For definiteness, let the total endowment of the economy (the sum of 1's and 2's endowments) be $(1, 1)$. Then an allocation — i.e., a "point" x in the Edgeworth box — is of the form $(x^1, x^2, 1-x^1, 1-x^2)$, where the superscripts index the commodities. In particular, each a_k, since it is on the diagonal, has the form

$$a_k = (\alpha_k, \alpha_k, 1 - \alpha_k, 1 - \alpha_k),$$

and similarly w has the form

$$w = (\omega, \omega, 1 - \omega, 1 - \omega).$$

Denote the utility functions of agents 1 and 2 in M^* by u^* and v^* respectively. Let f be a C^2 function from the real line onto itself, with $f' > 0$ everywhere, $f'(\alpha) < \alpha$ when $\omega < \alpha < 1$, and $f(\alpha) = \alpha$ otherwise; such a function will be called *admissible*. Define functions u and v by

(4) $$\begin{cases} u(x^1, x^2) = u^*(f(x^1), f(x^2)) \\ v(1 - x^1, 1 - x^2) = v^*(1 - f(x^1), 1 - f(x^2)) \end{cases}$$

when $(x^1, x^2, 1-x^1, 1-x^2)$ is in the box, i.e., when $0 \leq x^1 \leq 1$, $0 \leq x^2 \leq 1$; for other values of x^1 and x^2, we will define u and v later. For appropriate choice of f, we will take the utilities in M to be u and v.

First we show that the contract curve in M remains the diagonal. To this end, note that for points $(\alpha, \alpha, 1-\alpha, 1-\alpha)$ on the diagonal, we have[3]

(5) $$\begin{cases} \nabla u(\alpha, \alpha) = f'(\alpha) \nabla u^*(f(\alpha), f(\alpha)), \\ \nabla v(1 - \alpha, 1 - \alpha) = f'(\alpha) \nabla v^*(1 - f(\alpha), 1 - f(\alpha)). \end{cases}$$

Letting g_u denote the normalized gradient, i.e.,

$$g_u(x_1, x_2) = \frac{\nabla u(x_1, x_2)}{\|\nabla u(x_1, x_2)\|},$$

and defining g_{u^*}, g_v, and g_{v^*} similarly, we deduce from (5) that

(6) $$\begin{cases} g_u(\alpha, \alpha) = g_{u^*}(f(\alpha), f(\alpha)) \\ g_v(1 - \alpha, 1 - \alpha) = g_{v^*}(1 - f(\alpha), 1 - f(\alpha)). \end{cases}$$

On the other hand, since the diagonal is the contract curve in M^*, we have

$$g_u(\alpha, \alpha) = g_v(1 - \alpha, 1 - \alpha),$$

which means that the diagonal remains the contract curve in M, as asserted.

We next wish to show that none of the points $(\alpha, \alpha, 1-\alpha, 1-\alpha)$ in I other than

---

[3] By $\nabla v(1-\alpha, 1-\alpha)$ we mean ∇v evaluated at $(1-\alpha, 1-\alpha)$.

w are Walras allocations in M. Since by construction these points *are* Walras allocations in M^*, it is sufficient for this to show that

(7) $$g_u(\alpha, \alpha) \neq g_{u^*}(\alpha, \alpha).$$

Indeed, when $(\alpha, \alpha, 1-\alpha, 1-\alpha) \in I - w$, then $\omega < f(\alpha) < \alpha$; hence $(f(\alpha), f(\alpha), 1-f(\alpha), 1-f(\alpha))$ is in $I - w$ and is different from $(\alpha, \alpha, 1-\alpha, 1-\alpha)$, and so both are Walras allocations in M^*. Hence

(8) $$g_{u^*}(f(\alpha), f(\alpha)) \neq g_{u^*}(\alpha, \alpha),$$

and combining this with (6), we get (7).

Before proceeding, it will be convenient to establish the following lemma:

LEMMA. *Let $\{\gamma_l\}$ be a sequence of numbers tending to ω, and let $\{\varepsilon_l\}$ be an arbitrary sequence of positive numbers. Then there is an admissible function f such that*

(9) $$|f(\gamma_l) - \gamma_l| \leq \varepsilon_l$$

for all l. Furthermore, if ε is an arbitrary positive number, then an admissible f can be found satisfying (9) and, in addition,

(10) $$|f(\alpha) - \alpha| < \varepsilon, \quad |f'(\alpha) - 1| < \varepsilon, \quad |f''(\alpha)| < \varepsilon$$

for all real α.

PROOF. Since we will define $f(\alpha) = \alpha$ whenever α is not in the open interval $(\omega, 1)$, we may assume w.l.o.g. (without loss of generality) that $\omega < \gamma_l < 1$ for all l. Since ω is the only limit point of the γ_l, we may then also assume w.l.o.g. that γ_l is decreasing in l (no more than a finite number of γ_l can coincide, since $\gamma_l \to \omega$; if, then, there are several equal γ_l, we can simply eliminate all but one and use the smallest of the corresponding ε_l). Finally, we may assume w.l.o.g. that $\varepsilon_l \leq 1$ for all l and that $\varepsilon_l \to 0$; for if not, we can substitute $\min(\varepsilon_l, 1/l)$ for ε_l.

Now let f_1 be a continuous function on $\mathbf{R}$ such that $1 \geq f_1(\alpha) > 0$ on $(\omega, 1)$,

(11) $$f_1(\alpha) = 0 \quad \text{outside of} \quad (\omega, 1),$$

and

(12) $$f_1(\gamma_l) = \varepsilon_l \quad \text{for all } l.$$

For example, such a function can be constructed by defining it by (11), (12), and the requirement that it be linear on each of the segments $(\gamma_1, 1)$ and (γ_{i+1}, γ_i), $i = 1, 2, \ldots$. Next, define

$$f_2(\alpha) = \int_0^\alpha f_1(t)dt$$

and

$$f_3(\alpha) = \int_0^\alpha f_2(t)dt.$$

Let f_4 be a C^2 function such that $1 \geq f_4(\alpha) > 0$ for $\alpha \in (\omega, 1)$ and $f_4(\alpha) = 0$ for $\alpha \geq 1$ (for example, $f_4(\alpha) = (\alpha - 1)^4$ when $\alpha \leq 1$, and $f_4(\alpha) = 0$ when $\alpha \geq 1$). Define $f_5 = f_4 f_3$; then $f_5 \in C^2$, $f_5(\alpha) > 0$ on $(\omega, 1)$, $f_5(\alpha) = 0$ outside of $(\omega, 1)$, and

$$f_5(\gamma_l) \leq f_3(\gamma_l) < f_1(\gamma_l) \leq \varepsilon_l$$

for all l.

Let $K_1 = \max f_5$, $K_2 = \max f_5'$, $K_3 = \max f_5''$, $K = \max(K_1, K_2, K_3)$, and define

$$f(\alpha) = \alpha - \left(\frac{\varepsilon}{2K}\right) f_5(\alpha).$$

Then f obeys all the requirements of the lemma, and so the proof of the lemma is complete.

We are now ready to show that when f is appropriately chosen, $a_k \in C_k$ for k sufficiently large. Since a_k is a Walras allocation in M^*, it follows from (3) and (4) that $a_k \in C_k$ if the mapping

$$(x^1, x^2, 1 - x^1, 1 - x^2) \longrightarrow (f(x^1), f(x^2), 1 - f(x^1), 1 - f(x^2))$$

moves the three points a_k, $h^k(a_k)$, and $h_k(a_k)$ sufficiently little. To state this formally, set

(13)
$$h^k(a_k) = (\beta_k^1, \beta_k^2, 1 - \beta_k^1, 1 - \beta_k^2)$$
$$h_k(a_k) = (\beta_k^3, \beta_k^4, 1 - \beta_k^3, 1 - \beta_k^4).$$

Since a_k is a Walras allocation in M^*, we have

(14)
$$u^*(\beta_k^1, \beta_k^2) < u^*(\alpha_k, \alpha_k) > u^*(\beta_k^3, \beta_k^4)$$
$$v^*(1 - \beta_k^1, 1 - \beta_k^2) < v^*(1 - \alpha_k, 1 - \alpha_k) > v^*(1 - \beta_k^3, 1 - \beta_k^4).$$

From the continuity of u^* and v^* it follows that for each k there is a positive number η_k such that if we change each of the five numbers α_k, β_k^1, β_k^2, β_k^3, β_k^4 by at most η_k, then the strict inequalities in (14) will continue to hold. From this it follows that if f is constructed such that for all k,

(15) $\quad |f(\alpha_k) - \alpha_k| < \eta_k \quad$ and $\quad |f(\beta_k^i) - \beta_k^i| < \eta_k \quad$ for $\quad i = 1, 2, 3, 4$,

then we will have $a_k \in C_k$ for all k.

From (2) and (13) we deduce

(16) $\qquad\qquad\qquad \beta_k^i \longrightarrow \omega \quad$ as $\quad k \longrightarrow \infty$

for $i = 1, 2, 3, 4$. We can therefore combine the α_k and the β_k^i into a single sequence $\{\gamma_l\}$ defined by $\gamma_{5k+i} = \beta_k^i$ for $i = 1, 2, 3, 4$ and $\gamma_{5k+5} = \alpha_k$. Setting $\varepsilon_{5k+i} = \eta_k$ ($i = 1, 2, 3, 4, 5$) and applying the Lemma (in particular (10)), we conclude that it is possible to construct f so that (15) is satisfied, as desired.

Finally, we must show that u and v can be constructed so that they are everywhere C^2 and have positive first derivatives and positive Gaussian curvature.

Let q be a C^2 function from $\mathbf{R}$ to itself such that $q(\alpha)=1$ for $\alpha \leq 1$ and $q(\alpha)=0$ for $\alpha \geq 2$. For arbitrary $(x^1, x^2) \in R_+^2$, define

$$u(x^1, x^2) = u^*(q(x^2)f(x^1) + (1 - q(x^2))x^1, q(x^1)f(x^2) + (1 - q(x^1))x^2).$$

When $(x^1, x^2, 1-x^1, 1-x^2)$ is in the box, i.e., when $0 \leq x^1 \leq 1$ and $0 \leq x^2 \leq 1$, this coincides with the definition at (4). When $x^1 > 2$ or $x^2 > 2$, then $u(x^1, x^2) = u^*(x^1, x^2)$, in which case the required smoothness properties are inherited from M^*. We must therefore only concern ourselves with the compact set $\{(x^1, x^2): 0 \leq x^1 \leq 2 \text{ and } 0 \leq x^2 \leq 2\}$. In this set, u^* and its first and second derivatives are bounded, and the first derivatives of u^* and the Gaussian curvatures are bounded away from zero. Hence if we change the first and second derivatives by sufficiently little, the smoothness properties will continue to hold.[4] A straightforward calculation of the derivatives, together with an application of the Lemma (in particular (10)), then yields the desired result.

Similar considerations apply to v. Define $\hat{f}(\alpha) = 1 - f(1-\alpha)$, and for $(y^1, y^2) \in R_+^2$, set

$$v(y^1, y^2) = v^*(q(y^2)\hat{f}(y^1) + (1 - q(y^2))y^1, q(y^1)\hat{f}(y^2) + (1 - q(y^1))y^2).$$

Again, when $0 \leq y^1 \leq 1$ and $0 \leq y^2 \leq 1$, this coincides with the definition at (4). If we note that the inequalities (10) for $\hat{f}$ follow from the same inequalities for f, then the argument for v may be completed like that for u.

3. REMARKS AND CONCLUSION

The reader who has followed the construction through will realize that it can be carried out at no additional cost with C^∞ utility functions. Furthermore, the example can be constructed so that it has a unique Walras equilibrium. Thus unlike in the Shapley example, there is absolutely nothing pathological here, nothing overt that one might have guessed would lead to "trouble." The conclusion is that though Debreu's theorem shows that a slow-converging core is a rarity, it nevertheless can and does occur in the smoothest of environments, and can apparently not be ruled out by any kind of economically reasonable a priori requirement.

The measure with which we chose to gauge the speed of convergence was simply the Euclidean distance in the "equal treatment allocation space" E^{lm}. Of course, any other norm on E^{lm} would have done as well. But one might ask whether the example would go away if one used some economically more meaningful measure, such as difference in value at, say, the equilibrium prices. Again, the answer is no; the value at *any* price will yield the same slow rate of convergence, since the core is laid out along the diagonal, which is transversal to any budget line.

Finally, it is useful to examine our example in the light of the result of Dierker

---

[4] The Gaussian curvature is a continuous function of the first and second derivatives of u.

[1975],[5] which assumes no differentiability or even continuity. Roughly speaking, this result says that no matter how large an economy may be, the total deviation of a member of its core from "competitiveness", summed over all agents in the economy, will not exceed some constant (which is independent of the size of the economy). Translated to the replication context, it implies that for each k and each allocation $x=(x_1,\ldots, x_m)$ in C_k, there is a normalized price vector p such that for each agent i, $|p(e_i-x_i)|=O(1/k)$ and $|p(y_i-x_i)|=O(1/k)$, where e_i is i's endowment and y_i is the bundle of minimal value, at prices p, on the indifference surface of x_i. If in these two equations, $O(1/k)$ would be replaced by 0, then x would be a Walras allocation; this, therefore, strongly suggests the $O(1/k)$ convergence. The reason that such a conclusion would be erroneous is that p depends on x and k and need not itself be competitive (i.e., Walras); indeed, our example (and Shapley's) shows that convergence of p to the competitive price may be arbitrarily slow.

Intuitively, though, Dierker's theorem makes it clear that $O(1/k)$ is the "right" rate, even when the actual convergence is slower. Though p is not within $O(1/k)$ of a competitive price, it is a price that is within $O(1/k)$ of being competitive. The distinction here is similar to that between being near a fixed point of a mapping and being at a point that is nearly fixed. It is, of course, the second kind of approximation that has been the object of the approximation algorithms pioneered by Scarf [1973], and that seems, intuitively, to be the more significant concept.

REFERENCES

ANDERSON, R. M., "An Elementary Core Equivalence Theorem," *Econometrica*, 46 (1978), 1483–1487.
ARROW, K. J. AND F. H. HAHN, *General Competitive Analysis* (San Francisco: Holden-Day, 1971).
DEBREU, G., "The Rate of Convergence of the Core of an Economy," *Journal of Mathematical Economics*, 2 (March, 1975), 1–7.
DEBREU, G. AND H. SCARF, "A Limit Theorem on the Core of an Economy," *International Economic Review*, 4 (September, 1963), 235–246.
DIERKER, E., "Gains and Losses at Core Allocations," *Journal of Mathematical Economics*, 2 (June–September, 1975), 119–128.
EDGEWORTH, F. Y., *Mathematical Psychics* (London: Kegan Paul, 1881).
SCARF, H., *The Computation of Economic Equilibria* (New Haven: Yale University Press, 1973).
SHAPLEY, L. S., "An Example of a Slow-Converging Core," *International Economic Review*, 16 (June, 1975), 345–351.

---

[5] See Arrow and Hahn [1971, p. 188ff] for a weaker version of this result, and Anderson [1978] for a simpler proof.

X COALITIONAL GAMES: ECONOMIC AND POLITICAL APPLICATIONS OF THE SHAPLEY VALUE

The *value* of a game is an a priori measure of what a player can expect from playing it, when one abstracts away from the particular game, and takes into account only its coalitional form. First defined in 1953 for TU games by Lloyd Shapley,[1] its definition was extended[2] by Shapley in 1969 to NTU games, which are often more appropriate than TU games for economic modeling. The chapters in this group apply the value to five economic issues, most of which also have important political aspects: exchange, redistribution, public goods, fixed prices, and link formation between players. In the first eight there are many players, each individually negligible. In chapters 51–56, this is modeled as a nonatomic continuum (an "ocean"), like in chapters 46 and 47; chapter 57 uses an asymptotic approach. All but two of the chapters (54 and 59) use NTU models.

Chapter 51 does for the NTU value what chapter 45 did for the core—it shows that in markets (i.e., pure exchange economies) with a nonatomic continuum of traders, the NTU values coincide with the competitive allocations. Shapley and I had previously proved a TU version of this result in our book *Values of Non-Atomic Games*,[3] and at about the same time as chapter 49, P. Champsaur[4] proved an asymptotic, finite-type version of the NTU result (analogous to the Debreu–Scarf[5] core convergence theorem).

Chapters 52 and 53, coauthored with M. Kurz, apply the NTU value to the analysis of a politico-economic model in which society votes on taxation and redistribution. The analysis in chapter 52 is in terms of a single commodity—"money"—whereas chapter 53 is a multicommodity analysis, which takes into account the effects of taxation and redistribution on consumption. These papers broke away from what had been the traditional view of the taxing authority (the government) as a benevolent entity wishing to maximize "social welfare." Instead, the view here is that government policy is determined by the voters, each of whom is out to maximize his own welfare; that taxation and redistribution is the result of a power struggle between these voters. One of our conclusions is that the tax depends on attitudes toward risk-taking. Specifically, an individual's net tax (payments less receipts from government entitle-

---

1. "A value for *n*-person games," in *Contributions to the Theory of Games II*, H. W. Kuhn and A. W. Tucker, editors, *Ann. of Math. Studies* 28, (Princeton: Princeton University Press), 343–359.
2. "Utility comparison and the theory of games," in *La Decision* (Paris: Editions du CNRS), 251–263.
3. Princeton, Princeton University Press, 1974.
4. "Cooperation versus competition," *J. Econ. Th.* 11 (1975), 393–417.
5. "A limit theorem on the core of an economy," *Int. Econ. Rev.* 4 (1963), 235–246.

ments) depends on his *fear of ruin*;[6] the more fearful he is of large losses, the larger his tax.

Chapter 52 is the more intuitive and less formal of the two. The precise proofs are in chapter 53.

Taken together, chapters 46 and 51 imply that in an exchange economy with a nonatomic continuum of traders, the core and the value coincide (since both coincide with the competitive outcome). Chapter 54, coauthored with R. J. Gardner and R. W. Rosenthal, shows, by a simple example, that this equivalence does not extend to the case of public goods. The example is set in a TU framework, but since this is a counterexample rather than a theorem, it applies also to NTU games (as TU games are special cases of NTU games).

Chapters 55 and 56, coauthored with M. Kurz and A. Neyman, continue the analysis of power in politico-economic models begun in chapters 52 and 53. The difference is that now the taxes are used to provide public goods[7] rather than for redistribution. The chief conclusion of both papers is that unlike in the case of redistribution, the vote matters for very little in the determining which public goods are to be produced; the important source of power in this case is economic, not political. For an intuitive explanation of the result, see section 6a of chapter 55, and/or chapter 4.

Chapter 57 was begun in 1978, when I spent a few months' leave from the Hebrew University at CORE[8]. Jacques Dreze, the founding father and guiding spirit of CORE, had, during the seventies, developed a theory of economic equilibrium under a regime of fixed prices; the theory was motivated by the reality of widespread unemployment in the western world, which was thought to be related to wage rigidities. When I arrived at CORE that winter, Jacques and I asked ourselves what the NTU value could say about this problem. This led to a decade of some of the deepest, most challenging and frustrating, yet most rewarding work in which I have ever been involved.

It turns out that the NTU value predicts the endogenous emergence of a rationing scheme that brings the fixed price economy back into equilibrium. The analysis is unusual because, unlike in the above studies, the mass of individually "small" agents cannot be modeled by a nonatomic continuum. To understand why, recall that the NTU value assigns endogenous weights to the agents' utilities. As is well known, fixed prices

---

6. To be distinguished from the risk aversion, which measures aversion to small risks.
7. The public goods here are assumed *nonexclusive*—that is, they may be enjoyed by all players.
8. The Center for Operations Research and Econometrics at the Catholic University of Louvain.

often result in discrepancies between supply and demand. Agents whose endowments are in oversupply at the fixed prices—like unskilled workers in an economy with unemployment—contribute very little to Society. On the average, though, they do contribute something; *some* workers of this kind *are* needed. As a result, the weight of each such agent is small—but not 0!—as compared to that of an agent in short supply. When the economy is large, the weights of agents on the "long" side (i.e., in oversupply) become infinitesimal as compared to those of agents on the "short" side. Such phenomena cannot be modeled by a nonatomic continuum, where the weight of an agent's utility must be either zero or positive. Instead, one must use an asymptotic model, in which one looks at limits as the number n of agents goes to infinity. The agents on the long side then get weights that tend to 0 as $n \to \infty$, whereas on the short side, the weights remain bounded away from 0.

Chapter 58 is a succinct summary of the six NTU papers in this group (51–53 and 55–57), giving the main results as well as brief outlines of the proofs. It was presented orally as part of the NATO Advanced Study Institute that took place in July 1991 at the State University of New York at Stony Brook. The excellent written version is due to J.-F. Mertens.

Much of the social, economic, and political activity that we observe consists of jockeying for position. "Connections" play a crucial role: "It's not what you know, it's whom you know." Chapter 59 examines the formation of such connections between players, and where it is likely to lead. The starting point is work of R. Myerson[9] that generalizes the Shapley value to situations where, in addition to a TU game v, there is also given a network[10] G of links between the players; only coalitions S that form a connected subgraph of G can achieve the payoff $v(S)$. In chapter 59 we start with a TU game v, and ask which network G is likely to form endogenously, assuming that any two players who wish to form a link can do so. The viewpoint is forward-looking: in deciding whether to form a link, players consider not whether their prospects are improved by the formation of this link alone, but whether it is improved by all the subsequent link forming activity to which the formation of this link may lead. Here the "prospects" of a player are measured by his "Myerson value"—Myerson's above-mentioned generalization of the Shapley value.

9. "Graphs and cooperation in games," *Math. Oper. Res.* 2 (1977), 225–229.
10. Technically, a graph whose vertices are the players.

51 Values of Markets with a Continuum of Traders[1]

1. INTRODUCTION

THE "EQUIVALENCE PRINCIPLE" for the core[2] states that the core of a perfectly competitive market coincides with the set of its competitive allocations. It is the object of this paper to establish a similar equivalence principle for the Shapley value [5, 21, and 23], a game theoretic concept quite different from the core. Since the Shapley value can be interpreted in terms of "marginal worth", it is perhaps *the* game theoretic concept most closely related to traditional economic ideas.

By a "perfectly competitive" market we mean one in which there are many traders, with each trader holding only a negligible proportion of the total resources of the economy. This situation may be modelled by a continuum of traders—specifically, by a non-atomic continuum, as in Aumann [1].

We will be dealing with ordinary Walrasian pure exchange economies; no assumption of transferable utility will be made. The value was originally defined [21] for games with transferable utility only, but was later [23] adapted to apply also to games without transferable utility. It is this adaptation, itself adapted to non-atomic markets, that we shall use in this paper. Values are ordinarily expressed in units of utility; since we are dealing with markets that are defined a priori in terms of preferences rather than in terms of utilities, we shall wish to refer directly to allocations rather than to utilities. Allocations corresponding to values in such a set-up will be called "value allocations"; a precise definition will be given in Section 5 below.

To establish the equivalence principle for the value, we will have to make certain differentiability assumptions on the traders' preferences, assumptions that are not necessary for the core equivalence principle. Preferences satisfying these assumptions will be called "uniformly smooth." This notion is closely related to Debreu's "smooth preferences" [10]. We might say that the preferences in a market are uniformly smooth if and only if they are smooth in Debreu's sense and the properties embodying this smoothness hold uniformly over all the traders in the market. This, too, will be spelled out precisely below (Sections 4 and 5).

[1] Text of the Walras-Bowley Lecture delivered at the winter meeting of the Econometric Society held in Toronto in December, 1972. This work was supported in part by National Science Foundation Grant GS-40104 at the Institute for Mathematical Studies in the Social Sciences, Stanford University, and in part by CORE, the Center for Operations Research and Econometrics at the Catholic University of Louvain.

[2] For a comprehensive treatment of the core equivalence principle, see Hildenbrand [14].

This chapter originally appeared in *Econometrica* 43 (1975): 611–646. Reprinted with permission.

Recall that an allocation x is called *competitive* if there is a price vector p such that (p, x) is a competitive equilibrium. We are now ready for the statement of our main theorem:

THEOREM 1 (Value Equivalence Theorem): *In a non-atomic market with uniformly smooth preferences, the set of value allocations coincides with the set of competitive allocations.*

It should be noted that this result depends strongly on the non-atomicity of the market, i.e., on its perfectly competitive nature. The theorem is not true *in either direction* for finite markets.[3] This is to be contrasted with the situation for the core: competitive allocations are in the core for any market, even a finite one; there the continuum is needed only to establish the converse, that all core allocations are competitive.

The Value Equivalence Theorem can be interpreted in terms of social welfare functions, and the concept of an individual's contribution to society. Given utility functions for the traders, we may, for any allocation x and coalition S, consider the sum of the utilities of the members of S for the bundles assigned to them under x. Let us think of this sum as the "welfare" of the coalition S under the allocation x; in other words, assume the existence of an additively separable social welfare function consistent with individual preferences. The "marginal contribution" of an individual trader is then the amount of welfare added to society by his presence —i.e., the difference between the maximum levels of welfare that society as a whole can achieve, with that trader and without him. It can be shown (see the Appendix) that, under the conditions of Theorem 1, *the marginal contribution of a trader is almost surely the same as his "true" contribution*: If the traders are ordered at random, then the marginal contribution of a trader almost surely equals the amount added by him to the maximum welfare achievable by the coalition of traders preceding him in the random order. This justifies our reference to the marginal contribution of a trader simply as his "contribution." Our theorem then makes two assertions: First, that an allocation is competitive if its utility to each trader is precisely his contribution; and conversely, that for each allocation x, individual utility functions can be chosen (consistent with the preferences), so that the contribution of each trader is precisely the utility of the bundle assigned to him by x.

In brief, *an allocation is competitive if and only if for some choice of the individual utility functions, it yields to each trader precisely his contribution to the welfare of society.*

Among the precursors of this result are results of Shapley [22], Champsaur [6], and Aumann and Shapley [5, Theorem J and Proposition 31.7]. The literature will be more thoroughly discussed in Section 12.

The next two sections are devoted to a review of the Shapley value, first for finite games and then for games with a continuum of players. Section 4 is devoted to

---

[3] Counter-examples may be found toward the end of Section 11.

the relevant definitions of differentiability and smoothness. In Section 5 the market model is presented, "value allocation" is defined, and the main results are stated. In Section 6 we again discuss the intuitive content of the main theorem, in a more leisurely manner than above. Section 7 is devoted to an illustrative example, and Section 8 to a suggestive but unrigorous demonstration of the main results, based on the calculus of infinitesimals. In Section 9 we discuss the intuitive meaning of the differentiability and smoothness conditions, and Section 10 is devoted to a similar discussion of the uniformity conditions. In Section 11 we discuss the extraordinary relation, implicit in our results, between the basically cardinal concept of Shapley value and the ordinal concept of competitive equilibrium. Section 12 is devoted to a discussion of the literature, and Sections 13 through 15 to proofs. An appendix clarifies the precise relationship between the marginal and "true" contributions of a trader.

Readers who wish to skim through the paper should perhaps read only Sections 5 and 8, and the intuitive discussion immediately following the statement of the Value Equivalence Theorem in Section 1.

2. THE VALUE IN FINITE GAMES

Let N be a finite set; call the members of N *players* and the subsets of N *coalitions*. A *game on N* is a function v that associates with each coalition S a real number $v(S)$ (the *worth* of S), such that $v(\emptyset) = 0$. A *payoff vector on N* is a measure on the subsets of N. Intuitively, a payoff vector is simply a function that assigns a real number (the payoff) to each player; such functions are in an obvious one-to-one correspondence with the measures on N, and for reasons that are both technical and conceptual,[4] it is convenient to treat a payoff vector as a measure.

A *null player* in a game v is a player i such that $v(S \cup \{i\}) = v(S)$ for all coalitions S. Players i and j are called *substitutes* if $v(S \cup \{i\}) = v(S \cup \{j\})$ whenever S contains neither i nor j. For a fixed player set N, a *value* is a function φ that associates with each game v a payoff vector φv satisfying the following conditions:

CONDITION 2.1 (Additivity): $\varphi(v + w) = \varphi v + \varphi w$.

CONDITION 2.2 (Symmetry): $(\varphi v)(\{i\}) = (\varphi v)(\{j\})$ *whenever i and j are substitutes*.

CONDITION 2.3 (Efficiency): $(\varphi v)(N) = v(N)$.

CONDITION 2.4 (Null player condition): $(\varphi v)(\{i\}) = 0$ *whenever i is a null player*.

PROPOSITION 2.1: *For each finite player set N there is one and only one value φ; it is given by the formula $(\varphi v)(\{i\}) = E(v(S_i \cup \{i\}) - v(S_i))$, where S_i is the set of all players preceding i in a random order on N, and E is the expectation operator when all $|N|!$ such orders are assigned equal probability.*

---

[4] We wish to think of the payoff to a coalition, not only to an individual.

For a proof of Proposition 2.1 that is even simpler than the original proof [21], see [5, Appendix A].

3. THE VALUE IN CONTINUOUS GAMES

In this section we take for the underlying player space a measurable space $(T, \mathscr{C})$, i.e., a set T together with a σ-field $\mathscr{C}$ of subsets of T. The members of $\mathscr{C}$ are *coalitions*, but one should not think of the points t of T as individual players; rather, one should think of an individual player as an infinitesimal subset dt of T (cf. [5, Section 29]).

A *game on* $(T, \mathscr{C})$ is a function v from $\mathscr{C}$ to the real numbers such that $v(\varnothing) = 0$. There are various ways of extending the definition of value from the finite games studied in Section 2 to the more general situation we have here; they are studied in detail in [5]. Here we will adopt a definition due to Kannai [15] (see also [5, Section 17]).

Let v be a game and S a coalition; we wish to define $(\varphi v)(S)$. Intuitively, one proceeds by dividing both S and $T \backslash S$ into a large but finite number of "small" sets W_i, say n in all (see Figure 1). If one considers each of these small sets as an individual player, then one gets a finite game by restricting v to unions of the W_i. In this finite game the coalition S—or more precisely the coalition of those W_i whose union is S—has a value, which we denote $\varphi_n(S)$. Now if we let $n \to \infty$ and the W_i shrink, $\varphi_n(S)$ may or may not tend to a limit. If it does, and if this limit is independent of the various choices that must be made, then we denote the limit $(\varphi v)(S)$, and call it the "asymptotic value of S."

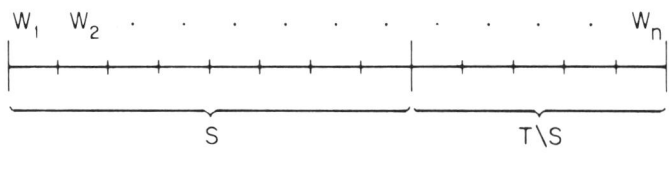

FIGURE 1

Formally, let $\Pi_1, \Pi_2, \ldots$ be a sequence of partitions of T into measurable sets (i.e., members of $\mathscr{C}$), such that each Π_m refines the previous partition Π_{m-1}. Assume moreover that the sequence is *separating*, i.e., that if s and t are distinct points in T, then for m sufficiently large, s and t are in different members of Π_m; this is a precise expression of the idea that the sets in the partitions shrink. For example, if Π_m is the partition of the unit interval $[0, 1]$ into the subintervals $[0, 1/2^m], (1/2^m, 2/2^m], \ldots, (1 - 1/2^m, 1]$, then the sequence $\{\Pi_m\}$ is separating. For each m, let v_m be the finite game on Π_m defined by $v_m(\Xi) = v(\bigcup_{W \in \Xi} W)$, and let φv_m be its value. Let $\Pi_1 = \{S, T \backslash S\}$, and for each m, let $S_m = \{W \in \Pi_m : W \subset S\}$; S_m is the coalition in the finite game v_m corresponding to S in the original game v. Now let $m \to \infty$; if $(\varphi v_m)(S_m)$ has a limit, and if this limit is independent of the choice of the sequence $(\Pi_1, \Pi_2, \ldots)$ (when chosen in accordance with the above conditions), then the limit is denoted $(\varphi v)(S)$. If, moreover, this is the case of all

$S \in \mathscr{C}$, then φv is called the *asymptotic value* of v. Note that the asymptotic value, if it exists, is a finitely additive measure on $(T, \mathscr{C})$. Note also that when T is finite, the asymptotic value coincides with the finite value.

The asymptotic value obeys conditions analogous to Conditions (2.1) through (2.4); see [5, Sections 2 and 17, in particular Theorem F]. Specifically, we note the *efficiency condition*:

(3.1) $\quad (\varphi v)(T) = v(T).$

4. DIFFERENTIABILITY, SMOOTHNESS, AND UNIFORM SMOOTHNESS

We will be working in a Euclidean space E^n. Superscripts will be used to denote coordinates. For x and y in E^n we write $x > y$ when $x^i > y^i$ for all i, and $x \geqslant y$ when $x^i \geqslant y^i$ for all i. The symbol $x \cdot y$ will denote the scalar product $\Sigma_i x^i y^i$. We will also use standard matrix notation, so that $x \cdot y = xy^*$, where both x and y are considered row vectors and $*$ denotes "transpose." The symbol 0 denotes the origin of E^n as well as the number 0; no confusion will result. The non-negative orthant $\{x \in E^n : x \geqslant 0\}$ in E^n is denoted Ω.

Unless otherwise specified, the word "function" will mean "real valued function."

A function u on Ω is *continuously differentiable* (on Ω) if it can be extended to an open neighborhood (in E^n) of Ω on which it is continuously differentiable. An equivalent definition [29] is that u is continuous on Ω and continuously differentiable in the interior of Ω, and each of the partial derivatives of u can be continuously extended from the interior of Ω to Ω. In that case, the row vector of the partial derivatives of u is denoted u'; on the boundary of Ω, u' is defined to be the unique continuous extension of u' in the interior. A function on Ω with values in a finite dimensional Euclidean space is *continuously differentiable* (C^1 for short) if each of its components is continuously differentiable. A real-valued function u on Ω is *twice continuously differentiable* (C^2 for short) if both u and u' are continuously differentiable; the matrix of second derivatives is denoted u''.

A function u on Ω is called *quasi-concave* if for all $\alpha \in [0, 1]$ and all x and y in Ω, $u(\alpha x + (1 - \alpha)y) \geqslant \min(u(x), u(y))$.

If u is a C^2 function on Ω, define

(4.1) $\quad u^H = \begin{vmatrix} u'' & u'^* \\ u' & 0 \end{vmatrix},$

where $*$ denotes "transpose." This is the *bordered Hessian* of u, i.e., the $(n + 1) \times (n + 1)$ determinant in which the $n \times n$ matrix u'' is bordered by the row vector u', by the column vector u'^*, and by the number 0. When $u' > 0$, define

(4.2) $\quad u^c = \dfrac{-u^H}{\|u'\|_2^{n+1}},$

where $\| \; \|_2$ is the Euclidean norm. It may be verified [10] that u^c depends only on the level (or indifference) surfaces of u, and not on u itself. The *Gaussian curvature* of the level surface of u at x is then defined as $u^c(x)$.

When $n = 2$, then $u^c(x) = 1/r$, where r is the radius of the circle that "best fits" the level curve of u at x. Thus if $u^c(x) = 0$, then $r = \infty$, so that the "circle" that best fits the level curve is a straight line. Further discussion of the Gaussian curvature can be found in Section 9 and also in [10].

Let $(T, \mathscr{C})$ be a measurable space. For each $t \in T$, let w_t be a function from Ω to some Euclidean space E^l (possibly $l = 1$). The w_t are *uniformly bounded* if there is a point b in E^l such that $w_t(x) < b$ for all t in T and x in Ω. They are *uniformly bounded in compact sets* if for each compact subset C of Ω, there is a point b in E^l such that $w_t(x) < b$ for all t in T and x in C. They are *uniformly positive (or bounded away from zero) in compact sets* if for each compact subset C of Ω, there is a point $b > 0$ in E^l such that $w_t(x) > b$ for all t in T and x in C. They are *simultaneously measurable* if $w_t(x)$ is measurable as a function of (t, x), i.e., measurable in the product field $\mathscr{C} \times \mathscr{B}$, where $\mathscr{B}$ is the σ-field of Borel sets in Ω.

A family $\{u_t\}$ of functions on Ω is called *bounded differentiable* if the u_t are simultaneously measurable, uniformly bounded, and continuously differentiable on Ω, and the derivatives u'_t are, in compact sets, uniformly bounded and uniformly positive. Finally, $\{u_t\}$ will be called *uniformly smooth* if it is bounded differentiable and the following conditions are satisfied:

CONDITION 4.1: *The u_t are quasiconcave and twice continuously differentiable on Ω.*

CONDITION 4.2: *The Gaussian curvatures u^c_t are uniformly positive in compact sets.*

CONDITION 4.3: *The second derivatives u''_t are uniformly bounded in compact sets.*

5. DESCRIPTION OF MODEL AND STATEMENT OF RESULTS

The market model is basically the same as in [1]. It consists of: (i) a measurable space $(T, \mathscr{C})$, called the *trader space*, together with a σ-additive, non-negative measure μ on $\mathscr{C}$, with $\mu(T) < \infty$; (ii) the non-negative orthant Ω—called the consumption set—of a Euclidean space E^n, whose dimension n represents the number of different goods being traded in the market; (iii) a measurable function a from T to Ω called the *initial allocation*; and (iv) for each t in T, a relation $\succ_t$ on Ω, called the *preference relation* of t. The system consisting of these four items is called a *market*. The market is called *non-atomic* if μ is non-atomic.

We will assume that the measurable space $(T, \mathscr{C})$ is *standard*, i.e., that it is finite, denumerable, or isomorphic[5] to the unit interval $[0, 1]$ with the Borel sets. This assumption is less restrictive than it sounds; any Borel subset of any Euclidean space (or indeed, of any complete separable metric space) has this property.

---

[5] Two measurable spaces are *isomorphic* if there is a one-to-one transformation from one onto the other that preserves measurability in both directions.

If f is a function on T, we often write $\int_S f$ instead of $\int_S f(t)\mu(dt)$, and $\int f$ instead of $\int_T f$. We will make the following assumptions:

ASSUMPTION 5.1: $\int a > 0$.

ASSUMPTION 5.2: *a is uniformly bounded.*[6]

Assumption 5.1 is fairly standard; it means that each commodity is actually present in the market, though not all traders need initially hold a positive amount of it. Assumption 5.2 is less familiar; actually we make it only as a kind of normalization, and at the expense of certain complications below, could dispense with it. This matter will be discussed further in Section 10.

In the non-atomic case, it is best to view an individual trader as an "infinitesimal subset" dt of T, rather than as a single point in T. Individual bundles are also infinitesimal; thus the initial bundles of the individual trader dt is $a(t)\mu(dt)$, whereas the point $a(t)$ itself is to be thought of as dt's initial bundle *density* (initial bundle of dt divided by his measure). Similarly, $x \succ_t y$ means that dt prefers $x\mu(dt)$ to $y\mu(dt)$, i.e., he prefers the bundle density x to the bundle density y. This point of view is expounded at some length in [5, Section 29]; see also [26, Footnote 12].

A *utility* for a relation $\succ$ on Ω is a real-valued function u on Ω such that $u(x) > u(y)$ if and only if $x \succ y$. The preferences $\succ_t$ are called *uniformly smooth* if there exists a uniformly smooth family of utilities for them. An intuitive discussion of uniform smoothness will be given in Section 10.

Let $U = \{u_t\}$ be a bounded differentiable[7] family of utilities for the preferences. Define a game v_U on $(T, \mathscr{C})$ by

(5.1) $\quad v_U(S) = \max \left\{ \int_S u_t(x(t))\mu(dt) : \int_S x = \int_S a \text{ and } x(t) \in \Omega \text{ for all } t \text{ in } S \right\}$

Intuitively, $v_U(S)$ is the maximum total utility that S can guarantee to itself by redistributing its total initial bundle $\int_S a$ among its members. We remark that the maximum in the definition of v_U is attained [4], and the game v_U has an asymptotic value φv_U [5, Proposition 31.7].

An *allocation* is a measurable function x from T to Ω such that $\int x = \int a$. It is called a *value allocation* if there exists a bounded differentiable family U of utilities for the preferences such that for all t,

(5.2) $\quad (\varphi v_U)(dt) = u_t(x(t))\mu(dt)$,

i.e., such that for all S,

(5.3) $\quad (\varphi v_U)(S) = \int_S u_t(x(t))\mu(dt)$.

In words, (5.2) says that x yields each trader precisely the utility that he is assigned by the value. An allocation x is called *competitive* if there exists a p in Ω with

---
[6] That is, there is a point b in Ω such that $a(t) < b$ for all t.
[7] See Section 4.

$p \neq 0$ (called the vector of *prices*) such that for almost all t in T, $x(t)$ is maximal with respect to $\succ_t$ in the *budget set* $\boldsymbol{B}_p(t) = \{x \in \Omega : p \cdot x \leqq p \cdot \boldsymbol{a}(t)\}$.

This completes the presentation of the basic definitions. The first result is:

PROPOSITION 5.1: *In a non-atomic market, every value allocation is competitive.*

Conversely, we have:

PROPOSITION 5.2: *In a non-atomic market, assume that either* (i) *the preferences are uniformly smooth, or* (ii) *there exists a bounded differentiable family of concave utilities for the preferences. Then every competitive allocation is a value allocation.*

Proposition 5.2 requires either smoothness or concavity; for Proposition 5.1 neither requirement is needed. But Proposition 5.1 does require differentiability, which is implicit in the definition of "value allocation".

The Value Equivalence Theorem—stated in the introduction—is, of course, an immediate consequence of Propositions 5.1 and 5.2.

We mention also that if in the definition of value allocation we replace "asymptotic value" by "mixing value" [5, Section 14, p. 115], then the Value Equivalence Theorem and Propositions 5.1 and 5.2 remain true without any change. The proofs are as in Sections 13 through 15; one merely has to replace the word "asymptotic" by the word "mixing".

6. INTERPRETATION

The simplest interpretation is essentially the one sketched in the introduction (after the statement of the Value Equivalence Theorem). It involves an additively separable social welfare function, defined by a family $U = \{u_t\}$ of utilities. It is best to think of trader dt's utility for the bundle $x\mu(dt)$ as $u_t(x)\mu(dt)$. Then $\int_T u_t(\boldsymbol{x}(t))\mu(dt)$ is society's level of welfare under the allocation $\boldsymbol{x}$; and $v_U(S)$ is the maximum level of welfare that the coalition S can achieve by its own efforts. Thus $v_U(S)$ measures the power or bargaining strength of the coalition S; and *given the bargaining strengths*, φv_U may be considered the "equitable" payoff distribution. Hence a value allocation (see (5.2)) is one that yields each trader precisely his "equity" in the market; i.e., an allocation $\boldsymbol{x}$ whose utility $u_t(\boldsymbol{x}(t))\mu(dt)$ to each trader dt is precisely $(\varphi v_U)(dt)$.

Here we used the word "equity" in describing the value, whereas in the introduction we referred to the traders' "contributions". The assertion that the value constitutes an equitable payoff distribution is based on its axiomatic definition (see Conditions 2.1 through 2.4). That it also may be thought of in terms of the players' contributions is a consequence of Proposition 2.1 (see also Step 5 of Section 7, and the Appendix). These two interpretations are simply two sides of the same coin—two ways of looking at the same concept.

It should be noted that for a given family U of utilities, there will in general[8] be at most one such x, and usually none; to obtain the set of all value allocations, one must allow U to vary. An illustrative example will be analyzed in Section 7.

There is a certain conceptual analogy between the process of determining a value allocation and that of determining a competitive allocation. To find a competitive allocation, one assigns prices to the commodities. Given the initial allocation, these prices generate individual demands. If the demands thus generated are feasible—i.e., if they can be obtained by reallocating the total initial supply—then we have a competitive allocation.

To find a value allocation, one assigns a utility function to each trader. Given the initial allocation, these utility functions generate a payoff distribution—the value. If this payoff distribution is feasible—i.e., if it can be obtained by reallocating the total initial supply—then we have a value allocation. The utility functions here play a role similar to that of the prices in the case of competitive allocations. Prices enable us to evaluate bundles of goods; utility functions enable us to evaluate coalitions of players. Workers in general equilibrium theory have in the past been loath to compare utilities; but there is some reason to believe that such comparisons play a significant role in economic activity. Workers in areas such as growth, income distribution, and taxation have recognized this and have been quite willing to assume social welfare functions and the utility comparisons they imply. The fact that there are certain conceptual difficulties involved in such comparisons does not permit us, ostrich-like, to ignore them. The analysis outlined here suggests that society may be generating utility comparisons by means of a mechanism similar to the familiar supply and demand mechanism that generates prices; more bluntly, that "social welfare" is a function of power rather than an ethical construct.

The above interpretation of the Value Equivalence Theorem—i.e., in terms of social welfare functions—is open to a certain objection; though not crippling, it bears discussion. The games v discussed in Sections 2 and 3 are usually interpreted as "side payment" or "transferable utility" (TU) games. Essentially, this means that the payoffs are assumed to be in a single, desirable commodity; that $v(S)$ is the *total* payoff that S can get for its members; and that S can get for its members *any* payoff totalling $v(S)$, as is for example the case when the players can freely transfer the single good among each other. The markets defined in Section 6, on the other hand, cannot be considered TU games: the payoff is in terms of the personal utilities u_t, which are not "commodities", and are certainly not transferable. How, then, can TU theory be relevant to these markets?

Naturally, one can answer that though the TU view of a game v is the usual one, it is not the only possible one, and that in the current application the social welfare view is preferred. But it is also possible to justify the notion of value allocation directly in terms of the TU view. One can think of markets in terms of production, as follows: there are n raw materials, not in themselves consumable, and one desirable consumer commodity, which for simplicity we shall call *food*. Each

---

[8] For example, whenever the u_t are strictly concave.

trader dt starts out with a bundle $a(t)\mu(dt)$ of raw materials, and with no food. The traders may exchange raw materials. If, after the exchange, dt holds a bundle $x\mu(dt)$ of raw materials, he can produce from it an amount $u_t(x)\mu(dt)$ of food. Transfers of food are forbidden.

A little reflection will convince the reader that this is conceptually entirely equivalent to the usual view of a market.

Since food transfers are forbidden, this cannot be viewed as a TU game. But suppose now that food transfers are legalized.[9] Then each coalition S can produce an amount $v_U(S)$ of food, which it can divide among its members in any way it pleases. Therefore the situation is well represented by the TU game v_U.

In general, the value φv_U of v_U is not achievable without actual transfers of food (cf. Section 7). Specifically, we will have

(6.1) $\quad (\varphi v_U)(dt) = u_t(x(t))\mu(dt) + \zeta(dt),$

where x is an allocation of resources that achieves $v_U(T)$, i.e., such that $v_U(T) = \int_T u_t(x(t))\mu(dt)$, and $\zeta(dt)$ is the net intake of food by dt resulting from food transfers (rather than from his own production). These transfers clearly sum to 0, i.e., $\zeta(T) = \int_T \zeta(dt) = 0$; but for individual traders they will usually not vanish, i.e., the measure ζ will not vanish identically.

Suppose now that ζ does happen to vanish identically; then (6.1) reduces to (5.2), i.e., x is a value allocation. This means that the value φv_U is achievable directly by means of the resource allocation x, without any transfers of food. In this case x is called a *value allocation*.

If x is a value allocation, one can think of the traders going to market under the impression that food transfers will be permitted. This leads them to trade so as to arrive at the value of the associated game v_U; i.e., they will reach the allocation x, but will undertake no subsequent food transfers. After going home, they hear that food transfers have been forbidden. Fortunately, nobody wanted to transfer food anyway. Thus, to arrive at a value allocation, one must only think about food transfers—imagine them—they don't actually have to be possible. The "transferable utility" does not really have to be transferable, because in the end nobody wants to transfer it.

Basically, this argument uses what is sometimes called the principle of independence of irrelevant alternatives (cf. [18]). We agree that when food may be transferred, then among all possible outcomes, an outcome that is in some sense best is to give to player dt the allocation $x(t)\mu(dt)$. But this allocation does not require the transfer of food, and so a fortiori it should be best when such transfers are forbidden.

We close this section with a few words about the notion of "equity" and its connection with the Shapley value. We stated above that the value of a game v is calculated by taking into account the bargaining strengths of players and coalitions, as reflected in the worths $v(S)$. Thus the more a player can get for himself, acting alone or in concert with other players, the more the value will

---

[9] Economies of this kind (i.e., with legalized food transfers) are called *TU markets*.

assign to him. Some readers may feel that this smacks more of power politics—of "might makes right"—than of what is commonly considered equity or fairness. Wouldn't it be more appropriate to adopt a point of view similar to that of Rawls [20] or of some of the recent literature on income redistribution,[10] which ignore the power structure of society and treat people in a basically egalitarian way?

We think not. Criteria of equity that do not take power considerations into account have a pleasing air of symmetry and abstract perfection; they appeal to our philosophical, ethical senses. But what would make society adopt such criteria? And even if adopted, what chance do they have of surviving the attacks of pressure groups, large and small? For better or for worse, economic realities are dictated not by ethical considerations, but by the exercise of economic power. We are not suggesting that the more egalitarian considerations are not worthy of study; but we do think that the kind of criterion exemplified by the Shapley value—which some readers might prefer to call a "reasonable compromise" rather than an "equitable solution"—is more relevant to economic reality.

7. AN EXAMPLE

This section is devoted to a trivial but nonetheless instructive example. Consider an economy in which there is only one commodity, which is desired by all traders (i.e., for all t, $x \succ_t y$ if and only if $x > y$). Let the initial bundle density be 1 for half the traders (say the set S_1), and 3 for the other half (say $S_3 = T \setminus S_1$). Of course one would not expect any trading to take place in such a situation, but let us follow through our formal analysis and look for value allocations. Let u be any strictly concave function on the nonnegative reals that is bounded, differentiable, and strictly increasing (e.g., $1 - 1/(x + 1)$). Set $u_t = u$ for all t; for simplicity, let $\mu(T) = 1$. Then $v_U(S_1) = (1/2)u(1)$, $v_U(S_3) = (1/2)u(3)$, and $v_U(T) = u(2)$. Since u is strictly concave, it follows that $v_U(T) > v_U(S_1) + v_U(S_3)$; in other words, trading is socially beneficial. To obtain the full benefit from trading, society must choose an allocation that actually yields a total utility of $v_U(T) = u(2)$. The only such allocation is $x_0(t) \equiv 2$. But this cannot be a value allocation, as it is highly inequitable: it does not differentiate at all between the players in S_1 and S_3, and in fact assigns to traders in S_3 bundles smaller than their initial bundles! Hence there is no value allocation at all associated with this choice of the u_t—no allocation that is both socially optimal, and divides the benefits from trading equitably among the traders.

If one calculates[11] φv_U, one finds that it assigns densities $u(2) - u'(2)$ and $u(2) + u'(2)$ to the players in S_1 and S_3, respectively[12] (see Figure 2). But to achieve such a payoff distribution one would have to start by assigning a density of 2 units of the single commodity to all traders, and then one would have to "transfer utility"

---

[10] See, for example, Sheshinski [25] and the papers quoted there.
[11] Step 5 of Section 8 or Lemma 13.2.
[12] When $u(x) = 1 - 1/(x + 1)$, then $u(2) = u'(2) = 5/9 = u(1) + 1/18$, and $u(2) + u'(2) = 7/9 = u(3) + 1/36$. Thus the value dictates a gain in utility (over that of the initial bundle) for all traders, but a smaller gain for S_3 than for S_1. "Equitable" is not necessarily "equal".

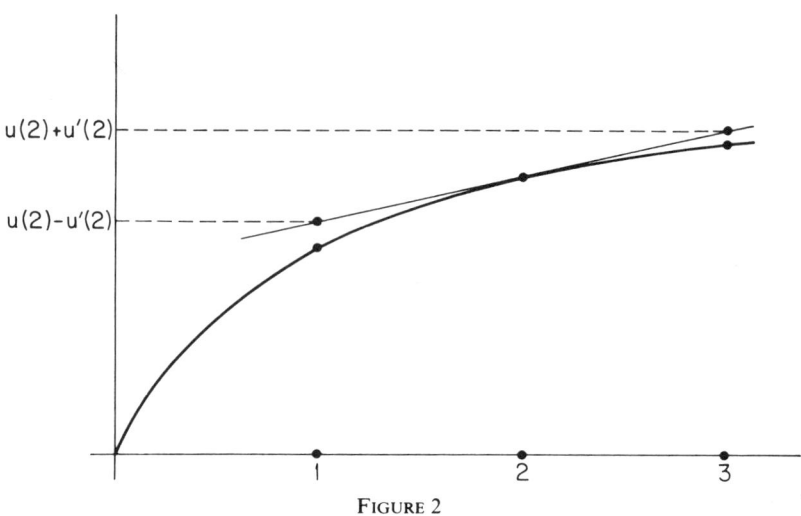

FIGURE 2

in the amount of $u'(2)$ from S_3 to S_1—an impossible (indeed meaningless) proceeding.

Since this choice of U does not lead to a value allocation, the question arises, which choice does? A trivial answer is provided by taking $u_t = w$ for all t, where $w(x) = x$ in an interval including $[1, 3]$. In that case, there is no social benefit from trading, and so the initial allocation is also a (in fact, *the*) value allocation.

But there is also a more instructive answer. Consider the u_t in our first example, i.e., $u_t(x) = 1 - 1/(x + 1)$. If we multiply these u_t by some $\alpha > 1$ for $t \in S_3$, and leave them unchanged for $t \in S_1$, then we are assigning a higher social benefit to consumption by S_3 than by S_1—consumption by S_3 is considered "more valuable".[13] As a result, maximum social welfare will be attained by assigning a higher commodity density to S_3 than to S_1. In particular, if α is sufficiently high, maximum social welfare will be attained by transferring goods from S_1 to S_3. We have already seen that if $\alpha = 1$, then maximum social welfare is attained by transferring from S_3 to S_1. For an appropriate intermediate α, therefore, maximum social welfare is attained precisely at the initial allocation, i.e., without any transfers. Of course this does not necessarily imply that the value is then also achieved without any transfers, but it may be verified that that is in fact the case.

When there is more than one commodity, value allocations will in general involve transfers of commodities; but the utilities must be chosen so that no "transfer of utility" is required.

8. AN UNRIGOROUS DEMONSTRATION OF THE RESULTS

The proof of our results relies heavily on the theory of non-atomic games developed in Aumann-Shapley [5]; when spelled out, it is long and involved. On the other hand, it is comparatively easy to "demonstrate" the theorem in

---

[13] Because the members of S_3 are more powerful, in that their initial bundles are greater.

terms of the calculus of infinitesimals and one or two other rough conceptual tools; that is what we shall do in this section. Such a demonstration, though unrigorous, gives insight into the situation and shows why one would expect the theorem to hold.

In Steps 1 through 5 of the demonstration, $U = \{u_t\}$ is a fixed bounded differentiable family of utilities, and x is an allocation at which $v_U(T)$ is achieved, i.e., such that

$$(8.1) \quad v_U(T) = \int_T u_t(x(t))\mu(dt).$$

The universal quantifier ("for all t") is to be understood in each one of these steps.

STEP 1: $(x(t), u_t(x(t)))$ *is on the boundary of the convex hull of the graph of* u_t.

When $n = 1$, this asserts that the situation cannot be as pictured in Figure 3. Indeed, if it were, we could split dt into two players, dt_1 and dt_2, and give them densities x_1 and x_2, respectively. The total utility that dt would get in this way would be $y\mu(dt)$ rather than $u_t(x(t))\mu(dt)$. Thus dt, and so also T as a whole, would

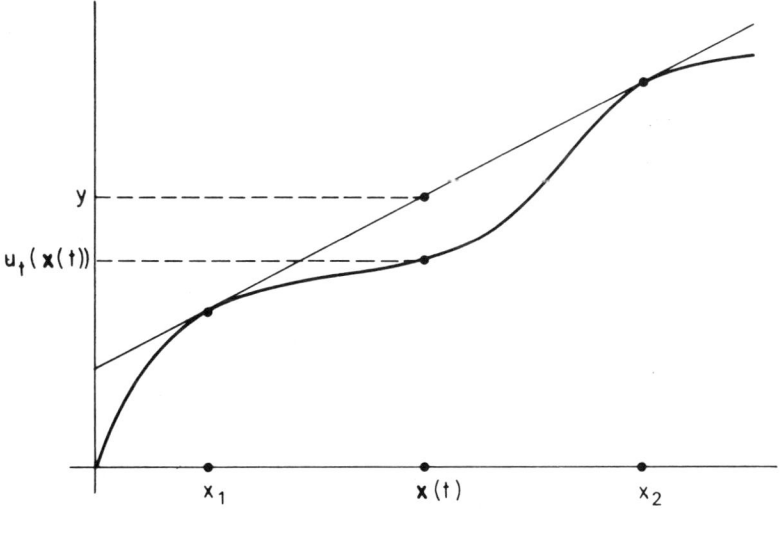

FIGURE 3

benefit from the split; this contradicts the efficiency of the value. The argument for arbitrary n is similar, except that it may be necessary to split dt into n traders $dt_1, \ldots, dt_n$. We are of course making strong use of the non-atomicity of the market in this step.

STEP 2: $u'_t(x(t)) = u'_s(x(s))$ *for all s and t.*

Since the demonstration is in any case unrigorous, we assume without twinges of conscience that we are in the interior,[14] i.e., that $x(t) > 0$, $x(s) > 0$. Let $n = 1$. If the assertion were false, say $u'_t(x(t)) > u'_s(x(s))$, then society could gain by moving $x(t)$ to the right and $x(s)$ to the left, in contradiction to the efficiency of the value (see Figure 4). The argument is similar for general n.

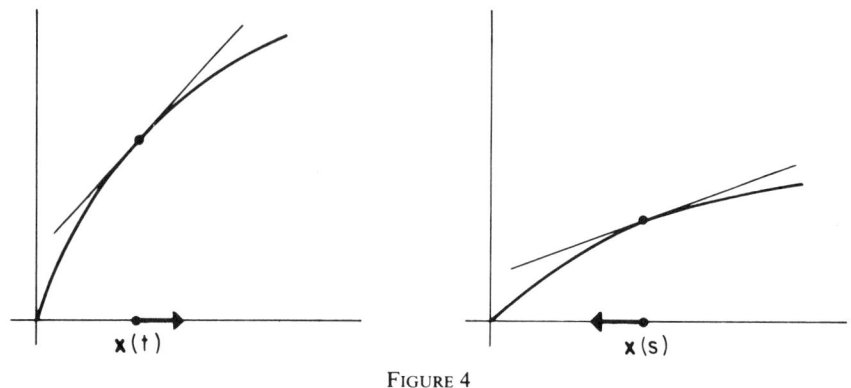

FIGURE 4

Set $p = u'_t(x(t))$; by Step 3, p does not depend on t.

STEP 3: *For all x in Ω, $u_t(x(t)) - u_t(x) \geqq p \cdot (x(t) - x)$.*

From Step 2 it follows that the hyperplane with slope p through the graph of u_t at $x = x(t)$ is tangent to the graph there; from Step 1 that this hyperplane supports the graph. This is precisely what is asserted here (see Figure 5).

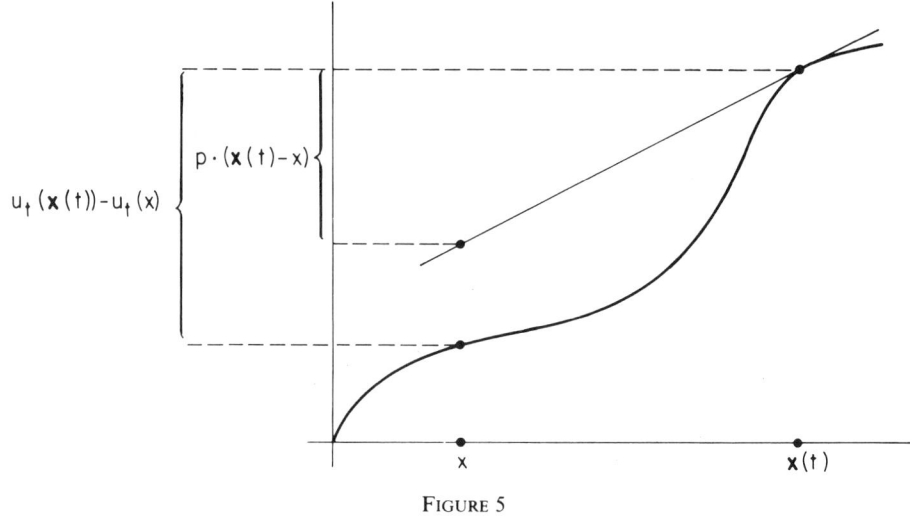

FIGURE 5

[14] Of course no such assumption is used in the rigorous proof. Even the statement of the step is false without the "interior" assumption, and must be replaced by a more complicated statement.

STEP 4: $x(t)$ is maximal with respect to the preference relation $\succ_t$ in $\{x : p \cdot x \leqslant p \cdot x(t)\}$.

Indeed, if $p \cdot x \leqslant p \cdot x(t)$, then from Step 3 it follows that $u_t(x) \leqslant u_t(x(t))$.

Step 4 asserts that the prices p are *efficiency prices*. When the u_t are concave, Steps 2, 3, and 4 are well-known even in the finite case; the p^i are the "shadow prices" in the maximization that defines $v_U(T)$. The non-atomicity enables the establishment of Step 1, and so the extension to non-concave u_t.

The following step is crucial:

STEP 5: $(\varphi v_U)(dt) = (u_t(x(t)) + p \cdot (a(t) - x(t)))\mu(dt)$.

To demonstrate this, let us first calculate the marginal contribution $v_U(T) - v_U(T \backslash dt)$ of dt to society. Since $v_U(T)$ is achieved at x, trader dt will, after joining the rest of society, consume $x(t)\mu(dt)$, whereas he has contributed $a(t)\mu(dt)$. Therefore his net contribution of commodities to society is $(a(t) - x(t))\mu(dt)$. This is distributed among all other traders, each of whom will receive only an infinitesimal portion of this (in itself infinitesimal) bundle. Therefore the bundle *density* of each trader will change by only an infinitesimal amount, and so by Step 2, the utility of each trader will change by p times the increment in his bundle. Hence the total increment in utility to all members of T due to the distribution of $(a(t) - x(t))\mu(dt)$ among them is $p \cdot (a(t) - x(t))\mu(dt)$. But over and above his contribution of goods to society, dt himself will get utility out of his consumption of $x(t)\mu(dt)$, and this additional utility, which is $u_t(x(t))\mu(dt)$, also must be taken into account when calculating the marginal contribution of dt. We conclude that

(8.2) $\quad v_U(T) - v_U(T \backslash dt) = (u_t(x(t)) + p \cdot (a(t) - x(t)))\mu(dt)$.

Now let S be the set of traders up to dt in a random ordering of all the traders. Because of the many traders, the randomly chosen coalition S will almost surely be an almost perfect sample of the population of all traders (insofar as the distribution of utilities and initial bundles is concerned). Hence the contribution $v_U(S) - v_U(S \backslash dt)$ of dt to S is almost surely the same as his marginal contribution $v_U(T) - v_U(T \backslash dt)$ to all of society. Hence the expectation of his contribution to S—which is precisely $(\varphi v_U)(dt)$—is also the same as his marginal contribution. Step 5 thus follows from (8.2).

STEP 6: *Every value allocation is competitive.*

A value allocation x necessarily satisfies (8.1), since by (5.3) and the efficiency of the value (3.1) we have $\int_T u_t(x(t))\mu(dt) = (\varphi v_U)(T) = v_U(T)$. From (5.4) and Step 5 we then obtain $p \cdot (a(t) - x(t)) = 0$ for all t. Thus $p \cdot a(t) = p \cdot x(t)$, and so from Step 4 it follows that (p, x) is a competitive equilibrium.

Step 6 is identical with Proposition 5.1, and thus the demonstration of the latter is complete.

STEP 7: *If the preferences have concave utilities, then every competitive allocation is a value allocation.*

Let $\{w_t\}$ be a family of concave utilities, and let (p, x) be a competitive equilibrium. From the concavity it follows that for all t and x, $w_t(\mathbf{x}(t)) - w_t(x) \geqq w'_t(\mathbf{x}(t)) \cdot (\mathbf{x}(t) - x)$, and hence, as in Step 4 above, that the hyperplane through $\mathbf{x}(t)$ orthogonal to $w'_t(\mathbf{x}(t))$ supports the indifference surface of u_t through $\mathbf{x}(t)$. Hence $w'_t(\mathbf{x}(t))$ must be proportional to p. So if we define a family $U = \{u_t\}$ of different utilities for the same preferences by multiplying each w_t by an appropriate constant (which may depend on t), then we obtain $p = u'_t(\mathbf{x}(t))$ for all t. Hence $u_t(\mathbf{x}(t)) - u_t(x) \geqq p \cdot (\mathbf{x}(t) - x)$ for all t. From this it follows that if $\mathbf{y}$ is another allocation, then $\int_T u_t(\mathbf{x}(t))\mu(dt) - \int_T u_t(\mathbf{y}(t))\mu(dt) \geqq p \cdot \int (\mathbf{x} - \mathbf{y}) = p \cdot \int (\mathbf{a} - \mathbf{a}) = 0$, so that $v_U(T)$ is achieved at $\mathbf{x}$, i.e., (8.1) holds. Hence, by Step 5, $(\varphi v_U)(dt) = (u_t(\mathbf{x}(t)) + p \cdot (\mathbf{a}(t) - \mathbf{x}(t)))\mu(dt)$. Since (p, x) is a competitive equilibrium, $p \cdot \mathbf{a}(t) = p \cdot \mathbf{x}(t)$, and hence $(\varphi v_U)(dt) = u_t(\mathbf{x}(t))\mu(dt)$. But this is precisely the definition (5.2) of value allocation, and so the proof of Step 7 is complete.

STEP 8: *Uniformly smooth preferences can be represented by concave utilities.*

For a time it was thought likely that all preferences with quasi-concave utilities could be represented also by concave utilities; that this is not the case was shown by examples of de Finetti [7] and Fenchel [12], which will be considered in more detail in Section 9. At this point we cannot "demonstrate" this step, but perhaps this matter will be clearer to the reader after he has read the discussion in Section 9.

We add that in independent work, Mas-Collel [17] and Kannai [16] have shown that all preferences having quasi-concave utilities can be approximated by preferences having concave utilities, so that Step 8 is "approximately" true even without the smoothness condition.

Proposition 5.2 is, of course, an immediate consequence of Steps 7 and 8. Since Proposition 5.1 has already been established (see Step 6), the demonstration of the Value Equivalence Theorem is complete. The most crucial steps in this demonstration are probably 5 and 8.

9. DISCUSSION OF SMOOTHNESS AND SOME COUNTER-EXAMPLES

This section is devoted to the differentiability and smoothness conditions for individuals; the uniformity conditions will be discussed in the next section.

If the utilities u_t are not differentiable, the TU game v_U will in general not have an asymptotic value,[15] so one cannot even begin to speak of a value allocation. Moreover, though φv_U may exist under certain circumstances even when the utilities are not differentiable,[16] value theory for the game v_U is in this case quite

---

[15] See [5, Example 33.9], especially the end of the discussion of that example.
[16] See Footnote 15.

undeveloped, and we would not be able to apply the basic tools from [5] that we need for the proof of our theorem.[17]

So much for the first-order differentiability. Next, we discuss smoothness, which is needed to ensure that the preferences have concave utilities;[18] the importance of this will soon become apparent.

Smoothness has three basic ingredients: quasi-concavity, positive Gaussian curvature, and positive gradients. We discuss the three ingredients separately.

Without quasi-concavity—e.g., for a preference order like that pictured in Figure 6—we will obviously not be able to find concave utilities (the preferred sets $\{x:u(x) \geq y\}$ are convex if u is concave). Moreover, it is then easy to find an example with a competitive allocation that is not a value allocation. Let all traders have the same preferences, and the same initial bundle $a(t) = a$, as pictured in

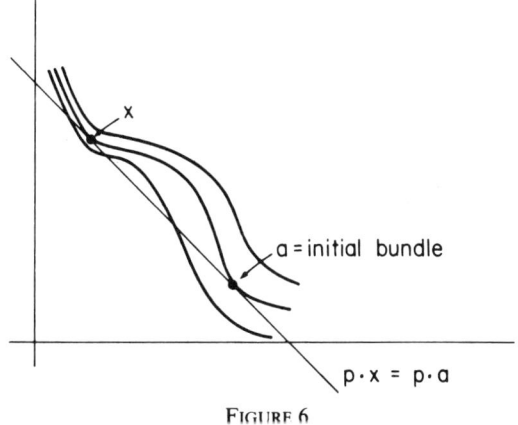

FIGURE 6

Figure 6. Then there is a unique competitive allocation, namely a; the prices will of course be the p indicated in Figure 6. But now we will argue that a cannot be a value allocation. Indeed, if it were, then there would, in particular, have to exist a bounded differentiable family $U = \{u_t\}$ such that the maximum in the definition of $v_U(T)$ is achieved when $x = a$. By the argument of Step 1 of Section 7 (which can easily be made precise), it would follow that $(a, u_t(a))$ would have to be on the boundary of the convex hull of the graph of u_t. If that were the case, there would exist a supporting hyperplane to the graph through $(a, u_t(a))$; and it would necessarily pass through $(x, u_t(x))$. Since the hyperplane supports the graph and passes through both these points on the graph, it follows that $u'_t(a) = u'_t(x)$. But we have constructed the preferences so that though $u'_t(x)$ has the same direction as $u'_t(a)$, it must have a larger magnitude (the indifference curves are closer). Hence a cannot be a value allocation. Indeed, since every value allocation is competitive, and a is the only competitive allocation, there will be no value allocation at all.

[17] Specifically, Lemma 13.2. On the basis of the available evidence, it seems reasonable to conjecture that without differentiability, something like Proposition 5.1 will still go through, but Proposition 5.2 will not.

[18] More precisely, utilities that are concave over the relevant part of Ω.

The requirement of positive Gaussian curvature says that the indifference surfaces should never be too "flat"—at any point in any direction. Specifically, for x in Ω, consider the hyperplane H_x tangent at x to the indifference surface I_x through x. Let G_x be the line orthogonal to H_x at x (i.e., the line spanned by the gradient at x of a utility function), and let L be any line in H_x through x. We may take G_x and L as coordinate axes in the plane $P_{x,L}$ spanned by them. The situation is pictured in Figure 7. Let f be the function (of distance along L) whose graph

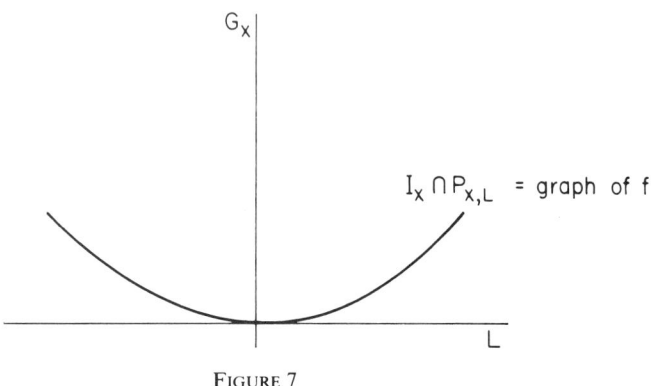

FIGURE 7

is the intersection of the indifference surface I_x with the plane $P_{x,L}$. Then the requirement of positive Gaussian curvature says[19] that the second derivative f'' be positive at the origin of $P_{x,L}$; i.e., that the indifference surface not be "too flat" at x in the direction L, and that this be true for all x and all L through x in H_x.

In economic terms, positive Gaussian curvature is equivalent[20] to finite elasticity of substitution.[21] Another familiar characterization is that the quadratic form $xu''x^*$ be positive definite when restricted to the plane $u' \cdot x = 0$.

The following example of an otherwise well-behaved preference relation with vanishing Gaussian curvature that has no concave utility is due to W. Fenchel [12]. Consider the relation pictured in Figure 8, with straight indifference curves that *are not parallel*.[22] If u is any C^1 utility for those preferences, the gradient of u at x has the same direction as the gradient at a; but it has a larger magnitude, since the indifference curves are closer at x. These facts enable us to apply an argument exactly as in our discussion of quasi-concavity, and to deduce that u cannot be concave; and moreover, that in the market M in which all the traders have the preference relations of Figure 8 and initial bundles $a(t) = a$, the competitive allocation a is not a value allocation and in fact there is no value allocation.[23]

[19] Given quasi-concavity.
[20] See Footnote 19.
[21] I am indebted for this observation to C. C. von Weiszäcker.
[22] In Figure 8 the indifference curves, when extended, all pass through a single point outside (and to the left of) Ω. But this feature is not crucial; what is needed is only that the indifference curves be straight and closer at x than at y.
[23] In this case there is more than one competitive allocation x, since $x(t)$ may be essentially anywhere on the indifference curve of a as long as $\int x = \int a$. However, the argument that *none* of the competitive allocations can be value allocations—and hence that there is no value allocation—is not essentially different from the one we had before.

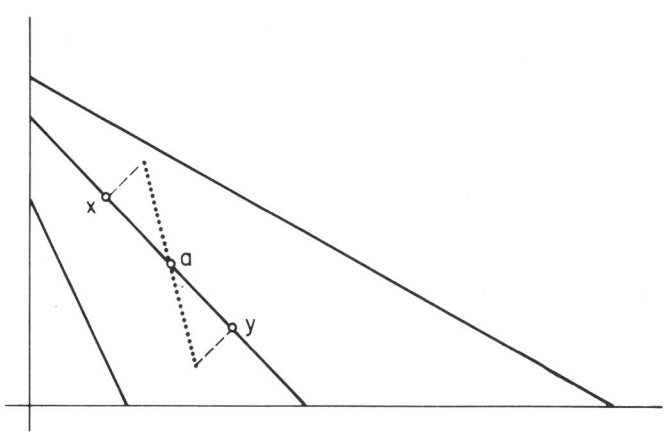

FIGURE 8

There is another argument for these facts that is of some interest. Let us think of Figure 8 as a contour map, the "height" being the differentiable utility u. The contour lines are closer at x than at y, which means the ground is steeper there. In particular, the ground rises faster along the dashed line perpendicular to the indifference curve at x than it falls along the parallel dashed line starting at y. So if one were to string a telegraph line along the dotted line and pull it taut, it would pass *over* the indifference curve containing x and y. This means that u cannot be concave. Moreover, in the market M defined in the previous paragraph, our telegraph line and an argument similar to that of Step 1 of Section 7 show that a cannot be a value allocation and in fact that there is no value allocation. Moreover, it is fairly clear that this argument goes through even under strict quasi-concavity,[24] as long as the indifference curves are still "very flat" at a, provided, of course, that the tangents to the indifference curves near a are "markedly non-parallel", e.g., that they all pass through the same point (outside of Ω).

We come now to the requirement that the gradients u' be positive. This prevents the "de Finetti phenomenon" [7]: the presence of two points x and a such that

(9.1) $\quad u(x) = u(a), \quad u'(a) = 0, \quad \text{and} \quad u'(x) > 0$

(see Figure 9). If (9.1) holds for some C^1 utility u, it holds for all such utilities; what it means is that the indifference curves are infinitely more bunched up near x than near a. Under (9.1) u cannot be concave, because that, together with $u'(a) = 0$, would imply that the maximum of u along the line perpendicular to the indifference surface (dotted in the figure) is reached at a; this is impossible if the preferences are monotone.

Consider finally a market, all of whose traders have the same utility obeying (9.1), and the same initial bundle $a(t) = a$. If the utilities are strictly quasi-concave. then the only competitive allocation is $x = a$. If a were a value allocation, there would exist a family $U = \{u_t\}$ of utilities such that $v_U(T)$ is achieved at a. This would imply[25] that $(a, u_t(a))$ is on the boundary of the convex hull of the graph

---

[24] That is, $u(\alpha x + (1 - \alpha)y) > \min(u(x), u(y))$ for $\alpha \in (0, 1)$.
[25] Step 1 of Section 8.

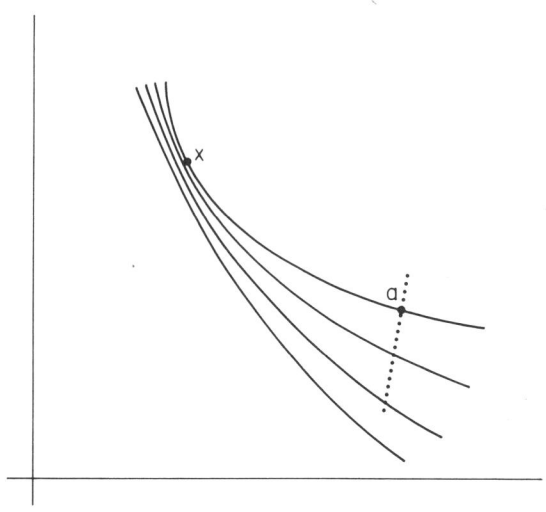

FIGURE 9

of u_t for each t. But since $u'_t(a) = 0$, this supporting hyperplane would be flat, and this would contradict the monotonicity of u_t. Thus a is not a value allocation and indeed, there is no value allocation.

10. DISCUSSION OF THE UNIFORMITY CONDITIONS

An individual preference relation is *smooth* [10] if it has a quasi-concave C^2 utility function with an everywhere strictly positive gradient and indifference curves that have everywhere positive Gaussian curvature. This implies that in each compact subset of Ω, the gradient u' is bounded and bounded away from 0, the Gaussian curvature is bounded away from 0, and the second derivative u'' is bounded. *Uniform* smoothness of a family $\{u_t\}$ requires that the bounds involved in these three statements be independent of t (though they may, of course, depend on the compact subset of Ω). It also requires[26] that the utilities u_t be bounded over all of Ω (i.e., not only compact subsets), uniformly in t.

Of these four requirements, one (about Gaussian curvatures) is stated directly in terms of preferences; but the others are stated in utility terms, and it would be desirable to see what they say about the preferences. The last one—about a uniform bound for the u_t—places no restriction on the preferences; if u_t are utilities satisfying the other restrictions but possibly not uniformly bounded, then $(u_t - u_t(0))/(1 + u_t - u_t(0))$ will satisfy *all* the requirements. The requirement relating to upper and lower positive bounds for the gradient u'_t has two aspects: that relating to the direction and that relating to the magnitude. That relating to the direction has an obvious interpretation in terms of the indifference surfaces (in the two-dimensional case, it simply says that the indifference curves have

---

[26] This requirement could be replaced by the far weaker requirement that $u_t(x) = o(\|x\|)$, integrably in t (see [4] or [5, Section 31, especially (31.2)]).

uniformly bounded slopes). To understand the requirement relating to the magnitudes of the gradients, note that whereas these magnitudes of course depend on the choice of a utility, the *ratio* between the magnitudes at two points x and y on the same indifference surface depends only on the preferences. This ratio is required to be bounded and bounded away from zero.

As for the uniform boundedness of the second derivatives, this is intuitively less clear, but it too can presumably be interpreted in terms of the preferences.

It is perhaps worthwhile to state that a sufficient condition for the family $\{u_t\}$ to be uniformly smooth is that (i) each of the u_t be smooth, (ii) the u_t be bounded over Ω, uniformly in t, and (iii) $\{u_t\}$ be contained in a compact subset of the space $\mathcal{U}$ of C^2 utility functions on Ω, endowed with the topology of C^2 uniform convergence on compact sets.[27]

Assumption 5.2, that a is uniformly bounded, is merely a normalization: if it fails, it is always possible to make it hold simply by replacing the underlying measure μ by the measure ν defined by $\nu(S) = \int_S \Sigma_{i=1}^n a_i(t)\mu(dt)$. In the new representation, all initial bundle densities would be $\leqslant (1,\ldots,1)$. But such a transformation would change also the utilities, and this would lead to complications in the statement of the uniformity conditions. Thus if we wanted to dispense with Assumption 5.2, we would have to replace the fixed compact sets in Conditions 4.2 and 4.3 by sets varying with t in such a way so that the maximal distance from the origin is a bounded multiple of $\Sigma_{i=1}^n a^i(t)$. In other words, the "size" of a trader is essentially determined by the size of his initial bundle (not by his measure), and when speaking about bounds of various kinds we must really scale them to the size of the trader. It is in order to avoid these complications that we made the normalization in Assumption 5.2.

11. THE THEORY IN A CARDINAL CONTEXT

The definition of the market model in this paper is an *ordinal* one; that is, it involves only the preferences (and initial bundles) of the traders, and not any specific choice of utility functions u_t. In the definition of "value allocation", the u_t are not fixed, but are allowed to range over *all* appropriate utility functions that represent the given preferences. Hence the concept of value allocation is also an ordinal one; i.e., whether or not a given x is a value allocation depends only on the preferences (and initial bundles) of the traders.

A purely ordinal model of this kind is, however, not the traditional setting for a value theory. In this section we will consider the Shapley value in a model in which it should feel more at home—one that specifies the *cardinal* utilities of the traders, rather than just their ordinal preferences. By this we mean a model in which the traders have an intrinsic numerical scale on which the intensity of their preferences can be measured. Thus, it becomes meaningful to say not only that x

---

[27] A sub-basis for this topology consists of sets of the form $W(w, \varepsilon, C)$, defined for each w in $\mathcal{U}$, $\varepsilon > 0$, and compact subset C of Ω to be the set of all u in $\mathcal{U}$ such that for all x in C, $|u(x) - w(x)| < \varepsilon$, $\|u'(x) - w'(x)\| < \varepsilon$, and $\|u''(x) - w''(x)\| < \varepsilon$. It is weaker than the Whitney topology discussed in [27], so that if a set is compact in the Whitney topology, it is a fortiori compact in this topology.

is preferred to y, and y to z, as in an ordinal context; one can go further and say, for example, that y is relatively close to x on the preference scale (e.g., $\frac{9}{10}$ of the way between z and x). One can lend operational significance to this kind of statement by the method of von Neumann and Morgenstern [28], i.e., by referring to lotteries with prizes x, y, and z. But this is by no means the only possible interpretation. For example, cardinal utilities can be defined by the method of Debreu [9] in the case of independent commodities, or when the set of commodities can be partitioned into independent "factors" (i.e., sets of commodities); or one can define them by the method of "just noticeable differences" (cf. Fishburn [13, p. 18ff. and p. 80ff.]). If one wishes, one can also simply accept intensity of preferences as a fact of life that is not to be ignored, and so assume that cardinal utilities are given a priori. In any case, in our application it makes no difference how the cardinal utilities are derived, as long as they are in fact given for all traders.

Formally, call two functions u and v on Ω *equivalent* if there are real numbers α and β, with $\alpha > 0$, such that $u = \alpha v + \beta$; and define a *preference scale* on Ω to be an equivalence class of functions on Ω. A member of the equivalence class is called a *cardinal utility* for that preference scale. Preference scales play the same role in cardinal contexts that preference relations play in ordinal contexts. A cardinal utility is to a preference scale what a utility is to a preference relation; i.e., a representation by means of a numerical function.

It should be particularly noted that though specifying preference scales for all traders enables each trader to compare the intensities of his own preferences, it allows no interpersonal comparisons—no comparison of the intensities of preference of different traders. Our definition of preference scale embodies the familiar statement that "cardinal utilities are determined only up to an additive and a positive multiplicative constant". Thus if we are given cardinal utilities for all traders we can multiply them by arbitrary constants that are different for the different traders, without changing the preference scales, so that no consistent comparisons can be based on such utilities.

To stress the ordinal character of the markets and utilities defined in Section 5, we will henceforth in this section refer to them as *ordinal markets* and *ordinal utilities*.

Define a *cardinal market* in exactly the same way as an ordinal market, except that the preference relations $\succ_t$ are replaced by preference scales $\wp_t$. Adopt Assumptions 5.1 and 5.2; define *non-atomic*, *allocation*, and *competitive* allocation as in an ordinal market. Define a *cardinal value allocation* to be an allocation x such that there is a bounded differentiable family $U = \{u_t\}$ of cardinal utilities for the preference scales $\wp_t$ for which (5.3) holds[28] for all S in $\mathscr{C}$, where the game v_U is defined by (5.1). All this is the same as in Section 5, except that we have replaced the preference relations by preference scales.

There are of course relationships between the ordinal and the cardinal concepts. A preference scale $\wp$ and a preference relation $\succ$ are called *associated* if the cardinal utilities for $\wp$ are ordinal utilities for $\succ$. There is exactly one preference relation

---

[28] Or equivalently, (5.2) holds for all t in T.

associated with each preference scale, but generally many scales with each relation. A cardinal and an ordinal market are *associated* if the ordinal market is obtained from the cardinal market by replacing each of the scales $\wp_t$ by the associated relation $\succ_t$. It is easy to see that the value allocations in an ordinal market are precisely those allocations x that are cardinal value allocations in some associated cardinal market.

As an illustration, consider a cardinal market $\mathcal{M}_c$ with two traders, say $T = \{1, 2\}$. Let u_1 and u_2 be cardinal utilities for $\wp_1$ and $\wp_2$, and let us assume that they are strictly concave. An *outcome* can be viewed as a pair $(u_1(y(1)), u_2(y(2)))$, where y is an allocation; the set X of all outcomes is compact and convex. There is exactly one cardinal value allocation x, and it is such that $(u_1(x(1)), u_2(x(2)))$ is the solution[29] to the "Bargaining Problem" of Nash [18], with outcome set X and conflict payoff $(u_1(a(1)), u_2(a(2)))$.

Next, consider the ordinal market $\mathcal{M}_0$ associated with $\mathcal{M}_c$. From an argument of Shapley [23, Sec. 3], it follows that *every* Pareto optimal allocation x with $x(1) \succ_1 a(1)$ and $x(2) \succ_2 a(2)$ is a value allocation in $\mathcal{M}_0$; in fact, every such x is the *unique* cardinal value allocation in some cardinal market associated with $\mathcal{M}_0$. Thus, in two-trader markets, the cardinal value allocations depend strongly on the preference scales of the traders, not only on their preference relations; which is of course what one would expect. A priori, moreover, there is every reason to expect that the situation is similar also when there are more than two traders, and also in the case of continuous markets. But exactly here we are in for a surprise.

PROPOSITION 11.1: *In a non-atomic cardinal market, assume that there is a bounded differentiable family of concave cardinal utilities for the preference scales. Then the set of cardinal value allocations coincides with the set of competitive allocations.*

The surprising aspect of this is that it asserts equality between the ordinal concept of "competitive allocation" and the cardinal concept of "cardinal value allocation." Non-atomicity turns the concept of "cardinal value allocation" into an ordinal concept—it does not depend on the preference scales, but only on the preference relations. It is to be noted that this is so only when we confine ourselves to preference scales with concave cardinal utilities; even then, though, it is a startling and somewhat mysterious phenomenon.

On the intuitive, unrigorous level, we can demonstrate Proposition 11.1 by the methods of Section 8—in particular, it follows from Step 6 and the demonstration of Step 7. A rigorous proof will be given in Section 14.

Like the Value Equivalence Theorem, Proposition 11.1 (which might be called the "Cardinal Value Equivalence Theorem") depends strongly on the non-atomicity: for finite markets, it is false in both directions. Indeed, we can easily find a two-trader market $\mathcal{M}_c$ of the type discussed above which has a unique

---

[29] That is, the maximum of $(z_1 - u_1(a(1)))(z_2 - u_2(a(2)))$ over $z \in X$ is achieved when $z = (u_1(x(1)), u_2(x(2)))$.

competitive allocation and a unique cardinal value allocation, which are different from each other.

The ordinal market $\mathcal{M}_0$ associated with such an $\mathcal{M}_c$ also shows that the Value Equivalence Theorem is false in finite markets, since in $\mathcal{M}_0$ there are many value allocations that are not competitive. It is also false in the converse direction: there are finite markets with competitive allocations that are not value allocations. For an example, take $T = \{1, 2, 3\}$, $n = 2$, $u_t(x) = \sqrt{x^1 + 1} + \sqrt{x^2 + 1}$ for all t, $a(1) = (2, 0)$, $a(2) = (1, 1)$, $a(3) = (0, 2)$. There is a unique competitive allocation x, given by $x(1) = x(2) = x(3) = (1, 1)$. Note that $x(2) = a(2)$, so that at x, trader 2 gets only his "personal minimum" (what he can guarantee to himself without trading). But any value allocation must yield trader 2 more than his personal minimum, since for any choice of utility functions, he contributes at least his personal minimum in all orderings of the players, and contributes more than his personal minimum in the orderings (1, 2, 3) and (3, 2, 1). Therefore x is not a value allocation.

In game theoretic terms, a cardinal market is an example of a *non-transferable utility (NTU) game*,[30] or what is more commonly called a "game without side payments" (see [2 or 3]). Shapley [23] has adapted the value notion to NTU games; our definition of "cardinal value allocation" is essentially a specialization to markets of Shapley's NTU value. There is absolutely no reason to expect the NTU value to be invariant under monotonic concave (but non-linear) transformations of the utilities; but Proposition 11.1 asserts that this is precisely what happens in a large class of NTU games, namely the non-atomic cardinal markets. It appears that the game theoretic condition underlying this phenomenon is the "homogeneity" of non-atomic markets: the payoffs available to the members of a coalition depend only on its composition, not on its size; and one may vary the size of a coalition at will, almost without affecting its composition.[31]

12. DISCUSSION OF THE LITERATURE

Though the material of this paper has many roots in previous work, we will here only touch on some of the highlights. Perhaps the first general theorem connecting the value and the competitive equilibrium was established by Shapley [22], using the so-called "transferable utility (TU) markets"[32] in an asymptotic model with a fixed finite number of types of traders, in which the number of traders of each type is allowed to tend to ∞. In a TU market with concave differentiable utilities there is a unique competitive payoff,[33] and Shapley proved that the values

---

[30] NTU games are generally defined in terms of a specific choice of cardinal utilities, not in terms of preference scales. However, since all concepts associated with such games are usually required to be "invariant" under monotonic linear transformations of the utilities, we are essentially dealing with games defined in terms of preference scales.

[31] Cf. [5, Chapter V]. A more familiar term for homogeneity is "constant returns to scale," as applied to coalitions rather than production processes; increasing or decreasing the scale of a coalition will not affect the possibilities open to its members.

[32] Cf. Footnote 9.

[33] That is, payoff corresponding to a competitive equilibrium.

of the TU games corresponding to the finite markets converge to this payoff. See also Shapley and Shubik [24] for a discussion of this work.

Recently, Aumann and Shapley [5] studied the value problem in the context of a TU non-atomic market. Assuming differentiability but not concavity, they showed that a TU non-atomic market has a unique competitive payoff, that the corresponding TU game has a value, and that the two coincide. No assumption of finite number of types was used.

In a parallel development Champsaur [6] studied the value problem in an asymptotic finite-type framework, but using general Walrasian markets (the kind studied here) rather than TU markets. He considered a special class of cardinal value allocations, namely those obtained by assigning to traders of the same type not only the same preference scale, but actually the same cardinal utility.[34] In that case, traders of the same type get the same utility payoffs; thus if m is the number of types and k the number of traders of each type, then for each k we get a vector $w_k \in E^m$ representing the utility payoffs of the various types. Champsaur [6] showed that as $k \to \infty$, every limit point of $\{w_k\}$ must be a competitive payoff. However, he did not prove the converse, that every competitive payoff is such a limit point. Concavity was assumed, but not differentiability.

Thus the situation can be summed up in Table I.

TABLE I

	TU	Walrasian
Asymptotic, finitely many types	Shapley [22]	Champsaur [6]
Non-atomic continuum	Aumann and Shapley [5]	This paper

Finally, we mention the important paper of Negishi [19], who was the first to use TU markets to prove a theorem about NTU markets; many of the underlying ideas of the current paper will be seen to be closely related to those of Negishi.

13. PROOF OF PROPOSITION 5.1

We start out by recalling a number of definitions, results, and notations from Aumann-Shapley [5].

Let us be given a non-atomic market as described in Section 5, and a family $U = \{u_t\}$ of utilities. Define v_U by (5.1).

---

[34] No such assumption is made in our treatment. Every trader may be of a different type; and even if there are many traders of the same type, they may be assigned different cardinal utilities, and in the ordinal case, even different preference scales. Nevertheless, there is implicit in our model a situation that is in principle similar to that of Champsaur. No matter how one assigns the cardinal utilities, one can, because of the non-atomicity, partition practically all the traders (i.e., all but a set of arbitrarily small measure) into a finite number of sets of equal measure, in each of which all the traders will have *approximately* the same cardinal utilities and initial bundles. In some sense, it is this that is responsible for the value equivalence phenomenon. (Compare the discussion of the core equivalence phenomenon at the top of [1, p. 49]; there the Debreu–Scarf asymptotic model plays a role similar to that played by the Champsaur model here.)

LEMMA 13.1: *If the family U is bounded differentiable, then the maximum in the definition of $v_U(S)$ is attained for all S, and the asymptotic value φv_U of v_U exists.*

Proof: This is an immediate consequence of [5, Proposition 31.7], which asserts these conclusions (among others) under conditions on U that are considerably weaker than bounded differentiability.

If x is an allocation, write $u(x)$ for the function on T whose value at t is $u_t(x(t))$. A *transferable utility competitive equilibrium* (TUCE) for U is a pair (x, p), where x is an allocation and $p \in \Omega$, such that for all $t \in T$, $u_t(x) - p \cdot (x - a(t))$ attains its maximum (over $x \in \Omega$) at $x = x(t)$.

LEMMA 13.2: *If (x, p) is any TUCE for U, then for all $S \in \mathcal{C}$,*

$$\int_S (u(x) - p \cdot (x - a)) = (\varphi v_U)(S).$$

Proof: This is an immediate consequence of [5, Proposition 32.3] (see the remark following the statement of the proposition), which asserts this conclusion (as well as some others) under Assumption 5.1 and conditions on U that are considerably weaker than those that define bounded differentiability.

Write $x \geqslant y$ if $x \geqslant y$ and $x \neq y$. The function u_t is called *increasing* if $x \geqslant y$ implies $u_t(x) > u_t(y)$.

LEMMA 13.3: *Let (x, p) be a TUCE for U, such that for almost all t in T, $p \cdot (x(t) - a(t)) = 0$. Assume that the u_t are increasing. Then x is a competitive allocation.*

PROOF: From the fact that the u_t are increasing it follows that $p \neq 0$. Since x is by definition an allocation, we must therefore only show that for almost all t, the function u_t is maximized over the budget set $\{x \in \Omega : p \cdot x \leqslant p \cdot a(t)\}$ by $x = x(t)$. Suppose, indeed, that it were not. Then for a set of t's of positive measure there is a y in Ω such that $u_t(y) > u_t(x(t))$ and $p \cdot y \leqslant p \cdot a(t)$. For almost all of those t's we will have $u_t(y) - p \cdot (y - a(t)) \geqslant u_t(y) > u_t(x(t)) = u_t(x(t)) - p \cdot (x(t) - a(t))$, in contradiction to the assumption that (x, p) is a TUCE.

PROOF OF PROPOSITION 5.1: Let x be a value allocation, and let U be the bounded differentiable family of utilities for the market such that for all $S \in \mathcal{C}$,

(13.1) $\quad (\varphi v_U)(S) = \int_S u(x).$

In particular, $v_U(T) = (\varphi v_U)(T) = \int_T u(x)$, i.e., the maximum in the definition of $v_U(T)$ is attained at x. Therefore by [5, Proposition 32.1], there is a p such that (x, p) is a TUCE with respect to U. Hence, by Lemma 13.2, for all $S \in \mathcal{C}$ we have

(13.2) $\quad (\varphi v_U)(S) = \int_S (u(x) - p \cdot (x - a)).$

Combining (13.1) and (13.2), we see that $\int_S p \cdot (x - a) = 0$ for all $S \in \mathscr{C}$. Hence $p \cdot (x(t) - a(t)) = 0$ almost everywhere. But members of a bounded differentiable family are by definition increasing, since $u'_t > 0$. Hence by Lemma 13.3, x is competitive. Thus the proof of Proposition 8.1 is complete.

14. TOWARD THE PROOF OF PROPOSITION 5.2

In this section we prove Proposition 5.2 under Hypothesis (ii), and also Proposition 11.1.

Let $x \in \Omega$. A continuously differentiable function u on Ω is called *concave over Ω at x* (or simply *concave at x*) if for all y in Ω, $u(y) - u(x) \leqslant u'(x) \cdot (y - x)$. It should be noted that this concept is not a local one; it depends on the behavior of u on all of Ω, not only near x.

REMARK 14.1: *A concave continuously differentiable function u on Ω is concave over Ω at each x in Ω.*

PROOF: First let x be the interior of Ω. For $0 \leqslant \theta \leqslant 1$, define $f(\theta) = u(\theta y + (1 - \theta)x)$. Then f is concave and continuously differentiable on $[0, 1]$ and hence $u(y) - u(x) = f(1) - f(0) \leqslant f'(0) = u'(x) \cdot (y - x)$. If x is on the boundary of Ω, one obtains the result by a limiting argument from the interior case.

Let b in Ω be a uniform bound on the initial allocation a, i.e., $a(t) < b$ for all t; such a bound exists by Assumption 5.2.

For $x \in E^n$, set $\|x\| = \max_i |x^i|$.

LEMMA 14.1: *Let x be a competitive allocation, with prices p. Let $\{w_t\}$ be a bounded differentiable family of utilities such that the w_t are concave at each x with $\|x\| \leqslant \max_i b^i / \min_i p^i$. Then there is a measurable function α on T, bounded and bounded away from 0, such that for the family $U = \{u_t\}$ of utilities defined by $u_t = \alpha(t)w_t$, we have $(\varphi v_U)(S) = \int_S u(x)$ for all $S \in \mathscr{C}$. In particular, x is a value allocation.*

PROOF: We will find an α such that (x, p) is a TUCE for U, i.e., such that for almost all t in T, we have that for all y in Ω,

(14.1) $\quad u_t(y) - u_t(x(t)) \leqslant p \cdot (y - x(t))$.

Since x is a competitive allocation with prices p, and the u_t are increasing, it follows that $p > 0$, and that

(14.2) $\quad p \cdot a(t) = p \cdot x(t) \quad$ for almost all t.

Hence for almost all t we have that for all y in Ω,

(14.3) $\quad w_t(y) > w_t(x(t)) \Rightarrow p \cdot y > p \cdot x(t)$.

For the time being, consider a fixed t satisfying (14.3) for all y in Ω; let $w = w_t$ and $x = x(t)$. Note that for each i, $p^i x^i \leqslant p \cdot x = p \cdot a(t) \leqslant p \cdot b \leqslant \max_i b^i$. Hence

$$(14.4) \quad \|x\| \leqslant \frac{\max_i b^i}{\min_i p^i},$$

and so w is concave at x. Assume first that $x \neq 0$; let $j = j(t)$ be such that $x^j > 0$. Set

$$(14.5) \quad \alpha = \alpha(t) = \frac{p^j}{w^j(x)},$$

where w^j is the jth derivative of w. We wish to prove (14.1). Without loss of generality assume $j = 1$.

We first show that for all i,

$$(14.6) \quad p^i \geqslant \alpha w^i(x).$$

Indeed, suppose $p^i < \alpha w^i(x)$. By (14.5) and $j = 1$, it follows that $p^i/p^1 < w^i(x)/w^1(x)$. Let θ be such that

$$(14.7) \quad \frac{p^i}{p^1} < \theta < \frac{w^i(x)}{w^1(x)},$$

and consider a point of the form $x_s = x + (-\theta s, 0, \ldots, 0, s, 0, \ldots, 0)$, where the term s occurs in the ith coordinate; if $s > 0$ is sufficiently small, then $x_s \in \Omega$, since $x^1 > 0$. Letting $s \to 0+$, we find $dw(x_s)/ds = -\theta w^1(x) + w^i(x) > 0$, by (14.7). Hence for $s > 0$ sufficiently small, $w(x_s) > w(x)$. But then from (14.3) it follows that $p \cdot x_t \geqslant p \cdot x$, i.e., $-p^1 \theta s + p^i s > 0$, contradicting (14.7). This proves (14.6).

Next, let i be such that $x^i > 0$. Then we claim that

$$(14.8) \quad p_i \leqslant \alpha w^i(x).$$

Indeed, suppose $p^i > \alpha w^i(x)$. Reasoning as in the proof of (14.6), we find a θ such that $(p^i/p^1) > \theta > (w^i(x)/w^1(x))$. Letting $x_s = x + (\theta s, 0, \ldots, 0, -s, 0, \ldots, 0)$, we note that $x_s \in \Omega$ for small s, since $x^i > 0$. Consideration of $dw(x_s)/ds$ then leads to a contradiction as before, so (14.8) is proved.

Let $L = \{i : x^i > 0\}$ and $M = \{i : x^i = 0\}$. From (14.6) and (14.8) it follows that $p^i = \alpha w^i(x)$ for $i \in L$ and $p^i \geqslant \alpha w^i(x)$ for $i \in M$. If $y \in \Omega$, then for $i \in M$ we have $y^i - x^i = y^i \geqslant 0$; hence $\alpha w^i(x)(y^i - x^i) \leqslant p^i(y^i - x^i)$, with equality holding for $i \in L$. From the concavity of w at f it then follows that $u_t(y) - u_t(x) = \alpha(w(y) - w(x)) \leqslant \alpha w'(x) \cdot (y - x) \leqslant p \cdot (y - x)$, which is what (14.1) asserts. Thus $\alpha(t)$ is defined and (14.1) is established when $x(t) \neq 0$.

Suppose next that $x = x(t) = 0$. Set $\alpha = \alpha(t) = \min_i p^i/\max_i w^i(0)$. The concavity of w at 0 then gives $u_t(y) - u_t(0) = \alpha(w(y) - w(0)) \leqslant \alpha w'(0) \cdot y \leqslant p \cdot y$, which is again what (14.1) asserts.

The measurability of α is easily verified. To show that α is bounded and bounded away from 0, note that p is a constant > 0, and w'_t is by assumption bounded and bounded away from 0 in compact sets; in particular, therefore, it is bounded and bounded away from 0 in the compact set

$$\left\{y \in \Omega : \|y\| \leqslant \frac{\max\limits_i b^i}{\min\limits_i p^i}\right\},$$

to which, by (14.4), all the $x(t)$ belong.

Thus we have shown that U is bounded differentiable and (x, p) is a TUCE. From Lemma 13.2 and (14.2), we then deduce that for all $S \in \mathscr{C}$, $(\varphi v_U)(S) = \int_S (u(x) - p \cdot (x - a)) = \int_S u(x)$. This completes the proof of Lemma 14.1.

From Remark 14.1 and Lemma 14.1 it follows that if there exists a bounded differentiable family of concave utilities for the preferences, then every competitive allocation is a value allocation. This yields Proposition 5.2 under Hypothesis (ii), and together with Proposition 5.1 (proved in the previous section), also Proposition 11.1.

15. PROOF OF THE VALUE EQUIVALENCE THEOREM

In view of Proposition 5.1 and Lemma 14.1, it remains only to prove that if the preferences are uniformly smooth, then we can find a bounded differentiable family of utilities that is concave over Ω in an arbitrarily large set. Specifically, we will prove

LEMMA 15.1: *Let $U = \{u_t\}$ be a uniformly smooth family of utilities for the given preferences. Then for every $\gamma > 0$ there exists a bounded differentiable family $\{w_t\}$ of utilities such that the w_t are concave over Ω at each x with $\|x\| \leqslant \gamma$.*

REMARK: Necessary and sufficient conditions for a preference relation to possess a concave utility have been obtained by Fenchel [12]. His conditions imply, as a special case, that under appropriate conditions, smooth preferences have concave utilities. Here, however, we have found it more convenient not to use his result, but to prove what is needed "from scratch." This is done for a number of reasons: First, we are interested only in a comparatively simple sufficient condition; whereas the mere statement of Fenchel's necessary and sufficient condition is complex and lengthy. Second, we need a uniform result for a family of preferences, whereas Fenchel's result applies to a single preference relation; therefore even if we used his result we would have to reprove it in the "uniformized" version. Finally, his conditions do not apply exactly here; for example, he considers an open rather than a closed domain of definition. Nevertheless, we will lean heavily on his methods, and there is no question that the principle that "smooth preferences have concave utilities" is to be attributed to Fenchel.

PROOF: Let $u^i_t = \partial u_t/\partial x_i$. Since u_t is bounded differentiable, u^i_t is uniformly bounded and uniformly positive in compact sets. Hence there is a $\delta > 0$ such that for all t and all x in Ω with $\|x\| \leq \gamma$, we have

$$(15.1) \quad \min_i \left(\frac{u_t^i}{\sum_{i=1}^{n} u_t^i} \right) \geq \delta.$$

Obviously, $\delta \leq 1/n$. Let $\Gamma = \{x \in \Omega : x \leq 2\gamma/\delta\}$, and let Δ be the interior of Γ. Set $h(\theta) = -e^{-\xi\theta}$ where $\xi > 0$ is a parameter that will be fixed later. We first wish to show that when ξ is sufficiently large, $h \circ u_t$ is concave on Δ for all t (i.e., $h \circ u_t|\Delta$ is concave for all t). For simplicity of notation, we will for the time being suppress the subscript t, i.e., we will write $u_t = u$.

It is well known (see, for example, [11, p. 51]) that a C^2 function w on a convex open set is concave if and only if the matrix $w'' = w''(x)$ of its second partial derivatives is negative semidefinite for each x, i.e., $yw''y^* \leq 0$ for all y in E^n, where * denotes "transpose." A straightforward calculation yields $(h \circ u)'' = \xi e^{-\xi u}(u'' - \xi u'^* u')$, where u' is a row vector. Thus we must show that $u'' - \xi u'^* u'$ is negative semidefinite for appropriate ξ. We will in fact show that for ξ sufficiently large,

$$(15.2) \quad u'' - \xi u'^* u' \text{ is negative definite at each } x \text{ in } \Delta,$$

i.e., that $y(u'' - \xi u'^* u')y^* < 0$ for all $y \neq 0$.

Let $H(=H_{x,t})$ be the hyperplane $\{y : y \cdot u' = 0\}$. Then we assert

$$(15.3) \quad y \in H \Rightarrow yu''y^* \leq 0.$$

Indeed, for $y \in H$, consider the expression $u(x + \theta y)$ as a function of the real variable θ; it is defined for θ in some neighborhood of 0, since x is in the interior of Ω. The point $x + \theta y$ is in the hyperplane through x orthogonal to u'; by its quasiconcavity, u achieves its maximum in this hyperplane at x. Hence $u(x + \theta y)$, as a function of θ, achieves its maximum at 0, and so

$$0 \geq \frac{d^2}{d\theta^2} u(x + \theta y) = yu''y^*;$$

this proves (15.3).

Next we assert

$$(15.4) \quad y \in H \Rightarrow yu''y^* < 0, \quad \text{or} \quad y = 0;$$

in other words, not only is u'' negative semidefinite on H, which is what (15.3) asserts, but it is in fact negative definite on H. Suppose it were not; let $y_0 \in H$ be such that $y_0 \neq 0$ and $y_0 u'' y_0^* = 0$ (the possibility $y_0 u'' y_0^* > 0$ is ruled out by (15.3)). By (15.3), $yu''y^*$ attains its maximum subject to $yu'^* = 0$ at $y = y_0$. Therefore there is a Lagrange multiplier λ such that $2u''y_0^* + \lambda u'^* = 0$ and $u'y_0^* = 0$. Hence the system

$$\begin{pmatrix} u'' & u'^* \\ u' & 0 \end{pmatrix} z^* = 0$$

of linear equations in $z = (z^1, \ldots, z^{n+1})$ has the distinct solutions $z = 0$ and

$z = (y_0, \lambda/2)$, and so the determinant of the matrix defining the system vanishes. But this violates the positivity of the Gaussian curvature (see Condition 4.2 and Equations (4.1) and (4.2)), so (15.4) is proved.[35]

According to a known theorem,[36] (15.4) implies (15.2) for sufficiently large ξ; but we must still show that the choice of ξ may be made independent of x in Δ, and of t.

Let $Q_\xi = u'' - \xi u'^* u'$, let $S^{n-1} = \{y \in E^n : \|y\| = 1\}$, and let $g(\xi)$ be the maximum of $y Q_\xi y^*$ over all y in S^{n-1}. Clearly g is non-increasing. Moreover, since $y Q_\xi y^*$ is continuous in y and ξ, and S^{n-1} is compact, it follows that g is continuous. Hence if ξ_0 is the infimum of all ξ such that $g(\xi) < 0$, then $g(\xi_0) = 0$; in other words, $y Q_\xi y^*$ is negative definite for $\xi > \xi_0$, whereas for $\xi = \xi_0$, it is negative semidefinite but not negative definite. Now it is well known that any quadratic form Q can be diagonalized, i.e., there is a non-singular matrix F such that FQF^* is a diagonal matrix; if Q is negative definite then all the diagonal entries are negative, and it follows that the determinant $|Q|$ does not vanish. If Q is negative semidefinite but not definite, then at least one of the diagonal entries vanishes and hence $|Q| = 0$. Thus $|Q_\xi| \neq 0$ for $\xi > \xi_0$ and $|Q_\xi| = 0$ for $\xi = \xi_0$, i.e.,

(15.5) $\quad \xi_0 = \max\{\xi : |u'' - \xi u'^* u'| = 0\}$.

Now letting i and j run from 1 to n, we have

$$|Q_\xi| = |u'' - \xi u'^* u'| = |u^{ij} - \xi u^i u^j| = \begin{vmatrix} u^{ij} - \xi u^i u^j & u^i \\ 0 & 1 \end{vmatrix}$$

$$= \begin{vmatrix} u^{ij} & u^i \\ u^j & 1 \end{vmatrix} = |u^{ij}| + \xi \begin{vmatrix} u^{ij} & u^i \\ u^j & 0 \end{vmatrix} = |u''| + \xi u^H.$$

From this and (15.5) it follows that $\xi_0 = -|u''|/u^H$. Now making the dependence of u on t explicit once more, and applying (4.2), we obtain

(15.6) $\quad \xi_0 = \dfrac{|u_t''| \, \|u_t'\|_2^{n+1}}{u_t^c}.$

By Conditions 4.2 and 4.3, and the bounded differentiability of the u_t, there is a uniform bound for the right side of (15.6) over all x in Δ and all t. If $\xi > 0$ is chosen greater than that bound, then $u_t'' - \xi u_t'^* u_t'$ is negative definite for all x in Δ and all t, and hence $h \circ u_t$ is concave on Δ. From a simple continuity argument it then follows that $h \circ u_t$ is concave also on the closure Γ of Δ.

It remains to demonstrate the concavity over Ω, for $\|x\| \leqslant \gamma$. Let r be a real function of a real variable θ with the following properties:

PROPERTY 15.1: $r \in C^2$ and for all θ, $r'(\theta) > 0$.

---

[35] I am indebted to Kenneth Arrow for this elegant proof of (15.4), which replaces a more awkward proof, using an explicit diagonalization, that I previously had.

[36] For an elegant proof in four lines, see Debreu [8, p. 296, Theorem 3].

PROPERTY 15.2: $r(\theta) \leqslant \min(\theta, 1)$.

PROPERTY 15.3: $r(\theta) = \theta$ for $\theta \leqslant 0$.

For each t, set $v_t = h \circ u_t$, and define q_t by

$$q_t(\theta) = (\beta_t - \alpha_t)r((\theta - \alpha_t)/(\beta_t - \alpha_t)) + \alpha_t,$$

where $e = (1, \ldots, 1)$, $\beta_t = v_t(2\gamma e)$, and $\alpha_t = v_t(\gamma e)$. Properties 15.1 through 15.3 then yield:

(15.7) $\quad q_t \in C^2$ and for all θ, $q_t'(\theta) > 0$.

(15.8) $\quad q_t(\theta) \leqslant \min(\theta, v_t(2\gamma e))$.

(15.9) $\quad q_t(\theta) = \theta$ for $\theta \leqslant v_t(\gamma e)$.

Now define $w_t = q_t \circ v_t$. We wish to demonstrate that w_t is concave on $\{x \in \Omega : \|x\| \leqslant \gamma\}$ over Ω. This means that for all x with $\|x\| \leqslant \gamma$, and for all y in Ω,

(15.10) $\quad w_t(y) \leqslant w_t(x) + w_t'(y) \cdot (y - x)$.

To prove (15.10), we again suppress the subscript t, writing w for w_t, v for v_t, and u for u_t. First note that because $\|x\| \leqslant \gamma$, it follows from (15.9) that $w(x) = v(x)$ and $w'(x) = v'(x)$. If $y \in \Gamma$, then from (15.8) and the concavity of v on Γ, we get $w(y) = q(v(y)) \leqslant v(y) \leqslant v(x) + v'(x) \cdot (y - x) = w(x) + w'(x) \cdot (y - x)$, proving (15.10) in this case. If $y \notin \Gamma$, then for at least one coordinate i, we have $y^i \geqslant 2\gamma/\delta$. Hence by (15.1),

$$\frac{u'(x)}{\sum_{i=1}^{n} u^i(x)} \cdot y \geqslant 2\gamma = \frac{u'(x)}{\sum_{i=1}^{n} u^i(x)} \cdot 2\gamma e.$$

Hence $u'(x) \cdot y \geqslant u'(x) \cdot 2\gamma e$, and so $v'(x) \cdot y \geqslant v'(x) \cdot 2\gamma e$ (since $v'(x) = h'(u(x))u'(x)$ and $h' > 0$ always). Combining this with (15.9) and the concavity of v on Γ we obtain $w(y) = q(v(y)) \leqslant v(2\gamma e) \leqslant v(x) + v'(x) \cdot (2\gamma e - x) \leqslant v(x) + v'(x) \cdot (y - x)$, as was to be shown. Thus (15.10) is established.

Since $\{w_t\}$ is easily seen to be bounded differentiable, the proof of Lemma 15.1 —and with it of the Value Equivalence Theorem—is complete.

The Hebrew University of Jerusalem

APPENDIX

Marginal and "True" Contributions

In the introduction we stated that "the marginal contribution of a trader to society is almost surely the same as his 'true' contribution." This assertion can be demonstrated on two levels: intuitively, and in a rigorous fashion. The intuitive demonstration was already given in Step 5 of Section 7, where we showed that if the coalition S is randomly chosen, then the contribution $v_U(S) - v_U(S \setminus dt)$ of dt to S is almost surely the same as his marginal contribution $v_U(T) - v_U(T\setminus dt)$. We will now state this assertion precisely, and prove it rigorously.

Let us be given a non-atomic market, and let $U = \{u_t\}$ be a bounded differentiable family of utilities for the preferences. Set $v = v_U$. Let $\mathscr{P} = (\Pi_1, \Pi_2, \ldots)$ be a separating sequence of partitions of T into measureable sets, such that each Π_m refines Π_{m-1}. For each m, let $\mathscr{R}_m$ be a random order[37] on Π_m. For each B in Π_m, denote by $Q = Q_B^m$ the union of the members of Π_m up to and including B in the order $\mathscr{R}_m$. For each S with $B \subset S$ let $\Delta_B v(S) = v(S) - v(S\setminus B)$; $\Delta_B v(S)$ is the *contribution of B to S*.

PROPOSITION A.1: *For every $\varepsilon > 0$ there is an m_0 such that if $m > m_0$ then for every $B \in \Pi_m$ we have*

(A.1) $\qquad |\Delta_B v(T) - (\varphi v)(B)| < \varepsilon \mu(B)$

and

(A.2) $\qquad prob\ \{|\Delta_B v(Q) - (\varphi v)(B)| < \varepsilon \mu(B)\} > 1 - \varepsilon.$

This says that in a random order on a sufficiently fine partition Π of the player space, with high probability the incremental worth ("true" contribution) of an element B of Π is $(\varphi v)(B) + o(\mu(B))$, and also that the marginal contribution is $(\varphi v)(B) + o(\mu(B))$. In particular, the "true" and marginal contribution differ by a term that is $o(\mu(B))$. Note that $(\varphi v)(B)$ is in general of the order of magnitude of $\mu(B)$ (see Lemma 13.2).

We will make extensive and free use of the notation, terminology, and results of [5]. The reader should note that T in this paper plays the same role as I in [5], and that $u_t(x)$ is also denoted $u(x, t)$ in [5]; also that it is assumed that $u_t(0) = 0$ for all t, an assumption we can make without loss of generality.

There is one respect in which our notation will differ from that of [5]. In [5], both random variables and functions of t were denoted by bold face letters; no confusion could result, since these two kinds of objects were never considered at the same time. Here confusion might result, and so we shall indicate random variables by a wiggly underline, and functions of t by bold face.

Before proceeding with the proof, we shall introduce some auxiliary definitions that will enable us to avoid unnecessary repetition of tiresome "epsilon-delta" phraseology. Let $F = F_B^m$ be a sentence depending on m and on the choice of an element B of Π_m (an example is Formula (A.2)). We shall say that F *eventually* holds if there is an m_0 such that F_B^m holds for $m > m_0$ and all $B \in \Pi_m$. If $\underline{f} = \underline{f}_B^m$ is a random variable depending on m and the choice of a B in Π_m (for example $\underline{v}(\underline{Q})$), then $\underline{f}$ is *eventually probably small* if for each $\varepsilon > 0$, $prob\ \{|\underline{f}| < \varepsilon\} > 1 - \varepsilon$ eventually holds (if $\underline{f}$ is multidimensional, replace $|\underline{f}|$ by $\|\underline{f}\|$). If $\underline{f} - g$ is eventually probably small, we shall say that $\underline{f}$ is *eventually probably close* to g. Proposition A.1 may be paraphrased by saying that (A.1) holds and that $(\Delta_B v(\underline{Q}) - \Delta_B v(T))/\mu(B)$ is eventually probably small.

A number of lemmas will be required. Let $r = r_m = |\Pi_m|$ and let $\underline{h} = \underline{h}_B^m$ be the number of elements of Π_m included in $\underline{Q}$, i.e., the number of elements up to and including B in the order $\mathscr{R}_m$.

LEMMA A.1: *Let $\xi \in NA^+$. Then $\xi(\underline{Q})$ is eventually probably close to $(\underline{h}/r)\xi(T)$.*

PROOF: Set $\delta^m = \max\ \{\xi(A) : A \in \Pi_m\}$. Since $\underline{h}$ takes on each integer value between 1 and r with the same probability, we have

$$E\left(\xi(\underline{Q}\setminus B) - \frac{\underline{h}}{r}\xi(T\setminus B)\right)^2 = \frac{1}{r}\sum_{h=1}^{r} E\left(\left(\xi(\underline{Q}\setminus B) - \frac{h}{r}\xi(T\setminus B)\right)^2 \Big| \underline{h} = h\right).$$

Referring to the proof of [5, Lemma 18.7], we find that each of the summands in the last sum is $\leq \delta^m$, and that $\delta^m \to 0$. Hence $E(\xi(\underline{Q}\setminus B) - (\underline{h}/r)\xi(T\setminus B))^2 \leq \delta^m \to 0$. Since $\xi(B) \leq \delta^m \to 0$, the lemma follows.

LEMMA A.2: *Let $\xi \in NA^+$, $\xi(T) > 0$. Then for each $\theta > 0$, eventually $prob\ \{\xi(\underline{Q}) > \theta\xi(T)\} > 1 - 2\theta$.*

PROOF: Since $\underline{h}$ takes on each integer value between 1 and r with equal probability, it follows that for every $\theta > 0$, $prob\ \{\underline{h}/r > \theta\} \to 1 - \theta$ as $m \to \infty$. Lemma A.2 then follows from Lemma A.1.

LEMMA A.3: *$u'_{\underline{Q}}(\int_{\underline{Q}} a)$ is eventually probably close to $u'_T(\int a)$.*

---

[37] A random variable whose values are orders on Π_m, where each of the $|\Pi_m|!$ possible orders is assigned the same probability.

PROOF: Let $\varepsilon > 0$ be given. Set $\alpha = \Sigma \int a$. By [5, Proposition 38.14], there is a $\delta > 0$ such that for all S with $\mu(S) \geq \varepsilon$, all b in $A(\varepsilon, \alpha)$, and all δ-approximations $\hat{u}$ to u, we have

(A.3) $\qquad \|\hat{u}'_S(b) - u'_S(b)\| < \varepsilon$.

Let $\hat{u}$ be a δ-approximation to u that is of finite type [5, Proposition 35.6]. Define an n-dimensional vector measure ζ by $\zeta(S) = \int_S a$. Then

(A.4) $\qquad \left\| u'_Q\!\left(\int_Q a\right) - u'_T\!\left(\int a\right) \right\| \leq \|u'_Q(\zeta(Q)) - \hat{u}'_Q(\zeta(Q))\| + \|\hat{u}'_Q(\zeta(Q)) - \hat{u}'_T(\zeta(T))\|$

$$+ \|\hat{u}'_T(\zeta(T)) - u'_T(\zeta(T))\| = D_1 + D_2 + D_3.$$

Applying Lemma A.2 (with $\xi = \mu$ and $\theta = \varepsilon$) and (A.3), we find that eventually

(A.5) $\qquad \text{prob}\{D_1 < \varepsilon\} > 1 - 2\varepsilon$

and

(A.6) $\qquad D_3 < \varepsilon$.

To estimate D_2, let k be the number of different "types" in $\hat{u}$, i.e., the number of different $\hat{u}_t$. Denote the nonnegative orthant of E^k by Ξ. Then there is a continuous function g on $\Xi \times \Omega$ and a k-dimensional vector η of NA^+ measures on T, such that

(A.7) $\qquad \hat{u}_S(z) = g(\eta(S), z)$

for all $S \in \mathscr{C}$ and $z \in \Omega$, $\eta(T) > 0$, and g is homogeneous of degree 1 on $\Xi \times \Omega$ and continuously differentiable in the interior of $\Xi \times \Omega$ [5, (39.17), Propositions 39.8 and 39.13, and Lemma 39.9]. Set $q = n + k$, $\varepsilon_1 = \varepsilon/3q$, and $v = (\eta, \zeta)$. Since $\varepsilon_1 v(T) > 0$, we may find a compact neighborhood U of the straight line segment connecting $\varepsilon_1 v(T)$ to $v(T)$, such that g' is uniformly continuous in U. Hence there is a δ_1, with $0 < \delta_1 < \varepsilon_1$, such that

(A.8) $\qquad x, y \in U \quad \text{and} \quad \|x - y\| < \delta_1 \Rightarrow \|g'(x) - g'(y)\| < \varepsilon$.

Now by Lemma A.2, eventually prob $\{v(Q) > \varepsilon_1 v(T)\} > 1 - 2q\varepsilon_1$, and by Lemma A.1, $v(Q)$ is eventually probably close to $(h/r)v(T)$. Hence by (A.8), eventually prob $\{\|g'(v(Q)) - g'(h/rv(T))\| < \varepsilon\} > 1 - 3q\varepsilon_1 = 1 - \varepsilon$. But $g'((h/r)v(T)) = g'(v(T))$, because g is homogeneous. Hence by (A.7), eventually prob $\{\|\hat{u}'_Q(\zeta(Q)) - \hat{u}'_T(\zeta(T))\| < \varepsilon\} > 1 - \varepsilon$. Combining this with (A.5) and (A.6), we find that eventually prob $\{D_1 + D_2 + D_3 < 3\varepsilon\} > 1 - 3\varepsilon$, and hence by (A.4), the proof of Lemma A.3 is complete.

LEMMA A.4: *For each $\varepsilon > 0$ there is a bounded function η such that if $\mu(S) \geq \varepsilon$, b in Ω satisfies $\Sigma b \leq \Sigma \int a$, and $u_S(b)$ is attained at y, then $y(s) \leq \eta(s)e$ for almost all $s \in S$.*

PROOF: First note that in [5, Lemma 37.1] we may choose η_δ bounded; indeed, if the uniform bound on u is γ, then we may choose $\eta_\delta(s) \equiv \max(\gamma/\delta, 1/\delta^2)$. Lemma A.4 then follows immediately from [5, Lemma 37.8], when one refers to the first paragraph of the proof of that lemma.

LEMMA A.5: *Let $\xi \in NA^+$. Then $\max\{\xi(A) : A \in \Pi_m\} \to 0$ as $m \to \infty$.*

PROOF: This follows from the fact that $\mathscr{P}$ is separating. For details, refer to the proof of [5, Lemma 18.7].

COROLLARY A.1: *Let $v(Q)$ be achieved at y. Then $\int_B y$ is eventually probably small.*

PROOF: Let $\varepsilon > 0$ be given, and let η correspond to ε in accordance with Lemma A.4. Let γ be a common bound for η and for Σa (see Assumption 5.2). If $\mu(Q) \leq \varepsilon$, then $\int_B y \leq \int_Q y = \int_Q a \leq \gamma \varepsilon$. If $\mu(Q) \geq \varepsilon$, then by Lemma 1.4, $\int_B y \leq \gamma \mu(B)e$. But by Lemma A.5, $\mu(B)$ is eventually smaller than ε. Thus the Corollary is proved.

PROOF OF PROPOSITION A.1: First we demonstrate that eventually (A.1) holds. Suppose $v(T)$ is attained at x. From [5, Propositions 32.3 and 38.5] it follows that

(A.9) $\qquad (\varphi v)(B) = \int_B \left[u(x) + u'_T\!\left(\int a\right)(a - x) \right]$.

From [5, Lemma 40.1], it follows that

(A.10) $\quad \Delta_B v(T) = v(T) - v(T\backslash B) = \int_B [u(x) + u'_{T\backslash B}(\underline{c})(a - x)],$

where $\underline{c}$ is on the straight line segment connecting $\int_{T\backslash B} x$ to $\int_{T\backslash B} a$, i.e., (since $\int x = \int a$) $\underline{c} = \int a - \int_B (\theta x + (1-\theta)a)$, where $0 \leqslant \theta \leqslant 1$. Set $\underline{b} = \int a - \int_B x$. Then $\underline{b} - \underline{c} = (1-\theta)\int_B (a-x)$, and hence from [5, Proposition 38.7] and Lemma A.5, it follows that for each $\varepsilon_1 > 0$, eventually

(A.11) $\quad \|u'_{T\backslash B}(\underline{b}) - u'_{T\backslash B}(\underline{c})\| < \varepsilon_1.$

Again using [5, Lemma 40.1], we find

(A.12) $\quad u'_{T\backslash B}(\underline{b}) = u'_T\left(\int a\right),$

on condition that $\int_{T\backslash B} a > 0$; but this follows from Assumption 5.1 and Lemma A.5 for sufficiently large m, so that (A.12) eventually holds. Formulas (A.9), (A.10), (A.11), and (A.12) yield eventually $|\Delta_B v(T) - (\varphi v)(B)| \leqslant \varepsilon_1 \Sigma \int_B (a + x)$; together with Assumption 5.2 and Lemma A.4, this yields that (A.1) holds eventually.

The proof of (A.2) is quite similar. Suppose $v(Q)$ is achieved at y; [5, Lemma 40.1] yields

(A.13) $\quad \Delta_B v(Q) = v(Q) - v(Q\backslash B) = \int_B [u(y) + u'_Q(\underline{c})(a - y)],$

where $\underline{c} = \int_Q a + \theta \int_B (a - y)$ and $0 \leqslant \theta \leqslant 1$. Set $\underline{b} = \int_Q a - \int_B y$. Then $\underline{b} - \underline{c} = (1-\theta)\int_B (a-y)$, and hence from [5, Proposition 38.7], Corollary A.1, and Lemma A.2 (with $\xi = \mu$ and θ small), it follows that $u'_{Q\backslash B}(\underline{b})$ is eventually probably close to $u'_{Q\backslash B}(\underline{c})$. By [5, Lemma 40.1], $u'_{Q\backslash B}(\underline{b}) = u'_Q(\int_Q a)$ whenever $\int_{Q\backslash B} a > 0$. By Lemma A.2, for each $\varepsilon_2 > 0$, eventually prob $\{\int_{Q\backslash B} a > 0\} > 1 - \varepsilon_2$. Hence $u'_{Q\backslash B}(\underline{b})$ is eventually probably close to $u'_Q(\int_Q a)$, and by Lemma A.3, this is eventually close to $u'_T(\int a)$. Hence

(A.14) $\quad u'_Q\left(\int_Q a\right)$ and $u'_{Q\backslash B}(\underline{c})$ are both eventually probably close to $u'_T\left(\int a\right).$

Now by [5, Proposition 38.5], we have $u_t(y(t)) - u_t(x(t)) \leqslant u'_T(\int a)(y(t) - x(t))$. Hence by (A.9) and (A.13),

(A.15) $\quad \Delta_B v(Q) - (\varphi v)(B) \leqslant \left(u'_Q(\underline{c}) - u'_T\left(\int a\right)\right)\int_B (a - y).$

Again by [5, Proposition 38.5], $u_t(x(t)) - u_t(y(t)) \leqslant u'_{Q\backslash Q} a)(x(t) - y(t))$, and hence by (A.9) and (A.13)

(A.16) $\quad (\varphi v)(B) - \Delta_B v(Q) \leqslant \left(u'_Q\left(\int_Q a\right) - u'_T\left(\int a\right)\right)\int_B (x - y) - \left(u'_T\left(\int a\right) - u'_Q(\underline{c})\right)\int_B (y - a).$

Now let $\varepsilon_3 > 0$ be given, and let η correspond to ε_3 in accordance with Lemma A.4. Let γ be a common bound for η and Σa (see Assumption 5.2). By Lemma A.2, eventually $\mu(Q) \geqslant \varepsilon_3$ with probability $\geqslant 1 - 2\varepsilon_3$. Set $\varepsilon_4 = \varepsilon_3/2\gamma m$. Hence by (A.14), (A.15), and (A.16), eventually, with probability $\geqslant 1 - 2\varepsilon_3$, $|\Delta_B v(Q) - (\varphi v)(B)| \leqslant 4\varepsilon_4 \gamma m \mu(B) = 2\varepsilon_3 \mu(B)$. This completes the proof of Proposition A.1.

REFERENCES

[1] AUMANN, R.: "Markets with a Continuum of Traders," *Econometrica*, 32 (1964), 39–50 [Chapter 46].

[2] ———: "A Survey of Cooperative Games without Side Payments," in *Essays in Mathematical Economics in Honor of Oskar Morgenstern*, ed. by M. Shubik. Princeton, N.J.: Princeton University Press, 1967, 3–27 [Chapter 41].

[3] AUMANN, R., AND B. PELEG: "von Neumann-Morgenstern Solutions to Cooperative Games Without Side Payments," *Bulletin of the American Mathematical Society*, 66 (1960), 173–179 [Chapter 38].

[4] AUMANN, R., AND M. PERLES: "A Variational Problem Arising in Economics," *Journal of Mathematical Analysis and Applications*, 11 (1965), 488–503 [Chapter 70].

[5] AUMANN, R., AND L. SHAPLEY: *Values of Non-Atomic Games.* Princeton, N.J.: Princeton University Press, 1971.
[6] CHAMPSAUR, P.: "Cooperation vs. Competition," *Journal of Economic Theory*, 11 (1975), 394–417.
[7] DE FINETTI, B.: "Sulle Stratificazioni Convesse," *Annali di Matematica*, 4 (1949), 173–183.
[8] DEBREU, G.: "Definite and Semi-Definite Quadratic Forms," *Econometrica*, 20 (1952), 295–299.
[9] ———: "Topological Methods in Cardinal Utility Theory," in *Mathematical Methods in the Social Sciences*, ed. by K. Arrow, S. Karlin, and P. Suppes. Stanford: Stanford University Press, 1959, 16–26.
[10] ———: "Smooth Preferences," *Econometrica*, 40 (1972), 603–616.
[11] EGGLESTON, H. G.: *Convexity.* Cambridge: Cambridge University Press, 1958.
[12] FENCHEL, W.: "Uber konvexe Funktionen mit vorgeschriebenen Niveaumannigfaltigkeiten," *Mathematische Zeitschrift*, 63 (1956), 496–506.
[13] FISHBURN, P.: *Utility Theory for Decision Making*, New York: Wiley, 1970.
[14] HILDENBRAND, W.: *Core and Equilibria of Large Economies.* Princeton, N.J.: Princeton University Press, 1974.
[15] KANNAI, Y.: "Values of Games with a Continuum of Players," *Israel Journal of Mathematics*, 4 (1966), 54–58.
[16] ———: "Approximation of Convex Preferences," *Journal of Mathematical Economics*, 1 (1974), 101–106.
[17] MAS-COLLEL, A.: "Continuous and Smooth Consumers: Approximation Theorems," *Journal of Economic Theory*, 8 (1974), 305–336.
[18] NASH, J. F.: "The Bargaining Problem," *Econometrica*, 18 (1950), 152–162.
[19] NEGISHI, T.: "Welfare Economics and Existence of an Equilibrium for a Competitive Economy," *Metroeconomica*, 12 (1960), 92–97.
[20] RAWLS, J.: *A Theory of Justice.* Cambridge, Mass.: Harvard University Press, 1971.
[21] SHAPLEY, L.: "A Value for *n*-Person Games," in *Contributions to the Theory of Games, Vol. II*, ed. by H. W. Kuhn and A. W. Tucker, Princeton, N.J.: Princeton University Press, 1953, 307–317.
[22] ———: "Values of Large Games VII: A General Exchange Economy with Money," Rand RM-4248-PR, 1964.
[23] ———: "Utility Comparison and the Theory of Games," in *La Decision*, ed. by G. Th. Guilbaud. Paris: Editions du CNRS, 1969.
[24] SHAPLEY, L., AND M. SHUBIK: "Pure Competition, Coalitional Power, and Fair Division," *International Economic Review*, 3 (1969), 337–362.
[25] SHESHINSKI, E.: "An Example of Income Tax Schedules which are Optimal for the Maximin Criterion," unpublished manuscript.
[26] SHITOVITZ, B.: "Oligopoly in Markets with a Continuum of Traders," *Econometrica*, 41 (1973), 467–501.
[27] SMALE, S.: "Global Analysis and Economics IIA: Extension of a Theorem of Debreu," *Journal of Mathematical Economics*, 1 (1974), 1–14.
[28] VON NEUMANN, J., AND O. MORGENSTERN: *Theory of Games and Economic Behavior.* Princeton, N.J.: Princeton University Press, 1944.
[29] WHITNEY, H.: "Functions Differentiable on Boundaries of Regions," *Annals of Mathematics*, 35 (1934), 482–485.

52 Power and Taxes
with M. Kurz

1 Introduction

Perhaps the most fundamental element in the theory of the public sector is the view that the government is an exogenous, benevolent economic agent. The benevolence of the government is often expressed by assuming it to maximize a "social welfare" function of the form $\int u_t(x(t))\mu(dt)$, where the u_t are utility functions, x is an allocation of consumption bundles and μ is a distribution of agent types; in other words, simply the sum of individual utilities. With such a social welfare function Arrow and Kurz [2] were able to derive optimal investment and taxation programs while Mirrlees [13], Sheshinski [19] and others were able to derive optimal tax policies for a population with heterogeneous endowment.

We do not think that this view is without merit. There are perhaps some public issues with regard to which a consensus may be reached, and then such an approach may suffice to explain the behavior of the government. But more often, redistributive effects are a central issue; and then the actions of the government, and in particular its tax policies, can be understood only as an endogenous consequence of the political forces that enable it to maintain power. For this reason one should investigate the connection between tax policies and the political forces that shaped those policies to begin with.

We propose to regard income distribution, taxation, the production of public goods and other actions of the public sector as determined by a political process simultaneously with the economic process of exchange and production. This means that we propose to study an economic-political equilibrium where the power of each individual is reflected both in the political and the economic spheres.

The present is our first paper on this subject,[1] where we formulate the basic structure and motivate it. All of our work in this paper deals with a world in which there is only one commodity ("money"), to be thought of as an aggregate of all real commodities. This enables us to focus on the purely redistributive aspect of government policy. In another paper [4] we treat a more complex and realistic multi-commodity model in which we

---

This chapter originally appeared in *Econometrica* 45 (1977): 1137–1161. Reprinted with permission.

1. This research was supported by U.S. National Science Foundation Grant GS-40104 at the Institute for Mathematical Studies in the Social Sciences at Stanford University, and by a grant from the Israel National Council for Research and Development at the Hebrew University of Jerusalem. We are greatly indebted to Kenneth Arrow for an extremely helpful conversation on the subject of this paper; the interpretation of δ_t in terms of "fear of ruin" (Section 6) is an outcome of that conversation.

discuss such matters as the relations between taxes and prices for various goods.

The basic tool that deals with the conflict part of our theory comes from game theory; it is the "Harsanyi–Shapley–Nash Value for Non-Transferable Utility Games" (Shapley [18]). The highlights of our results are as follows:

a. If a democratic power structure (majority vote) is assumed, then an income tax emerges which can be progressive, regressive or neutral.

b. The size of the tax depends upon attitudes toward risking large losses. We shall introduce a new measure for the attitude toward such risks, which we shall call "the Fear of Ruin." This may be contrasted with the Arrow–Pratt [1, 16] measures of risk aversion, which measure attitudes towards small risks only. The connection with risk is somewhat startling, since there is no overt element of risk in the model; however, a closer examination of the situation, which will be undertaken below, shows that considerations of risk do enter naturally.

c. The tax structure involves a personal support level and negative taxation for low income people. (The "support" is what the individual receives from the government in the absence of any income.)

d. The marginal tax rate is always between 50 per cent and 100 per cent.

The treatment in this paper will be on the conceptual level, and will not be entirely rigorous from the mathematical viewpoint. A rigorous treatment, with complete proofs, may be found in [4]. The "heaviest" parts of this paper from the mathematical viewpoint are Sections 4 and 9; readers who are willing to forego the precise definition of the Shapley value, and of the corresponding concept of value allocation, may skip these sections.

2   The Income Redistribution Game

In the model of this paper there is only one commodity, to be thought of as an aggregate of all "real" commodities; for convenience we shall call it "money."[2] There is a set T of agents ("society"), each one of whom has an initial endowment of money ("gross income"). It is assumed that there are many agents, and that each individual one is "small;" that is, remov-

---

2. We use the term "money" where often economists use the term "income." We wish to keep the ideas behind these terms distinct: in our terminology, the commodity "money" provides the units in which wealth or income are measured. Thus we can talk of "income," "gross income," and "net income" as specific quantities of money received by individuals (or groups) in various circumstances.

ing him would not appreciably affect society. By a political procedure to be specified below, the agents in T will be taxed, and the resulting revenue will be distributed among them; what is left to an agent after that is called his "net income" (it may exceed gross income). We shall be interested only in the relation between gross and net income—i.e., in the total effect of both taxation and redistribution—and not in the two processes separately.

The political procedure is majority rule. This means that any coalition containing more than half the agents can impose taxes and then redistribute them in any way it pleases. In particular, it may divide among its own members the taxes collected from the minority.

It might be objected that in a democracy, one aims at "uniform" tax laws, i.e. laws under which people with the same income are taxed in the same way; this would preclude taxing two people differently just because one is in the ruling coalition while the other is not. But in fact, tax laws are not uniform, as witness the myriads of different rules for different kinds of taxpayers and different kinds of income—rules that are often deliberately slanted to favor various pressure groups.[3] Moreover, it must be recalled that what we are discussing here is not just taxation, but rather the total effect of government activity on incomes. Though theoretically one might require uniform taxation, there can be no such requirement on government spending; and spending can easily be determined by the majority to wipe out any undesired effects of uniform taxation requirements. Finally, we are even willing to concede that total income redistribution may in fact exhibit certain aspects of uniformity, if the concept is interpreted broadly. But we would argue that this is the *result* of democratic forces at work—a compromise between clashing pressure groups—rather than a precondition for democracy. In formal terms, we would expect such uniformity as an *outcome* of the theory; a conclusion, not a hypothesis.

Since the majority can, in principle, tax the minority at 100 per cent and return nothing to it, it would seem at first sight that the total endowment of society becomes available to whomever is in the majority. Thus individual endowments would lose their significance, and we would be led to a completely egalitarian analysis. While this kind of analysis is not without interest, it appears too extreme to be considered relevant to the problem of income distribution in a democracy.

We now appear to be caught in a dilemma. On the one hand, majority rule appears to lead immediately to the extreme of total egalitarianism.

---

3. No adverse value judgment is intended. Pressure groups are what democracy is all about—they are as essential to healthy politics as competition is to a healthy economy.

On the other hand, if one ignores politics and does not allow the majority to redistribute income, one is simply left with the initial income distribution. How, then, can one account politically for the type of taxation scheme that is observed, in which some redistribution takes place but net income still remains related to gross income?

To answer this question, we will make an assumption that we consider a basic ingredient of a democratic society, namely that

every agent can, if he wishes, destroy part or all of his endowment.

It goes without saying that the part that is destroyed cannot be taxed. If one thinks of one's endowment as labor, then the above means that there is *no forced labor*: an individual may, if he wishes, "destroy" his labor, by simply working less (or not at all).[4]

It may not be immediately clear why this assumption changes anything—after all, who would want to destroy his endowment—what good would it do to anybody? The answer is that it gives the minority very considerable *threat power*—power that is vital in determining taxes. There are numerous cases in history where farmers' revolts against the tax collector were associated with the destruction of crops; this was recently demonstrated by the French farmers of Normandy who destroyed their crops on the roads against President De Gaulle. Also, a strike involves the destruction of endowment (in the form of labor services) in the face of what are considered unfavorable terms of trade or excessive taxation. Though the minority certainly cannot guarantee to itself any part of its endowment, nevertheless it can say to the majority "it will be neither mine nor thine."[5] This is a powerful threat, which can force the majority to compromise. And it is this compromise that underlies the delicate balance between individual rights and Society's needs inherent in all tax schemes.

There is another element that enters the description of our model. Every agent is assumed to have a von Neumann–Morgenstern (N–M) utility [20] for money. The reader may ask, of what possible relevance can N–M utility functions—which really measure attitudes toward risk—be in a situation that contains no overt elements of risk? The answer is that on the contrary, we are dealing with a situation that is replete with risky elements. The bargaining associated with entry into (or exclusion from) a ruling coalition is a risky business, as are the threats of the minority and of the majority once coalitions have been formed. One's

---

4. In this interpretation no positive utility is assigned to leisure. See the discussion in Section 10, Subsection b.

5. I Kings 3, 26.

attitude toward these risks is a decisive factor in how well one can do in the bargaining, even though in the end no random mechanism is used to determine the payoff, i.e. no risks are taken by anyone.

To sum up, the *Income Redistribution Game* is played as follow: each agent starts out with an endowment and a utility function. Redistribution decisions are made by majority vote, but each agent has the right to destroy some or all of his endowment.

Our results turn out to be insensitive to the exact strategic description of the game (e.g. whether destruction decisions are made before, after, or simultaneously with announcement of tax laws) and therefore we will leave these matters unspecified.

3   The Formal Description of Society

Society—the set of "agents" or "players"—will be denoted T. We will model our assumption that society consists of many individually small agents by taking T to be a *continuum*[6] (like a line or a region) rather than a finite set. In such a model it is best to think of an agent as an infinitesimal subset dt of T; however, we will usually still find it convenient to label agents by points t in T, where the label t is to be thought of as a "typical" point in the infinitesimal set[7] dt.

Next, we require some formal way of measuring the size of coalitions (i.e., sets of agents), in order to be able to say when they are in the majority. If T were finite one could do this simply by counting. Since T is a continuum, counting will not do, and one must specify the measure of size exogenously. We therefore assume given a nonnegative measure[8] μ on the coalitions, with $\mu(T) = 1$ (the *population measure*). Intuitively, one should think of $\mu(S)$ as the proportion of traders in S; thus S is in the majority if and only if $\mu(S) > 1/2$. Note that the measure $\mu(dt)$ of a single agent can be thought of as the reciprocal of the number of people in society.

Finally, for each t in T, there is given a nonnegative number $e(t)$ (t's *endowment*) and a function u_t on the nonnegative numbers (t's *utility*). The endowment function e is to be thought of as a density function, like

---

6. Compare Aumann and Shapley [5], especially Section 29, or Hildenbrand [10]. Technically T is taken as a measurable space, isomorphic to the unit interval with the Borel subsets. Only measurable sets are to be thought of as coalitions.

7. One may think of T as an interval that is cut up into a large number of small subintervals dt, each of which is labelled by a point t in it.

8. A *measure* is a function μ on the coalitions such that $\mu(S \cup W) = \mu(S) + \mu(W)$ when S and W are disjoint, and more generally, $\mu(\bigcup_{i=1}^{\infty} S_i) = \sum_{i=1}^{\infty} \mu(S_i)$ when the S_i are disjoint.

the density functions that occur in probability theory; the actual endowment of an agent t is $e(t)\mu(dt)$, and the total endowment of a coalition S is $\int_S e(t)\mu(dt)$. If x is a function on T, denote $\int_S x(t)\mu(dt)$ simply by $\int_S x$, and $\int_T x(t)\mu(dt)$ by $\int x$. Thus the total endowment of a coalition S is $\int_S e$.

We shall require some assumptions. The first is:

The population measure μ is nonatomic. (3.1)

Nonatomicity means that T can be cut up into coalitions all of which have μ-measure as small as we like; intuitively, it means that the agents are individually negligible. We shall also assume:

The u_t are increasing, concave, continuously differentiable at positive values of the argument, and continuous at 0. (3.2)

$$\infty > \int e > 0.$$ (3.3)

The u_t are uniformly bounded,[9] and $u_t(1)$ is uniformly positive.[10] (3.4)

$$u_t(0) = 0.$$ (3.5)

Assumptions (3.1) and (3.2) are substantive, and without them we could not prove our result. The first half of (3.3)—that $\infty > \int e$—merely says that e is integrable, i.e. that the total wealth of society is finite, whereas the second half—that $\int e > 0$—merely says that some significant set of agents has a positive endowment, without which the whole situation becomes trivial. Assumption (3.4) is also of a technical nature, and it might be possible to dispense with it or at least weaken it. Assumption (3.5) is, of course, merely a normalization.

There are also some very technical measurability assumptions, which we will not spell out here. The interested reader is referred to [4].

As usual, an *allocation* is a nonnegative function x on T such that $\int x = \int e$; here, too, we must view x as a density function. If x is a nonnegative function on T, we shall abuse our notation slightly by writing $u(x)$ for the function on T whose value at t is $u_t(x(t))$; the notation $u'(x)$ is to be interpreted analogously, u'_t being the first derivative of u_t. Since all individual quantities are scaled down to infinitesimal size, it is useful to think even of the utilities as densities; i.e., to think of t's utility for an allocation x as $u_t(x(t))\mu(dt)$.

---

9. $\sup_{x,t} u_t(x) < \infty$.
10. $\inf_t u_t(1) > 0$. This implies that $\inf_t u_t(x) > 0$ for all $x > 0$.

4 The Solution Concept

The game described in Section 2 is "played" on two levels. First, there is maneuvering to determine which players will be in the majority and minority coalitions, respectively; and then, there is bargaining between the two coalitions that actually form, involving threats by both sides. Our aim is to find an outcome that avoids conflict and assigns to each player a payoff reflecting his opportunities on both levels of play; in short, a reasonable compromise. It is our view that the notion most suitable for this purpose is the concept of value introduced by Shapley [17], as adapted to variable threat games by Harsanyi [9], to nontransferable utility (NTU) games by Shapley [18], and to games with a continuum of players by Kannai [11] and Aumann and Shapley [5]. The underlying ideas date back to the Nash variable threat model [15]. This history explains why we will refer to the solution concept adopted here as the Harsanyi–Shapley–Nash NTU value.

To explain this solution concept, we start with finite games. Recall that a *finite coalitional game* (or simply *game*) consists of a finite set T (the "players") together with a function v that associates with each subset S of T ("coalition") a real number $v(S)$ (the "worth" of S), such that $v(\emptyset) = 0$. A *payoff vector* is a measure ξ on the coalitions; intuitively, $\xi(S)$ represents the sum of the payoffs to all members of S. Since T is finite, ξ is determined by its values $\xi(\{t\})$ on one-player sets, so that it may be thought of as the vector of payoffs to the individual players.

A *value* [17] is an operator that associates with each game v a payoff vector ϕv satisfying certain plausible axioms. Here we will quote only the *efficiency* axiom, according to which

$$(\phi v)(T) = v(T), \tag{4.1}$$

and the *additivity* axiom, according to which
$$\phi(v + w) = \phi v + \phi w. \tag{4.2}$$

Shapley [17] also showed that the value is given by the formula

$$(\phi v)(\{t\}) = E(v(S_t^{\mathscr{R}} \cup \{t\}) - v(S_t^{\mathscr{R}})), \tag{4.3}$$

where $S_t^{\mathscr{R}}$ is the set of players preceding t in a random order $\mathscr{R}$ on the set of all players, and E is the expectation operator when all orders on T are assigned equal probability. Formula (4.3) means that the value of a player is the expectation of his contribution to the worth of the players preceding him in a random order of all players.

The concepts of coalitional game and value discussed above have been extended from a finite player set to a continuum of players; see [5, 11]. As in the finite case, the value associates to a coalitional game v a payoff vector ϕv, where by "payoff vector" we mean a measure on coalitions.

We turn now to the problem of defining a coalitional game corresponding to the income redistribution game described in Section 2. This means that we need to specify the "worth" $v(S)$ of each coalition S. In attempting to do that we are faced with two basic difficulties: First, the utilities of the players are not "comparable," so that it is meaningless to speak of one players getting, say, twice as much utility as another; on the other hand, any measure of the worth of a coalition must obviously involve some kind of aggregation of the utilities of the individuals, and it is difficult to see how such an aggregation can be carried out in the absence of comparability. The second difficulty is that even if we could somehow aggregate utilities to determine how much each outcome is worth to each coalition, we still would not know how to determine the worth of a coalition. This is because even after coalitional lines are drawn, the majority and the minority are faced with a fairly complex strategic situation in which threats, counterthreats, and compromises play a crucial role. It is not at all clear a priori how these strategic considerations would interact and to what outcome they would lead.

Recall that since u_t is an N–M utility for t, it follows that for any positive constant λ, λu_t is also an N–M utility. In formal terms, the problem of comparability is that the constants λ can be chosen entirely arbitrarily, with no restriction that they must be the same for different players. Solving this problem means finding a criterion of comparability—i.e., somehow fixing a "weight" $\lambda = \lambda(t)$ for each agent t. If that were done, then one could define the aggregate utility of a coalition S for an allocation x by the expression[11] $\int_S \lambda(t) u_t(x(t)) \mu(dt) = \int_S \lambda u(x)$.

Let us for the moment postpone discussing the choice of weights λ, and proceed at once to the second question, namely that of determining the worth of a coalition in view of the strategic possibilities open to it and its complement. Suppose, then, that the function λ is given; we wish to determine the worth $v_\lambda(S)$ of a coalition S. Consider first the case $S = T$. This is easier because the assumption that the coalition T of all agents has "formed" means that all agents are cooperating; hence it is only a question of maximizing total aggregate utility $\int \lambda u(x)$ subject to $\int x = \int e$, and there is no question of threats, counterthreats, and compromises. To achieve this maximum, the players will presumably reallocate the initial endowment between them in some way; the resulting distribution of net

---

11. This notion of "aggregate utility" is of course not without conceptual difficulties; but its intuitive meaning in connection with the NTU value has been discussed by Shapley [18] and Aumann [3], and we will not repeat the discussion here.

income will be an allocation, i.e. $\int x = \int e$, since total net income must equal total gross income (remember that we are only discussing redistribution of income). The aggregate utility of T will then be $\int \lambda u(x)$. Since the coalition T will act to maximize this aggregate utility, we may conclude that[12]

$$v_\lambda(T) = \max\left\{\int \lambda u(x) : \int x = \int e\right\}. \tag{4.4}$$

Suppose next that $S \neq T$. We will think of $v_\lambda(S)$ as being the aggregate utility of S if it "forms" and bargains as a unit with the agents outside of S. In analyzing this situation, we use a simplified version of the model of Nash [15]. Suppose both S and its complement $T\setminus S$ commit themselves to carrying out certain threats—let's call them σ and τ respectively—if an accomodation between them is not reached. Let's say that carrying out these threats will yield to S and $T\setminus S$ aggregate utilities of, say, $f = f_\lambda(\sigma, \tau)$ and $g = g_\lambda(\sigma, \tau)$ respectively. Thus after the threats are made, the surplus utility over which they are bargaining (and that somehow has to be split between them) is $v_\lambda(T) - f - g$, where $v_\lambda(T)$ is given by (4.4). Under these circumstances, symmetry conditions (cf. Nash [14]) indicate a compromise in which this surplus will be evenly divided between them, so that the resulting aggregate utility to S should be

$$f + \frac{v_\lambda(T) - f - g}{2} = \tfrac{1}{2}[v_\lambda(T) + (f - g)], \tag{4.5}$$

and similarly, to $T\setminus S$ it should be

$$\tfrac{1}{2}[v_\lambda(T) - (f - g)]. \tag{4.6}$$

Of course each of the sides will try to choose its threat in a way that is most advantageous to its final payoff. Hence, if we set

$$H = H_\lambda^S(\sigma, \tau) = f - g = f_\lambda(\sigma, \tau) - g_\lambda(\sigma, \tau),$$

then we may conclude from (4.5) and (4.6) that S will wish to maximize, and $T\setminus S$ to minimize, the expression H. By the minimax theorem[13] for 2-person 0-sum games, there is a number $w = w_\lambda(S)$ such that S can guarantee that H will be at least w, and $T\setminus S$ can guarantee that it will be at most w. Presumably S and $T\setminus S$, if formed would act to "cash

---

12. Fine points such as questions of existence of the maximum will be steadfastly ignored in this paper.
13. We remind the reader that we are ignoring "fine points" such as verifying that the conditions for the minimax theorem hold.

in" on these guarantees, so that one could then expect an outcome that yields to S and $T\setminus S$ aggregate utilities of $1/2(v_\lambda(T) + w_\lambda(S))$ and $1/2(u_\lambda(T) - w_\lambda(S))$ respectively. Thus given the function λ, we may define the worth $v_\lambda(S)$ by

$$v_\lambda(S) = \tfrac{1}{2}(v_\lambda(T) + w_\lambda(S)). \tag{4.7}$$

We are interested in the value ϕv_λ of the game v_λ.

But there is a difficulty here. To see it note first the value ϕv_λ calls for a "redistribution of utilities," since the worths $v_\lambda(S)$ defined in (4.4) are in units of "aggregate utility." In fact, we cannot redistribute utility, but only income; thus it is possible that the value ϕv_λ is not attainable—i.e., there is no redistribution of income (or allocation) that yields each agent the utility that he is assigned by the payoff vector ϕv_λ. In fact, we might remark parenthetically that in general there is not even any allocation that will yield S and $T\setminus S$ the amounts (4.5) and (4.6); that is because (4.5) and (4.6) involve an even split of surplus aggregate *utility*; and in our model, utility, though one may assume it interpersonally comparable, is certainly not transferable—only money can be transferred. Thus the compromises implicit in (4.5) and (4.6), as well as the larger compromise implicit in the value ϕv_λ, may simply be infeasible.

And as if that were not sufficiently worrisome, we must remind the reader that we have not yet resolved the problem of determining the weights λ.

Fortunately, these two difficulties cancel each other out. We will show that there is a unique[14] function λ such that ϕv_λ is feasible; thus the feasibility problem and the problem of determining λ solve each other. Formally, we define a *value allocation* to be an allocation x such that there exists a λ for which

$$(\phi v_\lambda)(dt) = \lambda(t) u_t(x(t)) \mu(dt) \tag{4.8}$$

for all agents t (or equivalently, $(\phi v_\lambda)(S) = \int_S \lambda u(x)$ for all S). This means that the value assigns to each agent precisely the utility he receives under the allocation x. Thus the value is feasible without any "transfers of utility;" the only transfers are those of money.

The Harsanyi–Shapley–Nash value just defined, and the corresponding concept of value allocation, have been thoroughly discussed elsewhere [3, 15, 18], and we will not enter into another such discussion here. A few words about the concept may, however, be in place. Integrating (4.8) and

---

14. This is true only in the income redistribution game described here; in general one will not get uniqueness of the λ in an NTU game, though the set of appropriate λ may be expected to be in some sense "small."

using the efficiency axiom (4.1), we find

$$v_\lambda(T) = (\phi v_\lambda)(T) = \int \lambda u(x).$$

From this and (4.4) it follows that the maximum that defines $v_\lambda(T)$ is achieved at the value allocation x; hence the marginal utilities $\lambda(t)u'_t(x(t))$ must be equal for all agents t with positive income. This means that although utility itself is not transferable, if one scales the utilities u_t by multiplying them by $\lambda(t)$, then in the neighborhood of a value allocation, utility is "locally transferable;" i.e., small amounts of utility can effectively be transferred by transferring money.

Of course for any allocation x one can find λ's with this property—it is only necessary to choose $\lambda(t) = 1/u'_t(x(t))$. But when one then proceeds to calculate ϕv_λ, one will in general find that it is infeasible. Thus the value allocation is uniquely determined by requiring utility comparisons to be made in such a way so that utility is locally transferable, *and* that the allocation be a reasonable compromise, from the point of view of the Shapley value.

5 The Main Theorem: Statement and Preliminary Discussion

MAIN THEOREM   *The income redistribution game has a unique value allocation x. This allocation satisfies*

$$\frac{u_t(x(t))}{u'_t(x(t))} + x(t) = c + e(t) \tag{5.1}$$

for all t, where c is a positive constant.

Note that

$$c = \int \frac{u(x)}{u'(x)}; \tag{5.2}$$

this follows from integrating (5.1) and using $\int x = \int e$.

We will demonstrate the Main Theorem informally in Section 9. A formal proof is given in [4].

Let us explore some implications of the main theorem. First, we have

IMPLICATION 5.3   *For agents with the same utility function, net income is an increasing function of gross income $e(t)$.*

This follows from the fact that by Assumption (3.2), the left side of (5.1) is an increasing function of $x(t)$.

Next, note that

t's income tax[15] $= e(t) - x(t)$. (5.4)

Hence, the *marginal income tax rate* $M(t)$ is the derivative of $e - x$ with respect to e at $e = e(t)$, when x is implicitly defined as a function of e by the equation

$$\frac{u_t(x)}{u'_t(x)} + x = c + e.$$

Implicit differentiation of this equation yields

IMPLICATION 5.5 *Assume that u_t is twice continuously differentiable. Then*

$$M(t) = 1 - \frac{1}{2 + (-u''_t u_t / u'^2_t)},\quad (5.6)$$

where u_t and its derivatives are calculated at $x(t)$.

In particular,

$$\tfrac{1}{2} \leq M(t) < 1, \quad (5.7)$$

i.e., the marginal rate is always at least 50 per cent! In Section 10 we return to this matter of the surprisingly high marginal rate.

Let us now interpret the main theorem in terms of the total rather than the marginal tax. From (5.1) and (5.4) we obtain

$$t\text{'s income tax} = \frac{u_t(x(t))}{u'_t(x(t))} - c. \quad (5.8)$$

Since $x(t)$ is the allocation of income to agent t and since $u'_t(x(t))$ is his marginal utility of income, $\lambda(t) = 1/u'_t(x(t))$ is the *money price* of a unit of t's utility. Thus in the final compromise, agent t receives utility with money valuation $V(t)$ defined by

$$V(t) = \lambda(t) u_t(x(t)) = \frac{u_t(x(t))}{u'_t(x(t))}. \quad (5.9)$$

Combining (5.1), (5.2), and (5.9), we find

IMPLICATION 5.10 t*'s income tax* $= V(t) - \bar{V}$, *where $\bar{V}$ is the average*[16] *of the $V(t)$.*

---

15. It should be recalled that we are really discussing the net effect of taxation *and* redistribution; use of the word "tax" to describe the combination of both processes is merely a matter of convenience. Note also that the decrement in t's income is actually $(e(t) - x(t))\mu(dt)$; thus $e(t) - x(t)$ is really "tax density" rather than "tax."

16. $\bar{V}$ is of course the same as the constant c of (5.1) and (5.2); we use the notation $\bar{V}$ only to emphasize the interpretation of this constant as a mean of the $V(t)$.

This means that a person's tax is the surplus of the dollar value of his utility over the average of the dollar values of everybody's utilities. Thus we get a tax structure which is negative on the lower part of the scale, i.e. provides for positive public transfer to low income agents. In fact if we denote the elasticity of utility by

$$\eta(t) = \frac{u'_t(x(t))}{u_t(x(t))} x(t),$$

then (5.1) can be solved to read

$$x(t) = \frac{\eta(t)}{1 + \eta(t)} (c + e(t)).$$

If agent t has no income, i.e. $e(t) = 0$, then this yields a social "support" (or negative tax) to agent t of size $[\eta(t)/(1 + \eta(t))]c$. As his income increases the agent pays a positive marginal income tax $M(t)$ until the support is exhausted at which time the agent begins to pay a positive amount of total taxes.

6 The Fear of Ruin

In the discussion above (at (5.8)), we saw that

$$t\text{'s income tax} = \frac{u_t(x(t))}{u'_t(x(t))} - c,$$

where c may be viewed as a constant tax credit. Let us now interpret the term $u_t(x(t))/u'_t(x(t))$, in terms of behavior under uncertainty.

For simplicity, set $x = x(t), u = u_t(x), u' = u'_t(x)$. The expression u/u' will be called t's *fear of ruin* at x. To understand why, let us consider its reciprocal u'/u, which will be called t's *boldness* at x. Suppose that t is considering a bet in which he risks his entire fortune x against a possible gain of a small amount h. The probability q of ruin would have to be very small in order for him to be indifferent between such a bet and retaining his current fortune. Moreover, the more unwilling he is to risk ruin, the smaller q will be. Thus q is an *inverse* measure of t's aversion to risking ruin, and a direct measure of boldness; obviously q tends to 0 as the potential winnings h shrink. We assert that the boldness is the *probability of ruin per dollar of potential winnings* for small potential winnings, i.e., it is the limit of q/h as $h \to 0$. To see this, note that for indifference we must have

$$u(x) = (1 - q)u(x + h) + qu(0) = (1 - q)u(x + h).$$

Hence

$$\frac{q}{h} = \frac{(u(x+h) - u(x))/h}{u(x+h)}$$

and as $h \to 0$, this tends to u'/u.

We conclude that *the tax equals the fear of ruin at the net income, less a constant tax credit*. Thus the more fearful a person, the higher he may expect his tax to be.

Next, set $u'' = u_t''(x)$, and let us examine the term

$$\delta_t = -uu''/u'^2 \qquad (6.1)$$

appearing in the expression (5.6) for the marginal tax rate. We have

$$\delta_t = \frac{-u''/u'}{u'/u}.$$

The denominator of the right side is the boldness; its numerator is the measure of *absolute risk aversion* (of t at x), as defined by Arrow [1] and Pratt [16]. To interpret this concept, let us suppose that t considers an even-money bet in a small amount. Since he is risk averse (i.e. $u'' < 0$), the probability p for success would have to be greater than $1/2$ in order for him to be indifferent between this bet and simply retaining his current fortune x. The probability premium $p - 1/2$ is a measure of his aversion to risk at x; because u is differentiable, it tends to 0 as the size of the bet shrinks. The measure of absolute risk aversion is the probability premium per dollar of bet size for small bets, i.e., it is the limit of $(p - (1/2))/h$ as $h \to 0$, where h is the size of the bet.

Conceptually, there are two components that enter into the boldness coefficient. One is t's attitude toward risking his fortune; the other is his attitude toward winnings. We know that he is risk averse, i.e., that even at even money he requires a probability premium for entering into a risk. Obviously this aversion is a factor in determining the probability q that entered into the calculation of boldness. If we wish to measure his attitude "purely" toward risking his fortune, we must somehow cancel out the component that measures his aversion to small risks. This motivates us to define the *pure boldness* as the ratio between boldness and absolute risk aversion, i.e. as $-(u'/u)/(u''/u')$. The *pure fear of ruin* is defined as the reciprocal of the pure boldness, i.e. the δ_t in formula (6.1).

The absolute boldness has the same dimensions as the absolute risk aversion, namely 1/dollar. Therefore the pure boldness and pure fear of ruin are both dimensionless, i.e., invariant under changes in the unit of utility and/or the unit of money.

Returning now to the taxation picture, we see from (5.6) that *the marginal tax rate depends only on the pure fear of ruin δ_t and is directly related to it*: the higher the pure fear of ruin, the higher is the marginal tax rate.

The concept of fear of ruin introduced here is of interest also in an entirely different context, namely that of the Nash Bargaining Problem [14]. Consider a case in which two players are bargaining over the division of a fixed amount R of money, and suppose that their utility functions are u_1 and u_2 respectively, where $u_i(x)$ represents the utility of i for receiving the amount x out of this bargaining process. Assume that the u_i are concave and differentiable and $u_i(0) = 0$. Then the Nash solution calls for maximizing $u_1(x_1)u_2(x_2)$ subject to $x_1 + x_2 = R$. This maximum is attained at the unique point (x_1^*, x_2^*) for which $x_1^* + x_2^* = R$ and

$$\frac{u_1'(x_1^*)}{u_1(x_1^*)} = \frac{u_2'(x_2^*)}{u_2(x_2^*)}. \tag{6.2}$$

We thus find that the Nash solution calls for that compromise which makes the two players *equally fearful of ruin*, where ruin is here taken to mean disagreement. This provides an alternative interpretation of the Nash solution, which has sometimes been ignored by economists because of its strongly cardinal nature. (For another interpretation, see Harsanyi [8].)

7 Illustrations

In this section we will assume that the endowment function $e(t)$ is bounded. From (5.1) it follows that $x(t) \leq c + e(t)$. Since e is bounded, it follows than in any particular example net income of all individuals must lie in a fixed finite interval, which we shall call the "relevant range." The behavior of the utility functions outside of this interval is irrelevant—the results do not change if we change the utility functions outside of the relevant range. In the illustrations of this section, we will describe the utilities only in the relevant range; outside of that, they can be chosen arbitrarily as long as they remain bounded and sufficiently differentiable. For example, when we refer to "linear utilities," we mean "utilities that are linear in the relevant range;" the utilities cannot, of course, be linear on the entire real line, since that would violate the boundedness condition.

Example 7.1 Linear Utilities. One's first impression is that the threat possibilities make the rich (i.e. high endowment) players extremely powerful: not only do they have a higher endowment, but also more to threaten with. It might even be thought that they would end up richer than before. But this is not the case. From the linearity of the u_t one obtains $(u_t(x)/u_t'(x)) + x = 2x$, and hence $x = (1/2)(c + e)$. Integrating, one obtains $c = e(T)$; thus

$$x(t)\mu(dt) = \tfrac{1}{2}\left(\int e\right)\mu(dt) + \tfrac{1}{2}e(t)\mu(dt). \tag{7.2}$$

The expression $(\int e)\mu(dt)$ represents t's "equal share" of Society's endowment, i.e., what he would get if Society's endowment $(\int e)$ were divided equally among all agents. Of course, $e(t)\mu(dt)$ and $x(t)\mu(dt)$ are t's gross and net income respectively. Thus (7.2) says that the net income distribution is a 50–50 compromise between the gross income distribution and an entirely egalitarian distribution—a far cry from our above guess that the rich will get richer! What is happening is that by compromising after a threat to destroy its endowment, the minority can hold on to half of its endowment, while the other half goes to the majority. Thus the term $(1/2)e(t)\mu(dt)$ represents that part of one's income that is not taxed; whereas the term $(1/2)(\int e)\mu(dt)$ represents that part that results from the fact that one has a priori a 50–50 chance of being in the ruling coalition.

In tax terms, we have a tax rate of 50 per cent, with an exemption in the amount of the equal share; this results in a support in the amount of half the equal share, since the exemption can lead to negative taxable income. Alternatively, one could say that everybody gets taxed at a straight 50 per cent (with no support), and that the resulting revenue is redistributed *equally* among the entire population.

Example 7.3 $u_t(x) = x^\alpha$, $0 < \alpha < 1$. Let us try again to guess the result beforehand. Here we have decreasing marginal utility; the less a person has, the greater his marginal utility of income. This kind of situation usually favors the rich, since they are less prone to threats. For example, suppose two people with utility function $\sqrt{x}$ who are worth $0 and $10,000 respectively must agree on how to share an additional $10,000, or else lose the additional amount entirely. Then the Harsanyi–Shapley–Nash value (which in this case boils down to the bargaining solution of Nash [14]) dictates that the richer man will receive approximately $6,404, while the poorer one will receive only $3,596 (see (6.2)). Thus, it seems safe to guess that in Example 7.3 the rich will be relatively better off than in Example 7.1.

Unfortunately, we are wrong again. Here

$$\frac{u_t(x)}{u_t'(x)} + x = \frac{\alpha+1}{\alpha}x,$$

and we find

$$x(t) = \frac{1}{\alpha+1}\left(\int e\right) + \frac{\alpha}{\alpha+1} e(t).$$

Again, we get a compromise between the endowment and the equal share, but in a more egalitarian ratio (when $\alpha = 1/2$ it is 2/3 to 1/3). The smaller α—i.e., the more intensely the poor suffer—the more egalitarian the outcome: an effect exactly the opposite of what we had thought!

The explanation is simple. A rich man who finds himself in the minority is subject to ruin (i.e. 0 utility) just as much as a poor man; and therefore in the minority, he is as prone to threats as the poor man. Thus the increased fear of ruin (x/α as compared with x in the linear case) works to the advantage not of the rich man, but of the majority. Since everybody has an equal chance of being in the majority, this tips the scale away from the initial endowment and towards the egalitarian outcome.

In tax terms, we get a tax rate of $1/(\alpha+1)$ (which may be anywhere between 1/2 and 1), an exemption in the amount of the equal share, and a possibility of negative tax. Alternatively, one could say that everybody gets taxed at the rate $1/(\alpha+1)$, with no exemption, and that the resulting revenue is distributed equally among the entire population.

Example 7.4 $u_t(x) = x^{a(t)}$, $0 < a(t) \leq 1$. Here the fear of ruin varies from agent to agent. One would expect that the more fearful agent—the one with lower $a(t)$—is penalized in the tax structure, since he is more prone to threats. This is indeed the case. We have

$$\frac{u_t(x)}{u'_t(x)} + x = \frac{a(t)+1}{a(t)} x$$

and hence

$$x(t) = \frac{a(t)}{a(t)+1}\left[\left(\int \frac{e}{a+1} \bigg/ \int \frac{a}{a+1}\right) + e(t)\right].$$

The tax is given by

$$e(t) - x(t) = \frac{1}{a(t)+1}\left[e(t) - a(t)\left(\int \frac{e}{a+1} \bigg/ \int \frac{a}{a+1}\right)\right].$$

This means that the agent who is more fearful of ruin has both a higher tax rate and a lower support.

Example 7.5 $u_t(x) = \log(1+x)$. From (5.6) we get a marginal tax rate of $1 - 1/(2 + \log(1+x(t)))$, which rises with $x(t)$, and hence also with $e(t)$. Thus in the case of this very classical utility function, the tax is progressive.

Example 7.6 $u_t(x) = x + \sqrt{x}$. We have

$$\frac{-u_t'' u_t}{(u_t')^2} = \frac{1}{2(1+2\sqrt{x})} + \frac{1}{2(1+2\sqrt{x})^2},$$

which decreases as x increases. Hence by (5.6) the marginal tax rate decreases as $x(t)$ increases, and hence also when $e(t)$ does. The tax is thus regressive; the rate goes from near 2/3 when $x(t)$ is small, to near 1/2 when it is large. What is happening is that when x is close to 0, u_t behaves very much like $\sqrt{x}$, and as it moves up, it behaves more and more like x. The results then follow the pattern of Examples 7.1 and 7.3.

In the next section we will see that there are examples in which the tax is progressive, neutral, and regressive at different points in the range of income.

8 Further Discussion of the Marginal Tax Rate

We have seen (5.7) that the marginal rate is always between 1/2 and 1. In Section 7 we saw that the tax may be progressive, regressive, or neutral, i.e. that the marginal rate may be increasing, decreasing, or constant as a function of wealth. This section is devoted to the question of just what we can say about the behavior of the marginal rate, i.e. to characterizing those functions of wealth that can appear as marginal tax rates.

Call a function u *admissible* if it is bounded, concave, increasing, twice continuously differentiable at positive values of the argument, continuous at 0 and satisfies $u(0) = 0$. Let u be admissible, and for $x > 0$, let

$$m(x) = 1 - \frac{1}{2 + (-u''(x)u(x)/u'(x)^2)}; \qquad (8.1)$$

the marginal tax rate of an individual with utility function u and net income x is precisely $m(x)$. Clearly

$$1/2 \leq m(x) < 1. \qquad (8.2)$$

We can also say something about the marginal tax rates of the very rich; we have

$$\limsup_{x \to \infty} m(x) = 1. \qquad (8.3)$$

Indeed, because of the concavity of u,

$$(x/2)u'(x) \leq u(x) - u(x/2).$$

Because u is bounded, the right side approaches 0, and hence $\lim_{x\to\infty} xu' = 0$. Arrow [1] has proved that when utility is bounded, $\lim_{x\to\infty} \inf(xu''/u') \geq 1$. Hence

$$\lim \inf(u''u/u'^2) = \lim \inf \frac{xu''/u'}{xu'/u} = \frac{\lim \inf(xu''/u')}{\lim xu'/\lim u} = \infty.$$

Hence $\lim \sup m(x) = 1$, as claimed.

Formula (8.3) says that for arbitrarily large net incomes, the marginal tax rate comes close to 1. On the other hand, it cannot come *too* close too often; letting R^+ be the nonnegative part of the real axis, we have

$1/(1-m)$ is integrable over any interval in R^+. (8.4)

Indeed, when the interval has positive end points, this follows from

$$\frac{1}{(1-m)} = 2 - (u''u/u'^2) = (x + (u/u'))';$$

when the left end point is 0, it follows from a limiting argument using $u/u' \to 0$ as $x \to 0$.

The question now arises whether anything can be said about m other than (8.2), (8.3), and (8.4), i.e. whether any function m satisfying these three conditions is the marginal tax rate associated with some admissible utility function. The answer is no; $u = \log(1 + x)$ is not admissible but the corresponding m (see Example 7.5) satisfies all three conditions. However, we have

Remark 8.5   Let m_0 be any continuous function on the positive real numbers satisfying (8.2) and (8.4). Then for each $K > 0$ there is an admissible u whose m coincides with m_0 for $x \leq K$.

This says that (8.2) and (8.4) are enough to characterize marginal tax rates if we restrict ourselves to bounded net incomes. Of course, (8.3) is irrelevant when one is looking only at a bounded set of x.

To prove Remark 8.5 we can simply set

$$u(x) = \exp\left(\int_0^x \left(1 - y + \int_0^y (1 - m_0(z))^{-1} dz\right)^{-1} dy\right) - 1$$

when $x \leq K$; when $x > K$, we adjust u smoothly so that it is bounded. Such a u satisfies all the requirements.

Remark 8.5, together with (8.2) and (8.4), provides a characterization of those functions that can appear as marginal tax rates when one restricts oneself to a bounded set of net incomes.

9 An Informal Demonstration of the Main Theorem

Though the proof of the main theorem is somewhat technical, it is possible, by using the calculus of infinitesimals, to outline the underlying ideas rather quickly. This we will now do. A completely rigorous proof will be presented in [4].

Write

$$r_\lambda(S) = \sup\left\{\int_S \lambda u(x) : \int_S x = \int_S e\right\}, \tag{9.1}$$

$$q_\lambda(S) = \begin{cases} r_\lambda(S) & \text{if } \mu(S) \geq 1/2, \\ 0 & \text{if } \mu(S) < 1/2. \end{cases} \tag{9.2}$$

Intuitively, $r_\lambda(S)$ is the maximum total utility (when weighted by the $\lambda(t)$) that S can get for itself by reallocating precisely its own endowment among its members, i.e., neither taking anything from other people nor having anything taken away from it nor destroying anything. On the other hand, $q_\lambda(S)$ is the maximum aggregate utility that S can *assure* itself.

In this demonstration we assume that $r_\lambda(S)$ is finite and that the sup is actually attained. These assumptions can be removed.

Step 1 We have

$$w_\lambda(S) = q_\lambda(S) - q_\lambda(T \setminus S). \tag{9.3}$$

An optimal pair of strategies in the threat game H_λ^S is for the majority to take from the minority everything that the minority does not destroy (i.e., to tax at 100 per cent), and for the minority to destroy its entire endowment.[17]

Demonstration First let S be in the majority. By taxing at 100 per cent and reallocating its own endowment, S can assure that its own payoff will be at least $r_\lambda(S)$, and that of its complement 0. Therefore it assures itself a payoff of $r_\lambda(S)$ in the game H_λ^S. On the other hand, by destroying its endowment, the minority can assure that the majority's payoff in H_λ^S will not be more than $r_\lambda(S)$. Hence the strategies described are indeed optimal, and when $\mu(S) > 1/2$, (9.3) is proved. The argument when S is in the minority is similar, using the fact that $T \setminus S$ is then in the majority.[18]

---

17. Of course these threats are not carried out in the compromise that is finally reached.
18. In this heuristic treatment we ignore the case in which $\mu(S)$ is exactly 1/2. In the rigorous treatment of Aumann-Kurz [4] it is of course taken into account.

Step 2 $\phi v_\lambda = \phi q_\lambda$.

Demonstration Define the game $q_\lambda^\#$ dual[19] to q_λ by $q_\lambda^\#(S) = q_\lambda(T) - q_\lambda(T \setminus S)$. By reversing orderings in (3.3), one can easily show that $\phi q_\lambda^\# = \phi q_\lambda$. On the other hand, from (4.7) and (9.3) it follows that $v_\lambda = (1/2)q_\lambda + (1/2)q_\lambda^\#$. The result then follows from the additivity axiom for the value (4.2).

Step 3 Suppose $v_\lambda(T)$ is attained at x. Let p be the shadow price associated with the maximization, i.e. $p = \lambda(t)u_t'(x(t))$ when $x(t) > 0$. Then

$$(\phi q_\lambda)(dt) = \tfrac{1}{2}r_\lambda(T)\mu(dt) + \tfrac{1}{2}(\lambda(t)u_t(x(t)) - p(x(t) - e(t)))\mu(dt).$$

Demonstration Let us fix attention on a particular agent; in this demonstration it will be convenient to refer to him directly as dt, rather than by the label t. Denote by S the set of all agents up to and including dt in a random order on all the agents; the value $(\phi q_\lambda)(dt)$ is the expectation of the contribution $q_\lambda(S) - q_\lambda(S \setminus dt)$ of dt to S (see (4.3)). The probability is $1/2$ that S is in the minority, in which case dt contributes nothing. With probability $\mu(dt)$ (i.e. the reciprocal of the number of players), S is in the majority with dt and in the minority without him, i.e. dt is "pivotal." Because the ordering is random and there are many agents, S is almost surely an almost perfect sample of all the agents (insofar as utilities, endowments and the $\lambda(t)$ are concerned). That is, S is just like T, but is operating at half the scale (because $\mu(S) = 1/2$). In this case, therefore, dt's contribution is $r_\lambda(S) = (1/2)r_\lambda(T)$.

Finally, with probability $1/2$, S is in the majority with and without dt. Again because of the random order, S is almost surely an almost perfect sample of the population T of all agents, and dt's contribution $r_\lambda(S) - r_\lambda(S \setminus dt)$ is the same as if he were the last agent, i.e., the same as $r_\lambda(T) - r_\lambda(T \setminus dt)$. But now a simple computation (see for example, Aumann [3, Section 8, Step 5]) shows that the latter contribution is just

$$(\lambda(t)u_t(x(t)) - p(x(t) - e(t)))\mu(dt). \tag{9.4}$$

Summing up, dt contributes nothing with probability $1/2$, $(1/2)r_\lambda(T)$ with probability $\mu(dt)$, and (9.4) with probability $1/2$. So his expected contribution $(\phi q_\lambda)(dt)$ is precisely as asserted.

Step 4 *If x is a value allocation, then there is a constant c satisfying (5.1).*

Demonstration That x is a value allocation means that $u(x)$ is a value, i.e.,

---

19. Compare Aumann and Shapley [5, p. 140], or Milnor and Shapley [12].

$$\lambda(t)u_t(x(t))\mu(dt) = (\phi v_\lambda)(dt)$$

(see (4.8)). Combining this with Steps 3 and 2, we get

$$\lambda(t)u_t(x(t)) = \tfrac{1}{2}r_\lambda(T) + \tfrac{1}{2}(\lambda(t)u_t(x(t)) - p(x(t) - e(t))),$$

whence

$$\frac{1}{p}\lambda(t)u_t(x(t)) + x(t) = \frac{1}{p}r_\lambda(T) + e(t). \tag{9.5}$$

From this it follows that $x(t) > 0$, and hence $p = \lambda(t)u'_t(x(t))$. Inserting this into (9.5) and setting $c = r_\lambda(T)/p$, we deduce (5.1).

Step 5   *If x is an allocation satisfying (5.1), then it is a value allocation.*

Demonstration   First note that $x(t) > 0$, since $c > 0$. Hence $u'_t(x(t))$ is defined and is positive. Set $\lambda(t) = 1/u'_t(x(t))$, and set $p = 1$; then (5.1) may be written in the form

$$\lambda(t)u_t(x(t)) = c + p(e(t) - x(t)). \tag{9.6}$$

Integrating and recalling that x is an allocation, we obtain

$$\int \lambda u(x) = c.$$

Because $\lambda(t)u'_t(x(t)) = 1$, the integral on the left actually achieves the maximum defining $r_\lambda(T)$, i.e. $c = r_\lambda(T)$. Inserting this in (9.6) and rearranging, we find

$$\lambda(t)u_t(x(t)) = \tfrac{1}{2}r_\lambda(T) + \tfrac{1}{2}(\lambda(t)u_t(x(t)) - p(x(t) - e(t))).$$

Hence by Steps 3 and 2,

$$\lambda(t)u_t(x(t))\mu(dt) = (\phi v_\lambda)(dt). \tag{9.7}$$

Since x is an allocation, it follows from (9.7) and (4.8) that it is a value allocation, as was to be shown.

Step 6   *There are precisely one allocation x and one real constant c satisfying (5.1), and this constant is necessarily positive.*

Demonstration   Define

$$g_t(x) = \frac{u_t(x)}{u'_t(x)} + x;$$

g_t is increasing and continuous, is defined for all nonnegative numbers, vanishes at 0, and tends to infinity as $x \to \infty$. Hence its inverse g_t^{-1} is defined and has the same properties. Moreover, $g_t(x) \geq x$, and so

$$g_t^{-1}(y) \leq y, \tag{9.8}$$

with strict inequality holding when $y > 0$. For each nonnegative number γ, define

$$f(\gamma) = \int_T g_t^{-1}(\gamma + e(t))\mu(dt); \tag{9.9}$$

by (9.8) and Assumption (3.3) the integral is finite. Again using (9.8) we get

$$f(0) < \int e.$$

Next, if we go to the limit under the integration sign[20] in (9.9) and use the fact that $g_t^{-1}(x) \to \infty$ as $x \to \infty$, then we deduce that for sufficiently large γ,

$$f(\gamma) > \int e.$$

Moreover f is strictly increasing (since the g_t^{-1} are) and continuous.[21] Hence there is one and only one γ with $f(\gamma) = \int e$; denote this γ by c and set

$$x(t) = g_t^{-1}(c + e(t)).$$

By construction x is an allocation, and together with c satisfies (5.1). Conversely, if x and c satisfy (5.1), then integrating (5.1) yields $c = \int u(x)/u'(x) > 0$, and then inverting (5.1) yields (9.10). Integrating (9.10) yields $\int e = \int_T g_t^{-1}(c + e(t))\mu(dt)$, and hence c and x can only be those already found. This completes Step 6.

Demonstration of the Main Theorem We have shown in Steps 4 and 5 that an allocation x is a value allocation if and only if there is a constant c satisfying (5.1). By Step 6, there are precisely one x and c satisfying (5.1); and $c > 0$.. Hence there is precisely one value allocation, and it satisfies (5.1), which is what the Main Theorem asserts.

---

20. This may be justified by a standard theorem like Fatou's lemma or the monotone convergence theorem.

21. This follows from the fact that $|g_t^{-1}(y_2) - g_t^{-1}(y_1)| < |y_2 - y_1|$, which in turn follows from the definition of g_t.

10 Additional Discussion of the Results

In the development above we passed over issues that are important but perhaps somewhat controversial. This we did for the sake of continuity; we now wish to address ourselves to some of these issues.

(a) Aggregation

Our purpose in this paper is to try to expose the effect of political power on income redistribution. Taxation has other important aspects, such as its effect on relative prices in a multicommodity economy and aspects relating to the production of private and public goods. The reader will agree, however, that an important element of taxation is the redistributive one, even in the case in which there is no overt cash redistribution, but all revenue is used for the provision of public goods. The relatively simple model of this paper deals with the redistributive element only; other, more complex issues are left for subsequent studies.

(b) Pareto Optimality, Second Best and the Problem of Leisure

An argument may be made in connection with the fact that one of the axioms underlying the Shapley value is Pareto optimality. This means that the value allocation x is Pareto optimal, in spite of the fact that we are introducing an explicit income tax into the system. It is known that the introduction of income taxation may in theory lead individuals to take more leisure than they would take in the absence of taxation, and so may lead to outcomes that are not Pareto optimal.

Certain aspects of the nonoptimality of taxation are implicit in the fact that any agent may destroy his endowment. If we think of one component of the endowment as "labor services," then within the context of the threat game, coalitions may threaten to withhold their labor services from the market. The problem arises with regard to the final compromise embodied in the value allocation x. What we are saying is that as part of the compromise, each individual promises to offer the same amount of labor services as before the bargaining process. Thus if $e(t)$ was the endowment of agent t, this same endowment will be offered to the market at the end. The difficulty with this is that in a decentralized, democratic society an individual may choose to work as much as he wishes. He thus may accept the compromise tax function but then work less and end up with a new $e^\tau(t)$ which is the taxable income arising from a tax schedule τ. The usual argument is that $e^\tau(t)$ and $e(t)$ may not be the same and the allocation based on $e^\tau(t)$ may not be Pareto optimal.

Although we think there may be possibilities of reformulating our model so as to take into account these "incentive effects" of taxation, we hesitate to do so because of the rather mild empirical evidence in support of "incentive effects." Both historical and cross-section analysis indicate a negligible wage elasticity of labor supply for males and a slightly higher elasticity for females (see, for example, Hall [7] and Boskin [6]). Thus our assumption appears to be supported by the empirical evidence.

(c) **Individualized Tax Schedules**

Our Main Theorem indicates that the tax rate depends upon the agent's utility function as well as on his gross income; thus the solution calls for individualized tax rates. The same phenomenon arises in all welfare theoretic treatments that seek optimal income distribution. For example, the criterion of "equal marginal utility of income" entails individualized taxation. Naturally we are aware of the fact that taxes ought to be "uniform;" this question was discussed already in Section 2, in a somewhat different context (that of majority power in the threat game). As we said there, the extremely complex tax laws that one observes are in fact designed to provide some degree of individualization. Moreover, it must be remembered that we are discussing redistribution as well as taxation; though taxation may be required to be "uniform," certainly there is no way to prevent individualized redistribution.

(d) **Cardinality**

The cardinal nature of the results may disturb some economists, who prefer the ordinal concepts that are familiar from general equilibrium theory. The point is that in the context of a power struggle, where threats are involved, it is to be expected that intensity of preference will play a determining role. Even on a purely intuitive level, it seems clear that in a bargaining situation, the more fearful side, or the side that is more "intense"—more interested in the outcome—is also the weaker one. "Fear" and "intensity" are cardinal concepts, and it would seem that they must be explicitly taken into account in any situation involving bargaining. For this reason we would expect value theories to be necessarily cardinal; on the contrary, it is surprising that such results can be obtained without interpersonal comparisons. For a more detailed analysis, see Shapley [18].

Our cardinal results should be contrasted with those of Aumann [3], who investigated value allocations in societies using the economic mechanism of exchange only, without any political voting mechanism. One of the surprising conclusions of that study was that though cardinal utilities enter in an essential manner into the description of the model, the outcomes are independent of the choice of cardinal utilities, and depend on ordinal preferences only. The reason for this phenomenon is that in a nonatomic market situation, the marginal contribution of an agent is essentially the same almost whatever coalition he joins, so that the result of the coalitional maneuvering is a foregone conclusion and there is little or no risk. But here that is far from being the case, as there is an enormous difference between being in the majority and being in the minority. The willingness to risk being in the minority—and to have the minority make threats—is therefore of vital importance. This explains in particular why the fear of ruin is decisive, since the optimal threat of the minority is precisely to ruin its members, and only in that way can it get a fair deal from the majority.

(e) **The High Tax Rates**

An eyebrow-raising result of this study is that the marginal tax rate must be at least 50 per cent. This result is due to our assumption of absolute majority rule. It can be shown that if the voting rules are altered to give some weight to wealth, the tax rates tend to be smaller.[22]

In fact, the power structure of existing societies does generally give weight to wealth; i.e., wealthier individuals have more say in decision making. In a representative democracy, the extra weight given to wealth arises from the fact that in their votes, the representatives are more responsive to the stronger pressure groups: in addition to the fact that the elected representatives come from the wealthier classes, the need to use resources in running for office enables the owners of wealth to form more effective pressure groups. The result is a lower marginal tax than is indicated in this study, which might be considered the extreme case of "pure democracy."

(f) **Concavity and Boundedness of the Utility Functions**

Concavity is merely an assertion of general risk averseness. Boundedness is a natural assumption in the context of von Neumann–Morgenstern utilities, since unbounded utilities lead to the St. Petersburg Paradox.

(g) **Interpretation of the Weights λ**

The weight $\lambda(t)$ should be considered a measure of the importance of t's utility, arising endogenously out of the given power structure; indirectly,

---

22. In the extreme case in which decisions are reached by a majority vote of the wealth rather than the population (i.e. "money votes") there are no taxes at all; i.e. the tax rate is 0.

it is an index of t's power. Society behaves as if it were maximizing the expression $\int \lambda u(x)$, i.e., the sum of individual utilities weighted by their relative importance. This expression resembles the social welfare function mentioned in the introduction; but the underlying idea is quite different, since the equilibrium values of λ change if we change the initial distribution of wealth or if we change the utilities. But apart from that we eschew all the paternalistic, ethical connotations of the phrase "social welfare," and so prefer here to call $\int \lambda u(x)$ a *social power function*. Society maximizes total weighted utility, where the more powerful individuals are considered more important—i.e., are given more weight. Of course we are not advocating this (or any other) procedure; but most observers will agree that in fact, society takes much less account of "equity" or "fairness" than of power.

This may at first sound cynical, but when one examines its implications more carefully it turns out to be quite the opposite: a reaffirmation of the importance of democracy. Since social decision making is a function of power, it follows that to achieve equity in the outcome, one must build equity into the political institutions. A system that concentrates power in the hands of the few, even with the most idealistic intentions, must in the end benefit the few. It has long been known that income distributions in countries with rightist totalitarian regimes are more skewed than in the western democracies; but as Wiles [21] has shown,[23] this is to some extent true also for leftist totalitarian regimes (once one removes the cosmetic layers under which official figures are buried). So it is not ideology that is decisive, but the power structure; and if we want a more equitable society, we must develop institutions that spread the political power as thinly and evenly as possible.

References

1. Arrow, K. J.: *Aspects of the Theory of Risk-Bearing*. Helsinki: Yrjö Jahnssonin Säätiö, 1965.

2. Arrow, K. J., and M. Kurz: *Public Investment, the Rate of Return, and Optimal Fiscal Policy*. Baltimore: Johns Hopkins Press, 1970.

3. Aumann, R. J.: "Values of Markets with a Continuum of Traders," *Econometrica*, 43 (1975), 611–646 [Chapter 51].

4. Aumann, R. J., and M. Kurz: "Power and Taxes in a Multi-Commodity Economy," *Israel Journal of Mathematics*, 27 (1977), 185–234 [Chapter 53].

5. Aumann, R. J., and L. S. Shapley: *Values of Non-Atomic Games*. Princeton: Princeton University Press, 1974.

23. We are indebted to E. Sheshinski for this reference.

6. Boskin, M. J.: "The Economics of Labor Supply" in *Income Maintenance and Labor Supply*, ed. by G. G. Cain and H. W. Watts. New York: Academic Press, 1973.

7. Hall, R. E.: "Wages, Income, and Hours of Work in the U.S. Labor Force" in *Income Maintenance and Labor Supply*, ed. by G. G. Cain and H. W. Watts. New York: Academic Press, 1973.

8. Harsanyi, J. C.: "Approaches to the Bargaining Problem Before and After the Theory of Games: A Critical Discussion of Zeuthen's, Hicks's and Nash's Theories," *Econometrica*, 24 (1956), 144–157.

9. ———: "A Bargaining Model for the Cooperative n-Person Game" in *Contributions to the Theory of Games IV*, ed. by R. D. Luce and A. W. Tucker. Princeton: Princeton University Press, 1959.

10. Hildenbrand, W.: *Core and Equilibria of a Large Economy*. Princeton: Princeton University Press, 1974.

11. Kannai, Y.: "Values of Games with a Continuum of Players," *Israel Journal of Mathematics*, 4 (1966), 54–58.

12. Milnor, J. W., and L. S. Shapley: "Values of Large Games: Oceanic Games," *Mathematics of Operations Research*, 3 (1978), 290–307.

13. Mirrlees, J. A.: "An Exploration in the Theory of Optimum Income Taxation," *Review of Economic Studies*, 38 (1971), 175–208.

14. Nash, J. F.: "The Bargaining Problem," *Econometrica*, 18 (1950), 155–162.

15. ———: "Two Person Cooperative Games," *Econometrica*, 21 (1953), 128–140.

16. Pratt, J. W.: "Risk Aversion in the Small and in the Large," *Econometrica*, 32 (1964), 122–136.

17. Shapley L. S.: "A Value for *n-Person Games"* in *Contributions to the Theory of Games II*, ed. by H. W. Kuhn and A. W. Tucker. Princeton University Press, 1953.

18. ———: "Utility Comparisons and the Theory of Games," in *La Décision*. Paris: Editions du Centre National de la Recherche Scientifique, 1969.

19. Sheshinski, E.: "The Optimal Linear Income-Tax," *Review of Economic Studies*, 39 (1972), 297–302.

20. Von Neumann, J., and O. Morgenstern: *Theory of Games and Economic Behavior*. Princeton: Princeton University Press, 1944.

21. Wiles, P.: *Distribution of Income: East and West*. Amsterdam: North-Holland, 1974.

53 Power and Taxes in a Multi-Commodity Economy
with M. Kurz

> Taxation and redistribution in a democratic majority-rule society are analyzed, using the Harsanyi–Shapley non-transferable utility value. The context is that of a multi-commodity pure exchange economy. Two approaches are treated: one in which taxes are in kind and exchange takes the form of barter; and one in which taxes are in money, exchange takes the form of sale and purchase, and prices are determined by a process of supply and demand. It is shown that in the presence of a non-atomic continuum of agents, the two approaches are equivalent, but that this is not so when there are only finitely many agents. It is also shown that the value exists under both approaches, and a characterization is found in the non-atomic case.

Most of modern economic theory treats the public sector as a "benevolent" social agent who behaves so as to maximize some social welfare function. In a different paper [2] we propose an alternative view in which the public sector with its fiscal structure emerges as an endogenous consequence of the power structure of society. We thus propose to regard taxation and the redistribution of wealth as determined simultaneously with the process of exchange, resulting in an outcome that is in equilibrium in both the economic and political spheres.

In the earlier paper we considered a relatively simple, one-commodity (i.e. "money") economy, where the entire struggle focused on income redistribution. Here we extend the characterization to a general l-commodity exchange economy. The classical competitive equilibrium in this economy is replaced by a game theoretic equilibrium, based on the Shapley value, that takes into account not only the economic function of exchange but also the political functions of voting and majority rule.

This work was supported by U.S. National Science Foundation Grant SOC74–11446 at the Institute for Mathematical Studies in the Social Sciences, Stanford University, and by a grant from the Israel National Council for Research and Development at the Hebrew University of Jerusalem.
Received June 24, 1976

This chapter originally appeared in *Israel Journal of Mathematics* 27 (1977): 185–234. Reprinted with permission.

Within this framework, two different approaches will be investigated. In the first, called the "Commodity Redistribution" approach, the entire economic-political system is considered as a single game, and the equilibrium we propose is simply the non-transferable-utility Shapley value of this one game (see [14]). In the second, called the "Income Redistribution" approach, the economic side of the model — consumption and exchange — is assumed to take place in a normal competitive environment, whereas the political side — the redistribution of income — is assumed governed by game theoretic considerations; specifically, by the Shapley value.

One of our major results (Theorem B) is that when the space of agents is a non-atomic continuum — representing the idea of many agents, each individually insignificant — then these two approaches lead to the same result. We will find that this is emphatically *not* true when the number of agents is finite. Our other major results are a characterization of the resulting allocations in the non-atomic case, and a general existence theorem that covers both approaches, for both non-atomic and finite populations.

Sections 2, 3, and 4 are devoted to game theoretic preliminaries and definitions. In Section 5 we present the basic exchange model. Sections 6 and 7 describe in detail the two approaches we have just outlined. The major results are stated in Section 8 and discussed in Section 9. Sections 10 and 19 are devoted to examples, and the remainder of the paper to proofs.

2. Values of finite coalitional games

Let T be a finite set; call the members of T *players* and the subsets of T *coalitions*. A *coalitional game* on T (or simply *game*) is a function v that associates with each coalition S a real number $v(S)$ (the *worth* of S), such that $v(\emptyset) = 0$. A *payoff vector* on T is a measure on the subsets of T. Intuitively, a payoff vector is simply a function that assigns a real number (the payoff) to each player; such functions are in an obvious 1-1 correspondence with the measures on T. The number of members of T is denoted $|T|$.

A *null player* in a game v is a player i such that $v(S \cup \{i\}) = v(S)$ for all coalitions S. Players i and j are called *substitutes* if $v(S \cup \{i\}) = v(S \cup \{j\})$ whenever S contains neither i nor j. For a fixed player set T, a *value* is a function ϕ that associates with each game v a payoff vector ϕv satisfying the following conditions:

(2.1) *Additivity*: $\phi(v + w) = \phi v + \phi w$.

(2.2) *Symmetry*: $(\phi v)(\{i\}) = (\phi v)(\{j\})$ *whenever i and j are substitutes.*

(2.3) *Efficiency*: $(\phi v)(T) = v(T)$.

(2.4) *Null player condition*: $(\phi v)(\{i\}) = 0$ whenever i is a null player.

The definition of value is due to Shapley [13], who also proved the following basic characterization of values:

PROPOSITION 2.5. *For each finite player set T there is one and only one value ϕ; it is given by the formula*

(2.6) $$(\phi v)(\{i\}) = E(v(S_i \cup \{i\}) - v(S_i)),$$

where S_i is the set of all players preceding i in a random order on T, and E is the expectation operator when all $|T|!$ such orders are assigned equal probability.

For a proof of Proposition 2.5 that is even simpler than the original proof (Shapley [13]), see appendix A of [4].

Define the *dual*[1] v^* of a finite game v by $v^*(S) = v(T) - v(T \setminus S)$. By reversing orders in Proposition 2.5, it is easy to see that

(2.7) $$\phi v^* = \phi v.$$

3. Values of non-atomic games

In much of this paper we shall be working with a non-atomic continuum of players, and for this reason it is important to examine the extension of the above model to the non-atomic case.

Let $(T, \mathscr{C}, \mu)$ be a non-atomic measure space; i.e. a set T, together with a σ-field $\mathscr{C}$ of subsets of T, and a non-atomic, non-negative measure μ on $\mathscr{C}$ with $\mu(T) = 1$. One should not think of the points t of T as individual players; rather, one should think of an individual player as an infinitesimal subset dt of T. The measure μ is the population measure, i.e. $\mu(S)$ represents the proportion of the total population in S.

A *coalitional game* (or simply *game*) on the measurable space $(T, \mathscr{C})$ is a function v from $\mathscr{C}$ to the real numbers such that $v(\emptyset) = 0$. There are various ways of extending the definition of value from the finite games above to the situation we have here; see [4]. Here we will adopt a variant of a definition due to Kannai [9]. In this variant a special role is played by the family $\mathscr{C}'$ of S in $\mathscr{C}$ with $\mu(S)$ rational; these S will be called *coalitions*.

---

[1] Cf. [10], or [4, p. 140].

Let v be a game and S a coalition; we wish to define $(\phi v)(S)$. Intuitively, one proceeds by dividing both S and $T\setminus S$ into a large but finite number of "small" sets W_i, all of equal μ-measure; let there be n in all (see Fig. 1).

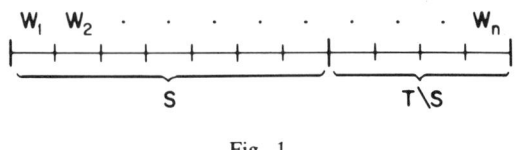

Fig. 1.

If one considers each of these small sets as an individual player, then one gets a finite game, by restricting v to unions of the W_i. In this finite game the coalition S — or more precisely the coalition of those W_i whose union is S — has a value, which we denote $\phi_n(S)$. Now if we let $n \to \infty$ and the W_i shrink, $\phi_n(S)$ may or may not tend to a limit. If it does, and if this limit is independent of the various choices that must be made, then we denote the limit $(\phi v)(S)$, and call it the "μ-value of S."

Formally, let $\Pi_1, \Pi_2, \cdots$ be a sequence of partitions of T into measurable sets (i.e. members of $\mathscr{C}$) of equal μ-measure, such that S is a union of members of Π_1, and each Π_m refines the previous partition Π_{m-1}. Assume moreover that the sequence is *separating*, i.e. that if s and t are distinct points in T, then for m sufficiently large, s and t are in different members of Π_m. For example, if Π_m is the partition of the unit interval $[0,1]$ into the subintervals $[0, 1/2^m]$, $(1/2^m, 2/2^m], \cdots, (1 - 1/2^m, 1]$, then the sequence $\{\Pi_m\}$ is separating. For each m, let v_m be the finite game on Π_m defined by $v_m(\Xi) = v(\bigcup_{W \in \Xi} W)$, and let ϕv_m be its value. Further, let $S_m = \{W \in \Pi_m : W \subset S\}$; S_m is the coalition in the finite game v_m corresponding to S in the original game v. Now let $m \to \infty$; if $(\phi v_m)(S_m)$ has a limit, and if this limit is independent of the choice of the sequence $(\Pi_1, \Pi_2 \cdots)$ (when chosen in accordance with the above conditions), then the limit is denoted $(\phi v)(S)$. If this is the case for all coalitions S, then the function ϕv is called the μ-*value* of v.

The game model of this section differs from the standard model of non-atomic games [4] in that in addition to T, $\mathscr{C}$ and v, we are here given an underlying measure μ. The μ-value is a variant of the asymptotic value [4, pp. 126–127]; it is obtained from the latter by considering only partitions of T into sets of equal μ-measure. Obviously if v has an asymptotic value, then it also has a μ-value, and it equals the asymptotic value; but the converse is false.[2]

---

[2] Example 19.2 of [4] has a λ-value but no asymptotic value.

In the applications below we deal with voting games in which we stress the democratic "one man—one vote" principle. In this context it is natural to use finite approximations in which the elements of the partitions are assigned equal amounts of the population measure, and this leads directly to the μ-value defined above.

As in the case of finite games, we define the *dual* v^* of a game v by $v^*(S) = v(T) - v(T\setminus S)$. Then by using a limiting argument and (2.7), one can easily show that v^* has a μ-value if and only if v does, and in that case

(3.1) $$\phi v^* = \phi v.$$

The μ-value satisfies conditions analogous to the axioms (2.1)–(2.4) defining the finite value. For future reference we quote here only the efficiency axiom

(3.2) $$(\phi v)(T) = v(T),$$

which follows easily from the efficiency axiom (2.3) for finite games.

It is sometimes convenient to treat finite and non-atomic games in the same context, and in particular to refer to the "μ-value" of a game that may either be non-atomic or finite. In that case, the μ-value of a finite game will be taken to be simply the value.

4. Threats

A *strategic game* Γ consists of

i) A measure space $(T, \mathscr{C}, \mu)$ (T is the *player space*, and μ the *population measure*; S in $\mathscr{C}$ with $\mu(S)$ rational are *coalitions*).

ii) For each coalition S, a set X^S (the *strategies* of S).

iii) For each coalition S, each strategy σ of S, each strategy τ of $T\setminus S$, and each t in T, a number $h^S_{\sigma\tau}(t)$ (the *payoff* to t), such that $h^S_{\sigma\tau}(t) = h^{T\setminus S}_{\tau\sigma}(t)$.

When T is finite, this is something very similar to the traditional definition of "games in normal form" [17]. However, it is not entirely the same; the point is that here strategies are assigned to coalitions, and not just to individual players. Though formally it is easy to derive coalitional strategies from individual strategies and vice versa, there may be certain strategies that are more naturally described in terms of coalitions than individuals, e.g. those involved in the imposition of taxes. Moreover, in the non-atomic case, coalitional strategies enable us to bypass the technical complexities that would arise from the need to define payoffs to "infinity-tuples" of individual strategies. For these two reasons we prefer the definition as given.

Strategic games will be assumed to satisfy the following four conditions, which are not substantive but merely a matter of convenience:

(4.1) $h_{\sigma\tau}^S(t) \geq 0$.

(4.2) $\mu(T) = 1$.

(4.3) $h_{\sigma\tau}^S$ is measurable in t for all S, σ, and τ.

(4.4)   *The empty coalition $\emptyset$ has exactly one strategy.*

In view of (4.4), we shall write h_σ^T instead of $h_{\sigma\tau}^T$. The meaning of (4.4) is that $\emptyset$ has no real choice of strategies and cannot affect the payoff.

All strategic games treated here will either have finite T, or non-atomic μ. If T is finite, we assume that $\mathscr{C}$ consists of all subsets of T, and $\mu(\{t\}) = 1/|T|$ for all t in T.

Before we go further let us agree on some notational conventions. The family of coalitions (i.e. S in $\mathscr{C}$ with $\mu(S)$ rational) is denoted $\mathscr{C}'$. If f is a (vector or scalar) function on T, we shall write $\int_S f$ for $\int_S f(t)\mu(dt)$, and $\int f$ for $\int_T f$. The set of non-negative real numbers will be denoted R^+. W.r.t. means "with respect to"; w.l.o.g. means "without loss of generality"; integrable means "μ-integrable."

For $S \in \mathscr{C}'$ and $(\sigma, \tau) \in X^S \times X^{T\setminus S}$, write

(4.5) $$H^S(\sigma, \tau) = \int_S h_{\sigma\tau}^S - \int_{T\setminus S} h_{\sigma\tau}^S.$$

We may view H^S as a 2-person 0-sum game, the players being the coalitions S and $T\setminus S$. If this game has a saddle point (σ_0, τ_0) — i.e., if[3]

(4.6) $$H^S(\sigma, \tau_0) \leq H^S(\sigma_0, \tau_0) \leq H^S(\sigma_0, \tau)$$

for all σ in X^S and τ in $X^{T\setminus S}$ — then we shall denote the minmax value $H^S(\sigma_0, \tau_0)$ of this game by $w(S)$. Note that

(4.7) $$w(T) = \max\left\{\int h_\sigma^T : \sigma \in X^T\right\}.$$

Finally, write

(4.8) $$v(S) = \frac{1}{2}w(T) + \frac{1}{2}w(S)$$

---

[3] Implicit in formula (4.6) is the assumption that the three expressions appearing therein are defined as extended real numbers, i.e., that none of them is of the form $\infty - \infty$.

for all S in $\mathscr{C}'$. We call v the *Harsanyi coalitional form* of the strategic game Γ, and say that it is *defined* whenever all the games H^S have saddle points.

Intuitively, the games H^S, w and v are meaningful only when utility is "transferable." In that case one can meaningfully speak of $\int_S h_{\sigma\tau}^S$ as the total payoff to S, since S can divide that sum in an arbitrary way among its members.

If a coalition S acts in concert in such a situation, what total payoff can it expect? The answer given by von Neumann and Morgenstern [17] is

$$(4.9) \qquad \min_{\tau \in X^{T\setminus S}} \max_{\sigma \in X^S} \int_S h_{\sigma\tau}^S.$$

This has been criticized as "too pessimistic" because it assumes the worst possible, as if the only object of the complementary coalition $T\setminus S$ is to minimize the payoff to S. A more sophisticated answer, based on a subtle interplay of threats, counterthreats, and compromises, was suggested by Harsanyi [6], who followed up the pioneering work of Nash [12] on the subject. Suppose the members of S have decided to act in concert, as have the members of $T\setminus S$. Presumably S and $T\setminus S$ will eventually wish to cooperate so that they will jointly receive the maximum payoff $w(T)$ (see (4.7)); the only question is how this amount is to be divided between them. Before deciding on this, each side will wish to put itself in as good a bargaining position as possible vis-à-vis its opponent. To this end, each side makes a threat — i.e., announces a strategy to be carried out in case of disagreement. If the threats are σ and τ, the payoffs in case of disagreement will be $\int_S h_{\sigma\tau}^S$ to S, and $\int_{T\setminus S} h_{\sigma\tau}^S$ to $T\setminus S$. Thus the total payoff is $\int h_{\sigma\tau}^S$, which is in general less than the maximum amount $w(T)$ that T can achieve. It therefore seems reasonable for the sides to compromise by splitting the difference between $w(T)$ and $\int h_{\sigma\tau}^S$, and adding this amount to the disagreement payoff of each side. The final payoff to S will then be $(w(T) + H^S(\sigma, \tau))/2$, and to $T\setminus S$, it will be $(w(T) - H^S(\sigma, \tau))/2$. Thus if in choosing their threats, the sides take into account their effect on the final outcome of bargaining, then S will try to maximize, and $T\setminus S$ to minimize, the quantity $H^S(\sigma, \tau)$. Hence the final amount that S can expect to obtain is precisely $v(S)$.

Suppose now that v is defined and has a μ-value ϕv. Then ϕv is called the *Harsanyi-Shapley transferable utility (TU) value*, or simply *TU value*, of the strategic game Γ.

Next, let us refer to a positive real-valued measurable function on $(T, \mathscr{C})$ by the term "comparison function." If λ is a comparison function, denote by $\lambda\Gamma$ the strategic game obtained from Γ by multiplying the payoffs $h_{\sigma\tau}^S(t)$ by $\lambda(t)$, and by v_λ the Harsanyi coalitional form of $\lambda \Gamma$. If v_λ is defined and has a μ-value ϕv_λ, and if there is a σ in X^T such that

$$(4.10) \qquad \int_S \lambda h_\sigma^T = (\phi v_\lambda)(S)$$

for all S in $\mathscr{C}'$, then h_σ^T is called a *Harsanyi–Shapley non-transferable utility* (*NTU*) *value*, or simply a *value*, of the strategic game Γ.

The reasoning behind this definition of value may be briefly described as follows: if side payments were permitted with the "exchange rates" λ, then the value for the resulting coalitional game would dictate paying each player dt an amount $(\phi v_\lambda)(dt)$, where v_λ is the Harsanyi coalitional form of $\lambda \Gamma$. Formula (4.10) may be rewritten

$$(4.11) \qquad \lambda(t) h_\sigma^T(t) \mu(dt) = (\phi v_\lambda)(dt);$$

that means that at the exchange rates λ, the value ϕv_λ is achievable without any "transfers of utility." This may therefore be taken as a value even in the absence of transferability, since no transfers are called for.

A more thorough discussion may be found in Shapley[4] [14], and in [1, section 6]. It is worthwhile to note that for $|T| = 2$, the Harsanyi–Shapley NTU value coincides with the bargaining solution of Nash [12] for two-person cooperative games.

The value notion differs in a fundamental way from solution notions such as the core, the von Neumann–Morgenstern solution, and the bargaining set, which are based on the concept of domination. Recall that an outcome x *dominates* an outcome y if there is a coalition S that prefers x to y and can achieve *for itself* an outcome at least as good as x. Thus domination expresses dissatisfaction on the part of a coalition because *it can do better by itself*. But the value, though it is based partly on this kind of consideration, also takes into account the opportunities of a coalition to cause harm to players outside it — i.e. to threaten. Thus the core, von Neumann–Morgenstern solution and bargaining set take into account arguments of the form "I should get more because I do not need you to do better"; whereas the value takes into account not only this kind of argument, but also the kind that says "I should get more because you need me to get what you're getting."

---

[4] Shapley permits some (but not all) of the exchange rates $\lambda(t)$ to vanish. Vanishing exchange rates are awkward to interpret; it seems best to avoid them when possible. A value in our sense (with non-vanishing $\lambda(t)$) is of course also a value in Shapley's sense.

5. The market model

A *market*[5] M consists of

i) A measurable space $(T, \mathscr{C})$ (the space of *agents*[6]) together with a σ-additive, non-negative measure μ on $\mathscr{C}$ with $\mu(T) = 1$ (the *population measure*[7]).

ii) The non-negative orthant Ω — called the *consumption set* — of a Euclidean space E^l (l represents the number of different commodities in the market).

iii) An integrable function e from T to Ω (the *endowment function* or *initial allocation*).

iv) For each t in T, a function u_t on Ω (the *utility function* of t).

A market is called *finite* if T is finite, and *non-atomic* if μ is non-atomic. In this paper we will assume that every market is either finite or non-atomic.[8]

We will assume that the measurable space $(T, \mathscr{C})$ is finite or isomorphic[9] to the unit interval $[0, 1]$ with the Borel sets. This assumption is less restrictive than it sounds; any non-denumerable Borel subset of any Euclidean space (or indeed, of any complete separable metric space) is isomorphic to $[0, 1]$.

If x and y are in E^l, we write $x \geqq y$ if $x^i \geqq y^i$ for all i, $x > y$ if $x^i > y^i$ for all i, and $x \geq y$ if $x \geqq y$ and $x \neq y$. The origin of E^l, as well as the number zero, are denoted 0. A function u on Ω is *increasing* if $x \geq y$ implies $u(x) > u(y)$. The partial derivative $\partial u / \partial x^i$ of a function u on Ω is denoted u^i, and the gradient $(u^1, \cdots, u^l)$ is denoted u'. The following assumptions will be made throughout:

(5.1) For each t, u_t is increasing, concave, and continuous on Ω.

(5.2) $u_t(0) = 0$.

(5.3) $u_t(x)$ is simultaneously measurable[10] in t and x.

(5.4) $\int e > 0$.

---

[5] The terminology differs somewhat from that of [1]. There a "market" is defined by preference relations rather than utility functions.

[6] "Agents" are the same as "players"; we prefer the former in the current politico-economic context, the latter in purely game theoretic contexts.

[7] In [1], μ was not interpreted as a population measure.

[8] This excludes the case in which μ has a denumerable infinity of atoms, as well as the "mixed" case, in which μ has some atoms as well as a non-atomic part.

[9] Two measurable spaces are *isomorphic* if there is a one-one transformation from one onto the other that preserves measurability in both directions.

[10] I.e. measurable in the product field $\mathscr{B} \times \mathscr{C}$, where $\mathscr{B}$ is the Borel field on Ω.

(5.5) *For each t and i, the partial derivative $u_t^i(x)$ exists and is continuous at each x in Ω with $x^i > 0$.*

A market is called *bounded* if

(5.6) *u_t is uniformly bounded,*[11]

and

(5.7) *$u_t(1, \cdots, 1)$ is uniformly positive.*[12]

We close this section by describing some notation and terminology that will be used throughout. For S in $\mathscr{C}$, we will sometimes write $e(S)$ for $\int_S e$. An *S-allocation* is a measurable function x from S to Ω with $\int_S x = e(S)$. An *allocation* is a T-allocation. If x is a function from T to Ω then we will sometimes write $u(x)$ for the function on T whose value at t is $u_t(x(t))$. A *price vector* is a member p of Ω with $p > 0$; it is called *normalized* if $\Sigma_{i=1}^l p^i = 1$. If p is a price vector and $x \in E^l$, then $\Sigma_{i=1}^l p^i x^i$ is denoted px. The expressions "almost everywhere" (a.e.), "almost all" (a.a.), and so on will refer to the measure μ. If λ is a comparison function, denote by λM the market obtained from M by multiplying each utility function u_t by $\lambda(t)$.

6. Commodity redistribution

We wish to define a game that embodies the redistribution process. First we describe the game verbally. Taxation[13] decisions are reached by majority vote. This means that any coalition S with $\mu(S) > 1/2$ can impose taxes in any way it pleases. In particular, it may divide among its own members the taxes collected from the minority. The taxes discussed in this section are imposed on commodities and paid in commodities.

It might be argued that in a democracy, the tax laws must be uniform, so that two people cannot be taxed differently just because one is a member of a ruling coalition while the other is not. But it must be recalled that what we are discussing here is not merely taxation but rather net income redistribution due to government activity. Though taxation is required to be uniform, there is no such requirement on government spending, nor would such a requirement be feasible. The net effect of this is that the majority can tax the minority as heavily as it wishes, and distribute the proceeds among its own members.

---

[11] $\text{Sup}\{u_t(x): t \in T, x \in \Omega\} < \infty$.

[12] $\text{Inf}\{u_t(1, \cdots, 1): t \in T\} > 0$.

[13] Actually, we are studying the net effect of both taxation and redistribution.

This being the case, it would seem at first sight that the entire income $e(T)$ of society becomes available to whoever is in the majority, since the majority can, in principle, tax the minority at 100% and return nothing to it. If one calculates the resulting value, one gets an allocation in which no account is taken of the fact that different individuals may have different endowments. Though the calculation is not without interest, the result appears too extreme to be considered relevant to the problem of the distribution of wealth in a democracy.

We now appear to be caught in a dilemma. If the majority rules, one gets an entirely egalitarian outcome. If, on the other hand, one ignores the political structure and does not allow the majority to redistribute, one is simply left with the initial allocation. Neither case seems realistic. How, then, can one account for the type of taxation scheme that one observes?

The answer lies in what we said at the end of Section 4. *The power of the minority lies in its threat possibilities.* We are going to assume that there is no "forced labor" — i.e. that the minority can, if it wishes, destroy part or all of its endowment. This is a powerful threat, which can force the majority to compromise.

The formal treatment starts out with a market M. Given M, define a strategic game $\Gamma = \Gamma(M)$, called the *redistribution game*, as follows: T, $\mathscr{C}$, and μ are as in the market M. As for the strategy spaces and payoff functions, we will not describe these fully, because that would lead to irrelevant complications; but we will make three assumptions about them, which suffice to characterize completely the games v_λ and their values.

The first of the three assumptions is:

(6.1) *If $\mu(S) > 1/2$, then for each S-allocation x there is a strategy σ of S such that for each strategy τ of $T\setminus S$,*

$$h^S_{\sigma\tau}(t) \begin{cases} \geq u_t(x(t)), & t \in S \\ = 0, & t \notin S. \end{cases}$$

This means that a coalition in the majority can force every member outside of it down to the zero level, while reallocating to itself its initial bundle in any way it pleases.

Next, we assume

(6.2) *If $\mu(S) \geq 1/2$, then there is a strategy τ of $T\setminus S$ such that for each strategy σ of S, there is an S-allocation x such that*

$$h^S_{\sigma\tau}(t) \leq u_t(x(t)), \quad t \in S.$$

This means that a coalition in the minority can prevent the majority from making use of any endowment other than its own (the majority's). Finally, we assume

(6.3) *If $\mu(S) = 1/2$, then for each S-allocation x there is a strategy σ of S such that for each strategy τ of $T\backslash S$ there is a $T\backslash S$-allocation y such that*

$$h^S_{\sigma\tau}(t) \begin{cases} \geq u_t(x(t)), & t \in S \\ \leq u_t(y(t)), & t \in T\backslash S. \end{cases}$$

This simply means that if neither S nor its complement are in the majority, then each side can divide its endowment in any way it pleases, while at the same time not giving anything to the other side.

This completes the definition of the redistribution game. Values for all strategic games have been defined in Section 4, and the definition applies in particular to this game. More interesting than the values themselves, though, are the allocations x such that $u(x)$ is a value of $\Gamma(M)$. These are the commodity redistributions to which we are most directly led by value considerations; we call them *commodity tax allocations* for M.

7. Income redistribution

In the Redistribution Game of Section 6, it was assumed that in taxing an individual, Society could take specific cognizance of the *vector* of his endowments. It is possible to take a different approach, in which Society would only be allowed to tax income — i.e. the monetary worth of the endowment vector at prevailing prices. That is what we will do in this section.

Given a price vector p, define the *indirect utility function* u^p_t of trader dt to be the function from R^+ to itself given by

(7.1) $$u^p_t(y) = \max\{u_t(x): x \in \Omega \quad \text{and} \quad px \leq y\}.$$

Intuitively, $u^p_t(y)$ is the highest utility dt can attain by buying goods at prices p with a maximum expenditure of y.

If all traders are assured that they can always trade at the fixed prices p, then the given l-good economy becomes an economy with only one good, namely money. In this economy the initial endowments are $pe(t)$, and the utility functions are u^p_t; this may be analyzed as a redistribution game with only one commodity, and this analysis yields a certain taxation-redistribution system.

Taxation and redistribution in this one-good "money" economy, as well as the ordinary incentives for trading, will in general create a situation in which at prices p, the supply and demand in the original l-good economy are out of balance. One therefore would like to know whether there exists a price vector p such that if the above procedure is carried out, supply and demand for each of the l goods will match *after* the taxation and redistribution are carried out, where the same price vector p is used both in assessing the endowment for purposes of taxation and in trading after taxes have been collected. Such a p, together with the resulting tax scheme, is called a "competitive tax equilibrium."

Formally, given a market M and a price vector p, define a market M^p, called the *market derived from M at prices p* (or simply the *derived market*) as follows: $(T, \mathscr{C}, \mu)$ is as in M; the number of commodities is 1; the initial allocation is pe; and the utility functions are the indirect utilities u_t^p defined in (7.1). We will see below (Lemma 16.1) that M^p does indeed satisfy all the conditions required of markets as defined in Section 5 ((5.1) through (5.5)). A *competitive tax equilibrium* in M is a pair consisting of an allocation x and a price vector p such that

(7.2) $x(t)$ a.e. maximizes u_t over $\{x \in \Omega : px \leq px(t)\}$,

and

(7.3) px is a commodity tax allocation in M^p.

If (x, p) is a competitive tax equilibrium, then x will be called an *income tax allocation*. Note that in M^p there is just one "commodity," namely money, and that the quantities $px(t)$ appearing in (7.3) are in units of that "commodity." Note also that in any market in which $l = 1$ — i.e. in which there is just one commodity — the commodity and income tax allocations are the same. Hence when $l = 1$, we will sometimes refer simply to *tax allocations*.

8. Statement of major results

Three basic results are proved in this paper. Theorem A is a general existence theorem that covers both finite and non-atomic markets, and both commodity and income tax allocations. Theorem B is an equivalence theorem, which asserts that in non-atomic markets, the commodity and income redistribution approaches lead to the same result; this result does *not* hold for finite markets. Theorem C provides a characterization of tax allocations in one-commodity non-atomic markets, and asserts that in such markets, there is a unique tax allocation. Theorems B and C together yield a characterization of commodity tax allocations in many-commodity non-atomic markets: By Theorem B, every such allocation is associated with a tax allocation in a derived market M^p; and these, in turn, are characterized by Theorem C (see Proposition 9.14).

A market is called *trivial* if there are just two agents ($|T| = 2$) and one of them has an endowment vector equal[14] to 0.

THEOREM A (Existence). *Every non-trivial bounded*[15] *market has a commodity tax allocation and an income tax allocation.*

THEOREM B (Equivalence). *In a non-atomic bounded market, the commodity tax allocations coincide with the income tax allocations.*

THEOREM C (Characterization). *A non-atomic bounded market with a single commodity ($l = 1$) has a unique tax allocation x. This allocation is characterized*[16] *by a.e. $x(t) > 0$ and*

(8.1) $$x(t) + \frac{u_t(x(t))}{u'_t(x(t))} = e(t) + \int \frac{u(x)}{u'(x)}.$$

Theorem A is proved in Section 18, Theorem B in Section 16, and Theorem C in Section 15. A counter-example to Theorem B when T is finite is given in Section 19.

9. Interpretation, discussion, further results

We start out by recalling some basic facts about "efficient" allocations. An allocation x in a market M is called *efficient* if there is no allocation y such that a.e. $u_t(y(t)) > u_t(x(t))$. With each efficient allocation there is associated an essentially[17] unique pair (λ, p), consisting of a measurable function λ from $(T, \mathscr{C})$ to R^+, and a price vector p, such that

(9.1) *the maximum of $\lambda(t)u_t(x) - px$ over Ω is a.e. achieved when $x = x(t)$.*

This (λ, p) is called an *efficiency pair* for x; we will call it *normalized* if p is normalized. From (9.1) it follows that p is an *efficiency price vector*, i.e. that

(9.2) *the maximum of $u_t(x)$ over $\{x \in \Omega : px \leq px(t)\}$ is a.e. achieved when $x = x(t)$,*

and that

---

[14] When $|T| > 2$, a vanishing endowment does not make an agent powerless, because he still has his vote. But when $|T| = 2$, the vote plays no role because neither player alone has a majority, and so lacks the power to tax the other agent.

[15] I.e. obeying (5.6) and (5.7).

[16] I.e., an allocation x is a tax allocation if and only if a.e. $x(t) > 0$ and (8.1).

[17] Unique up to multiplication by a positive constant.

(9.3) *the maximum of $\int \lambda u(y)$ over all allocations y is achieved when $y = x$.*

Conversely, any one of the three statements (9.1), (9.2), or (9.3) implies that x is efficient; moreover, given an efficient x, the p satisfying (9.2) is essentially unique, as is the λ satisfying (9.3). From (9.1) it also follows that a.e.

(9.4) $\quad\quad\quad\quad \lambda(t)u_t^i(x(t)) \leq p^i \text{ for all } i,$

and

(9.5) $\quad\quad\quad\quad \lambda(t)u_t^i(x(t)) = p^i \text{ when } x^i(t) > 0.$

Note, incidentally, that (9.2) and (7.2) are the same.

From (9.5) it follows that a.e.

$$x(t) \neq 0 \Rightarrow \lambda(t) > 0.$$

In particular, if $x(t) \neq 0$ a.e., then λ is a comparison function (i.e. $\lambda(t) > 0$ a.e.). The converse, however, is false: if λ is a comparison function, x may still vanish at a set of positive measure.

The existence of an efficiency price vector p is well known[18]; its essential uniqueness follows from the differentiability of the u_t. One can then use (9.5) to define λ, and deduce (9.1) (cf. the proof of lemma 14.1 of [1]). Our other assertions follow without difficulty from these considerations.

Given an efficient allocation x, the efficiency comparison function λ can be thought of as providing "coefficients of importance" for the players[19]: the redistribution x would result if one would want to maximize total utility when the individual utilities u_t are weighted by $\lambda(t)$.

In case $l = 1$, we may w.l.o.g. take $p = 1$. If x is a tax allocation, then by Theorem C, $x(t) > 0$ a.e., so by (9.5), $\lambda(t) = 1/u_t'(x(t))$. Thus (8.1) becomes

(9.6) $\quad\quad\quad\quad \lambda(t)u_t(x(t)) - \int \lambda u(x) = e(t) - x(t).$

The right side of (9.6) represents the net[20] taxes of dt. The left side is the excess of dt's final (i.e. after-tax) utility over the average final utility of all agents, when the utilities are compared using λ. Thus (9.6) says that one's taxes are

---

[18] See e.g. Hildenbrand [7].

[19] Or rather for the players' utilities. If $\lambda(t) = 2\lambda(s)$, a unit of dt's utility is considered equivalent to two of ds's. Of course any rescaling of the u_t's involves a corresponding rescaling of the $\lambda(t)$'s.

[20] I.e., the net decrement in dt's worth after both taxation and redistribution are taken into account. In fact, both sides of (9.6) are densities; to get actual amounts of tax, one should multiply by $\mu(dt)$.

proportional to how much one is better off than the average person, when utilities are weighted by one's "importance."

An alternative interpretation of (8.1) can be given in terms of marginal utilities. Since $u'_t(x(t))$ is the marginal utility of income, its inverse $1/u'_t(x(t))$ is the money worth of a unit of utility. Hence $u_t(x(t))/u'_t(x(t))$ is the monetary worth of dt's final utility when evaluated at the marginal rate. Thus (8.1) says that dt's taxes are equal, in dollars, to the amount by which the monetary worth of his utility exceeds the average monetary worth of everybody's utility.

Yet another interpretation of (8.1), in terms of the "fear of ruin," was discussed in [2]; we will not repeat this discussion here.

We note that

(9.7) *if x is an allocation satisfying* (8.1) *a.e., then* $x(t) > 0$ *a.e.*;

this follows from the fact that $\int x = \int e$, hence x cannot a.e. vanish, and hence $\int u(x)/u'(x) > 0$. Thus (8.1) alone is sufficient for an allocation x to be a tax allocation. Note also that if c is any constant such that for some allocation x, we have

(9.8) $$u_t(x(t))/u'_t(x(t)) = e(t) - x(t) + c$$

a.e., then x must be a tax allocation, since (9.8) implies $c = \int u(x)/u'(x)$.

Next, we show how Theorems B and C can be combined to yield a characterization of commodity tax allocations when there are many commodities ($l \geq 1$). We require two lemmas.

LEMMA 9.9. *If a market M is bounded, so are all its derived markets M^p.*

This will be proved below (Lemma 16.1).

LEMMA 9.10. *Let x be an efficient allocation in a market M, with efficiency pair (λ, p). Then*

(9.11) px *is an efficient allocation in M^p, with efficiency pair $(\lambda, 1)$*;

(9.12) $u(x) = u^p(px)$; *and*

(9.13) $\lambda(t) = 1/(u^p)'(px(t))$ *whenever $x(t) \neq 0$.*

PROOF. (9.12) follows from (9.2) and (7.1). To prove (9.11), let $y \in R^+$, and let the maximum in the definition of $u^p(y)$ be achieved at x. Then $y = px$ and $u^p(y) = u(x)$; since (λ, p) is an efficiency pair for x, we deduce, using (9.12), that

$$\lambda(t)u_t^p(y) - y = \lambda(t)u_t(x) - px \leqq \lambda(t)u_t(x(t)) - px(t)$$
$$= \lambda(t)u_t^p(px(t)) - px(t).$$

But this means precisely that $(\lambda, 1)$ is an efficiency pair for px. (9.13) follows immediately from (9.11) and (9.5) applied to M^p.

The characterization of commodity tax allocations for $l \geqq 1$ is as follows:

PROPOSITION 9.14. *In a non-atomic bounded market M, an allocation x is a commodity tax allocation if and only if it is efficient and a.e.*

(9.15) $\qquad \lambda(t)u_t(x(t)) - \int \lambda u(x) = p(e(t) - x(t)),$

where (λ, p) is an efficiency pair for x.

Note that (9.15) is subject to exactly the same interpretations as (9.6), except that the taxes are expressed in "dollar" (rather than commodity) terms, the prices being simply the efficiency prices for x.

To derive Proposition 9.14 from Theorems B and C, first let x be a commodity tax allocation. Then x is efficient. By Theorem B, x is an income tax allocation, so there is a normalized p satisfying (7.2) and (7.3). By (7.3) and Theorem C, $px(t) > 0$ a.e., and hence $x(t) \neq 0$ a.e. By (7.2), p is the normalized efficiency price vector for x, so there is a comparison function λ such that (λ, p) is the normalized efficiency pair for x. By (7.3), Lemma 9.9, and Theorem C, we have a.e.

(9.16) $\qquad \dfrac{u_t^p(px(t))}{u_t^{p\prime}(px(t))} + px(t) = pe(t) + \int \dfrac{u^p(px)}{u^{p\prime}(px)}.$

Together, (9.12), (9.13) and (9.16) yield (9.15).

Conversely, let x be an efficient allocation, with normalized efficiency pair (λ, p) satisfying (9.15). Since $\int x = \int e$, x cannot vanish a.e., hence $\int \lambda u(x) > 0$, hence by (9.15) a.e. $x(t) \neq 0$, and hence a.e. $px(t) > 0$. From (9.12), (9.13) and (9.15) we then deduce (9.16), and so from Lemma 9.9 and Theorem C it follows that (7.3) holds. (7.2) follows from the fact that p is an efficiency price vector. Hence x is an income tax allocation, and so by Theorem B a commodity tax allocation. This completes the derivation of Proposition 9.14 from Theorems B and C.

In *presenting* our results, we stated Theorems B and C first and derived Proposition 9.14 from them. Our strategy of *proof* will be the reverse. First we will prove Proposition 9.14 (characterization in the many commodity case), in

Sections 11 through 14. This is the longest and deepest part of the paper; a reader seeking an intuitive overview of this part is referred to [2], Section 9, Steps 1 through 5.[21] From this we derive Theorem C (existence, uniqueness, and characterization in the non-atomic case) in Section 15, and Theorem B (equivalence in the non-atomic case) in Section 16. Theorem B and the second part of Theorem C — i.e. the characterization (8.1) — are derived rather easily from Proposition 9.14, in much the same way that we just did the reverse derivation. The first part of Theorem C (existence and uniqueness in the non-atomic one-commodity case), however, requires a separate, though not particularly difficult argument. Theorem A (existence in the many-commodity case) is then proved in two parts. The non-atomic case is done in Section 17: One finds an income tax allocation by applying Theorem C to calculate demand in the derived markets M^p, allowing p to vary, and using Debreu's lemma [5, p. 82]; by Theorem B, the income tax allocation thus found is also a commodity tax allocation. In the finite-player case there is no equivalence; we apply in two different ways a basic lemma of Shapley [14] about the existence of NTU values, to find both kinds of tax allocations (Section 18).

Additional discussion of these results, and especially of Theorem C, will be found in [2].

10. Examples

Several one-dimensional examples were discussed in [2]. In this section we wish to examine a class of multidimensional examples. We will confine ourselves to the non-atomic case.

We start with a lemma that will simplify the calculations.

LEMMA 10.1. *Let x be a commodity (or equivalently, income) tax allocation, with efficiency pair (λ, p). Set $c = \int \lambda u(x)$. Assume $x(t) > 0$ a.e., and set*[22]

$$\varepsilon^i(t) = u_i^i(x(t))x^i(t)/u_t(x(t)),$$

$$\varepsilon(t) = \sum_{i=1}^{l} \varepsilon^i(t).$$

Then a.e.

(10.2) $$p^i x^i(t)/\varepsilon^i(t) = c + p(e(t) - x(t))$$

---

[21] [2] treats only $l = 1$; but the intuitive considerations for general l are in many respects similar, though somewhat more complex.

[22] $\varepsilon^i(t)$ is the elasticity of t's utility w.r.t. x^i at $x^i(t)$.

and a.e.

(10.3) $$px(t) = \frac{\varepsilon(t)}{1+\varepsilon(t)}(c + pe(t)).$$

If ε is a constant — say $\varepsilon(t) = \varepsilon$ — then

(10.4) $$c = \int pe/\varepsilon.$$

PROOF. (10.2) follows from (9.15) and (9.5). Multiplying both sides of (10.2) by $\varepsilon^i(t)$, summing over i, and dividing by $1 + \varepsilon(t)$, we obtain (10.3). To obtain (10.4), we integrate (10.3) and remember that $\int x = \int e$. This completes the proof of the lemma.

Let u be a utility function that is homogeneous of degree α, where $0 < \alpha \leq 1$. For our first example, we would like to choose $u_t = u$ for all t. Because u is homogeneous, the induced preferences are homothetic; hence all efficient allocations consist of bundles lying on the ray from the origin through the aggregate initial endowment $\int e$, and we may write

(10.5) $$p = u'(\int e).$$

Let x be a commodity tax allocation; then x is efficient, and hence $x(t)$ is a scalar multiple of $\int e$ for each t. From this and the homogeneity we obtain $\varepsilon^i(t) = u^i(\int e)\int e^i/u(\int e)$. By Euler's formula, it follows that

$$\varepsilon(t) = \sum_{i=1}^{l} \varepsilon^i(t) = \sum_{i=1}^{l} (\int e^i) u^i(\int e)/u(\int e) = \alpha.$$

From (10.3) and (10.4) we obtain

(10.6) $$px(t) = \frac{1}{1+\alpha} \int pe + \frac{\alpha}{1+\alpha} pe(t),$$

i.e. the net income of each agent is a mix of his gross income and the average gross income of all agents, in the ratio $\alpha:1$. For the tax we get

(10.7) $$pe(t) - px(t) = \frac{1}{1+\alpha}(pe(t) - \int pe),$$

and this means that we have a linear tax with a rate of at least 50% (since $\alpha \leq 1$). The bundle $x(t)$ itself (as distinguished from its worth $px(t)$) may be calculated by recalling that it must be a scalar multiple of $\int e$; hence by (10.6),

(10.8) $$x(t) = \frac{px(t)}{p\int e}\int e = \left(\frac{1}{1+\alpha} + \frac{\alpha}{1+\alpha}\frac{pe(t)}{\int pe}\right)\int e,$$

where $p = u'(\int e)$.

There is, however, a difficulty; we cannot take $u_t = u$, because homogeneous utility functions are not bounded. To get around this, assume e is bounded, say $e(t) \leq b \in \Omega$ for all t. Define u_t by $u_t(x) = f(u(x))$ for all x, where f is bounded and $f(y) = y$ for $y \leq u((pb/\int pe)\int e)$, where $p = u'(\int e)$. Though the u_t are no longer homogeneous, all agents still have the same homothetic preferences; hence all efficient allocations are on the ray through $\int e$ and the efficiency prices are $p = u'(\int e)$. Since by Theorem C the derived market M^p has only one tax allocation, it follows that there is exactly one income (and hence commodity) tax allocation in M. But by (10.8), $x(t) \leq (pb/\int pe)\int e$, hence u_t and its derivatives are the same as u and its derivatives at $x(t)$, and so from Proposition 9.14 it follows that the x of (10.8) is the unique commodity tax allocation in the given market.

For a more specific example, we may let $a_1, \cdots, a_k$ be l-dimensional vectors, and define

(10.9) $$u(x) = (a_1 x)^{\alpha_1}(a_2 x)^{\alpha_2} \cdots (a_k x)^{\alpha_k},$$

where $\alpha_j > 0$ and $\sum_{j=1}^{k} \alpha_j \leq 1$. If $k = l$ and a_j is the j-th unit vector, this is a Cobb–Douglas utility; such utilities are however excluded, since they are not monotonic on the boundary of Ω. If $a_j > 0$ (i.e. all components are positive) for all j, monotonicity is restored. In this case the degree α of homogeneity is simply $\sum_{j=1}^{k} \alpha_j$.

Another example is given by

(10.10) $$u(x) = \sum_{j=1}^{k}(a_j x)^\alpha;$$

here it is only required that $\sum_{j=1}^{k} a_j > 0$ and $a_j \geq 0$ for all j.

To some extent the above method may be used even when the u_t are different, as long as they are homogeneous, all with the same degree of homogeneity α, in the "relevant" part of Ω (i.e. up to an appropriately chosen bound). This would be the case, for example, if the right side of (10.9) were replaced by $(a_1(t)x)^{\alpha_1(t)} \cdots (a_k(t)x)^{\alpha_k(t)}$, subject to the restriction $\sum_{j=1}^{k} \alpha_j(t) = \alpha$; or if the right hand side of (10.10) were replaced by $\sum_{j=1}^{k}(a_j(t)x)^\alpha$. In this kind of situation the elasticities $\varepsilon'(t)$ may be different for different t, but their sum $\varepsilon(t)$ will always equal α. From (10.3) and (10.4) we can then deduce (10.6) and (10.7), i.e. we can calculate the net income and the tax in terms of gross income, average gross income, and the degree α of homogeneity. But calculating the prices p and the actual allocation x is quite another matter, since we can no longer say that $x(t)$ must be on the ray through $\int e$. We will not go further into this matter here.

As before, we must modify the utilities to make them bounded; the appropriate way to do this is in this case a little complex, and it may be necessary to assume that the normalized gradient $u'_i(x)/\Sigma_{i=1}^l u'_i(x)$ is bounded away from 0 (as will be the case for utility functions of type (10.9), but not for those of type (10.10) if the a_j are permitted to have vanishing components). Unlike before, we have no way of knowing what prices in the modified economy will look like. Therefore, though we can be sure that there is a commodity tax allocation satisfying (10.6) and (10.7), we cannot be sure that there are no others.

11. The optimal threat

Let M be a market, $\Gamma = \Gamma(M)$ the corresponding Redistribution Game, H^S, v, and w as in Section 4; in particular, v is the Harsanyi coalitional form of Γ. In this section we shall show that the minority's optimal threat — i.e., optimal strategy in H^S — is actually to destroy its entire endowment.[23] We shall then show that the value of v is the same as that of a coalitional game q in which $q(S)$ is the total utility of S when this threat is carried out.

Define coalitional games $r = r_M$ and $q = q_M$ by

(11.1) $$r(S) = \sup\left\{\int_S u(x): \int_S x = e(S)\right\}$$

and

(11.2) $$q(S) = \begin{cases} r(S) & \text{if} & \mu(S) \geq \frac{1}{2} \\ 0 & \text{if} & \mu(S) < \frac{1}{2}. \end{cases}$$

Note that if $r(T)$ is finite then $r(S)$ is finite for all S. We shall say that $r(S)$ is *attained* if it is finite and there is an S-allocation x with $r(S) = \int_S u(x)$.

PROPOSITION 11.3. *Assume that v is defined. Then $r(S)$ is attained whenever $\mu(S) > 1/2$, and for all S we have*

$$w(S) = q(S) - q(T\setminus S).$$

PROOF. That v is defined means that all the H^S have saddle points. Let (σ_0, τ_0) be a saddle point for H^S. Suppose first that $\mu(S) > 1/2$. Using the τ of (6.2), we

---
[23] Of course this threat is not carried out in the final outcome.

find that there is an S-allocation x such that $h^S_{\sigma_0\tau}(t) \leq u_t(x(t))$ for all $t \in S$. Hence from (4.6) and (4.5) we get

$$H^S(\sigma_0, \tau_0) \leq H^S(\sigma_0, \tau) \leq \int_S h^S_{\sigma_0\tau} \leq \int_S u(x).$$

Applying (6.1) to this same S-allocation x, we get a strategy σ of S such that

$$h^S_{\sigma\tau_0}(t) \begin{cases} \geq u_t(x(t)) & \text{if} \quad t \in S \\ = 0 & \text{if} \quad t \notin S. \end{cases}$$

Again using (4.6) and (4.5) we get

$$H^S(\sigma_0, \tau_0) \geq H^S(\sigma, \tau_0) = \int_S h^S_{\sigma\tau_0} - \int_{T\setminus S} h^S_{\sigma\tau_0} \geq \int_S u(x).$$

Hence

(11.4) $$w(S) = H^S(\sigma_0, \tau_0) = \int_S u(x).$$

Suppose now that $r(S)$ is not attained at x, i.e. that there is an S-allocation x' with $\int_S u(x') > \int_S u(x)$. If σ' corresponds to this x' in accordance with (6.1), then again using (4.6) and (4.5) we get

$$H^S(\sigma_0, \tau_0) \geq H^S(\sigma', \tau_0) = \int_S h^S_{\sigma'\tau_0} - \int_{T\setminus S} h^S_{\sigma'\tau_0} \geq \int_S u(x') > \int_S u(x),$$

contradicting (11.4). Hence when $\mu(S) > 1/2$, $r(S)$ is attained at x, and

(11.5) $$q(S) - q(T\setminus S) = r(S) = \int_S u(x) = w(S).$$

Next, note that $H^S(\sigma, \tau) = -H^{T\setminus S}(\tau, \sigma)$. Hence (σ_0, τ_0) is a saddle point of $H^{T\setminus S}$, and therefore

(11.6) $$w(S) = -w(T\setminus S).$$

This shows that when $\mu(S) < 1/2$,

$$q(S) - q(T\setminus S) = -r(T\setminus S) = -w(T\setminus S) = w(S).$$

Consider finally the case $\mu(S) = 1/2$. We have already shown that $r(T)$ is attained and in particular is finite, and hence $r(S)$ is also finite. Given $\varepsilon > 0$, let x be an S-allocation with $\int_S u(x) > r(S) - \varepsilon$. By (6.3), there is a strategy σ of S and a $T\setminus S$-allocation y such that

$$h^s_{\sigma\tau_0}(t) \begin{cases} \geq u_t(x(t)), & t \in S \\ \leq u_t(y(t)), & t \in T\setminus S. \end{cases}$$

Hence

(11.7) $\quad w(S) = H^s(\sigma_0, \tau_0) \geq H^s(\sigma, \tau_0) \geq \int_S u(x) - \int_{T\setminus S} u(y)$

$> r(S) - \varepsilon - r(T\setminus S) = q(S) - q(T\setminus S) - \varepsilon.$

Since also $\mu(T\setminus S) = 1/2$, we deduce

$$w(T\setminus S) > q(T\setminus S) - q(S) - \varepsilon;$$

hence by (11.6),

(11.8) $\quad w(S) < q(S) - q(T\setminus S) + \varepsilon.$

Letting $\varepsilon \to 0$ in (11.7) and (11.8), we obtain the desired result.

COROLLARY 11.9. *Assume that v is defined. Then v has a μ-value if and only if q does, and in that case the μ-values are equal.*

PROOF. From Proposition 11.3 and (4.8) we obtain

(11.10) $\quad v(S) = \tfrac{1}{2}[q^*(S) + q(S)],$

and the result then follows from (3.1).

PROPOSITION 11.11. *Assume that all the $r(S)$ are attained. Then all the games H^s have saddle points, i.e. v is defined.*

PROOF. First let $\mu(S) > 1/2$. Let $r(S)$ be attained at x. By (6.1), there is a strategy σ_0 of S such that for all strategies τ of $T\setminus S$, $\int_S h^s_{\sigma_0\tau} \geq \int_S u(x)$ and $\int_{T\setminus S} h^s_{\sigma_0\tau} = 0$; hence

$$H^s(\sigma_0, \tau) \geq \int_S u(x).$$

On the other hand, by (6.2) there is a strategy τ_0 of $T\setminus S$ such that for all strategies σ of S,

$$\int_S h^s_{\sigma\tau_0} \leq \max\left\{\int_S u(y): \int_S y = e(S)\right\} = r(S) = \int_S u(x)$$

and hence

$$H^s(\sigma, \tau_0) \leq \int_S u(x).$$

Hence (σ_0, τ_0) is a saddle point of H^s. Since $H^{T\setminus S}(\tau, \sigma) = -H^s(\sigma, \tau)$, we have completed the proof when $\mu(S) > 1/2$ or $\mu(S) < 1/2$.

Assume now that $\mu(S) = 1/2$. Then also $\mu(T\setminus S) = 1/2$. Then by (6.3), there is a strategy σ_0 of S such that for all strategies τ of $T\setminus S$,

$$H^s(\sigma_0, \tau) \geq \int_S u(x) - r(T\setminus S) = r(S) - r(T\setminus S).$$

Similarly, there is a strategy τ_0 of $T\setminus S$ such that for each strategy σ of S,

$$H^{T\setminus S}(\tau_0, \sigma) \geq r(T\setminus S) - r(S)$$

and hence

$$H^s(\sigma, \tau_0) \leq r(S) - r(T\setminus S).$$

Hence (σ_0, τ_0) is a saddle point. This completes the proof of Proposition 11.11.

PROPOSITION 11.12. *A necessary and sufficient condition that an allocation* x *in M be a commodity tax allocation is that there exists a comparison function* $\boldsymbol{\lambda}$ *such that* v_λ *is defined and has a μ-value given by*

(11.13) $$(\phi v_\lambda)(S) = \int_S \lambda u(x).$$

PROOF. The necessity follows immediately from the definitions. To prove the sufficiency, note that by (6.1), there is a strategy σ of T such that

(11.14) $$\boldsymbol{h}_\sigma^T(t) \geq u_t(x(t)) \quad \text{for all } t.$$

Multiplying (11.14) by $\boldsymbol{\lambda}(t)$, integrating over T, and using (3.2), we obtain

(11.15) $$\int \boldsymbol{\lambda h}_\sigma^T \geq \int \lambda u(x) = (\phi v_\lambda)(T) = v_\lambda(T).$$

But by applying (4.8) to the game $\boldsymbol{\lambda}\Gamma$, we may deduce $v_\lambda(T) = w_\lambda(T)$; and then applying (4.7) to $\boldsymbol{\lambda}\Gamma$, we deduce $v_\lambda(T) \geq \int \boldsymbol{\lambda h}_\sigma^T$. Combining this with (11.15), we deduce $\int \boldsymbol{\lambda h}_\sigma^T = \int \lambda u(x)$, and hence equality holds a.e. in (11.14). Hence

$$\int_S \boldsymbol{\lambda h}_\sigma^T = \int_S \lambda u(x) = (\phi v_\lambda)(S)$$

for all S in $\mathscr{C}$, and hence $\boldsymbol{h}_\sigma^T$ is a value; since equality holds a.e. in (11.14), it follows that $u(x)$ is a value and so x a commodity tax allocation. This completes the proof of Proposition 11.12.

The major conclusion of this section is Corollary 11.9, which enables us to replace the rather complex strategic game Γ, for purposes of calculating the value, by the relatively transparent coalitional game q. What enables us to do this is the explicit calculation, in Proposition 11.3, of the optimal threat strategies σ_0 and τ_0 of the majority and the minority: namely, for the majority to tax at 100%, and for the minority to destroy its entire endowment.

12. Preliminaries on non-atomic games

In proving our results, we shall make extensive use of the theory of non-atomic games developed in [4]. This section is devoted to reviewing some of the relevant results of that theory.

The *asymptotic value* of a game ([9], [4, section 17]) is defined in exactly the same way as the μ-value (Section 3), except that the W_i need not have equal μ-measure, and the $\mu(S)$ need not be rational. If under these conditions, $\phi_n(S)$ still tends to a limit $(\phi v)(S)$ and the limit is independent of the choice of the sequence of partitions, and if this is true for all S in $\mathscr{C}$, then ϕv is called the *asymptotic value* of v. Obviously we have

REMARK 12.1. *If the asymptotic value of v exists, so does the μ-value, and they are equal.*

Throughout this section, r will refer to a general (coalitional) game, not necessarily of the form (11.1). A game r is said to be *monotonic* if $S \supset U$ implies $r(S) \geq r(U)$. It is of *bounded variation* if it is the difference of monotonic games; the linear space of games of bounded variation on $(T, \mathscr{C})$ is denoted BV. The *variation norm* (or simply *norm*) on BV is defined by

$$\|r\| = \sup \sum_{i=1}^{n} |r(S_i) - r(S_{i-1})|,$$

where the supremum is over all chains of coalitions $\emptyset = S_0 \subset \cdots \subset S_n = T$.

The bounded games on $(T, \mathscr{C})$ form a linear space called BS; clearly $BS \supset BV$. The *supremum* (or *sup*) norm on BS is defined by

$$\|r\|' = \sup |r(S)|,$$

where the supremum is over all S in $\mathscr{C}$.

The non-atomic σ-additive measures form a subspace of BV called NA; the subset of NA consisting of non-negative measures ν with $\nu(T) = 1$ is called NA^1. The set of all linear combinations of positive integer powers of NA^1

measures is called P; games in P are called *measure polynomials*. The variation closure of P in BV is called pNA, and the sup closure of P in BS is called pNA'. Clearly $pNA \subset pNA'$.

The set of measurable functions from $(T, \mathscr{C})$ to $[0, 1]$ is denoted $\mathscr{I}$. It is useful to think of a member of $\mathscr{I}$ as an "ideal" subset of T; the number $f(t)$ is the "degree" to which the point t belongs to the ideal set f. Ordinary sets correspond to functions whose value is either 1 or 0, which is interpreted as meaning that the point either "completely belongs" or "completely fails to belong" to the set. If $S \in \mathscr{C}$, we denote by χ_S the characteristic function, defined by $\chi_S(t) = 1$ if $t \in S$, and $\chi_S(t) = 0$ if $t \notin S$. An *ideal game* is a function from $\mathscr{I}$ to the real numbers that vanishes at 0. The linear subspace of all bounded ideal games is denoted IBS; on it we define the *supremum* (or *sup*) norm by

$$\|r\|' = \sup\{|r(f)|: f \in \mathscr{I}\}.$$

PROPOSITION 12.2. *There is a unique linear, sup-continuous mapping that associates with each game r in pNA' an ideal game r^*, so that*

(12.3) $$(\nu^k)^* = (\nu^*)^k, \quad \text{and}$$

(12.4) $$\nu^*(f) = \int_T f(t)\nu(dt)$$

for all measures ν in NA^1, positive integers k, and ideal sets f.

This is proposition 22.16 of [4]. The operator $r \to r^*$ is called the *extension operator*; this name is justified by the following proposition:

PROPOSITION 12.5. *If $r \in pNA'$, then $r^*(\chi_S) = r(S)$.*

PROOF. If $r \in NA^1$, the proposition follows from (12.4); hence by (12.3), it follows for all powers of NA^1 measures; by linearity of the extension operator, for all measure polynomials; and finally, by sup-continuity, for all of pNA'.

If $r \in pNA'$, $t \in (0, 1)$, and $S \in \mathscr{C}$, denote

$$\partial r^*(t, S) = \lim_{\tau \to 0} \frac{r^*(t\chi_T + \tau\chi_S) - r^*(t\chi_T)}{\tau}.$$

At this point we are of course making no claim about the existence of this limit, we are merely introducing a notation. If $\nu \in NA^1$ and $r = \nu^k$ for some positive integer k, then

(12.6) $$\partial r^*(t, S) = kt^{k-1}\nu(S).$$

Define *DIAG* as the linear space of all r in *BV* for which

(12.7) there is a positive integer k, a k-dimensional vector ζ of measures in NA^1, and a neighborhood U in E^k of the diagonal $[(0,\ldots,0),(1,\ldots,1)]$ such that if $\zeta(S) \in U$ then $r(S) = 0$.

Motivation for this definition may be found in [4], on p. 252. Define *pNAD* to be the variation closure of *pNA + DIAG* (or equivalently, of *P + DIAG*).

PROPOSITION 12.8. *Let $r \in pNAD \cap pNA'$; then r has an asymptotic value ϕr. Furthermore, for each coalition S, the derivative $\partial r^*(t, S)$ exists for almost all t in $[0, 1]$ and is integrable over $[0, 1]$ as a function of t; and*

(12.9) $$(\phi r)(S) = \int_0^1 \partial r^*(t, S)\,dt.$$

Finally,

(12.10) $$\|r\| \geq \int_0^1 |\partial r^*(t, S)|\,dt.$$

PROOF. The three sentences are, respectively, corollary 43.12, proposition 44.22 and formula (44.23) of [4].

PROPOSITION 12.11. *Suppose r in $BV \cap pNA'$ is of the form $r_1 + r_2$, where $r_1 \in DIAG$. Then $\int_0^1 |\partial r^*(t, S)|\,dt \leq \|r_2\|$, and $|r^*(\alpha\chi_T)| \leq \|r_2\|$ for all α between 0 and 1.*

PROOF. The first assertion follows from formula[24] (44.23) and lemma 44.14 of [4]; the second from lemma 44.14 of [4].

A game r in *pNA'* is called *homogeneous of degree* 1 if for all α in $[0, 1]$ and all S in $\mathscr{C}$, we have $r^*(\alpha\chi_S) = \alpha r(S)$.

PROPOSITION 12.12. *Let r in $pNAD \cap pNA'$ be homogeneous of degree 1. Then for all S in $\mathscr{C}$ and all t in $(0, 1)$,*

$$\partial r^*(t, S) = (\phi r)(S).$$

PROOF. Existence of $\partial r^*(t, S)$ for almost all t follows from Proposition 12.8. Then reasoning precisely as in the proof of lemma 27.2 of [4] we deduce that $\partial r^*(t, S)$ exists and is the same for all t in $(0, 1)$. The fact that it must be ϕr then follows from (12.9).

[24] The δ-norm $\|\ \|_\delta$ appearing in this formula is defined on p. 262 of [4].

13. The value of truncated games

Corollary 11.9 indicates the importance of the game q defined in (11.2), which we might say is derived from r by "truncating" below $\mu(S) = 1/2$. In this section we prove a proposition that relates the value of a "truncated" game q to that of the corresponding untruncated game r in a more general context, one in which r is not necessarily derived from a market.

PROPOSITION 13.1. *Let* $\mu \in NA^1$, *let* $r \in pNAD \cap pNA'$, *and let* $0 < \alpha < 1$. *Define*

$$(13.2) \qquad q(S) = \begin{cases} r(S) & \text{if } \mu(S) \geq \alpha, \\ 0 & \text{otherwise}. \end{cases}$$

Then q has a μ-value, given by

$$(13.3) \qquad (\phi q)(S) = r^*(\alpha \chi_T)\mu(S) + \int_\alpha^1 \partial r^*(t, S)dt.$$

The proof is along lines similar to those used in section 18 of [4]. Let $S \in \mathscr{C}'$, and let $\mathscr{P} = \{\Pi_1, \Pi_2, \cdots\}$ be a separating sequence of partitions of T into coalitions of equal μ-measure, such that S is a union of members of Π_1. Let $\mathscr{R}_m$ be a *random order* on Π_m, i.e. a random variable whose values are orders on Π_m, in which all of the $|\Pi_m|!$ possible orders have the same probability. Let $\boldsymbol{B}_h^m$ be the h-th member of Π_m in the order $\mathscr{R}_m$, and set

$$\boldsymbol{Q}_h^m = \boldsymbol{B}_1^m \cup \cdots \cup \boldsymbol{B}_h^m.$$

The following lemma is proved in [4, p. 132, corol. 18.10]:

LEMMA 13.4. *Let ζ be a vector of NA^1 measures. Let U be a neighborhood of the "diagonal" in the range of ζ, i.e. the line segment with end points $(0, \cdots, 0)$ and $(1, \cdots, 1)$. Then for every ε, there is an m_0, such that for $m \geq m_0$,*

$$\text{Prob}\{\zeta(\boldsymbol{Q}_h^m) \in U \quad \text{for all} \quad h \quad \text{with} \quad 1 \leq h \leq |\Pi_m|\} > 1 - \varepsilon.$$

In words, this says that for sufficiently fine partitions, the sequence $\zeta(\boldsymbol{Q}_h^m)$ will with high probability remain in an arbitrarily small neighborhood of the diagonal.

We now proceed with the

PROOF OF PROPOSITION 13.1. Fix m, set $\Pi = \Pi_m$, $\mathscr{R} = \mathscr{R}_m$, $\boldsymbol{B}_h = \boldsymbol{B}_h^m$, $\boldsymbol{Q}_h = \boldsymbol{Q}_h^m$, and $n = |\Pi|$. For each h between 1 and n, let

$$z_h = \begin{cases} 1 & \text{if } \boldsymbol{B}_h \subset S, \\ 0 & \text{otherwise.} \end{cases}$$

Let h^* be the smallest integer not smaller than αn. Set

(13.5)
$$\Delta = \Delta^m(r) = \sum_{h=1}^{n} [q(\boldsymbol{Q}_h) - q(\boldsymbol{Q}_{h-1})] z_h$$
$$= r(\boldsymbol{Q}_{h^*}) z_{h^*} + \sum_{h > h^*} [r(\boldsymbol{Q}_h) - r(\boldsymbol{Q}_{h-1})] z_h,$$

where $\boldsymbol{Q}_0 = \emptyset$. The expression Δ is the total contribution of the players in S_m (i.e. the players of the finite game q_m who are included in S) to q_m, when Π is ordered according to $\mathcal{R}$; thus

(13.6)
$$(\phi q_m)(S_m) = E\Delta^m(r).$$

Set

$$\theta r = r^*(\alpha \chi_T) \mu(S) + \int_\alpha^1 \partial r^*(t, S) dt;$$

we wish to prove that $E\Delta^m(r) \to \theta r$ as $m \to \infty$.

Consider first the special case in which $r = \nu^k$, where $\nu \in NA^1$ and k is a positive integer. Setting $f(x) = x^k$, we obtain

$$r(\boldsymbol{Q}_h) - r(\boldsymbol{Q}_{h-1}) = f(\nu(\boldsymbol{Q}_{h-1}) + \nu(\boldsymbol{B}_h)) - f(\nu(\boldsymbol{Q}_{h-1})) = \nu(\boldsymbol{B}_h) f'(x_h),$$

where

(13.7)
$$\nu(\boldsymbol{Q}_{h-1}) \leq x_h \leq \nu(\boldsymbol{Q}_h);$$

thus (13.5) becomes

(13.8)
$$\Delta = f(\nu(\boldsymbol{Q}_{h^*})) z_{h^*} + \sum_{h > h^*} \nu(\boldsymbol{B}_h) z_h f'(x_h).$$

Let $\eta(\varepsilon)$ be the minimum of the moduli of uniform continuity of f and f' on $[0, 1]$. Let $\zeta = (\mu, \nu)$, and for each $\varepsilon > 0$ define a neighborhood U of the diagonal in the range of ζ by $U = \{(x, y): |x - y| < \eta(\varepsilon)/2\}$. From Lemma 13.4 it then follows that for m sufficiently large, we have with probability $> 1 - \varepsilon$ that for all h,

$$|\nu(\boldsymbol{Q}_h) - h/n| = |\nu(\boldsymbol{Q}_h) - \mu(\boldsymbol{Q}_h)| < \eta(\varepsilon)/2.$$

From this and (13.7) it follows that for m sufficiently large,

$$|x_h - h/n| < \eta(\varepsilon)/2 + 1/n < \eta(\varepsilon).$$

Hence if throughout (13.8), we replace $\nu(Q_h)$ and x_h by h/n, then we are with probability $> 1 - \varepsilon$ making an error $< \varepsilon$. Thus we may write

$$\Delta = f(h^*/n)z_{h^*} + \sum_{h > h^*} \nu(B_h) z_h f'(h/n) + \psi$$

for m sufficiently large, where $|\psi| < \varepsilon$ with probability $> 1 - \varepsilon$ and $|\psi|$ is bounded; the bound — let us call it C — is the maximum of $|f'|$ on $[0, 1]$. Since the B_h have equal μ-measure, it follows that the number of B_h such that $B_h \subset S$ is precisely $n\mu(S)$; from this we conclude that $E z_{h^*} = \mu(S)$ and $E(\nu(B_h)z_h) = \nu(S)/n$ for all h. Hence

$$E(\Delta) = f(h^*/n)\mu(S) + \sum_{h > h^*} f'(h/n)\nu(S)/n + \psi,$$

where $|\psi| < 2\varepsilon C$ for m sufficiently large. The fact that $|\psi|$ becomes arbitrarily small for m sufficiently large means that $\psi = \psi^m \to 0$ as $m \to \infty$; from the definition of the Riemann integral it then follows that

(13.9) $$\lim_{m \to \infty} E\Delta^m(r) = f(\alpha)\mu(S) + \int_\alpha^1 f'(t)\nu(S)dt.$$

But by (12.3), $f(\alpha) = \alpha^k = (\nu^*(\alpha \chi_T))^k = r^*(\alpha \chi_T)$; applying this and (12.6) to (13.9), we deduce that when $r = \nu^k$, then indeed $E\Delta^m(r) \to \theta r$. From this and the additivity of $E\Delta(r)$ as a function of r, it follows that

(13.10) $$E\Delta^m(r) \to \theta r \quad \text{for} \quad r \in P,$$

i.e. when r is a measure polynomial.

Suppose now that r is an arbitrary member of $pNAD \cap pNA'$. Let $\varepsilon > 0$ be given. Since $pNAD$ is the variation closure of $P + DIAG$, we have $r = r_0 + r_1 + r_2$, where $r_0 \in P$, $r_1 \in DIAG$, and $\|r_2\| < \varepsilon$. Let ζ and U correspond to r_1 in accordance with (12.7). Then by Lemma 13.4, for m sufficiently large we have with probability $> 1 - \varepsilon$ that $\zeta(Q_h) \in U$ for all h, and hence $\Delta^m(r_1) = 0$. From Proposition 12.11 it follows that $|\theta(r_1 + r_2)| \leq 2\|r_2\|$ (note that $r_1 + r_2 = r - r_0 \in pNA'$). Hence by (13.10) and Proposition 12.11, we have that for m sufficiently large,

$$|E\Delta^m(r) - \theta r| \leq |E\Delta^m(r_0) - \theta r_0| + |E\Delta^m(r_1)| + |E\Delta^m(r_2)| + |\theta(r_1 + r_2)|$$

$$< \varepsilon + 0 + \|r_2\| + 2\|r_2\| < 4\varepsilon.$$

Hence $E\Delta^m(r) \to \theta r$ as $m \to \infty$, and so by (13.6), $(\phi q_m)(S_m) \to \theta r$ as $m \to \infty$. This completes the proof of Proposition 13.1.

COROLLARY 13.11. *Let r in $pNAD \cap pNA'$ be homogeneous of degree 1. Define q by (13.2). Then q has a μ-value, given by*

(13.12) $$(\phi q)(S) = \alpha r(T)\mu(S) + (1-\alpha)(\phi r)(S).$$

PROOF. Propositions 13.1 and 12.12.

14. Characterization of commodity tax allocations in the many-commodity non-atomic case: Proof of Proposition 9.14

Let us be given a non-atomic market M, and let $\Gamma = \Gamma(M)$ be the corresponding commodity redistribution game. Boundedness of M will be assumed only when specified.

The market M will be called *integrably sublinear*[25] if for each $\varepsilon > 0$ there is a μ-integrable function η on T such that $u_t(x) \leq \varepsilon \|x\|$ whenever $\|x\| > \eta(t)$. A *transferable utility competitive equilibrium* (t.u.c.e.) in M (see Aumann and Shapley [4], section 32) is a pair (x, p) where x is an allocation and p a price vector, such that a.e. $u_t(x) - px$ attains its maximum over x in Ω at $x = x(t)$. If λ is a comparison function, then clearly (x, p) is a t.u.c.e. in λM if and only if (λ, p) is an efficiency pair for x in M.

Define the coalitional games v, w, r and q by (4.8), (4.7), (11.1), and (11.2).

PROPOSITION 14.1. *Let M be integrably sublinear. Then*

(14.2) $\qquad\qquad r(S)$ *is attained for all S, and*

(14.3) $\qquad\qquad v$ *is defined.*

Furthermore, if (x, p) is a t.u.c.e. for M, then v has a μ-value ϕv, given by

(14.4) $$(\phi v)(S) = \frac{1}{2}\int_S [r(T) + u(x) - p(x-e)].$$

PROOF. (14.2) is a special case of the main theorem of [3] (see also [4, prop. 36.1]). Assertion (14.3) follows from (14.2) and Proposition 11.11. Next, we note that $r \in pNAD \cap pNA'$ and is homogeneous of degree 1; this follows from [4, corol. 45.8 and prop. 45.10]. Hence by Corollary 13.11 with $\alpha = 1/2$, the game q has a μ-value, given by

(14.5) $$(\phi q)(S) = \tfrac{1}{2}r(T)\mu(S) + \tfrac{1}{2}(\phi r)(S).$$

---

[25] Shapley and Shubik [15] use the term "sublinear" to describe a function that is $o(\|x\|)$ as $\|x\| \to \infty$. The concept of *integrable* sublinearity was introduced by Aumann and Perles [3], though they used somewhat different terminology ($u_t(x) = o(\|x\|)$, *integrably* in t). For a discussion of the concept, see [4, p. 183].

Proposition 32.3 of [4] asserts that r has an asymptotic value, given by

(14.6) $$(\phi r)(S) = \int_S [u(x) - p(x - e)].$$

By (14.3), v is defined, and so from Corollary 11.9 we deduce that v has a μ-value ϕv equal to ϕq. Formula (14.4) then follows from (14.5) and (14.6). This completes the proof of Proposition 14.1.

Note that $\Gamma(\lambda M) = \lambda \Gamma(M)$ for all comparison functions λ. Recall that v_λ is the Harsanyi coalitional form of $\lambda \Gamma$. Similarly, define $r_\lambda = r_{\lambda M}$ and $q_\lambda = q_{\lambda M}$; explicitly,

(14.7) $$r_\lambda(S) = \sup\left\{\int_S \lambda u(x): \int_S x = e(S)\right\},$$

(14.8) $$q_\lambda(S) = \begin{cases} r_\lambda(S), & \text{when } \mu(S) \geq \frac{1}{2}, \\ 0, & \text{otherwise.} \end{cases}$$

PROPOSITION 14.9. *Let M be bounded, and assume that $r_\lambda(T) < \infty$. Then λM is integrably sublinear.*

PROOF. Let $\theta = \min\{1, \int e^1, \cdots, \int e^l\}$ and $\delta = \inf_t u_t(1, \cdots, 1)$; by (5.4), $\theta > 0$, and by (5.7), $\delta > 0$. Define an allocation y by $y(t) = \int e$ for all t. Then

$$u_t(y(t)) \geq u_t(\theta, \cdots, \theta) \geq \theta u_t(1, \cdots, 1) + (1 - \theta)u_t(0, \cdots, 0) \geq \theta\delta.$$

Hence $\theta\delta \int \lambda \leq \int \lambda u(y) \leq r_\lambda(T)$, and so λ is integrable.

Let β be the uniform bound on u_t provided by (5.6). Then for each $\varepsilon > 0$,

$$\|x\| > \lambda(t)\beta/\varepsilon \Rightarrow \lambda(t)u_t(x) < \varepsilon\|x\|.$$

Since $\lambda(t)\beta/\varepsilon$ is integrable, it follows that λM is integrably sublinear.

PROPOSITION 14.10. *Let M be bounded, let x be an allocation in M, and let λ be a comparison function. Then a necessary and sufficient condition that v_λ be defined and have a μ-value given by (11.13) is that there exists a price vector p satisfying (9.1) (the definition of efficiency pair) and (9.15).*

PROOF. First assume (9.1) and (9.15) a.e. From (9.1) we obtain

(14.11) $$r_\lambda(T) = \int \lambda u(x),$$

and in particular $r_\lambda(T) < \infty$. Hence by Proposition 14.9, λM is integrably sublinear. Hence by Proposition 14.1, v_λ has a μ-value ϕv_λ, given by

(14.12) $$(\phi v_\lambda)(S) = \frac{1}{2} \int_S [r_\lambda(T) + \lambda u(x) - p(x-e)].$$

On the other hand, by integrating (9.15) over S and rearranging, we find

(14.13) $$\int_S \lambda u(x) = \frac{1}{2} \int_S \left[\left(\int \lambda u(x) \right) + \lambda u(x) - p(x-e) \right].$$

Combining this with (14.11) and (14.12), we deduce (11.13).

Conversely, assume that v_λ is defined and has a μ-value given by (11.13). Then by Proposition 11.3, $r_\lambda(T)$ is attained and in particular $r_\lambda(T) < \infty$. Hence by Proposition 14.9, λM is integrably sublinear. From (11.13) and (3.2) it follows that $\int \lambda u(x) = v_\lambda(T)$. From (4.8) and Proposition 11.3 it follows that $v_\lambda(T) = r_\lambda(T)$, and hence (14.11) holds. Hence there is a p such that (λ, p) is an efficiency pair for x, i.e. such that (x, p) is a t.u.c.e. in λM. From the integrable sublinearity of λM and (14.4) we then obtain (14.12). Combining this with (11.13) and rearranging, we get

$$\int_S \lambda u(x) = \int_S [r_\lambda(T) - p(x-e)]$$

for all S. Hence a.e.

$$\lambda(t) u_t(x(t)) = r_\lambda(T) - p(x(t) - e(t)).$$

Combining this with (14.11), we deduce (9.15). This completes the proof of Proposition 14.10.

Proposition 9.14 follows immediately from Propositions 11.12 and 14.10.

15. Non-atomic one-commodity markets: Proof of Theorem C

Since $l = 1$, we may in Proposition 9.14 take $p = 1$. For any allocation x we have $\int x = \int e > 0$, hence $\int \lambda u(x) > 0$; so if x satisfies (9.15), then a.e. $x(t) > 0$. Then applying (9.5), we deduce (8.1). Conversely, if $x(t) > 0$ a.e. and (8.1) holds, then from (9.5) we get (9.15). Hence we have established the second sentence of Theorem C, i.e. that the tax allocations x are precisely those for which $x(t) > 0$ a.e. and (8.1) holds. It remains to prove that there is precisely one such x.

Define

$$g_t(x) = \frac{u_t(x)}{u'_t(x)} + x;$$

g_t is increasing and continuous, is defined for all non-negative numbers, vanishes at 0, and tends to infinity as $x \to \infty$. Hence its inverse g_t^{-1} is defined and has the same properties. Moreover, $g_t(x) > x$ when $x > 0$, and so

(15.1) $$g_t^{-1}(y) < y \quad \text{when} \quad y > 0.$$

For each non-negative number γ, define

(15.2) $$f(\gamma) = \int_T g_t^{-1}(\gamma + e(t))\mu(dt);$$

by (15.1) the integrand is integrable, so there is no difficulty with the finiteness of the integral. Again using (15.1) we get

$$f(0) < \int e.$$

By the monotone convergence theorem we may go to the limit under the integration sign in (15.2); then using the fact that $g_t^{-1}(x) \to \infty$ as $x \to \infty$, we deduce that for sufficiently large γ,

$$f(\gamma) > \int e.$$

Moreover f is strictly increasing (since the g_t^{-1} are); and it is continuous, since by the definition of g_t, $|g_t^{-1}(y_2) - g_t^{-1}(y_1)| < |y_2 - y_1|$. Hence there is one and only one γ with $f(\gamma) = \int e$; denote this γ by c and set

(15.3) $$x(t) = g_t^{-1}(c + e(t)).$$

By construction x is an allocation, and satisfies $x(t) > 0$ for all t. From (15.3) we obtain

$$x(t) + \frac{u_t(x(t))}{u_t'(x(t))} = e(t) + c,$$

and so integrating and using the fact that x is an allocation, we deduce $c = \int u(x)/u'(x)$. Thus there is an allocation satisfying the conditions of the corollary. Conversely, if x is an allocation satisfying (8.1), then it also satisfies (15.3) with $c = \int u(x)/u'(x)$. Integrating (15.3) yields

$$\int e = \int_T g_t^{-1}(c + e(t))\mu(dt) = f(c).$$

Thus c is the same as the one previously found, and so x also is. This completes the proof of Theorem C.

16. Equivalence: Proof of Theorem B

Let M be a market. First we settle an old debt: the proof that the derived markets M^p are indeed markets, and in particular satisfy the differentiability condition.

LEMMA 16.1. *For each price vector p, the derived market M^p satisfies conditions (5.1) through (5.5); and if M is bounded, so is M^p.*

PROOF. The verification of (5.1) through (5.4) is straightforward. To prove the differentiability condition (5.5), write $u_t = u$, and let $y > 0$. Since u^p is concave, it possesses a right derivative $D^+u^p(y)$ and a left derivative $D^-u^p(y)$ at y, and

$$(16.2) \qquad D^+u^p(y) \leq D^-u^p(y).$$

Let the maximum in the definition of $u^p(y)$ be achieved at x, i.e. $u^p(y) = u(x)$ and $px = y$. Let a_i be the i-th unit vector in Ω, i.e. $a_i^i = 1$, $a_i^j = 0$ for $j \neq i$. For $\delta > 0$ we then have $u^p(y + p^i\delta) \geq u(x + \delta a_i)$, hence

$$[u^p(y + p^i\delta) - u^p(y)]/\delta \geq [u(x + \delta a_i) - u(x)]/\delta,$$

and hence letting $\delta \to 0$, we deduce

$$(16.3) \qquad p^i D^+u^p(y) \geq u^i(x),$$

where $u^i = \partial u/\partial x^i$. If, moreover, $x^i > 0$, then for $0 < \delta < x^i$ we have $u^p(y - p^i\delta) \geq u(x - \delta a_i)$, hence

$$[u^p(y - p^i\delta) - u^p(y)]/(-\delta) \leq [u(x - \delta a_i) - u(x)]/(-\delta),$$

and hence letting $\delta \to 0$, we deduce

$$(16.4) \qquad p^i D^-u^p(y) \leq u^i(x).$$

Since $y > 0$, there must be an i with $x^i > 0$; for this i we may combine (16.2), (16.3) and (16.4) to deduce $D^+u^p(y) = D^-u^p(y)$, i.e. u^p is differentiable at y. The continuity of the derivative is then a consequence of the concavity of u^p.

If M is bounded, then the verification of (5.6) for M^p is straightforward. To prove (5.7), write Σp for $\Sigma_{i=1}^l p^i$. Then

$$u^p(\Sigma p) = \max\{u(x): px = \Sigma p\} \geq u(1, \cdots, 1).$$

If $\Sigma p \geq 1$, then by the concavity of u^p it follows that

$$u^p(1) \geq \frac{1}{\Sigma p} u^p(\Sigma p) + \left(1 - \frac{1}{\Sigma p}\right) u^p(0) \geq \frac{1}{\Sigma p} u(1, \cdots, 1).$$

If $\Sigma p \leq 1$, then by the monotonicity of u^p it follows that

$$u^p(1) \geq u^p(\Sigma p) \geq u(1, \cdots, 1).$$

Since $u(1, \cdots, 1) = u_t(1, \cdots, 1)$ is uniformly positive by assumption, it follows that so is $u^p(1) = u_t^p(1)$, and the proof of (5.7) — and so of Lemma 16.1 — is complete.

For future reference we note the following corollary.

COROLLARY 16.5. *Let the maximum in the definition of $u_t^p(y)$ be achieved at x. Then $(u_t^p)'(y) \geq u_t^i(x)/p^i$; and if $x^i > 0$, then $(u_t^p)'(y) = u_t^i(x)/p^i$.*

PROOF. Follows from (16.3), (16.4), and the differentiability of u_t^p.

PROOF OF THEOREM B. First let x be a commodity tax allocation in M. Then x is efficient, say with efficiency pair $(\boldsymbol{\lambda}, p)$, and obeys (9.15) (by Proposition 9.14). Hence $x(t) \neq 0$ a.e., and so from (9.12) and (9.13) we deduce (9.16). By Lemma 16.1, M^p is bounded, and so Theorem C applies to it; thus (9.16) shows that px is a tax allocation in M^p, i.e. (7.3) is satisfied. The other condition for (x, p) to be a competitive tax equilibrium — Condition (7.2) — follows from the fact that p is an efficiency price vector for x. Hence x is an income tax allocation.

Conversely, let x be an income tax allocation. Then there is a p satisfying (7.2) and (7.3), and also a $\boldsymbol{\lambda}$ such that $(\boldsymbol{\lambda}, p)$ is an efficiency pair for x. By (7.3) and Theorem C we have (9.16), hence a.e. $x(t) \neq 0$, hence (9.12) and (9.13) yield (9.15), and hence by Proposition 9.14, x is a commodity tax allocation. This completes the proof of Theorem B.

17. Existence in the non-atomic case

The idea of the proof is as follows: For each price vector p, consider the market M^p derived from M at prices p. By Theorem C, M^p has a unique tax allocation, which assigns to each trader a certain income. This income generates a certain demand, and so we get a certain total excess demand. We can then apply Debreu's lemma [5, p. 82], and deduce that there is a p for which the excess demand contains 0. But this yields a competitive tax equilibrium, hence an income tax allocation in M, and so by Theorem B a commodity tax allocation in M.

Let $\Delta = \{p \in \Omega : \Sigma_{i=1}^l p^i = 1\}$. Let Int Δ denote the relative interior of Δ; every point in Int Δ is a price vector. Let $\partial \Delta = \Delta \setminus \text{Int } \Delta$. We will make use of the following variant of Debreu's lemma, proved in [8, p. 150, lemma 1]:

LEMMA 17.1. *Let Z be an uppersemicontinuous,[26] compact-valued correspondence[27] from Int Δ to E^l that is bounded from below,[28] such that*

(17.2) $$pz = 0 \quad \text{for all} \quad z \in Z(p),$$

and such that

(17.3) *if the sequence $\{p_n\}$ in Int Δ converges to p_0 in $\partial \Delta$ then*
$$\inf\left\{\sum_{i=1}^{l} z^i : z \in Z(p_n)\right\} > 0 \text{ for } n \text{ large enough.}$$

Then there is a p in Int Δ such that 0 is in the convex hull of $Z(p)$.

We proceed in a series of lemmas.

LEMMA 17.4. *For all t in T, $y > 0$ and p in Int Δ, $u_t^p(y)$ and $(u_t^p)'(y)$ are continuous in (p, y).*

PROOF. Fix t and write $u_t = u$, $u_t^p = u^p$. The continuity of $u^p(y)$ follows from the definition (7.1). To prove the continuity of $(u^p)'(y)$, let $p_n \to p$, $y_n \to y$, and let the maximum in the definition of $u^{p_n}(y_n)$ be attained at x_n, i.e. $u^{p_n}(y_n) = u(x_n)$ and $p_n x_n = y_n$. Then $\limsup x_n^i \leq y/p^i$ for all i, so that $\{x_n\}$ is bounded; let x be a limit point of $\{x_n\}$, w.l.o.g. the limit. Then $px = y$ and $u^p(y) = u(x)$ (by the continuity of $u^p(y)$). Since $px = y$, there is an i with $x^i > 0$; then for n sufficiently large, also $x_n^i > 0$. Applying Corollary 16.5 and the continuity of u^i, we deduce

$$(u^{p_n})'(y_n) = \frac{u^i(x_n)}{p_n^i} \to \frac{u^i(x)}{p^i} = (u^p)'(y);$$

this completes the proof of the lemma.

LEMMA 17.5. *Suppose that for each p in Int Δ, h^p is an increasing continuous function from R^+ onto R^+, such that $h^p(y)$ is continuous in (p, y). Then the inverse functions $(h^p)^{-1}$ are defined, increasing, continuous, and take R^+ onto R^+; and $(h^p)^{-1}(z)$ is continuous in (p, z).*

PROOF. Except for the continuity of $(h^p)^{-1}(z)$ in (p, z), the lemma is readily verified. To prove the continuity, let $(p_n, z_n) \to (p_0, z_0) \in \text{Int } \Delta \times R^+$. Set $y_n = (h^{p_n})^{-1}(z_n)$ for $n = 0, 1, 2, \cdots$. We wish to show $y_n \to y_0$. Since

---

[26] By this we mean that each p has a neighborhood in which the graph of Z is closed.
[27] A *correspondence* is a set-valued function with non-empty values.
[28] I.e., there is a y in E^l such that $x \geq y$ for all p and all x in $Z(p)$.

$$h^{P_n}(y_0+1) \to h^{P_0}(y_0+1) > h^{P_0}(y_0) = z_0$$

and $z_n \to z_0$, it follows that $z_n < h^{P_n}(y_0+1)$ for sufficiently large n. Hence

$$y_n = (h^{P_n})^{-1}(z_n) < y_0 + 1,$$

and so $\{y_n\}$ has a limit point y'. Allowing $n \to \infty$ through values for which $y_n \to y'$, we obtain from the continuity of $h^p(y)$ in (p, y) that

$$z_0 = \lim z_n = \lim h^{P_n}(y_n) = h^{P_0}(y').$$

Hence $y' = (h^{P_0})^{-1}(z_0) = y_0$. Thus $\{y_n\}$ has a limit point, and every such limit point is y_0, which means $y_n \to y_0$. This completes the proof of Lemma 17.5.

For $p \in \text{Int } \Delta$, let y^p be the unique tax allocation of M^p provided by Theorem C.

LEMMA 17.6. $y^p(t)$ is a.e. continuous in p.

PROOF. We closely follow the proof of Theorem C (Section 15), keeping track of the parameter p. Set

(17.7) $$g_t^p(y) = y + \frac{u_t^p(y)}{(u_t^p)'(y)}.$$

Then g_t^p satisfies the hypotheses of Lemma 17.5, and so its inverse $(g_t^p)^{-1}$ is defined, continuous and increasing, takes R^+ onto R^+, and $(g_t^p)^{-1}(z)$ is continuous in (p, z); moreover from (17.7) we get

(17.8) $$(g_t^p)^{-1}(z) < z \quad \text{for} \quad z > 0.$$

Defining $f^p(\gamma) = \int_T (g_t^p)^{-1}(\gamma + pe(t)) \mu(dt)$, we deduce from Lebesgue's dominated convergence theorem (using (17.8)) that $f^p(\gamma)$ is continuous in (p, γ). Moreover f^p is strictly increasing, and from the monotone convergence theorem it follows that $f^p(\gamma) \to \infty$ as $\gamma \to \infty$. If we define $h^p(\gamma) = f^p(\gamma) - f^p(0)$, then h^p obeys all the conditions of Lemma 17.5, and so $(h^p)^{-1}$ is defined, increasing, continuous, takes R^+ onto R^+, and $(h^p)^{-1}(\beta)$ is continuous in (p, β). From (17.8) and (5.4) it follows that $h^p(0) = 0 < \int pe - f^p(0)$. Hence $(h^p)^{-1}(\int pe - f^p(0))$ is defined and continuous in p; call it c^p. Then $f^p(c^p) = h^p(c^p) + f^p(0) = \int pe$. From (15.3) we then deduce

(17.9) $$y^p(t) = (g_t^p)^{-1}(c^p + pe(t)),$$

and so from the continuity properties of $(g_t^p)^{-1}$ and c^p we deduce the assertion of our lemma.

COROLLARY 17.10. *For each compact subset C of Int Δ, there is a constant c such that $y^p(t) \leq c + pe(t)$ for all t and all p in C.*

PROOF. From (17.8) and (17.9) we get $y^p(t) \leq c^p + pe(t)$. From the continuity of c^p and the compactness of C it follows that c^p is bounded in C, which proves the corollary.

Let

(17.11) $$B^p(t) = \{x \in \Omega : px \leq y^p(t)\},$$

and let $D^p(t)$ be the set of elements of $B^p(t)$ at which u_t is maximized; $D^p(t)$ is t's (after tax) *demand* at prices p. Let $D(p) = \int D^p$, and let $Z(p) = D(p) - \int e$; $D(p)$ is aggregate demand, and $Z(p)$ aggregate excess demand.

We note for future reference that

(17.12) $$\frac{u_t^p(y)}{(u_t^p)'(y)} \geq y$$

for all p, y, and t; this follows from the concavity of u_t^p.

LEMMA 17.13. *If the sequence $\{p_n\}$ in Int Δ converges to p_0 in $\partial \Delta$, then a.e.*

$$\inf \left\{ \sum_{i=1}^{l} x^i : x \in D^{p_n}(t) \right\} \to \infty.$$

PROOF. Let $\delta = \min_i \int e^i$; by (5.4), $\delta > 0$. From (17.12) and the fact that y^p is an allocation in M^p it follows that for all p in Int Δ,

$$\int \frac{u^p(y^p)}{(u^p)'(y^p)} \geq \int y^p = \int pe \geq \delta.$$

Hence from Theorem C it follows that for all p in Int Δ, a.e.

(17.14) $$y^p(t) + \frac{u_t^p(y^p(t))}{(u_t^p)'(y^p(t))} = pe(t) + \int \frac{u^p(y^p)}{(u^p)'(y^p)} \geq \delta.$$

If the lemma is false, then there is a set U of positive measure such that for each t in U there is a compact set $C(t)$ such that

(17.15) $$D^{p_n}(t) \cap C(t) \neq \emptyset \quad \text{for arbitrarily large } n.$$

W.l.o.g. we can choose U so that (17.14) holds for *all* t in U (not only a.e.) whenever p is one of the p_n. Fix a t in U, and set $u = u_t$, $u^p = u_t^p$, $y^p = y^p(t)$, $D^p = D^p(t)$, $C = C(t)$; thus we have

(17.16) $$y^{p_n} + \frac{u^{p_n}(y^{p_n})}{(u^{p_n})'(y^{p_n})} \geq \delta$$

for all $n \geq 1$. We next assert that there are *positive* numbers β and δ' such that

(17.17) $$u(x) \leq \beta$$

for all x in C, and

(17.18) $$u^i(x) \geq \delta'$$

for all i and all x in C at which $u^i(x)$ exists. Assertion (17.17) follows immediately from (5.6), or alternatively from the continuity of u and the compactness of C. To prove (17.18), define $\xi^i = \max\{x^i : x \in C\} + 1$ for each i, and let

$$C^i = \{z \in \Omega : z^i = \xi^i, z^j \leq \xi^j \text{ for } j \neq i\}.$$

Then u^i is defined and continuous on C^i, which is compact; hence u^i attains its minimum, which we call δ^i, on C^i. But for each x in C and each i there is a z in C^i with $x^i < z^i$ and $x^j = z^j$ for $j \neq i$. From the concavity of u it then follows that $u^i(x) \geq u^i(z)$ when $u^i(x)$ exists (which is always the case when $x^i > 0$). Since $u^i(z) \geq \delta^i$, we may set $\delta' = \min_i \delta^i$, and thus (17.18) is proved.

By Corollary 16.5, $(u^p)'(y^p) \geq u^i(x)/p^i$ for all i and all x in D^p. Hence from (17.15) and (17.18) we get

(17.19) $$(u^p)'(y^p) \geq \delta'/p^i$$

for all i. From (17.15) and (17.17) we get $u^p(y^p) \leq \beta$. Combining this with (17.16) and (17.19), we get

(17.20) $$y^p \geq \delta - p^i \beta/\delta'$$

for all i and all p in Int Δ.

Since $p_0 \in \partial \Delta$, there is a coordinate of p_0 that vanishes; w.l.o.g. let $p_0^1 = 0$. Then $p_n^1 \to 0$, and so from (17.20) we deduce

(17.21) $$y^{p_n} \geq \delta/2$$

for n sufficiently large.

For each n, let $x_n \in D^{p_n} \cap C$; by (17.15), there is such an x_n. The sequence $\{x_n\}$ has a limit point x_0; w.l.o.g. let it be the limit. By (17.21), $p_n x_n = y^{p_n} \geq \delta/2$ for all sufficiently large n; hence $p_0 x_0 \geq \delta/2$. Hence there is an i with $p_0^i > 0$ and $x_0^i > 0$; w.l.o.g. let $i = 2$. Let $C' = \{x \in C : x^2 \geq x_0^2/2\}$, and let $\beta' = \max\{u^2(x) : x \in C'\}$; this max is achieved, because u^2 is continuous on the compact set C'. For sufficiently large n, we have $x_n^2 > x_0^2/2$, and hence $x_n \in C'$ and $u^2(x_n) \leq \beta'$. From Corollary 16.5 we get for these n that

$$u^2(x_n) = p_n^2(u^{p_n})'(y^{p_n}) \geq p_n^2 \delta'/p_m^1$$

hence $p_n^2 \leq p_n^1 \beta'/\delta'$. Since $p_n^1 \to 0$ it follows that $p_n^2 \to 0$, and hence $p_0^2 = 0$. This contradiction proves Lemma 17.13.

LEMMA 17.22. *There is a price vector p such that $0 \in Z(p)$.*

PROOF. We will use Lemma 17.1. To show that the hypotheses of Lemma 17.1 hold, note first that from Lemma 17.6 and $p \in \text{Int}\,\Delta$ it follows that $B^p(t)$ is continuous (in the set sense) in p for a.a. t. Hence from a known theorem [5, p. 19, comment 4], it follows that

(17.23) $D^p(t)$ *is uppersemicontinuous in* p *for a.a.* t.

Next, from the definition of D^p it follows that $px \leq y^p(t)$ for all $x \in D^p(t)$. Hence from Corollary 17.10 it follows that if C is a compact subset of Int Δ, then there is a constant c such that for all i,

$$0 \leq x^i \leq \left(c + \sum_{j=1}^{l} e^j(t)\right) \Big/ \min\{p^i : p \in C, \, 1 \leq i \leq l\}$$

whenever $p \in C$ and $x \in D^p(t)$. Thus

(17.24) D^p *is integrably bounded throughout* C,

i.e. there is an integrable function h such that $\|x\| \leq h(t)$ whenever $x \in D^p(t)$ and $p \in C$. Now it is known (see e.g. [8, p. 73, prop. 8]) that the integral of an integrably bounded uppersemicontinuous set-valued function is uppersemicontinuous. Thus (17.23) and (17.24) yield the uppersemicontinuity of $D(p)$ and hence of $Z(p)$ throughout C. Since every point in Int Δ has a compact neighborhood in Int Δ, it follows that

(17.25) Z *is uppersemicontinuous in* Int Δ.

The values of an uppersemicontinuous set-valued function are necessarily closed, so that Z is closed-valued. From (17.24) it follows that each $Z(p)$ is also bounded; hence

(17.26) Z *is compact-valued.*

Moreover $x \in Z(p) \Rightarrow x \geq -\int e$, whence

(17.27) $\quad\quad\quad\quad\quad\quad Z$ is bounded from below.

To establish the non-emptiness of $Z(p)$, we will use a measurability argument. Let's use the phrase *a.e. Borel-measurable* for a function (point- or set-valued) that differs from one with Borel graph on a set of measure 0 only. Then since y^p is an allocation it is a.e. Borel-measurable. A fairly standard argument then shows that $\boldsymbol{D}^p$ is also a.e. Borel-measurable. Then from the selection theorem of von Neumann [16] it follows that $\boldsymbol{D}^p$ has a measurable selection. Since by (17.24), $\boldsymbol{D}^p$ is integrably bounded, this measurable selection is integrable, and its integral is a member of $D(p)$. Hence $D(p)$, and so also $Z(p)$, are non-empty, i.e.

(17.28) $\quad\quad\quad\quad\quad\quad Z$ is a correspondence.

Next, from the monotonicity of u_t it follows that $px = y^p(t)$ for all p and t and all $x \in \boldsymbol{D}^p(t)$. Since $\int y^p = \int pe$, it follows that $px = p\int e$ for all $x \in D(p)$, whence

(17.29) $\quad\quad\quad\quad pz = 0 \quad \text{for all} \quad z \in Z(p).$

It remains only to establish the boundary condition (17.3). Let $p_n \to p_0$ where $p_n \in \text{Int } \Delta$ and $p_0 \in \partial \Delta$. Let $x_n \in D(p_n)$; it is sufficient to prove that $\sum_{i=1}^{l} x_n^i \to \infty$. Since $D(p_n) = \int \boldsymbol{D}^{p_n}$, there is a sequence $\{x_n\}$ such that $x_n(t) \in \boldsymbol{D}^{p_n}(t)$ for all t and $\int x_n = x_n$. Then by Fatou's lemma and Lemma 7.13,

$$\liminf \sum_{i=1}^{l} x_n^i = \liminf \int \sum_{i=1}^{l} x_n^i \geq \int \liminf \sum_{i=1}^{l} x_n^i = \infty.$$

This proves (17.3). Hence by Lemma 17.1, there is a p in Int Δ such that 0 is in the convex hull of $Z(p)$. But the integral of any set-valued function is convex [8, p. 62, theor. 3]. Hence $D(p)$ is convex, and hence so is $Z(p)$. Hence $Z(p)$ is its own convex hull, and so $0 \in Z(p)$. This completes the proof of Lemma 17.22.

Let p be such that $0 \in Z(p)$. Then there is an x such that $x(t) \in \boldsymbol{D}^p(t)$ for all t, and $\int x = \int e$. But this means precisely that x is an allocation satisfying (7.2) and (7.3). Thus (x, p) is a competitive tax equilibrium, hence x is an income tax allocation, and so by Theorem B a commodity tax allocation. This completes the proof of Theorem A in the non-atomic case.

18. Existence for finite T: Completion of the proof of Theorem A

Suppose the agent space T is finite. Then there is no equivalence theorem, and the existence of commodity tax allocations and income tax allocations must be established separately. But though the two proofs are different, they use similar ideas and many of the same tools. We proceed first to develop these tools.

Let E^T be the set of all real valued functions on T; since T is finite, E^T is a Euclidean space of dimension $|T|$. Define a *generalized comparison function* to be a non-negative (rather than positive) valued function on T, that does not vanish identically; the set of all such functions is denoted Λ. One can think of Λ as the non-negative orthant of E^T, excluding the origin. Notations for generalized comparison functions are similar to those for comparison functions. Thus if $\lambda \in \Lambda$ and M is a market with agent space T, then the coalitional games r_λ and q_λ are given by (14.7) and (14.8).

At the basis of both proofs lies the following lemma, proved by L. S. Shapley [14, p. 261], and used by him to establish the existence of NTU values in a general context.

PROPOSITION 18.1. *Let C be a convex and compact subset of E^T. Let $\lambda \to \psi_\lambda$ be a continuous mapping*[29] *from Λ to E^T such that for all λ,*

$$(18.2) \qquad \sum_{t \in T} \psi_\lambda(t) = \max \left\{ \sum_{t \in T} \lambda(t) z(t) : z \in C \right\},$$

and for all t and λ,

$$(18.3) \qquad \lambda(t) = 0 \Rightarrow \psi_\lambda(t) \geq 0.$$

Then there is a λ in Λ such that

$$(18.4) \qquad \psi_\lambda \in C.$$

Let M be a fixed market with the finite agent space T. In both of our applications of Proposition 18.1, we will choose[30]

$$(18.5) \qquad C = \{z \in E^T : \text{there is an allocation } x \text{ with } u(x)/|T| = z\}$$

(recall that $u(x)$ is the function on T whose value at t is $u_t(x(t))$). The convexity of C follows from the concavity of the u_t, and the compactness of C from the fact that the set of all allocations is compact and the u_t are continuous, so that C is the continuous image of a compact set.

To prove the existence of a commodity tax allocation, set $\psi_\lambda(t) = (\phi q_\lambda)(\{t\})$, where ϕ is the value. To apply Proposition 18.1, we shall have to establish the continuity of $\lambda \to \psi_\lambda$, (18.2), and (18.3).

By (14.7), for all $S \subset T$ we have

---

[29] The proposition remains true if the mapping is defined on the unit simplex only.
[30] We divide $u(x)$ by $|T|$ because we wish to think of the payoff to an individual agent t as $u_t(x(t))\mu(\{t\})$, just as in the non-atomic case we think of the payoff to dt as $u_t(x(t))\mu(dt)$.

$$r_\lambda(S) = \sup\left\{\int_S \lambda u(x) : \int_S x = e(S)\right\}.$$

Note that because of the finiteness of T, the integrals appearing on the right are in fact sums. The sup is over all S-allocations x; this is a compact subset of Euclidean space of dimension $l|S|$, and so the sup is attained and is a continuous function of λ in Λ. Hence by (14.8), $q_\lambda(S)$ is also a continuous function of λ. Hence by Proposition 2.5, $\psi_\lambda(t) = (\phi q_\lambda)(\{t\})$ is indeed continuous in λ.

The right side of (18.2) is simply $r_\lambda(T)$; thus (18.2) merely is the efficiency axiom (2.3) for the value, since $q_\lambda(T) = r_\lambda(T)$.

Next, the game q_λ is monotonic, i.e. $S \supset U \Rightarrow q_\lambda(S) \geq q_\lambda(U)$. Together with Proposition 2.5, this shows that $\psi_\lambda(t) = (\phi q_\lambda)(\{t\}) \geq 0$ for all λ and t, and this in particular implies (18.3).

From Proposition 18.1 we then deduce (18.4). Taking into account the definition (18.5) of C, we find that (18.4) says that there is a generalized comparison function λ and an allocation x such that

(18.6) $\qquad (\phi q_\lambda)(\{t\}) = \lambda(t)u_t(x(t))/|T|$

for all t in T.

We show next that the λ obeying (18.6) is in fact a comparison function, i.e. $\lambda(t) > 0$ for all t. Recall that given an ordering of the agents and an agent t, S_t represents the set of agents preceding t in the ordering. Assume first that $|T| > 2$. Let t_1 be an agent with $e(t_1) \neq 0$; there must be such an agent since $\int e > 0$ (Assumption (5.4)). Since λ does not vanish identically, either $\lambda(t_1) > 0$, or there is an agent t_0 different from t_1 with $\lambda(t_0) > 0$. In the latter case, consider an order on the agents in which t_0 is first, and t_1 is the first t such that $\mu(S_t \cup \{t\}) \geq 1/2$; i.e., t_1 is number $|T|/2$ in the order if $|T|$ is even, number $(|T|+1)/2$ if $|T|$ is odd. Then $q_\lambda(S_{t_1}) = 0$ and

$$q_\lambda(S_{t_1} \cup \{t_1\}) = r_\lambda(S_{t_1} \cup \{t_1\}) \geq \lambda(t_0)u_{t_0}(e(t_1))/|T| > 0.$$

Hence $q_\lambda(S_{t_1} \cup \{t_1\}) - q_\lambda(S_{t_1}) > 0$ for this order, and so $(\phi q_\lambda)(\{t_1\}) > 0$. Hence by (18.6), $\lambda(t_1) > 0$ in this case as well.

Suppose now t_2 is any agent other than t_1. Consider the order in which t_1 is first and t_2 is the first t such that $\mu(S_t \cup \{t\}) \geq 1/2$. Then $q_\lambda(S_{t_2}) = 0$ and

$$q_\lambda(S_{t_2} \cup \{t_2\}) = r_\lambda(S_{t_2} \cup \{t_2\}) \geq \lambda(t_1)u_{t_1}(e(t_1))/|T| > 0.$$

Hence $q_\lambda(S_{t_2} \cup \{t_2\}) - q_\lambda(S_{t_2}) > 0$ for this order, and so $(\phi q_\lambda)(\{t_2\}) > 0$. Hence by (18.6), $\lambda(t_2) > 0$. Since t_2 was chosen to be any agent, we have proved

(18.7) $$\lambda(t) > 0$$

when λ satisfies (18.6) and $|T| > 2$.

The case $|T| = 2$ requires special treatment since the first t in the order is also the first t such that $\mu(S_t \cup \{t\}) \geq 1/2$. Let $T = \{t_1, t_2\}$; w.l.o.g. $\lambda(t_1) > 0$. Since M is non-trivial (see Section 8), we must have $e(t_2) \neq 0$. Hence $q_\lambda(\{t_1, t_2\}) - q_\lambda(\{t_1\}) > 0$, hence $(\phi q_\lambda)(\{t_2\}) > 0$ and so by (18.6), $\lambda(t_2) > 0$. When $|T| < 2$, (18.7) is trivial. Thus we have shown (18.7) whenever λ satisfies (18.6); in other words, there is a comparison function (not only a generalized comparison function) obeying (18.6).

We have already noted that the sup in the definition of $r_\lambda(S)$ is attained for all λ in Λ and all $S \subset T$, and so in particular for $S = T$ and for the λ obeying (18.6). Since this λ is a comparison function, it follows from Proposition 11.11 that v_λ is defined, and hence from Corollary 11.9 that $\phi q_\lambda = \phi v_\lambda$. Thus from (18.6) we get $(\phi v_\lambda)(\{t\}) = \lambda(t) u_t(x(t))/|T|$ for all t in T. Because of the finiteness of T this implies (11.13), and so from Proposition 11.12 it follows that x is a commodity tax allocation.

We turn next to the proof of existence of an income tax allocation. This is in many respects similar to the proof just completed, and we will on several occasions refer the reader to an argument in "the previous case," rather than going through it in detail again.

Given a λ in Λ, let x be an allocation at which $r_\lambda(T)$ is attained, i.e. such that $\int \lambda u(x) = r_\lambda(T)$. Such an allocation is efficient, and so there is associated with it a unique normalized efficiency price vector p. We now show that though there may be more than one x associated with λ, all these different x's associated with the same λ will have the same normalized efficiency price vector p, which we may call $p(\lambda)$.

Indeed, suppose $r_\lambda(T)$ is attained at two different allocations, x_1 and x_2. Let (λ, p_1) and (λ, p_2) be efficiency pairs for x_1 and x_2 respectively. For a given i, let t be such that $x_1^i(t) > 0$. Applying (9.1) first with $x = x_1$ and $p = p_1$, and then with $x = x_2$ and $p = p_2$, we find

$$p_1(x_1(t) - x_2(t)) \leq \lambda(t)(u_t(x_1(t)) - u_t(x_2(t))) \leq p_2(x_1(t) - x_2(t)).$$

Integrating (i.e. summing) over t yields 0 both on the extreme right and on the extreme left, since $\int x_1 = \int x_2 = \int e$. Hence we must have equality throughout for each t, and so in particular

$$\lambda(t) u_t(x_1(t)) - p_2 x_1(t) = \lambda(t) u_t(x_2(t)) - p_2 x_2(t).$$

Thus the maximum of $\lambda(t)u_t(x) - p_2 x$ is taken on at $x_1(t)$ as well as at $x_2(t)$. From this it follows that $\lambda(t)u'_t(x_1(t)) = p'_2$. But by (9.5), $\lambda(t)u'_t(x_1(t)) = p'_1$; hence $p'_1 = p'_2$. Since this holds for each i we conclude[31] that $p_1 = p_2$. From this it follows that the *normalized* efficiency price vectors associated with x_1 and x_2 are also the same, so that $p(\lambda)$ is indeed well defined. Note that $p(\lambda) > 0$ even when λ has some vanishing components.

Now if p is any price vector, let r_λ^p and q_λ^p be the coalitional games defined by

(18.8)
$$r_\lambda^p(S) = \sup\left\{\int_S \lambda u^p(px): \int_S px = pe(S)\right\}$$
$$= \sup\left\{\int_S \lambda u(x): \int_S px = pe(S)\right\},$$

$$q_\lambda^p(S) = \begin{cases} r_\lambda^p(S) & \text{if } \mu(S) \geq \frac{1}{2} \\ 0 & \text{if } \mu(S) < \frac{1}{2}. \end{cases}$$

We may think of r_λ^p and q_λ^p as being associated with λM^p in the same way that r and q are associated with M; but it should be remembered that λM^p is not really a market, since $\lambda(t)$ may vanish for some t, and those t will not have increasing utilities in λM^p. Set $\psi_\lambda(t) = \phi q_\lambda^{p(\lambda)}(\{t\})$. To apply Proposition 18.1, we must, as in the previous case, prove the continuity of $\lambda \to \psi_\lambda$, (18.2), and (18.3).

First we show that $p(\lambda)$ is continuous in λ. Let $\lambda_k \to \lambda_0$. Let $r_{\lambda_k}(T)$ be attained at x_k, and let x_0 be a limit point of x_k, which w.l.o.g. we may take to be the limit; then $r_{\lambda_0}(T)$ is attained at x_0. For each i, there is a t such that $x_0^i(t) > 0$; this follows from $\int x_0 = \int e > 0$. Hence for sufficiently large k, which w.l.o.g. we may take to be all k, we have $x_k^i(t) > 0$. Define

(18.9)
$$p_k^i = \lambda_k(t)u'_t(x_k(t))$$

for such t; note that $\lambda_k(t)$ cannot vanish, since $x_k(t) = 0$ if it does. Then by (9.5), p_k is an efficiency price vector for x_k for $k = 0, 1, 2, \cdots$, i.e.

(18.10)
$$p(\lambda_k) = p_k \Big/ \sum p_k,$$

where $\sum p_k = \sum_{i=1}^{l} p_k^i$. From (18.9), $x_k \to x_0$, $\lambda_k \to \lambda_0$, and the continuity of the derivative u'_t, we deduce $p_k \to p_0$. Hence by (18.10), $p(\lambda_k) \to p(\lambda_0)$, which proves that $p(\lambda)$ is continuous in λ.

---

[31] For a similar argument, see [4, p. 190].

From (18.8), it follows that $r_\lambda^p(S)$ is simultaneously continuous as a function of λ and p (recall that $p > 0$ always). Hence $r_\lambda^{p(\lambda)}(S)$ is continuous in λ, hence so is $q_\lambda^{p(\lambda)}(S)$, and hence so is $\psi_\lambda(t) = \phi q_\lambda^{p(\lambda)}(\{t\})$. To prove (18.2), note that on the right side we have $r_\lambda(T)$, which is the same as $r_\lambda^{p(\lambda)}(T)$ because $p(\lambda)$ is an efficiency price vector for the x at which $r_\lambda(T)$ is attained. On the left side we have $\Sigma_{t \in T}(\phi q_\lambda^{p(\lambda)})(\{t\}) = (\phi q_\lambda^{p(\lambda)})(T)$, which because of the efficiency condition (2.3) for values, $= q_\lambda^{p(\lambda)}(T) = r_\lambda^{p(\lambda)}(T)$. Thus (18.2) is proved. (18.3) follows by a monotonicity argument exactly as in the previous case.

From Proposition 18.1 we then deduce (18.4). Taking into account the definition (18.5) of C, we find that (18.4) says that there is a generalized comparison function λ and an allocation x such that

(18.11) $$(\phi q_\lambda^{p(\lambda)})(\{t\}) = \lambda(t) u_t(x(t))/|T|$$

for all t in T.

We next claim that the λ obeying (18.11) is in fact a comparison function, i.e. $\lambda(t) > 0$ for all t. This proof is exactly the same as in the previous case. We must only replace u_t by $u_t^{p(\lambda)}$ and e by $p(\lambda)e$; since $p^i(\lambda)$ never vanishes, we have $p(\lambda)e = 0$ if and only if $e = 0$, and so the proof goes through as before.

Finally, arguing as in the previous case — but in $M^{p(\lambda)}$ rather than M — we find that the existence of a comparison function λ and an allocation x satisfying (18.11) implies that the Harsanyi coalitional form $v_\lambda^{p(\lambda)}$ of the redistribution game $\Gamma(M^{p(\lambda)})$ is defined, and that $(\phi v_\lambda^{p(\lambda)})(S) = \int_S \lambda u(x)$ for all S. Since $p(\lambda)$ is an efficiency price vector for x, we have $u(x) = u^{p(\lambda)}(p(\lambda)x)$. Hence $(\phi v_\lambda^{p(\lambda)})(S) = \int_S \lambda u^{p(\lambda)}(p(\lambda)x)$ for all S, and so by Proposition 11.12, $p(\lambda)x$ is a tax allocation in $M^{p(\lambda)}$. Since $p(\lambda)$ is an efficiency price vector for x, it follows that $(x, p(\lambda))$ is a competitive tax equilibrium, and hence x is an income tax allocation. This completes the proof of Theorem A.

19. Non-equivalence for finite T

When there are finitely many agents and at least 2 commodities, the commodity tax allocations are in general not the same as the income tax allocations. We present an example, from which it will be clear that for finite T, equivalence is the exception rather than the rule.

Define M by $T = \{1, 2\}$, $e(1) = (8, 0)$, $e(2) = (0, 27)$, and $u_1(x) = u_2(x) = f((\sqrt[3]{x^1} + \sqrt[3]{x^2})^3)$, where f is an increasing C^1 concave bounded function that is differentiable and is the identity on the "relevant" part of the line ($f(s) = s$ for $s \leq 125$ is sufficient). We need f only to satisfy the boundedness condition on the u_t.

The example is relatively transparent because of the homothetic preferences. Any Pareto optimal allocation — including any commodity tax allocation and any income tax allocation — consists exclusively of bundles lying on the "diagonal," i.e. the straight line segment connecting the origin to the aggregate endowment $(8, 27)$ of the economy (see Fig. 2). Therefore the efficiency price vector associated with any Pareto optimal allocation must be orthogonal to the indifference curve at $(8, 27)$, which means that it is proportional to $(9, 4)$.

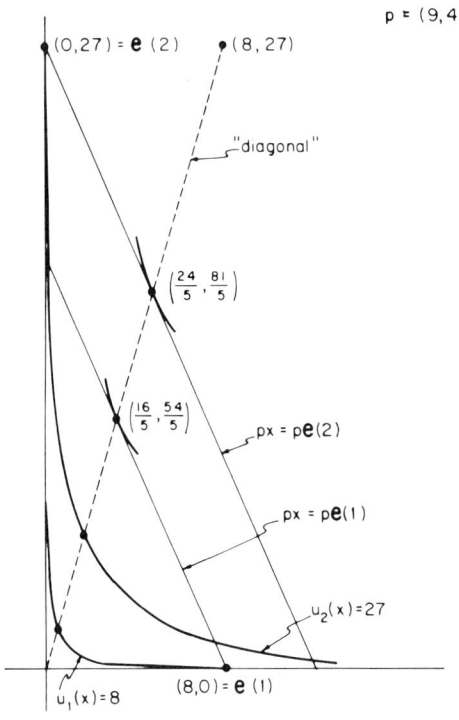

Fig. 2.

Let us first calculate the commodity tax allocations. From $|T| = 2$ it follows that $q_\lambda = r_\lambda$, and hence by Proposition 11.11 and Corollary 11.9 that $\phi v_\lambda = \phi r_\lambda$. Thus we must find an allocation x and a comparison function λ for which $\lambda(t) u_t(x(t)) = (\phi r_\lambda)(\{t\})$ for $t = 1, 2$. By a remark of Shapley [14], this is equivalent to solving a Bargaining Problem in the sense of Nash [11]; namely, the problem in which the disagreement payoffs are $r(\{1\}) = u_1(e(1)) = 8$ and $r(\{2\}) = u_2(e(2)) = 27$, and the set of feasible contracts is the set of all utility pairs $(u_1(x(1)), u_2(x(2)))$, where x ranges over all allocations. We have already seen that all Pareto optimal allocations consist of bundles on the diagonal; furthermore, the utility functions are equal to each other and are homogeneous of degree 1, hence linear on the diagonal. Hence the Pareto optimal surface of the feasible set is the line $u_1 + u_2 = 125$. In this case the Nash solution is to "split the surplus," the surplus being $125 - (8 + 27) = 90$. Hence agents 1 and 2 end up with utilities of $8 + 45 = 53$ and $27 + 45 = 72$ respectively. There is exactly one commodity tax allocation, and it is given by $x(1) = (8 \cdot 53/125, 27 \cdot 53/125) \approx (3.4, 11.4)$ and $x(2) = (8 \cdot 72/125, 27 \cdot 72/125) \approx (4.6, 15.6)$ (see Fig. 2).

We turn next to the competitive tax equilibria (x, p). Here the price vector p is an efficiency price vector for the income tax allocation x (see (7.2)); therefore by the homotheticity, we may take $p = (9, 4)$, as we saw above. In the market M^p, each trader t can guarantee to himself $u_t^p(pe(t))$. The maximum of u_t on the line $px = pe(t)$ is taken on when that line crosses the "diagonal" (see Fig. 2), which is at $(16/5, 54/5)$ for $t = 1$ and at $(24/5, 81/5)$ for $t = 2$. Hence $u_1^p(pe(1)) = 50$ and $u_2^p(pe(2)) = 75$. Reasoning as before but in the market M^p, we obtain a Nash Problem with disagreement payoffs of 50 and 75 and a Pareto optimal line given by $u_1 + u_2 = 125$. Here, therefore, there is no surplus, so the agents end up with utilities of 50 and 75 respectively. There is exactly one income tax allocation, given by $x(1) = (16/5, 54/5) = (3.2, 10.8)$ and $x(2) = (24/5, 81/5) = (4.8, 16.2)$.

The reason for the discrepancy between the commodity tax allocation and the income tax allocation is, of course, that though the feasible payoffs in M and M^p are the same, the disagreement payoffs are quite different and in fact larger in M^p than in M.

References

1. R. J. Aumann, *Values of markets with a continuum of traders*, Econometrica **43** (1975), 611–646 [Chapter 51].
2. R. J. Aumann and M. Kurz, *Power and taxes*, Econometrica **45** (1977), 1137–1161 [Chapter 52].
3. R. J. Aumann and M. Perles, *A variational problem arising in economics*, J. Math. Anal. Appl. **11** (1965), 488–503 [Chapter 70].
4. R. J. Aumann and L. S. Shapley, *Values of Non-Atomic Games*, Princeton University Press, Princeton, 1974.
5. G. Debreu, *Theory of Value*, John Wiley and Sons, New York, 1959.
6. J. C. Harsanyi, *A bargaining model for the cooperative n-person game*, in *Contributions to the Theory of Games IV* (R. D. Luce and A. W. Tucker, eds.), Princeton University Press, Princeton, 1959, pp. 325–355.
7. W. Hildenbrand, *Pareto optimality for a measure space of economic agents*, International Economic Review **10** (1969), 363–372.
8. W. Hildenbrand, *Core and Equilibria of a Large Economy*, Princeton University Press, Princeton, 1974.
9. Y. Kannai, *Values of games with a continuum of players*, Israel J. Math. **4** (1966), 54–58.
10. J. W. Milnor and L. S. Shapley, *Values of Large Games II: Oceanic Games*, Math. Oper. Res. **3** (1978), 290–307.

11. J. F. Nash, *The bargaining problem*, Econometrica **18** (1950), 155–162.

12. J. F. Nash, *Two person cooperative games*, Econometrica **21** (1953), 128–140.

13. L. S. Shapley, *A value for n-person games*, in *Contributions to the Theory of Games II* (H. W. Kuhn and A. W. Tucker, eds.), Princeton University Press, Princeton, 1953, pp. 307–317.

14. L. S. Shapley, *Utility comparisons and the theory of games*, in *La Décision*, Editions du Centre National de la Recherche Scientifique, Paris, 1969, pp. 251–263.

15. L. S. Shapley and M. Shubik, *Quasi-cores in a monetary economy with nonconvex preferences*, Econometrica **34** (1966), 805–827.

16. J. von Neumann, *On rings of operators. Reduction theory*, Ann. of Math. **50** (1949), 401–485.

17. J. von Neumann and O. Morgenstern, *Theory of Games and Economic Behavior*, Princeton University Press, Princeton, 1944.

DEPARTMENT OF MATHEMATICS
  HEBREW UNIVERSITY OF JERUSALEM
    JERUSALEM, ISRAEL

AND

DEPARTMENT OF ECONOMICS
  STANFORD UNIVERSITY
    STANFORD, CALIFORNIA, USA

54 Core and Value for a Public Goods Economy: An Example
with R. J. Gardner and R. W. Rosenthal

1. Introduction

One can approach the problem of distribution in an economy in at least three different ways: competitive equilibrium; core; and Shapley value. In an economy with a continuum of agents, these three approaches are equivalent in that they lead to the same imputation of utility [1, 2]. The concept of Lindahl equilibrium [7] has been advanced to fill the role of competitive equilibrium when public goods are present in the economy, the other approaches remaining the same. Foley [4] has shown that the Lindahl equilibrium is in the core, regardless of the size of the economy. Muench [8] has given an example of a continuum economy whose core and Lindahl equilibrium are not equivalent, but whose value and Lindahl equilibrium are. Gardner [5] and Rosenthal [9] have given examples in which the value and Lindahl equilibrium are different, but both are in the core. This leaves open the question as to whether the value must always be in the core. This note answers that question in the negative, by presenting an example in which the core consists of a single point, which is different from the value.

2. The Example

The economy consists of two types of agents, of measure 1 each, one public good and one private good. Formally, the space of agents is the measure space $(T, \mathbf{B}, \mu)$, where $T = [0, 2]$, $\mathbf{B}$ consists of the Borel subsets of T, and μ is Lebesgue measure. The types are $T_1 = [0, 1]$ and $T_2 = (1, 2]$. Each infinitesimal agent dt in T_1 has utility function $u_i(x, y) = 2x - x^2$, where x is the aggregate amount of the public good produced, and $y\mu(dt)$ is the amount of private good consumed by dt (i.e., y is his consumption density). The utility functions for the agents in T_2 are identically zero. Agents in T_1 have no initial endowment, whereas the endowment of an agent dt in T_2 is $\mu(dt)$ (i.e., his endowment density is 1). The public good can be produced from the private good in the ratio 1 : 1; i.e., if each agent dt

This chapter originally appeared in *Journal of Economic Theory* 15 (1977): 363–365. © Academic Press. Reprinted with permission.

inputs the amount $\mathbf{y}(t)\, \mu(dt)$ into the production process, then a total of $\int \mathbf{y}(t)\, \mu(dt)$ of the public good will be produced. Side payments are permitted (as is well known, this is equivalent to assuming the existence of another private good with a linear utility that enters additively into the utility function of both types). Note that the utilities of both types are nondecreasing over the set of possible consumption vectors.

For each coalition S, denote by $\mu_i(S)$ the measure of traders in S of type i. Consider a coalition S with $\mu_1(S) = a$, $\mu_2(S) = b$. Since the private good is useless except for producing the public good, this coalition will produce an amount b of the public good. Each agent of type 1 has utility $2b - b^2$ for the amount of public good, so that the total utility of S is $\int_s (2b - b^2)\, d\mu_1 = (2b - b^2)a$. Thus if v is the characteristic function of this game, then

$$v(S) = (2\mu_2(S) - \mu_2^2(S))\, \mu_1(S).$$

The Shapley value of this game may be calculated by using the diagonal formula [3, Theorem B]. It turns out to be $\tfrac{2}{3}\mu_1 + \tfrac{1}{3}\mu_2$. As for the core, we assert that it consists precisely of μ_1. Indeed, it is clear that μ_1 is in the core. Suppose another measure, μ_3, also were. We must have $\mu_3(T) = 1$, and μ_3 must be nonatomic [3, Proposition 27.8]. Since $\mu_3 \neq \mu_1$ there is a coalition U with $\mu_3(U) < \mu_1(U)$. Set $\alpha_i = 1 - \mu_i(U)$; then $\alpha_3 > \alpha_1$. By Lyapunov's theorem, there is for each small positive ϵ a coalition S such that $\mu_i(S) = (1 - \epsilon) + \epsilon\mu_i(U) = 1 - \alpha_i\epsilon$. Hence

$$v(S) - \mu_3(S) = (1 - \alpha_2^2\epsilon^2)(1 - \alpha_1\epsilon) - (1 - \alpha_3\epsilon)$$
$$= (\alpha_3 - \alpha_1)\epsilon - \alpha_2^2\epsilon^2 + \alpha_2^2\alpha_1\epsilon^3.$$

The linear term is positive for all ϵ under consideration, and hence the whole expression is positive for ϵ sufficiently small. That means $v(S) > \mu_3(S)$, and so μ_3 is not in the core.

The interpretation is that the core gives everything to type 1, whereas the value gives twice as much to type 1 as to type 2.

3. Conclusion

In the past, strong claims have been made for both the core [10] and Lindahl equilibrium [7] as principles of just taxation. Yet on intuitive grounds, in the example we have considered the value appears rather more just. Now one may wonder whether the special nature of the utility functions or the initial endowments is responsible for the result; but it is easy to adjust this example so that all endowments are positive and every agent in the economy gets positive utility from the public good and still the value is not in the core. Again, one may wonder whether the von Neumann–Morgenstern characteristic function, with its conservative treatment of threats, is responsible for the result. If one tries the Harsanyi characteristic function [6], one gets a different value, but still a positive imputation to type 2 individuals.

References

1. R. J. AUMANN, Markets with a continuum of traders, *Econometrica* **32** (1964), 39–50 [Chapter 46].
2. R. J. AUMANN, Values of markets with a continuum of traders, *Econometrica* **43** (1975), 611–646 [Chapter 51].
3. R. J. AUMANN AND L. S. SHAPLEY, "Values of Non-Atomic Games," Princeton Univ. Press, Princeton, N. J., 1974.
4. D. K. FOLEY, Lindahl's solution and the core of an economy with public goods, *Econometrica* **38** (1970), 66–72.
5. R. J. GARDNER, "Studies in the Theory of Social Institutions," Ph.D. Thesis, Cornell University, Ithaca, New York, 1975.
6. J. C. HARSANYI, A bargaining model for the cooperative *n*-person game, *in* "Contributions to the Theory of Games IV" (R. D. Luce and A. W. Tucker, Eds.), pp. 325–356, Princeton Univ. Press, Princeton, N.J., 1959.
7. E. LINDAHL, Just taxation—a positive solution, *in* "Classics in the Theory of Public Finance" (R. Musgrave and A. Peacock, Eds.), pp. 168–176, Macmillan, New York, 1958.
8. T. MUENCH, The core and the Lindahl equilibrium of an economy with a public good: An example, *J. Econ. Theory* **4** (1972), 241–255.
9. R. W. ROSENTHAL, Lindahl's solution and values for a public goods example, *J. Math. Econ.* **3** (1976), 37–41.
10. K. WICKSELL, A new principle of just taxation, *in* "Classics in the Theory of Public Finance" (R. Musgrave and A. Peacock, Eds.), pp. 72–119, Macmillan, New York, 1958.

R. J. AUMANN[*]
Department of Mathematics
Hebrew University
Jerusalem, Israel

R. J. GARDNER
Department of Economics
Iowa State University
Ames, Iowa

R. W. ROSENTHAL
Bell Laboratories
Murray Hill, New Jersey 07974

[*] Research supported by National Science Foundation grant SOC 74-11446 at the Institute for Mathematical Studies in the Social Sciences, Stanford University.

55 Power and Public Goods
with M. Kurz and A. Neyman

1. Introduction

Taxation has two related but distinct primary[1] purposes: redistribution and the provision of public goods. Most treatments of taxation have assumed a "benevolent" central government, that seeks to maximize some social welfare function. A contrasting game-theoretic model, in which the government responds to pressures from its constituents, was introduced in [2, 3]. That analysis was set in the context of private goods, so that the main issue was redistribution. In the current study we apply a similar analysis to public goods. We assume that exclusion is ruled out; i.e., once produced, a public good is available to all.

This work was supported by National Science Foundation Grant SES80-06654 at the Institute for Mathematical Studies in the Social Sciences, Stanford University, and by the Institute for Advanced Studies of the Hebrew University of Jerusalem.

[1] Secondary purposes include providing incentives or disincentives of various kinds, controlling inflation, and so on.

This chapter originally appeared in *Journal of Economic Theory* 42 (1987): 108–127. © Academic Press. Reprinted with permission.

Our main conclusion is that the distribution of voting rights has little or no relevance to the choice of public goods. This result is qualitatively different from those of the previous study, where it was shown that the vote is of central importance in redistributional questions.

The framework within which we work is that of a *public goods economy*, defined by a set of agents, a collection of public goods, a collection of nonconsumable resources, and a technology (enabling public goods to be produced from resources); moreover, each agent has a utility function for public goods, an initial endowment of resources, and a voting weight. It will be assumed that the agents form a nonatomic continuum, i.e., that there are many agents, each of whom is individually insignificant.

Private consumption goods are totally absent from this model. It represents the opposite extreme from the model of [2, 3] and indeed from all the classical general equilibrium models (e.g., Debreu [6]), in which *public* consumption goods are totally absent.[2]

The explicit introduction of voting weights into an economic model of this kind is a new feature of this study. It enables one to consider situations in which certain agents, while perhaps possessing economic power, are not citizens; or in which some agents have more influence on the political process than others.

The question at issue is which bundle of public goods will be produced; this depends on the decision process, or constitution, that is in force. One simple and natural framework is the *voting game*, studied in [4], in which any coalition (set of agents) with a majority of the vote may produce any bundle of public goods that it wishes, using its own resources only; minority coalitions may produce nothing. It is, however, desirable to have a more general framework. For example, we may wish to allow the minority to produce certain public goods (or bads) on its own; or we may wish to allow the majority to expropriate certain of the minority's resources, such as land. To this end, consider the class of economies obtainable from a given public goods economy by varying the system v of voting weights, but keeping all the other specifications fixed. A given coalition S may thus be a majority in some economies of the class, a minority in others. In defining a *public goods game*, we do not specify the rules, like in the voting game. Rather, we postulate that to each coalition S there is available a set X_v^S of strategies; the choice of a pair of strategies by a coalition and its complement determines which public goods are produced, which are then enjoyed by all agents. Two assumptions are made: First, a monotonicity assumption according to which any coalition has at least those strategic options available to any of its sub-coalitions (i.e., $S \supset U$ implies $X_v^S \supset X_v^U$). Second, that the only way in which the voting weights

---

[2] See Section 6b.

can affect the strategies available to a coalition S is in determining whether or not S is a majority (i.e., if v and ζ are different weighting systems, and S is a majority under both v and ζ, or a minority under both v and ζ, then $X_v^S = X_\zeta^S$).

To this game we shall apply the solution notion known as the *Harsanyi–Shapley Non-Transferable-Utility* (NTU) *Value*;[3] the resulting outcomes (i.e., bundles of public goods produced) will be called *value outcomes*. We obtain the following:

THEOREM. *In any public goods game, the value outcomes are independent of the voting weights.*

As an illustration, think of a public goods economy in which there are two groups of people, one preferring public libraries and the other television. It seems almost obvious that if the television fans may vote but the book lovers may not, then one may expect more television programming and fewer books than when the voting rights situation is reversed. But this is not what value theory predicts; by our theorem, the value outcomes in the two situations are identical. An intuitive discussion of this phenomenon is found in Section 6.

The paper is organized as follows. In Section 2 we formally describe public goods economies and set forth our assumptions. Section 3 contains the formal description of public goods games. Section 4 specifies the variant of the Harsanyi–Shapley NTU value used in this paper, thus completing the formal specification of all terms appearing in the above statement of the theorem. In Section 5 we demonstrate the theorem informally, stressing the intuitive background. Section 6 is devoted to intuitive discussion, Section 7 to the formal proof of the theorem, and Section 8 to some technical comments.

The paper is constructed so that readers who are not interested in the formal treatment can avoid it entirely. Such readers, after completing the Introduction, should go immediately to Sections 5 and 6. Conversely, readers interested *only* in the formal proofs may omit Sections 5, 6, and 8.

This paper is a companion piece to [4], which was devoted to the voting game mentioned above. The voting game is an instance of a public goods game, and its particular nature enabled us to obtain a result that is more specific (stronger) than the general result obtained here. The proof in [4] relies heavily on the proof in this paper (cited in [4] as AKN); this paper, however, may be read independently. Though the papers are related, there

---

[3] Sometimes called the "λ-transfer" value. See Shapley [15]; also Sections 4 of [2] or [3]. There is a considerable literature on the NTU value, including discussion, applications, and critical evaluation. Of course, like any other single solution concept, the NTU value does not capture all the strategic aspects of the games under consideration.

is little overlap between them. Thus, readers interested in additional background material and comment, or in numerical illustrations and examples, are referred to [4].

2. PUBLIC GOODS ECONOMIES

The real line is represented by $\mathbb{R}$, the euclidean space of dimension n by E^n, its nonnegative orthant by E^n_+ (i.e., $E^N_+ = \{x \in E^n: x^j \geq 0 \text{ for all } j\}$).

A nonatomic *public goods economy* consists of

(i) A measure space $(T, \mathscr{C}, \mu)$ (T is the space of *agents* or *players*, $\mathscr{C}$ the family of *coalitions*, and μ the *population measure*); we assume that $\mu(T) = 1$ and that μ is σ-additive, nonatomic and nonnegative.

(ii) Positive integers l (the number of different kinds of resources) and m (the number of different kinds of public goods).

(iii) A correspondence G from E^l_+ to E^m_+ (the *production correspondence*).

(iv) For each t in T, a member $\mathbf{e}(t)$ of E^l_+ ($\mathbf{e}(t)\,\mu(dt)$ is dt's *endowment* of resources).

(v) For each t in T, a function $u_t\colon E^m_+ \to \mathbb{R}$ (dt's von Neumann–Morgenstern *utility*).

(vi) A σ-additive, nonatomic, nonnegative measure v on $(T, \mathscr{C})$ (the *voting measure*); we assume $v(T) = 1$.

Note that the total endowment of a coalition S—its input into the production technology if it wishes to produce public goods by itself—is $\int_S \mathbf{e}(t)\,\mu(dt)$; for simplicity, this vector is sometimes denoted $e(S)$. A public goods bundle is called *jointly producible* if it is in $G(e(T))$, i.e., can be produced by all of society.

We assume that the measurable space $(T, \mathscr{C})$ is isomorphic[4] to the unit interval $[0, 1]$ with the Borel sets. This assumption is less restrictive than it sounds; any non-denumerable Borel subset of any euclidean space (or, indeed, of any complete separable metric space) is isomorphic to $[0, 1]$. We also assume:

(2.1) $u_t(y)$ is Borel measurable simultaneously in t and y, continuous in y for each fixed t, and bounded uniformly in t and y.

(2.2) G has compact and nonempty values.

[4] An *isomorphism* is a one-to-one correspondence that is measurable in both directions.

3. Public Goods Games

Recall that a strategic game[5] with player space $(T, \mathscr{C}, \mu)$ is defined by specifying, for each coalition S, a set X^S of *strategies*, and for each pair (σ, τ) of strategies belonging respectively to a coalition S and its complement $T\backslash S$, a payoff function $\mathbf{h}^S_{\sigma\tau}$ from T to $\mathbb{R}$.

In formally defining public goods games, we shall describe pure strategies only; but it is to be understood that arbitrary mixtures of pure strategies are also available to the players. The pure strategies we shall describe will have a natural Borel structure, and mixed strategies should be understood as random variables whose values are pure strategies.

As indicated in the Introduction, we consider a class of public goods economies in which all the specifications except for the voting measure v are fixed. We assume that the set X^S_v of pure strategies of a coalition S in the economy with voting measure v is a compact metric space, such that

$$S \supset U \quad \text{implies} \quad X^S_v \supset X^U_v \tag{3.1}$$

and

$$(v(S) - \tfrac{1}{2})(\zeta(S) - \tfrac{1}{2}) > 0 \quad \text{implies} \quad X^S_v = X^S_\zeta \tag{3.2}$$

(the intuitive meaning of (3.1) and (3.2) is given in the Introduction). From (3.2), it follows that X^T_v is independent of v, so that $X^T_v = X^T$; and from (3.1), it follows that $X^S_v \subset X^T$ for all v, i.e., X^T contains all strategies of all coalitions. Now we assume given a continuous function that associates with each pair (σ, τ) in $X^T \times X^T$ a public goods bundle $y(\sigma, \tau)$ in E^m_+. Intuitively, if a coalition S chooses σ and its complement $T\backslash S$ chooses τ, then the public goods bundle produced is $y(\sigma, \tau)$. (It may of course happen that a (σ, τ) in $X^T \times X^T$ is not "feasible" as a strategy pair of a coalition and its complement, i.e., there are no S and v with $(\sigma, \tau) \in X^S_v \times X^{T\backslash S}_v$; in that case it simply does not matter how $y(\sigma, \tau)$ is defined.) Finally, we define

$$\mathbf{h}^S_{\sigma\tau}(t) = u_t(y(\sigma, \tau)) \tag{3.3}$$

for each S, v, σ in X^S_v, and τ in $X^{T\backslash S}_v$.

Note that the *feasible* public goods bundles—those that can actually arise as outcomes of a public goods game—are contained in a compact set (the image of $X^T \times X^T$ under the mapping $(\sigma, \tau) \to y(\sigma, \tau)$), and hence constitute a bounded set.

---
[5] See Section 4 of [3].

4. VALUE OUTCOMES

As in [4], we shall be working here with the asymptotic value,[6] an analogue of the finite-game Shapley value for games with a continuum of players, obtained by taking limits of finite approximants. Let Γ be a public goods game. A *comparison function* is a nonnegative valued μ-integrable function λ on T that is positive on a set of agents of positive measure;[7] the corresponding *comparison measure* λ is defined by $\lambda(dt) = \lambda(t)\mu(dt)$, i.e., $\lambda(S) = \int_S \lambda(t)\mu(dt)$. A *value outcome* in Γ is then a random bundle of public goods associated with the Harsanyi–Shapley NTU value based on ϕ; i.e., a random variable $\underline{y}$ with values in $G(e(T))$, for which there exists a comparison function λ such that the Harsanyi coalitional form[8] v_λ^Γ of the game $\lambda\Gamma$ is defined and has an asymptotic value, and

$$(\phi v_\lambda^\Gamma)(S) = \int_S Eu_t(\underline{y})\,\lambda(dt) \qquad \text{for all} \quad S \in \mathscr{C}, \tag{4.1}$$

where $Eu_t(\underline{y})$ is the expected utility of $\underline{y}$.

5. AN INFORMAL DEMONSTRATION OF THE THEOREM

We start by motivating and describing in more detail than above the concept of "value outcome." Let us use the word *outcome* for a bundle of public goods.[9] Let λ be a comparison measure, i.e., a nonnegative measure on the agent space; $\lambda(dt)$ is interpreted as an infinitesimal "exchange rate" that enables comparison of agent dt's utility u_t with that of other agents. For each coalition S and each outcome y, write

$$U^y(S) = \int_S u_t(y)\,\lambda(dt).$$

$U^y(S)$ represents the "total" payoff to S when the exchange rates $\lambda(dt)$ are used and y is the total bundle of public goods produced by all coalitions;

---

[6] Kannai [9]; see also [5, Sect. 18], or [3, Sect. 3].

[7] This is a slight variant of previous definitions; see Section 8c.

[8] The Harsanyi coalitional form was first explicitly defined (for finite games and $\lambda \equiv 1$) by Selten [13, p. 592], who called it "the $\frac{1}{2}$-characteristic of Γ" and denoted it $c_{1/2}$; but its roots go back further (see Footnote 11). The formal definition of v_λ^Γ is reviewed at the beginning of Section 7, and an informal description and motivation is given in the next section.

[9] In general, the outcomes arising in the analysis of a public goods game are "mixed," i.e., random variables whose values are pure outcomes; but for simplicity, the informal discussion of this section is restricted to pure outcomes. For the general case, one need only replace pure outcomes y by mixed outcomes $\underline{y}$, and the utilities $u_t(y)$ by expected utilities $Eu_t(\underline{y})$.

this follows from the fact that all agents can enjoy all public goods produced by anybody. Note that $U^y(S)$ is a measure (i.e., additive) in S for each fixed y.

Now let Γ be a public goods game. Suppose that each coalition S announces a "threat" strategy ξ^S, representing what it would do if it forms. Then if a specific coalition S and its complement $T\setminus S$ actually form, their threats ξ^S and $\xi^{T\setminus S}$ will lead to a specific outcome $y(\xi^S, \xi^{T\setminus S})$; total payoff to S will then be

$$V(S) = U^{y(\xi^S, \xi^{T\setminus S})}(S).$$

Presumably, each agent dt will wish each coalition S to which he belongs to announce its threat ξ^S in such a way that his value $(\phi V)(dt)$ in the resulting game V is maximized. A priori, it would seem likely that different members of the same coalition S would wish S to announce different threats. It can be shown, though [3, Sect. 4], that this is not the case, i.e., that all members of S wish S to announce the same strategy ξ_0^S; and that for each S, the pair $(\xi_0^S, \xi_0^{T\setminus S})$ of "optimal" threats constitutes a saddle point of $H^{y(\xi^S, \xi^{T\setminus S})}(S)$, where[10]

$$H^y(S) = U^y(S) - U^y(T\setminus S) = 2U^y(S) - U^y(T) \tag{5.1}$$

for all outcomes y. Denoting by V_0 the game V resulting from the announcement of the optimal threats ξ_0^S, we define

$$q_\lambda^\Gamma(S) = V_0(S) = U^{y(\xi_0^S, \xi_0^{T\setminus S})}(S).$$

In words, $q_\lambda^\Gamma(S)$ is the total payoff to S when the coalitions S and $T\setminus S$ form and use their optimal threats.[11] A value outcome in the game Γ is a feasible outcome y for which there exists a comparison measure λ such that for all agents dt,

$$(\phi q_\lambda^\Gamma)(dt) = u_t(y)\, \lambda(dt), \tag{5.2}$$

where ϕ is the Shapley value; in words, it is a feasible outcome whose utility for each player, in terms of the exchange rates $\lambda(dt)$, is precisely his value in the coalitional game q_λ^Γ.

---

[10] $H^{y(\xi^S, \xi^{T\setminus S})}(S)$ here corresponds to $H^S(\xi_S, \xi_{T\setminus S})$ in [3, Sect. 4].

[11] $q_\lambda^\Gamma(S)$ was first defined (for finite games) by Harsanyi [7, p. 354], where it was denoted (for $\lambda \equiv 1$) by α^S.

Define

$$w_\lambda^\Gamma(S) = \max_{\xi^S} \min_{\xi^{T\setminus S}} H^{y(\xi^S, \xi^{T\setminus S})}(S)$$

$$= H^{y(\xi_0^S, \xi_0^{T\setminus S})}(S)$$

$$= q_\lambda^\Gamma(S) - q_\lambda^\Gamma(T\setminus S). \tag{5.3}$$

The formal definition of value outcomes, at the end of Section 4, is in terms of the Harsanyi coalitional form v, defined by

$$v_\lambda^\Gamma(s) = \frac{(w_\lambda^\Gamma(S) + w_\lambda^\Gamma(T))}{2}$$

$$= \frac{(q_\lambda^\Gamma(S) + q_\lambda^\Gamma(T) - q_\lambda^\Gamma(T\setminus S))}{2}. \tag{5.4}$$

From (5.4) it follows (cf. [2, p. 1154, Step 3]) that $\phi v_\lambda^\Gamma = \phi q_\lambda^\Gamma$, so that (4.1) and (5.2) are equivalent. The v_λ^Γ are mathematically better behaved, and therefore more convenient for the demonstration in this section; whereas the q_λ^Γ are conceptually more transparent, and so were used in discussing the examples in [4, Section 6]. Briefly, $w_\lambda^\Gamma(S)$ measures S's bargaining strength, or ability to threaten; $v_\lambda^\Gamma(S)$, the total utility that S can expect from the resulting efficient compromise. Cf. [3, Section 4], and [2, p. 1144].

Intuitively, it is not surprising that the coalition S must choose its threat to maximize the *difference* $H^y(S) = U^y(S) - U^y(T\setminus S)$ between its own payoff and that of its complement, and not its own payoff. One must remember that once the final compromise is agreed upon, these threats are not carried out; their only function is to influence the final compromise. And for *this* purpose, lowering the conflict payoff of the threatened party is as important as raising that of the threatener.

We proceed now to demonstrate our result. For each θ with $0 \leqslant \theta \leqslant 1$, let θT denote a "diagonal coalition of size θ." Intuitively, θT may be thought of as a perfect sample of the population T of all agents, containing a proportion θ of the agents. A precise formal definition may be given in terms of "ideal coalitions;" see [5, Chap. IV]. The crucial property of diagonal coalitions is that $\xi(\theta T) = \theta\xi(T)$ for nonatomic measures ξ. In particular, for any outcome y,

$$U^y(\theta T) = \int_{\theta T} u_t(y)\, \lambda(dt) = \theta \int_T u_t(y)\, \lambda(dt) = \theta U^y(T). \tag{5.5}$$

The importance of diagonal coalitions stems from the formula for the

Shapley value of a player t in a coalitional game r with a finite number n of players. This may be written[12]

$$(\phi r)(t) = \sum_{j=0}^{n-1} E(r(\underline{S}_j \cup \{t\}) - r(\underline{S}_j))\frac{1}{n}, \quad (5.6)$$

where E stands for "expectation," and $\underline{S}_j$ is a coalition chosen at random from among all those not containing t and having exactly j members, all such coalitions receiving the same probability. If now r is a nonatomic game and dt an infinitesimal agent, then by passing to the limit in (5.6) we obtain[13]

$$(\phi r)(dt) = \int_0^1 (r(\theta T \cup dt) - r(\theta T))\, d\theta. \quad (5.7)$$

Let v and ζ be two different vote measures; denote the corresponding variants of Γ by Γ^v and Γ^ζ, and write v_λ^v for v_λ^{Γ}, etc. By the definition of value outcome, it is sufficient to demonstrate

$$\phi v_\lambda^v = \phi v_\lambda^\zeta$$

for all comparison measures λ. In the remainder of the section, therefore, we consider λ fixed, and suppress the subscript throughout (i.e., write v^v for v_λ^v, etc.).

Let $r = v^v - v^\zeta$; we wish to demonstrate that $\phi r = 0$. For given dt, set $\delta = \max(v(dt), \zeta(dt))$. Let us call a coalition *even* if it is either a majority under both v and ζ, or a minority under both v and ζ. If S is even, then its complement $T \setminus S$ is also even; therefore the strategic options of S and $T \setminus S$ are the same under v and under ζ, therefore $v^v(S) = v^\zeta(S)$, and so $r(S) = 0$. Now θT is even for all θ, and $\theta T \cup dt$ is even for $\theta > \frac{1}{2}$ and $\theta < \frac{1}{2} - \delta$. Hence (5.7), (5.4) and $w^v(T) = w^\zeta(T)$ yield

$$(\phi r)(dt) = \int_{(1/2)-\delta}^{1/2} r(\theta T \cup dt)\, d\theta$$

$$= \int_{(1/2)-\delta}^{1/2} (v^v(\theta T \cup dt) - v^\zeta(\theta T \cup dt))\, d\theta$$

$$= \int_{(1/2)-\delta}^{1/2} \tfrac{1}{2}(w^v(\theta T \cup dt) - w^\zeta(\theta T \cup dt))\, d\theta. \quad (5.8)$$

[12] Shapley [14, Formula (13)].

[13] A precise formulation and proof of (5.7) are given in [5], Theorem H, for a restricted class of games; and for a much wider class in Mertens [10].

Let $\frac{1}{2} - \delta < \theta < \frac{1}{2}$. Suppose $w^v(\theta T \cup dt)$ is achieved at the outcome y, i.e.,

$$w^v(\theta T \cup dt) = H^y(\theta T \cup dt) = 2U^y(\theta T) + 2U^y(dt) - U^y(T)$$
$$= (2\theta - 1) U^y(T) + 2u_t(y) \lambda(dt)$$

(by (5.1) and (5.5)). Now $2\theta - 1$ is infinitesimal, since $|2\theta - 1| < 2\delta$; also $\lambda(dt)$ is infinitesimal. Hence $w^v(\theta T \cup dt)$ is infinitesimal, and by similar reasoning, so is $w^\zeta(\theta T \cup dt)$. But then the bottom line of (5.8) is the integral of an infinitesimal over an infinitesimal range, which is an infinitesimal of the second order and so may be ignored. Thus $(\phi r)(dt) = 0$, as was to be shown.

The demonstration hinges on the fact that $w^v(\theta T \cup dt)$ is infinitesimal when $\frac{1}{2} - \theta$ is infinitesimal; i.e., that $w^v(S)$ is small when S is near[14] $\frac{1}{2}T$, that w^v *is continuous at* $\frac{1}{2}T$ (where it vanishes). Indeed, if S is near $\frac{1}{2}T$, then it is also near its complement $T \setminus S$; hence no matter what the outcome is, S and $T \setminus S$ must have approximately the same utility, since all agents enjoy the same public goods. Note specifically that this breaks down in the case of redistribution of private goods [2, 3], or in a public goods voting game in which the majority may exclude the minority from enjoying the public goods [4, p. 682, Example 2]. In both cases, the minority may be prevented from consuming anything, whereas the majority can at least use its own resources; so even when S has only a slight v-majority, $w^v(S)$ will in general be very far from 0. The theorem is actually false in these cases.

6. Conceptual Discussion

a. *The Result*

To focus our thoughts, consider again the public goods economy discussed at the end of the Introduction, consisting of television fans and book lovers; by our theorem, the outcome is independent of who has the vote. To understand this intuitively, consider first the transferable utility (TU) context, in which we allow individuals to trade their votes for money if they wish. Thus the process of vote trading will continue while alternative proposals of television programming and libraries are debated. When an equilibrium of the vote trading is achieved, the winning public goods program is voted in. The power and influence of every individual thus has two elements: his material endowment (money or resources), and the market value of his vote. The effect that different voting rights have must be understood by studying the market price of a vote in equilibrium.

For concreteness, assume that there are as many television fans as book

---

[14] Two coalitions are "near" each other if they have similar profiles of characteristics, are *statistically* not too different. In particular, both their average utilities and their sizes are close, and so also their total utilities.

lovers, but only television fans may vote; that there is only one resource, of which all agents have the same endowment; that the utility functions are strictly concave and are "mirror images" of each other (i.e., $u_1(y^1, y^2) = u_2(y^2, y^1)$, where u_1 and u_2 are the utilities of book lovers and television fans, respectively, and y^1 and y^2 are quantities of libraries and television programming, measured in some appropriate units); that from an amount x of the resource one can produce any combination of public goods that total x; that the total amount of resource is 1; that the decision rule is as in the "voting game" of the Introduction; and that the marginal utility of money is 1.

An outcome of the TU game has two components: A vector y of public goods, and a schedule of accompanying side payments. We wish to make out an intuitive case for our contention that the vector of public goods "actually" produced is $(\frac{1}{2}, \frac{1}{2})$, and that there will be no accompanying side payments. Indeed, suppose first that more television programming than libraries are produced, as one might expect from the voting rights situation; for example, let us consider the vector $(\frac{1}{3}, \frac{2}{3})$. The strict concavity of the utilities implies that such an outcome cannot be Pareto optimal: the book lovers, by appropriate side payments, could make it worthwhile for all the television fans to vote for more books and less television. Indeed, they could make it worthwhile for them to vote for $(\frac{1}{2}, \frac{1}{2})$, since what is gained by the book lovers in going from $(\frac{1}{3}, \frac{2}{3})$ to $(\frac{1}{2}, \frac{1}{2})$ is more than what is lost by the television fans $(u_1(\frac{1}{2}, \frac{1}{2}) - u_1(\frac{1}{3}, \frac{2}{3}) > u_2(\frac{1}{3}, \frac{2}{3}) - u_2(\frac{1}{2}, \frac{1}{2}))$, by the "mirror symmetry" and the strict concavity).

In fact, though it is worthwhile for the book lovers to make sidepayments to *all* television fans to get them to vote for $(\frac{1}{2}, \frac{1}{2})$, this is not at all necessary; only slightly more than half the television fans are needed. It is obvious that this enables the book lovers to cut their expenses for side payments by almost 50%. Only slightly less obvious, though, is the fact that the book lovers can get much more out of the situation: they can play the television fans off against each other. The television fans know that the book lovers will pick some 51% of them and get them to vote for $(\frac{1}{2}, \frac{1}{2})$ by appropriate side payments. Naturally, they would like to be among the 51% who receive side payments, not among the 49% who do not. So they will bid against each other, offering to accept lower and lower amounts; and in the end, the equilibrium side payment will be practically zero.

The argument can be summed up as follows:

(1) The actual outcome must be Pareto optimal.

(2) The only Pareto optimal outcome involves a production of $(\frac{1}{2}, \frac{1}{2})$, possibly with an accompanying schedule of side payments from book lovers to television fans.

(3) As only slightly more than half the television fans are needed to

approve this outcome, competition between them will drive the side payments down to zero.

In brief, by playing the television fans off against each other, the book lovers can achieve parity in public goods for only a pittance in bribes.

This argument sounds almost too simple, and we would like to examine it from several viewpoints. First, when the issue is redistribution rather than choice of public goods, as in [2, 3], then the vote has a very important effect. Suppose, for example, that there are two groups of people, citizens and noncitizens; each is endowed with a single unit of a consumption good, but only citizens are allowed to vote. Any redistribution of the good may be decided on by a vote of the majority, but the minority has the right to destroy some or all of its goods. Then the citizens will always get a considerably larger share of the pie; when the utilities are linear (in the relevant range), citizens will get three times as much as noncitizens.

The vote also has an important effect when one is dealing with public goods with exclusion allowed. Specifically, consider the above television–library game, modified by the proviso that the majority may exclude the minority from use of any or all of the public goods produced. In that case, again, the value allocation calls for considerably more television than libraries.

When this result was presented at Professor E. Malinvaud's seminar in Paris, Professor K. Shell asked, why doesn't the argument about Pareto optimality and competition for side payments work in the cases of redistribution and exclusive public goods? What makes these cases different from that of nonexclusive public goods?

The answer is as follows. In the voting game with nonexclusive public goods, each voter will enjoy the public goods eventually produced no matter how he votes. He is a *free rider*; therefore he sells his vote for whatever it will fetch, producing cutthroat competition which drives its market value down to zero.

However, in the case of redistribution, the payoff of each player depends on how he votes. The man who votes with the majority will usually get all of his own initial bundle plus a part of the minority's; whereas the man who votes with the minority will not even get all of his own initial bundle. In the case of exclusive public goods, enjoyment of the public goods depends on voting the right way. The ride is not free in either of these cases; and a man will not sell his vote as lightly as in the voting game of this paper, as voting "wrong" is liable to cost him personally dearly.

As this contrast is crucial to an understanding of our results, we would like to dwell on it a little longer. It is not true that in our voting game, there is no cost at all to switching one's vote. Every player's vote does have some influence, since when all is said and done, all the players together do determine which public goods are produced. If there are 10^8 television fans, then roughly speaking, each one's vote can be expected to affect the amount of television programming by something of the order of magnitude of 10^{-8}, on the average. So to him, the cost of selling his vote is approximately one one-hundred millionth of his television viewing, a quite negligible amount. Competition therefore drives the price of the vote to 0. But in the case of redistribution or exclusive public goods, the cost of vote switching to the individual voter may be all of his television viewing, by all odds a considerable cost; and so the price of his vote will also be considerable.

The arguments adduced up to the present appear to depend critically on the TU assumption. However, the theorem of this paper is set in an NTU (nontransferable utility) context. Can we make economic sense of our result in that context as well?

To understand the situation, note that the TU assumption has two levels. The first may be called the *ordinal* level—simply that side payments are permitted, that an agent can act so that others gain while he loses. In our context, this is achieved if there are nonexpropriable, desirable private goods that may be transferred among the agents at will; specifically, if $u_t(y, x)$ is monotonic in t's private goods bundle x, where u_t is t's utility, and y is the public goods bundle.

The second level is the *cardinal* one—that an agent can act so that (for an appropriate choice of utilities), the total utility gained by others precisely equals the utility lost by him. This is achieved if there is just one private good, and $u_t(y, x) = u_t(y, 0) + x$.

To express a game in coalitional[15] form, as in (5.4), one needs the cardinal TU assumption. This is sometimes considered excessively strong, e.g., because it implies a complete lack of income effects. The NTU ("λ-transfer") value was developed for situations in which this strong assumption does not hold. In particular, it is needed whenever only the weak, ordinal form of the TU assumption holds.

The arguments in this section can be modified so as to use only this weak, ordinal form. At this level the TU assumption is intuitively very plausible; in most real situations at least some small amount of private goods is available for transfer. Perhaps it would have been methodologically preferable to put these private goods explicitly into the model. We did not do so because it would have cluttered up both the description and the analysis of the model without adding much insight. Indeed, the result is the same as before: both the choice of public goods, and the schedule of transfers (if any), are independent of the voting weights. In the television–library game, under appropriate symmetry conditions, the outcome remains $\frac{1}{2} - \frac{1}{2}$, and there are no side payments.

---

[15] Or "characteristic function."

To sum up: Strictly speaking, in our NTU result there are no private goods; intuitively, it can be considered the limit of what happens when the amount of private goods goes to 0. More broadly, the same intuitive ideas, and a similar theorem, apply when transfers of private goods are possible in any amount.

b. *The Model*

In the third paragraph of this paper we made the point that a public goods economy represents an extreme, idealized model of a certain politico-economic phenomenon. Studying such phenomena in isolation is typical of economic theory.[16] Pure exchange economies, pure monopolies, purely constant returns to scale, pure competition, etc., are all idealizations. We study them because they are associated with certain phenomena that are found (or sought) in a mixed, attenuated manner in the "real world"; they enable us to try to see some order, some regularity, in the chaos.

An example of a "real" system similar to the model of this paper is a commune like a kibbutz. But this paper is not meant to be about communes; rather, it is about the public goods *aspect* of complex politico-economic systems.

7. Formal Proof of the Theorem

Throughout this section, the word "measure" means "signed σ-additive measure on (T, C)." The symbol $\|\ \|$ denotes the max norm ($\|x\| = \max_i |x_i|$) when applied to points x in a euclidean space, and the variation norm ($\|\xi\| = \max_{S \in \mathscr{C}}(|\xi(S)| + |\xi(T \setminus S)|)$) when applied to measures ξ. Sets of measures are always endowed with the metric induced by the variation norm. K denotes a uniform bound on $|u_t(y)|$ (Assumption 2.1), and C an m-dimensional hypercube containing all feasible public goods bundles (see the end of Sect. 3).

Let Γ be a public goods game and λ a comparison measure. "Value outcomes" for Γ were defined in Section 4, in terms of the Harsanyi coalitional form v_λ^Γ of the game $\lambda\Gamma$, whose explicit definition we now recall. For each public goods bundle y and coalition S, define

$$U_\lambda^y(S) = U^y(S) = \int_S u_t(y)\,\lambda(dt), \tag{7.1}$$

$$H_\lambda^y(S) = H^y(S) = U^y(S) - U^y(T \setminus S), \tag{7.2}$$

$$w_\lambda^\Gamma(S) = w^\Gamma(S) = \min \max EH^{y(\sigma,\tau)}(S) = \max \min EH^{y(\sigma,\tau)}(S), \tag{7.3}$$

$$v_\lambda^\Gamma(S) = v^\Gamma(S) = (w^\Gamma(S) + w^\Gamma(T))/2; \tag{7.4}$$

---

[16] And of much theory in the physical sciences as well.

in (7.3), E is the expectation operator, the max is over mixed strategies σ of the coalition S (i.e., random variables with values in the pure strategy space X^S), and similarly the min is over mixed strategies τ of $T\backslash S$. To see that the min max in (7.3) is attained and equals the max min, note that Lebesgue's dominated convergence theorem and the continuity of $y(\sigma, \tau)$ in pairs (σ, τ) of pure strategies imply that $H^{y(\sigma,\tau)}(S)$ is continuous in (σ, τ), and then use the minimax theorem on arbitrary compact strategy spaces.

LEMMA 7.5. *Let Γ and Δ be two public goods games with the same player space $(T, \mathscr{C}, \mu)$, the same utilities u_t, and the same set $G(e(T))$ of jointly producible public goods bundles. Assume that for every comparison function λ, the asymptotic value of $v_\lambda^\Gamma - v_\lambda^\Delta$ exists and vanishes identically. Then Γ and Δ have the same value outcomes.*

Proof. Follows from the definition of value outcome.

In what follows, λ will be a fixed integrable comparison function, λ the corresponding measure. We usually suppress the subscript λ, e.g., write v^Γ for v_λ^Γ. Also, we assume, as we may, that $\lambda(T) = 1$.

Before proceeding it is useful to recall some definitions. A coalitional game v is *monotonic* if $v(S) \geq v(T)$ whenever $S \supset T$. The difference of two monotonic games is *of bounded variation*; the linear space of all such games (on $(T, \mathscr{C})$) is denoted BV. A nondecreasing sequence Ω of coalitions $S_1 \subset S_2 \subset \cdots \subset S_k$ is a *chain*. The *variation* of a coalitional game v over the chain Ω is defined by $\|v\|_\Omega = \sum_{i=1}^{k-1} |v(S_{i+1}) - v(S_i)|$; of course $\|v\|_\Omega$ is a seminorm on BV. For $\mathscr{D} \subset \mathscr{C}$ the seminorm $\|v\|_\mathscr{D}$ is defined by $\|v\|_\mathscr{D} = \sup \|v\|_\Omega$ where the sup is taken over all chains $S_1 \subset S_2 \subset \cdots \subset S_k$ in $\mathscr{D}$, i.e., with all $S_i \in \mathscr{D}$. If $\varepsilon > 0$ and Ψ is a collection of nonatomic probability measures on $(T, \mathscr{C})$, define $\mathscr{U}(\Psi, \varepsilon)$ to consist of all coalitions S such that $(\psi(S) - \psi'(S)) < \varepsilon$ whenever ψ and ψ' are in Ψ. A *diagonal neighborhood* is a family of coalitions that includes some $\mathscr{U}(\Psi, \varepsilon)$ in which Ψ is finite (i.e., is essentially a finite-dimensional vector measure).

LEMMA 7.6. *Given a game r in BV, suppose that for every $\varepsilon > 0$ there is a diagonal neighborhood $\mathscr{D}$ such that $\|r\|_\mathscr{D} < \varepsilon$. Then the asymptotic value of r exists and vanishes identically.*

Proof. For given ε, let $\mathscr{D}$ be as in the hypothesis. Let Π be a partition of the player space. Construct a chain Ω by taking successive unions of the elements of Π, one at a time and in a random order. Corollary 18.10 in [5] asserts that if Π is sufficiently "fine," the entire chain Ω will with arbitrarily high probability be in $\mathscr{D}$, and therefore $\|r\|_\Omega < \varepsilon$. But $\|r\|_\Omega$ is bounded even if Ω is not in $\mathscr{D}$, since $v \in BV$; since the residual probability may be made arbitrarily small, it follows that $E\|r\|_\Omega < 2\varepsilon$ (where E stands for "expectation"). Hence the Shapley value of the finite approximant to r corresponding to Π has (variation) norm $<2\varepsilon$. Since ε can be made arbitrarily small, the lemma follows from standard arguments.

LEMMA 7.7. *Let C be a compact subset of the euclidean space E^m, and let $g: T \times C \to \mathbb{R}$ be a strictly positive uniformly bounded measurable function, such that for any fixed t in T, $g(t, y)$ is continuous in y. For each y in C, define a measure ψ^y by*

$$\psi^y(S) = \int_S g(t, y) \, \lambda(dt)$$

and a probability measure $\hat{\psi}^y$ by

$$\hat{\psi}^y(S) = \psi^y(S)/\psi^y(T).$$

Then the set $\{\hat{\psi}^y: y \in C\}$ is compact.

Proof. If $y_n \to y$, then by Lebesgue's dominated convergence theorem,

$$\|\psi^{y_n} - \psi^y\| = \int_T |g(t, y_n) - g(t, y)| \, \lambda(dt) \to 0;$$

hence the mapping $y \to \psi^y$ is continuous. Since g is strictly positive, $\psi^y(T) > 0$ for each y in C; hence $\psi^y \to \hat{\psi}^y$ is continuous. Hence $y \to \hat{\psi}^y$ is continuous, and so Lemma 7.7 follows from the compactness of C.

LEMMA 7.8. *If Ψ is a compact set of nonatomic probability measures, then for every $\varepsilon > 0$, $\mathcal{U}(\Psi, \varepsilon)$ is diagonal neighborhood.*

Proof. Since Ψ is compact, it has a finite subset Ψ' such that for every ψ in Ψ there is a ψ' in Ψ' with $\|\psi - \psi'\| < \varepsilon/3$. Then $\mathcal{U}(\Psi', \varepsilon/3) \subset \mathcal{U}(\Psi, \varepsilon)$, completing the proof of Lemma 7.8.

LEMMA 7.9. *Let Γ be a public goods game in which the utilities u_t are non-negative. Then v^Γ is monotonic.*

Proof. Assume $Q \supset S$. As $X^Q \supset X^S$ and $X^{T\setminus Q} \subset X^{T\setminus S}$ (see (3.1)), it is enough to show that for every y in $G(e(T))$,

$$I \equiv \int_T u_t(y)[(\chi_Q - \chi_{T\setminus Q}) - (\chi_S - \chi_{T\setminus S})] \, \lambda(dt) \geq 0;$$

but $(\chi_Q - \chi_{T\setminus Q}) - (\chi_S - \chi_{T\setminus S}) = 2\chi_{Q\setminus S}$ and therefore $I \geq 0$. This completes the proof.

Proof of the Theorem. Let Γ^ν and Γ^ζ be two variants of a public goods game corresponding to voting measures ν and ζ, respectively. Let $\varepsilon > 0$ be given. Let Ψ consist of all the measures $\hat{U}^y = U^y/U^y(T)$ with y in C (see (7.1)), together with the two voting measures ν and ζ. Set $\mathscr{D} = U(\Psi, \varepsilon)$.

For the moment, assume that $u_t(y)$ is strictly positive for all t and y; this assumption will be removed later. Then by Lemma 7.7, Ψ is compact, and hence by Lemma 7.8, $\mathscr{D}$ is a diagonal neighborhood.

Set $v^\nu = v^{\Gamma^\nu}$, $v^\zeta = v^{\Gamma^\zeta}$, $r = v^\nu - v^\zeta$, and let $\varnothing = S_0 \subset \cdots \subset S_k = T$ be a chain in $\mathscr{D}$, which we call Ω. Let i_1 be the greatest index for which $\max(\nu(S_i), \zeta(S_i)) < \frac{1}{2}$, and i_2 the smallest index for which $\min(\nu(S_i), \zeta(S_i)) > \frac{1}{2}$; clearly $i_1 < i_2$. Let Ω_1 be the chain $S_0 \subset S_1 \subset \cdots \subset S_{i_1}$, Ω_2 the chain $S_{i_1} \subset S_{i_1+1}$, Ω_3 the chain $S_{i_1+1} \subset \cdots \subset S_{i_2-1}$, Ω_4 the chain $S_{i_2-1} \subset S_{i_2}$ and Ω_5 the chain $S_{i_2} \subset \cdots \subset S_k = T$. Clearly $\|r\|_\Omega = \sum_{i=1}^5 \|r\|_{\Omega_i}$. From (3.2) it follows that for $i \leq i_1$ as well as for $i \geq i_2$, $v_\lambda^\nu(S_i) = v_\lambda^\zeta(S_i)$ and therefore $\|r\|_{\Omega_1} = \|r\|_{\Omega_5} = 0$.

Next, let $i_1 < i < i_2$. The definition of $\mathscr{D}$ then implies that $|\hat{U}^y(S_i) - \frac{1}{2}| < 2\varepsilon$ for each y, and similarly $|\hat{U}^y(T \setminus S_i) - \frac{1}{2}| < 2\varepsilon$ for each y; hence $|U^y(S_i) - U^y(T \setminus S_i)| < 4\varepsilon U^y(T) \leq 4\varepsilon K$ for all y. Hence by (7.2), $|H^y(S_i)| < 4\varepsilon K$ for all y, and hence $|w^\nu(S_i)| < 4\varepsilon K$, where $w^\nu = w^{\Gamma^\nu}$ (see (7.3)). Hence by the monotonicity of v^ν (Lemma 7.9), and by (7.4),

$$\|v^\nu\|_{\Omega_3} = v^\nu(S_{i_2-1}) - v^\nu(S_{i_1+1}) = w^\nu(S_{i_2+1}) - w^\nu(S_{i_1+1}) < 8\varepsilon K,$$

and similarly $\|v^\zeta\|_{\Omega_3} < 8\varepsilon K$. Hence

$$\|r\|_{\Omega_3} < \|v^\nu\|_{\Omega_3} + \|v^\zeta\|_{\Omega_3} < 16\varepsilon K.$$

Finally, setting $w^\zeta = w^{\Gamma^\zeta}$, we have

$$\|r\|_{\Omega_2} \leq |r(S_{i_1})| + |w^\nu(S_{i_1+1})| + |w^\zeta(S_{i_1+1})| \leq 0 + 4\varepsilon K - 8\varepsilon K,$$

and similarly $\|r\|_{\Omega_4} < 8\varepsilon K$. Summing up, we obtain

$$\|v^\nu - v^\zeta\|_\Omega = \|r\|_\Omega < 0 + 8\varepsilon K + 16\varepsilon K + 8\varepsilon K + 0 = 32\varepsilon K.$$

Hence by Lemmas 7.5 and 7.6, the proof of the theorem is complete when $u_t(y)$ is strictly positive.

In the general case, one may modify the utility functions by adding the constant $K+1$ to them; they will then be strictly positive. The corresponding game r is not changed by the modification; since it has vanishing asymptotic value with the modified utilities, the same holds for the original utilities, and so by Lemma 7.5 the proof is complete.

8. Technical Comments

a. *Redistribution as a Public Good*

When this result was presented at a seminar at the London School of Economics, Professor William Gorman pointed out that technically, a redistribution plan [2] may be viewed as a public good for which different agents have different utilities. This presents a paradox, since the vote counts heavily in determining redistribution but not at all in the choice of public goods.

To resolve the paradox, note that the dimension of a redistribution plan is one less than the number of agents; thus with a continuum of agents one would need an infinite-dimensional public goods space to accommodate all feasible redistribution plans, whereas the model presented in Section 2 limits the number of public goods to the finite number m.

Unfortunately, this does not quite resolve the paradox. The proof of our theorem still works when the public goods space E_+^m is replaced by any separable metric space Y, as long as the strategy spaces X_ν^S are compact (which implies that the feasible bundles are in a compact subspace of Y). The crucial point is not finite dimensionality, but compactness.

Intuitively, a compact topological space is one that can be approximated by one with finitely many points. Thus the compactness of the space of public goods bundles means that there cannot be too many dissimilar feasible outcomes, where "similar" outcomes are those considered similar by all agents. With a continuum of agents, there is a continuum of dissimilar redistribution plans, and that is the reason that the results of this paper do not work for redistribution.[17]

To sum up, for the vote not to matter, we need a large number of individually insignificant agents (the continuum), but a relatively restricted choice of feasible outcomes (the compact outcome space).

b. *Beyond the Asymptotic Value*

Our theorem continues to hold when the asymptotic value is replaced by the μ-value [8, 2], or by a partition value [12]. Like the asymptotic value, these are obtained by taking limits of values of finite approximants to the given game. Unlike the asymptotic value, they do not cover *all* finite approximants; the μ-value, for example, looks only at approximating games in which all players have the same "size," when measured by the population measure μ. It follows that these values are "stronger" than the asymptotic value, in the sense that they exist and equal the asymptotic

---

[17] Of course, even with a continuum of agents one could restrict oneself to a finitely parametrized family of redistribution plans. The outcome space then really is compact, and the theorem of this paper does apply. Intuitively, what is happening in that case is that the individual voter is restricted in his effectiveness because he can benefit himself only if he simultaneously benefits others.

value whenever the latter exists, but also exist for many more games. Therefore Lemma 7.6 applies to these values as well, and the rest of the proof follows without change.

The advantage of using one of these stronger values is that it may well provide value outcomes in public goods games in which there are no value outcomes based on the asymptotic value; we have proved no general existence theorem, only an equivalence theorem.

Similar remarks apply to certain other values, such as the values obtained by Mertens [10, 11], or the mixing value [5, Chap. II]. While these are not necessarily stronger than the asymptotic value, they do appear to satisfy Lemma 7.6, and so our proof carries over to them as well.

c. Comparison Functions and Measures

In Section 4, we defined a comparison function to ba a μ-integrable nonnegative function on $(T, \mathscr{C})$ that is positive on a set of positive measure. This is in line with the original definition of Shapley [15], but differs slightly from the definition in [1] and in [3], in which $\lambda(t) > 0$ for all t, and λ is measurable but not necessarily integrable. In our case, $\lambda(t)$ may vanish for some t even under the simplest of circumstances. As for the integrability, conceptually it involves no loss of generality. Indeed, by applying appropriate linear transformations, we may obtain a bundle y in $G(e(T))$ with uniformly positive utilities (i.e., $\inf_t u_t(y) > 0$); then non-integrable λ lead to undefined (in fact, infinite) $v_\lambda^r(T)$, and hence cannot correspond to value outcomes. In [3] the uniform positivity assumption was made explicit (5.7); but there it had substantive content, since the assumptions of $u_t(0) = 0$ (not made here) and uniform boundedness do not in general permit further linear adjustment to obtain uniform positivity.

Perhaps most natural would be to dispense altogether with the comparison function λ, and define value outcomes directly in terms of the comparison measure λ. In that case one would have to start out by proving that value outcomes can correspond only to non-atomic comparison measures; this offers no particular difficulty, but is in any case avoided under our approach. In our approach, the comparison measures are in fact absolutely continuous w.r.t. μ, since $\lambda(ds) = \lambda(t)\,\mu(ds)$. Like non-atomicity, absolute continuity can be *proved* in the alternative approach; unlike non-atomicity, it is not needed to prove our results.

d. Representative Democracy and Other Voting Schemes

When this result was presented at the Berkeley–Stanford Value Theory conference in 1981, Professor Lloyd Shapley inquired whether it also applies to other voting schemes, e.g., when the voting is by district, and a majority of the districts is required. (Technically, this is represented by a finite number of nonatomic vote measures $v_1, ..., v_k$, where a coalition S "wins" if and only if more than $k/2$ of the $v_i(S)$ are $> v_i(T)/2$.)

The answer is "yes." The theorem applies whenever there are finitely many nonatomic measures $v_1,..., v_k$, such that S "wins" if all the $v_i(S)$ are $> v_i(T)/2$. Roughly, this condition means that a coalition that is both a good sample of the population and a majority always wins. Any two voting schemes satisfying this condition will lead to the same choice of a public goods bundle.

The condition is of more general applicability than may at first appear. Assuming that a legislator's vote reflects the wishes of a majority of his constituents, it is satisfied even for the process of amending the constitution of the United States, which requires majorities of both houses of Congress, and of each of the legislatures of $\frac{3}{4}$ of the states; and for that of removing the president of the United States, which requires a majority of the House of Representatives and $\frac{2}{3}$ of the Senate.[18]

REFERENCES

1. R. J. AUMANN, Values of markets with a continuum of traders, *Econometrica* **43** (1975), 611–646 [Chapter 51].
2. R. J. AUMANN AND M. KURZ, Power and taxes, *Econometrica* **45** (1977), 1137–1161 [Chapter 52].
3. R. J. AUMANN AND M. KURZ, Power and taxes in a multi-commodity economy (updated) *J. Pub. Econom.* **9** (1978), 139–161 [Chapter 53].
4. R. J. AUMANN, M. KURZ, AND A. NEYMAN, Voting for public goods, *Rev. Econom. Stud.* **50** (1983), 677–693 [Chapter 56].
5. R. J. AUMANN AND L. S. SHAPLEY, "Values of Non-Atomic Games," Princeton Univ. Press, Princeton, N.J., 1974.
6. G. DEBREU, "Theory of Value," Wiley, New York, 1959.
7. J. C. HARSANYI, A bargaining model for the cooperative n-person game, *in* "Contributions to the Theory of Games IV" (A. W. Tucker and R. D. Luce, Eds.), Ann. of Math. Studies, No. 40, pp. 325–355, Princeton Univ. Press, Princeton, N.J., 1959.
8. S. HART, Measure-based values of market games, *Math. Oper. Res.* **5** (1980), 197–228.
9. Y. KANNAI, Values of games with a continuum of players, *Israel J. Math.* **4** (1966), 54–58.
10. J. F. MERTENS, Values and derivatives, *Math. Oper. Res.* **5** (1980), 523–552.
11. J. F. MERTENS, The Shapley value in the non-differentiable case, *Internat. J. Game Theory* **17** (1988), 1–65.
12. A. NEYMAN AND Y. TAUMAN, The partition value, *Math. Oper. Res.* **4** (1979), 236–264.
13. R. SELTEN, Valuation of n-person games, *in* "Advances in Game Theory" (M. Dresher, L. S. Shapley, and A. W. Tucker, Eds.), Ann. of Math. Studies, No. 52, pp. 577–626, Princeton Univ. Press, Princeton, N.J., 1964.
14. L. S. SHAPLEY, A value for n-person games, *in* "Contributions to the Theory of Games II" (H. W. Kuhn and A. W. Tucker, Eds.), Ann. of Math. Studies, No. 28, pp. 307–317, Princeton Univ. Press, Princeton, N.J., 1953.
15. L. S. SHAPLEY, Utility comparisons and the theory of games, *in* "La Décision," pp. 251–263, Editions du Centre National de la Recherche Scientifique, Paris, 1969.

---

[18] On the other hand, there *are* voting schemes for which the theorem is false; for example, if $\frac{2}{3}$ of the entire population is required in order to "win." In that case w_λ^v is discontinuous at the point $\frac{2}{3}T$ on the diagonal, and the jump at that point makes the vote measure important.

56 Voting for Public Goods
with M. Kurz and A. Neyman

1 Introduction

Several years ago, a game theoretic model that took explicit account of power relationships was introduced to analyse taxation in a democratic society (Aumann and Kurz, 1977a, b). Those analyses were set in the context of private goods, so that only redistribution and exchange were at issue. In the current paper we apply a similar analysis to public goods.

The framework within which we work is that of a public goods economy, defined by a set of agents, a collection of public goods, a collection of non-consumable resources, and a technology (enabling public goods to be produced from resources); moreover, each agent has a utility function for public goods, an initial endowment of resources, and a voting weight. It will be assumed that the agents form a non-atomic continuum, i.e. that there are many agents, each of whom is individually insignificant.

We consider two games, the voting game and the non-voting game. In the voting game, any coalition (i.e. set of agents) with a majority of the vote may produce public goods, using its own resources only; once produced, the public goods may be enjoyed by all agents.[1] In the non-voting game, any coalition, irrespective of its size, may produce public goods, using its own resources only; public goods produced by different (disjoint) coalitions may be enjoyed by all. (For example, if we are discussing television, any programme produced by any coalition may be viewed by any agent.)

To these games we shall apply the solution notion known as the Harsanyi–Shapley Non-Transferable-Utility (NTU) Value;[2] the resulting outcomes (i.e. bundles of public goods produced) will be called value outcomes. We obtain the following:

THEOREM *The voting game has the same value outcomes as the non-voting game.*

In Aumann, Kurz and Neyman (1987) (henceforth AKN) we prove a related result, namely that the value outcomes in the voting game are independent of the voting weights. This follows from the current theorem, since obviously the voting weights cannot affect the outcome of the non-

---

This chapter originally appeared in *Review of Economic Studies* 50 (1983): 677–694. Reprinted with permission.

1. See Section 9a for an alternative description of the voting game, parallel to that used for the redistribution games of Aumann and Kurz (1977a, b) (i.e. allowing for expropriation of the minority's resources by the majority, and their destruction by the minority).
2. Shapley (1969).

voting game. But the theorem of AKN was proved under weaker conditions, under which the voting and non-voting games can actually have different value outcomes (cf. Examples 6 and 7). Briefly, in AKN, it is the voting weights that turn out irrelevant; here, the whole institution of voting turns out to be irrelevant.

The paper is organized as follows. In Section 2 we formally describe public goods economies and set forth our assumptions. Section 3 contains the formal description of our games. Section 4 specifies the variant of the Harsanyi–Shapley NTU value used in this paper, thus completing the formal specification of all terms appearing in the above statements of the theorems. In Section 5 we demonstrate the theorem informally, stressing the intuitive background. Section 6 contains illustrations and counterexamples, and Section 7 the formal proof of the theorem. Section 8 is devoted to discussion.

The paper is constructed so that readers who are not interested in the formal treatment can avoid it entirely. Such readers, after completing the introduction, should go immediately to Section 5, then peruse the informal part of Section 6, and then read Section 8. Conversely, readers interested *only* in the formal proofs may omit Sections 5, 6, and 8.

2 Public Goods Economies

The real line is represented by $\mathbb{R}$, the euclidean space of dimension n by E^n, its non-negative orthant by E^n_+ (i.e. $E^n_+ = \{x \in E^n : x^j \geq 0 \text{ for all } j\}$).

A non-atomic public goods economy consists of

i. A measure space $(T, \mathscr{C}, \mu)$ (T is the space of agents or players, $\mathscr{C}$ the family of coalitions, and μ the population measure); we assume that $\mu(T) = 1$ and that μ is σ-additive, non-atomic and non-negative.

ii. Positive integers l (the number of different kinds of resources) and m (the number of different kinds of public goods).

iii. A correspondence G from E^l_+ to E^m_+ (the production correspondence).

iv. For each t in T, a member $e(t)$ of E^l_+ ($e(t)\mu(dt)$ is dt's endowment of resources).

v. For each t in T, a function $u_t : E^m_+ \to \mathbb{R}$ (dt's von Neumann–Morgenstern utility).

vi. A σ-additive, non-atomic, non-negative measure v on $(T, \mathscr{C})$ (the voting measure); we assume $v(T) = 1$.

Note that the total endowment of a coalition S—its input into the production technology if it wishes to produce public goods by itself—is

$\int_S e(t)\mu(dt)$; for simplicity, this vector is sometimes denoted $e(S)$. A public goods bundle is called jointly producible if it is in $G(e(T))$, i.e. can be produced by all of society.

We assume that the measurable space $(T, \mathscr{C})$ is isomorphic[3] to the unit interval [0, 1] with the Borel sets. This assumption is less restrictive than it sounds; any non-denumerable Borel subset of any euclidean space (or indeed, of any complete separable metric space) is isomorphic to [0, 1]. We also assume the following (as usual $x \leq y$ means $x^j \leq y^j$ for all j):

ASSUMPTION 1 $u_t(y)$ is Borel measurable simultaneously in t and y, continuous in y for each fixed t, and bounded uniformly in t and y.

ASSUMPTION 2 G has compact and non-empty values.

ASSUMPTION 3   If $x \leq y$, then $G(x) \subset G(y)$ and $u_t(x) \leq u_t(y)$ for all t.

ASSUMPTION 4 $0 \in G(0)$.

ASSUMPTION 5   *Either* (i) u_t is C^1 (continuously differentiable) on[4] E_+^m and the derivatives $\partial u_t / \partial y^i$ are strictly positive and uniformly bounded, *or* (ii) there are only finitely many different utility functions u_t.

Assumption 3 may be called "monotonicity of production and utility" or "free disposal of resources and of public goods". The theorem is actually false without this assumption; see Example 6. Assumption 4 says that the technology is capable of producing nothing from nothing. In Assumption 5, we assume that either the utility functions are smooth, or that there are only finitely many "utility types" (though perhaps a continuum of "endowment types"). The situation is reminiscent of that in Geometric Topology, where to avoid "wild imbeddings" one may assume either that all maps are piecewise linear, or that they are differentiable. We require Assumption 5 for the proof, but we do not know whether the theorem is actually false without it; see Section 8d.

The other assumptions are of a technical nature. Note that conceptually, uniform boundedness involves no loss of generality. Indeed, in each of the games we will consider, the set of feasible public goods bundles is contained in a compact set (see the end of Section 3); by changing the u_t outside this set, we can make them bounded without really affecting anything. Uniform boundedness can then be obtained by applying (possibly different) positive linear transformations to each of the u_t. Omitting the assumption altogether might however cause technical diffi-

---

3. An isomorphism is a one-to-one correspondence that is measurable in both directions.
4. A function is C^1 on a closed set A if it can be extended to a C^1 function on an open set containing A.

culties, since transformations of this kind might affect the integrability of the utility functions $u_t(y)$ when weighted by the comparison function $\lambda(t)$ (see Section 4). While the difficulties may be circumventable, it did not seem worthwhile to expend our energy—or the readers'—in removing this conceptually harmless assumption.

3 The Games

For a verbal description of the games we shall define here, see Section 1.

Recall that a strategic game[5] with player space $(T, \mathscr{C}, \mu)$ is defined by specifying, for each coalition S, a set X^s of strategies, and for each pair (σ, τ) of strategies belonging respectively to a coalition S and its complement $T \setminus S$, a payoff function $h^s_{\sigma\tau}$ from T to $\mathbb{R}$.

In formally defining the games, we shall describe pure strategies only; but it is to be understood that arbitrary mixtures of pure strategies are also available to the players. The pure strategies we shall describe will have a natural Borel structure, and mixed strategies should be understood as random variables whose values are pure strategies.

In the non-voting game, a pure strategy for S is simply a member x of $G(e(S))$, i.e. a choice of a public goods bundle which can be produced from the total resource bundle $e(S)$. If S has chosen $x \in G(e(S))$ and $T \setminus S$ has chosen $y \in G(e(T \setminus S))$, then the payoff to any t is $u_t(x+y)$.

In the voting game, a strategy for a coalition S in the majority $(v(S) > 1/2)$ is again a member x of $G(e(S))$. Minority coalitions $(v(S) < 1/2)$ have only one strategy (essentially "doing nothing"). If a majority coalition S chooses $x \in G(e(s))$ and $T \setminus S$ chooses its single strategy (as it must), then the payoff to any t is $u_t(x)$. The definition of strategies and payoffs for coalitions with exactly half the vote is not important, as these coalitions play practically no role in the analysis; the reader may define them in any way he considers appropriate.

Note that the set of feasible public goods bundles—those that can actually arise as outcomes of one of our games—is precisely the compact set $G(e(T))$.

4 Value Outcomes

We shall be working here with the asymptotic value,[6] an analogue of the finite-game Shapley value for games with a continuum of players,

---

5. See Section 4 of Aumann and Kurz (1977b).
6. Kannai (1966).

obtained by taking limits of finite approximants. Let Γ be the voting or the non-voting game. A comparison function is a non-negative valued μ-integrable function λ on T that is positive on a set of agents of positive measure; the corresponding comparison measure $\tilde{\lambda}$ is defined by $\tilde{\lambda}(dt) = \lambda(t)\mu(dt)$, i.e. $\tilde{\lambda}(S) = \int_S \lambda(t)\mu(dt)$. A value outcome in Γ is then a random bundle of public goods associated with the Harsanyi-Shapley NTU value based on ϕ; i.e. a random variable $\underset{\sim}{y}$ with values in $G(e(T))$, for which there exists a comparison function $\tilde{\lambda}$ such that the Harsanyi coalitional form[7] v_λ^Γ of the game $\lambda\Gamma$ is defined and has an asymptotic value, and

$$(\phi v_\lambda^\Gamma)(S) = \int_S Eu_t(\underset{\sim}{y})\tilde{\lambda}(dt) \quad \text{for all } S \in \mathscr{C}, \tag{1}$$

where $Eu_t(\underset{\sim}{y})$ is the expected utility of $\underset{\sim}{y}$.

5 An Informal Demonstration of the Theorems

We start by briefly reviewing Section 5 of AKN; the reader is referred to there for a more comprehensive treatment. Let us use the word outcome for a bundle of public goods.[8] Let λ be a comparison measure, i.e. a non-negative measure on the agent space; $\lambda(dt)$ is interpreted as an infinitesimal "exchange rate" that enables comparison of agent dt's utility u_t with that of other agents. For each coalition S and each outcome y, write

$$U^y(S) = \int_S u_t(y)\lambda(dt).$$

$U^y(S)$ represents the "total" payoff to S when the exchange rates $\lambda(dt)$ are used and y is the total bundle of public goods produced by all coalitions; this follows from the fact that all agents can enjoy all public goods produced by anybody.

Denote the non-voting and voting games by A and B respectively, and let Γ be either A or B. Define

$$w_\lambda^\Gamma(S) = \max \min[U^y(S) - U^y(T\setminus S)], \tag{2}$$

---

7. v_λ^Γ is formally defined in Section 7 of AKN; for an informal definition, see (3).

8. In general, the outcomes arising in the analysis of the non-voting game are "mixed", i.e. random variables whose values are pure outcomes; but for simplicity, the informal discussion of this section is restricted to pure outcomes. For the general case, one need only replace pure outcomes y by mixed outcomes $\underset{\sim}{y}$, and the utilities $u_t(y)$ by expected utilities $Eu_t(\underset{\sim}{y})$. The voting game always leads to pure outcomes.

where the max and the min are over the strategies of S and $T\setminus S$ respectively. Set

$$v_\lambda^\Gamma(S) = \frac{(w_\lambda^\Gamma(S) + w_\lambda^\Gamma(T))}{2}. \tag{3}$$

Briefly, $w_\lambda^\Gamma(S)$ measures S's bargaining strength, or ability to threaten; $v_\lambda^\Gamma(S)$, the total utility that S can expect from the resulting efficient compromise.

Recall that an outcome y in Γ is a value outcome iff

$$(\phi v_\lambda^\Gamma)(dt) = u_t(y)\lambda(dt), \tag{4}$$

where ϕ is the Shapley value; i.e., iff it is feasible, and its utility for each infinitesimal agent dt, in terms of the exchange rates $\lambda(dt)$, is precisely his value in the coalitional game v_λ^Γ.

For each θ with $0 \leq \theta \leq 1$, let θT denote a "diagonal coalition of size θ". Intuitively, θT may be thought of as a perfect sample of the population T of all agents, containing a proportion θ of the agents. If r is a non-atomic game and dt an agent, then (AKN, (5.7)),

$$(\phi r)(dt) = \int_0^1 (r(\theta T \cup dt) - r(\theta T))d\theta. \tag{5}$$

Let S be a perfect—or almost perfect—sample of the population T with a clear majority; i.e. a coalition of the form αT, $\alpha T \cup dt$, or $\alpha T \setminus dt$, where α is larger than $1/2$ by more than an infinitesimal. Then in the non-voting game, the optimal threat of the minority is not to produce any public goods. This is because any public goods produced by the minority will also be enjoyed by the majority. Both are perfect—or almost perfect—samples, so the per capita rise in utility from such production is about the same in the two coalitions; but the majority is larger than the minority, so its total utility rises by more. Thus in the difference $U^y(S) - U^y(T\setminus S)$ between the payoffs to the two coalitions—which is the criterion for defining the optimal threats (see (2))—any enjoyment by the minority is more than offset by the corresponding enjoyment of the majority. The upshot is that in the voting game, the minority *may* not produce; in the non-voting game, it chooses not to produce; in any case, it does not produce. Therefore the outcome is the same in the two cases, namely what the majority chooses to produce; thus suppressing the subscript λ, we conclude that

$$v^A(S) = v^B(S) \quad \text{and} \quad v^A(T\setminus S) = v^B(T\setminus S). \tag{6}$$

Set $r = v^A - v^B$; we wish to show that $\phi r = 0$. By (5), the relevant coalitions U are those of the form θT or $\theta T \cup dt$. If $\theta - (1/2)$ is non-infinitesimal, then each such coalition is either of the form S considered in the previous paragraph, or is the complement of such a coalition. Hence by (6), r vanishes on all such coalitions.

If $\theta - (1/2)$ is infinitesimal, then U is "near"[9] $(1/2)T$, and so also near its complement $T \setminus U$; hence no matter what the outcome is, U and $T \setminus U$ must have approximately the same utility, since all agents enjoy the same public goods. Hence $w^\Gamma(U)$ is infinitesimal, and since $w^A(T) = w^B(T)$, it follows from (3) that $r(U)$ is infinitesimal.

Summing up, the integrand in (5) vanishes when $\theta - (1/2)$ is not infinitesimal, and is infinitesimal when $\theta - (1/2)$ *is* infinitesimal. Hence ϕr is the integral of an infinitesimal over an infinitesimal range, i.e. an infinitesimal of the second order, which may be ignored. Thus indeed $\phi r = 0$. Hence $\phi v_\lambda^A = \phi v_\lambda^B$ for all λ, and so by (4), A and B have the same value outcomes. This completes the demonstration of our theorem.

The second part of the argument (θ near 1/2) breaks down in the case of redistribution of private goods (Aumann and Kurz 1977a, b), or when the majority may exclude the minority from enjoying the public goods. In both cases, the minority may be prevented from consuming anything, whereas the majority can at least use its own resources; so even when S has only a slight v-majority, $w^B(S)$ will in general be very far from 0. The theorem is actually false in these cases (see Examples 1 and 2).

The first part of the argument (θ not near 1/2) works only if utilities are monotonic (Assumption 3); for otherwise, the minority may threaten to lower the majority's total utility by producing public bads, more than it lowers its own (Example 6).

6 Illustrations and Counterexamples

As above, $w(S)$ denotes the difference between total payoff to S and that to $T \setminus S$ when they minimax the difference. We denote by $q(S)$ the total payoff to S under the same circumstances (i.e. when the coalitions minimax the difference), so that

$$w(S) = q(S) - q(T \setminus S).$$

As above, $v(S)$ denotes $(w(S) + w(T))/2$ and λ denotes a comparison measure. It may be verified that $\phi v = \phi q$.

---

9. i.e., has similar characteristics, is statistically similar.

In the examples of this section, we will describe the utilities only in the "relevant range"—the compact set of feasible public goods bundles (see the end of Section 3). Outside of the relevant range they can be chosen arbitrarily, as long as they satisfy our assumptions. For example, the phrase "linear utilities" means "utilities that are linear in the relevant range"; the utilities cannot, of course, be linear throughout E_+^m, since that would violate the boundedness condition.

In the first four examples of this section, the comparison measure λ and the population measure μ coincide. The subscript λ is omitted in these examples.

To provide contrast and perspective, we first consider two games *different* from those forming the main subject of this paper.

Example 1 A Redistribution Game. This is a variant of one of the games discussed by Aumann and Kurz (1977a).[10] There is one kind of commodity, serving both as resource and consumption good; it is private, in the sense that any amount consumed by one agent cannot also be consumed by another agent. Each agent dt is endowed with an amount $e(dt) = e(t)\mu(dt)$ of this commodity, μ being the population measure. Utilities are linear; specifically $u_t(x) = x$ for all t and x. The vote measure v may be different from μ. A coalition with a majority of the vote may redistribute its own resources in any way it pleases among its own members; a minority coalition may consume nothing.[11] We assume $e(T) = v(T) = 1$.

It may be seen that $q(S) = e(S)$ or 0 according as $v(S) > 1/2$ or $<1/2$. In particular, q is a function of the vector measure (e, v). For diagonal coalitions, we get $q(\theta T) = 0$ or θ according as $\theta < 1/2$ or $\theta > 1/2$ (see Figure 1). For the value, (5) yields

$$(\phi q)(dt) = \left[\int_0^{1/2-v(dt)} + \int_{1/2-v(dt)}^{1/2} + \int_{1/2}^1\right](q(\theta T \cup dt) - q(\theta T))d\theta. \quad (7)$$

In the first integral, both $\theta T \cup dt$ and θT are minorities, so the integrand vanishes identically. In the second integral, $\theta T \cup dt$ is a majority whereas θT is a minority; hence the integrand $\approx ((1/2)T) = 1/2$, and so the integral is $\approx (1/2)v(dt)$. In the last integral, both are majorities; hence the integrand is $e(\theta T \cup dt) - e(\theta T) = e(dt)$, and the integral is $(1/2)e(dt)$. Summing up, we get $\phi q = (v + e)/2$. Thus the vote measure v is an important component of the value, so that our theorem does not hold here.

---

10. Example 7.1 there; it differs from this example only in that there, the vote and population measures are the same.

11. This formulation of the strategic game is different from, but equivalent to, that of Aumann and Kurz (1977a). See the discussion in Section 8a.

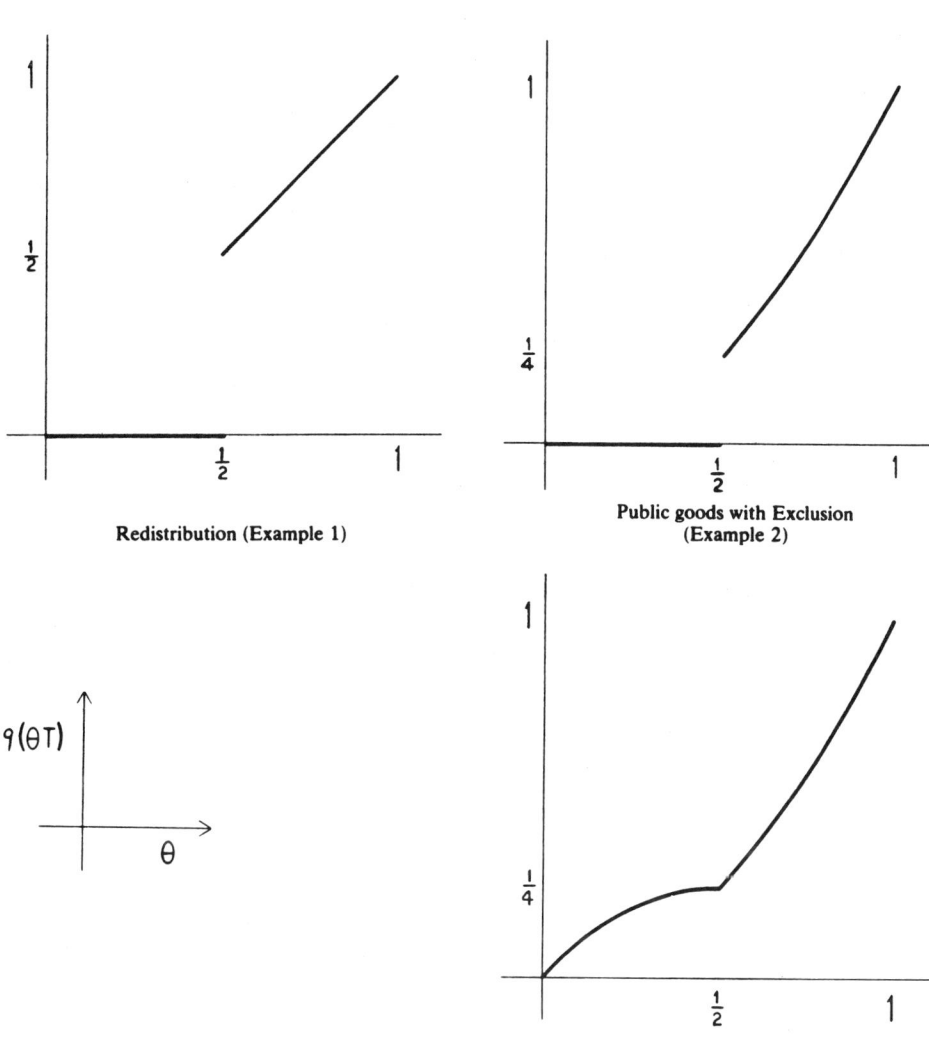

Figure 1
Three voting games.

Example 2 Public Goods With Exclusion. Here we are back in a public goods context, but we allow the coalition producing the goods to exclude others from using it. To keep the example simple, we assume one kind of resource ($l = 1$), which is non-consumable; one public good ($m = 1$); one unit of the public good may be produced from each unit of the resource ($G(x) = [0, x]$); agent dt is endowed with $e(dt) = e(t)\mu(dt)$ units of the resource; and $u_t(x) = x$. Only coalitions with a majority of the vote measure v may produce public goods, using their own resources only; they may (and therefore in the optimal threat do) prevent non-members from using the output. We assume $\mu(T) = e(T) = v(T) = 1$.

If S is a majority coalition, it will produce $e(S)$, and hence its payoff $q(S)$ is

$$\int_S u_t(e(S))\mu(dS) = \mu(S)e(S).$$

If S is a minority, then $q(S) = 0$. For diagonal coalitions, we get $q(\theta T) = 0$ or θ^2 according as $\theta < (1/2)$ or $\theta > (1/2)$ (see Figure 1). Applying (7), we find that the first integral vanishes and the second one yields $v(dt)/4$. The integrand in the third integral is

$$\mu(\theta T \cup dt)e(\theta T \cup dt) - \mu(\theta T)e(\theta T),$$

which works out to $\theta(e(dt) + \mu(dt))$ plus an infinitesimal of higher order. Since $\int_{1/2}^{1} \theta d\theta = 3/8$, we conclude that

$$\phi q = (1/4)v + (3/8)e + (3/8)\mu.$$

Thus the vote measure v is again an important component of the value.

Example 3 A Non-Voting Game. With this example, we return to the main subject of this paper. Suppose that the specifications are precisely as in the previous example, except that all may produce public goods, and all enjoy any produced. A coalition will produce if and only if it gains more out of such production than its complement; since all the utilities are the same, this means simply that it is larger than its complement. Thus a coalition will produce what it can if it is in the majority, and otherwise will produce nothing; all coalitions will enjoy whatever is produced. Therefore

$$q(S) = \begin{cases} \mu(S)e(S) & \text{if } \mu(S) > \tfrac{1}{2} \\ \mu(S)(1 - e(S)) & \text{if } \mu(S) < \tfrac{1}{2}. \end{cases}$$

For diagonal coalitions, we get $q(\theta T) = \theta \max(\theta, 1 - \theta)$ (see figure 1). Again using (7), we find that this time the middle integral is an infinitesimal of the second order, and so may be ignored; the other two integrals yield

$$\phi q = (3/4)\mu + (1/4)e.$$

Of course, the vote measure v does not figure in the expression for the value, since it does not figure in the description of the game.

Example 4 A Voting Game. Like the previous example, except that coalitions with a minority of the vote measure v may not produce public goods. This time a coalition will produce if and only if it wants to *and* may; i.e. iff $v(S) > \frac{1}{2}$ and $\mu(S) > \frac{1}{2}$. Denoting the q of the previous example by q^A, we find that

$$q(S) = \begin{cases} q^A(S) & \text{if } (\mu(S) - \frac{1}{2})(v(S) - \frac{1}{2}) > 0 \\ 0 & \text{otherwise} \end{cases}$$

(see figure 2). Since the diagonal is wholly within the area in which $q = q^A$, we get the same worth for diagonal coalitions as before (see figure 1).

By the theorem, the value ϕq is the same as in the previous example,[12] i.e. $(3/4)\mu + (1/4)e$.

The characteristic feature of the first two examples, which is absent from the last two, is the jump in $q(\theta T)$ at $\theta = \frac{1}{2}$. It is because of this jump that the middle integral in (7) contributes non-negligibly to $(\phi q)(dt)$. The contribution is precisely $v(dt)$ times the size of the jump; only here does the vote measure put in an appearance. Thus the discontinuity in $q(\theta T)$ at $\frac{1}{2}$ is intimately associated with the relevance of the vote measure. A little thought will convince the reader that this makes economic sense as well. An individual's vote is only significant because it may pivot, i.e. turn a minority into a majority; if nothing much happens to anybody even when pivoting occurs, the vote can't be very important.

Example 4 bears further examination because though $q(\theta T)$ is continuous, q itself has an essential discontinuity at $(1/2)T$. To enable the discussion to take place in two dimensions, let us take $e(t) \equiv 1$, i.e. $e(S) = \mu(S)$. If in Figure 2 one approaches the centre $(1/2)T$ of the diagonal from the southwest or northeast, then $q \to 1/4$, whereas $q \to 0$ if it is approached from the northwest or southeast. If one considers v instead of q, one finds

---

12. The perspicacious reader will have observed that (5) (or (7)) applied directly to this q yields a result different from $(3/4)\mu + (1/4)e$, in fact one that is obviously "wrong" in that it does not satisfy the efficiency axiom for values $((\phi q)(T) = q(T))$. This demonstrates once more that rough, intuitive methods have their limitations, and are no substitute for careful proofs.

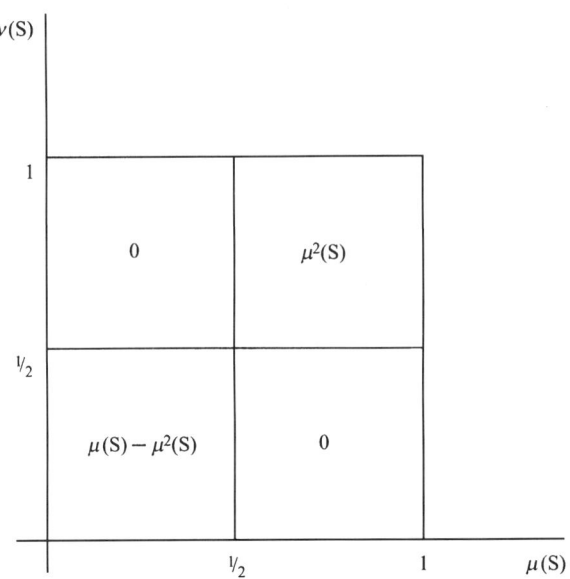

Figure 2
$q(S)$ in Example 4 when $e \equiv 1$

$$v(S) = \begin{cases} (2\mu^2(S) - \mu(S) + 1)/2 & \text{if } \mu(S) > \tfrac{1}{2} \text{ and } v(S) > \tfrac{1}{2} \\ (\mu(S) - 2\mu^2(S) + 1)/2 & \text{if } \mu(S) < \tfrac{1}{2} \text{ and } v(S) < \tfrac{1}{2} \\ \tfrac{1}{2} & \text{otherwise.} \end{cases}$$

Note that whereas $v(S)$, too, is discontinuous along almost all of the "voting line" $v(S) = 1/2$, it is continuous along the line $\mu(S) = 1/2$; in particular, the discontinuity at $(1/2)T$ itself has disappeared.

In a sense, $q(S)$ is an unnatural object, because it is the payoff to S when S is acting with a different objective in mind: not its own payoff, but the difference between the payoffs to it and to its complement. This accounts for q's bad behaviour. However, it is not this bad behaviour that causes the discontinuities in $q(\theta T)$; in each of our three examples, $v(\theta T)$ has the same jump (or lack of it) at $\theta = 1/2$ as $q(\theta T)$.

Example 5 The Optimal Quantity of a Public Good. Examples 3 and 4 are instructive from the point of view of understanding the TU analysis in this paper, which underlies the NTU analysis. But the NTU analysis itself is in these particular games trivial, since there is only one public good, so that in the end all agents will agree to produce a maximum amount of it. To obtain a non-trivial NTU example, we require at least two public goods, so that there can be some difference of opinion as to how much of each one should be produced. Since the comparison measure will now be endogenous, we abandon the convention that $\lambda = \mu$. We do adopt the normalization

$$\lambda(T) = 1. \tag{8}$$

Our public goods economy has a single resource ($l = 1$). There are two kinds of public goods, and the utility functions are of the form

$$u_t(y) = f_t(y^1) + y^2,$$

where f_t is concave, differentiable, increasing on the non-negative reals, and satisfies $f'_t(0) > 1 > f'_t(1/2)$ ($f'_t(0) = \infty$ is not excluded); it will be convenient to call the first good concave, and the second one linear. Given an amount z of resource, any combination of public goods totalling up to z may be produced; i.e.

$$G(z) = \{y \in E_+^2 : y^1 + y^2 \leqq z\}.$$

The total endowment $e(T)$ of the resource is 1.

Intuitively, one may think of the concave good as representing some particular government activity on which one wishes to focus attention, say defence; the problem is to determine its budget. The linear good represents an amalgam of all other government activity, and the resource z an amalgam of all resources; one may think of z in units of money. We consider the voting and the non-voting games, which as we know have the same value outcomes.

The point of the example is that the agents differ in their assessment of the concave good. Since $f'_t(0) > 1$, all agents would like some of this good to be produced, but they differ as to how much. If dt could decide by himself what is to be done with the total amount of resource available to Society (namely 1), he would produce that amount y^1_t of the concave good for which $f'_t(y^1_t) = 1$; the remaining resources would be allocated to production of the linear good. Since $f'_t(1/2) < 1$, it follows that $y^1_t < 1/2$.

Let λ be given. We have

$$(\phi v_\lambda)(dt) = (\phi q_\lambda)(dt) = \int_0^1 (q_\lambda(\theta T \cup dt) - q_\lambda(\theta T)) d\theta.$$

From Section 5 we know that if $\theta > 1/2$, then the optimal strategy for θT is to produce a y that will maximize

$$\int_{\theta T} u_t(y) \lambda(dt) = \theta \int_T (f_t(y^1) + y^2) \lambda(dt) = \theta y^2 + \theta F_\lambda(y^1),$$

where

$$F_\lambda(y^1) = \int_T f_t(y^1)\lambda(dt).$$

From $f'_t(0) > 1 > f'_t(1/2)$ and $\lambda(T) = 1$ it follows that $F'_\lambda(0) > 1 > F'_\lambda(1/2)$; hence there is a y^1_λ between 0 and 1/2 with $F'_\lambda(y^1_\lambda) = 1$. If $\theta > 1/2$, then the optimal strategy for θT is to produce precisely y^1_λ of the concave good; the remainder of the resource will be used for producing the linear good, yielding $\theta - y^1_\lambda$ of the linear good. The optimal strategy for the complementary coalition $(1 - \theta)T$ is to produce nothing.

Consider next the coalition $\theta T \cup ds$, where still $\theta > 1/2$. The arrival of ds places additional resources in the amount of $e(ds)$ at the disposal of the coalition. Since $y^1_\lambda < 1/2$, the coalition is already producing all it wants of the concave good, so that this additional amount will be used to produce the linear good. Each agent dt in θT will therefore obtain an additional utility of $e(ds)\lambda(dt)$; all together, they obtain an additional utility of

$$e(ds)\lambda(\theta T) = \theta e(ds)\lambda(T) = \theta e(ds),$$

by the normalization condition (8). By joining the coalition θT, though, ds causes an increase in utility in another way, namely by adding his own utility for the entire bundle now present. This is

$$(f_s(y^1_\lambda) + \theta - y^1_\lambda)\lambda(ds).$$

Hence we conclude that for $\theta > 1/2$,

$$q(\theta T \cup ds) - q(\theta T) = \theta e(ds) + (f_s(y^1_\lambda) + \theta - y^1_\lambda)\lambda(ds). \tag{9}$$

Suppose next that $\theta < 1/2$. The optimal strategy for θT is to produce nothing; and for its complement $(1 - \theta)T$, it is to produce $(y^1_\lambda, (1 - \theta) - y^1_\lambda)$, which all agents will enjoy, including those in θT. If ds joins θT, then the resources of the complement *decrease* by $e(ds)$. Since $1 - \theta > 1/2$, we are in the interior of the range where the linear good is being produced, so that the loss in utility to θT by having ds join it is

$$e(ds)\lambda(\theta T) = \theta e(ds),$$

i.e. a gain of $-\theta e(ds)$. On the other hand, there is a *gain* to θT of ds's own utility, given by

$$(f_s(y^1_\lambda) + (1 - \theta) - y^1_\lambda)\lambda(ds).$$

Hence we conclude that for $\theta < 1/2$,

$$q(\theta T \cup ds) - q(\theta T) = -\theta e(ds) + (f_s(y_\lambda^1) + (1 - \theta) - y_\lambda^1)\lambda(ds). \tag{10}$$

The analysis of Section 5 shows that the immediate neighbourhood of $(1/2)T$ may be ignored. Hence using (9) and (10), we find

$$(\phi v_\lambda)(ds) = \int_0^1 (q(\theta T \cup ds) - q(\theta T))d\theta$$

$$= e(ds)\left(\int_0^{1/2} (-\theta)d\theta\right) + \left(\int_{1/2}^1 \theta d\theta\right) + [f_s(y_\lambda^1) + (1 - y_\lambda^1)]\lambda(ds)$$

$$- \lambda(ds)\left(\int_0^{1/2} \theta d\theta + \int_{1/2}^1 (1 - \theta)d\theta\right)$$

$$= [f_s(y_\lambda^1) + (1 - y_\lambda^1)]\lambda(ds) + \tfrac{1}{4}e(ds) - \tfrac{1}{4}\lambda(ds). \tag{11}$$

Suppose now that we are at a value outcome. This means that we have found a λ such that the utility to each ds of the optimal outcome produced by the all-player set T is precisely equal to the value of ds. The optimal outcome produced by T is $(y_\lambda^1, 1 - y_\lambda^1)$, and its utility to ds is

$$[f_s(y_\lambda^1) + (1 - y_\lambda^1)]\lambda(ds).$$

Equating this with $(\phi v_\lambda)(ds)$ and using (11), we find that for all s,

$$\tfrac{1}{4}e(ds) - \tfrac{1}{4}\lambda(ds) = 0,$$

which means $e = \lambda$. Thus in this example, the weights turn out to be the initial resources; there is a unique value outcome, found by maximizing $\int_T u_t(y)e(dt)$. More precisely, the value outcome is $(y_e^1, 1 - y_e^1)$, where y_e^1 is that amount of the concave good for which $\int_T f_t'(y_e^1)e(dt) = 1$; i.e. a sort of average of the y_t^1, weighted by the $e(dt)$ and taking the utilities into account.

For an instance of an explicit calculation, take

$$f_t(z) = w(t)\log(1 + z),$$

where $1 < w(t) < 3/2$ (this comes from $f_t'(0) > 1 > f_t'(1/2)$). Then $y_t^1 = w(t) - 1$, and the unique value outcome is given by $y_e^1 = \int_T y_t^1 e(dt)$.

In this example the utility functions are separable, with a linear utility for one of the goods. This is reminiscent of the utility function traditionally used to get a TU effect in an NTU pure exchange private goods economy (see Aumann and Shapley (1974), Sections 30 and 34, and the sources quoted there). The resemblance is, however, superficial. The presence of a private good with a separable linear utility is tantamount to having side payments, and in particular λ must coincide with μ. Here the linear good is public, not private, and so cannot be used for side payments; and we have already seen that λ need not coincide with μ.

Example 6 Public Bads. Without the monotonicity assumption, Assumption 3, our theorem fails. To show this, we modify Example 5 by adding a public good y^0 with a negative utility (called a "bad"), and a resource z^0 that can be used only to produce the bad. Formally,

$$u_t(y^0, y) = f_t(y^1) + y^2 - y^0$$

$$G(z^0, z) = \{(y^0, y) \in E_+^3 : y^1 + y^2 \leq z, y^0 \leq z^0\},$$

where y, z, f_t, and the endowment e of the "old" resource z are exactly as in Example 5. The total endowment $e^0(T)$ of the new (or "nuisance") resource is 1, but no particular relationship between e and e^0 is assumed.

In both the voting and the non-voting games, majority coalitions near the diagonal will produce no bads, and will produce public goods in exactly the same amounts as in Example 5. But a minority coalition near the diagonal will behave differently in the voting and non-voting games. In the voting game, it has no licence to produce, and must simply consume what is produced by the majority. But in the non-voting game, it will produce all it can of the public bad, since it suffers less than the majority from it (because it is smaller than the majority), and its aim is to minimize the difference between the majority's payoff and its own; and as in Example 5, it will produce nothing of the original public "goods" y^1 and y^2.

In the voting game, therefore, the value outcome is exactly the same as in Example 5. In the non-voting game, we obtain for $\theta > 1/2$

$$q(\theta T \cup ds) - q(\theta T) = \theta e(ds) + \theta e^0(ds) + (f_s(y_\lambda^1) + \theta - y_\lambda^1 - (1 - \theta))\lambda(ds).$$

The right side here differs from that in (9) in two places: first, in the term $\theta e^0(ds)$, which is the total increment caused to the entire coalition by having ds deny its "bad" resource $e^0(ds)$ to the minority opposition coalition; and second, in the term $-(1 - \theta)$ which now appears in the terms describing the utility of ds for the entire bundle now present, and which is due to the production of public bad by the minority coalition. In a similar manner, we obtain

$$q(\theta T \cup ds) - q(\theta T) = -\theta e(ds) - \theta e^0(ds) + \lambda(s)(f_s(y_\lambda^1)$$
$$+ (1 - \theta) - y_\lambda^1 - \theta)\lambda(ds)$$

when $\theta < 1/2$. Proceeding as in (11), we obtain

$$(\phi v_\lambda)(ds) = [f_s(y_\lambda^1) + (1 - y_\lambda^1)]\lambda(ds) + \tfrac{1}{4}e(ds) + \tfrac{1}{4}e^0(ds) - \tfrac{1}{2}\lambda(ds). \tag{12}$$

Since a value outcome is Pareto optimal, no public bad is produced in the end; therefore the expression for the value outcome, is, as in Example 5, of the form $(0, y_\lambda^1, 1 - y_\lambda^1)$, and hence its utility to ds is given by

$$[f_s(y_\lambda^1) + (1 - y_\lambda^1)]\lambda(ds). \tag{13}$$

(The value outcome itself is of course not as in Example 5, since λ is different, as we shall soon see.) Equating (13) with $(\phi v_\lambda)(ds)$ and using (12), we deduce $\lambda = (e + e^0)/2$; this is quite different from the comparison measure obtained for the voting game, namely $\lambda = e$. Specifically, in the logarithmic example calculated at the end of Example 5, we obtain

$$y_\lambda^1 = \begin{cases} \int_T y_t^1 e(dt) & \text{in the voting game} \\ \int_T y_t^1 (e(dt) + e^0(dt))/2 & \text{in the non-voting game.} \end{cases}$$

Example 7 Our theorem also fails if we modify the voting game so that the majority is allowed to expropriate some of the resources of the minority against its will. To show this, we change Example 6 slightly, by making the "bad" y^0 into a "good," i.e., setting

$$u_t(y^0, y) = f_t(y^1) + y^2 + y^0,$$

and specifying that the resource z^0 is expropriable against the minority's will. The production function G and all other features of the example are as before.

If he wishes, the reader may think of z as labour (or "time"), and z^0 as land. A person can "destroy" his productive time simply by refusing to work, and thus avoid taxation; but land cannot be destroyed (compare Section 8a).

In the non-voting game, land and labour play similar roles; the calculations are like those of Examples 5 and 6, and yield $\lambda = (e + e^0)/2$. In the modified voting game, though, private ownership of the land is essentially meaningless, since the majority can—and therefore will—always expropriate the land. The calculations then yield $\lambda = e$. Thus the result is as in Example 6; the formula for the value outcome y_λ^1 in the logarithmic case is of course also the same.

Both this example and the previous one are examples of "public goods games" in the sense of AKN. This implies that the outcomes are independent of the voting weights, as indeed is apparent from the results (v

does not enter the formulas). But as these examples show, the stronger theorem of this paper fails for these games. In a sense, resources that can be expropriated by the majority against the will of the minority are to all intents and purposes public property, and cannot enter the calculations like privately held resources.

7 Formal Proof

The proof of our theorem uses much of the machinery developed in connection with the proof of the main theorem of AKN, in Section 7 of that paper. Rather than reviewing all this material, we simply assume that it is before us, and continue the development from there. For convenience, we use a separate numeration of formulas in this section, starting from (7.10); Formulas (7.1) through (7.9) are in AKN.

Denote the non-voting and voting games by A and B respectively. Set $r = v^A - v^B$. As in Section 7 of AKN, we may assume that $u_t(y)$ is strictly positive. Let $\varepsilon > 0$ be given.

First assume (i) in Assumption 5. For each j with $1 \leq j \leq m$ and each y in C, define

$$U_j^y(S) = \int_S u_t^j(y)(dt),$$

where $u_t^j(y) = \partial u_t(y)/\partial y^j$. Note that $U_j^y(T) > 0$, since $u_t^j(y) > 0$ for all j; define a probability measure $\hat{U}_j^y$ by $\hat{U}_j^y = U_j^y/U_j^y(T)$. Let Ψ consist of all the measures $\hat{U}_j^y$, together with the $\hat{U}^y$ defined in Section 7 of AKN, and the voting measure v; let $\mathscr{D} = U(\Psi, \varepsilon)$. By Lemma 7.7, Ψ is compact, and hence by Lemma 7.8, $\mathscr{D}$ is a diagonal neighbourhood.

Before proceeding, we note that

$$\frac{\partial}{\partial y^j} U^y(S) = U_j^y(S); \tag{7.10}$$

this follows from Lebesgue's dominated convergence theorem and the mean value theorem (which implies that the difference ratios tending to $\partial u_t(y)/\partial y^j$ are uniformly bounded). Again using the mean value theorem, we obtain from (7.2) and (7.10) that if x, y, and $x+y$ are in C, then for any S there is a point z on the line segment connecting x to $x+y$ such that

$$H^{x+y}(S) - H^x(S) = \sum_{j=1}^m y^j(U_j^z(S) - U_j^z(T\setminus S)). \tag{7.11}$$

Now let $\emptyset = S_0 \subset \cdots \subset S_k = T$ be a chain in $\mathscr{D}$, which we call Ω. Let

i_1 be the greatest index for which $v(S_1) < 1/2 - \varepsilon$, and i_2 the smallest index for which $v(S_i) > 1/2 + \varepsilon$. Define five chains $\Omega_1, \ldots, \Omega_5$ exactly as in Section 7 of AKN. First, let $i \geq i_2$. From the definition of $\mathscr{D}$ it follows that $\hat{U}_j^z(S_i) > 1/2$ for all z in C and all j; hence $\hat{U}_j^z(T \setminus S_i) < 1/2$, and therefore $U_j^z(S_i) - U_j^z(T \setminus S_i) > 0$. Thus if x, y, and $x + y$ are in C, then (7.11) yields

$$H^{x+y}(S_i) \geq H^x(S_i). \tag{7.12}$$

Since $v(S_i) > 1/2$, the coalition S_t—but not its complement $T \setminus S_i$—may produce public goods in the voting game B, and hence

$$w^B(S_i) = \max_{x \in G(e(S_i))} H^x(S_i). \tag{7.13}$$

In the non-voting game A, both S_1 and $T \setminus S_i$ may produce; hence by (7.12), by $0 \in G(e(T \setminus S_i))$ (Assumptions 3 and 4), and by (7.13),

$$w^A(S_i) = \max_{x \in G(e(S_i))} \min_{y \in G(e(T \setminus S_i))} H^{x+y}(S_i)$$
$$= \max_{x \in G(e(S_i))} H^x(S_i) = w^B(S_i). \tag{7.14}$$

Hence $v^A(S_i) = v^B(S_i)$, hence $r(S_i) = 0$, and so $\|r\|_{\Omega_5} = 0$. If $i \leq i_1$, the same proof applied to $T \setminus S_i$ instead of S_i shows that $r(S_i) = -r(T \setminus S_i) = 0$, whence $\|r\|_{\Omega_1} = 0$.

Next, proceeding exactly as in Section 7 of AKN (using v^A and v^B instead of v^ν and v^ς), one shows that $\|r\|_{\Omega_3} < 16\varepsilon K$, $\|r\|_{\Omega_2} < 8\varepsilon K$, and $\|r\|_{\Omega_4} < 8\varepsilon K$. Thus

$$\|v^A - v^B\|_\Omega = \|r\|_\Omega < 0 + 8\varepsilon K + 16\varepsilon K + 8\varepsilon K + 0 = 32\varepsilon K.$$

Hence by Lemmas 7.5 and 7.6, the proof of our theorem under the differentiability assumption, Assumption 5(i) is complete.

Finally, assume (ii) in Assumption 5, i.e. that there are only finitely many utility types $T_1 \ldots, T_h$; thus all agents in a fixed T_j have the same utility function u_j, and $\bigcup_{j=1}^{h} T_j = T$. Define

$$\lambda_j(S) = \begin{cases} \lambda(S \cap T_j)/\lambda(T_j) & \text{if } \lambda(T_j) > 0, \\ 0 & \text{if } \lambda(T_j) = 0. \end{cases}$$

Let Ψ consist of the voting measure v and all the λ_j, and let $\mathscr{D} = U(\Psi, \varepsilon)$. Since ψ is finite, $\mathscr{D}$ is by definition a diagonal neighbourhood. Let $\emptyset = S_0 \subset \cdots \subset S_k = T$ be a chain in $\mathscr{D}$, which we call Ω. Let i_1 be the greatest index for which $v(S_i) < 1/2 - \varepsilon$, and i_2 the smallest index for which $v(S_i) > (1/2) + \varepsilon$. Define five chains $\Omega_1, \ldots, \Omega_5$ exactly as in Section 7 of AKN.

First let $i > i_2$. From the definition of $\mathscr{D}$ it follows that when $\lambda(T_j) > 0$, then $\lambda_j(S_i) > 1/2$, hence $\lambda_j(T \setminus S_i) < 1/2$, and hence $\lambda_j(S_i) - \lambda_j(T \setminus S_i) > 0$. Hence for all x and y in E_+^m, the monotonicity of the utilities (Assumption 3) yields

$$H^{x+y}(S_i) = U^{x+y}(S_i) - U^{x+y}(T \setminus S_i)$$

$$= \sum_{j=1}^{h} \lambda(T_j)u_j(x+y)(\lambda_j(S_i) - \lambda_j(T \setminus S_i))$$

$$\geq \sum_{j=1}^{h} \lambda(T_j)u_j(x)(\lambda_j(S_i) - \lambda_j(T \setminus S_i)) = H^x(S_i).$$

Now this is precisely (7.12), and the remainder of the proof is as in the differentiable case.

8 Discussion

a Description of the Voting Game

In describing the voting game, we specified that only majority coalitions were permitted to produce, using their own resources only. This appears different from the corresponding description in Aumann and Kurz (1977a), in which it was specified that the majority may expropriate resources from the minority, but that faced with expropriation, the minority may destroy part or all of its resources.

In fact, the descriptions are equivalent. Since the minority may destroy its own resources, and the majority may also effectively destroy the minority's resources (simply by refusing to use them), destruction of the minority's resources is an option available to either side. But the zero-sum nature of the threat game implies that for some pair of optimal strategies, any option available to both sides will be taken up by at least one of them, since at least one of the sides will gain (or at least not lose) by doing so. Thus there is no loss of generality in specifying that the minority's resources will in fact be destroyed.

b The Coase Theorem and Related Issues

The "Theorem" of Coase (1960) asserts, among other things, that property rights do not affect the level at which economic activities that generate externalities are performed.[13] For example, suppose steam

---

13. The assertion concerns rational economic agents who are permitted to trade in their property rights.

locomotives emit sparks that damage crops on nearby land. The Coase Theorem states that whether the train runs is independent of whether the railroad must recompense the farmer for lost crops. If profits from running the train exceed the value of the lost crops, it will run; otherwise it won't. Property rights determine the level (and direction) of side payments, but not that of the actual activities.

This is subject to a lot of "ifs": no transaction costs, no income effects,[14] and so on. In effect, one must assume a transferable utility (TU) context. But in that case ordinary Pareto optimality dictates that the protagonists always engage in whatever activities are necessary to maximize the sum of the payoffs, regardless of the capabilities (or "rights") of individuals or subcoalitions (who can always be compensated by transfers). Thus the Coase Theorem is simply an expression of Pareto optimality in the context of externalities.[15]

Public goods are a classic instance of externalities, and the vote may be considered analogous to property rights. Thus our result, which implies that the choice of public goods is independent of who has the vote, sounds like a version of the Coase Theorem. But the resemblance is superficial; our result goes much further.

To clarify this issue, note first that as stated, our model permits no side payments. When side payments are impossible, the Coase Theorem predicts only Pareto optimality, and, of course, there are in general many Pareto optimal outcomes. The choice among these outcomes may very well be affected by property rights; if the railroad is prevented from compensating the farmer, his property rights may be decisive in determining whether the train will run. Similarly, one would expect that when side payments are impossible, the vote *does* affect the choice of public goods. But our theorem says that it does not.

Consider next the TU version of our model, in which it is possible to make side payments so that the utility lost by the payor always precisely equals that gained by the payee. This yields a model like the one in this paper, but with an exogenous comparison measure λ that expresses actual rates of exchange. In that case the Coase principle (i.e. the principle of Pareto optimality) leads us to expect a choice of public goods that is independent of the voting weights, but with compensation between the agents that *does* depend on the voting weights. Here again our theorem goes much further: it says that the choice of public goods *and* the schedule of compensations between the agents is independent of the vote.

---

14. Dolbear (1967).
15. We are referring to the principle expounded by Coase (1960), not to later developments that discuss the formation of markets in externalities.

Finally, consider the possibility of making side payments that cause the payor to lose and the payee to gain utility, but not necessarily in equal amounts.[16] In spite of appearances this is essentially an NTU (non-transferable utility) situation, much like that in which no side payments at all are permitted. While we have not considered this model explicitly, our methods can be applied to it in a straightforward fashion; the conclusion is that as in the cardinal TU case, both the choice of public goods *and* the schedule of side payments are independent of the vote. This is to be contrasted with the Coase Theorem, which with similar side payments allows both the level of externality-producing activity and the schedule of side payments to depend on property rights, stipulating only that the overall outcome be Pareto optimal.

The question arises, what economic factors that are absent in the Coase case account for our strong results. Part of the answer is that we deal with a large number of individually insignificant agents—"a perfect competition" context, so to speak; whereas Coase considers any number of agents, and in fact most of his examples have just two. There is more to it than that, though; the reader is invited to consult Section 6 of AKN.

c Existence of Value Outcomes

This paper has concentrated on equivalence results—on relationships between value outcomes of different games—and has avoided questions of existence. We have proved a statement of the form "two games have the same set of value outcomes," it being understood that the set may be empty. Here we briefly address the existence problem.

This problem divides naturally into two parts: (i) Existence of an asymptotic value for the game v_λ with given λ, and (ii) given a positive solution to (i), finding equilibrium λ (i.e. solving (1)).

As far as (i) is concerned, it is likely that a differentiability assumption in the spirit of Assumption 5(i) would be sufficient to ensure existence of an asymptotic value for the non-voting and voting games v_λ^A and v_λ^B. The proof would perhaps use methods similar to those used for exchange economies (Aumann and Shapley (1974), Chapters VI and VII), and the finite dimensionality of the outcome space would further simplify matters.[17]

Without differentiability matters become more problematic. Even with a finite type assumption like Assumption 5(ii), the asymptotic value in

---

16. We called this "ordinal TU" in Section 9 of AKN, to distinguish it from the previous case, which we called "cardinal TU" there, and which is the plain "TU" of most of the literature.

17. Exchange economies have infinite dimensional outcome spaces.

general fails to exist in the case of exchange economies (op. cit., Section 19), and presumably for our public goods economies as well. Hart (1977) has demonstrated the existence of an asymptotic value for non-differentiable exchange economies obeying a certain symmetry condition; perhaps a similar result could be proved here.

For (ii), the main problem would be continuity of the value ϕv_λ as a function of λ. With differentiability, this would probably be OK. But without differentiability, it would cause difficulties even when (i) is satisfactorily resolved; the value, which is a kind of average derivative, would not in general be continuous in λ when one passes over a kink.

Summing up, it appears that an appropriate differentiability condition is probably sufficient to ensure existence of a value outcome, and that for a general existence theorem, one cannot get away with much less. On the other hand, a *generic* existence theorem may well be provable without any kind of differentiability condition.

We stress that we do not have any existence proof; the remarks in this subsection should be considered conjectures.

d The Role of Differentiability

The proof of the theorem depends on Formula (7.14), which asserts that $w^A(S) = w^B(S)$ for any coalition S that is "close to the diagonal" and has a "considerable" majority; more precisely, that

for any $\delta > 0$ there is a diagonal neighbourhood $\mathscr{D}$ such that

$$w^A(S) = w^B(S) \text{ whenever } S \in \mathscr{D} \text{ and } v(S) > \tfrac{1}{2} + \delta. \tag{13}$$

Assumption 5—that the utilities are either differentiable or are of finitely many different types—is essential to prove (13): It is possible[18] to construct a public goods economy satisfying Assumptions 1 through 4—but not 5—and violating (13). Thus Assumption 5 is essential for our line of proof. It is not known whether the theorem is actually false without it.

Acknowledgments

This work was supported by National Science Foundation grant SES80-06654 at the Institute for Mathematical Studies in the Social Sciences, Stanford University, and by the Institute for Advanced Studies of the Hebrew University of Jerusalem.

---

18. See Example 6.18, p. 36 of "Public Goods and Power," by R. Aumann, M. Kurz, and A. Neyman, TR 273 (revised) of the Institute for Mathematical Studies in the Social Sciences (Economics), Stanford University, September 1980.

References

Aumann, R. J. and Kurz, M. (1977*a*), "Power and Taxes," *Econometrica*, 45, 1137–1161 [Chapter 52].

Aumann, R. J. and Kurz, M. (1977*b*), "Power and Taxes in a Multi-Commodity Economy," *Israel Journal of Mathematics*, 27, 185–234 [Chapter 53].

Aumann, R. J., Kurz, M. and Neyman, A. (1987), "Power and Public Goods," *Journal of Economic Theory*, 42, 108–127 [Chapter 55].

Aumann, R. J. and Shapley, L. S. (1974), *Values of Non-Atomic Games* (Princeton: Princeton University Press).

Coase, R. H. (1960), "The Problem of Social Cost," *The Journal of Law and Economics*, 3, 1–44.

Dolbear, F. T., Jr. (1967), "On the Theory of Optimum Externality," *American Economic Review*, 57, 90–103.

Hart, S. (1977), "Values of Non-Differentiable Markets with a Continuum of Traders," *Journal of Mathematical Economics*, 4, 103–116.

Kannai, Y. (1966), "Values of Games with a Continuum of Players," *Israel Journal of Mathematics*, 4, 54–58.

Shapley, L. S. (1969). "Utility Comparisons and the Theory of Games," in *La Decision*, Editions due Centre National de la Recherche Scientifique, Paris, 251–263.

57 Values of Markets with Satiation or Fixed Prices
with J. Dreze

To Gerard Debreu on his sixty-fifth birthday, with admiration and affection.

In markets with satiation, competitive equilibria may fail to exist, because no matter what the prices are, the satiation points of some traders may be in the interiors of their budget sets. Thus some traders will be using less than the maximum budget available to them, creating a total budget excess. This suggests a revision of the equilibrium concept that allows the budget excess to be divided among all the traders, as dividends. Each trader's budget is then the sum of his dividend and the market value of his endowment. A given system of dividends and prices defines a dividend equilibrium if it generates equal supply and demand.

This in itself is not satisfactory because it is too broad: Every Pareto optimal allocation is sustained by some system of dividends and prices. However, the Shapley value yields much more specific information. We prove that, when there are many individually insignificant agents, every Shapley value allocation is generated by a system of dividends and prices in which all dividends are nonnegative and depend only on the net trade sets of the agents, not on their utilities. Moreover, the dependence is monotonic; the larger the net trade set, the higher the dividend.

The same result holds for markets with fixed prices, which can be analyzed formally as a special case of markets with satiation.

On a more technical level, our analysis has some unusual features. We use a finite-type asymptotic model, rather than a nonatomic continuum. Surprisingly, the results are qualitatively different. (The continuum is too rough a tool for our problem, and leads to inconclusive results.) Also, small coalitions play a critical role in our analysis. (We are led to equations in which the first-order terms cancel; the second-order terms, which take events of small probability into account, become decisive.)

KEYWORDS: Coupons equilibrium, exchange economy, fixed prices, game theory, satiation, Shapley value, unemployment.

1. INTRODUCTION

PURE EXCHANGE ECONOMIES, or *markets*, in which the preferences satisfy conditions of monotonicity and nonsatiation have been studied thoroughly in the past. In this paper we investigate the opposite situation: the utility functions need not be monotonic, and do have absolute maxima. The resulting theory has significant new qualitative features.

This study[1] is not motivated by an abstract desire to remove as many assumptions as possible. It originated in the analysis of *fixed price* economies, which have been used extensively in the past decade[2] to model market failures such as unemployment. In such economies, all trade is restricted to take place at

---

[1] The work of R. J. Aumann was supported by CORE at Université Catholique de Louvain, by the Institute for Advanced Studies at the Hebrew University of Jerusalem, and by the Institute for Mathematical Studies in the Social Sciences (Economics) at Stanford University under a grant from the U.S. National Science Foundation. This work is part of the Projet d'Action Concertée on "Applications of Economic Decision Theory" sponsored by the Belgian Government under Contract No. 80/85-12. We are grateful to Jean-François Mertens for carefully reading the manuscript and suggesting significant improvements.

[2] See, e.g., the survey by Drazen (1980).

This chapter originally appeared in *Econometrica* 54 (1986): 1271–1318. Reprinted with permission.

exogenously fixed prices $\bar{p}$. In effect, this limits each trader t to his *fixed price hyperplane*, i.e. the set of all bundles x in his original consumption set for which $\bar{p} \cdot x = \bar{p} \cdot e(t)$, where $e(t)$ is t's endowment; under the usual assumptions, t's utility has an absolute maximum on this set, and is not monotonic there.[3]

In general, price rigidities prevent a market from "clearing" (i.e., supply from matching demand); various quantity constraints or rationing schemes have been proposed to bring the situation back into equilibrium. In the more traditional markets, without fixed prices, there is a close relationship between competitive equilibria and game theoretic concepts such as the core[4] and the Shapley value;[5] thus one may expect game theory also to be helpful in suggesting equilibria for fixed price economies. It turns out that the core is not well suited to this purpose (see Section 11). But we shall find that the Shapley value allocations in fixed price economies correspond to a natural extension of competitive equilibria, closely related to the concept of coupons equilibrium defined by Drèze and Müller (1980).

To describe our results, let us return to the more general context of markets with satiation. The reason that competitive equilibria may fail to exist in such markets is that no matter what the prices[6] are, the satiation points of some traders may be in the interiors of their budget sets.[7] Thus some traders will be using less than the maximum budget available to them, creating a total budget excess. This suggests a revision of the equilibrium concept that allows the budget excess to be divided among all the traders, say as *dividends*: Each trader's budget is then the sum of his dividend and the market value of his endowment at the market prices. A given system of dividends and prices is in equilibrium if it generates equal supply and demand.

This in itself is not satisfactory because it is too broad: Drèze and Müller showed that the fundamental proposition of welfare economics continues to apply here, i.e., that *every* Pareto optimal allocation is generated by some system of dividends and prices. However, the Shapley value yields much more specific information. Our main result says that when there are many individually insignificant agents, every Shapley value allocation is generated by a system of dividends and prices in which all dividends are nonnegative and depend only on the net trade sets[8] of the agents, not on their utilities. Thus the income allocated to each agent—over and above the market value of his endowment—depends only on his trading *opportunities*; on what he is *able* to offer, not on what he wants to offer. Moreover, the dependence is monotonic; the larger the net trade set, the higher the dividend.

---

[3] Indeed, monotonicity is meaningless in this context, since there is no natural partial order on the fixed price hyperplane.

[4] Cf., e.g., Hildenbrand (1982).

[5] Cf., e.g., Aumann (1975) or Hart (1977b).

[6] We are here discussing endogenous market prices q, which should not be confused with the exogenous prices $\bar{p}$ in fixed price economies. See Section 10.

[7] See Section 3 for an example.

[8] The net trade set of agent t is $C(t) - e(t)$, where $C(t)$ is his consumption set, and $e(t)$ his endowment.

Two brief illustrations may clarify this point. When a bond issue is oversubscribed, bonds are normally rationed to the subscribers in proportion to the amount requested. Under complete information, this procedure has no equilibrium; the subscribers will always request more than they really want, this will be taken into account by the other subscribers, and so on. But in the rationing scheme implied by the Shapley value, the maximum that a subscriber may buy is based not on what he requests, which is subject to manipulation, but on what he *could* buy; on his net worth, say.

The second illustration deals with unemployment in a fixed wage context. Various rationing schemes that involve cutting down on working hours have been proposed. In the scheme suggested by the Shapley value, the maximum work week for any particular worker would depend on how much time he has. Thus a youngster who must by law attend school, or a kidney patient undergoing time-consuming dialysis, would be assigned a quota smaller than the average, *even though he might be able to fill the average quota.*

Economic models have two basic components, the objective and the subjective. The first consists of the physical opportunities or *abilities* of the agents: resources, technologies, constraints on consumption, and so on. The second consists of the utilities or *preferences*. In a market, the objective component is completely described by the net trade sets of the agents. Outcomes of economic models usually depend on both components, often quite intricately.

Competitive equilibria "decouple" the two components. Each agent optimizes over an endogenous choice set, his budget set; in equilibrium, the choices mesh, they "clear" the market. The optimization, of course, is subjective; it depends on the agent's preferences. But the choice set itself does not; it depends only on his net trade set, i.e., on purely objective factors. Our result implies that the dividend equilibria to which the Shapley value leads also decouple in this way.

On a more technical level, our analysis has several unusual features. Though we are dealing with a large number of individually insignificant agents, we do not model it with a nonatomic continuum; rather, we use a finite-type asymptotic model of the Debreu and Scarf (1963) genre. Asymptotic and continuous results may differ in various ways,[9] but usually, the results are qualitatively similar. Here they are not. The continuum is too rough a tool; it obliterates the fine structure of the problem, and so leads to inconclusive results. The matter will be discussed further in Section 11.2.

Another unusual feature, related to the first, is the critical importance of small coalitions. The Shapley value of a player is the expectation of his "contribution to Society" when the players are ordered at random; the probability that he is second or third in the order is small, and is usually ignored. Here, we are led to equations in which the first-order terms cancel, and the second-order terms, which take events of small probability into account, become decisive. When there is an excess supply of labor, the length of the work week allocated to a given worker

---

[9] E.g. in ease and transparency of the formulation, in the generality of the results, in the methods of proof, and in the discussion of errors and rates of convergence. Compare Aumann and Shapley (1974, Section 34, 208-210).

depends on his expected contribution when he arrives on the scene; unless he is very early, this is negligible.

The plan of the paper is as follows. In Sections 2-5, we present the model and state our main result; it is proved in Sections 7-9. Section 10 contains the application to fixed prices, and Section 11 is devoted to a discussion of some alternative approaches. In Section 12 we state some additional results of a more quantitative nature, including properties of the dividends that go beyond mere monotonicity. These results enable the calculation of some numerical examples in Section 13. Finally, Section 14 discusses open problems. Appendices A and B establish two mathematical propositions that are needed for the proof of the main result. Appendix C contains the proofs of the results stated in Section 12.

Since the proof of the main theorem is rather complex, we offer three aids to its understanding. Section 6 contains a summary of the underlying economic ideas. Section 9.2 gives a brief outline of the mathematical ideas. Finally, the flow charts in Appendix D provide an overview of the relationships between the various lemmas and propositions constituting the proof.

2. MARKETS WITH SATIATION

A (finite) *market with satiation* is defined by:

(2.1)   a finite set T (the *trader space*);

(2.2)   a positive integer d (the number of *commodities*);

(2.3)   for each trader t, a compact convex subset X_t of R^d, whose interior is nonempty and contains the origin 0 (t's *net trade set*); and

(2.4)   for each trader t, a concave continuous function u_t on X_t (t's *utility function*).

Because X_t is compact, the continuous function u_t must attain its maximum; denote by B_t the set of all points in X_t at which the maximum is attained (the *satiation* or *bliss* set of trader t), and note that it is compact and convex. To avoid trivialities, assume $0 \notin B_t$, i.e., the initial bundle never satiates.

A few matters of terminology and notation: the inner product of two vectors q and x is denoted $q \cdot x$; "w.r.t." means "with respect to" and "w.l.o.g." means "without loss of generality"; R^k, R^k_+, and R^k_{++} denote, respectively, Euclidean k-space, its (closed) nonnegative orthant, and its (open) strictly positive orthant; int and bd denote "interior" and "boundary" respectively.

3. DIVIDEND EQUILIBRIA

An *allocation* in a market with satiation is a vector $x = (x_t)_{t \in T}$, where x_t is a feasible net trade for trader t (i.e., $x_t \in X_t$), and $\sum_{t \in T} x_t = 0$. A *price vector* is any member of R^d other than 0; since utilities need not be monotonic, one cannot confine oneself to nonnegative prices. The classical notion of *competitive equilibrium* is defined for markets with satiation just as it is for ordinary markets: it consists of a price vector q and an allocation x such that for all t, x_t maximizes u_t over the *budget set* $\{x \in X_t: q \cdot x \leq 0\}$.

Competitive equilibria do not in general exist in markets with satiation. Consider, for example,[10] a market with two agents, 1 and 2, and one commodity; suppose that the satiation points are on opposite sides of the origin, e.g., $u_1(x) = -(x-1)^2$, $u_2(x) = -(x+2)^2$. W.l.o.g. the price vector is ± 1; in either case one agent receives 0 and the other his satiation point, and these do not sum to 0.

The example is not due to any pathologies associated with the low dimension.[11] Consider a market with two commodities and three agents having the same net trade set, and with indifference maps as illustrated in Figure 1. No matter what the price vector is, the satiation point of at least one trader must be in the interior of the budget set,[12] so that his utility will be maximized there over the budget

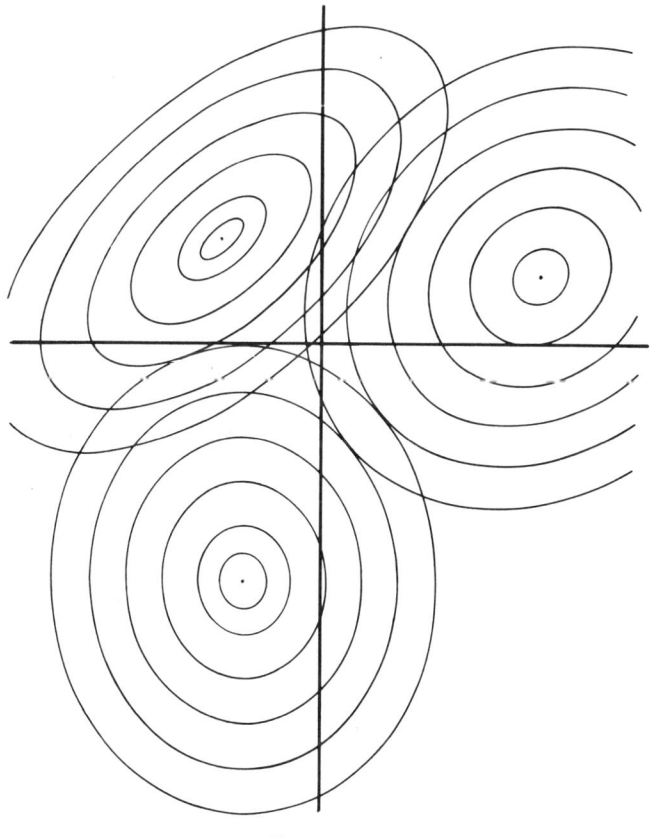

FIGURE 1

[10] This example appears in Drèze and Müller (1980).
[11] Such as the disconnectedness of the set of price vectors.
[12] This holds as long as 0 is in the interior of the convex hull of the three satiation points.

set; whereas the utilities of the remaining agents will be maximized on the budget line. The resulting three points cannot sum to 0.

What is happening is that at least one trader creates a surplus by refusing to make use of his entire budget; but the definition of competitive equilibrium does not permit the other agents to use this surplus, so an imbalance results.[13]

To overcome this problem, define a *dividend* to be a vector $c = (c_t)_{t \in T}$ whose components c_t are real numbers. A *dividend equilibrium* consists of a price vector q, a dividend c, and an allocation x, such that for all t, x_t maximizes u_t over the *dividend budget set*

$$\{x \in X_t : q \cdot x \leq c_t\}.$$

A dividend may be thought of as a cash allowance added to the budget of each trader; its function is to distribute among the unsatiated agents the surplus created by the failure of the satiated agents to use their entire budget.

A dividend c is *nonnegative* if all the c_t are nonnegative; *monotonic in the net trade sets*, if $X_t \supset X_s$ implies $c_t \geq c_s$. Occasionally, the word dividend will also be used for a component c_t of c.

4. VALUE ALLOCATIONS

A *comparison vector* on the trader space T is a vector $\boldsymbol{\lambda} = (\lambda_t)_{t \in T}$, where each λ_t is a positive real number. For each comparison vector $\boldsymbol{\lambda}$ and coalition[14] S, define

(4.1) $\quad v_\lambda(S) := \text{Max} \left\{ \sum_{t \in S} \lambda_t u_t(x_t) : \sum_{t \in S} x_t = 0 \text{ and } x_t \in X_t \text{ for all } t \text{ in } S \right\}.$

In words, $v_\lambda(S)$ is the maximum total utility that S can get for itself by redistributing its endowment among its members, when the utilities u_t are weighted by the λ_t. A *value allocation* (in a given market with satiation) is an allocation x for which there exists a comparison vector $\boldsymbol{\lambda}$ such that for all traders t,

(4.2) $\quad (\phi v_\lambda)(t) = \lambda_t u_t(x_t),$

where ϕv_λ denotes the Shapley value of the game v_λ; we say that $\boldsymbol{\lambda}$ and x are *associated* with each other. (Recall that

(4.3) $\quad (\phi v_\lambda)(t) = E(v_\lambda(S \cup t) - v_\lambda(S)),$

where E denotes "expectation", and S is the set of traders preceding t in a random order on all traders.[15])

At a value allocation, the weighted utility each player receives is equal to his Shapley value in the game v_λ. In other words, if transfers of utility are permitted

---

[13] If free disposal were permitted, the example would go away; but this is not a reasonable assumption in the absence of monotonicity. Moreover, "disposal" is not really possible in the fixed-price application (Section 10).

[14] A *coalition* is a subset of T.

[15] Shapley's definition (1953) of ϕ is in terms of a set of axioms, from which (4.3) is derived. See Roth (1977) for an interesting discussion of Shapley's axioms.

with exchange rates λ_t, then the Shapley value of the resulting game is achieved, *without transfers of utility*, at the allocation x. Compare Shapley (1969) and Aumann (1975).

5. THE MAIN THEOREM

Let M^1 be a market with satiation; denote the traders by $1, \ldots, k$, the utility functions by $u_1, \ldots, u_k$, the net trade sets by $X_1, \ldots, X_k$, and the satiation sets by $B_1, \ldots, B_k$. The *n-fold replication* M^n of M^1 is the market with satiation in which there are nk traders, n of each of the k "types" in M^1; i.e. the trader space T^n in M^n is the union of k disjoint sets $T_1^n, \ldots, T_k^n$ (the *types*), such that $u_t = u_i$ and $X_t = X_i$ whenever $t \in T_i^n$. We assume as follows:

ELBOW ROOM ASSUMPTION: *For each $J \subset \{1, \ldots, k\}$,*

$$(5.1) \quad 0 \notin \operatorname{bd}\left[\sum_{i \in J} B_i + \sum_{i \notin J} X_i\right].$$

In words, for any coalition J in M^1, if it is at all possible simultaneously to satiate all agents in J, then this can also be done "with room to spare," i.e. when all other agents are restricted to the interiors of their net trade sets.[16]

This assumption holds generically in a certain very natural sense. The left side of (5.1) represents the total endowment of the market, whereas its right side is the boundary of a convex set in R^d, hence at most $(d-1)$-dimensional. Since there are only finitely many J, the assumption holds for all but a $(d-1)$-dimensional set of total endowments. A formal genericity statement can be made in terms of translates of the X_i; translating the net trade set is equivalent to varying the endowment. For details, see Section 11.3.

Note also that since the J's are sets of types, the number of conditions (5.1) is fixed at 2^k, and does not vary with n.

Call an allocation $\hat{x}$ in M^n *equal treatment* if it assigns the same net trade to traders of the same type. Such an allocation $\hat{x}$ defines a k-tuple x of net trades, one for each type; x is an allocation in M^1, which is said to *correspond* to $\hat{x}$.

MAIN THEOREM: *For each n, let x^n be an allocation in the unreplicated market M^1, corresponding to an equal treatment value allocation in the n-fold replication M^n. Let x^∞ be a limit point[17] of $\{x^n\}$. Then there is a nonnegative dividend c that is monotonic in the net trade sets, and a price vector q, such that (q, c, x^∞) is a dividend equilibrium in M^1.*

From the monotonicity it follows that the dividends are determined by the net trade sets, i.e., $X_i = X_j$ implies $c_i = c_j$. Thus, the theorem says that all value allocations in large markets with satiation approximate dividend equilibrium

---
[16] See Section 10 for an interpretation in the fixed price case.
[17] Limit of a subsequence.

allocations, where the dividends depend only on the net trade sets, and monotonically so.[18] In particular, call a dividend equilibrium (q, c, x) *uniform*[19] if all the c_t are the same. Then we have the following:

COROLLARY 5.3: *Under the conditions of the theorem, assume that all traders have the same net trade set. Then (q, c, x^∞) is a uniform dividend equilibrium.*

The following existence result, for which (5.1) is not needed, gives substance to the main theorem:

PROPOSITION 5.4: *For every n, there is an equal treatment value allocation in the n-fold replication M^n.*

Further results will be stated in Section 12.

6. AN INFORMAL DEMONSTRATION OF THE MAIN THEOREM

Let $\boldsymbol{\lambda}^n = (\lambda_1^n, \ldots, \lambda_k^n)$ be a comparison vector associated[20] with x^n. Normalize $\boldsymbol{\lambda}^n$ so that the sum of its coordinates is 1. For simplicity,[21] assume for each i that x_i^n and λ_i^n converge as $n \to \infty$; that each of the x_i^n as well as their limit x_i^∞ is interior to the net trade set X_i; and that the utility function u_i is strictly convex and continuously differentiable on X_i.

In the classical context of monotonicity and nonsatiation, λ_i tends to a positive limit for all types i (Champsaur, 1975). But as we shall see below, in our context some of the λ_i^n may tend to 0. This is the crucial difference between the two contexts, and it is this that leads to the positive dividends.

The situation may in fact be quite complicated; there may be differences in order of magnitude even between those λ_i^n that tend to 0. In this section, though, we will assume for simplicity that those λ_i^n that do tend to 0 all have the same order of magnitude.[22] If there are such types, they are called *lightweight*, the others *heavyweight*.[23]

By definition, the allocation x^n is optimal[24] for the all-trader set T^n. This implies that all the (weighted) utility gradients $\lambda_i^n u_i'(x_i^n)$ are equal; otherwise

---

[18] Of course, the dividends are endogenously determined by all the data of the market, including all the utilities. Yet, in any given market, traders with the same net trade set have the same dividend.

[19] This concept is due to Drèze and Müller (1980), who showed, using fixed point methods, that uniform dividend equilibria always exist. They worked in a fixed price context, using "uniform coupons equilibrium" for what we call "uniform dividend equilibrium."

[20] More precisely, $\boldsymbol{\lambda}^n$ is a k-dimensional vector corresponding to an nk-dimensional equal treatment comparison vector $\hat{\boldsymbol{\lambda}}^n$ associated with the equal treatment allocation $\hat{x}^n$ corresponding to x^n. There must always be such a $\hat{\boldsymbol{\lambda}}^n$; cf. footnote 33.

[21] Use of the phrase "for simplicity" means that the restriction involved is for purposes of the informal demonstration in this section only, and is not needed for the rigorous treatment in the ensuing sections.

[22] More precisely, $\lim_{n \to \infty} \lambda_i^n / \lambda_j^n$ exists and is positive whenever both λ_i^n and λ_j^n tend to 0.

[23] The formal definition of "lightweight" in Section 9 is slightly different, but yields the same result.

[24] An allocation is called *optimal* for a coalition S if it achieves the maximum total utility for S when the utilities u_i are multiplied by the weights λ_i^n.

transfers could lead to gains in total utility. Denote the common value of all these gradients by q^n; thus

(6.1) $\quad \lambda_i^n u_i'(x_i^n) = \lambda_j^n u_j'(x_j^n) = q^n$

for all i and j. If we let $n \to \infty$ and set $\lambda_i^\infty := \lim_{n \to \infty} \lambda_i^n$, $q^\infty := \lim_{n \to \infty} q^n$, we get

(6.2) $\quad \lambda_i^\infty u_i'(x_i^\infty) = \lambda_j^\infty u_j'(x_j^\infty) = q^\infty$.

Our demonstration is based on (4.3), which says that the Shapley value of a given trader t is his expected contribution to a randomly chosen coalition; more precisely, to the coalition S of traders before t in a randomly chosen order on all traders.

If S is large, it is very likely to be a good sample of all the traders, i.e. to have approximately equal numbers of traders of each type. The allocation that is optimal for S is then approximately the same as the allocation x^n that is optimal for the all-trader coalition T^n.

Adding t to S will not change this optimal allocation by much; each trader will be allocated approximately the same net trade as before. In particular, if t is of type i, he will be allocated approximately x_i^n. Since all the net trades must sum to 0, the net trade of t must somehow be divided among all the traders, with each trader subtracting a small part of x_i^n from his net trade. Since all the utility gradients are q^n, the utility of each trader is decreased by q^n times that small part. Altogether, this causes a change in total utility of $-q^n \cdot x_i^n$. To this must be added the utility $\lambda_i^n u_i(x_i^n)$ that t himself now enjoys. Thus t's contribution to S (the worth of $S \cup t$ less the worth of S) is given approximately by

(6.3) $\quad \Delta := \lambda_i^n u_i(x_i^n) - q^n \cdot x_i^n$.

All this is valid only when S is reasonably large. Otherwise—e.g. when S has no more than a fixed finite number of traders (such as one or a hundred or a thousand)—the reasoning breaks down. Denote by P^n the probability that S is "small," so that t's contribution is not measured by Δ. This event is perhaps not very well defined, but in this section we are making no attempt at precision. It is in any case clear that $P^n \to 0$ as $n \to \infty$, and that the order of P^n is at least $1/n$ (obtained already when S has only one trader).

Letting δ denote the conditional expectation of t's contribution given that S is small,[25] we conclude that

(6.4) $\quad (\phi v_{\lambda^n})(t) \approx (1 - P^n) \Delta + P^n \delta$.

Note that we are ignoring the probability that S is large but nevertheless not a good sample of the entire population; this probability is very small indeed, much smaller than P^n, and in fact has no influence on the value. Note that δ is uniformly bounded; this follows, e.g., from the continuous differentiability of the utilities u_i on the compact sets X_i.

The definition (4.2) of the value stipulates that

$$(\phi v_{\lambda^n})(t) = \lambda_i^n u_i(x_i^n).$$

---

[25] Both Δ and δ depend on n; we suppress the corresponding superscript to keep our notation as uncluttered as possible.

Together with (6.4) and the definition of Δ, this yields

(6.5) $\quad q^n \cdot x_i^n \approx \varepsilon^n(\delta - \lambda_i^n u_i(x_i^n))$,

where $\varepsilon^n := P^n/(1 - P^n) \to 0$.

The case in which none of the λ_i^n tend to 0 is the simplest, and we dispose of it first. With strict convexity, the elbow room assumption (5.1) rules out the possibility of simultaneously satiating all traders, a situation that is in any case rather uninteresting. Hence $u_i'(x_i^\infty) \neq 0$ for at least one i; and since $\lambda_i^\infty > 0$ for all i, it follows from (6.2) that x^∞ satiates no one, that $q^\infty \neq 0$, and that for all i, the gradient of u_i at x_i^∞ is in the direction of q^∞. Letting $n \to \infty$ in (6.5), we obtain $q^\infty \cdot x_i^\infty = 0$; hence the net trade x_i^∞ maximizes u_i over the budget set $\{x \in X_i: q \cdot x \leq 0\}$. Thus x^∞ is an ordinary competitive allocation, and hence trivially part of a dividend equilibrium that satisfies the appropriate conditions.

Up to now the analysis has been as in the classical context of monotonicity and nonsatiation, where limits of value allocations are always competitive (Champsaur, 1975). But as we saw in Section 3, in our context there are situations in which competitive allocations need not even exist. By the above analysis, then, there must be some lightweights; of course there must also be heavyweights, since the sum of the k weights λ_i^n is normalized to be 1. This is the case of central interest in this paper, to which we now turn.

If we take i lightweight in (6.2), we find $q^\infty = 0$; hence $u_j'(x_j^\infty) = 0$ for heavyweight j, and hence x^∞ satiates all heavyweights. Suppose now that t is a lightweight trader of type i. Since $q^\infty = 0$, letting $n \to \infty$ in (6.5) as before would simply yield $0 = 0$. For something more informative, we must look at the fine structure, at the second order effects. This is done by dividing (6.5) by $\|q^n\|$, which, like q^n, tends to 0 as $n \to \infty$.

Assume for simplicity that $q^n/\|q^n\|$ actually tends to a limit q; note that $\|q\| = 1$. We shall see below that t's expected contribution δ to small coalitions is of larger order of magnitude than the term $\lambda_i^n u_i(x_i^n)$ on the right side of (6.5). Hence dividing (6.5) by $\|q^n\|$ and letting $n \to \infty$, we find

(6.6) $\quad q \cdot x_i = \lim_{n \to \infty}(\varepsilon^n/\|q^n\|)\delta =: c_i;$

the limit exists on the right because it exists on the left.

By definition, q^n is proportional to the unweighted utility gradient $u_i'(x_i^n)$; therefore its direction $q^n/\|q^n\|$ is equal to the direction of $u_i'(x_i^n)$. Letting $n \to \infty$, we deduce that q is the direction of $u_i'(x_i^\infty)$ whenever x_i^∞ does not satiate t. In that case, therefore, (6.6) says that x_i^∞ maximizes t's utility over the dividend budget set defined by prices q and dividend c_i. Of course, when x_i^∞ does satiate t, his utility is maximized globally, and a fortiori over his dividend budget set.

For lightweight i, it remains therefore only to show that c_i depends monotonically on the net trade set X_i, and in particular is independent of the utility u_i. To see this, let us examine the contribution of t when joining a fixed small coalition S. This may be divided into three components:

    (i) t's own utility after joining;

    (ii) the change in the total utility of the lightweight traders in S due to t's joining; and

(iii) the change in the total utility of the heavyweight traders in S due to t's joining.

In the first two of these three components, the utilities have weights tending to 0; in the third they do not. Thus for large n, the contribution of t to himself and to other lightweights is negligible; the importance of his contribution to S comes from what he can do to improve the lot of the heavyweights in S. Therefore he should distribute his resources so as to maximize the heavyweights' gain in utility, paying no attention to his own. His ability to do this is limited only by his net trade set, and has nothing to do with his utility. Moreover, the larger his net trade set, the more he can do, and this yields the monotonicity.

The reasoning works only when S is a fixed coalition of relatively small size. In that case the heavyweights in S cannot, in general, all be simultaneously satiated; since S is small, they will then be a significant distance from satiation.[26] When t joins, he brings in resources (not utility!) that could be used significantly to improve the lot of at least one heavyweight trader, perhaps even to bring that one all the way to satiation; that would be a good deal more worthwhile than using the resources for himself or for other lightweight traders, whose utilities have weights tending to 0. A more even handed distribution of the resources among the heavyweights would yield still more, but giving it all to one gives us a lower bound on t's contribution, and indicates that it is of larger order than λ_i^n.

If, however, S is large, it is probably a good sample of all agents, and then all types j will be close to x_j^∞; in particular, the heavyweights will already be satiated, even before t joins. Thus by joining, t cannot improve the heavyweights by much. The upshot is that no matter how t uses his resources—whether for himself, for his lightweight colleagues, or for the heavyweights—the increment in total utility will be the same; in the first two cases the utilities are weighted by small weights of the order λ_i^n, and in the last, the increase in the utility u_j is small.

We come finally to the case in which the type i of the additional trader t is heavyweight. In calculating the contribution δ to small coalitions, the significant components are now (i) and (iii), rather than just (iii); component (ii) remains negligible. Note, though, that on the right side of (6.5) we now have not δ, but something close to $\delta - \lambda_i^\infty u_i(x_i^\infty)$. Since $\lambda_i^\infty u_i(x_i^\infty)$ is the absolute maximum that t can get, component (i) of δ is certainly at least cancelled out, and very likely more than cancelled out. Thus what is left is at most component (iii). The rest of the argument is as before, with (6.6) modified to read:[27]

(6.7) $\quad q \cdot x_i^\infty \leq \lim_{n \to \infty} (\varepsilon^n / \|q^n\|)[\text{component (iii) of } \delta] =: c_i$.

---

[26] In principle, the equality of marginal utilities expressed by (6.1) should still hold when x_i^n and x_j^n are replaced by y_i^n and y_j^n, where y^n is optimal for an arbitrary (fixed) S. In fact, when S is small and n large, y_i^n is very likely to be on the boundary of X_i, so that we have a corner situation, in which marginal utilities need not be equal. We therefore cannot deduce that y_j^n is close to satiation for heavyweight j, and indeed it will usually not be.

[27] For technical reasons, the definition of the dividend c_i for heavyweight types i that we use in Section 9 is a little different from (6.7). Since x_i^∞ is generally in the interior of the dividend budget set when i is heavyweight, there is sometimes a little leeway in defining the dividend. Of course, if $X_i = X_j$ for some lightweight j, then we must have $c_i = c_j$, so the leeway disappears.

Since i is heavyweight, x_i^∞ satiates; hence we must only show that it satisfies the budget inequality, which (6.7) indeed shows.

We end with a word of caution. The argument in this section is meant only to be indicative, and cannot easily be made rigorous. The difficulties we will encounter in the rigorous treatment below are intrinsic; they are not due to the generality of the treatment. Assuming differentiability, strict concavity, and so on enabled a simplified presentation in this section, but it would not help appreciably in the treatment below.

7. THE EXISTENCE PROOF

In this section we prove Proposition 5.4. Define a *generalized comparison vector* on the trader space T of a market with satiation to be a vector $\boldsymbol{\lambda} = (\lambda_t)_{t \in T}$ of nonnegative real numbers not all of which vanish. A *generalized value allocation* is defined like a value allocation, except that the comparison vector is replaced by a generalized comparison vector.

In 1969, Shapley proved that every game has a nontransferable utility value, if generalized comparison vectors are admitted. In this proof, the only properties of the Shapley value that are used are continuity in the comparison weights, Pareto optimality, and individual rationality (see the theorem on p. 261 of Shapley (1969)); any function of the comparison vector enjoying these properties will be called a *pseudo-value*.

Fix the replication index n. To each generalized comparison vector $\boldsymbol{\lambda}$ on $\{1, \ldots, k\}$, there corresponds a generalized comparison vector $\hat{\boldsymbol{\lambda}}$ on T^n, which assigns weight $\hat{\lambda}_t = \lambda_i$ to each of the n traders t of type i. Define $v_{\hat{\lambda}}$ on the subsets of T^n as in (4.1). By the symmetry of the Shapley value ϕ, traders of the same type in T^n are assigned equal values by $\phi v_{\hat{\lambda}}$; thus $\phi v_{\hat{\lambda}}$ defines a k-dimensional vector ψ_λ, whose ith coordinate is $(\phi v_{\hat{\lambda}})(t)$ for any t of type i in T^n. Then the function $\boldsymbol{\lambda} \to \psi_\lambda$ is a pseudo-value, and so by the theorem of Shapley quoted above, there is an allocation x in M^1 and a generalized comparison vector $\boldsymbol{\lambda}$ such that $(\psi_\lambda)_i = \lambda_i u_i(x_i)$ for all $i = 1, \ldots, k$. Now define an allocation $\hat{x}$ in M^n by $\hat{x}_t = x_i$ whenever t in T^n is of type i; then

(7.1) $\quad (\phi v_{\hat{\lambda}})(t) = (\psi_\lambda)_i = \lambda_i u_i(x_i) = \hat{\lambda}_t u_t(\hat{x}_t),$

and so $\hat{x}$ is an equal treatment generalized value allocation in M^n.

It remains only to show that $\boldsymbol{\lambda}$ is in fact a comparison vector, i.e., that $\lambda_i > 0$ for all i. W.l.o.g. $\lambda_1 \geq \lambda_2 \geq \cdots \geq \lambda_k$; hence $\lambda_1 > 0$. Suppose $\lambda_k = 0$. Let $b \in B_1$; since 0 does not satiate, $u_1(b) > u_1(0)$. Since $0 \in \text{int } X_k$, we have $-\theta b \in X_k$ for $\theta > 0$ sufficiently small. Hence if t is a type k trader and S a coalition in M^n consisting of a single type 1 trader, then

(7.2) $\quad v_\lambda(S \cup t) - v_\lambda(S) \geq \lambda_1 u_1(\theta b) + \lambda_k u_k(-\theta b) - \lambda_1 u_1(0)$

$$= \lambda_1(u_1(\theta b) - u_1(0)) \geq \lambda_1 \theta(u_1(b) - u_1(0)) > 0,$$

by the concavity of u_1. On the other hand, for any $S \subset T^n$ not containing t, the superadditivity of v_λ yields

(7.3) $\quad v_\lambda(S \cup t) - v_\lambda(S) \geq v_\lambda(t) = \lambda_k u_k(0) = 0.$

Combining (7.2) and (7.3) with (4.3), and noting that S is as in (7.2) with positive probability, we deduce $(\phi v_{\hat{\lambda}})(t) > 0$. Since $\hat{\lambda}_t = \lambda_k$, (7.1) then yields $\lambda_k > 0$ after all. Hence $\lambda_i > 0$ for all i. $\hfill$ Q.E.D.

8. SOME TOOLS

If f is a concave function on a convex set X in a Euclidean space R^d, define the *superdifferential*[28] $\partial f(x)$ of f at a point x in X by[29]

(8.1) $\quad \partial f(x) = \{p \in R^d: f(z) - f(x) \leq p \cdot (z - x) \text{ for all } z \in X\}.$

Note that $\partial f(x)$ is always nonempty, closed, and convex, and that when x is interior and f is differentiable, it consists of the gradient only.

As in the foregoing, for $i = 1, \ldots, k$, let u_i be a concave continuous function on a compact convex subset[30] X_i of R^d. Throughout the rest of the paper, write $\sum$ for $\sum_{i=1}^k$ or $\sum_{j=1}^k$. Define a closed convex cone U in R_+^k by

(8.2) $\quad U = \{y \in R_+^k: 0 \in \sum y_i X_i\};$

note that U coincides with R_+^k whenever each X_i contains 0. Define a real-valued function w on U by

(8.3) $\quad w(y) = \text{Max } \{\sum y_i u_i(x_i): x_i \in X_i \text{ for all } i, \text{ and } \sum y_i x_i = 0\};$

the maximum is attained because of the compactness of the X_i. Note that

(8.4) $\quad w$ is 1-homogeneous,[31] concave, and superadditive

$\quad$ (i.e. $w(y + y') \geq w(y) + w(y')$).

The function w plays a vital role in the sequel. The following proposition, related to the core equivalence theorem, characterizes the superdifferentials of w in terms of the superdifferentials of u_i (which play the role of prices). It is proved in Appendix A.

PROPOSITION 8.5: *Let y in R_{++}^k be such that $0 \in \text{Int} \sum y_i X_i$, and let $w(y)$ be attained at $(x_1, \ldots, x_k)$. Then $p \in \partial w(y)$ if and only if there is an element q of $\bigcap_{i=1}^k \partial u_i(x_i)$ such that for each i,*

(8.6) $\quad p_i = u_i(x_i) - q \cdot x_i.$

---

[28] Cf. Rockafellar (1970, p. 215), where this notation is used for *sub*differentials. The superdifferential $\partial f(x)$ is denoted $P(x; f)$ by Aumann and Shapley (1974, p. 216 ff.).
[29] The symbol p in this and the next section bears no relation whatever to the fixed price vector $\bar{p}$ in Sections 1 and 10.
[30] In this section, the X_i need not have interiors, nor contain 0, nor satisfy $0 \notin B_i$, nor satisfy (5.1).
[31] Homogeneous of degree 1.

The second tool presented in this section is of an explicitly game-theoretic nature. As in the foregoing, let T^n be a set with nk members (the *players*), divided into disjoint subsets $T_1^n, \ldots, T_k^n$ (the *types*) with n members each. For each $S \subset T^n$, let $\eta(S) \in R_+^k$ denote the *profile* of S, i.e.,

(8.7) $\quad \eta_i(S) = |S \cap T_i^n|$,

where $|\cdot|$ denotes cardinality; in words, $\eta_i(S)$ is the number of type i players in S. Let ϕ denote the Shapley value.

PROPOSITION 8.8: *Let $w_1, w_2, \ldots$ be 1-homogeneous continuous functions on R_+^k that are uniformly bounded on $[0, 1]^k$. Assume that on a convex neighborhood of $(1, \ldots, 1)$, the w_n are concave and pointwise approach a concave function*[32] *w_∞. Define a game v_n on the subsets of T^n by $v_n = w_n \circ \eta$, and define p^n in R^k by*

(8.9) $\quad p_i^n = (\phi v_n)(T_i^n)/n \qquad (i = 1, \ldots, k)$.

Then p^n is bounded, and every limit point of $\{p^n\}$ as $n \to \infty$ is in $\partial w_\infty(1, \ldots, 1)$.

Propositions 8.5 and 8.8 constitute the technical foundation on which the proof of the main theorem rests. Proposition 8.8, proved in Appendix B, tells us that as the number of traders increases, the value approaches the superdifferential of the appropriate w-function; and Proposition 8.5 tells us that this superdifferential can be interpreted in terms of prices in the original market.

9. PROOF OF THE MAIN THEOREM

9.1. *Preliminaries*

We return now to the situation described in Sections 2 through 5. Assume w.l.o.g. that $u_i(0) = 0$ for all i. Let $\hat{x}^n$ be an equal treatment value allocation in the n-fold replication M^n, let $\hat{\lambda}^n$ be an equal treatment comparison vector[33] associated with $\hat{x}^n$, let x^n be the allocation in M^1 corresponding to $\hat{x}^n$, and let λ^n be the comparison vector on $T^1 = \{1, \ldots, k\}$ corresponding to $\hat{\lambda}^n$. Assume w.l.o.g. that $\sum \lambda_i^n = 1$. Then there is a sequence of positive integers, called the *convergence indicator*,[34] such that if $n \to \infty$ over members of this sequence only, then $x^n \to x^\infty$, and also $\{\lambda^n\}$ converges. From now on, all finite values of n will be in the convergence indicator; in particular, $n \to \infty$ means that n tends to infinity over members of the convergence indicator only.

By possibly taking a smaller convergence indicator, we may assume w.l.o.g. that λ_i^n/λ_j^n approaches a (finite or infinite) limit as $n \to \infty$, for all i and j. We will assume the types arranged in the limiting order of size of the weights λ_i^n; that

---

[32] The w_n and w_∞ are not necessarily related to the w of (8.3).

[33] One that assigns equal weight to traders of the same type. That there is one such associated with $\hat{x}^n$ follows from the fact that $\hat{x}^n$ is itself equal treatment. Indeed, if $\hat{\mu}^n$ is any comparison vector associated with $\hat{x}^n$, then we may define an equal treamtent $\hat{\lambda}^n$ associated with $\hat{x}^n$ by taking $\hat{\lambda}_t^n$, for each trader t, to be the average of the weights $\hat{\mu}_s^n$ over traders s of t's type.

[34] We are grateful to Lloyd Shapley for suggesting this name.

is, $\lim \lambda_{i+1}^n/\lambda_i^n \le 1$ for all $i < k$. W.l.o.g. x^∞ does not satiate all types;[35] let ℓ be the "heaviest" type not satiated by x^∞ (i.e., $\ell = \min\{i: x_i^\infty \notin B_i\}$). Call type i *lightweight* ($i \in L$) if $\lambda_i^n = 0(\lambda_\ell^n)$, *heavyweight* ($i \in H$) otherwise.

Define w^n on R_+^k as in (8.3), except that u_i is replaced by $\lambda_i^n u_i$. Define an open convex cone U_L in R_{++}^k by

$$U_L = \left\{ y \in R_{++}^k : 0 \in \text{Int}\left(\sum_{i \in H} y_i B_i + \sum_{i \in L} y_i X_i\right)\right\}.$$

For each lightweight trader i, define

$$\xi_i = \lim_{n \to \infty} \lambda_i^n / \lambda_\ell^n.$$

For $y \in U_L$, define $w_L(y)$ as in (8.3), except that for heavyweight i, X_i is replaced by B_i and u_i by 0; and for lightweight i, u_i is replaced by $\xi_i u_i$ (that the constraint set is nonempty follows from $y \in U_L$). For $n < \infty$, define w_H^n on R_+^k as in (8.3), except that u_i is replaced by $\lambda_i^n u_i$ for heavyweight i, and by 0 for lightweight i. Also for $n < \infty$, define w_L^n on R_+^k by

(9.1) $\quad w^n = w_H^n + \lambda_\ell^n w_L^n.$

Recall that $\eta(S)$ denotes the profile of the coalition S (see (8.7)). On T^n, define games v^n, v_L^n, and v_H^n by

$$v^n = w^n \circ \eta, \quad v_L^n = w_L^n \circ \eta, \quad v_H^n = w_H^n \circ \eta.$$

From the concavity of the u_i it follows that v^n is the same as the v_{λ^n} defined on $S \subset T^n$ as in (4.1). Hence by (4.2),

(9.2) $\quad (\phi v^n)(T_i^n) = n\lambda_i^n u_i(x_i^n) \qquad (i = 1, \ldots, k).$

9.2. Outline

Before proceeding, we briefly outline the proof; the reader may also wish to consult the flow charts in Appendix C. The thought that first comes to mind is to apply the value convergence theorem (Proposition 8.8) directly to the v^n. But this would wipe out all information about all types i for which $\lambda_i^n \to 0$; i.e., about all but the heaviest of the heavyweights. We therefore decompose w^n as in (9.1), and deduce

(9.3) $\quad \phi v^n = \phi v_H^n + \lambda_\ell^n \phi v_L^n.$

After showing $w_L^n \to w_L$ (Lemma 9.8), we apply Proposition 8.8 to deduce $\phi v_L^n \to p \in \partial w_L(1, \ldots, 1)$. We then use Proposition 8.5 to express p in terms of prices q in the commodity space. By plugging all this back into (9.3), and applying (9.2), we obtain our result.

---

[35] Otherwise the theorem is immediate.

Intuitively, $v_H^n(S)$ represents what the coalition S can do for the heavyweight traders in it; i.e., the result of ignoring completely the utilities of the lightweights in S. What is left over for the lightweights (possibly negative!) is of order[36] λ_ℓ^n, since that is the order of all their utilities. Dividing by λ_ℓ^n to get $v_L^n(S)$ is a way of looking at the situation through a microscope, so to speak on the scale of the lightweight traders themselves; it yields something of order 1. If $v_L := w_L \circ \eta$, then $v_L(S)$ represents what the lightweights in S, when assigned their limiting "relative" weights ξ_i, can do for themselves (again, on their own "scale"), after first satiating all the heavyweights in S; it is defined only if the heavyweights can indeed be satiated.

If S is large, then its profile is likely to be close to the "diagonal" (i.e., to a multiple of $(1, \ldots, 1)$); hence the heavyweights can be satiated (Lemma 9.11). Then $v_L^n(S)$ will for large n be close to $v_L(S)$ (Lemma 9.8), and hence an additional trader of type i contributes approximately p_i to v_L^n by joining S, where $p = \partial w_L(1, \ldots, 1) \approx \partial w_L(\eta(S))$. To v_H^n he contributes nothing if he is lightweight, since the heavyweights are already satiated. Any contribution to v_H^n by lightweights therefore comes from the coalitions with few traders, which stand a reasonable chance of not being able to satiate all their heavyweights.

Figure 2 depicts the path of the profile of a coalition Q that grows as traders are added to it in a random order (cf. (4.3)). When Q is small, there is a good chance that $\eta(Q)$ is outside of the cone U_L. During this initial period, lightweight traders of type i make heavyweight contributions, which sum to $(\phi v_H^n)(T_i^n)$; but it is a short period, and relative to n it tends to zero. (There is also a lightweight contribution during this period, but it is negligible.) Afterwards, $\eta(Q)$ is in the cone U_L and in fact close to the diagonal; lightweight traders of type i are no longer making heavyweight contributions, but altogether their contributions are over a much longer period, and the total, which is $\approx n\lambda_\ell^n p_i$, has the same order of magnitude as $(\phi v_H^n)(T_i^n)$. By the definition of the NTU value, all the contributions together must add up to the total (weighted) utility of type i; this yields

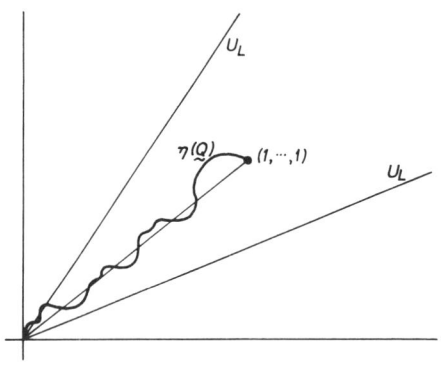

FIGURE 2

---

[36] By Corollary 9.26, $\xi_i > 0$ for all $i \in L$; hence the lightweight i are those with λ_i^n of the lowest order of magnitude (that of λ_k^n).

(9.25). The v_L^n term gives us the prices q, as in ordinary markets without satiation (Cf. Champsaur (1975)); the v_H^n term gives us the dividends. The analysis is similar when i is heavyweight.

We now proceed with the formal proof, which we divide into three parts.

9.3. w_L^n Satisfies the Hypotheses of Proposition 8.8

If $y \in R_+^k$, a *y-allocation* is a k-tuple $x = (x_1, \ldots, x_k)$, with $x_i \in X_i$ for all i, and $\sum y_i x_i = 0$. An *allocation* is simply a $(1, \ldots, 1)$-allocation, i.e., an allocation in M^1. A y-allocation *satiates* type i if $x_i \in B_i$. Denote

$$\mu_i = \text{Max}\{u_i(x): x \in X_i\}.$$

LEMMA 9.4. *Let $y \in U_L$, and for $n < \infty$, let $w^n(y)$ be attained at $x^n(y)$. Let $x^\infty(y)$ be a limit point of $\{x^n(y)\}$. Then w_L and w_H^n are attained at $x^\infty(y)$; that is, $x^\infty(y)$ is a y-allocation that satiates all heavyweight types,*

(9.5) $\quad w_L(y) = \sum_{i \in L} y_i \xi_i u_i(x_i^\infty(y)),$

and

(9.6) $\quad w_H^n(y) = \sum_{i \in H} y_i \lambda_i^n u_i(x_i^\infty(y)) = \sum_{i \in H} y_i \lambda_i^n \mu_i.$

Finally, if $x^n(y) \to x^\infty(y)$ for some sequence of n, then for n in that sequence and for all i, we have, as $n \to \infty$,

(9.7) $\quad \lambda_i^n u_i(x_i^n(y)) = \lambda_i^n u_i(x_i^\infty(y)) + o(\lambda_\ell^n).$

PROOF: We restrict attention to a subsequence of the convergence indicator for which $x^n(y) \to x^\infty(y)$. Since the $x^n(y)$ are y-allocations, so is $x^\infty(y)$. Suppose that there is a heavyweight i with $x_i(y) \notin B_i$. Since $y \in U_L$, there is a y-allocation z that satiates all j in H. Now

$$(\sum y_j \lambda_j^n u_j(z_j)) - w^n(y) = \sum y_j \lambda_j^n (u_j(z_j) - u_j(x_j^n(y))).$$

Since $z_j \in B_j$ for all j in H, each term on the right with $j \in H$ is nonnegative. Moreover, for n in the subsequence we have chosen, the ith term is of the order λ_i^n. But the terms with $j \in L$ are of smaller order of magnitude; so for n sufficiently large, the right side is positive. Thus z satisfies the constraints in the definition of $w^n(y)$ and yields a larger value than $x^n(y)$, a contradiction. Hence indeed $x^\infty(y)$ is a y-allocation that satiates all heavyweight types.

Now let z be any other such y-allocation. Then

$$\sum_{i \in H} y_i \lambda_i^n \mu_i + \sum_{i \in L} y_i \lambda_i^n u_i(x_i^n(y))$$

$$\geq \sum y_i \lambda_i^n u_i(x_i^n(y)) = w^n(y)$$

$$\geq \sum y_i \lambda_i^n u_i(z_i) = \sum_{i \in H} y_i \lambda_i^n \mu_i + \sum_{i \in L} y_i \lambda_i^n u_i(z_i).$$

Cancelling $\sum_{i \in H} y_i \lambda_i^n \mu_i$, dividing by λ_ℓ^n, and allowing $n \to \infty$, we obtain

$$\sum_{i \in L} y_i \xi_i u_i(x_i^\infty(y)) \geq \sum_{i \in L} y_i \xi_i u_i(z_i),$$

which yields (9.5).

As for (9.6), this follows from the fact that $x^\infty(y)$ satisfies the constraints in the definition of $w_H^n(y)$ and yields the maximum value for each of the u_i appearing in this definition.

Finally, to prove (9.7), note that

$$\sum y_i \lambda_i^n u_i(x_i^\infty(y)) \leq w^n(y) = \sum y_i \lambda_i^n u_i(x_i^n(y)).$$

Hence

$$\sum_{i \in H} y_i \lambda_i^n (u_i(x_i^\infty(y)) - u_i(x_i^n(y)))$$

$$\leq \sum_{i \in L} y_i \lambda_i^n (u_i(x_i^n(y)) - u_i(x_i^\infty(y))) = o(\lambda_\ell^n).$$

Since each summand on the left is nonnegative, and since each y_i is positive, (9.7) follows. Q.E.D.

LEMMA 9.8: w_L^n *is concave and 1-homogeneous on* U_L, *and for each y in* U_L,

$$w_L^n(y) \to w_L(y) \quad \text{as} \quad n \to \infty.$$

PROOF: The concavity and 1-homogeneity of w_L^n follow from its definition (9.1), from the concavity and 1-homogeneity (8.4) of w^n, and from the linearity (9.6) of w_H^n on U_L. The convergence follows from

(9.9) $\quad w^n(y) = w_H^n(y) + \lambda_\ell^n w_L(y) + o(\lambda_\ell^n)$

for each y in U_L. To demonstrate (9.9) for a sequence of n for which $x^n(y)$ converges, write $w^n(y) = \sum y_i \lambda_i^n u_i(x_i^n(y))$ and apply (9.7), (9.6), (9.5), and $\lambda_i^n = \lambda_\ell^n \xi_i + o(\lambda_\ell^n)$ for $i \in L$; the truth of (9.9) in general follows from its truth for those special sequences. Q.E.D.

LEMMA 9.10: $w_L^n(y) = O(1)$ *as* $n \to \infty$, *uniformly for y in* $[0, 1]^k$.

PROOF: Let $w^n(y)$ and $w_H^n(y)$ be attained at the y-allocations $x^n(y)$ and z^n respectively. Then

$$w^n(y) = \sum_{i \in H} y_i \lambda_i^n u_i(x_i^n(y)) + \sum_{i \in L} y_i \lambda_i^n u_i(x_i^n(y))$$

$$\leq w_H^n(y) + O(\lambda_\ell^n),$$

where the O is uniform because the u_i are uniformly bounded. Next,

$$w_H^n(y) = \sum y_i \lambda_i^n u_i(z_i^n) - \sum_{i \in L} y_i \lambda_i^n u_i(z_i^n)$$

$$\leq w^n(y) + O(\lambda_\ell^n),$$

where again the O is uniform. Q.E.D.

LEMMA 9.11: U_L is a convex neighborhood of $(1, \ldots, 1)$.

PROOF: U_L is open and convex by definition. By the definition of H, the allocation x^∞ satiates all heavyweight types; hence $0 \in \sum_{i \in H} B_i + \sum_{i \in L} X_i$, and hence by[37] the elbow room assumption (5.1), $0 \in \text{Int} (\sum_{i \in H} B_i + \sum_{i \in L} X_i)$; i.e., $(1, \ldots, 1) \in U_L$. Q.E.D.

9.4. Monotonicity Properties of v_H^n

In a finite coalitional game v, a player s is called *at least as desirable as*[38] a player t, written $s \geqslant t$, if $v(S \cup s) \geqslant v(S \cup t)$ for each[39] coalition S that contains neither s nor t.

PROPOSITION 9.12: *If $s \geqslant t$, then $(\phi v)(s) \geqslant (\phi v)(t)$.*

PROOF: Suppose there are h players in all. The orders $\mathcal{R}$ on the players can be divided into $h!/2$ pairs, in each of which the two orders are the same, except that the positions of s and t are interchanged. Let $\mathcal{R}$ and $\mathcal{R}^*$ constitute such a pair; suppose s precedes t in $\mathcal{R}$. Let S_1 be the set of players preceding s in $\mathcal{R}$, and let S_2 be the set of players strictly between s and t. Setting $S_3 = S_1 \cup S_2$, we find that

$$\begin{cases} \phi^{\mathcal{R}}(s) = v(S_1 \cup s) - v(S_1), \\ \phi^{\mathcal{R}}(t) = v(S_3 \cup s \cup t) - v(S_3 \cup s), \end{cases}$$

where $\phi^{\mathcal{R}}(s)$ and $\phi^{\mathcal{R}}(t)$ are the contributions of s and t respectively in the order $\mathcal{R}$. Similarly,

$$\begin{cases} \phi^{\mathcal{R}^*}(s) = v(S_3 \cup s \cup t) - v(S_3 \cup t), \\ \phi^{\mathcal{R}^*}(t) = v(S_1 \cup t) - v(S_1). \end{cases}$$

Using $s \geqslant t$, we conclude that

$$\phi^{\mathcal{R}}(s) + \phi^{\mathcal{R}^*}(s) \geqslant \phi^{\mathcal{R}}(t) + \phi^{\mathcal{R}^*}(t),$$

and the proposition follows. Q.E.D.

COROLLARY 9.13: *Suppose $i, j \in L$, and $X_i \subset X_j$. Then $(\phi v_H^n)(T_i^n) \leqslant (\phi v_H^n)(T_j^n)$.*

If S is a coalition in T^n, an *S-allocation* is a vector $x = (x_t)_{t \in S}$ such that $x_t \in X_i$ whenever t is of type i, and $\sum_{t \in S} x_t = 0$. Whereas every $\eta(S)$-allocation may be viewed as an S-allocation, the converse is false; an S-allocation allows traders of the same type to have different net trades, which an $\eta(S)$-allocation does not.

LEMMA 9.14: *Suppose $i \in H$, $j \in L$, and $X_i \subset X_j$. Then*

$$(\phi v_H^n)(T_i^n) \leqslant (\phi v_H^n)(T_j^n) + n\lambda_i^n \mu_i.$$

---

[37] (5.1) is used only here.
[38] Cf. Maschler and Peleg (1966).
[39] $\{s\}$ and $\{t\}$ are abbreviated by s and t.

PROOF: Let t and s be type i and j traders respectively. First we show that

(9.15) $\quad v_H^n(S \cup t) \leq v_H^n(S \cup s) + \lambda_i^n \mu_i$

for all coalitions S containing neither t nor s. Indeed, suppose $v_H^n(S \cup t)$ is attained at the $(S \cup t)$-allocation x. Now transfer the amount x_t from t to s. Since $X_i \subset X_j$, this yields an $(S \cup s)$-allocation z; z coincides with x on S, and obeys $z_s = x_t$. Since lightweight traders have utilities 0 in the definitions of w_H^n and v_H^n, it follows that

$$v_H^n(S \cup s) \geq \sum_{r \in S} \hat{\lambda}_r^n u_r(x_r),$$

where the comparison function $\hat{\lambda}^n$ on T^n corresponds to λ^n on T. Hence

$$v_H^n(S \cup t) = \sum_{r \in S \cup t} \hat{\lambda}_r^n u_r(x_r) \leq v_H^n(S \cup s) + \hat{\lambda}_t^n u_t(x_t),$$

and (9.15) follows. Proceeding from (9.15) as in the proof of Proposition 9.12, we find

$$\phi^{\mathcal{R}}(t) + \phi^{\mathcal{R}^*}(t) \leq \phi^{\mathcal{R}}(s) + \phi^{\mathcal{R}^*}(s) + 2\lambda_i^n \mu_i.$$

Summing over all $(nk)!/2$ pairs $(\mathcal{R}, \mathcal{R}^*)$ and dividing by $(nk)!$, we obtain

$$(\phi v_H^n)(t) \leq (\phi v_H^n)(s) + \lambda_i^n \mu_i,$$

and the lemma follows. Q.E.D.

9.5. Derivation of Prices and Dividends

By restricting the convergence indicator, we may assume that $(\phi v_L^n)(T_i^n)/n$ has a (finite or infinite) limit for each i; denote

(9.16) $\quad p_i = \lim_{n \to \infty} (\phi v_L^n)(T_i^n)/n.$

By Lemma 9.11, U_L is a convex neighborhood of $(1, \ldots, 1)$; by Lemma 9.8, on U_L the w_L^n are concave and 1-homogeneous, and converge to the concave function w_L; the w_L^n are 1-homogeneous throughout R_+^k by definition, and by Lemma 9.10 are uniformly bounded on $[0, 1]^k$. Therefore by Proposition 8.8, all p_i are finite, and setting $p = (p_1, \ldots, p_k)$, we have

(9.17) $\quad p \in \partial w_L(1, \ldots, 1).$

Now construct a vector q corresponding to p in accordance with Proposition 8.5. Here w is replaced by w_L, i.e. u_i is replaced by $\xi_i u_i$ for lightweight i, and by 0 for heavyweight i; and X_i is replaced by B_i for heavyweight i. Applying Lemma 9.4 to $y = (1, \ldots, 1)$, we see that $w_L(1, \ldots, 1)$ is achieved at x^∞; and $0 \in \text{Int} (\sum_{i \in H} B_i + \sum_{i \in L} X_i)$ by Lemma 9.11. Hence (8.6) yields

(9.18) $\quad p_i = \begin{cases} \xi_i u_i(x_i^\infty) - q \cdot x_i^\infty, & i \in L, \\ -q \cdot x_i^\infty, & i \in H. \end{cases}$

As for q, the condition in Proposition 8.5 implies that

(9.19) $\quad q \in \partial \xi_i u_i(x_i^\infty), \quad i \in L.$

Since $\ell \in L$, $\xi_\ell = 1$, and $u_\ell(x_\ell^\infty) < \mu_\ell$, it follows from (9.19) that

(9.20) $\quad q \neq 0$

(choose $z \in B_\ell$ in the definition (8.1) of superdifferential).

Note that v_H^n is monotonic,[40] since $u_i(0) = 0$ for all i; hence by (4.3),

(9.21) $\quad (\phi v_H^n)(T_i^n) \geq 0, \quad \text{all } i.$

LEMMA 9.22: *If i is lightweight, then*

(9.23) $\quad q \cdot x_i^\infty = \dfrac{(\phi v_H^n)(T_i^n)}{n \lambda_\ell^n} + o(1).$

If i is heavyweight, j lightweight, and $X_i \subset X_j$, then

(9.24) $\quad q \cdot x_i^\infty = \dfrac{(\phi v_H^n)(T_i^n) - n \lambda_i^n \mu_i}{n \lambda_\ell^n} + o(1).$

PROOF: By (9.2) (which is part of the definition of the λ_i^n), (9.1), and (9.16),

(9.25) $\quad n \lambda_i^n u_i(x_i^n) = (\phi v^n)(T_i^n) = (\phi v_H^n)(T_i^n) + \lambda_\ell^n (\phi v_L^n)(T_i^n)$

$\qquad = (\phi v_H^n)(T_i^n) + n \lambda_\ell^n p_i + o(n \lambda_\ell^n)$

for all i. Moreover, $x^n \to x^\infty$ and the continuity of u_i yield

$$u_i(x_i^n) = u_i(x_i^\infty) + o(1).$$

Hence if i is lightweight, then $\lambda_i^n = \lambda_\ell^n \xi_i + o(\lambda_\ell^n)$, together with (9.25) and (9.18), yield

$n \lambda_\ell^n \xi_i u_i(x_i^\infty) + o(n \lambda_\ell^n) = n \lambda_i^n u_i(x_i^n)$

$\qquad = (\phi v_H^n)(T_i^n) + n \lambda_\ell^n \xi_i u_i(x_i^\infty) - n \lambda_\ell^n q \cdot x_i^\infty + o(n \lambda_\ell^n).$

Cancelling, transposing, and dividing by $n \lambda_\ell^n$, we obtain (9.23).

To prove the second part of the lemma, note that $\mu_i = u_i(x_i^\infty)$, since i is heavyweight. Hence (9.7), (9.25), and (9.18) yield

$n \lambda_i^n u_i(x_i^\infty) + o(n \lambda_\ell^n) = n \lambda_i^n u_i(x_i^n) = (\phi v_H^n)(T_i^n) + n \lambda_\ell^n p_i + o(n \lambda_\ell^n)$

$\qquad = (\phi v_H^n)(T_i^n) - n \lambda_\ell^n q \cdot x_i^\infty + o(n \lambda_\ell^n);$

transposing and dividing by $n \lambda_\ell^n$, we obtain (9.24). Q.E.D.

---

[40] $v^n(S \cup t) \geq v^n(S)$.

COROLLARY 9.26: $\xi_i > 0$ *for all lightweight* i.

PROOF: If $\xi_i = 0$, then by (9.19), the definition (8.1) of superdifferential, $0 \in \text{Int } X_i$ (see (2.3)), and $q \neq 0$ (see (9.20)), we have

$$q \cdot x_i^\infty = \min \{q \cdot x : x \in X_i\} < 0;$$

by (9.21), this contradicts (9.23). Q.E.D.

We now define the dividend c. First, define

(9.27) $\quad c_i := \lim_{n \to \infty} \dfrac{(\phi v_H^n)(T_i^n)}{n \lambda_\ell^n}, \quad \text{when} \quad i \in L;$

the limit exists by (9.23). When $i \in H$, define $J(i) := \{j \in L : X_i \subset X_j\}$, and

(9.28) $\quad c_i := \min \{c_j : j \in J(i)\}, \quad \text{when} \quad i \in H \quad \text{and} \quad J(i) \neq \varnothing.$

If $J(i)$ is empty, then, as usual for the minimum over the empty set, c_i should be taken as $+\infty$. Since, however, we want c_i finite for all i, we define it as a sufficiently large finite number; specifically,

(9.29) $\quad c_i := \max \{q \cdot x : x \in \sum X_j\}, \quad \text{when} \quad i \in H \quad \text{and} \quad J(i) = \varnothing.$

With these definitions it follows from (9.21) that the dividends are nonnegative, and from Corollary 9.13 that they are monotonic in the net trade sets. By (9.23),

(9.30) $\quad q \cdot x_i^\infty = c_i \quad \text{when} \quad i \in L;$

and by (9.23), (9.24), and Lemma 9.14,

$$q \cdot x_i^\infty \leq c_i \quad \text{for all } i,$$

i.e., x_i^∞ is in the dividend budget set $\{x \in X_i : q \cdot x \leq c_i\}$ (note that q is a price vector by (9.20)). If i is heavyweight, then x_i^∞ is a global maximum of u_i, and a fortiori maximizes u_i over the dividend budget set. If i is lightweight, then from (9.19), (9.30), and $\xi_i > 0$ (Corollary 9.26), it follows that x_i^∞ maximizes u_i over the dividend budget set. This completes the proof of the main theorem.

10. FIXED PRICE MARKETS

A (finite) fixed price market is defined by:

(10.1) a finite set T (the *trader space*);

(10.2) an integer $d + 1$ that is at least 2 (the *number of commodities*);

(10.3) for each trader t, a point e_t in R_{++}^{d+1} (t's *endowment*);

(10.4) for each trader t, a concave continuous function u_t^* on R_+^{d+1}

(t's *utility function*); and

(10.5) a point $\bar{p}$ in R_{++}^{d+1} (the *fixed price vector*).

A fixed price market is just like an ordinary market (without satiation), except that all trade is constrained to take place at the exogenously given prices $\bar{p}$. In effect, this means that each trader t can only consume bundles in his *fixed price hyperplane*

$$H_t := \{y \in R_+^{d+1}: \bar{p} \cdot y = \bar{p} \cdot e_t\}.$$

The utility u_t^* is defined on the entire orthant R_+^{d+1} only for convenience; all that is actually used is its restriction to H_t.

An *allocation* in a fixed price market is a vector $v = (y_t)_{t \in T}$, where $y_t \in H_t$ for each t (trading takes place at prices $\bar{p}$ and all consumptions are nonnegative), and $\sum_{t \in T} y_t = \sum_{t \in T} e_t$ (trading does not affect the total quantity of each good). A *coupons price vector* is a member q^* of R^{d+1} not proportional to $\bar{p}$ (i.e., unequal to $\alpha \bar{p}$ for any real α). A *coupons endowment* for agent t is a real number c_t. A *coupons equilibrium* consists of a coupons price vector q^*, a vector $c = (c_t)_{t \in T}$ of coupons endowments, and an allocation y, such that for all traders t, y_t maximizes u_t^* over the *coupons budget set*

$$\{y \in R_+^{d+1}: \bar{p} \cdot y = \bar{p} \cdot e_t \text{ and } q^* \cdot y \leq q^* \cdot e_t + c_t\}.$$

The notion of coupons equilibrium is due to Drèze and Müller (1980). If the traders maximize their utilities subject only to the fixed prices $\bar{p}$, then in general the market will not clear. To obtain market clearing, one introduces an auxiliary currency, *in addition* to the ordinary currency in which the fixed prices $\bar{p}$ are stated. This auxiliary currency may be thought of as rationing "coupons"; each transfer of commodities must be paid for both in ordinary money, at prices $\bar{p}$, and in coupons, at prices q^*. Coupons may not be exchanged for ordinary money.

The coupons endowments c_t are called *monotonic in the commodity endowments* if $e_t \geq e_s$ (coordinatewise) implies $c_t \geq c_s$; and *uniform* if all the c_t are the same.

In a $(d+1)$-commodity fixed price market M^*, the spaces $H_t - e_t =: X_t$ of net trades are compact convex subsets of the d-dimensional subspace Q of R^{m+1} that is orthogonal to $\bar{p}$; M^* may be viewed as a d-commodity market M with satiation. If (q, c, x) is a dividend equilibrium in this market, then q is a linear functional on Q, and x is in Q^k; extending q in an arbitrary way to a linear functional q^* on all of R^{d+1} yields a coupons equilibrium $(q^*, c, x+e)$ in M^*, where e is the initial allocation in M^*. In brief, dividend equilibria in M correspond naturally to coupons equilibria in M^* (cf. Drèze and Müller, 1980, p. 133). Hence our main theorem implies that in fixed price markets with k types, limiting value allocations are associated with coupons equilibria enjoying the appropriate monotonicity properties.

Clearly, monotonicity in the net trade sets is equivalent to monotonicity in the endowments:

(10.6) $X_t \supset X_s$ if and only if $e_t \geq e_s$ (coordinatewise).

Thus, *all value allocations in large fixed price markets approximate coupons equilibrium allocations, where the coupons endowments depend only on the commodity endowments, and monotonically so.* In particular, if all commodity endowments are the same, we are led to uniform coupons equilibria.

Note that in the context of fixed prices, the elbow room assumption is satisfied if there is no set J of types whose aggregate demand for some good h precisely exhausts the total supply of that good.[41]

So much for the technical treatment. We end this section with some remarks of a more conceptual nature, which relate this work to other work on fixed prices.

As mentioned in the introduction, our interest in markets with satiation arose from the desire to discover what kind of allocations the Shapley value would generate, in markets with fixed prices. The study of these markets in economic theory has been mostly oriented towards equilibria with one-sided, market-by-market rationing. The specific features of the equilibrium concept are either imposed directly, as in Drèze (1975), or derived from more basic assumptions (no involuntary trading, efficient recourse to a set of admissible trades, ...), as in Malinvaud and Younès (1977). By contrast, the analysis presented here imposes no conditions on the problem or its solution, beyond the constraint that all trading should take place at exogenously given prices.[42] Rather, we apply a general solution concept (the Shapley value) to the problem. The equilibrium concept (coupons equilibrium) and its specific features (nonnegative coupons endowments monotonically geared to initial resources) are an output of the analysis, not an input.

Whatever further properties Shapley value allocations may be found to possess, these properties will also emerge from the problem formulation, kept here to essentials. By this we mean in particular keeping out of the problem formulation elements like "market-by-market rationing," which make the solution set depend upon inessentials like the definition of commodities.[43]

To clarify this point, note that the formal description of a market specifies for each trader a set (the consumption set), a real function on it (the utility function), and a point in it (the endowment). The set must also be endowed with an additive structure (to enable us to describe transfers between traders). Nothing more is required to describe a market from the economic viewpoint; the above structure completely specifies the opportunities as well as the incentives.

This suggests that one might want an economic "solution" (such as an equilibrium concept) to depend only on this structure, to be invariant under "inessential" changes, changes in the specification of the situation that leave this basic structure invariant. Familiar examples of such "inessential" changes are changing the units of the commodities, or using different commodities that are utility substitutes, or permuting the commodities. But the principle of invariance under inessential changes applies equally well in less familiar cases, e.g. for rotations or other affine transformations.[44]

---

[41] That is, $\sum_{i \in J} y_i^h = \sum e_i^h$, where y_i maximizes u_i^* over H_i.

[42] This is not the place to discuss the rationale for studying markets with fixed prices.

[43] Indeed, in applied work, identifying specific "commodities" is often quite difficult.

[44] For example, suppose that each of two mutual funds is composed of stock in the same two companies, but in different proportions. Suppose that the companies themselves are not public, but that the funds are: in essence, therefore, one can buy into the companies in any proportion between those offered by the two funds. Then it should make no difference whether the "commodities" are defined to be company stock or fund stock.

The Walras equilibrium is invariant in this sense; so are the kinds of dividend equilibrium and coupons equilibrium defined in this paper.[45] But the "market-by-market" rationing equilibria mentioned above are not; they depend on identifying specific "commodities," which are not present in the opportunities or the incentives.

All this is reflected in the game-theoretic treatment. Game theory gets at the essence; it separates the intrinsic from the conventional. Thus it is not surprising that the game-theoretic analysis leads to a "commodity-free" solution.

When institutional aspects (like market-by-market rationing) are deemed important, they should perhaps be introduced exogenously into the problem formulation. This brings us to the basic methodological dichotomy, whether economic theory should be concerned with "explaining" the genesis of institutions, or with "predicting" their consequences. In the end, each of these activities has its own validity.

11. ALTERNATIVE APPROACHES

11.1 *The Core*

The core of a market is the set of all allocations that cannot be improved upon by any coalition S. "Improved upon" has two possible meanings:

(a) Some members of S are better off and none worse off.

(b) All members of S are better off.

In classical markets these two meanings lead to the same core, but here they do not. Neither core is very interesting; the first is too small, the second is too large.

Let M^1 be a market with satiation, M^n its n-fold replication. For simplicity we assume that the utilities are strictly convex and that not all traders can be simultaneously satiated; the elbow room assumption (5.1) is, however, unnecessary here.

Under (a), the Debreu-Scarf Theorem (1963) applies; the proof goes through without difficulty. Specifically, the a-core of M^n enjoys the equal treatment property. Hence it may be represented by a set C_a^n of allocations in the unreplicated market M^1. Then $C_a^{n+1} \subset C_a^n$, and the limiting a-core, $C_a^\infty := \bigcap_{n=1}^\infty C_a^n$, coincides with the set of competitive allocations.

As we saw in Section 3, markets with satiation often have no competitive allocations; the limiting a-core is then empty. This is what we meant by "too small".

Under (b), the core of M^n may be very large, and does not even enjoy the equal treatment property. If we nevertheless confine ourselves to equal treatment allocations, we are, as above, led to a set C_b^n of allocations in M^1, where again $C_b^{n+1} \subset C_b^n$. We are interested in $C_b^\infty := \bigcap_{n=1}^\infty C_b^n$.

For each allocation x in M^1, let $G_i(x) := \{x \in X_i : u_i(x) > u_i(x_i)\}$ be the set of net trades preferred by i to x. Then C_b^∞ consists precisely of those allocations x

---

[45] The monotonicity condition for coupons equilibria ostensibly involves the commodities, but (10.6) shows that it is merely a restatement of an "invariant" condition.

for which 0 is not in the convex hull of the union of the preferred sets $G_i(x)$; since the preferred set is empty for satiated traders, we may take the union over unsatiated i only. For example, any dividend equilibrium allocation with nonnegative dividends is in the limiting b-core, even if the dividends are not in any sense monotonic; they may even be different for identical[46] types. Any individually rational allocation x at which only one type is unsatiated will also be in the limiting b-core. What is happening is that the satiated traders are useless as partners in an improving scheme; the unsatiated must fend for themselves, and they may well lack the resources for this. This makes the b-core very large.

We note for the record that the limiting value allocations are in the limiting b-core; but in general they constitute only a small subset. This fits in well with our experience in other market contexts with cores. For example, in large transferable utility markets with nondifferentiable utilities the core may be quite big, but if it has a center of symmetry, then the value is that center of symmetry (Hart, 1977a). More generally (asymmetric core, nontransferable utility), the value allocations in a large nondifferentiable market often constitute a small, "central" subset of a relatively big core (Hart, 1977b, 1980; Mertens, 1988; Tauman, 1981).

11.2. The Continuum Approach

There is no difficulty in defining nonatomic markets with satiation. One simply replaces the trader space T by a nonatomic measure space $(T, \mathscr{C}, \mu)$ with $\mu(T) = 1$, and requires that the net trade sets X_t and the utility functions u_t be measurable in an appropriate sense, and the u_t uniformly bounded. An *allocation* is now a measurable function x from T to R^d with $x(t) \in X_t$ for all t and $\int_T x = 0$. As before, we assume that no allocation satiates all traders, but do not require anything like the "elbow room" assumption (5.1). The definition of competitive equilibrium remains literally unchanged.

A *generalized comparison function* is an integrable function from T to R^1_+ with a positive integral; if it is to R^1_{++}, it is a *comparison function*. A *coalition* is a measurable subset of T (i.e. a member of $\mathscr{C}$). Given a generalized comparison function λ, define a nonatomic game v_λ by

$$v_\lambda(S) := \max\left(\int_S \lambda(t) u_t(x(t)) \mu(dt) : \int_S x = 0 \quad \text{and } x(t) \in X_t \text{ for all } t \in S\right)$$

for each coalition S. A (*generalized*) *value allocation* is an allocation x for which there exists a (generalized) comparison function λ such that

(11.1) $(\phi v_\lambda)(dt) = \lambda(t) u_t(x(t)) \mu(dt)$

for all "infinitesimal" agents dt, where ϕ is an appropriate[47] value operator. (A more formal statement of (11.1) is that $(\phi v_\lambda)(S) = \int_S \lambda(t) u_t(x(t)) \mu(dt)$ for all coalitions S.)

---

[46] Having the same utilities and net trade sets. Confining ourselves to equal treatment allocations in M^n does not mean that identical types get the same net trade; it only means that in M^n, different replicas of the same trader in M^1 get the same net trade.

[47] See, e.g., Kannai (1966), Aumann and Shapley (1974), Hart (1980), Mertens (1980, 1988), or Neyman and Tauman (1979). What is needed here is a value with the "diagonal" property.

So much for the definitions. Unfortunately, the results are rather disappointing. All we can say is:

(11.2) *Every value allocation is competitive.*

(11.3) *An allocation is a generalized value allocation if and only if it is competitive or satiates some agents.*

We have already noted that markets with satiation often have no competitive equilibria; in that case there are *no* value allocations in the continuum approach. The *generalized* value allocations, on the other hand, constitute a very large set even then, consisting of all allocations satiating at least one agent. There is no restriction at all on what nonsatiated agents get; they may even be assigned individually irrational net trades.

To demonstrate these results, assume for simplicity that the u_t are continuously differentiable and strictly convex, and that the allocations in question are interior (i.e., $x(t) \in$ int X_t for all t). Let x be a generalized value allocation. As at (6.1), there is a vector q such that

(11.4) $\lambda(t) u'_t(x(t)) = q$,

for all t. Moreover, the value $(\phi v_\lambda)(dt)$ is the average contribution of dt to a coalition S. In the continuum case, "almost all" coalitions are large; hence as at (6.3),

$$(\phi v_\lambda)(dt) = \lambda(t) u_t(x(t)) \mu(dt) - q \cdot x(t) \mu(dt),$$

and together with (11.1), we obtain that for all t,

(11.5) $q \cdot x(t) = 0$.

If x is a value allocation, i.e., $\lambda(t) > 0$ for all t, then $u'_t(x(t))$ is either equal to 0 for all t or is unequal to 0 for all t. If it is equal to 0 for all t, then all traders are simultaneously satiated, which we have ruled out. Hence it is unequal to 0 for all t, whence $q \neq 0$; together, (11.4) and (11.5) then assert precisely that (q, x) is a competitive equilibrium, whence x is a competitive allocation.

If x is not a value allocation, i.e., $\lambda(t) = 0$ for some t, then $q = 0$. Hence for those t for which $\lambda(t) \neq 0$, we must have $u'_t(x(t)) = 0$, and hence these t are satiated. Conversely, if x is any allocation that satiates some traders, let λ be a generalized comparison function that assigns weight 0 to all traders unsatiated at x. The value $(\phi v_\lambda)(dt)$ is the average contribution of dt to a diagonal[48] coalition θT, where θ ranges from 0 to 1. The case $\theta = 0$ has no effect on the average and may be ignored. As soon as $\theta > 0$, there are enough unsatiated traders in θT to supply all the resources desired by the satiated traders in θT. Thus dt contributes nothing if he is unsatiated, and only his own utility $\lambda(t) u_t(x(t)) \mu(dt)$ if he is satiated, which means that (11.1) is satisfied. Thus any such x is a generalized value allocation.

---

[48] See Aumann and Shapley (1974, Chapter III).

The reader will have realized that what prevents the continuum approach from achieving a more satisfactory result is that "small" coalitions play no role. Coalitions either have positive measure, in which case they behave like "large" coalitions, or have measure 0, in which case they are ignored. The crucial coalitions in the limit approach[49] (the one used in this paper) are those whose size is positive but of smaller order of magnitude than that of the all-trader set; the continuum approach is not equipped to take account of such coalitions.

11.3. The Elbow Room Assumption in a Model with Explicit Endowments

To show that (5.1) holds generically, we must reformulate the definition of a market with satiation so that the endowments appear explicitly. Accordingly, define a *market with satiation and explicit endowments* to consist of a finite set T (the *trader space*), a positive integer m (the number of *commodities*), and for each trader t,

a compact convex subset X_t^0 of R^d (t's *consumption set*);

a concave continuous function u_t^0 on X_t^0 (t's *utility function*); and

a point e_t in the interior of X_t^0 (t's *endowment*).

To regain from this a market with satiation as in Section 2, simply define $X_t := X_t^0 - e_t$ (algebraic subtraction!) and $u_t(x) := u_t^0(x + e_t)$. The remainder of the treatment is then exactly as before.

In this formulation, the elbow room assumption is equivalent to the following: for each $J \subset \{1, \ldots, k\}$,

$$(11.6) \quad \sum_{i=1}^{k} e_i \notin \mathrm{bd}\left[\sum_{i \in J} B_i^0 + \sum_{i \notin J} X_i^0\right],$$

where e_i is the endowment of a type i trader, and B_i^0 is the set of points in X_i^0 at which u_i^0 is maximized. Note that for each J, the right side of (11.6) represents the boundary of a compact convex set in R^d that is independent of the endowments; such sets are closed and of measure 0. The left side is simply the total endowment. Since there are only finitely many different possible choices of J, it follows that the elbow room assumption holds for all total endowment vectors except for a closed set of measure 0 in R^d; hence also for all k-tuples $(e_1, \ldots, e_k)$ of endowments except for a closed set of measure 0 in R^{dk}.

Conceptually, the situation here is perhaps a little different from that of other generic theorems in the literature. The exceptional set is entirely explicit and has transparent geometric and economic meanings; in any given market one can, so to speak, "see at a glance" whether or not the elbow room assumption is satisfied.

---

[49] Of course, the value operator ϕ may itself be defined by a limit approach, even when it is applied to nonatomic games. Nevertheless, the kind of second-order effect that is crucial in the proof of the main theorem does not obtain then.

12. FURTHER RESULTS

The main theorem (Section 5) shows that the dividends depend only on the net trade sets, and monotonically so. In fact, we know much more about them, both qualitatively and quantitatively. This section discusses additional properties of the dividends, as well as the order of magnitude of the weights λ_i^n, and some results concerning the differentiable case and that of a single commodity. The next section applies these results to computing some examples.

We will maintain here the terminology, notations, and conventions introduced previously, especially in Sections 5, 8, and 9. The allocations x^n, the comparison vectors λ^n, and the limiting allocation x^∞ will be fixed throughout; so will the price vector q and the dividend c, which are taken to be as in the proof (Section 9) of the main theorem. We will call a type i *satiated* (*unsatiated*) if it is satiated (unsatiated) at x_i^∞.

The competitive case is from our point of view less interesting, and it is convenient to exclude it; thus throughout this section, we assume

(12.1) (q, x^∞) *is not a competitive equilibrium*.

A summary of the results is as follows. The dividends c_i are all strictly positive (rather than just nonnegative); in addition to monotonicity, they satisfy a concavity condition; and there is an "explicit" formula for them. The order of magnitude of the weights λ_i^n of each lightweight type i is exactly $1/n$. Lightweight types whose utilities are differentiable, and whose maxima are interior, are unsatiated; hence if all utilities are differentiable and all maxima interior, then the heavyweights are precisely the satiated types, the lightweights precisely the unsatiated. In the case of one commodity ($d = 1$), all traders on the "short" side are satiated.

We start the formal treatment by discussing concavity of the dividends. Let J be a set of types. A dividend c in M^1 is called *monotonically concave in the net trade sets of the types in J*, if

(12.2) $c_j \geq \sum_{i \in J} \alpha_i c_i$

for all j in J and constants $\alpha_i \geq 0$ such that $\sum_{i \in J} \alpha_i \leq 1$ and

(12.3) $X_j \supset \sum_{i \in J} \alpha_i X_i$.

This implies that if a convex combination of net trade sets X_i is itself a net trade set X_j, then the dividend c_j is at least as great as the corresponding combination of the dividends c_i. As a function of the net trade sets, so to speak, the dividends are concave.

Actually, the condition involves somewhat more, in two directions. First, note that in (12.3) we have inclusion, not only equality. This expresses a kind of amalgamation of concavity and monotonicity. We have already seen that "ordinary" concavity is implied; note that by taking $\alpha_i = 1$ for one of the i in (12.3), we also get ordinary monotonicity.

Second, note that we demand only that $\sum_{i \in J} \alpha_i \leq 1$, not $\sum_{i \in J} \alpha_i = 1$. Intuitively, one can think of this as if we were talking about ordinary convex combinations ($\sum_{i \in J} \alpha_i = 1$), but adding the additional "virtual" element $X_0 := \{0\}$, with $c_0 := 0$.

Denote by $\mathscr{C}_J$ the set of all convex compact subsets of linear combinations of the X_i (for $i \in J$) that contain 0; note that $\mathscr{C}_J$ itself is a convex set. M. Perles has shown (private communication) that c is concave over J if and only if there is a concave monotonic function on $\mathscr{C}_J$ that vanishes at $\{0\}$ and coincides with c_i at each X_i with $i \in J$.

We now state our first results.

THEOREM 12.4: $c_i > 0$ for all i.

THEOREM 12.5: c is monotonically concave in the net trade sets of the lightweight types.

In particular, since by definition, all unsatiated types are lightweight, it follows that c is monotonically concave in the net trade sets of the unsatiated types.

THEOREM 12.6: If i is lightweight, then λ_i^n has order of magnitude $1/n$ exactly.

By possibly restricting the convergence indicator, we may assume that $\lim_{n \to \infty} n\lambda_\ell^n$ exists; denote it by λ_*. The above theorem asserts that

(12.7) $\quad 0 < \lambda_* < \infty.$

We come next to the explicit formula. Define

(12.8) $\quad \lambda_i^\infty := \lim_{n \to \infty} \lambda_i^n.$

Define w_H on R_+^k as in (8.3), except that u_i is replaced by $\lambda_i^\infty u_i$ for all i. Consider a pool of traders containing infinitely many of each of the k types. If S is a finite coalition chosen from this pool, let $v_H(S)$ be the maximum total utility that S can achieve by trading within itself, with utilities weighted by the λ_i^∞; formally, $v_H := w_H \circ \eta$, where $\eta(S)$ is S's profile ($\eta_i(S)$ is the number of type i traders in S). Consider next an infinite sequence of independent random choices from this pool, which pick traders from the k types with probability $1/k$ each. Denote by Q_m the coalition of traders chosen up to stage m, and let $\rho_i^m = 1$ or 0 according as to whether or not i is chosen at stage m. Let t_i denote a separate trader of type i, not included in the pool.

THEOREM 12.9: For lightweight i,

$$c_i = \frac{1}{\lambda_*} \sum_{m=1}^{\infty} E((v_H(Q_m) - v_H(Q_{m-1}))\rho_i^m)$$

$$= \frac{1}{k\lambda_*} \sum_{m=0}^{\infty} E(v_H(Q_m \cup t_i) - v_H(Q_m)).$$

In words, the first line says that the dividend is the expected total contribution of all type i traders in the infinite sequence, divided by λ_*. The second expression is the sum (over m) of the expected contributions made by a type i trader to the coalition chosen up to stage m, divided by $k\lambda_*$. The two expressions are equal because at each stage, a type i trader is picked with probability $1/k$ exactly.

Another alternative expression, directly in terms of w_H, will be given in Appendix C (Proposition C.15). There we will also discuss analogous formulas for heavyweight types (Corollary C.19).

A noteworthy feature of this formula is that it involves *independent* choices (like in sampling with replacement), which are much easier to work with than the more complicated permutations (sampling without replacement) that are usually associated with values. For example, it is the independence that enables the proof of concavity of the dividends (Theorem 12.5). Another noteworthy feature is that the limiting order represented here—a discrete infinite sequence—is quite different both from the continuous interval associated with the familiar diagonal formula,[50] and from the denumerable dense order type (the order type of the rationals) introduced into value theory by N. Z. Shapiro.[51]

The last two results are shallower than the others, but are useful to keep in mind, particularly when computing examples.

PROPOSITION 12.10: *Let i be a lightweight type with $B_i \subset \text{Int } X_i$, and u_i differentiable at each point of B_i. Then i is unsatiated.*

The final result deals with the case of one commodity ($d = 1$), which is of considerable special interest, both theoretically—because of its relative simplicity—and in the applications, e.g. to unemployment.

In this case, each net trade set X_i is a compact interval on the real line, that contains 0 in its interior. For each i, the bliss set B_i is also a compact interval; we have assumed (Section 2) that it does *not* contain 0. If $0 \in \sum B_j$, then all agents can be simultaneously satiated. Since value allocations are Pareto optimal, it follows that x^∞ satiates all agents, a trivial case that was excluded in Section 9. Thus, $0 \notin \sum B_j$. Since $\sum B_j$ is a compact interval, it is included in either the strictly positive or the strictly negative half-line. We shall say that Type i is on the *short (long) side of the market* if B_i is on the opposite (same) side of 0 from $\sum B_j$.

Intuitively, $\sum B_j$ represents total demand. If $\sum B_j$ is, say, in the positive half-line, then there is excess demand for the single good;[52] the good is scarce. A type i with B_i in the negative half-line wants to supply this scarce good, which is what we mean by "being on the short side."

PROPOSITION 12.11: *If $d = 1$, then $\lambda_i^\infty > 0$ for all types i on the short side of the market; in particular, they are satiated.*

---

[50] See Footnote 47.
[51] See Shapley (1962).
[52] This is the convention in the examples below; it is convenient because then $q > 0$. In applications, the case of interest is often that of excess supply. Practically, there is no difference; demand for positive amounts is the same as supply of negative amounts.

All results stated in this section will be proved in Appendix C. Also, some useful generalizations and strengthenings of these results may be found there.

13. EXAMPLES

EXAMPLE 13.1: Unless x^∞ is competitive, there are at least two orders of magnitude for the weights: $1/n$ for the lightweights, and 1 for at least some of the heavyweights. Here we show that there may be more than two orders of magnitude, by bringing an example with three. We will not compute the third one explicitly, but will use theoretical arguments to show that it is different from both 1 and $1/n$. Indeed, we do not know what it is.

Let $d = 1$, $k = 3$, $X_1 = X_2 = X_3 = [-3, 3]$,
$$u_1(x) = 4 - (x+2)^2, \quad u_2(x) = 1 - (x-1)^2, \quad u_3(x) = 4 - (x-2)^2.$$

The maxima are at -2, 1, and 2 respectively. The short side consists of Type 1 only, and so by Proposition 12.11, Type 1 is satiated, and indeed $\lambda_1^\infty > 0$. Since all net trade sets are equal, all dividends are equal, which means that the maxima of 2 and 3 over their dividend budget sets are achieved at 1. At this point, 3 is unsatiated and so must be lightweight. Type 2 is satiated and so, by Proposition 12.10, must be heavyweight; that is, λ_2^n has order of magnitude greater than $1/n$. If $\lambda_2^\infty > 0$, then by Lemma C.28, $1 = x_2^\infty < x_3^\infty = 1$. Hence $\lambda_2^\infty = 0$; that is, λ_2^n has order of magnitude less than 1. Q.E.D.

EXAMPLE 13.2: In this example, the net trade sets are not all the same, and we use the "explicit formula" (Theorem 12.9) to calculate the dividends. Let $d = 1$, $k = 3$,
$$X_2 = [-1, 1], \quad X_1 = X_3 = [-2, 2],$$
$$u_1(x) = 1 - |x+1|, \quad u_2(x) = u_3(x) = 1 - (x-1)^2.$$

The short side consists of Type 1 only, so $\lambda_1^\infty > 0$ (Proposition 12.11), and $q > 0$. Because $c_i > 0$ for all i (Theorem 12.4), both long types, 2 and 3, have a positive consumption. Therefore if either one is satiated, the market does not clear; there is excess demand. Hence they are both unsatiated, hence lightweight, and so Theorem 12.9 applies to them. Together, they consume the supply of Type 1, and hence
$$1 = x_2^\infty + x_3^\infty = \frac{c_2}{q} + \frac{c_3}{q}.$$

Hence it suffices to calculate c_2/c_3, which we can do from Theorem 12.9 without knowing λ_*. The normalization $\sum \lambda_i^\infty = 1$ yields $\lambda_1^\infty = 1$.

Denote Q_m's profile by $\underline{y}^m := (y_1^m, y_2^m, y_3^m)$. Set
$$f(\underline{y}) := y_1 - y_2 - 2y_3.$$

If $f(y^m) \leq 0$, then the long types together can satiate Type 1, and so an additional long trader t_i contributes nothing to v_H. If $f(y^m) = 1$, then the whole long side together falls short by just one unit of trading capacity to be able to satiate Type 1. This unit of capacity is supplied by t_i whether $i = 2$ or 3, and so in either case, t_i contributes exactly 1. If $f(y^m) \geq 2$, then t_i contributes 1 if $i = 2$, and 2 if $i = 3$. The expected contribution of t_i is therefore

(13.2.1) $\text{prob}\{f(y^m) = 1\} + (i-1) \text{prob}(f(y^m) \geq 2)$.

Since $f(y^m)$ is the sum of m i.i.d. r.v.'s distributed like $f(y^1)$, the generating function of $f(y^m)$ is $g(x) := ((x + x^{-1} + x^{-2})/3)^m$. The first probability in (13.2.1) is the coefficient of x in $g(x)$, and the second is the sum of all coefficients starting with that of x^2. Using this we find

(13.2.2) $x_2^\infty = c_2/q = .414$,

$x_3^\infty = c_3/q = .586$,

to within .002. Of course, $x_1^\infty = -1$.

Several aspects of this example are worthy of special note. First, the limiting value allocation x^∞ is unique, though there are no symmetries that would lead to this conclusion in any obvious way. Second, our method does *not* involve calculating x^n for large n, say by some fixed point method. On the contrary, we use a formula that is valid only "in the limit." This formula enables a precise calculation of x^∞, to within errors with definite, theoretically proven bounds; but it does *not* enable even an approximate calculation of x^n, for any finite n.

Third, the outcome is fairly insensitive to the utility functions on the long side (Types 2 and 3). For simplicity, let each bliss set B_i consist of a unique point b_i. The above reasoning certainly continues to apply when u_2 and u_3 are changed only cardinally,[53] e.g., if $u_2(x) = x$, $u_3(x) = 1 - (x-1)^4$. But even if there are ordinal changes, i.e., the bliss points are changed, x^∞ will not change if b_2 and b_3 remain sufficiently large; e.g., if $b_2 > \frac{1}{2}$ and $b_3 > \frac{2}{3}$ (see C.19 and C.20). Of course, if they are sufficiently small—e.g., $b_2 \leq .412$ or $b_3 \leq .584$—then either 2 or 3 must be satiated, and the result necessarily changes. (We are assuming that u_1 and the X_i do not change.) Again, all this applies to x^∞ only; each x^n may, and in general will, change considerably, even if u_2 or u_3 are only changed cardinally. Indeed, the x^n may perhaps be nonunique, even though x^∞ is unique.

On the other hand, the outcome *is* quite sensitive to changes in u_1, even if they are only cardinal. If u_1 is strictly concave, then the fact that Type 3 has more capacity than Type 2 matters less than when u_1 is linear. Specifically, if $u_1 = 1 - (x+1)^\alpha$, then $c_3/c_2 \to 1$ when $\alpha \to \infty$, and hence $x_3^\infty \to \frac{1}{2}$, $x_2^\infty \to \frac{1}{2}$. Indeed, when α is large, the lion's share of the contribution to Type 1 comes on the very first occasion when a trader on the long side joins the market, and it comes with the first unit that that trader contributes. When α is moderate, additional units remain more significant.

---

[53] This is as in ordinary markets, without satiation. In such markets, of course, *all* agents are unsatiated; and indeed, under appropriate smoothness conditions, the limiting value allocations coincide with the competitive ones (Mas-Colell, 1977), and so are ordinally invariant.

Thus on one end of the scale, when the satiated traders are very risk averse, the unsatiated traders get dividends that are almost independent of their trading capacity. One might have thought that on the other end of the scale, when the satiated traders are risk neutral, the dividends of the satiated will be proportional to their capacities. Our example shows that this is not so. Type 3 traders have 100% more capacity than type 2, but their dividends are only about 40 per cent larger. This is because of the possibility that the shortfall, $f(y^m)$, in the capacity to satiate Type 1 is exactly 1, in which case the extra capacity of Type 3 is useless.[54]

Intuitively, it may seem strange that what the unsatiated traders get depends on the cardinal utility of the satiated traders, but not on their own cardinal utility. The reason is that because the unsatiated traders are on the "long" side, they contribute very little to Society. When there is unemployment, an additional employer is a lot more welcome than an additional worker. This is reflected in the weights, which are much higher for the employers; the workers are judged not by what they can do for themselves, but by what they can do for the employers.

It should not puzzle the reader that the dividends may depend on cardinal utilities, in spite of our statement that they depend only on the net trade sets. The latter statement refers to different types within the same market; such types do get the same dividend as long as they have the same net trade sets, even if their utilities are different. But if one changes the utility of one type, everybody's dividend may change. The situation is similar in an ordinary market, without satiation. Incomes depend only on endowments, not on utilities, in the sense that agents with the same endowment have the same incomes, even if their utilities are different. But prices, and therefore incomes, are determined by everybody's utilities, and may change when a utility changes.

EXAMPLE 13.3: In this example, there are two types on the long side with identical utilities (but different net trade sets), one of whom is satiated and the other not. The satiated one of the two has $\lambda_i^\infty > 0$.

Let $k=3$, $d=1$, the X_i and u_1 as in the previous example, and

$$u_2(x) = u_3(x) = (.55)^2 - (x-.55)^2.$$

As above, the short side consists of Type 1 only; therefore Type 1 is satiated ($x_1^\infty = -1$), and indeed $\lambda_1^\infty > 0$. Type 2 cannot be satiated, because then by monotonicity, Type 3 is also satiated, and then the market does not clear (there is excess demand). Hence $\lambda_2^\infty = 0$. If also $\lambda_3^\infty = 0$, then v_H is as in the previous example, and hence $x_2^\infty < .416 < .45$. Hence $c_3/q \geq x_3^\infty > .55$. But then Type 3 is not maximizing over its budget set, since the maximum is at .55. Hence $\lambda_3^\infty > 0$, and hence $x_3^\infty = .55$, $x_2^\infty = .45$.

---

[54] It is worthy of note that this happens in a significant proportion of the cases, not only when m is small, but also when it is large. When $m = 40$, the probability is about .0003 that Type 1 cannot be satiated. In 37 per cent of these cases, the shortfall is exactly 1, though it ranges up to 40. This phenomenon is closely related to the exponential decay in Lemma B.9; it lies at the root of our results.

14. OPEN PROBLEMS

Foremost among the open problems is that of the converse. To what extent are the necessary conditions that we have found for limiting value allocations also sufficient? In the case of ordinary markets, without satiation, this is related to smoothness: see Mas-Colell (1977) and Hart (1977b). It is quite likely that smoothness is relevant here too.

Another interesting task is to dispense with the finite type assumption. As we have seen, one cannot simply use a continuum; what is called for is a limiting approach, in which the limit is a continuum of different types. There is a large literature on this type of model in connection with the core equivalence principle; cf. Hildenbrand's book (1974) and survey article (1982). Another approach that could conceivably be helpful for this purpose is that of nonstandard analysis (cf. Brown and Robinson, 1972).

One might also like to explore the consequences of dispensing with the equal treatment restriction, the elbow room assumption, or the assumption that 0 is in the interior of each net trade set.

Perhaps most interesting at this stage would be to derive additional qualitative properties of the solution in particular contexts. In the case $d = 1$ (one commodity), for example, what happens to the dividends when the capacity of the long side is much larger than the supply of the short side? When they are almost equal? Can this kind of result, once obtained, be generalized to $d > 1$? The "explicit formula" (12.9) gives us a powerful tool for investigating these and other questions arising in particular contexts.

Institute of Mathematics, Hebrew University, 91904 Jerusalem, Israel
and
CORE, Universite Catholique de Louvain, B-1348 Louvain-la-Neuve, Belgium

Manuscript received June, 1984; final revision received February, 1986.

APPENDIX A

PRICE CHARACTERIZATION OF $\partial w(y)$

In this Appendix we prove Proposition 8.5.

LEMMA A.1: *Let g be concave and 1-homogeneous on a convex cone V, and let $y \in V$. Then $p \in \partial g(y)$ if and only if*

(A.2) $\quad p \cdot y = g(y),$

and

(A.3) $\quad p \cdot y' \geq g(y') \quad \text{for all } y' \text{ in } V.$

PROOF: By definition, $p \in \partial g(y)$ if and only if

(A.4) $\quad g(y') - g(y) \leq p \cdot (y' - y) \quad \text{for all } y'.$

Obviously (A.2) and (A.3) imply (A.4). Conversely, applying (A.4) to $y' = 2y$ and $y' = 0$ yields $g(y) \leq p \cdot y$ and $g(y) \geq p \cdot y$ respectively, and therefore $p \cdot y = g(y)$. Again applying (A.4), we then obtain $p \cdot y' \geq g(y')$. Q.E.D.

PROOF OF PROPOSITION 8.5: Intuitively, Lemma A.1 shows that one can view the superdifferential $\partial w(y)$ as the core of a transferable utility market with a nonatomic continuum of traders of types $1, \ldots, k$, in which type i has measure y_i. Proposition 8.5 is then simply an expression of the core equivalence principle. It needs to be reproved because the published versions use different assumptions (e.g., they do not permit satiation), but the proof follows known ideas.

Suppose $q \in \bigcap_{i=1}^{k} \partial u_i(x_i)$, and define p by (8.6). Let $y' \in R_+^k$, and let $w(y')$ be attained at $(x_1', \ldots, x_k')$. From $\sum y_i x_i = 0 = \sum y_i' x_i'$ it follows that

$$\sum y_i' q \cdot (x_i' - x_i) = \sum (y_i' - y_i)(-q \cdot x_i).$$

Hence

$$w(y') - w(y) = \sum (y_i' u_i(x_i') - y_i u_i(x_i))$$
$$= \sum [(y_i' - y_i) u_i(x_i) + y_i'(u_i(x_i') - u_i(x_i))]$$
$$\leq \sum [(y_i' - y_i) u_i(x_i) + y_i' q \cdot (x_i' - x_i)]$$
$$= \sum (y_i' - y_i)(u_i(x_i) - q \cdot x_i) = p \cdot (y' - y),$$

i.e., $p \in \partial w(y)$.

Conversely, let $p \in \partial w(y)$. Let G_i denote the "strict subgraph" of $u_i(x) - p_i$, i.e.,

$$G_i = \{(x, \theta) : x \in X_i \text{ and } \theta < u_i(x) - p_i\}.$$

We assert that

(A.5) $\quad 0$ is not in the convex hull of $\bigcup_i G_i$.

Indeed, suppose 0 *is* in this convex hull. Then there are nonnegative $y_1', \ldots, y_k'$ summing to 1, and (x_i', θ_i) in G_i, such that $\sum y_i'(x_i', \theta_i) = 0$, i.e.

(A.6) $\quad 0 = \sum y_i' x_i'$

and

(A.7) $\quad 0 = \sum y_i' \theta_i.$

From (A.6) it follows that

$$\sum y_i' u_i(x_i') \leq w(y');$$

hence from $\theta_i < u_i(x_i') - p_i$ and (A.7) we obtain

$$0 < \sum [y_i' u_i(x_i') - y_i' p_i] \leq w(y') - p \cdot y',$$

in contradiction to (A.3). Thus (A.5) is established.

From (A.5) it follows that there is a hyperplane through 0 that supports all the G_i; i.e., a nonzero vector $(-q, q_0)$ in R^{d+1} such that

(A.8) $\quad -q \cdot x + q_0 \theta \leq 0$

whenever $\theta < u_i(x) - p_i$, and hence also whenever $\theta \leq u_i(x) - p_i$. Choosing θ to be a negative number with a large absolute value shows that q_0 cannot be negative. If $q_0 = 0$, then $q \neq 0$ and $q \cdot x \geq 0$ for all x in all X_i; but this contradicts $0 \in \text{Int} \sum y_i X_i$. Thus $q_0 > 0$, so w.l.o.g. $q_0 = 1$. Setting $\theta = u_i(x) - p_i$ in (A.8) we then obtain

(A.9) $\quad -q \cdot x + u_i(x) - p_i \leq 0 \quad \text{for all} \quad x \in X_i,$

and in particular when $x = x_i$. If strict inequality would hold in (A.9) for one of the x_i, then since $y \in R_{++}^k$,

$$0 > \sum y_i(-q \cdot x_i + u_i(x_i) - p_i) = -q \cdot \sum y_i x_i + w(y) - p \cdot y = 0,$$

where the last equation follows from $\sum y_i x_i = 0$ and (A.2). Therefore

(A.10) $\quad p_i = u_i(x_i) - q \cdot x_i$

for all i. From (A.10) and (A.9) we obtain $q \in \partial u_i(x_i)$ for all i, and together with (A.10), this is what was to be proved.

APPENDIX B

THE VALUE CONVERGENCE THEOREM

In this Appendix we prove Proposition 8.8. Results that are in many respects similar appear in Champsaur (1975) and Hart (1977a); but they do not contain quite what we need here,[55] and it is simpler and quicker to prove what we need directly than to derive it from those results.

Lemmas B.9 and B.10 constitute the probabilitistic foundations of the paper. The proofs of these lemmas do not depend on what goes before.

Let U be the convex neighborhood of $(1, \ldots, 1)$ mentioned in the proposition;[56] since the w_i are 1-homogeneous, we may take U to be an open cone. Denote by w an arbitrary concave 1-homogeneous function on U, and by $w'(y, z)$ its directional derivative $\lim_{\theta \to 0+} [w(y + \theta z) - w(y)]/\theta$ in the direction z.

LEMMA B.1: *Let* $y_n \to y \in U$ *and* $z_n \to z \in R^k$. *Then*

$$\liminf_{n \to \infty} w'_n(y_n; z_n) \geq w'_\infty(y; z).$$

PROOF: Rockafellar (1970, Theorem 2.4.5, p. 233).   Q.E.D.

For $z \in R^k$, set $\|z\| = \max_i z_i$.

COROLLARY B.2: *Let* $y \in U$ *and* $z \in R^k$. *Then for any* $\varepsilon > 0$ *there is a* $\delta > 0$ *such that*

$$w'_n(\hat{y}; \hat{z}) \geq w'_\infty(y; z) - \varepsilon$$

whenever

$$\|\hat{y} - y\| < \delta, \quad \|\hat{z} - z\| < \delta, \quad \text{and} \quad n > 1/\delta.$$

PROOF: Immediate from Lemma B.1.

LEMMA B.3: *Let* $y \in U$ *and* $z \in R^k$. *Then*

(B.4) $w'(y; z) \geq w(y + z) - w(y)$ *if* $y + z \in U$;

(B.5) $-w'(y; -z) \geq w'(y; z)$;

(B.6) $w(y) - w(y - z) \geq w'(y; z)$ *if* $y - z \in U$;

(B.7) $w'(y; z)$ *is 0-homogeneous*[57] *in* y *and* 1-*homogeneous in* z.

PROOF: The concavity of w yields (B.4). For (B.5), see Rockafellar (1970, Theorem 2.3.1, p. 214). For (B.6), substitute $-z$ for z in (B.4) and apply (B.5). The definition of $w'(y; z)$ yields (B.7) directly.   Q.E.D.

LEMMA B.8: *Let* $y \in U$, *and let* p *in* R^k *be such that* $p \cdot z \geq w'(y; z)$ *for all* z *in* R^k_+, *and* $p \cdot y = w(y)$. *Then* $p \in \partial w(y)$.

PROOF: Let $y' \in U$. Then $y' \in R^k_{++}$, since U is an open cone in R^k_+. Hence $\theta y' = y$ for sufficiently large positive θ, and hence there is a z in R^k_+ such that $y + z = \theta y'$. Hence by (B.4),

$$p \cdot z \geq w'(y; z) \geq w(y + z) - w(y) = w(y + z) - p \cdot y;$$

hence by the homogeneity of w,

$$p \cdot (\theta y') = p \cdot (y + z) \geq w(y + z) = w(\theta y') = \theta w(y');$$

hence $p \cdot y' \geq w(y')$, and hence by Lemma A.1, $p \in \partial w(y)$.   Q.E.D.

LEMMA B.9: *Let* $x_1, x_2, \ldots$ *be independent identically distributed random variables with mean* 0, *all bounded in absolute value by* 1. *Set* $\sigma_m = x_1 + \cdots + x_m$, *and let* $0 < \delta < 1$. *Then for all* m,

$$\text{prob}\left\{\left|\frac{\sigma_m}{m}\right| \geq \delta\right\} \leq 2e^{-(\delta^2/4)m}.$$

---

[55] For example, we do not have concavity on the entire nonnegative orthant, but only on a conical neighborhood of the diagonal.
[56] Not the U of (8.2).
[57] Homogeneous of degree 0.

REMARK: This provides an explicit bound for the residual probability in the weak law of large numbers. The particular expression $\delta^2/4$ is of no importance; for our purposes, it can be replaced by any positive constant depending on δ only (but not on m or on the distribution of the x_i). We are grateful to B. Weiss for providing the following elegant direct proof.

PROOF: Let $0 < \lambda < 1$. Since $E\underline{x}_i = 0$, we have

$$\text{prob}\{\underline{\sigma}_m \geq m\delta\} e^{\lambda m \delta} \leq E e^{\lambda \underline{\sigma}_m} = (E e^{\lambda \underline{x}_1})^m$$

$$= \left(1 + \lambda(E\underline{x}_1) + \frac{\lambda^2}{2}(E\underline{x}_1^2) + \cdots\right)^m$$

$$\leq \left(1 + \frac{\lambda^2}{2} + \frac{\lambda^3}{3!} + \cdots\right)^m \leq (1+\lambda^2)^m \leq e^{m\lambda^2}.$$

Setting $\lambda = \delta/2$, we deduce

$$\text{prob}\{\underline{\sigma}_m \geq m\delta\} \leq e^{-m\delta^2/4}.$$

By substituting $-\underline{x}_i$ for $\underline{x}_i$, we deduce that also

$$\text{prob}\{\underline{\sigma}_m \leq -m\delta\} \leq e^{-m\delta^2/4}.$$
Q.E.D.

We now proceed to calculate ϕv_n, using (4.3). Pick at random an order $\underline{\mathcal{R}}^n$ on T^n, assigning equal probability to all $(nk)!$ orders. For $0 \leq m \leq nk$, denote by $\underline{Q}^n_m$ the coalition[58] consisting of the first m agents in the order $\underline{\mathcal{R}}^n$.

LEMMA B.10: *For $1 > \delta > 0$ and $S \subset T^n$, we have*

$$\text{prob}\left\{\left|\frac{|\underline{Q}^n_m \cap S|}{m} - \frac{|S|}{nk}\right| \geq \delta\right\} \leq 2e^{-(\delta^2/4)m}.$$

REMARK: This is the weak law of large numbers for sampling without replacement, with an explicit bound as in the previous lemma.

PROOF: Fix S, and write $\underline{r}^n_m = |S \cap \underline{Q}^n_m|$. The sequence $\underline{r}^n_1, \ldots, \underline{r}^n_{nk}$ may be obtained by choosing traders at random, one at a time from the population of all nk traders, *without* replacement; if choosing an agent in S is considered as "success," then $\underline{r}^n_m$ is the number of successes in the first m trials. Now let $\underline{s}_m$ be the number of successes in m trials when traders are chosen at random from the same population, but *with* replacement (here m may be as large as we like); i.e., the number of successes in m independent Bernoulli trials, each with success probability $|S|/nk$. Note that both $\underline{r}^n_m$ and $\underline{s}_m$ have mean $m|S|/nk$. Our proof is based on the principle that "sampling without replacement is uniformly better than sampling with replacement."[59] Here "uniformly better" means that the probability of any given deviation from the mean is smaller; thus if

$$\underline{\mathcal{F}} = \left\{\left|\frac{\underline{r}^n_m}{m} - \frac{|S|}{nk}\right| \geq \delta\right\}, \quad \underline{\mathcal{G}} = \left\{\left|\frac{\underline{s}_m}{m} - \frac{|S|}{nk}\right| \geq \delta\right\},$$

then prob $\underline{\mathcal{F}} \leq$ prob $\underline{\mathcal{G}}$. Noting that $(\underline{s}_m/m) - (|S|/nk)$ is of the form $\underline{\sigma}_m/m$ in the previous lemma completes the proof.
Q.E.D.

PROOF OF PROPOSITION 8.8: Let e^i denote the ith unit vector $(0, \ldots, 0, 1, 0, \ldots, 0)$ in R^k, and let $e = (1, \ldots, 1)$. Let z be an arbitrary but, for the time being, fixed point in $[0, 1]^k/k$. Let $S^1, S^2, \ldots$ be a sequence of coalitions in $T^1, T^2, \ldots$ respectively such that

(B.11) $$\lim_{n \to \infty} \frac{\eta(S^n)}{nk} = z;$$

in fact, we choose S^n so that $\eta(S^n) \neq 0$ and in the maximum norm,

(B.12) $$\left\|\frac{\eta(S^n)}{nk} - z\right\| \leq \frac{1}{nk}.$$

---

[58] The wiggles are to indicate that $\underline{\mathcal{R}}^n$ and $\underline{Q}^n_m$ are random variables.
[59] Cf. e.g. Aumann and Shapley (1974, p. 135, Note 1), or Champsaur (1975, p. 415, 6.13).

Let $T_m^n = 1$ or 0 according as to whether the mth trader in the order $\mathcal{R}^n$ is or is not in S^n. Set

(B.13) $\quad y^{mn} = \eta(Q_m^n), \quad z^{mn} = \dfrac{\eta(S^n \cap Q_m^n)}{m}.$

For all $z \in R_+^k$, set $z_* = z/\sum z_i$ if $z \neq 0$, $z_* = 0$ if $z = 0$.

The value of a type i player in v_n is p_i^n, where p^n is as in (8.9); hence by the efficiency of the value,

$$p^n \cdot ne = p^n \cdot \eta(T^n) = (\phi v_n)(T^n) = v_n(T^n) = w_n(\eta(T^n)) = w_n(ne),$$

and so by the 1-homogeneity of w_n,

(B.14) $\quad p^n \cdot e = w_n(e) \to w_\infty(e) \quad \text{as} \quad n \to \infty.$

Next, by (4.3) we have

(B.15) $\quad p^n \cdot \eta(S^n) = (\phi v_n)(S^n) = E \sum_{m=1}^{nk} (v_n(Q_m^n) - v_n(Q_{m-1}^n)) T_m^n$

$$= \sum_{m=1}^{nk} E[(w_n(\eta(Q_m^n)) - w_n(\eta(Q_{m-1}^n))) T_m^n] = \sum_{m=1}^{nk} E\Delta_m^n,$$

the Δ_m^n being defined by the expressions in square brackets. Now Δ_m^n is either 0, or it is $w_n(y^{mn}) - w_n(y^{mn} - e^i)$ for some type i. The latter happens if and only if the mth trader in $\mathcal{R}^n$ is of type i and is in S^n; given Q_m^n, the probability of this is $\eta_i(S^n \cap Q_m^n)/m = z_i^{mn}$. Setting

(B.16) $\quad \Gamma_m^n = E(\Delta_m^n | Q_m^n),$

we thus conclude that

(B.17) $\quad \Gamma_m^n = \sum_{i=1}^{k} [w_n(y^{mn}) - w_n(y^{mn} - e^i)] z_i^{mn}.$

Now define the event $\mathcal{E}_m^n$ by

(B.18) $\quad \mathcal{E}_m^n = \{y^{mn} - [0,1]^k \subset U\}.$

If $\mathcal{E}_m^n$ obtains, then the concavity of w_n on U, (B.6), and (B.7) yield

(B.19) $\quad \Gamma_m^n = \left(\sum_{i=1}^{k} [w_n(y^{mn}) - w_n(y^{mn} - e^i)] z_{*i}^{mn} \right) \sum z_i^{mn}$

$$\geq (w_n(y^{mn}) - w_n(y^{mn} - z^{mn})) \sum z_i^{mn}$$

$$\geq w_n'(y^{mn}; z^{mn}) \sum z_i^{mn}$$

$$= w_n'(y^{mn}; z^{mn}) = w_n'\left(\dfrac{y^{mn}}{m}; z^{mn} \right).$$

Now let $\varepsilon > 0$ be given. Choose δ in accordance with Corollary B.2, with $y = e/k$ and z as chosen at the beginning of the proof. Define events $\mathcal{D}_m^n$, $\mathcal{C}_m^n$, and $\mathcal{B}_m^n$ by

$$\mathcal{D}_m^n = \left\{ \left\| \dfrac{y^{mn}}{m} - \dfrac{e}{k} \right\| < \delta \right\}, \quad \mathcal{C}_m^n = \left\{ \left\| z^{mn} - \dfrac{\eta(S^n)}{nk} \right\| < \dfrac{\delta}{2} \right\}, \quad \mathcal{B}_m^n = \mathcal{D}_m^n \cap \mathcal{C}_m^n.$$

By setting $S = T_i^n$ in Lemma B.10, we deduce

$$\text{Prob (not } \mathcal{D}_m^n) \leq 2k\, e^{-m\delta^2/4}.$$

Similarly, by setting $S = S^n \cap T_i^n$, we deduce

$$\text{prob (not } \mathcal{C}_m^n) \leq 2k\, e^{-m\delta^2/16},$$

and hence

(B.20) $\quad \text{prob (not } \mathcal{B}_m^n) \leq 4k\, e^{-m\delta^2/16}.$

Now from (B.12) we deduce that

(B.21) $\quad \mathcal{B}_m^n \subset \mathcal{C}_m^n \subset \{\|z^{mn} - z\| < \delta\} \quad \text{whenever} \quad n > 2/\delta.$

Moreover, we may assume w.l.o.g. that δ is chosen sufficiently small so that

$$\left\| y - \frac{e}{k} \right\| < 2\delta \;\Rightarrow\; y \in U;$$

as U is a cone, this yields

$$\left\| \frac{y}{m} - \frac{e}{k} \right\| < 2\delta \;\Rightarrow\; \frac{y}{m} \in U \;\Rightarrow\; y \in U.$$

Since $x \in [0,1]^k$ and $m > 1/\delta$ imply $\|x/m\| < \delta$, we deduce

(B.22) $\quad \mathcal{B}_m^n \subset \mathcal{D}_m^n \subset \mathcal{E}_m^n \quad$ whenever $\quad m > 1/\delta$.

Combining (B.19), (B.21), (B.22), Corollary B.2 and (B.7), we obtain that for $n > 2/\delta$ and $m > 1/\delta$

(B.23) $\quad E(\Gamma_m^n \mid \mathcal{B}_m^n) \geq w'_\infty\!\left(\frac{e}{k}; z\right) - \varepsilon = w'_\infty(e; z) - \varepsilon.$

Since $n \geq m/k$, (B.23) holds whenever $m > 2k/\delta$.

Let μ be a uniform bound on the $|w_n|$ in $[0,1]^k$. By (B.17), we always have

(B.24) $\quad |\Gamma_m^n| \leq 2k\mu m.$

By (B.16), $E(\Delta_m^n) = E(\Gamma_m^n)$. Hence by (B.15), (B.23), and (B.24), for $m > 2k/\delta$ we have

$$p^n \cdot \eta(S^n) = \sum_{m=1}^{nk} E(\Delta_m^n) = \sum_{m=1}^{nk} E(\Gamma_m^n)$$

$$= \sum_{m=1}^{nk} [E(\Gamma_m^n \mid \mathcal{B}_m^n)\,\text{prob}\,\mathcal{B}_m^n + E(\Gamma_m^n \mid \text{not } \mathcal{B}_m^n)\,\text{prob}\,(\text{not } \mathcal{B}_m^n)]$$

$$\geq nkw'_\infty(e; z) - nk\varepsilon - \sum_{m=1}^{nk} 4k\mu m \,\text{prob}\,(\text{not } \mathcal{B}_m^n);$$

and hence by (B.20)

(B.25) $\quad p^n \cdot \dfrac{\eta(S^n)}{nk} \geq w'_\infty(e; z) - \varepsilon - \left[\dfrac{1}{nk}\sum_{m=1}^{nk} m\,e^{-m\delta^2/16}\right] 16k^2\mu.$

Denote by c_n the expression in square brackets. Then c_n is the Cesaro mean of the sequence $\{m\,e^{-m\delta^2/16}\}$, which tends to 0 as $m \to \infty$; hence also $c_n \to 0$ as $n \to \infty$. In particular, for n sufficiently large, we have $16k^2\mu c_n < \varepsilon$, and then

(B.26) $\quad p^n \cdot \dfrac{\eta(S^n)}{nk} \geq w'_\infty(e; z) - 2\varepsilon.$

In the particular case in which $z = e^i$, we may choose $S^n = T_i^n$, and then (B.26) yields

$$p_i^n = p^n \cdot \frac{\eta(S^n)}{n} \geq kw'_\infty(e; e^i) - 2k\varepsilon.$$

Thus the p_i^n are bounded from below, and by (B.14), $\sum p_i^n$ is bounded from above. Hence $\{p^n\}$ is bounded. If p is a limit point of $\{p^n\}$, then (B.12) and (B.26) yield

$$p \cdot z \geq w'_\infty(e; z) - 2\varepsilon.$$

Since this is true for each ε, we deduce

(B.27) $\quad p \cdot z \geq w'_\infty(e; z).$

Now (B.27) holds for any z in $[0,1]^k/k$, and so by (B.7), for any z in R_+^k. By (B.14), $p \cdot e = w_\infty(e)$. Hence by Lemma B.8, $p \in \partial w_\infty(e)$, as was to be proved.

APPENDIX C

FURTHER PROPERTIES OF DIVIDENDS AND WEIGHTS

This Appendix is devoted to proving the results stated in Section 12.

W.l.o.g. x^∞ does not satiate all types (see footnote 35), which implies that there are lightweight types. If none of the λ_i^n tend to 0, it follows that all types are lightweight; hence by (9.27), $c_i = 0$ for all i, and hence (q, x^∞) is a competitive equilibrium, which we have excluded (12.1). Thus

(C.1)   *there are both heavyweight and lightweight types.*

Formulas (9.27) through (9.29) indicate that to gain information about the dividends, we should study $(\phi v_H^n)(T_i^n)$ for lightweight i. Define $\mathcal{R}^n$, Q_m^n and y^{mn} as in Appendix B (just before Lemma B.10, and (B.13)). Let $\tau_i^{mn} = 1$ or 0 according as whether the mth trader in the order $\mathcal{R}^n$ is or is not of Type i, and set

(C.2) $\quad \Delta_i^{mn} := (v_H^n(Q_m^n) - v_H^n(Q_{m-1}^n)) \tau_i^{mn}$
$\qquad\qquad = (w_H^n(y^{mn}) - w_H^n(y^{(m-1)n})) \tau_i^{mn}.$

Δ_i^{mn} represents the contribution of the mth player to v_H^n if he is of Type i; otherwise it is 0. Hence

(C.3) $\quad (\phi v_H^n)(T_i^n) = E \sum_{m=1}^{nk} \Delta_i^{mn} = \sum_{m=1}^{nk} E\Delta_i^{mn}.$

LEMMA C.4: *If i is lightweight and N a nonnegative integer, then uniformly in n,*

$$(\phi v_H^n)(T_i^n) = \sum_{m=1}^{N} E\Delta_i^{mn} + o(1).$$

REMARK: The error $o(1)$ is to be understood as a function of N. In words, the lemma says that the total contribution $(\phi v_H^n)(T_i^n)$ of *all* Type i traders to the heavyweights can be approximated arbitrarily closely by the contribution of the Type i traders that are among the first N traders of the order, where N is a fixed finite number that is independent of the number n of times that the market is replicated. We will in fact show that the error decreases exponentially, i.e. that there is a positive α, independent of n, such that the error is $O(e^{-\alpha N})$.

PROOF: By Lemma 9.11, we may choose δ sufficiently small so that $\delta < 1$ and

$$\left\| y - \frac{e}{k} \right\| < \delta \implies y \in U_L.$$

Define[60]

$$\mathcal{D}_m^n := \left\{ \left\| \frac{y^{mn}}{m} - \frac{e}{k} \right\| < \delta \right\}.$$

Using Lemma B.10 with $S = T_j^n$ for each j, we find

$\qquad$ prob (not $\mathcal{D}_m^n$) $\leq 2k\, e^{-m\delta^2/4}$,

$\qquad$ prob (not $\mathcal{D}_{m-1}^n$) $\leq 2k\, e^{-(m-1)\delta^2/4} \leq 3k\, e^{-m\delta^2/4}$.

Setting

$\qquad \mathcal{F}_m^n := \{y^{mn}$ and $y^{(m-1)n}$ are in $U_L\}$,

we deduce

(C.5) $\quad$ prob (not $\mathcal{F}_m^n$) $\leq 5k\, e^{-m\delta^2/4}$.

Since i is lightweight, it follows from (9.6) and (C.2) that

(C.6) $\quad \Delta_i^{mn} = 0 \quad$ whenever $\mathcal{F}_m^n$ obtains.

---

[60] This is as in Appendix B, with U_L instead of U.

Letting $\mu_i := \max u_i$ as in Section 9.3, and setting $\mu := \sum_{i \in H} \mu_i$, we find that $0 \leq w_H^n(y) \leq \mu$ whenever $y \in [0, 1]^k$. From (C.2) and the 1-homogeneity of w_H^n (see (8.4)), it follows that always

(C.7) $\quad 0 \leq \underline{\Delta}_i^{mn} \leq m\mu$.

Combining (C.5) through (C.7) with (C.3) and (C.2), we find

(C.8) $\quad (\phi v_H^n)(T_i^n) = \sum_{m=1}^{N} E\underline{\Delta}_m^n + \sum_{m=N+1}^{nk} E(\underline{\Delta}_i^{mn} | \text{not } \mathcal{F}_m^n) \text{ prob (not } \mathcal{F}_m^n)$

$$\leq \sum_{m=1}^{N} E\underline{\Delta}_i^{mn} + 5k\mu \sum_{m=N+1}^{nk} m e^{-m\delta^2/4}.$$

The second term in the last expression is part of the tail of a convergent series, and therefore tends to 0 uniformly as $N \to \infty$. This proves the lemma. The explicit bound on the error term in the remark follows from a slightly closer look at the tail. Q.E.D.

COROLLARY C.9: *If i is heavyweight and N is a nonnegative integer, then uniformly in n,*

$$(\phi v_H^N)(T_i^n) = \left(n - \frac{N}{k}\right)\lambda_i^n \mu_i + \sum_{m=1}^{N} E\underline{\Delta}_i^{mn} + o(1).$$

REMARK: As in Lemma C.4, the error is a function of N, and decreases exponentially. The difference between the two cases is in the first term on the right, which reflects the fact that a heavyweight trader makes a contribution to the heavyweights even when they are all satiated, simply by adding his own consumption to that of the coalition. As $n \to \infty$, this contribution of the heavyweights cannot be approximated by a fixed finite N.

PROOF: Follows the proof of Lemma C.3. Formula (C.5) must be replaced by

(C.10) $\quad \underline{\Delta}_i^{mn} = \lambda_i^n \mu_i$ whenever $\mathcal{F}_m^n$ obtains.

In (C.8), this leads to the additional term

$$\sum_{m=N+1}^{nk} E(\underline{\Delta}_i^{mn} | \mathcal{F}_m^n) \text{ prob } \mathcal{F}_m^n = \left(n - \frac{N}{k}\right)\lambda_i^n \mu_i + o(1),$$

where, as before, the term $o(1)$ decreases exponentially in N, uniformly in n. Q.E.D.

LEMMA C.11: *If i is lightweight, then as n varies, $(\phi v_H^n)(T_i^n)$ remains bounded and bounded away[61] from 0. If i is heavyweight, then as n varies, $(\phi v_H^n)(T_i^n) - n\lambda_i^n \mu_i$ remains bounded.*

PROOF: For the boundedness part, which holds in both cases, choose $N = 0$ in Lemma C.4 (for $i \in L$) and Corollary C.10 (for $i \in H$); the result follows since the error term in those results is uniform in n. To show the boundedness away from 0 for lightweight i, note that $\underline{\Delta}_i^{mn}$ is always ≥ 0, so it is enough to show that $\underline{\Delta}_i^{2n}$, say, is bounded away from 0. Now the probability is $> 1/k^2$ that the first trader in $\mathcal{R}^n$ is of Type 1 (which is heavyweight by (C.1)), and the second trader is of Type i. In that case

$$\underline{\Delta}_i^{2n} = \lambda_1^n \max\{u_1(x): x \in X_1 \cap (-X_i)\}.$$

Since 0 does not satiate 1, and is in the interior of both X_1 and $-X_i$ (see (2.3)), it follows that the max on the right is a fixed positive number, say $\zeta > 0$. The λ_i^n are all positive by definition, and since 1 is the "heaviest" type, they do not tend to 0; therefore they have a positive minimum, say $\beta > 0$. It follows that

$$(\phi v_H^n)(T_i^n) \geq E\underline{\Delta}_i^{2n} > \beta\zeta/k^2 > 0. \qquad \text{Q.E.D.}$$

---

[61] Greater than a positive constant.

PROOF OF THEOREM 12.6: W.l.o.g. restrict the convergence indicator so that $n\lambda_\ell^n$ converges or tends to infinity. By (9.23), if i is lightweight, then

(C.12) $(\phi v_H^n)(T_i^n) = n\lambda_\ell^n q \cdot x_i + o(n\lambda_\ell^n).$

By Lemma C.11, the left side is bounded away from 0, so

(C.13) $0 < \lim n\lambda_\ell^n.$

Suppose $\lim n\lambda_\ell^n = \infty$. Since the left side of (C.12) is bounded (Lemma C.11), it follows that $q \cdot x_i^\infty = 0$; this holds for all $i \in L$. If $i \in H$, then (9.24) yields

$$(\phi v_H^n)(T_i^n) - n\lambda_i^n \mu_i = n\lambda_\ell^n q \cdot x_i^\infty + o(n\lambda_\ell^n).$$

By Lemma C.11, the left side is bounded in n, and so again, since $\lim n\lambda_\ell^n = \infty$, we must have $q \cdot x_i^\infty = 0$. Thus $q \cdot x_i^\infty = 0$ both for $i \in L$ and for $i \in H$, i.e., for all i. But this implies that (q, x^∞) is a competitive equilibrium (cf. the end of Section 9), which we have ruled out (C.1). Hence $\lim n\lambda_\ell^n < \infty$, and so by (C.13),

$$0 < \lim n\lambda_\ell^n < \infty.$$

Since by Corollary 9.26, the λ_i^n have the same order of magnitude for all lightweight i, the theorem is proved. Q.E.D.

PROOF OF THEOREM 12.4: Suppose $i \in L$. By Lemma C.11, $(\phi v_H^n)(T_i^n)$ is bounded away from 0; by Theorem 12.6, proven above, $n\lambda_\ell^n$ is bounded. Hence $c_i > 0$ by (9.27). If $i \in H$, the result follows from (9.28), (9.29), and from its truth for $i \in L$. Q.E.D.

We come next to the explicit formula for the dividends, Theorem 12.9. Note that the ρ^m are i.i.d. r.v.'s, which take on each of the values $e^1, \ldots, e^k$ in R_+^k with equal probability (recall that $e^i = (0, \ldots, 0, 1, 0, \ldots, 0)$). Set

(C.14) $y^m := \rho^1 + \cdots + \rho^m.$

PROPOSITION C15: *For lightweight i,*

$$0 < \lambda_* c_i = \lim_{n \to \infty} (\phi v_H^n)(T_i^n) = \sum_{m=1}^\infty E((w_H(y^m) - w_H(y^{m-1}))\rho_i^m)$$

$$= \sum_{m=0}^\infty \frac{1}{k} E(w_H(y^m + e^i) - w_H(y^m)).$$

REMARK: In particular, it is asserted that the limit exists and is finite, and the series converge to a finite limit.

PROOF: The positivity, $0 < \lambda_* c_i$, follows from (12.7) and Theorem 12.4 (proven above). The equality of $\lambda_* c_i$ with the limit is (9.27). The equality between the two sums is straightforward. It remains only to prove that the limit equals the first sum.

First note that if $z^n \to z$, then

(C.16) $w_H^n(z) \to w_H(z).$

Let us now examine the behavior of $E\Delta_i^{mn}$ for fixed m, as $n \to \infty$. Each of the random variables y^{mn} and y^m has precisely k^m possible values; as $n \to \infty$, the distribution of y^{mn} approaches that of y^m. Noting that $\pi_i^{mn} = y_i^{mn} - y_i^{(m-1)n}$ and $\rho_i^m = y_i^m - y_i^{m-1}$, and using (C.16), we deduce that as $m \to \infty$,

(C.17) $E\Delta_i^{mn} \to E((w_H(y^m) - w_H(y^{m-1}))\rho_i^m).$

Now let $\varepsilon > 0$ be given. By Lemma C.4, there is an N_0 such that for all $N > N_0$, and for all n,

$$\left|(\phi v_H^n)(T_i^n) - \sum_{m=1}^N E\Delta_i^{mn}\right| \leq \varepsilon.$$

Letting $n \to \infty$ on the left while keeping N fixed, and using (C.17), yields

$$\left|\lim_{n \to \infty} (\phi v_H^n)(T_i^n) - \sum_{m=1}^N E((w_H(y^m) - w_H(y^{m-1}))\rho_i^m)\right| \leq \varepsilon$$

for all $N > N_0$. Since ε was chosen arbitrarily, the proof is complete. Q.E.D.

PROOF OF THEOREM 12.9: Follows from Proposition C.15, by noting that

(C.18) $y^m = \eta(Q_m), \quad w_H(y^m) = v_H(Q^m), \quad w_H(y^m + e^i) = v_H(Q^m \cup t_i).$ Q.E.D.

COROLLARY C.19: *For all heavyweight i,*

$$\lambda_* q \cdot x_i^\infty = \lim_{n \to \infty} ((\phi v_H^n)(T_i^n) - n\lambda_i^n \mu_i) = \frac{1}{k} \sum_{m=0}^\infty E(w_H(y^m + e^i) - w_H(y^m) - \lambda_i^\infty \mu_i)$$

$$= \frac{1}{k} \sum_{m=0}^\infty E(v_H(Q_m \cup t_i) - v_H(Q_m) - \lambda_i^\infty \mu_i).$$

PROOF: The first equality is (9.24). The second equality follows as in the proof of Proposition C.15, using Corollary C.9 instead of Lemma C.4. The third equality follows from (C.18). Q.E.D.

Before proceeding, we should point out that whereas v_H is formally defined as $w_H \circ \eta$, the two represent slightly different concepts. In $v_H(S)$, the maximum is over S-allocations that need not be equal treatment; that may give different bundles to traders of the same type (cf. (4.1)). This was used already in the proof of Lemma 9.14, and will be used again in the proof of Lemma C.20 below. Because the X_i are convex and the u_i concave, the maximum is in fact achieved at equal treatment allocations, and this enables us to express v_H by $w_H \circ \eta$.

Denote by T_i^∞ the infinite pool of Type i traders, and set $T^\infty = T_1^\infty \cup \cdots \cup T_k^\infty$. Recall that t_i denotes an additional Type i trader, outside of the pool. Like before, $u_t = u_i$, $X_t = X_i$, and $\lambda_t^\infty = \lambda_i^\infty$ whenever t is of Type i.

PROPOSITION C.20: *Let $\alpha_1, \ldots, \alpha_k$ be nonnegative numbers whose sum α is ≤ 1, let $S \subset T^\infty$, and let j be such that $\lambda_j^\infty = 0$ and*

(C.21) $\quad X_j \supset \sum \alpha_i X_i.$

Then

(C.22) $\quad v_H(S \cup t_j) - v_H(S) \geq \sum \alpha_i (v_H(S \cup t_i) - v_H(S) - \lambda_i^\infty \mu_i).$

PROOF: Assume first that $\alpha = 1$. Suppose that $v_H(S \cup t_i)$ is achieved at an $(S \cup t_i)$-allocation that assigns to each t in $S \cup t_i$ the bundle x_t^i; that is, $x_t^i \in X_t$,

(C.23) $\quad \lambda_i^\infty u_i(x_{t_i}^i) + \sum_{t \in S} \lambda_t^\infty u_t(x_t^i) = v_H(S \cup t_i),$ and

(C.24) $\quad x_{t_i}^i + \sum_{t \in S} x_t^i = 0.$

Now for $t \in S$, assign to t the bundle $x_t := \sum \alpha_i x_t^i$, and to t_j, assign the bundle $x_{t_j} := \sum \alpha_i x_{t_i}^i$. The concavity of X_t yields $x_t \in X_t$ when $t \in S$, (C.21) yields $x_{t_j} \in X_j$, and (C.24) yields $\sum_{t \in S \cup t_j} x_t = 0$. Thus x is an $(S \cup t_j)$-allocation, and hence using $\lambda_j^\infty = 0$, the concavity of the u_t, and (C.23), we get

(C.25) $\quad v_H(S \cup t_j) \geq \sum_{t \in S} \lambda_t^\infty u_t(x_t) + \lambda_j^\infty u_j(x_j)$

$$= \sum_{t \in S} \lambda_t^\infty u_t(x_t) \geq \sum_{t \in S} \lambda_t^\infty \sum \alpha_i u_t(x_t^i)$$

$$= \sum \alpha_i \sum_{t \in S} \lambda_t^\infty u_t(x_t^i) = \sum \alpha_i (v_H(S \cup t_i) - \lambda_i^\infty u_i(x_{t_i}^i))$$

$$\geq \sum \alpha_i (v_H(S \cup t_i) - \lambda_i^\infty \mu_i);$$

this yields the result for $\alpha = 1$.

When $\alpha < 1$, we add an additional type, with index 0, net trade set $X_0 = \{0\}$, utility defined by $u_0(0) = 0$, and weight $\lambda_0^\infty = 0$. This violates the conditions $0 \in \text{Int } X_i$ and $0 \notin B_i$, but in the above proof for the case $\alpha = 1$, no use was made of these conditions. Setting $\alpha_0 := 1 - \alpha$, and applying the previous case to this situation, yields (C.22) in this case as well. Q.E.D.

PROOF OF THEOREM 12.5: Since $\lambda_i^\infty \mu_i = 0$ for lightweight i, the theorem follows from Theorem 12.9 (proven above), (12.7), and Proposition C.20. Q.E.D.

LEMMA C.26: *If $q \cdot x_i^\infty \leq 0$, then $\lambda_i^\infty > 0$.*

PROOF: If $\lambda_i^\infty = 0$, then by (9.30), Proposition C.15, and Corollary C.19,

$$\lambda_* q \cdot x_i^\infty = \frac{1}{k} \sum_{m=0}^\infty E(w_H(y^m + e^i) - w_H(y^m)).$$

By the superadditivity (8.4) of w_H, each term on the right is nonnegative, and as in the proof of Lemma C.11, at least one must be positive. Hence $q \cdot x_i^\infty > 0$.

REMARK C.27: Let L' be the set of i with $\lambda_i^\infty = 0$. Theorem 12.5 says that the dividends are monotonically concave over L. We can replace L by L', thus strengthening the theorem, but only at the cost of revising the formulas for the dividends. Set $H' := H \setminus L'$. For $i \in L'$, define $c_i' := q \cdot x_i^\infty$. For $i \in H'$, define c_i' analogously to the definition of c_i for $i \in H$ ((9.28) and (9.29)); in words, c_i' is the smallest c_j' among those with $j \in L'$ and $X_j \supset X_i$, when there is such a j; otherwise, it is a sufficiently large finite number. By Lemma C.26, $x_i' > 0$ for i in L', and hence for i in H' as well. From Proposition C.20 it follows that c' is monotonically concave in the net trade sets of L'. It may, incidentally, be seen that c', like c, is monotonic over all the net trade sets. Q.E.D.

LEMMA C.28: *If $\lambda_i^\infty > 0$, $\lambda_j^\infty = 0$, and $X_i \subset X_j$, then $q \cdot x_i^\infty < q \cdot x_j^\infty$.*

PROOF: By (9.30), Proposition C.15, and Corollary C.19,

$$\text{(C.29)} \quad q \cdot x_i^\infty = \frac{1}{\lambda_* k} \sum_{m=0}^{\infty} E(v_H(Q_m \cup t_i) - v_H(Q_m) - \lambda_i^\infty \mu_i),$$

$$\text{(C.30)} \quad q \cdot x_j^\infty = \frac{1}{\lambda_* k} \sum_{m=0}^{\infty} E(v_H(Q_m \cup t_j) - v_H(Q_m)).$$

By Proposition (C.20) with $\alpha_i = 1$, each term on the right of (C.29) is less than or equal to the corresponding term in (C.30). In fact it is strictly less than, because there is a positive probability that Q_m consists of Type i traders only, in which case

$$v_H(Q_m \cup t_i) - v_H(Q_m) - \lambda_i^\infty \mu_i = -\lambda_i^\infty \mu_i < 0 \le v_H(Q_m \cup t_j) - v_H(Q_m).$$

Hence $q \cdot x_i^\infty < q \cdot x_j^\infty$.

PROOF OF PROPOSITION 12.10: If i is satiated at x_i^∞, then $x_i^\infty \in B_i$, so u_i is differentiable at x_i^∞, so $\partial u_i(x_i^\infty) = \{0\}$, so $\partial \xi_i u_i(x_i^\infty) = \{0\}$, so $q = 0$ (by (9.19)), contradicting (9.20). Q.E.D.

PROOF OF PROPOSITION 12.11: Since $d = 1$, the price vector q is a real number $\ne 0$, w.l.o.g. > 0 (otherwise reflect around 0). We say that i's demand is *positive* (*negative*) if $B_i \subset R_{++}^1 (B_i \subset -R_{++}^1)$; since all dividends are positive, the bliss sets B_i of types i with negative demand are included in i's dividend budget set, and hence $x_i^\infty \in B_i \subset -R_{++}^1$. Hence all unsatiated i have positive demand. The corresponding B_i cannot intersect i's dividend budget set, and so must be to the right of (greater than) x_i^∞. Thus for each unsatiated i, there is a point y_i in B_i with $y_i > x_i^\infty$. Setting $y_i := x_i^\infty \in B_i$ for unsatiated i, we find

$$0 = \sum x_i^\infty < \sum y_i \in \sum B_i,$$

the first equality being the feasibility of x^∞. Thus the short side of the market consists precisely of types with negative demand. But for such types i, we have already seen that $x_i^\infty \in -R_{++}^1$, hence $q \cdot x_i^\infty < 0$, and hence $\lambda_i^\infty > 0$ by Lemma C.26. Q.E.D.

APPENDIX D

OVERVIEW OF THE PROOF OF THE MAIN THEOREM

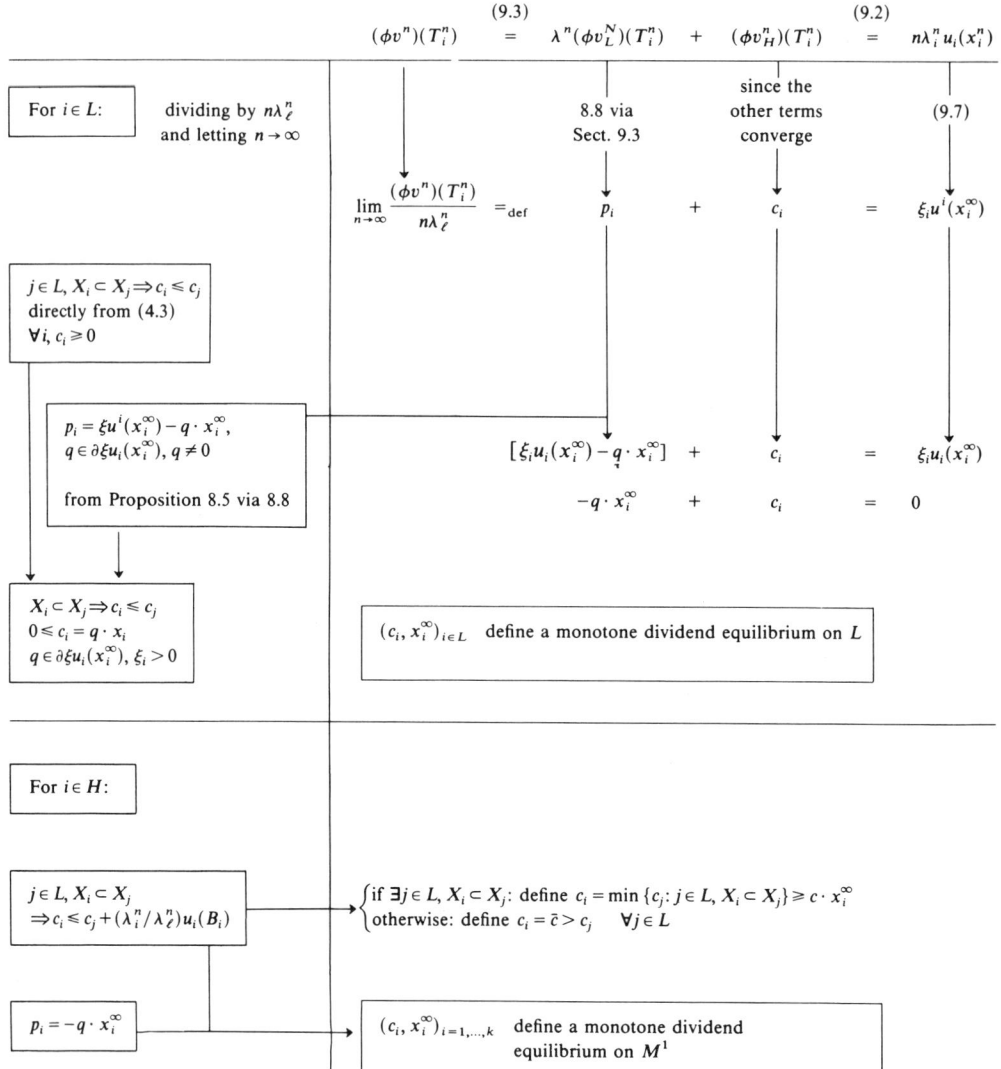

LOGICAL STRUCTURE OF THE PROOF OF THE MAIN THEOREM

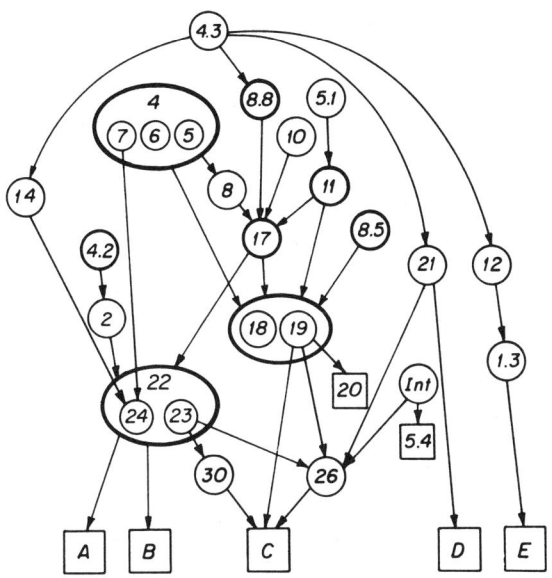

Key to flow chart

Results that form part of the conclusion are identified in a square.
Results that are grouped naturally together (e.g. different parts of the same lemma) are encircled together.
Results that are more "central" are enclosed in a heavy circle. Since everything is used, this is necessarily somewhat subjective.
Numbers without decimals refer to Section 9 (e.g., 20 means 9.20).
"Int" is the assumption that 0 is in the interior of each net trade set (see (2.3)).
Lettered results are *not* identified by letter in the text:
A. The dividends c_i are well defined.
B. x_i^∞ is in i's dividend budget set $(q \cdot x_i^\infty \le c_i)$.
C. i weakly prefers x_i^∞ to everything in his dividend budget set $(q \cdot x \le c_i \Rightarrow u_i(x_i^\infty) \ge u_i(x))$.
D. The dividends are nonnegative $(c_i \ge 0)$.
E. The dividends are monotonic in the net-trade sets $(X_i \subset X_j \Rightarrow c_i \le c_j)$.

REFERENCES

AUMANN, R. J. (1975): "Values of Markets with a Continuum of Traders," *Econometrica*, 43, 611–646 [Chapter 51].
AUMANN, R. J., AND L. SHAPLEY (1974): *Values of Non-Atomic Games*. Princeton, N.J.: Princeton University Press.
BROWN, D. J., AND A. ROBINSON (1972): "A Limit Theorem on the Cores of Large Standard Exchange Economies," *Proceedings of the National Academy of Sciences of the U.S.A.*, 69, 1258–1260.
CHAMPSAUR, P. (1975): "Cooperation versus Competition," *Journal of Economic Theory*, 11, 394–417.
DEBREU, G., AND H. SCARF (1963): "A Limit Theorem on the Core of an Economy," *International Economic Review*, 4, 225–246.
DRAZEN, A. (1980): "Recent Developments in Macroeconomic Disequilibrium Theory," *Econometrica*, 48, 283–306.
DRÈZE, J. H. (1975): "Existence of an Exchange Equilibrium under Price Rigidities," *International Economic Review*, 16, 301–320.
DRÈZE, J. H., AND H. MÜLLER (1980): "Optimality Properties of Rationing Schemes," *Journal of Economic Theory*, 23, 150–159.

HART, S. (1977a): "Asymptotic Value of Games with a Continuum of Players," *Journal of Mathematical Economics*, 4, 57-80.
—— (1977b): "Values of Non-Differentiable Markets with a Continuum of Traders," *Journal of Mathematical Economics*, 4, 103-116.
—— (1980): "Measure-Based Values of Market Games," *Mathematics of Operations Research*, 5, 197-228.
HILDENBRAND, W. (1974): *Core and Equilibria of a Large Economy*. Princeton, N.J.: Princeton University Press.
—— (1982): "Core of an Economy" in *Handbook of Mathematical Economics*, Volume II, ed. by K. J. Arrow and M. D. Intriligator. Amsterdam: North-Holland Publishing Company, Ch. 18, pp. 831-877.
KANNAI, Y. (1966): "Values of Games with a Continuum of Players," *Israel Journal of Mathematics*, 4, 54-58.
MALINVAUD, E., AND Y. YOUNÈS (1977): "Une nouvelle formulation générale pour l'étude de certains fondements microéconomiques de la macroéconomie," *Cahiers du Séminaire d'Econométrie*, 18, 63-112.
MAS-COLELL, A. (1977): "Competitive and Value Allocations of Large Exchange Economies," *Journal of Economic Theory*, 14, 419-438.
MASCHLER, M., AND B. PELEG (1966): "A Characterization, Existence Proof and Dimension Bounds for the Kernel of a Game," *Pacific Journal of Mathematics*, 18, 289-328.
MERTENS, J.-F. (1980): "Values and Derivatives," *Mathematics of Operations Research*, 5, 523-552.
—— (1988): "The Shapley Value in the Non-Differentiable Case," *International Journal of Game Theory*, 17, 1-65.
NEYMAN, A., AND Y. TAUMAN (1979): "The Partition Value," *Mathematics of Operations Research*, 4, 236-264.
ROCKAFELLAR, R. T. (1970): *Convex Analysis*. Princeton, N.J.: Princeton University Press.
ROTH, A. E. (1977): "The Shapley Value as a von Neumann-Morgenstern Utility," *Econometrica*, 45, 657-664.
SHAPLEY, L. (1953): "A Value for n-Person Games" in *Contributions to the Theory of Games*, Volume II, ed. by H. W. Kuhn and A. W. Tucker. Princeton, N.J.: Princeton University Press, pp. 307-317.
—— (1962): "Values of Games with Infinitely Many Players" in *Recent Advances in Game Theory* (Proceedings of a Princeton University Conference, October 4-6, 1961), ed. by M. Maschler. Philadelphia: Ivy Curtis Press, pp. 113-118.
—— (1969): "Utility Comparisons and the Theory of Games," in *La Décision*, ed. by G. Th. Guilbaud. Paris: Editions du CNRS, pp. 251-263.
SHAPLEY, L., AND M. SHUBIK (1969): "Pure Competition, Coalitional Power and Fair Division," *International Economic Review*, 10, 337-362.
TAUMAN, Y. (1981): "Value on a Class of Non-Differentiable Market Games," *International Journal of Game Theory*, 10, 155-162.

58 Economic Applications of the Shapley Value

1 Introduction

In the previous chapters, the concept of value was presented in a very abstract way. It has proved, however, to be a powerful tool in modelling some economic problems. In fact, since the Shapley value can be interpreted in terms of "marginal worth," it is closely related to traditional economic ideas. To illustrate this, we first present the Value Equivalence Theorem—the analogue of the Core Equivalence Theorem. Though other important applications exist, we then focus on three applications of the value concept to economic models other than the general equilibrium model. Each of them describes a way of departing from the market model environment. The first two are economic-political models dealing with taxation. Taxation has (at least) two purposes: redistribution and the raising of funds to finance public goods. The classical literature assumes that a benevolent government takes decisions so as to maximize some social utility function. On the contrary, analysing the government as subject to the influence of those who elected it brings new light on both aspects. Value appears to be a natural tool to deal with the voting games that are part of the two corresponding models. In the last section, we deal with economies with fixed prices.

All along this chapter, we will try to provide intuitions on why one would expect the results to hold (or not) rather than to give detailed proofs. For the real proofs, the reader is referred to the original papers.

2 The Value Equivalence Theorem

First, we present the Value Equivalence Theorem (see Hart [1994] and Aumann [1975]) similar to the Core Equivalence Theorem (see Allen & Sorin [1994] and Aumann [1964]); under certain assumptions, *in a competitive economy, every value allocation is a Walrasian allocation, and conversely.*

Define a competitive economy as $(T, \ell, e, (u_t)_{t \in T})$ where $T = [0, 1]$ stands for the set of traders, endowed with Lebesgue measure μ. $\mathbb{R}_+^\ell$ is the commodity space, $e : T \to \mathbb{R}_+^\ell$, integrable, is the initial allocation. An allocation is a (measurable) map $\mathbf{x} : T \to \mathbb{R}_+^\ell$ such that $\int_T \mathbf{x}_t \mu(dt) = \int_T e_t \mu(dt)$. $u_t : \mathbb{R}_+^\ell \to \mathbb{R}$ is the utility function of t.

An individual trader is best viewed as an "infinitesimal subset" dt of T. Hence, $e_t \mu(dt)$ is trader dt's initial endowment, and $\mathbf{x}_t \mu(dt)$ denotes what

---

This chapter originally appeared in *Game Theoretic Methods in General Equilibrium Analysis*, edited by J.-F. Mertens and S. Sorin, Kluwer Academic Publishers, Dordrecht, 1994, pp. 121–133. Reprinted with permission.

he gets under an allocation **x**, while $u_t(\mathbf{x}_t)\mu(dt)$ is the utility he derives from it. It is assumed that $u_t(0) = 0$.

Value Allocations

Consider a weight function $\lambda : T \to \mathbb{R}_+$ (integrable). The worth $v_\lambda(S)$ of coalition $S \subset T$ with respect to λ is the maximum "weighted total utility" it can get on its own; *i.e.* by properly reallocating the total initial endowment of members of S among themselves. In the previous setting,

$$v_\lambda(S) = \max\left\{\int_S \lambda_t u_t(\mathbf{x}_t)\mu(dt) \text{ such that } \int_S \mathbf{x}_t \mu(dt) = \int_S e_t \mu(dt)\right\}$$

The value of a trader is his average marginal contribution to the coalitions to which he belongs, where "average" means expectation with respect to a distribution induced by a random order of the players. Thus, the value of trader dt is

$$(\varphi v_\lambda)(dt) = E[v_\lambda(S_{dt} \cup dt) - v_\lambda(S_{dt})],$$

where S_{dt} is the set of players "before dt" in a random order (see Aumann [1994] and Neyman [1994]).

An allocation **x** is called a *value allocation* with respect to λ if

$$(\varphi v_\lambda)(dt) = \lambda_t u_t(\mathbf{x}_t)\mu(dt).$$

We want to prove that **x** is also a Walrasian allocation.

First, remark that, using $u'_t(x)$ for the gradient of u_t at x,

$$\forall (t_1, t_2) \in T^2, \lambda_{t_1} u'_{t_1}(\mathbf{x}_{t_1}) = \lambda_{t_2} u'_{t_2}(\mathbf{x}_{t_2});$$

otherwise society could always profitably reallocate its initial endowment. This would contradict the fact that **x** is a value allocation. Call p this common value.

Second, the same applies to S_{dt}. Further S_{dt} can be considered as a perfect sample of T, in the sense that in S_{dt} the distribution of players corresponds to that in T. Therefore, the common value of the gradients for S_{dt} is also p.

Hence, dt's contribution to S_{dt} is twofold: dt's weighted utility and the change in other traders' aggregate utility. Under the new optimal allocation of the initial endowment of $S_{dt} \cup dt$, dt gets $\mathbf{x}_t \mu(dt)$ and therefore its weighted utility is $\lambda_t u_t(\mathbf{x}_t)\mu(dt)$. Hence, $e_t \mu(dt) - \mathbf{x}_t \mu(dt)$ has to be distributed among the traders in S_{dt}. Their increase in utility is then $p \cdot [e_t - \mathbf{x}_t]\mu(dt)$.

Hence,

$$(\varphi v_\lambda)(dt) = p \cdot [e_t - \mathbf{x}_t]\mu(dt) + \lambda_t u_t(\mathbf{x}_t)\mu(dt)$$

and

$$p \cdot [e_t - \mathbf{x}_t] = 0$$

Now, $(\mathbf{x}_t, u_t(\mathbf{x}_t))$ is on the boundary of the convex hull of the graph of u_t. The idea is that otherwise, it would be possible to split trader dt (meaning his endowment) into several players so that the weighted sum of their utilities would be greater than dt's utility. This transformation taking place inside T, this would be contradictory to $\mathbf{x}$ being a value allocation. This result and the equality of gradients lead to:

$$\forall x \in \mathbb{R}_+^\ell, u_t(\mathbf{x}_t) - u_t(x) \geqslant p \cdot (\mathbf{x}_t - x).$$

Hence, $\mathbf{x}_t$ maximizes $u_t(x)$ under the constraint $p \cdot x \leqslant p \cdot \mathbf{x}_t$, and hence under the constraint $p \cdot x \leqslant p \cdot e(t)$.

So $\mathbf{x}_t$ is a Walrasian allocation corresponding to the price system p.

3 Taxation and Redistribution

Before 1977, perhaps the most fundamental element in the theory of the public sector was that the government was regarded as an exogenous benevolent economic agent who tried to maximize some social utility, usually the sum of individual utilities (see Arrow & Kurz [1970]). On the other hand, within a democratic system, a person can vote and try to influence the government's decision according to its own utility. This section, based on Aumann & Kurz [1977], aims at taking this idea into account in what concerns taxation and redistribution. It introduces a model in which each agent's power is reflected in two spheres: politics and economics. This *Income Redistribution Game* is very simple: each agent has an initial endowment and a utility function, a tax and redistribution policy is decided by majority voting but every agent can destroy part or all of his endowment. The idea is that while any majority can expropriate the corresponding minority, anyone can, for example, decide not to work so that the others get nothing from expropriating him. Though he does not feel better in this case (no utility of leisure is assumed), he can use this as a threat to make the majority compromise. This will influence the nature of the majority coalition formed and the tax policy it enforces.

We start from the previous model but with a single commodity. Since we want to accomodate for threats and for non-transferable utility among agents, we are using Harsanyi-Shapley NTU value. Suppose a weight function λ has been fixed somehow. Then the aggregate utility of T is

$$v_\lambda(T) = \max\left\{ \int_T \lambda_t u_t(\mathbf{x}_t)\mu(dt) \text{ such that } \int_T \mathbf{x}_t \mu(dt) = \int_T e_t \mu(dt) \right\}$$

Suppose now that two complementary coalitions S and $T\setminus S$ have formed. Think of $v_\lambda(S)$ as being the aggregate utility of S if it forms and bargains against $T\setminus S$. As in Nash [1953], suppose that the two parties can commit to carry out *threat strategies* if no satisfactory agreement is reached. If under these strategies S and $T\setminus S$ get respectively f and g, the two parties are bargaining for $v_\lambda(T) - f - g$ and, under the symmetry assumption, this is split evenly. Hence, S gets $1/2(v_\lambda(T) + f - g)$ and $T\setminus S$ gets $1/2(v_\lambda(T) + g - f)$ so that the derived game between S and $T\setminus S$ is a constant-sum game.

The optimal threat strategy for the majority coalition is 100% tax since it can at least ensure its own endowment while the optimal threat strategy for the minority is precisely to destroy all of its endowment so that the majority cannot ensure more than its endowment. Hence, the reduced game value is $q(S) =$

$$\begin{cases} \max\{\int_S \lambda_t u_t(\mathbf{x}_t)\mu(dt) \text{ s.t. } \int_S \mathbf{x}_t \mu(dt) = \int_S e_t \mu(dt)\} & \text{if } \mu(S) > \frac{1}{2} \\ 0 & \mu(S) < \frac{1}{2} \end{cases}$$

and we have

$$v_\lambda(S) = \tfrac{1}{2}[q_\lambda(T) + q_\lambda(S) - q_\lambda(T\setminus S)]$$

It can be shown that

$$(\varphi v)(dt) = (\varphi q)(dt) = E[q(S_{dt} \cup dt) - q(S_{dt})]$$

As in the previous section, S_{dt} is almost certainly a perfect sample of T. We can then use the self-explanatory notation $S_{dt} = \theta T$, with $\theta \in [0,1]$ being the size of the sample. So θ can also be viewed as the random time when dt enters the room, thus is uniformly distributed. Hence,

$$(\varphi q_\lambda)(dt) = \int_0^1 [q(\theta T \cup dt) - q(\theta T)] d\theta$$

We may have three different situations:

1. θT is a majority, $\theta T \cup dt$ is still majority;
2. $\theta T \cup dt$ is minority (and so is θT);
3. θT is minority, and $\theta T \cup dt$ is majority, i.e. dt is pivotal.

Suppose $q_\lambda(T)$ is achieved at $\mathbf{x}$. Then, for each case, we have dt's expected contribution—case 1 being as in the last section:

1. $\frac{1}{2}\lambda_t[u_t(\mathbf{x}_t) - u'_t(\mathbf{x}_t) \cdot (\mathbf{x}_t - e_t)]\mu(dt)$ $[= \int_{1/2}^{1}[q(\theta T \cup dt) - q(\theta T)]d\theta]$
2. $\frac{1}{2}(0 - 0)$ $[= \int_{0}^{(1/2)-\mu(dt)}(0)d\theta]$
3. $\frac{1}{2}q_\lambda(T)\mu(dt)$ $[= \int_{(1/2)-\mu(dt)}^{1/2}[q(\frac{1}{2}T) - 0]dt]$

By definition, $\varphi q_\lambda(dt) = \lambda_t u_t(\mathbf{x}_t)\mu(dt)$ and, as in the last section, $\lambda_t u'_t(\mathbf{x}_t)$ is constant in t, say p. We get thus

$$\lambda_t u_t(\mathbf{x}_t) = q_\lambda(T) - p \cdot (\mathbf{x}_t - e_t) \; \mu\text{-a.e., or (single commodity ...):}$$

$$e_t - \mathbf{x}_t = \frac{u_t(\mathbf{x}_t)}{u'_t(\mathbf{x}_t)} - C \quad \text{where } C = [q_\lambda(T)]/p \text{ is a constant.}$$

It can be shown that satisfying a condition of the kind

$$e_t - \mathbf{x}_t = \frac{u_t(\mathbf{x}_t)}{u'_t(\mathbf{x}_t)} - c$$

is a necessary and sufficient condition for $\mathbf{x}$ to be a value allocation (meaning that λ is allowed to vary). Moreover, there exists a single solution $(\mathbf{x}, c)$ and c is positive. This constitutes the main result in Aumann & Kurz [1977].

Let us now look for the economic interpretation. The left-hand side of the equality is the (signed) tax on t. Notice that under the assumption that utility functions u_t are increasing and concave, e_t is increasing in $\mathbf{x}_t$ or, more intuitively, $\mathbf{x}_t$ is increasing in e_t—but with slope $< 1/2$. That is to say, marginal tax rates are between 50% and 100%.

Despite that no explicit uncertainty was introduced in the model, call *fear of ruin* the ratio $u_t(\mathbf{x}_t)/u'_t(\mathbf{x}_t)$. In fact, consider the reciprocal u'/u. Suppose that some player is ready to play a game in which with probability $(1 - p)$ his initial fortune x is increased by a small amount ε and with probability p he is ruined and his fortune is 0. Certainly, if he is indifferent between playing the game or not, p/ε is a measure of his boldness (at x). Indifference implies:

$$u(x) = p \cdot 0 + (1 - p) \cdot u(x + \varepsilon)$$

Hence, when ε goes to zero, p/ε goes to $u'(x)/u(x)$.

Thus, the tax equals the fear of ruin at the net income, less a constant tax credit.

Example Let us end this section with an example where we can explicitly calculate the tax policy. Consider $T = [0, 1]$ with identical traders

$$u_t(x) = u(x) = x^\alpha \quad 0 < \alpha \leqslant 1$$

The fear of ruin is

$$\frac{u(x)}{u'(x)} = \frac{x^\alpha}{\alpha x^{\alpha-1}} = \frac{x}{\alpha}$$

By integrating we get

$$C = \int_T \frac{u(\mathbf{x})}{u'(\mathbf{x})} + 0 = \int_T \frac{\mathbf{x}}{\alpha} = \frac{\int_T e}{\alpha}$$

since $\int_T e = \int_T \mathbf{x}$. Hence,

$$\mathbf{x}_t = \frac{\alpha}{1+\alpha} \cdot e_t + \frac{\int_T e}{1+\alpha}$$

4 Public Goods without Exclusion

The second purpose of taxation is to raise funds to finance the production of public goods. As in the case of redistribution, considering government as subject to the influence of its electors instead of benevolently maximizing a given social welfare function sheds new light on the subject. A *Public Goods Economy* is modelled, where a continuum of agents, endowed with resources and voting rights, take part in the production of non-exclusive public goods. More precisely, when a coalition forms, it chooses one strategy amongst the available, which together with the complementary coalition's choice determines which bundle of public goods will be produced. A natural question is the dependence of the outcome of the game on the distribution of voting rights, which should at first sight exert a major influence. But the Harsanyi-Shapley NTU Value leads to the surprising new and pessimistic result that the distribution of voting rights has (little or) nothing to do with the choice. Aside from its proof, this result is reinforced by an economic argument based on the implicit price of a vote.

A non-atomic public goods economy is modelled by the set of agents $T = [0, 1]$, the space $\mathbb{R}_+^\ell$ of resources, and the public goods space $\mathbb{R}_+^m$. G is a correspondence from $\mathbb{R}_+^\ell$ to $\mathbb{R}_+^m$ representing the production function. G is supposed to take compact and nonempty values. $u_t : \mathbb{R}_+^m \to \mathbb{R}$ is t's utility function. v is a non-atomic voting measure with $v(T) = 1$.

Every person has resources that can be used to produce public goods. The voting measure v is not necessarily identical to the distribution of population μ. For example, it is possible that noncitizens do not have the right to vote.

A vote takes place and the majority decides which public good will be produced. The minority is not entitled to produce public goods but, as in the previous model, it has the right to destroy its resources. The important thing is that, once the public good has been produced, the minority may not be excluded from consuming it.

THEOREM 1 *In the voting game, the value outcomes are independent of the voting measure v.*

Consider an example with 2 public goods, TV and libraries. There are two kinds of people in two equally weighted sets, the ones fond of TV programmes and the others fond of books. Assume further that TV fans possess all the voting rights: you might expect TV programmes to be the only leisure available after voting, but it happens to be both TV and books in an equal manner! Whatever the voting rights, the same bundle of public goods will be chosen.

Sketch of Proof We first describe the definition of the Harsanyi-Shapley NTU value of a game, already used in the last section. As there, for a fixed weight function λ, every coalition S announces a threat strategy z_S it would carry out in case negotiations would break down with $T\setminus S$. Together with that of the complementary coalition, it yields S a total payoff: $V(S) = U(z_S, z_{T\setminus S})(S)$. After those announcements, players are thus in a fixed threat, TU game, with as solution the Shapley value. Players want thus the threats of the different coalitions they are members of to be chosen such as to maximize their own final payoff according to this Shapley value. This can be shown to imply that all members of any given coalition S unanimously want to maximize $H(S) = V(S) - V(T\setminus S)$. Since the same holds for $T\setminus S$, the *optimal threat strategies* z_S^* and $z_{T\setminus S}^*$ are the saddle points of the two-person zero-sum game $H(S)$. Let $q_\lambda(S) = U(z_S^*, z_{T\setminus S}^*)(S)$ be the total payoff to S when both S and $T\setminus S$ carry their optimal threat out. Define $w_\lambda(S) = q_\lambda(S) - q_\lambda(T\setminus S) = H(S)(z_S^*, z_{T\setminus S}^*)$. $w_\lambda(S)$ measures the bargaining power of S (its ability to threaten). Define also $v_\lambda(S) = 1/2(w_\lambda(S) + w_\lambda(T))$. As argued in the last section, $v_\lambda(S)$ is the total utility S can expect to result from an efficient compromise with $T\setminus S$. Observe that $\varphi v_\lambda = \varphi q_\lambda$ (since the difference is a game where every coalition gets the same as its complement), but whereas q_λ might depend on the particular choices of optimal threats z_S^*, v_λ no longer does. The function $v_\lambda(\cdot)$ is the Harsanyi coalitional form of the game with weight function $\lambda(\cdot)$, and we define a *value allocation* as a bundle y achieving the Harsanyi-Shapley NTU value:

$$\varphi v_\lambda(S) = \int_S \lambda_t u_t(y) \mu(dt) \text{ for every } S.$$

We know

$$(\varphi v^v)(dt) = \int_0^1 [v(\theta T \cup dt) - v(\theta T)]d\theta$$

We want to compare this expression for two different voting measures v and ξ.

$$(\varphi v^v - \varphi v^\xi)(dt) = \int_0^1 [(v^v - v^\xi)(\theta T \cup dt) - (v^v - v^\xi)(\theta T)]d\theta$$

We call a coalition S *even* if S is either a majority under both v and ξ or a minority under both v and ξ. If S is even, so is $T \setminus S$ and their strategic options are the same under v and ξ: $v^v(S) - v^\xi(S) = 0$. Every perfect sample of the whole population θT is even—it is determined by its size. For $\theta T \cup dt$, it is even if $\theta > 1/2$ or $\theta < 1/2 - \delta$ [with $\delta = \max(v(dt), \xi(dt))$]. The previous difference thus amounts to:

$$(\varphi v^v - \varphi v^\xi)(dt) = \int_{(1/2)-\delta}^{1/2} [(v^v - v^\xi)(\theta T \cup dt)]d\theta$$

$$= \frac{1}{2}\int_{(1/2)-\delta}^{1/2} [(w^v - w^\xi)(\theta T \cup dt)]d\theta$$

If $w^v(\theta T \cup dt)$ is achieved at the outcome y, then, by additivity of the integral, and homogeneity:

$$w^v(\theta T \cup dt) = H^y(\theta T \cup dt) = U^y(\theta T) + 2U^y(dt) - U^y((1-\theta)T)$$

$$= (2\theta - 1)U^y(T) + 2\lambda_t u_t(y)\mu(dt).$$

$(2\theta - 1)$ as well as $\mu(dt)$ are infinitesimal. Independently of the voting measure v, we have shown that in the relevant range ($\theta \in [1/2 - \delta, 1/2]$), $w^v(\theta T \cup dt)$ is infinitesimal: the idea is that, under those circumstances, both the coalition and its complement resemble $1/2T$, thus are close to each other, and whatever the outcome, they enjoy the same utility derived from the *common* consumption of the same public good. Nobody enjoys a real bargaining advantage, and the efficient compromise induced by the Shapley value leads to equal treatment. Going back to more technical arguments, assume $u_t(y) \leq K$ for all feasible y and all t.

Then $U^y(T) \leq K \int \lambda_t \mu(dt)$.

Given any coalition S, and $\delta > 0$, partition S into $S_1 \cup \cdots \cup S_n$, with $(v + \xi)(S_i) \leq \delta$, and $n \leq 2/\delta$.

Then we obtain $\varphi v^y(S) - \varphi v^\xi(S) =$

$$\frac{1}{2} \sum_{i=1}^{n} \int_{(1/2)-\delta}^{1/2} [(2\theta - 1)(U^{y_i^y}(T) - U^{y_i^\xi}(T))$$

$$+ 2 \int_{S_i} \lambda_t (u_t(y_i^y) - u_t(y_i^\xi)) \mu(dt)] d\theta$$

$$\leqslant \frac{1}{2} \cdot n \cdot \left| \int_{(1/2)-\delta}^{1/2} (2\theta - 1) d\theta \right| \cdot (2K) \int \lambda_t \mu(dt) + \frac{1}{2} \cdot 4K\delta \int_S \lambda_t \mu(dt)$$

$$\leqslant 2K\delta \left[\int_T \lambda_t \mu(dt) + \int_S \lambda_t \mu(dt) \right] \leqslant 4K\delta \int_T \lambda_t \mu(dt).$$

This being true for all δ, we obtain $\varphi v^y(S) = \varphi v^\xi(S)$.

This being true for any weight function λ, the result follows. □

Though counter-intuitive, this result might have been guessed from a similar analysis conducted in a transferable-utility (TU) context: we allow agents to trade their votes for money. The outcome of such a TU-game is made of a public goods vector and of a side-payments vector. In the book vs TV game, it can be derived easily from the following conditions:

1. It is Pareto-optimal.

2. Under sensible assumptions, the only Pareto-optimal situation involves a production of $(1/2, 1/2)$ with the accompanying schedule of side-payments from book-lovers to TV-lovers.

3. To have this outcome approved, as book fans need only 50% of the vote and thus can play TV-fans off against each other, they drive the price of a vote down to zero and can achieve the desired outcome without effective payments.

In the case of non-exclusive public goods, every agent endowed with a voting right is potentially a free-rider: he believes his personal vote not to influence the final outcome, which he may like or dislike, and is thus ready to sell it even at a low price. This is wrong in the case of redistribution or more generally of exclusive public goods, where we assumed, as here, voting not to be secret, and where the identify of the voters is crucial in eventually determining everyone's payoff.

There remains to stress the connection between the TU and NTU situations: the TU-games that we obtain corresponding to the latter are similarly public good economies, but with in addition a single desirable private good available for transfers.

5 Economies with Fixed Prices

5.1 Introduction

In economies with fixed prices all trading must take place at exogenously given prices, which determine (together with the positive orthant) a new consumption set—the net trade set—for each trader. This model has been used to describe market failures such as unemployment. In general, price rigidities prevent market clearance: a trader's consumption set is exogenous and, under the standard assumptions, his utility function has an absolute maximum, a satiation point, generally in the interior of the consumption set; it may be the case that for any price vector, at least one of the trader's utility functions is satiated; this trader uses less than the maximum budget available to him, creating a total budget excess.

Example Consider a fixed price exchange economy with one commodity and two traders. Let their net trade sets be $X_1 = X_2 = [-5, +5]$ and their utility functions be $u_1(x) = -(x-1)^2$ and $u_2(x) = -(x+2)^2$. The satiation points being respectively 1 and -2, if $p > 0$ then $x_1^* = 0$ and $x_2^* = -2$. Hence, the market does not clear. Similarly if $p < 0$ or if $p = 0$.

This has suggested a generalization of the equilibrium concept in the general class of markets with satiation: the total budget excess is divided among all the traders, as dividends, so that supply matches demand. However, Drèze & Müller [1980] extended the First Theorem of Welfare Economics to this equilibrium concept, proving it to be too broad: with appropriate dividends, one can obtain any Pareto-optimum.

In this respect, the Shapley value leads to more specific results: the income allocated to a trader depends only and monotonically on his trading opportunities and not on his utility function! This will be formally stated. Then, a sketch of proof will be given. A formulation in the particular context of fixed price economies will then be presented.

5.2 Dividend Equilibria

Define a market with satiation as $M^1 = (T; \ell; (X_t)_{t \in T}; (u_t)_{t \in T})$, where $T = \{1, \ldots, k\}$ is a finite set of traders, $\mathbb{R}^\ell$ is the space of commodities, $X_t \subset \mathbb{R}^\ell$ is trader t's net trade set, supposed to be compact, convex, with nonempty interior and containing 0, and u_t is trader t's utility function, assumed concave and continuous on X_t.

A price vector is any element in $\mathbb{R}^\ell$.

Let $B_t = \{x \in X_t | u_t(x) = \max_{y \in X_t} u_t(y)\}$ be the set of satiation points of trader t. B_t is nonempty *i.e.* every trader has at least one satiation point. For simplicity, traders such that $0 \in B_t$ may be taken out of the economy. They are fully satisfied with their initial endowment. Thus we will suppose that $\forall t, 0 \notin B_t$.

An *allocation* is a vector $\mathbf{x} \in \prod_{t \in T} X_t$ such that $\sum_{t \in T} \mathbf{x}_t = 0$.

As noted above, competitive equilibria may fail to exist since, whatever the price vector, a trader may well refuse to make use of his entire budget, thus preventing market clearance. The idea of dividends is to let the other traders use the excess budget.

A *dividend* is a vector $c \in \mathbb{R}^k$. A *dividend equilibrium* is a triplet constituted of a price q, a dividend c and an allocation $\mathbf{x}$ such that, for all t, $\mathbf{x}_t$ maximizes $u_t(x)$ on X_t sub $q \cdot x \leqslant c_t$.

5.3 Value Allocations

This is the "finite" version of the definition in section 2.

A *comparison vector* is a non-zero vector $\lambda \in \mathbb{R}_+^k$. For each λ and each coalition $S \subset T$, the *worth of S according to λ* is

$$v_\lambda(S) = \max\left\{\sum_{t \in S} \lambda_t u_t(\mathbf{x}_t) \text{ s.t. } \sum_{t \in S} \mathbf{x}_t = 0 \text{ and } \forall t \in S, \mathbf{x}_t \in X_t\right\}$$

$v_\lambda(S)$ is the maximum total utility that coalition S can get by internal redistribution when its members have weights λ_t.

An allocation is called a *value allocation* if there exists a comparison vector λ such that $\lambda_t u_t(\mathbf{x}_t) = \varphi v_\lambda(t)$ where φv_λ is the Shapley value of the game v_λ.

5.4 The Main Result

M^n, the n-fold replica of market M^1, is the market with satiation where every agent of M^1 has n twins. Formally stated:

$T^n = \cup_{i \in T} T_i^n$ — set of nk traders.

$\forall i \in T, |T_i^n| = n$ — there are n traders of type i.

$\forall t \in T_i^n, u_t = u_i$ and $X_t = X_i$.

ELBOW ROOM ASSUMPTION $\quad \forall J \subset \{1, \ldots, k\}$

$$0 \notin \text{bd}\left[\sum_{i \in J} B_i + \sum_{i \notin J} X_i\right]$$

To put this in words, if it is possible to satiate simultaneously all traders in any J then it is also possible to do so when they are restricted to the relative interior of their satiation sets, and the others to that of their net trade sets. Note that since the right-hand side is the boundary of a convex subset of $\mathbb{R}^\ell$, its dimension is at most $(\ell - 1)$. Since the possible J are finite in number, the assumption holds for all but an $(\ell - 1)$-dimensional set of total endowments. In that sense, it is generic.

An allocation $\hat{\mathbf{x}}$ of M^n is an *equal treatment* allocation if traders of the same type are assigned the same net trade. Trivially, there is then a corresponding allocation $\mathbf{x}$ in M^1.

THEOREM 2 *Consider a sequence* $(\mathbf{x}^n)_{n \in N}$ *where* $\mathbf{x}^n$ *is an allocation corresponding to an equal treatment value allocation in* M^n. *Let* $\mathbf{x}^\infty$ *be a limit of a subsequence of* $(\mathbf{x}^n)_{n \in N}$. *Then, there is a dividend (vector)* c *and a price vector* q *such that* $(q, c, \mathbf{x}^\infty)$ *is a dividend equilibrium where*

c *is nonnegative* i.e. $\forall i, c_i \geqslant 0$
c *is monotonic* i.e. $\forall i, X_i \subset X_j \Rightarrow c_i \leqslant c_j$.

What gives substance to the theorem is the following existence result.

PROPOSITION 1 *There exists an equal treatment value allocation for every* M^n.

Sketch of Proof of the Theorem   This proof is very informal. To make things simpler, suppose that:

- $\sum_i \lambda_i^n = 1$ (normalization);
- $\forall i, \mathbf{x}_i^n$ and λ_i^n converge;
- $\forall i, \mathbf{x}_i^n$ and $\mathbf{x}_i^\infty$ are in $\text{int}(X_{it})$;
- $\forall i, u_i$ is strictly concave and continuously differentiable on X_i.

Call *lightweight* those types i such that $\lim_{n \to \infty} \lambda_i^n = 0$ and *heavyweight* the rest. Suppose that all lightweight types' weights converge to 0 at the same speed.

We have

$$\forall (i,j), \forall n, \lambda_i^n u_i'(\mathbf{x}_i^n) = \lambda_j^n u_j'(\mathbf{x}_j^n) \; (:= q^n)$$

In the limit,

$$\forall (i,j), \lambda_i^\infty u_i'(\mathbf{x}_i^\infty) = \lambda_j^\infty u_j'(\mathbf{x}_j^\infty) \; (:= q^\infty)$$

Now, consider the contribution of a trader to a coalition S.

If S is "large enough," it is very likely to be a good sample of the population T^n. Thus an optimal allocation for S is approximately the optimal $\mathbf{x}^n$ for T^n. The first term of the new trader's contribution is what he gets for himself. Since he does not change the optimal allocation by much this is $\lambda_i^n u_i(\mathbf{x}_i^n)$. The second term is his influence on the other traders' utility. Since the net trade must equal zero and all gradients are equal, this is approximately $-q^n \cdot x_i^n$. Thus, the contribution is approximately $\Delta = \lambda_i^n u_i(x_i^n) - q^n \cdot x_i^n$.

If S is "too small," the previous considerations do not hold. However, a new trader's contribution to a small coalition is uniformly bounded. This follows from the continuous differentiability of utilities on compact net trade sets. Moreover, the probability P^n of S being a small coalition goes to zero as n goes to infinity. Denote by δ_i^n the expected contribution of t conditional on the coalition being small.

Now we have (very roughly):

$$\varphi v_\lambda^n(t) \approx (1 - P^n)\Delta + P^n \delta_i^n$$

Since $\varphi v_\lambda^n(t) = \lambda_i^n u_i(x_i^n)$ we have:

$$\lambda_i^n u_i(x_i^n) \approx (1 - P^n)\Delta + P^n \delta_i^n$$

Hence,

$$q^n \cdot x_i^n \approx \frac{P^n}{1 - P^n}(\delta_i^n - \lambda_i^n u_i(x_i^n)) \qquad (A)$$

Note that $P^n/(1 - P^n) \to 0$.

Suppose there is no lightweight type. If simultaneous satiation of all traders is possible, then this is the Shapley value for all n sufficiently large, since then $\lambda_i^n > 0 \; \forall i$. It is clear that the theorem holds then, for any price system q, and $c_i = C$ sufficiently large. Otherwise, for at least some $i, u_i'(x_i^\infty) \neq 0$. Hence, because of the equality of the gradients, $\forall j, u_j'(x_j^\infty) \neq 0$. Hence $q^\infty \neq 0$ and for all i, the gradient of u_i at x^∞ is in the direction of q^∞. Going to the limit in (A) gives $q^\infty \cdot x_i^\infty = 0$. Hence, x_i^∞ maximizes $u_i(x)$ on X_i sub $q^\infty \cdot x \leq 0$.

Hence, it is an ordinary competitive equilibrium and trivially a dividend equilibrium.

Suppose now that type i is lightweight. We have $q^\infty = 0$. Hence, for any heavyweight type j, $u_j'(x_j^\infty) = 0$ that is to say x_j^∞ satiates j. Before letting n go to infinity, divide equality (A) by $\|q^n\|$. We shall see that δ's order of magnitude is greater than that of $\lambda_i^n u_i(x_i^n)$. Assume, for simplicity, that the sequence $q^n/\|q^n\|$ converges to some point q. We get:

$$q \cdot x_i^\infty = \lim_{n \to \infty} \frac{P^n}{(1 - P^n)\|q^n\|}(\delta_i^n - \lambda_i^n u_i(x_i^n))$$

Denote this quantity c_i.

If $u_i'(x_i^\infty) = 0$ then x_i^∞ maximizes u_i over X_i. Hence it maximizes u_i over $\{x \in X_i, q \cdot x \leq c_i\}$. The same holds if $u_i'(x_i^\infty) \neq 0$, because then q is in the direction of $u_i'(x_i^\infty)$.

We claim that c_i is non-negative, depends monotonically on X_i, and not at all on u_i.

A lightweight trader t's joining a "small" coalition S contibutes to three utilities: (i) his own, (ii) that of other lightweight traders, and (iii) that of heavyweight traders.

Roughly (again), since we consider a small coalition, probably all the heavyweight traders are not simultaneously satiated. Since the weights in (i) and (ii) tend to zero when n goes large, when joining the coalition, trader t's resources are best used if distributed to unsatiated heavyweight traders.

His ability to give his resources depends only and monotonically on X_t, and not on u_t.

Though the optimal redistribution of t's resources may involve several heavyweight traders, giving it all to only one trader gives a lower bound to (iii): δ_i^n is of larger order than λ_i^n.

This establishes our claim about c_i.

If trader t is heavyweight, contributions (i) and (ii) in δ_i^n are at least cancelled out by $\lambda_i^n u_i(\mathbf{x}_i^n)$, since $\lambda_i^\infty u_i(\mathbf{x}_i^\infty)$ is the maximum t can get. Thus, the rest of the argument is as before with

$$q \cdot \mathbf{x}_i^\infty \leq \lim_{n \to \infty} \left(\frac{\varepsilon^n}{\|q^n\|} \right) \text{ [component (iii) of } \delta \text{]}$$

and c_i defined as the right side of this inequality.

Since i is heavyweight, $\mathbf{x}_i^\infty$ satiates. By the inequality above, it also satisfies the budget constraint. □

5.5 Concluding Remarks

A question has been eluded until now: "What about the core in an economy that allows satiation?" Remember that an allocation is in the core if it cannot be improved upon by any coalition. This can be understood either in a strong or a weak way. The strong version requires that no "weak improvement" is possible. That is to say that it is not possible that some members are strictly better off while others are not worse off. With this definition the Core Equivalence Theorem for replica economies applies. Since the set of competitive equilibria may be empty, the same holds for the core. On the other hand, the "weak" core may be too large and not even enjoy the equal treatment property.

Acknowledgments

The author is very grateful to J.-F. Mertens, who prepared this chapter on the basis of the author's oral presentation.

References

Allen, B. & S. Sorin [1994], "Cooperative games," in "Game–Theoretic Methods in General Equilibrium Analysis," (J.-F. Mertens & S. Sorin, eds.), 20–23, Kluwer, Dordrecht.

Arrow, K. J. & M. Kurz [1970], *Public Investment, the Rate of Return, and Optimal Fiscal Policy*, Baltimore: Johns Hopkins Press.

Aumann, R. J. [1964], "Markets with a Continuum of Traders," *Econometrica*, 32, 39–50 [Chapter 46].

Aumann, R. J. [1975], "Values of markets with a continuum of traders," *Econometrica*, 43, 611–646 [Chapter 51].

Aumann, R. J. [1994], "The Shapley Value," in "Game–Theoretic Methods in General Equilibrium Analysis," (J.-F. Mertens & S. Sorin, eds.), 61–66, Kluwer, Dordrecht [Chapter 64].

Aumann, R. J. & M. Kurz [1977], "Power and taxes," *Econometrica*, 45, 1137–1161 [Chapter 52].

Aumann, R. J. & M. Kurz [1977], "Power and taxes in a multi-commodity economy," *Israel Journal of Mathematics*, 27, 186–234 [Chapter 53].

Aumann, R. J., M. Kurz & A. Neyman [1983], "Voting for Public Goods," *Review of Economic Studies*, 50, 677–693 [Chapter 56].

Aumann, R. J., M. Kurz & A. Neyman [1987], "Power and Public Goods," *Journal of Economic Theory*, 42, 108–127 [Chapter 55].

Aumann, R. J. & J. H. Drèze [1986], "Values of Markets with Satiation or Fixed Prices," *Econometrica*, 54, 1271–1318 [Chapter 57].

Champsaur P. [1975], "Cooperation versus Competition," *Journal of Economic Theory*, 11, 394–417.

Drèze, J. H. & H. Müller [1980], "Optimality Properties of Rationing Schemes," *Journal of Economic Theory*, 23, 150–159.

Harsanyi, J. C. [1959], "A bargaining model for the cooperative n-person game," in "Contributions to the Theory of Games IV," (A. W. Tucker and R. D. Luce, eds.), Ann. of Math. Studies, 40, 325–355. Princeton University Press, Princeton, N.J.

Hart, S. [1994], "Value Equivalence Theorems: The TU and NTU Cases," in "Game–Theoretic Methods in General Equilibrium Analysis," (J.-F. Mertens & S. Sorin, eds.), 113–120, Kluwer, Dordrecht.

Nash, J. F. [1953], "Two Person Cooperative Games," *Econometrica*, 21, 128–140.

Neyman, A. [1994], "Value of Games with a Continuum of Players," in "Game–Theoretic Methods in General Equilibrium Analysis," (J.-F. Mertens & S. Sorin, eds.), 67–80, Kluwer, Dordrecht.

Shapley, L. S. [1953], "A value for n-person games," in "Contributions to the Theory of Games II," (H. W. Kuhn and A. W. Tucker, eds.), Ann. of Math. Studies, 28, 307–317, Princeton University Press, Princeton, N.J.

Shapley, L. S. [1969], "Utility Comparisons and the Theory of Games," in "La Décision," 251–263, Editions du Centre National de la Recherche Scientifique (CNRS), Paris.

Endogenous Formation of Links between Players and of Coalitions: An Application of the Shapley Value
with R. Myerson

1   Introduction

Consider the coalitional game v on the player set $\{1,2,3\}$ defined by

$$v(S) = \begin{cases} 0 & \text{if } |S| = 1, \\ 60 & \text{if } |S| = 2, \\ 72 & \text{if } |S| = 3, \end{cases} \qquad (1)$$

where $|S|$ denotes the number of players in S. Most cooperative solution concepts "predict" (or assume) that the all-player coalition $\{1,2,3\}$ will form and divide the payoff 72 in some appropriate way. Now suppose that P_1 (player 1) and P_2 happen to meet each other in the absence of P_3. There is little doubt that they would quickly seize the opportunity to form the coalition $\{1,2\}$ and collect a payoff of 30 each. This would happen in spite of its inefficiency. The reason is that if P_1 and P_2 were to invite P_3 to join the negotiations, then the three players would find themselves in effectively symmetric roles, and the expected outcome would be (24,24,24). P_1 and P_2 would not want to risk offering, say, 4 to P_3 (and dividing the remaining 68 among themselves), because they would realize that once P_3 is invited to participate in the negotiations, the situation turns "wide open"–anything can happen.

All this holds if P_1 and P_2 "happen" to meet. But even if they do not meet by chance, it seems fairly clear that the players in this game would seek to form pairs for the purpose of negotiation, and not negotiate in the all-player framework.

Research by Robert J. Aumann supported by the National Science Foundation at the Institute for Mathematical Studies in the Social Sciences (Economics), Stanford University, under Grant Number IST 85-21838.
Research by Roger B. Myerson supported by the National Science Foundation under grant number SES 86-05619.

This chapter originally appeared in *The Shapley Value: Essays in Honor of Lloyd S. Shapley*, edited by Alvin E. Roth, pp. 175–191, Cambridge University Press, Cambridge, 1988. Reprinted with permission.

The preceding example is due to Michael Maschler (see Aumann and Dreze 1974, p. 235, from which much of this discussion is cited). Maschler's example is particularly transparent because of its symmetry. Even in unsymmetric cases, though, it is clear that the framework of negotiations plays an important role in the outcome, so individual players and groups of players will seek frameworks that are advantageous to them. The phenomenon of seeking an advantageous framework for negotiating is also well known in the real world at many levels – from decision making within an organization, such as a corporation or university, to international negotiations. It is not for nothing that governments think hard – and often long – about "recognizing" or not recognizing other governments; that the question of whether, when, and under what conditions to negotiate with terrorists is one of the utmost substantive importance; and that at this writing the government of Israel is tottering over the question not of *whether* to negotiate with its neighbors, but of the *framework* for such negotiations (broad-base international conference or direct negotiations).

Maschler's example has a natural economic interpretation in terms of S-shaped production functions. The first player alone can do nothing because of setup costs. Two players can produce 60 units of finished product. With the third player, decreasing returns set in, and all three together can produce only 72. The foregoing analysis indicates that the form of industrial organization in this kind of situation may be expected to be inefficient.

The simplest model for the concept "framework of negotiations" is that of a *coalition structure,* defined as a partition of the player set into disjoint coalitions. Once the coalition structure has been determined, negotiations take place only within each of the coalitions that constitute the structure; each such coalition B divides among its members the total amount $v(B)$ that it can obtain for itself. Exogenously given coalition structures were perhaps first studied in the context of the bargaining set (Aumann and Maschler 1964), and subsequently in many contexts; a general treatment may be found in Aumann and Dreze (1974). Endogenous coalition formation is implicit already in the von Neumann–Morgenstern (1944) theory of stable sets; much of the interpretive discussion in their book and in subsequent treatments of stable sets centers around which coalitions will "form." However, coalition structures do not have a formal, explicit role in the von Neumann–Morgenstern theory. Recent treatments that consider endogenous coalition structures explicitly within the context of a formal theory include Hart and Kurz (1983), Kurz (1988), and others.

Coalition structures, however, are not rich enough adequately to capture the subtleties of negotiation frameworks. For example, diplomatic relations between countries or governments need not be transitive and, therefore, cannot be adequately represented by a partition; thus both Syria and Israel have diplomatic relations with the United States but not with each other. For another example, in salary negotiations within an academic department, the chairman plays a special role; members of the department cannot usually negotiate directly with each other, though certainly their salaries are not unrelated.

To model this richer kind of framework, Myerson (1977) introduced the notion of a *cooperation structure* (or *cooperation graph*) in a coalitional game. This graph is simply defined as one whose vertices are the players. Various interpretations are possible; the one we use here is that a link between two players (an edge of the graph) exists if it is possible for these two players to carry on meaningful direct negotiations with each other. In particular, ordinary coalition structures $(B_1, B_2, \ldots, B_k)$ (with disjoint B_j) may be modeled within this framework by defining two players to be linked if and only if they belong to the same B_j. (For generalizations of this cooperation structure concept, see Myerson 1980.)

Shapley's 1953 definition of the value of a coalitional game v may be interpreted as evaluating the players' prospects when there is full and free communication among all of them – when the cooperation structure is "full," when any two players are linked. When this is not so, the prospects of the players may change dramatically. For an extreme example, a player i who is totally isolated – is linked to no other player – can expect to get nothing beyond his own worth $v(\{i\})$; in general, the more links a player has with other players, the better one may expect his prospects to be. To capture this intuition, Myerson (1977) defined an extension of the Shapley value of a coalitional game v to the case of an arbitrary cooperation structure g. In particular, if g is the complete graph on the all-player set N (any two players are directly linked), then Myerson's value coincides with Shapley's. Moreover, if the cooperation graph g corresponds to the coalition structure $(B_1, B_2, \ldots, B_k)$ in the sense indicated here, then the Myerson value of a member i of B_j is the Shapley value of i as a player of the game $v|B_j$ (v restricted to B_j).

This chapter suggests a model for the endogenous formation of cooperation structures. Given a coalitional game v, what links may be expected to form between the players? Our approach differs from that of previous writers on endogenous coalition formation in two respects: First, we work with cooperation graphs rather than coalition structures, using the Myerson value to evaluate the pros and cons of a given cooperation structure for any particular player. Second, we do not use the usual myopic, here-and-now kind of equilibrium condition. When a player considers forming a link with another one, he does not simply ask himself whether he may expect to be better off with this link than without it, given the previously existing structure. Rather, he looks ahead and asks himself, "Suppose we form this new link, will other players be motivated to form further new links that were not worthwhile for them before? Where will it all lead? Is the *end result* good or bad for me?"

In Section 2 we review the Myerson value and illustrate the "look-ahead" reasoning by returning to the three-person game that opened the chapter. The formal definitions are set forth in Section 3, and the following sections are devoted to examples and counterexamples. The final section contains a general discussion of various aspects of this model, particularly of its range of application.

No new theorems are proved. Our purpose is to study the conceptual implications of the Shapley value and Myerson's extension of it to cooperation structures in examples that are chosen to reflect various applied contexts.

2   Looking ahead with the Myerson value

We start by reviewing the Myerson value. Let v be a coalitional game with N as player set, and g a graph whose vertices are the players. For each player i the *value* $\phi_i^g = \phi_i^g(v)$ is determined by the following axioms.

Axiom 1. If a graph g is obtained from another graph h by adding a single link, namely the one between players i and j, then i and j gain (or lose) equally by the change; that is,

$$\phi_i^g - \phi_i^h = \phi_j^g - \phi_j^h.$$

Axiom 2. If S is a connected component of g, then the sum of the values of the players in S is the worth of S; that is,

$$\sum_{i \in S} \phi_i^g(v) = v(S)$$

(Recall that a *connected component* of a graph is a maximal set of vertices of which any two may be joined by a chain of linked vertices.)

That this axiom system indeed determines a unique value was demonstrated by Myerson (1977). Moreover, he showed that if v is superadditive, then two players who form a new link never lose by it: The two sides of the equation in Axiom 1 are nonnegative. He also established[1] the following practical method for calculating the value: Given v and g, define a coalitional game v^g by

$$v^g(S) := \sum v(S_j^g), \qquad (2)$$

where the sum ranges over the connected components S_j^g of the graph $g|S$ (g restricted to S). Then

$$\phi_i^g(v) = \phi_i(v^g), \qquad (3)$$

where ϕ_i denotes the ordinary Shapley value for player i.

We illustrate with the game v defined by (1). If P_1 and P_2 happen to meet in the absence of P_3, then the graph g may be represented by

$$\begin{array}{c} \diagup^{\displaystyle 2} \\ 1 3 \end{array} \qquad (4)$$

with only P_1 and P_2 connected. Then $\phi^g(v) = (30,30,0)$; we have already seen that in this situation it is not worthwhile for P_1 and P_2 to bring P_3 into the negotiations, because that would make things entirely symmetric, so P_1 and P_2 would get only 24 each, rather than 30. But P_2, say, might consider offering to form a link with P_3. The immediate result would be the graph

$$(5)$$

This graph is not at all symmetric; the central position of P_2 – all communication must pass through him – gives him a decided advantage. This advantage is reflected nicely in the corresponding value, (14,44,14). Thus P_2 stands to gain from forming this link, so it would seem that he should go ahead and do so. But now in this new situation, it would be advantageous for P_1 and P_3 to form a link; this would result in the complete graph

$$\begin{array}{c} \diagup^{\displaystyle 2}\diagdown \\ 1 \text{_____} 3 \end{array} \qquad (6)$$

which is again symmetric and so corresponds to a payoff of (24,24,24). Therefore, whereas it originally seemed worthwhile for P_2 to forge a new link, on closer examination it turns out to lead to a net loss of 6 (he goes from 30 to 24). Thus the original graph, with only P_1 and P_2 linked, would appear to be in some sense "stable" after all.

Can this reasoning be formalized and put into a more general context? It is true that if P_2 offers to link up with P_3, then P_1 also will, but wouldn't P_1 do this anyway? To make sense of the argument, must one assume that P_1 and P_2 explicitly agree not to bring P_3 in? If so, under what conditions would such an agreement come about?

It turns out that no such agreement is necessary to justify the argument. As we shall see in the next section, the argument makes good sense in a framework that is totally noncooperative (as far as link formation is concerned; once the links are formed, enforceable agreements may be negotiated).

3 The formal model

Given a coalitional game v with n players, construct an auxiliary *linking game* as follows: At the beginning of play there are no links between any players. The game consists of pairs of players being offered to form links, the offers being made one after the other according to some definite rule; the rule is common knowledge and will be called the *rule of order*. To form a link, *both* potential partners must agree; once formed, a link cannot be destroyed, and, at any time, the entire history of offers, acceptances, and rejections is known to all players (the game is of perfect information). The only other requirements for the rule of order are that it lead to a finite game, and that after the last link has been formed, each of the $n(n-1)/2$ pairs must be given a final opportunity to form an additional link (as in the bidding stage of bridge). At this point some cooperation graph g has been determined; the payoff to each player i is then defined as $\phi_i^g(v)$.

Most of the analysis in the sequel would not be affected by permitting the rule of order to have random elements as long as perfect information is maintained. It does, however, complicate the analysis, and we prefer to exclude chance moves at this stage.

Note that it does not matter in which order the two players in a pair decide whether to agree to a link; in equilibrium, either order (with perfect information) leads to the same outcome as simultaneous choice.

In practice, the initiative for an offer may come from one of the players rather than from some outside agency. Thus the rule of order might give the initiative to some particular player and have it pass from one player to another in some specified way.

Because the game is of perfect information, it has subgame perfect equilibria (Selten 1965) in pure strategies.[2] Each such equilibrium is associated with a unique cooperation graph g, namely the graph reached at the end of play. Any such g (for any choice of the order on pairs) is called a *natural structure* for v (or a *natural outcome* of the linking game).

Rather than starting from an initial position with no links, one may start from an exogenously given graph g. If all subgame perfect equilibria of the resulting game (for any choice of order) dictate that no additional links form, then g is called *stable*.

4   An illustration

We illustrate with the game defined by (1). To find the subgame perfect equilibria, we use "backwards induction." Suppose we are already at a stage in which there are two links. Then, as we saw in Section 2, it is worthwhile for the two players who have not yet linked up to do so; therefore we may assume that they will. Thus one may assume that an inevitable consequence of going to two links is a graph with three links. Suppose now there is only one link in the graph, say that between P_1 and P_2 [as in (4)]. P_2 might consider offering to link up with P_3 [as in (5)], but we have just seen that this necessarily leads to the full graph [as in (6)]. Because P_2 gets less in (6) than in (4), he will not do so.

Suppose, finally, that we are in the initial position, with no links at all. At this point the way in which the pairs are ordered becomes important;[3] suppose it is 12, 23, 13. Continuing with our backwards induction, suppose the first two pairs have refused. If the pair 13 also refuses, the result will be 0 for all; if, on the other hand, they accept, it will be (30,0,30). Therefore they will certainly accept. Going back one step further, suppose that the pair 12 – the first pair in the order – has refused, and the pair 23 now has an opportunity to form a link. P_2 will certainly wish to do so, as otherwise he will be left in the cold. For P_3, though, there is no difference, because in either case he will get 30; therefore there is a subgame perfect equilibrium at which P_3 turns down this offer. Finally, going back to the first stage, similar considerations lead to the conclusion that the linking game has three natural outcomes, each consisting of a single link between two of the three players.

This argument, especially its first part, is very much in the spirit of the informal story in Section 2. The point is that the formal definition clarifies what lies behind the informal story and shows how this kind of argument may be used in a general situation.

5 Some weighted majority games

Weighted majority games are somewhat more involved than the one considered in the previous section, and we will go into less detail. We start with a fairly typical example. Let v be the five-person weighted majority game [4; 3,1,1,1,1] (4 votes are needed to win; one player has three votes, the other four have one vote each). Let us say that the coalition S has *formed* if g is the complete graph on the members of S (two players are linked if both are members of S). We start by tabulating the values for the complete graphs on various kinds of coalitions, using an obvious notation.

$\{1,1,1,1\}$ $\quad$ $[0,\frac{1}{4},\frac{1}{4},\frac{1}{4},\frac{1}{4}]$
$\{3,1\}$ $\quad$ $[\frac{1}{2},\frac{1}{2},0,0,0]$
$\{3,1,1\}$ $\quad$ $[\frac{2}{3},\frac{1}{6},\frac{1}{6},0,0]$
$\{3,1,1,1\}$ $\quad$ $[\frac{3}{4},\frac{1}{12},\frac{1}{12},\frac{1}{12},0]$
$\{3,1,1,1,1\}$ $\quad$ $[\frac{3}{5},\frac{1}{10},\frac{1}{10},\frac{1}{10},\frac{1}{10}]$

Intuitively, one may think of a parliament with one large party and four small ones. To form a government, the large party needs only one of the small ones. But it would be foolish actually to strive for such a narrow government, because then it (the large party) would be relatively weak *within* the government, the small party could topple the government at will; it would have veto power within the government. The more small parties join the government, the less the large party depends on each particular one, and so the greater the power of the large party. This continues up to the point where there are so many small parties in the government that the large party itself loses its veto power; at that point the large party's value goes *down*. Thus with only one small party, the large party's value is $\frac{1}{2}$; it goes up to $\frac{2}{3}$ with two small parties and to $\frac{3}{4}$ with three, but then drops to $\frac{3}{5}$ with four small parties, because at that point the large party itself loses its veto power within the government. Note, too, that up to a point, the fewer small parties there are in the government, the better for those that are, because there are fewer partners to share in the booty.

We proceed now to an analysis by the method of Section 3. It may be verified that any natural outcome of this game is necessarily the complete graph on some set of players; if a player is linked to another one indirectly, through a "chain" of other linked players, then he must also be linked to him directly. In the analysis, therefore, we may restrict attention to "complete coalitions"–coalitions within which all links have formed.

As before, we use backwards induction. Suppose a coalition of type

{3,1,1,1} has formed. If any of the "small" players in the coalition links up with the single small player who is not yet in, then, as noted earlier, the all-player coalition will form. This is worthwhile both for the small player who was previously "out" and for the one who was previously "in" (the latter's payoff goes up from $\frac{1}{12}$ to $\frac{1}{10}$). Therefore such a link will indeed form, and we conclude that a coalition of type {3,1,1,1} is unstable, in that it leads to {3,1,1,1,1}.

Next, suppose that a coalition of type {3,1,1} has formed. If any player in the coalition forms a link with one of the small players outside it, then this will lead to a coalition of the form {3,1,1,1}, and, as we have just seen, this in turn will lead to the full coalition. This means that the large player will end up with $\frac{3}{5}$ (rather than the $\frac{2}{3}$ he gets in the framework of {3,1,1}) and the small players with $\frac{1}{10}$ (rather than the $\frac{1}{6}$ they get in the framework of {3,1,1}). Therefore none of the players in the coalition will agree to form any link with any player outside it, and we conclude that a coalition of type {3,1,1} is stable.

Suppose next that a coalition of type {3,1} has formed. Then the large player does have an incentive to form a link with a small player outside it. For this will lead to a coalition of type {3,1,1}, which, as we have seen, is stable. Thus the large player can raise his payoff from the $\frac{1}{2}$ he gets in the framework of {3,1} to the $\frac{2}{3}$ he gets in the framework of {3,1,1}. This is certainly worthwhile for him, and therefore {3,1} is unstable.

Finally, suppose no links at all have as yet been formed. If the small players all turn down all offers of linking up with the large player but do link up with each other, then the result is the coalition {1,1,1,1}, and each one will end up with $\frac{1}{4}$. If, on the other hand, one of them links up with the large player, then the immediate consequence is a coalition of type {3,1}; this in turn leads to a coalition of type {3,1,1}, which is stable. Thus for a small player to link up with the large player inevitably leads to a payoff of $\frac{1}{6}$ for him, which is less than the $\frac{1}{4}$ he could get in the framework of {1,1,1,1}. Therefore considerations of subgame perfect equilibrium lead to the conclusion that starting from the initial position (no links), all small players reject all overtures from the large player, and the final result is that the coalition {1,1,1,1} forms.

This conclusion is typical for weighted majority games with one "large" player and several "small" players of equal weight. Indeed, we have the following general result.

Theorem A. In a superadditive weighted majority game of the form $[q; w,1, \ldots ,1]$ with $q > w > 1$ and without veto players, a cooperation structure is natural if and only if it is the complete graph on a minimal winning coalition consisting of "small" players only.

The proof, which will not be given here, consists of a tedious examination of cases. There may be a more direct proof, but we have not found it.

The situation is different if there are two large players and many small ones, as in [4; 2,2,1,1,1,] or [6; 3,3,1,1,1,1,1]. In these cases, either the two large players get together or one large player forms a coalition with *all* the small ones (not minimal winning!). We do not have a general result that covers all games of this type.

Our final example is the game [5; 3,2,2,1,1]. It appears that there are two types of natural coalition structure: one associated with coalitions of type {2,2,1,1}, and one with coalitions of type {3,2,1,1}. Note that neither one is minimal winning.

In all these games some coalition forms; that is, the natural graphs all are "internally complete." As we will see in the next section, that is not the case in general. For simple games, however, and in particular for weighted majority games, we do not know of any counterexample.

6 A natural structure that is not internally complete

Define v as the following sum of three voting games:

$$v := [2; 1,1,1,0] + [3; 1,1,1,0] + [5; 3,1,1,2].$$

That is, v is the sum of a three-person majority game in which P_4 is a dummy, a three-person unanimity game in which P_4 is again a dummy, and a four-person voting game in which the minimal winning coalitions are {1,2,3} and {1,4}. The *sum* of these games is defined as any sum of functions, so the worth $v(S)$ of a coalition S is the number of component games in which S wins. For example, $v(\{2,3\}) = 1$ and $v(\{1,2,4\}) = 2$.

The unique natural structure for this game is

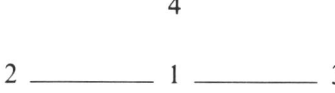

That is, P_1 links up with P_2 and P_3, but P_2 and P_3 do not link up with each other, and no player links up with P_4. The Myerson value of this game for this cooperation structure is $(\frac{5}{3},\frac{5}{6},\frac{5}{6},0)$.

The Shapley value of this game, which is also the Myerson value for the complete graph on all the players, is $(\frac{5}{4},\frac{3}{4},\frac{3}{4},\frac{1}{4})$. Notice that P_1, P_2, and P_3 all do strictly worse with the Shapley value than with the Myerson value for the natural structure described earlier. It can be verified that for any other graph either the value equals the Shapley value or there is at least one pair of players who are not linked and would do strictly better with the Shapley value. This implies inductively that if any pair of players forms a link that is not in the natural structure, then additional links will continue to form until every player is left with his Shapley value. To avoid this outcome, P_1, P_2, and P_3 will refuse to form any links beyond the two already shown.

For example, consider what happens if P_2 and P_3 add a link so that the graph becomes

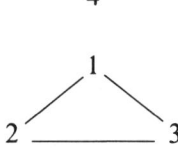

The value for this graph is $(1,1,1,0)$, which is better than the Shapley value for P_2 and P_3, but worse than the Shapley value for P_1. To rebuild his claim to a higher payoff than P_2 and P_3, P_1 then has an incentive to form a link with P_4.

Intuitively, P_1 needs both P_2 and P_3 in order to collect the payoff from the unanimity game $[3; 1,1,1,0]$. They, in turn, would like to keep P_4 out because he is comparatively strong in the weighted voting game $[5; 3,1,1,2]$, whose Shapley value is $(\frac{7}{12},\frac{1}{12},\frac{1}{12},\frac{3}{12})$. With P_4 out, all three remaining players are on the same footing, because all three are then needed to form a winning coalition. Therefore P_1 and P_2 may each expect to get $\frac{1}{3}$ from this game, which is $\frac{1}{4}$ more than the $\frac{1}{12}$ they were getting with P_4 in. On the other hand, excluding P_4 lowers P_1's value by $\frac{1}{4}$, from $\frac{7}{12}$ to $\frac{1}{3}$, and P_1 will therefore want P_4 in.

This is where the three-person majority game $[2; 1,1,1,0]$ enters the picture. If P_2 and P_3 refrain from linking up with each other, then P_1's centrality makes him much stronger in *this* game, and his Myerson value in it is then $\frac{2}{3}$ (rather than $\frac{1}{3}$, the Shapley value). This gain of $\frac{1}{3}$ more than makes up for the loss of $\frac{1}{4}$ suffered by P_1 in the game $[5; 3,1,1,2]$, so he is willing to keep P_4 out. On the other hand, P_2 and P_3 also gain thereby, because the $\frac{1}{4}$ each gains in $[5; 3,1,1,2]$ more than makes up for the $\frac{1}{6}$ each loses in the three-person majority game. Thus P_2 and P_3 are motivated to refrain from forming a link with each other, and all are motivated to refrain from forming links with P_4.

In brief, P_2 and P_3 gain by keeping P_4 isolated; but they must give P_1 the central position in the {1,2,3} coalition so as to provide an incentive for him to go along with the isolation of P_4, and a credible threat if he doesn't.

7 Natural structures that depend on the rule of order

The natural outcome of the link-forming game may well depend on the rule of order. For example, let u be the majority game [3; 1,1,1,1], let $w := [2; 1,1,0,0]$, and let $w' := [2; 0,0,1,1]$. Let $v := 24u + w + w'$. If the first offer is made to {1,2}, then either {1,2,3} or {1,2,4} will form; if it is made to {3,4}, then either {1,3,4} or {2,3,4} will form.

The underlying idea here is much like in the game defined by (1). The first two players to link up are willing to admit one more player in order to enjoy the proceeds of the four-person majority game u; but the resulting coalition is not willing to admit the fourth player, who would take a large share of those proceeds and himself contribute comparatively little. The difference between this game and (1) is that here each player in the first pair to get an *opportunity* to link up is positively motivated to seize that opportunity, which was not the case in (1).

The nonuniqueness in this example is robust to small changes in the game. That is, there is an open neighborhood of four-person games around v such that, for all games in this neighborhood, if P_1 and P_2 get the first opportunity to form a link then the natural structures are graphs in which P_1, P_2, and P_3 are connected to each other but not to P_4; but if P_3 and P_4 get the first opportunity to form a link, then the natural structures are graphs in which P_2, P_3, and P_4 are connected to each other but not to P_1. (Here we use the topology that comes from identifying the set of n-person coalitional games with euclidean space of dimension $2^n - 1$.)

Each example in this chapter is also robust in the phenomenon that it is designed to illustrate. That is, for all games in a small open neighborhood of the example in Section 4, the natural outcomes will fail to be Pareto optimal; and for all games in a small open neighborhood of the example in Section 6, the natural outcomes will not be complete graphs on any coalition.

8 Discussion

The theory presented here makes no pretense to being applicable in all circumstances. The situations covered are those in which there is a preliminary period that is devoted to link formation only, during which, for one reason or another, one cannot enter into binding agreements of any kind (such as those relating to subsequent division of the payoff, or even conditional link-forming, or nonforming, deals of the kind "I won't link up with Adams if you don't link up with Brown"). After this preliminary period one carries out negotiations, but then new links can no longer be added.

An example is the formation of a coalition government in a parliamentary democracy in which no single party has a majority (Italy, Germany, Israel, France during the Fifth Republic, even England at times). The point is that a government, once formed, can only be altered at the cost of a considerable upheaval, such as new elections. On the other hand, one cannot really negotiate in a meaningful way on substantive issues before the formation of the government, because one does not know what issues will come up in the future. Perhaps one does know something about some of the issues, but even then one cannot make *binding* deals about them. Such deals, when attempted, are indeed often eventually circumvented or even broken outright; they are to a large extent window dressing, meant to mollify the voter.

An important assumption is that of perfect information. There is nothing to stop us from changing the definition by removing this assumption — something we might well wish to try — but the analysis of the examples would be quite different. Consider, for example, the game [4; 3,1,1,1,1] treated at the beginning of Section 5. Suppose that the rule of order initially gives the initiative to the large player. That is, he may offer links to each of the small players in any order he wants; links are made public once they are forged, but rejected offers do not become known. This is a fairly reasonable description of what may happen in the negotiations for formation of governments in parliamentary democracies of the kind described here. In this situation the small players lose the advantage that was conferred on them by perfect information; formation of a coalition of type $\{3,1,1\}$ becomes a natural outcome. Intuitively, a small player will refuse an offer from the large player only if he feels reasonably sure that all the small players will refuse. Such a feeling is justified if it is common knowledge that all the others have already refused, and from there one may work one's way backward by induction. But the induction is broken if refused offers do not become known; and then the small players may become suspicious of each other — quite likely rightfully, as under imperfect information, mutual suspicion becomes an equilibrium outcome. We hasten to add that mutual trust — all small players refusing offers from the large one–remains in equilibrium; but unlike in the case of perfect information, where everything is open and aboveboard, it is no longer the *only* equilibrium. In short, secrecy breeds mistrust–justifiable mistrust.

Which model is the "right" one (i.e., perfect or imperfect information) is moot. Needless to say, the perfect information model is not being suggested as a universal model for all negotiations. But one may feel that the secrecy in the imperfect information model is a kind of irrelevant noise that muddies the waters and detracts from our ability properly to analyze power relationships. On the other hand, one may feel that the backwards induction in the perfect information model is an artificiality that overshadows and dominates the analysis, much as in the finitely repeated Prisoner's Dilemma, and again obscures the "true" power relationships. Moreover, the outcome predicted by the perfect information model in the game [4; 3,1,1,1,1] (formation of the coalition of all small players) *is* somewhat strange and anti-intuitive. On the contrary, one would have thought that the large player has a better chance than each individual small player to get into the ruling coalition; one might expect *him* to "form the government," so to speak.

In brief, there is no single "right" model. Each model has something going for it and something going against it. You pay your money, and you take your choice.

We end with an anecdote. This chapter is based on a correspondence that took place between the authors during the first half of 1977. That spring, there were elections in Israel, and they brought the right to power for the first time since the foundation of the state almost thirty years earlier. After the election, one of us used the perfect information model proposed here to try to predict which government would form. He was disappointed when the government that actually did form after about a month of negotiations did not conform to the prediction of the model, in that it failed to contain Professor Yigael Yadin's new "Democratic Party for Change." Imagine his delight when Yadin did after all join the government about four months later!

Appendix

We state and prove here the main result of Myerson (1977).

For any graph g, any set of players S, and any two players j and k in S, we say that j and k are connected in S by g if and only if there is a path in g that goes from j to k and stays within S. That is, j and k are connected in S by g if there exists some sequence of players $i_1, i_2, \ldots, i_M$ such that

$i_1 = j$, $i_M = k$, $\{i_1, i_2, \ldots, i_M\} \subseteq S$, and every pair (i_n, i_{n+1}) corresponds to a link in g. Let S/g denote the partition of S into the sets of players that are connected in S by g. That is,

$$S/g = \{\{k \mid j \text{ and } k \text{ are connected in } S \text{ by } g\} \mid j \in S\}.$$

With this notation, the definition of v^g from (2) becomes

$$v^g(S) = \sum_{T \in S/g} v(T) \tag{A1}$$

for any coalition S. Then the main result of Myerson (1977) is as follows.

Theorem. Given a coalitional game v, Axioms 1 and 2 (as stated in Section 2) are satisfied for all graphs if and only if, for every graph g and every player i,

$$\phi_i^g(v) = \phi_i(v^g), \tag{A2}$$

where ϕ_i denotes the ordinary Shapley value for player i. Furthermore, if v is superadditive and if g is a graph obtained from another graph h by adding a single link between players i and j, then $\phi_i(v^g) - \phi_i(v^h) \geq 0$, so the differences in Axiom 1 are nonnegative.

Proof: For any given graph g, Axiom 1 gives us as many equations as there are links in g, and Axiom 2 gives us as many equations as there are connected components of g. When g contains cycles, some of these equations may be redundant, but it is not hard to show that these two axioms give us at least as many independent linear equations in the values ϕ_i^g as there are players in the game. Thus, arguing by induction on the number of links in the graph (starting with the graph that has no links), one can show that there can be at most one value satisfying Axioms 1 and 2 for all graphs.

The usual formula for the Shapley (1953) value implies that

$$\phi_i(v^g) - \phi_j(v^g) = \sum_{S \subseteq N \setminus \{i,j\}} \frac{|S|!(|N|-|S|-2)!}{(|N|-1)!} (v^g(S \cup \{i\}) - v^g(S \cup \{j\})).$$

Notice that a coalition's worth in v^g depends only on the links in g that are between two players both of whom are in the coalition. Thus, when S does not contain i or j, the worths $v^g(S \cup \{i\})$ and $v^g(S \cup \{j\})$ would not be changed if we added or deleted a link in g between players i and j. Therefore, $\phi_i(v^g) - \phi_j(v^g)$ would be unchanged if we added or deleted a link in g between players i and j. Thus, (A2) implies Axiom 1.

Given any coalition S and graph g, let the games u^S and w^S be defined by $u^S(T) = v^g(T \cap S)$ and $w^S(T) = v^g(T\setminus S)$ for any $T \subseteq N$. Notice that S is a carrier of u^S, and all players in S are dummies in w^S. Furthermore, if S is a connected component of g, then $v^g = u^S + w^S$. Thus, if S is a connected component of g, then

$$\sum_{i \in S} \phi_i(v^g) = \sum_{i \in S} \phi_i(u^S) = u^S(S) = v^g(S),$$

and so (A2) implies Axiom 2.

Now suppose that the graph g is obtained from the graph h by adding a single link between players i and j. If v is superadditive and $i \in S$, then $v^g(S) \geq v^h(S)$, because S/g is either the same as S/h or a coarser partition than S/h. On the other hand, if $i \notin S$, then $v^g(S) = v^h(S)$. Thus, by the monotonicity of the Shapley value, $\phi_i(v^g) \geq \phi_i(v^h)$ if v is superadditive.

Q.E.D.

NOTES

1 These statements are proved in the appendix, and they imply the assertions about the Myerson value that we made in the introduction.
2 Readers unfamiliar with German and the definition of subgame perfection will find the latter repeated, in English, in Selten (1975), though this reference is devoted mainly to the somewhat different concept of "trembling hand" perfection (even in games of perfect information, trembling hand perfect equilibria single out only *some* of the subgame perfect equilibria).
3 For the analysis, not the conclusion.

REFERENCES

Aumann, R. J., and J. H. Dreze (1974), "Cooperative Games with Coalition Structures," *International Journal of Game Theory* 3, 217–37 [Chapter 44].

Aumann, R. J., and M. Maschler (1964), "The Bargaining Set for Cooperative Games," in Dresher, Shapley, and Tucker (1964), pp. 443–76 [Chapter 42].

Dresher, M., L. S. Shapley, and A. W. Tucker (1964), editors, *Advances in Game Theory,* Annals of Mathematics Studies No. 52, Princeton: Princeton University Press.

Hart, S., and M. Kurz (1983), "Endogenous Formation of Coalitions," *Econometrica* 51, 1047–64.

Kuhn, H. W., and A. W. Tucker (1953), editors, *Contributions to the Theory of Games,* Volume II, Annals of Mathematics Studies No. 28, Princeton: Princeton University Press.

Kurz, M. (1988), "Coalitional Value," in *The Shapley Value: Essays in Honor of Lloyd S. Shapley*, Roth, A. E. (ed.), Cambridge: Cambridge University Press, 155–173.

Myerson, R. B. (1977), "Graphs and Cooperation in Games," *Mathematics of Operations Research* 2, 225–9.

(1980), "Conference Structures and Fair Allocation Rules," *International Journal of Game Theory* 9, 169–82.

von Neumann, J., and O. Morgenstern (1944), *Theory of Games and Economic Behavior,* Princeton: Princeton University Press.

Selten, R. C. (1965), "Spieltheoretische Behandlung eines Oligopolmodells mit Nachfragetraegheit," *Zeitschrift fuer die gesamte Staatswissenschaft* 121, 301–24, 667–89.

(1975), "Reexamination of the Perfectness Concept for Equilibrium Points in Extensive Games," *International Journal of Game Theory* 4, 22–55.

Shapley, L. S. (1953), "A Value for *n*-Person Games," in Kuhn and Tucker (1953), pp. 307–17.

XI COALITIONAL GAMES: FOUNDATIONS OF THE NTU SHAPLEY VALUE

Unlike his 1953 definition of the TU value, Shapley's 1969 definition[1] of the NTU value is not axiomatic. Rather, it is a construction that uses weights to associate a TU game with a given NTU game, then determines the weights so that the value of the associated TU game is feasible in the original NTU game. During the early eighties, this definition came under critical fire (see chapters 61 and 62 below), in spite—or perhaps because—of the successes that the NTU value had enjoyed in applications. Inter alia, it was claimed that introduction of the weights somehow introduced a hidden TU assumption into the conceptual background of the NTU value.

This was part of the motivation for chapter 60, "An Axiomatization of the Nontransferable Utility Value," one of whose features is that the axioms do not mention weights. Shapley had already hinted at the possibility of such an axiomatization, and provided some of the underlying ideas, in his 1969 paper defining the value.

The next two items constitute the "NTU value controversy," which took place in the pages of *Econometrica* in the eighties. Chapter 61 has five parts. It starts with two pieces, by Al Roth (61a) and Wayne Shafer (61b), attacking the NTU value on the basis of two related examples in which, they assert, the NTU value behaves counterintuitively. There follows a rebuttal by me (61c), a reply by Roth (61d), and finally, a rejoinder by me (61e). Item 62 has two parts: an attack on the NTU value by Scafuri and Yannelis (62a), and a rebuttal by me (62b). The whole discussion is largely nontechnical, and has conceptual implications that go beyond the NTU value, touching on matters such as the meaning of solution concepts in game theory, and the significance of counterintuitive examples.

1. "Utility Comparisons and the Theory of Games," in "La Décision," 251–263, Editions du Centre National de la Recherche Scientifique (CNRS), Paris.

60 An Axiomatization of the Non-Transferable Utility Value[1]

1 Introduction

The NTU (Non-Transferable Utility) Value is a solution concept for multi-person cooperative games in which utility is not "transferable" (games "without side payments"). Introduced by Shapley in [12], it generalizes the Shapley value [11] for TU (Transferable Utility) games.[2] Many economic contexts are more naturally modelled by NTU than by TU games; and indeed, the NTU value has been applied with some success to a variety of economic and economic-political models.[3] Two well-known applications are Nash's solution for the bargaining problem [7] and for two-person cooperative games [8], both of which are instances of the NTU value.

The original definition [12] of the NTU value works roughly as follows: Given an NTU game V and a vector λ of "comparison weights" for the players, one derives a TU game v_λ, and calculates its value $\varphi(v_\lambda)$; if this value is feasible in the original NTU game V, then it is defined to be a value of V. A precise definition is given in Section 4.

Technically, the definition is reminiscent of that of the competitive equilibrium, with λ playing the role of prices, and $\varphi(v_\lambda)$ the role of the demand. Historically, it grew out of successive attacks by several investigators, notably J. Harsanyi [4, 5], on the value problem for NTU games. The bare definition may perhaps seem a little strange and unmotivated; but when one delves deeper [12, 1], one finds that it is quite natural. Nevertheless, it has been the object of controversy [9, 10, 6, 2].

In this paper, we offer an axiomatization of the NTU value. Like any axiomatization, it should enable us to understand the concept better, and hence to focus discussion. One can now view the NTU value as *defined* by the axioms, with the treatment in [12] serving as a formula or method of calculation. Thus the NTU value joins the ranks of the TU value and Nash's solution to the bargaining problem, each of which is defined by axioms, but usually calculated by a formula[4]—a formula whose intuitive significance is not, on the face of it, entirely clear.

This chapter originally appeared in *Econometrica* 53 (1985): 599–612. Reprinted with permission.

1. Research supported by a National Science Foundation Grant at the Institute for Mathematical Studies in the Social Sciences, Stanford University. Important conversations with M. Maschler, B. Peleg, and M. Perles are gratefully acknowledged.

2. I.e., games with side payments, representable by a coalitional worth ("characteristic") function.

3. See the references of [2].

4. The random order expected contribution formula for the TU value, and the maximum product formula for the Nash Bargaining Problem.

This work is an outgrowth of ideas that have been "in the air" for many years. The problem of axiomatizing the NTU value is a natural one; already in his original paper [12, p. 260], Shapley discusses "properties of our ... solution that ... could be used in the derivation of our definition." Our treatment owes much to that discussion, and to subsequent oral discussions with Shapley.[5]

Worthy of particular note is that the axioms refer to values as payoff vectors only—the comparison weights associated with a value make no explicit appearance in the axioms. This is important because the question of the intuitive significance of the comparison weights has often been raised in critical discussion. By contrast, Shapley's viewpoint is that his solution consists of both the payoff vector and the comparison weights [12, p. 259, line 20 ff.; p. 261, line 1], with the latter playing at least as important a role as the former.[6] Also worthy of note is the smoothness condition (3.1), which is indispensable for our approach (see Section 9).

The domain of the axioms—the family of games to which they apply—is described in Section 3; the axioms themselves are presented and discussed in Section 5. Section 6 is devoted to an alternative treatment, in which one of the axioms (Independence of Irrelevant Alternatives) is dropped. Proofs are presented in Sections 7 and 8. Section 10 discusses possible variations on the theme; it also contains a discussion of the implications of the axioms for our understanding of the intuitive content of the value solution.

2 Some Notation and Terminology

Denote the real numbers by $\mathbb{R}$. If N is a finite set, denote by $|N|$ the cardinality of N, and by R^N the set of all functions from N to $\mathbb{R}$. We will think of members x of R^N as $|N|$-dimensional vectors whose coordinates are indexed by members of N; thus when $i \in N$, we will often write x^i for $x(i)$. If $x \in R^N$ and $S \subset N$, write x^S for the restriction of x to S, i.e., the member of R^S whose ith coordinate is x^i. Write 1_s for the *indicator* of S, i.e., the member of R^N whose ith coordinate is 1 or 0 according as i is or is not in S. Call x *positive* if $x^i > 0$ for all i in N. If λ and x are in R^N, define λx in R^N by $(\lambda x)^i = \lambda^i x^i$, and denote the scalar product $\sum_{i \in N} x^i y^i$ by $x \cdot y$. Write $x \geqslant y$ if $x^i \geqslant y^i$ for all i in N. Denote the origin of R^N (the vector all of whose coordinates are 0) by O.

---

5. Specifically, the idea of adding the zero-game V^0 (see Section 8) to a given NTU game in order to obtain the induced transfer game is due to him.
6. The importance that Shapley attaches to the endogenous determination of comparison weights is evident from the title of his paper, as well as from its introduction.

Let $A, B \subset R^N$ and $\lambda, x \in R^N$. Write $A + B = \{a + b : a \in A \text{ and } b \in B\}$, $\lambda A = \{\lambda a : a \in A\}$, $A + x = A + \{x\}$, and $(1/2)A = \{(1/2)x : x \in A\}$. Denote the closure of A by $\bar{A}$, its complement by $\sim A$, and its frontier $\bar{A} \cap \overline{(\sim A)}$ by ∂A. If A is convex, call it *smooth* if it has a unique supporting hyperplane at each point of its frontier. Call A *comprehensive* if $x \in A$ and $x \geqslant y$ imply $y \in A$.

3 NTU Games

Let N be a finite set, which will henceforth be fixed; set $n = |N|$. The members of N are called *players*, its non-empty subsets *coalitions*; points in R^N are called *payoff vectors*. An *NTU game* on N (or simply *game*) is a function V that assigns to each coalition S a convex comprehensive non-empty proper subset $V(S)$ of R^S, such that

$V(N)$ is smooth; (3.1)

if $x, y \in \partial V(N)$ and $x \geqslant y$, then $x = y$; and (3.2)

for each coalition S there is a payoff vector x such that (3.3)

$V(S) \times \{0^{N \setminus S}\} \subset V(N) + x.$

Of these three conditions, only (3.1) is a substantive restriction from the intuitive viewpoint; the others are technical in nature. Condition (3.2) says that $\partial V(N)$ has no "level" segments, i.e., segments parallel to a coordinate hyperplane; it is a familiar regularity condition in game theory. Condition (3.3) says that if one thinks of $V(S)$ as embedded in R^N by assigning 0 to players outside S, then $V(S)$ is included in some translate of $V(N)$; it can be thought of as an extremely weak kind of monotonicity.

A *TU game* (on N) is a function v that assigns to each coalition S a real number $v(S)$. The NTU game V *corresponding* to a TU game v is given by

$$V(S) = \left\{ x \in R^S : \sum_{i \in S} x^i \leqslant v(S) \right\}.$$

If T is a coalition, define a *TU* game u_T by

$$u_T(S) = \begin{cases} 1 & \text{if } S \supset T, \\ 0 & \text{otherwise.} \end{cases} \quad (3.4)$$

The NTU game U_T corresponding to u_T is called the *unanimity game* on T.

Operations on games are defined like the corresponding operations on sets, for each coalition separately. Thus $(V + W)(S) = V(S) + W(S)$, $(\lambda V)(S) = \lambda^S V(S)$, $\overline{V}(S) = \overline{V(S)}$, and so on.

4 Shapley Values of NTU Games

Recall that the *value* of a TU game v is the vector $\varphi(v)$ in R^N given by

$$\varphi^i(v) = \frac{1}{n!} \sum_R [v(S_i^R \cup \{i\}) - v(S_i^R)], \qquad (4.1)$$

where R ranges over all $n!$ orders on N, and S_i^R denotes the set of players preceding i in the order R. The value is usually defined by a set of axioms, which are then shown [11] to lead to (4.1).

Let V be a game. For each positive λ in R^N, write

$$v_\lambda(S) = \sup\{\lambda^S \cdot x : x \in V(S)\}. \qquad (4.2)$$

We say that the TU game v_λ is *defined* if the right side of (4.2) is finite for all S. A *Shapley value* of V is a point y in $\overline{V}(N)$ such that for some positive λ in R^N, the TU game v_λ is defined, and $\lambda y = \varphi(v_\lambda)$. The set of all Shapley values of V is denoted $\Lambda(V)$. The set of games V for which $\Lambda(V) \neq \emptyset$—i.e., that possess at least one Shapley value—is denoted Γ^N or simply Γ. The correspondence from Γ to R^N that associates the set $\Lambda(V)$ to each game V is called the *Shapley Correspondence*.

5 The Axioms

A *value correspondence* is a correspondence that associates with each game V in Γ a set $\Phi(V)$ of payoff vectors, satisfying the following axioms for all games U, V, W in Γ:

Axiom 0—Non-Emptiness: $\Phi(V) \neq \emptyset$.

Axiom 1—Efficiency: $\Phi(V) \subset \partial V(N)$.

Axiom 2—Conditional Additivity: If $U = V + W$, then $\Phi(U) \supset (\Phi(V) + \Phi(W)) \cap \partial U(N)$.

Axiom 3—Unanimity: If U_T is the unanimity game on a coalition T, then $\Phi(U_T) = \{1_T/|T|\}$.

Axiom 4—Closure Invariance: $\Phi(\overline{V}) = \Phi(V)$.

Axiom 5—Scale Covariance: If λ in R^N is positive, then $\Phi(\lambda V) = \lambda \Phi(V)$.

Axiom 6—Independence of Irrelevant Alternatives: If $V(N) \subset W(N)$ and $V(S) = W(S)$ for $S \neq N$, then $\Phi(V) \supset \Phi(W) \cap V(N)$.

For a fixed value correspondence Φ, call x a *value* of V if $x \in \Phi(V)$. *Efficiency* says that all values are Pareto optimal. Suppose next that y and z are values of V and W respectively. We cannot in general expect $y + z$ to be a value of $V + W$, because it need not be Pareto optimal there. *Conditional additivity* says that if $y + z$ does happen to be Pareto optimal in $V + W$, then it is a value of $V + W$; i.e., that additivity obtains whenever it does not contradict efficiency. *Unanimity* says that the unanimity game on T has a *unique* value, which provides that the coalition T split the available amount equally. *Closure invariance* is a conceptually harmless technical assumption; we simply do not distinguish between a convex set and its closure. If the payoffs are in utilities, then *scale invariance* says that representing the same real outcome by different utility functions does not affect the value in real terms. *Independence of irrelevant alternatives* (IIA) says that a value y of a game W remains a value when one removes outcomes other than y ("irrelevant alternatives") from the wet $W(N)$ of all feasible outcomes, without changing $W(S)$ for coalitions S other than the all player coalition. (For a thorough discussion of this assumption, see the next section.)

These axioms are an amalgam of those that characterize the value for TU games [11] and those that characterize Nash's solution to the Bargaining Problem [7]. Axioms 1, 2, and 3 are fairly straightforward analogues of the TU value axioms, with the unanimity axiom combining the symmetry and dummy axioms. As we have noted, Axiom 4 is purely technical; and Axioms 5 and 6 are essentially the same as the corresponding axioms in Nash's treatment.

THEOREM A *There is a unique value correspondence, and it is the Shapley correspondence.*

6 An Axiomatic Treatment without IIA

IIA is perhaps the best-known of the axioms in the preceding section. This is partly due to its key role in Nash's work, and partly to its having stirred some controversy. In this section, after discussing the axiom, we offer an axiomatic treatment that avoids using it.

Whether or not IIA is reasonable depends on how we view the value. If we view it as an expected or average outcome, then IIA is not very convincing. By removing parts of the feasible set, we decrease the range of possible outcomes, and so the average may change even if it itself remains feasible. But in NTU games, viewing the value as an average is fraught with difficulty even without IIA, because the convexity of $V(N)$ implies that in general, an average will not be Pareto optimal.

An alternative is to view the value as a group decision or arbitrated outcome; i.e., a reasonable compromise[7] in view of all the possible alternatives open to the players. In that case IIA does sound quite convincing and even compelling. An anecdote—it happens to be a true one—may serve to illustrate its force. Several years ago I served on a committee that was to invite a speaker for a fairly prestigious symposium. Three candidates were proposed; their names would be familiar to many of our readers, but we will call them Alfred Adams, Barry Brown, and Charles Clark. A long discussion ensued, and it was finally decided to invite Adams. At that point I remembered that Brown had told me about a family trip that he was planning for the period in question, and realized that he would be unable to come. I mentioned this and suggested that we re-open the discussion. The other members looked at me as if I had taken leave of my senses. "What difference does it make that Brown can't come," one said, "since in any case we decided on Adams?" I was amazed. All the members were eminent theorists and mathematical economists, thoroughly familiar with the nuances of the Nash model. Not long before, the very member who had spoken up had roundly criticized IIA in the discussion period following a talk. I thought that perhaps he had overlooked the connection, and said that I was glad that in the interim, he had changed his mind about IIA. Everybody laughed appreciatively, as if I had made a good joke, and we all went off to lunch. The subject was never reopened, and Adams was invited.

Note that we are discussing a true game, not an individual decision problem. The members had different interests, coalitions could be formed, etc. Occasionally issues even came to a vote; and when they did not, the vote was definitely "there," in the background. If there ever was a situation in which IIA could be criticized, this was it.

Yet I think that the members were right to laugh off my suggestion. No matter how convincing such criticism may seem in the abstract, the concrete suggestion to reconsider the choice of Adams because Brown could not come sounded—and was—absurd.

Let us nevertheless examine the consequences of omitting this axiom. It turns out that IIA is not nearly as central here as in the Nash theory; something is lost, but less than might have been expected. The result is as follows:

THEOREM B *The Shapley correspondence is the maximal correspondence from Γ to R^N satisfying Axioms 0 through 5.*

---

7. We are purposely staying away from the word "fair," in order to avoid ethical connotations.

More explicitly:

Λ satisfies Axioms 0 through 5. (6.1)

If Φ satisfies Axioms 0 through 5, then $\Phi(V) \subset \Lambda(V)$
for all games V in Γ. (6.2)

At issue, of course, is the categoricity of the axioms. There may[8] be many correspondences Φ satisfying Axioms 0 through 5. With Axiom 6, there is only one; the system is fully determined. On a practical level, though, there isn't much difference. Many of the applications involve necessary conditions only: they assert that every value has a particular form (e.g., competitive equilibrium). This kind of result remains unchanged when Axiom 6 is omitted. The other kind of result—every outcome of a particular form is a value—is weakened; but if we interpret a "value" of V to mean a member of $\Phi(V)$ for *some* Φ (rather than for a particular, fixed Φ) satisfying the axioms, then this kind of result also remains true. Another kind of application in which dropping IIA changes nothing is when there is only one Shapley value ($|\Lambda(V)| = 1$); for example, this is the case for 2-person superadditive games, and in [3].

7 Proof that the Shapley Value Satisfies the Axioms

In the remainder of the paper, we abbreviate $\partial V(N)$ by ∂V. We call a member λ of R^N *normalized* if $\max_i |\lambda^i| = 1$.

Let V be a game, and let $y \in \partial V$. Since $V(N)$ is smooth (3.1), there is a unique supporting hyperplane to $V(N)$ at y. That means that there is a unique normalized λ in R^N such that $\lambda \cdot x$ is maximized over $\overline{V}(N)$ at $x = y$. By comprehensiveness and (3.2), this λ is positive; denote it $\delta(V, y)$.

LEMMA 7.1 $\Lambda(V) \subset \partial V$.

Proof Follows from the efficiency of the TU value.

LEMMA 7.2   Let $y \in \Lambda(V)$, and let $\lambda = \delta(V, y)$. Then the TU game v_λ is defined, and $\lambda y = \phi(v_\lambda)$.

Proof By the definition of the Shapley value (Section 4), there is a positive μ in R^N, which we may assume normalized, such that v_μ is defined

---

8. A referee asked for an example to show that IIA is really needed; i.e., for a correspondence other than Λ satisfying Axioms 0 through 5. We don't know of one. Thus at present, it is conceivable that Axioms 0 through 5 are already categoric.

and $\mu y = \varphi(v_\mu)$. By the efficiency of the TU value,

$$\mu \cdot y = \sum \mu^i y^i = \sum \varphi^i(v_\mu) = v_\mu(N) = \sup\{\mu \cdot x \colon x \in V(N)\}.$$

Hence $\mu \cdot x$ is maximized over $\bar{V}(N)$ at $x = y$, i.e., $\mu = \delta(V, y) = \lambda$, and the proof is complete.

PROPOSITION 7.3 *The correspondence Λ from Γ to R^N satisfies Axioms 0 through 6.*

Proof Axiom 0 follows from the definition of Γ. Axiom 1 is Lemma 7.1. To verify Axiom 2, let $y \in \Lambda(V)$, $z \in \Lambda(W)$, $y + z \in \partial U$; we wish to show $y + z \in \Lambda(U)$. Let $\lambda = \delta(U, y + z) = \delta(V + W, y + z)$. Then $\lambda \cdot x$ is maximized over $\bar{V}(N) + \bar{W}(N)$ at $x = y + z$, and hence over $\bar{V}(N)$ at $x = y$; hence $\lambda = \delta(V, y)$. Since $y \in \Lambda(V)$ it follows from Lemma 7.3 that the TU game v_λ is defined, and $\lambda y = \varphi(v_\lambda)$. Similarly the TU game w_λ (the notations w_λ and u_λ are analogous to v_λ) is defined, and $\lambda z = \varphi(w_\lambda)$. Hence the TU game u_λ is defined, and $u_\lambda = v_\lambda + w_\lambda$. Hence by the additivity axiom for the TU value,

$$\lambda(y + z) = \lambda y + \lambda z = \varphi(v_\lambda) + \varphi(w_\lambda) = \varphi(v_\lambda + w_\lambda) = \varphi(u_\lambda). \tag{7.4}$$

But $y + z \in \bar{V}(N) + \bar{W}(N) \subset \bar{U}(N)$; together with (7.4), this shows that $y + z$ is a Shapley value of U, as was to be shown. The remaining axioms are straightforward, and so the proof of the proposition is complete.

8 Proofs of the Theorems

Throughout this section, Φ is an arbitrary but fixed correspondence from Γ to R^N satisfying Axioms 0 through 5.

LEMMA 8.1 *If V is the game corresponding to a TU game v, then $\Phi(V) = \{\varphi(v)\}$.*

Proof Note first that Γ contains all games corresponding to TU games, so that we can apply our axioms to all these games at will.

Let V correspond to the TU game v. For any real number α, let V^α correspond to the TU game αv. Then V^0 corresponds to the TU game that is identically 0 (i.e., vanishes on all coalitions), and hence by Axioms 1, 2, and 3,

$$\Phi(V^0) + \{1_N/n\} = \Phi(V^0) + \Phi(U_N) \subset \Phi(V^0 + U_N) = \Phi(U_N) = \{1_N/n\}.$$

By Axiom 0, it follows that

$$\Phi(V^0) = \{O\}. \tag{8.2}$$

Hence by Axioms 1 and 2, $\Phi(V) + \Phi(V^{-1}) \subset \Phi(V + V^{-1}) = \Phi(V^0) = \{O\}$. By Axiom 0, it follows that each of $\Phi(V)$ and $\Phi(V^{-1})$ consists of a single point, and

$$\Phi(V^{-1}) = -\Phi(V). \tag{8.3}$$

If α is a positive scalar, then Axiom 5 with $\lambda = (\alpha, \ldots, \alpha)$ yields

$$\Phi(V^\alpha) = \alpha \Phi(V). \tag{8.4}$$

Combining this with (8.2) and (8.3), we deduce (8.4) for *all* scalars α, no matter what their signs are. From Axiom 3 and $\varphi(\alpha u_T) = \alpha \, 1_T/|T|$ we then deduce that

$$\Phi(U_T^\alpha) = \{\varphi(\alpha u_T)\} \tag{8.5}$$

for all coalitions T and all real numbers α.

Now each TU game v may be expressed in the form $v = \sum_T \alpha_T u_T$ where the α_T are real. Hence for the corresponding game V we have $V = \sum_T U_T^{\alpha_T}$. By (8.5), and Axioms 1 and 2, it follows that

$$\{\varphi(v)\} = \sum_T \{\varphi(\alpha_T u_T)\} = \sum_T \Phi(U_T^{\alpha_T}) \subset \Phi\left(\sum_T U_T^{\alpha_T}\right) = \Phi(V).$$

But we have already seen that $\Phi(V)$ consists of a single point. Hence $\{\varphi(v)\} = \Phi(V)$, and the proof of the lemma is complete.

LEMMA 8.6 $\Phi(V) \subset \Lambda(V)$ *for each V in Γ.*

Proof Let $y \in \Phi(V)$. By Axiom 1, $y \in \partial V$. Setting $\lambda = \delta(V, y)$, we deduce from (3.2) that λ is positive. By Axiom 5 (scale covariance) applied both to Φ and to Λ, we may assume without loss of generality that $\lambda = (1, \ldots, 1)$. If V^0 corresponds to the game that is identically 0, then by (3.3), $V + V^0$ is a game;[9] moreover, $y \in \partial(V + V^0)$, and $\overline{V + V^0}$ corresponds to the TU game v_λ (see (4.2)). Hence by Lemma 8.1, Axioms 2 and 4, and again Lemma 8.1, we have

$$\lambda y = y \in \Phi(V) \cap \partial(V + V^0) = (\Phi(V) + O)) \cap \partial(V + V^0)$$
$$= (\Phi(V) + \Phi(V^0)) \cap \partial(V + V^0) \subset \Phi(V + V^0) = \Phi(\overline{V + V^0})$$
$$= \{\varphi(v_\lambda)\}.$$

Hence $\lambda y = \varphi(v_\lambda)$, which means that $y \in \Lambda(V)$. This completes the proof of Lemma 8.6.

---

9. Condition (3.3) is needed to ensure that $(V + V^0)(S)$ does not fill all of R^S.

Theorem B follows from Proposition 7.3 and Lemma 8.6.

LEMMA 8.7 *If Φ is a value correspondence,[10] then $\Lambda(V) \subset \Phi(V)$ for all V in Γ.*

Proof Let $y \in \Lambda(V)$. Then $y \in \overline{V}(N)$, and there is a comparison vector λ such that the TU game v_λ (see (4.2)) is defined, and

$$\lambda y = \varphi(v_\lambda).$$

Let V_λ be the game corresponding to v_λ. Define a game W by

$$W(S) = \begin{cases} V_\lambda(N) & \text{when} \quad S = N, \\ \lambda \overline{V}(S) & \text{when} \quad S \neq N. \end{cases}$$

Then λy is a Shapley value of W, so $W \in \Gamma$, so $\Phi(W)$ is defined.

Let V^0 correspond to the TU game that vanishes on all coalitions. Then $V_\lambda = \overline{W + V^0}$ and $\partial(W + V^0) = \partial W$, and so by Lemma 8.1 and Axioms 4, 2, 1, and 0, we have

$$\{\varphi(v_\lambda)\} = \Phi(V_\lambda) = \Phi(\overline{W + V^0}) = \Phi(W + V^0)$$
$$\supset (\Phi(W) + \Phi(V^0)) \cap \partial(W + V^0)$$
$$= (\Phi(W) + O) \cap \partial W = \Phi(W) \neq \emptyset.$$

Hence

$$\Phi(W) = \{\varphi(v_\lambda)\} = \{\lambda y\}.$$

By definition, $W(N) = V_\lambda(N) \supset \lambda \overline{V}(N)$, and $W(S) = \lambda \overline{V}(S)$ for $S \neq N$. Moreover $y \in \overline{V}(N)$ yields $\lambda y \in \lambda \overline{V}(N)$. Hence by Axioms 6, 5 and 4, $\lambda y \in \Phi(\lambda \overline{V}) = \lambda \Phi(V)$. Hence $y \in \Phi(V)$, as was to be proved.

Theorem A follows from Proposition 7.3 and Lemmas 8.6 and 8.7.

9 Smoothness

The smoothness condition (3.1) is of the essence; without it, the Shapley correspondence fails to satisfy the conditional additivity axiom, and both our theorems become irreparably false.

To see how smoothness works, let y be a Shapley value of V. The associated "comparison vector" λ always defines a supporting hyperplane to $V(N)$ at y; because of smoothness, it is the *only* supporting hyper-

---

10. Satisfies Axiom 6 as well as 0 through 5.

An Axiomatization of the Non-Transferable Utility Value

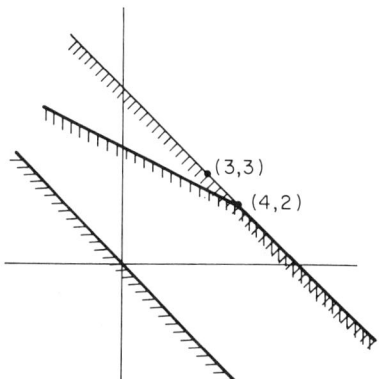

Figure 1
$V(N)$, $W(N)$, and $U(N)$ are, respectively, horizontally, vertically, and diagonally hatched.

plane. If now z is a Shapley value of W, then $y + z$ is efficient in $V + W$ if and only if the supporting hyperplanes at y and z are parallel; therefore, y and z must be associated with the *same* comparison vector, and then additivity follows from the additivity of the TU value.

Without smoothness, the reasoning breaks down. It is possible for $V(N)$ and $W(N)$ to have parallel supporting hyperplanes at y and z, by dint of which $y + z$ is efficient in $V + W$; but these need not be the hyperplanes defined by the comparison vectors that make y and z Shapley values. For example, let $N = \{1, 2\}$, let V correspond to the TU game given by $v(12) - v(1) = (2) = 0$, and define W by

$$W(1) = W(2) = (-\infty, 0],$$

$$W(12) = \{x \in R^N : x^1 + x^2 \leqslant 6 \text{ and } x^1 + 2x^2 \leqslant 8\}$$

(see Figure 1); setting $U = V + W$, we see that U corresponds to the TU game u given by $u(12) = 6$, $u(1) = u(2) = 0$. Then $\Lambda(V) = \{(0, 0)\}$, $\Lambda(W) = \{(4, 2)\}$, and $\Lambda(U) = \{(3, 3)\}$; $(0, 0) + (4, 2)$ is efficient in U, but it is not a value. What is happening is that $V(N)$ and $W(N)$ both have hyperplanes at the respective Shapley values that are orthogonal to $(1, 1)$; but the value $(4, 2)$ of W is associated with the comparison vector $(1, 2)$, not with $(1, 1)$.

Smoothness may be interpreted as local linearity, or, if one wishes, local TU; but note that it is needed for the all-player coalition only. In the guise of differentiability, it has played a significant role in several of the applications; so it is interesting that it makes an appearance on the foundational side as well.

10 Discussion

A. Vanishing Comparison Weights and the Non-Levelness Condition

Shapley's treatment [12] permits some of the comparison weights λ^i to vanish. Ours does not. Vanishing comparison weights are undesirable for several reasons. In the direct, non-axiomatic approach, their intuitive significance is murky; and in the axiomatic approach, they greatly complicate matters. In the applications,[11] vanishing λ^i have played no significant role; in most specific cases it can be shown that the λ^i must be positive, though the definition allows them to vanish.

Our definition of "Shapley value" explicitly takes λ positive; and the non-levelness condition (3.2) assures that whatever emerges from the axioms will be associated with a positive λ. A verbal statement of (3.2) is that weak and strong Pareto optimality are equivalent.

One can avoid the non-levelness condition by strengthening the efficiency axiom to read as follows:

Axiom 1*—Strong Efficiency : If $y \in \Phi(V)$, then
$\{x \in \partial V(N) : x \leqslant y\} = \{y\}$.

This is more than strong Pareto optimality; it says that y is in the relative (to $\partial V(N)$) interior of the strongly Pareto optimal set, or equivalently that $\delta(V, y)$ is positive.[12] If we replace Axiom 1 by Axiom 1*, then one can simply drop (3.2), and our theorems remain true.

B. The Domain

The domain Γ of the axioms is the set of all games that possess at least one Shapley value. This might be considered an esthetic drawback, since in this way the Shapley value enters into its own axiomatic characterization (albeit only via the domain). If one wishes to avoid this, one can replace Γ by any family Δ with the following properties:

Every game in Δ has a Shapley value. (10.1)

All games corresponding to TU games are in Δ. (10.2)

If $V \in \Delta$ and λ is a positive vector in R^N, then $\lambda V \in \Delta$. (10.3)

---

11. We are referring to the existing applications to economic and/or political models, not to isolated numerical examples.

12. In effect, (3.2) asserts that no part of the efficient surface is level, whereas 1* asserts this only on a neighborhood of the value.

The game obtained from any V in Δ by replacing $V(N)$ by
any one of its supporting half-spaces is also in Δ. (10.4)

$V \in \Delta$ if and only if $\bar{V} \in \Delta$. (10.5)

For example, we may take Δ to be the family of all games V such that for all coalitions S, the set of extreme points of $\bar{V}(S)$ is bounded (or equivalently, $\bar{V}(S)$ is the sum of a compact set and a cone).

That this Δ satisfies (10.2) through (10.5) is easily verified. To see that it satisfies (10.1), let $V \in \Delta$. Let C be the set of all positive λ in R^N for which v_λ is defined and $\sum_{i \in N} \lambda^i = 1$. Then C is convex and compact, and the mapping $\lambda \to v_\lambda(S)$ is continuous on C for each S; hence also $\lambda \to \varphi v_\lambda$ is continuous on C. For each λ in C, define x_λ in R^N by $\lambda x_\lambda = \varphi v_\lambda$; let y_λ be the point in $\bar{V}(N)$ closest to x_λ; and let λ' be the normal to the supporting hyperplane of $\bar{V}(N)$ at x_λ. By (3.3), $\lambda' \in C$; hence $\lambda \to \lambda'$ is a continuous mapping from C into itself, and so has a fixed point λ^*. Then x_{λ^*} is a Shapley value of V.

Thus Δ does satisfy (10.1) through (10.5), and so can replace Γ as the domain of the value correspondence. We adopted Γ as the domain because it is the largest on which the axioms "work," and is thus the most useful from the point of view of applications.

Note that the above existence proof may be modified so as to make no use of smoothness.

Finally, we mention that (10.1) is not gratuitous; there are indeed games not possessing any Shapley value. For example, let $N = \{1,2\}$, and define V by

$$V(1) = V(2) = (-\infty, 0],$$

$$V(12) = \{(\xi, \eta) \in R^N : \eta < 0 \text{ and } \xi\eta^2 \leq -1\}.$$

If $x = (\xi, \eta)$ were a value, then the tangent to $\partial V(N)$ at x would have a slope equal in magnitude (but opposite in sign) to the slope of the line connecting x with the origin; and this can never be, since the respective slopes are $-\eta/2\xi$ and η/ξ. The example is of course highly pathological, since each player can guarantee 0 to himself, but can never achieve this in $V(N)$; but it does show that one cannot simply take the domain to be the set of *all* games.

C. Conditional Additivity

The Conditional Additivity Axiom can be replaced by the following pair of axioms:

Axiom 2*a—Conditional Sure-Thing: *If* $U = \frac{1}{2}V + \frac{1}{2}W$, *then*

$\Phi(U) \supset \Phi(V) \cap \Phi(W) \cap \partial U(N)$.

Axiom 2*b—Translation Covariance: *For all x in R^N,*

$\Phi(V + x) = \Phi(V) + x$.

In Section 6, we suggested that the value of a game may be viewed as a group decision, compromise, or arbitrated outcome, that is reasonable in view of the alternatives open to the players and coalitions (rather than an outcome that is itself in some sense stable, such as a core point). In these terms, Axiom 2*a says the following: Suppose that y is a reasonable compromise both in the game V and in the game W. Suppose further that one of the games V and W will be played; at present it is not yet known which one, but it is common knowledge that the probabilities are half-half. Then y is a reasonable compromise in this situation as well,[13] unless the players can use the uncertainty to their mutual advantage.

D. Non-Uniqueness of the Value

Given the above view of the value as a reasonable compromise, some readers may be disturbed by the fact that a given game V may have more than one value.[14] Non-uniqueness, they may say, is all very well for stability or equilibrium concepts; but a theory of arbitration, of reaching reasonable compromises, should "recommend" a single point.

On closer examination, there seems to be no particular reason to accept such a view. Compromises may be based on many different kinds of principles and criteria. Such criteria are usually overlapping, in the sense that a given one applies to only a limited range of situations, and to a given situation several criteria may apply. This results in a multi-valued function—a correspondence.

A good analogy is to law; in fact, one can view civil law as society's way of reaching "reasonable compromises." Specific laws always have limited ranges; these ranges often overlap and yield contradictory results. An important function of a judge is to "resolve" such contradictions in each specific case brought before him, by selecting one of the applicable laws. It is no wonder that judgments are often overturned on appeal, and that different jurisdictions reach different opinions on identical cases. Law is multi-valued, not incoherent.

---

13. Compare [12, p. 261, IV].

14. I.e., that $\Phi(V)$ may contain several points. Of course, Theorem A guarantees that the *correspondence Φ is* unique.

In much the same way, a value correspondence is a coherent system. Its coherence is expressed by the axioms, by the way that they relate values of different games to each other. The axioms say, *if* you can decide such-and-such in case (a), *then* you can decide so-and-so in case (b). There is no reason to expect such a system to be single valued.

The original definition of the NTU value is an instance of this kind of system. Here a "criterion" is a vector λ of comparison weights, which the players (or the arbitrator) use to compare utilities. Given such a criterion, a "reasonable compromise" is the TU value $\varphi(v_\lambda)$; and the criterion "applies" to the NTU game V if $\varphi(v_\lambda)$ is feasible in V.

Our results say that *every* value system that is coherent, in the sense that it satisfies the axioms, *must* be of this specific kind.

References

1. Aumann, R. J.: "Values of Markets with a Continuum of Traders," *Econometrica*, 43 (1975), 611–646 [Chapter 51].
2. ———: "On the Non-transferable Utility Value: A Comment on the Roth-Shafer Examples," *Econometrica*, 53 (1985), 667–677 [Chapter 61c].
3. Aumann, R. J., and M. Kurz: "Power and Taxes," *Econometrica*, 45 (1977), 1137–1161 [Chapter 52].
4. Harsanyi, J. C.: "A Bargaining Model for the Cooperative *n*-Person Game," in *Contributions to the Theory of Games IV*, ed. by A. W. Tucker and R. D. Luce. Princeton: Princeton University Press, 1959, pp. 325–355.
5. ———: "A Simplified Bargaining Model for the *n*-Person Cooperative Game," *International Economic Review*, 4 (1963), 194–220.
6. ———: "Comments on Roth's paper, 'Values for Games without Side Payments'," *Econometrica*, 48 (1980), 477.
7. Nash, J. F.: "The Bargaining Problem," *Econometrica*, 18 (1950), 155–162.
8. ———: "Two-Person Cooperative Games," *Econometrica*, 21 (1953), 128–140.
9. Roth, A.: "Values for Games without Side Payments: Some Difficulties with Current Concepts," *Econometrica*, 48 (1980), 457–465 [Chapter 61a].
10. Shafer, W.: "On the Existence and Interpretation of Value Allocations," *Econometrica*, 48 (1980), 467–477 [Chapter 61b].
11. Shapley, L. S.: "A Value for *n*-Person Games," in *Contributions to the Theory of Games II*, ed. by H. W. Kuhn and A. W. Tucker. Princeton: Princeton University Press, 1953, pp. 307–317.
12. ———: "Utility Comparison and the Theory of Games," in *La Decision, Aggregation et Dynamique des Ordres de Preference*, Editions du Centre National de la Recherche Scientifique, Paris, 1969, pp. 251–263.

61a Values for Games without Side Payments: Some Difficulties with Current Concepts

Alvin E. Roth

1 Introduction

The game theory literature currently contains two well-developed solution concepts which can be applied to a cooperative game without sidepayments to select an outcome of the game as its "value." The first of these is due to Harsanyi [6, 7], and is developed as a generalization of Nash's [12] solution of the two-person bargaining problem. Harsanyi's procedure selects a "stable bargaining solution" which is interpreted to be the outcome which would be selected by rational (utility-maximizing) players who are aware of all of the possibilities in the game.

The second solution concept, called the "λ-transfer value," was first proposed by Shapley [19] as an extension of the Shapley value [18] for games with sidepayments. Shapley [19, p. 260] writes that the idea was first motivated as an attempt to approximate Harsanyi's bargaining solution in the context of market games with many players, and was then perceived to have virtues of its own. The λ-transfer value has subsequently been studied and developed, primarily in the context of market games, by Aumann [1] and others (e.g., Aumann and Kurz [2, 3]; Champsaur [5]; Hart [8]; Mas-Colell [9, 10, 11]; Neyman [13]). However, like Harsanyi's stable bargaining solution, the λ-transfer value is defined for essentially all cooperative games, and it is customarily justified and interpreted without reference to markets. Since the Shapley value for games with sidepayments can be interpreted either as a stable outcome of bargaining (e.g., Harsanyi [7, Chapter 11]) or as an expected outcome or expected utility of playing a game (e.g., Shapley [18]; Roth [15, 16]), it can be argued that the λ-transfer value should be interpreted in either way.

Regardless of which interpretation is used, however, both the λ-transfer value and the stable bargaining solution can yield predictions which are highly counterintuitive. In particular, Section 2 exhibits some games for which it seems that neither solution concept yields a result which is consistent with the hypothesis that the players of the game are rational utility-maximizers who are aware of all of the possible outcomes of the game.[1]

This paper originally appeared in *Econometrica* 48 (1980): 457–65. Reprinted with permission.

1. Since I wish to show that these solution concepts can yield counter-intuitive results, the discussion will depend, to some extent, on informal, intuitive reasoning about the appropriate *interpretation* of mathematical ideas. Although I have tried to make the arguments as compelling as possible, I recognize that this sort of discussion may leave room for disagreement.

Section 2 contains the examples and an analysis of them; Section 3 contains a brief description of the stable bargaining solution, and applies it to the examples; Section 4 briefly describes the λ-transfer value and applies it to the examples. Section 5 contains a discussion of these results. Readers who are not interested in how the stable bargaining solution and λ-transfer value are computed may wish to read Section 2 and then skip directly to Section 5.

2 The Examples

Consider a class of 3-player games defined by a single parameter p which varies between 0 and $1/2$. For a given $p \in [0, 1/2]$, let $G(p)$ be the game in which the players acting alone can assure themselves of achieving a utility of 0, players 1 and 2 acting together can achieve the outcome $(1/2, 1/2, 0)$, players 1 and 3 acting together can achieve the outcome $(p, 0, 1 - p)$, and players 2 and 3 together can achieve $(0, p, 1 - p)$. All three players acting together can achieve any convex combination of the vectors $(1/2, 1/2, 0)$, $(p, 0, 1 - p)$, and $(0, p, 1 - p)$. Assume that no sidepayments of any sort are feasible between players.

This is essentially a game in characteristic function form,[2] so it can be represented by the set $N = \{1, 2, 3\}$ of players, the set of feasible outcomes $H(p)$ equal to the convex hull of the point $\{(1/2, 1/2, 0), (p, 0, 1 - p), (0, p, 1 - p), (0, 0, 0)\}$, and the characteristic function[3] V_p such that

$V_p(i) = \{(u_1, u_2, u_3) | u_i \leq 0\}$ for $i \in N$,

$V_p(\{12\}) = \{(u_1, u_2, u_3) | (u_1, u_2) \leq (1/2, 1/2)\}$,

$V_p(\{13\}) = \{(u_1, u_2, u_3) | (u_1, u_3) \leq (p, 1 - p)\}$,

$V_p(\{23\}) = \{(u_1, u_2, u_3) | (u_2, u_3) \leq (p, 1 - p)\}$, and

$V_p(N) = \{u = (u_1, u_2, u_3) | u \leq y$ for some y in the convex hull of

$\{(1/2, 1/2, 0), (p, 0, 1 - p), (0, p, 1 - p)\}\}$.

---

2. That is, the payoffs available to a coalition are independent of the actions of the complementary coalition.

3. In writing the characteristic function I have adopted the usual convention (cf. Aumann and Peleg [4]) that if x is in $V(S)$ and $y \leq x$, then y is also in $V(S)$. In general, however, y need not be a feasible outcome of the game, and in the game $G(p)$, only outcomes in the set $H(p)$ are feasible. Thus the game $G(p)$ is formally described by the triple $(N, V_p, H(p))$. This description is included only to insure that there is no ambiguity in the definition of the game $G(p)$, and readers who are unfamiliar with the characteristic function can safely ignore it.

I claim that, for $p < 1/2$, the payoff vector $(1/2, 1/2, 0)$ is the unique outcome of the game consistent with the hypothesis that the players are rational utility maximizers. This is because, when $p < 1/2$, the outcome $(1/2, 1/2, 0)$ is *strictly* preferred by *both* players 1 and 2 to *every* other feasible outcome, and because the rules of the game permit players 1 and 2 to achieve this outcome without the cooperation of player 3. So in the game $G(p)$, for $p < 1/2$, there is really no conflict between players 1 and 2: their interests coincide in the choice of the outcome $(1/2, 1/2, 0)$, and the rules permit them to achieve this outcome.

This is perhaps clearest when $p = 0$, since in the game $G(0)$ players 1 and 2 have identical interests over the entire set of feasible outcomes. There is no pair of outcomes such that player 1 would choose one over the other, and player 2 would make the opposite choice. Furthermore, as far as players 1 and 2 are concerned, player 3 has nothing to offer in the game $G(0)$—his cooperation never offers either of them an increased reward. So, from the point of view of players 1 and 2, this game offers no prospect of reward different from the two-player game in which players 1 and 2 can achieve $(1/2, 1/2)$ if they agree, and $(0, 0)$ otherwise. The only rational outcome of this game, and consequently of the original game, is that players 1 and 2 should each receive a utility of $1/2$.

For $p < 1/2$, the outcome $(1/2, 1/2, 0)$ is not only the unique point in the core of the game $G(p)$, but is also the unique von Neumann–Morgenstern solution of the game.[4] It should be emphasized, however, that it is not this observation which prompts the conclusion that $(1/2, 1/2, 0)$ is the unique rational outcome of the game; the conclusion is due to the fact that players 1 and 2 agree on that outcome as preferable to all others.[5] (Note that this can never occur in a game with sidepayments, or

---

4. By the core and solution of the game $G(p)$, I mean the core and solution of the feasible set $H(p)$ under the domination relation induced by the characteristic function V_p (cf. Aumann and Peleg [4]).

5. Games without sidepayments are often described just by a set N of players and a characteristic function V. The set of feasible outcomes is then assumed to be equal to the set $V(N)$. Given the construction of the characteristic function (cf. footnote 4) this amounts to making the somewhat restrictive assumption that it is feasible to freely dispose of utility. Nevertheless, such an assumption would make no consequential difference in the analysis of the games $G(p)$ for $p < 1/2$. The principal effect of replacing the feasible set $H(p)$ by the set $V_p(N)$ would be that player 1, say, could achieve his maximum utility not only at the outcome $(1/2, 1/2, 0)$, but also at outcomes of the form $(1/2, q, 0)$ where $q < 1/2$, obtained by having player 2 dispose of some of the utility he could have obtained at the outcome $(1/2, 1/2, 0)$. Of course player 2 would prefer not to dispose of utility, and player 1 has no incentive to seek such a disposal of utility. (The hypothesis that the players are utility maximizers means player 1 doesn't want player 2 to dispose of utility in this way, since player 1 can't *want* anything which isn't captured by his utility function, and his utility remains constant at $1/2$.) Since each of players 1 and 2 can achieve his maximum utility only with the cooperation of the other, it is now straightforward to argue, as before, that $(1/2, 1/2, 0)$ must be the outcome of the game.

in any game in which it is possible to freely redistribute wealth between players.)

Finally, note that the conclusions reached in this section apply only when $p < 1/2$. When $p = 1/2$ the game $G(p)$ is completely symmetric with respect to the players, so it is no longer the case that cooperation with player 3 offers strictly less to players 1 or 2 than cooperation with one another.

3 Harsanyi's Stable Bargaining Solution

Let G be a game with a set $N = \{1, \ldots, n\}$ of players and a set H of feasible, individually rational payoffs. Harsanyi's model can be divided into two parts. The first part identifies one or more *bargaining solutions* of the game. In the event that more than one such bargaining solution is identified, the second part identifies a unique *stable bargaining solution*.[6]

A bargaining solution defined by the first part of the model consists of a feasible outcome u associated with a non-negative vector λ such that $\lambda \cdot u \geq \lambda \cdot x$ for any feasible outcome x. (That is, λ is the vector of coefficients of a hyperplane tangent to H at the point u.) To qualify as a bargaining solution, the payoff u_i to each player i must be the sum of dividends w_i^S which he receives from each coalition S of which he is a member, according to the rule that each coalition pays the maximum dividends that it can afford, subject to the restriction that $\lambda_i w_i^S = \lambda_j w_j^S$ for all $i, j \in S$. (The dividends w_i^S may be either positive or negative.)[7]

Let B be the (non-empty) set of bargaining solutions identified above. If B contains a unique element, then that is defined to be the unique stable bargaining solution of the game. Otherwise the unique stable bargaining solution is defined to be Nash's [12] solution of the pure bargaining game with outcome set H and disagreement point d such that $d_i = \min_{u \in B} u_i$. Thus the unique *stable bargaining solution*[8] is defined to be the outcome $u \in H$ such that $u \geq d$ and $\prod_{i \in N}(u_i - d_i) \geq \prod_{i \in N}(x_i - d_i)$ for all $x \in H$ such that $x \geq d$.

To see how this model applies to the games $G(p)$, first note that the plane $x_1 + x_2 + x_3 = 1$ is tangent to the set $H(p)$. When $\lambda = (1, 1, 1)$,

---

6. Only a brief description of the model will be given in this section. For a complete description, see Harsanyi [7, Chapter 12].

7. If $\lambda_i = 0$ then w_i^S is not uniquely determined, and may be taken to be any quantity which yields a feasible solution (cf. Harsanyi [7, p. 258]). Of course, not every vector of weights yields dividends which produce a bargaining solution, but Harsanyi has shown that at least one such solution always exists for a wide class of games.

8. Note that the stable bargaining solution need not be a bargaining solution.

the dividends which each coalition pays to determine a bargaining solution are $w_i^{\{i\}} = 0$ for $i \in N$, $w_1^{\{12\}} = w_2^{\{12\}} = 1/2$, $w_1^{\{13\}} = w_3^{\{13\}} = p$, $w_2^{\{23\}} = w_3^{\{23\}} = p$, and $w_1^N = w_2^N = w_3^N = -4p/3$. (Note that the grand coalition N must give a negative dividend, to preserve feasibility.) Summing these dividends, we find the bargaining solution for the game $G(p)$ corresponding to the weights $\lambda = (1,1,1)$ is the outcome $u = (1/2 - p/3, 1/2 - p/3, 2p/3)$.

It is easy to verify, however, that this is not the only bargaining solution of the game $G(p)$. In fact, each of the three extreme points of the set of Pareto optimal outcomes is the bargaining solution corresponding to some set of weights, for any game $G(p)$ with $p \geqslant 1/4$.

If a and c are positive numbers such that $pa = (1-p)c$, then the vector of weights $\lambda = (a, 0, c)$ yields the outcome $(p, 0, 1-p)$ as a bargaining solution of the game $G(p)$, since for these weights, the dividends $w_1^{\{13\}} = p$, $w_3^{\{13\}} = 1-p$, and all other dividends equal zero.[9] Similarly, if b and c are positive numbers such that $pb = (1-p)c$, then the bargaining solution corresponding to the weights $\lambda = (0, b, c)$ is the outcome $(0, p, 1-p)$, and the bargaining solution corresponding to $\lambda = (1, 1, 0)$ is the outcome $(1/2, 1/2, 0)$. Thus, for each player in the game, there is a bargaining solution at which he receives his minimum individually rational payoff.

Because there are multiple bargaining solutions, the second part of the model identifies a unique stable bargaining solution as follows.

PROPOSITION 3.1 *The stable bargaining solution for the game $G(p)$ $p \in [1/4, 1/2]$ is the outcomes $u = (1/3, 1/3, 1/3)$.*

Proof Since there is some bargaining solution at which each player receives 0, his minimum individually rational payoff, the disagreement point used to determine the stable bargaining solution is $d = (0, 0, 0)$. Thus the stable bargaining solution is the Nash solution of the three-person bargaining game with feasible set $H(p)$ and disagreement point d.

But the point $u = (1/3, 1/3, 1/3)$, which maximizes the product $\prod_{i \in N} x_i$ over the simplex $\{x \geqslant 0 | \Sigma x_i = 1\}$ is always feasible in the game $G(p)$ for $p \in [1/4, 1/2]$; i.e., $(1/3, 1/3, 1/3) \in H(p)$ for all $p \in [1/4, 1/2]$. Consequently u is also the Nash solution to the bargaining game with feasible set $H(p)$, so u is the unique stable bargaining solution of the game $G(p)$.

9. The requirement that $p \geqslant 1/4$ is needed in order that the plane defined by the equation $ax_1 + cx_3 = ap + c(1-p)$, which passes through the point $(p, 0, 1-p)$, be tangent to the set of feasible outcomes $H(p)$.

4 The λ-Transfer Value for the Games $G(p)$

Let $\lambda = (a, b, c)$ be a non-negative vector of weights, at least one of which is not zero. Define the *weighted game* $G_\lambda(p)$ to be the game in which each coalition S can achieve the payoffs (au_1, bu_2, cu_3), where (u_1, u_2, u_3) is a payoff which that coalition could achieve in the game $G(p)$. Define the weighted game with sidepayments $g_\lambda(p)$ to be the game in which a coalition can achieve any distribution of utility whose sum over members of the coalition does not exceed the sum of the utilities available to that coalition at some outcome in the game $G_\lambda(p)$. Since the games $G_\lambda(p)$ are in characteristic function form, so are the games $g_\lambda(p)$, which can be represented by the characteristic function $v_{\lambda p}$ given by

$v_{\lambda p}(i) = 0 \quad \text{for } i = 1, 2, 3;$

$v_{\lambda p}(12) = \dfrac{a+b}{2};$

$v_{\lambda p}(13) = pa + (1-p)c;$

$v_{\lambda p}(23) = pb + (1-p)c;$

$v_{\lambda p}(123) = \max\left\{\dfrac{a+b}{2},\, pa + (1-p)c,\, pb + (1-p)c\right\}.$

Consequently, the Shapley value $\phi(v_{\lambda p})$ is always well defined for the game $g_\lambda(p)$. In general, the Shapley value for the game $g_\lambda(p)$ need not be feasible for the game $G_\lambda(p)$. However if no utility transfers are required to achieve the Shapley value for the game $g_\lambda(p)$, then it is a feasible outcome of the game $G_\lambda(p)$, and in this case it is called a λ-transfer value for the game $G(p)$. Formally, a *λ-transfer value* for the game $G(p)$ is a feasible outcome $u = (u_1, u_2, u_3)$ of the game $G(p)$ such that $(au_1, bu_2, cu_3) = \phi(v_{\lambda p})$ for some vector of weights $\lambda = (a, b, c)$.

Although there are a number of ways to motivate this definition, the argument which has been most influential is made in two parts.[10] The first part considers the original game (i.e., the game $G(p) = G_\lambda(p)$ for $\lambda = (1, 1, 1)$) and the corresponding game with sidepayments. If the Shapley value of the sidepayment game is feasible in the original game, then it is justified as the value for the original game by invoking Nash's [12] principle of "independence of irrelevant alternatives." Aumann [1, Section 6][11] puts the argument in a picturesque way by supposing that

---

10. See Shapley [19] and Aumann [1] for a more complete account.
11. Aumann is actually considering a modified version of the value described here.

the players first negotiate under the assumption that utility is transferable, and only find that this is not the case after reaching an agreement. If the agreement requires no utility transfers to be made, however, then it could still be implemented.

Since this procedure may not yield a feasible outcome, the second part of the argument notes that, for any *strictly positive* vector of weights λ, the game $G_\lambda(p)$ is equivalent to $G(p)$ in the sense that the utility functions of the players are unique only up to positive linear transformations. So if some such λ can be found which yields a feasible outcome, the argument of the previous paragraph should still apply.[12]

In order to prove an existence theorem for a broad class of games (Shapley [19]) it is necessary to extend the definition to include vectors of weights λ which may be zero in some components. However the equivalence argument breaks down in this case, and this extension is customarily viewed as simply a technical expedient. In many cases of interest, existence can be obtained without resorting to weights of zero.

We are now in a position to study the λ-transfer value for the game $G(p)$.

PROPOSITION 4.1 *For any game $G(p)$ with $p \in [0, 1/2]$, the outcome $u = (1/3, 1/3, 1/3)$ is a λ-transfer value for $\lambda = (1, 1, 1)$.*

Proof Observe that when $\lambda = (a, b, c) = (1, 1, 1)$, then $v_{\lambda p}$ is independent of p and is the completely symmetric characteristic function representing 3-person majority-rule. Consequently, $\phi(v_{\lambda p}) = u = (1/3, 1/3, 1/3)$ for all p. Furthermore, the outcome $u = (1/3, 1/3, 1/3)$ is feasible in the game $G(p)$ for all $p \in [0, 1/2]$, since it lies on the line joining the outcomes $(1/2, 1/2, 0)$ and $(p/2, p/2, 1-p)$. Consequently u is the λ-transfer value for the game $G(p)$.

Of course, another vector λ of weights could yield another λ-transfer value. For instance, it is not difficult to verify that the weights $\lambda = (a, b, c) = (1, 1, 0)$ yield the λ-transfer value $u = (1/2, 1/2, 0)$ for the game $G(0)$. In view of the analysis of the game $G(p)$ given in Section 2, we wish to know in exactly what circumstances weights λ can be found such that the λ-transfer value gives player 3 a payoff of zero in games $G(p)$ for $p < 1/2$.

PROPOSITION 4.2 *If $u = (u_1, u_2, u_3)$ is a λ-transfer value for a game $G(p)$ $p \in (0, 1/2)$, then $u_3 > 0$.*

---

12. The vector λ can be interpreted as a vector of exchange rates according to which utility can be transferred in the sidepayment game—hence the name "λ-transfer value."

PROPOSITION 4.3 *If* $u = (u_1, u_2, u_3)$ *is a* λ-*transfer value for the game* $G(p)$, *and if* $\lambda = (a, b, c)$ *with* $c > 0$, *then* $u_3 > 0$ *for all* $p \in [0, 1/2]$.

Proof When $p > 0$ at least one of $v_{\lambda p}(13)$ or $v_{\lambda p}(23)$ must be positive, since not all of the components of λ may be equal to zero. When $c > 0$, both of these quantities must be positive. Consequently, under the conditions of both propositions, the Shapley value for the game with sidepayments gives player 3 a positive payoff for any λ. Since a λ-transfer value corresponds to the Shapley value of some sidepayment game, the result follows.

Thus the outcome (1/2, 1/2, 0) can be achieved as a λ-transfer value only in the game $G(0)$, and even then only by first multiplying the utility function of player 3 by zero.

5 Discussion

In order to simplify the discussion, consider a unique λ-transfer value for the games $G(p)$ by taking $\lambda = (1, 1, 1)$. (By Propositions 4.2 and 4.3, most of the discusson will apply as well to any other choice of weights λ.)

Both the stable bargaining solution and the λ-transfer value select the outcome $u = (1/3, 1/3, 1/3)$ for any game $G(p)$ with $p \in [1/4, 1/2]$. (In fact the λ-transfer value selects the outcome u for any $p \in [0, 1/2]$.) In view of the analysis presented in Section 2 of the games $G(p)$ for $p < 1/2$, it no longer seems tenable to interpret either solution concept as yielding a "stable" outcome of the game.[13] Perhaps something more needs to be said, however, about the idea that the outcome (1/3, 1/3, 1/3) might represent some sort of expected outcome or expected utility. These are two distinct ideas, and must be addressed separately.

To interpret $u = (1/3, 1/3, 1/3)$ as an expected outcome, it must be maintained either that u is the only outcome likely, in some sense, to occur, or else that there is a set of likely outcomes distributed in such a way as to make u their expectation. For the three-person majority game with sidepayments (or for the game $G(p)$ with $p = 1/2$), the complete symmetry of the game makes both of these positions defensible, since u is the only efficient outcome which is symmetric. For instance, any theory which depends only on the structure of the game and which finds the outcome (1/2, 1/2, 0) to be stable must also find the outcomes (1/2, 0, 1/2) and (0, 1/2, 1/2) to be stable, and u is the mean of these three outcomes.

---

13. In private correspondence, John Harsanyi has informed me that he is currently developing a solution concept which *does* yield the outcome (1/2, 1/2, 0) as the solution to the games $G(p)$ with $p < 1/2$.

But when $p > 1/2$, the game $G(p)$ is *not* symmetric (and in particular the outcomes $(1/2, 0, 1/2)$ and $(0, 1/2, 1/2)$ are not feasible). If u is to be interpreted as an expected outcome, then one must be prepared to argue that there are some "likely" outcomes which can be balanced against $(1/2, 1/2, 0)$ to yield u as an expected outcome. In view of the analysis presented in Section 2, this would seem to be a difficult position to defend. Note that I am essentially suggesting that the principle of "independence of irrelevant alternatives" is inappropriate under this interpretation—the notion of an "expected outcome" seems to require consideration of feasible outcomes other than the one chosen for the solution.

Finally, consider whether the vector $u = (1/3, 1/3, 1/3)$ might reasonably represent the expected utility of playing one of the positions in a game $G(p)$. Of course, any function which assigns numerical values to the positions in a game can be used as a utility function in the limited sense that it induces binary choices which are transitive and complete (since the ordering of the real numbers has this property). To decide whether a function is a reasonable utility function, it is necessary to consider what kind of preferences it reflects. The Shapley value for games with sidepayments was studied in this context in Roth [15, 16].

In order for the λ-transfer value (with $\lambda = (1, 1, 1)$) to represent an individual's utility for playing in the games $G(p)$, it must be that he is indifferent between playing in any position of any game $G(p)$ for $p \in [0, 1/2]$. In particular, he must be indifferent between playing position 1 or position 3 in any game $G(p)$ for $p < 1/2$, and indifferent between playing position 3 in the game $G(p)$ or in the game $G(1/2)$, since all of these prospects have a λ-transfer value of $1/3$. In view of the preceding discussion, these preferences are inconsistent with the notion that a rational player's preferences over games should be influenced by the payoff he might reasonably expect to achieve in those games.

Thus, for a simple family of superadditive games without sidepayments, both the stable bargaining solution and the λ-transfer value yield results which are difficult to justify. Unfortunately, the analysis of these games does not itself suggest any alternative theory for general cooperative games, since the arguments depend critically on the extreme simplicity of the games $G(p)$, which permitted us to analyze the games as in Section 2, essentially from first principles.

What this analysis does suggest is that, at the very least, some modifications are required in the existing theory.[14]

---

14. For other suggestive examples, see Owen [14, Example 2] or Shafer [17].

Acknowledgments

This research was supported by National Science Foundation grant SOC75-21820 to the Institute for Mathematical Studies in the Social Sciences, Stanford University and by National Science Foundation Grant SOC78-09928 to the University of Illinois. It is also a pleasure to acknowledge stimulating conversations on this topic with R. Aumann, T. Groves, J. Harsanyi, S. Hart, M. Kurz, A. Neyman, M. Osborne, L. Shapley, and R. Wilson.

References

1. Aumann, Robert J.: "Values of Markets with a Continuum of Traders," *Econometrica*, 43 (1975), 611–646 [Chapter 51].
2. Aumann, Robert J., and M. Kurz: "Power and Taxes," *Econometrica*, 45 (1977), 1137–1161 [Chapter 52].
3. ———: "Power and Taxes in a Multicommodity Economy," *Israel Journal of Mathematics*, 27 (1977), 185–234 [Chapter 53].
4. Aumann, Robert J., and B. Peleg: "Von Neumann–Morgenstern Solutions to Cooperative Games without Side Payments," *Bulletin of the American Mathematical Society*, 66 (1960), 173–179 [Chapter 38].
5. Champsaur, Paul: "Cooperation versus Competition," *Journal of Economic Theory*, 11 (1975), 393–417.
6. Harsanyi, John C.: "A Simplified Bargaining Model for the n-Person Cooperative Games," *International Economic Review*, 4 (1963), 194–220.
7. ———: *Rational Behavior and Bargaining Equilibrium in Games and Social Situations*. Cambridge: Cambridge University Press, 1977.
8. Hart, Sergiu: "Values of Non-Differentiable Markets with a Continuum of Traders," *Journal of Mathematical Economics*, 4 (1977), 103–116.
9. Mas-Colell, Andreu: "Competitive and Value Allocations of Large Exchange Economies," *Journal of Economic Theory*, 14 (1977), 419–438.
10. ———: "On the Asymptotic Equivalence Theorems," mimeo, Universität Bonn, March, 1977.
11. ———: "Remarks on the Game-Theoretic Analysis of a Simple Distribution of Surplus Problem," *International Journal of Game Theory*, 9 (1980), 125–140.
12. Nash, John F.: "The Bargaining Problem," *Econometrica*, 28 (1950), 155–162.
13. Neyman, Abraham: "Values for Non-Transferable Utility Games with a Continuum of Players," Technical Report No. 351, School of Operations Research and Industrial Engineering, Cornell University, 1977.
14. Owen, Guillermo: "Values of Games without Sidepayments," *International Journal of Game Theory*, 1 (1972), 95–108.
15. Roth, Alvin E.: "The Shapley Value as a von Neumann–Morgenstern Utility," *Econometrica*, 45 (1977), 657–664.
16. ———: "Bargaining Ability, the Utility of Playing a Game, and Models of Coalition Formation," *Journal of Mathematical Psychology*, 16 (1977), 153–160.
17. Shafer, Wayne J.: "On the Existence and Interpretation of Value Allocation," *Econometrica*, 48 (1980), 467–476 [Chapter 61b].
18. Shapley, Lloyd S.: "A Value for n-Person Games," in *Contributions to the Theory of Games*, II, ed. by H. W. Kuhn and A. W. Tucker. Princeton: Princeton University Press, 1953, 307–317.
19. ———: "Utility Comparisons and the Theory of Games," in *La Décision*, Editions du Centre National de la Recherche Scientifique, Paris, 1969, 261–263.

61b On the Existence and Interpretation of Value Allocation

Wayne J. Shafer

1 Introduction

The concept of a value allocation in an economy without transferable utility, due to Shapley [9], has provided a way of characterizing certain allocations in an exchange economy which apparently reflect the bargaining strength of the agents, analogous to the Shapley value in side payment games. The idea of a value allocation may be either in the context of cardinal utility, as in the formulation of Shapley, or in a purely ordinal framework as formulated by Aumann [1]. The purpose of this paper is basically to analyze some properties of value allocations by means of examples. First, a problem of existence of ordinal value allocations has been raised by an example, due to Kannai and Mantel [5], of an exchange economy in which for no choice of utility functions will the Pareto set be convex. This problem is resolved by means of an existence theorem, and an example of a value allocation in the Kannai–Mantel example is presented. Secondly, a series of examples, in which the value allocations can be interpreted in either a cardinal or ordinal sense, is presented which have the curious property that one agent receives more of all goods than he is initially endowed with, even if this agent has a zero initial endowment.

2 Definitions and Notation

The commodity space will always be the nonnegative orthant $\mathbb{R}^l_+$ of Euclidean l space. Given two vectors x, y in $\mathbb{R}^l_+$, xy denotes the inner product. The term *preference ordering* will always mean a complete preorder $\gtrsim$ of $\mathbb{R}^l_+$ such that the graph $\{(x,y) \in \mathbb{R}^l_+ \times \mathbb{R}^l_+ : x \gtrsim y\}$ is closed. $\mathscr{P}^*$ denotes the space of all such preference orderings endowed with the topology of closed convergence of Hausdorff (see Hildenbrand [3]). Given a $\gtrsim \in \mathscr{P}^*$, $x \gtrsim y$ means "x is as good as y," and $x \succ y$ is defined to mean $[x \gtrsim y$ and not $y \gtrsim x]$, i.e., "x is better than y." A *utility function* for a $\gtrsim \in \mathscr{P}^*$ is a continuous function $u : \mathbb{R}^l_+ \to \mathbb{R}$ such that $[u(x) \geq u(y) \Leftrightarrow x \gtrsim y]$. A $\gtrsim \in \mathscr{P}^*$ is *monotone* if $[x \geq y, x \neq y \Rightarrow x \succ y]$; *convex* if $\{y : y \gtrsim x\}$ is a convex set for each $x \in \mathbb{R}^l_+$; *concavifiable* if there exists a concave utility function for $\gtrsim$; *locally nonsatiated* if for each

This paper originally appeared in *Econometrica* 48 (1980): 467–477. Reprinted with permission.

$x \in \mathbb{R}_+^l$, and each neighborhood $N(x)$ of x, there exists a $y \in N(x)$ such that $y \succ x$.

An *exchange economy* $\mathscr{C} = (\succsim_a, \omega_a)_{a \in A}$ is a specification, for each point a in a finite set of agents A, of a $\succsim_a \in \mathscr{P}^*$ and an "initial endowment" vector $\omega_a \in \mathbb{R}_+^l$, such that $\sum_{a \in A} \gg 0$. A *game with side payments* $\Gamma = (A, v)$ consists of a finite set of agents A and a superadditive real valued function v defined on the subsets of A such that $v(\emptyset) = 0$. A $S \subset A$ is called a "coalition" and $v(S)$ is the "worth" of the coalition. The *Shapley value* (Shapley [8]) of a game $\Gamma = (A, v)$ is a rule which assigns to each agent a the "payoff" s_a according to the formula

$$s_a = \sum_{S \subset A} \frac{(|S|-1)!(|A|-|S|)!}{|A|!} [v(S) - v(S \setminus \{a\})].$$

The Shapley value is intended to be a measure of an individual's "bargaining power," and can be interpreted to be the agent's expected marginal contribution to a coalition S. The basic properties of the value that will be needed later are that $\sum_{a \in A} s_a = v(A)$ (efficiency), and $s_a \geq v(\{a\})$ for each $a \in A$ (individual rationality). Given an economy $\mathscr{C} = (\succsim_a, \omega_a)_{a \in A}$, and a vector $u = (u_a)_{a \in A}$ of utility functions, one for each $\succsim_a$, define a game (A, v_u) by

$$v_u(S) = \max_{x_a} \sum_{a \in S} u_a(x_a) \quad \text{subject to} \quad \sum_{a \in S} (x_a - \omega_a) = 0. \tag{1}$$

Then $\{x_a\}_{a \in A}$ is defined to be a *value allocation* for $\mathscr{C}$ if and only if $\sum_{a \in A}(x_a - \omega_a) = 0$ and there exists utility functions $u = (u_a)_{a \in A}$ for each $\succsim_a$ such that $(u_a(x_a))_{a \in A}$ is the Shapley value of the game (A, v_u). This is clearly an "ordinal" approach and is due to Aumann [1]. There is also a "cardinal" approach, due to Shapley [9] (see also Shapley and Shubik [10]); one specifies a utility function for each agent a priori as part of his characteristics. In this case utility is considered unique up to a linear affine transformation, i.e., if u_a is a utility for a, then v_a is also a utility for a if and only if $v_a = \alpha u_a + \beta$ for some numbers α and β such that $\alpha > 0$. $\{x_a\}_{a \in A}$ is a (cardinal) value allocation for $\mathscr{C}$ if and only if there exists nonnegative numbers $(\lambda_a)_{a \in A}$, not all zero, such that $(\lambda_a u_a(x_a))_{a \in A}$ is the Shapley value of $(A, v_{\lambda u})$, where $\lambda u = (\lambda_a u_a)_{a \in A}$. For a more detailed comparison of the two approaches, see Aumann [1]. Note that in both cases, the basic principle is the same; associate to $\mathscr{C}$ a side payment game in which utility is allowed to be transferable, but as a solution only consider those games in which the Shapley value does not require any transfers of utility to be made. This basic idea, which allows solution concepts designed for games with transferable utility to be applied to games without transferable utility, is called "the principle of irrelevant alternatives" by Shapley [9]. As a final remark for this section, Champsaur [2] and Mas-Colell [6] have shown that the cardinal value allocations converge to competitive equilibrium allocations if the economy $\mathscr{C}$ becomes "large" in an appropriate way; Aumann [1] demonstrated, in an economy with a nonatomic measure space of agents, that under certain conditions on preferences, ordinal value allocation, cardinal value allocations, and competitive equilibrium allocations all coincide. In general, however, in an economy with a finite set of agents, none of these three solution concepts will be the same.

3 Existence of Value Allocations

Shapley [9] has demonstrated that if the individual utility functions u_a have the property that the set of all feasible utility vectors for the economy (the "Pareto set") is convex and compact, then a cardinal value allocation will exist. This, for example, will always be true if the utility functions are concave. It is not difficult to construct examples where a cardinal value allocation fails to exist if this condition on the Pareto set is not met. Shapley's result can also be used to prove the existence of a value allocation in the ordinal case if it is possible to choose utility functions such that the Pareto set is convex, and if additional conditions are imposed to guarantee that each $\lambda_a > 0$ in the solution (for the latter, monotonicity and nonzero initial endowment vectors would suffice). Kannai and Mantel [5], however, have constructed examples of economies with monotone convex preference orderings for each agent, such that for no choice of utility functions will the Pareto set be convex. This poses the question as to whether the concept of an ordinal value allocation can be applied to economies without making the essentially cardinal restriction that the Pareto set can be made convex. It will be shown here that it can. The apparent difficulty in this case is that the utility functions cannot be chosen in advance, but must be computed along with the value allocation. This is handled by considering a class of utility functions for each agent: the "expenditure functions" of standard demand theory. The following lemma is implicit in much of the demand theory literature (see, for example, Hurwicz and Uzawa [4]), but does not appear in the following form:

LEMMA   Define $M: \mathscr{P}^* \times (\text{int } \mathbb{R}^l_+) \times \mathbb{R}^l_+ \to R$ by $M(\succsim, p, x) = \min_{x'} p \cdot x'$ subject to $x' \succsim x$. Then (i) M is continuous at each $(\succsim, p, x)$ for which $\succsim$ is locally nonsatiated. (ii) If $\succsim$ is locally nonsatiated and satisfies $[\forall x: x \succsim 0]$, then for each $p \in \text{int } \mathbb{R}^l_+$, $M(\succsim, p, \cdot)$ is a utility function for $\succsim$.

In other words, if one interprets p as a price vector, then $M(\succsim, p, x)$ measures the "utility" of x to be the least amount of money the agent would need to purchase a commodity vector at least as good as x.

Now economies $(\succsim_a, \omega_a)_{a \in A}$ will be considered for which each $\succsim_a$ is monotone and convex. Note that monotonicity guarantees that the hypothesis of parts (i) and (ii) of the lemma are satisfied. For each $p \in \text{int } \mathbb{R}_+^l$, consider the game $(A, v(p, \cdot))$ defined by

$$v(p, S) = \max_{x_a} \sum_{a \in S} M(\succsim_a, p, x_a) \quad \text{subject to} \quad \sum_{a \in S}(x_a - \omega_a) = 0. \qquad (2)$$

Let $Sh(p) = Sh_a(p))_{a \in A}$ denote the Shapley value of $(A, v(p, \cdot))$. Then an allocation $\{x_a\}_{a \in A}$ will be a value allocation as defined in Section 2, if there exists a $p \in \text{int } \mathbb{R}_+^l$ such that

$$M(\succsim_a, p, x_a) = Sh_a(p) \quad \text{for each } a \in A, \quad \text{and} \quad \sum_{a \in A}(x_a - \omega_a) = 0. \qquad (3)$$

Note that in such a value allocation two individuals with identical preferences will be assigned the same utility function. In fact, since M is continuous, heuristically, "similar agents" will receive "similar" utility functions. Thus such value allocations will have nice symmetry properties.

THEOREM Let $\mathscr{C} = (\succsim_a, \omega_a)_{a \in A}$ be an economy in which each $\succsim_a$ is monotone and convex. Then value allocations exist. In particular, there exist $p \in \text{int } \mathbb{R}_+^l$ and a $\{x_a\}_{a \in A}$ satisfying (3).

The technique of finding a value allocation by solving (3) has the advantage that it provides a means for *computing* the allocation and corresponding utility functions. This is illustrated by computing a value allocation for one of the Kannai–Mantel examples, in which the Pareto set is never convex.

Example 1 Three agents and two commodities, as follows:

$$u_i(x_i, y_i) = \frac{x_i}{2 - y_i}, \quad i = 1, 2 \quad (y_i < 2),$$

$$u_3(x_3, y_3) = x_3 + y_3,$$

$\omega = (1, 1)$ (the aggregate endowment vector).

The above is the simplest example in Kannai and Mantel [5]. To compute M, it is convenient to extend u_1 and u_2 to $\mathbb{R}_+^l$, as follows:

$$u_i(x_i, y_i) = \frac{x_i}{2 - y_i}, \quad y_i \leq \tfrac{3}{2},$$

$$u_i(x_i, y_i) = 2x_i, \quad y_i \geq \tfrac{3}{2}.$$

This does not affect the ordering induced by u_1 and u_2 on the attainable set. M can be computed for this example without difficulty; it has the following form, with $p = (p_x, p_y)$ and with $p_y \equiv 1$:

$$M(\succsim_i, p, (x_i, y_i)) = 2p_x \frac{x_i}{2 - y_i} \quad \text{if} \quad p_x \leq \frac{2 - y_i}{x_i},$$

$$= \tfrac{1}{2} p_x \frac{x_i}{2 - y_i} + \tfrac{3}{2} \quad \text{if} \quad p_x \geq \frac{2 - y_i}{x_i} \quad (i = 1, 2);$$

$$M(\succsim_3, p, (x_3, y_3)) = \min[p_x, 1](x_3 + y_3).$$

It will now be verified that $\bar{p}_x = 20/9$ and $(\bar{x}_1, \bar{y}_1) = (\bar{x}_2, \bar{y}_2) = (1/2, 0), (\bar{x}_3, \bar{y}_3) = (0, 1)$, is a value allocation in the case where the initial endowment vector for each individual is $(1/3, 1/3)$. Solving the equations (2) for $\bar{p}_x = 20/9$, one gets the characteristic function $v(\bar{p}, \cdot)$ as below:

$$v(\bar{p}, \{1\}) = v(\bar{p}, \{2\}) = \tfrac{8}{9},$$

$$v(\bar{p}, \{3\}) = \tfrac{2}{3},$$

$$v(\bar{p}, \{1, 2\}) = v(\bar{p}, \{1, 3\}) = v(\bar{p}, \{2, 3\}) = \tfrac{58}{27},$$

$$v(\bar{p}, \{1, 2, 3\}) = \tfrac{29}{9},$$

and the solution to (2) defining $v(\bar{p}, \{1, 2, 3\})$ is $(\bar{x}_3, \bar{y}_3) = (0, 1)$, $(\bar{x}_1, \bar{y}_1) = (\bar{x}_2, \bar{y}_2) = (1/2, 0)$.

The Shapley value of this game is $Sh_1(\bar{p}) = 10/9 = Sh_2(\bar{p})$, and $Sh_3(\bar{p}) = 1$, and thus equations (3) will be satisfied. Note that in this case the utility functions $M(\succsim_i, \bar{p}, \cdot)$ for 1 and 2 will fail to be differentiable at points in the attainable set, even though preferences have a differentiable representation. This failure of differentiability appears to be critical in overcoming the nonconvexity of the Pareto set.

4 Interpretation of Value Allocations: Examples

The hypotheses of the Theorem do not require that $\omega_a \neq 0$ for any particular agent. On the other hand, it is not clear from (1) and (3) that a solution will yield $Sh_a(p) = 0$ even if $\omega_a = 0$: this is because $Sh_a(p) > 0$ if just for one coalition S, $v(p, S) - v(p, S \setminus \{a\}) > 0$, and this may happen even if $\omega_a = 0$, since the maximand in the definition of $v(p, S)$ is different from that of $v(p, S \setminus \{a\})$. The question as to whether this could happen at a value allocation led to the following examples, which apply as well to the cardinal approach since all utility functions used are concave.

Example 2 Three agents 0, 1, 2 and two commodities, x and y, as follows:

$$u_0(x_0, y_0) = \left[\tfrac{1}{2}x_0^\beta + \tfrac{1}{2}y_0^\beta\right]^{1/\beta}, \quad \omega_0 = (\varepsilon, \varepsilon),$$

$$u_1(x_1, y_1) = \left[\tfrac{1}{2}x_1^p + \tfrac{1}{2}y_1^p\right]^{1/p}, \quad \omega_1 = (1-\varepsilon, 0),$$

$$u_2(x_2, y_2) = \left[\tfrac{1}{2}x_2^p + \tfrac{1}{2}y_2^p\right]^{1/p}, \quad \omega_2 = (0, 1-\varepsilon).$$

The cases where $0 \leqslant \varepsilon < 1$ and $0 \leqslant p < \beta \leqslant 1$ are considered (for $p = 0$, use $(xy)^{1/2} = \lim_{p \to 0} \left[\tfrac{1}{2}x^p + \tfrac{1}{2}y^p\right]^{1/p}$). The characteristic function v_u had the following form:

$$v_u(\{1\}) = v_u(\{2\}) = 2^{-1/p}(1-\varepsilon),$$

$$v_u(\{0\}) = \varepsilon,$$

$$v_u(\{1,2\}) = 1 - \varepsilon,$$

$$v_u(\{0,1\}) = v_u(\{0,2\}) = \left(\tfrac{1}{2} + \tfrac{1}{2}\varepsilon^\beta\right)^{1/\beta},$$

$$v_u(\{0,1,2\}) = 1,$$

and the solution can be attained by the following allocations: $(x_0, y_0) = (\theta, \theta)$ and $(x_1, y_1) = (x_2, y_2) = ((1-\theta)/2, (1-\theta)/2)$, for any θ such that $0 \leqslant \theta \leqslant 1$. The computation of v_u is quite easy if one uses that fact that $\beta > p$ implies

$$\left[\tfrac{1}{2}x^\beta + \tfrac{1}{2}y^\beta\right]^{1/\beta} \geqslant \left[\tfrac{1}{2}x^p + \tfrac{1}{2}y^p\right]^{1/p}$$

with equality if and only if $x = y$.

Note that agents 1 and 2 are treated symmetrically in the game described by v_u, so that if s_0 is the Shapley value of agent 0, then $(1-s_0)/2$ will be the common Shapley value for agents 1 and 2. Let $s_0(\varepsilon)$ denote the Shapley value for agent 0 of $(\{0,1,2\}, v_u)$ for a specific ε. Then direct computation shows

$$s_0(\varepsilon) = \left[\tfrac{2}{3} + \tfrac{1}{3}\left(\tfrac{1}{2}\right)^{1/p}\right]\varepsilon + \tfrac{1}{3}\left[\left(\tfrac{1}{2} + \tfrac{1}{2}\varepsilon^\beta\right)^{1/\beta} - \left(\tfrac{1}{2}\right)^{1/p}\right].$$

Thus $(x_0, y_0) = (s_0(\varepsilon), s_0(\varepsilon))$, and $(x_1, y_1) = (x_2, y_2) = ((1-s_0(\varepsilon))/2, (1-s_0(\varepsilon))/2)$ will be a value allocation. But $s_0(0) > 0$, and in fact $s_0(\varepsilon) > \varepsilon$ for each ε such that $0 \leqslant \varepsilon < 1$. That is, agent 0 will always receive more of both commodities than he is initially endowed with, even if his initial endowment is $(0, 0)$. This outcome depends critically on the assumption that $\beta > p$, i.e., that agent 0 has a higher elasticity of substitution than agents 1 and 2. Note that this is a value allocation in the ordinal sense, it is a value allocation in the cardinal sense ($\lambda_0 = \lambda_1 = \lambda_2 = 1$), and it is a solution to (3) for a $p = (p_x, p_y)$ such that $p_x = p_y = 1/2$, since with homothetic preferences $M(\succsim, p, \cdot)$ is the homogeneous of degree 1 utility function representing $\succsim$.[1]

The following two examples investigate this phenomenon for economies with a large number of agents.

Example 3 The above example can be extended to the case where there are l commodities and $l+1$ agents. Consider the simplest case, $p = 0$ and $\beta = 1$, i.e.,

$$u_0(x_0) = \frac{1}{l} \sum_{i=1}^{l} x_{0i} \quad \text{and} \quad \omega_0 = \varepsilon u,$$

$$u_j(x_j) = \left(\prod_{i=1}^{l} x_{ji} \right)^{1/l} \quad \text{and} \quad \omega_j = (1-\varepsilon) e_j \quad (j \geqslant 1, \ldots, l).$$

Here e_j is the jth unit vector and $u = \sum_j e_j$. In this case a value allocation can be calculated, and is for agent 0 $\bar{x}_0 = s_0(\varepsilon) u$, where $s_0(\varepsilon)$ is

$$s_0(\varepsilon) = \frac{1}{2} \left[\frac{l+3}{l+1} \varepsilon + \frac{l-1}{l+1} \right].$$

The outcome $s_0(\varepsilon) > \varepsilon$ still persists, and in fact $s_0(0)$ is approximately $1/2$ for l large.

Example 4 Fix an l in Example 3 and consider a replication of the economy, say m of each type. A value allocation for an agent of type 0 will still be of the form $s_0(\varepsilon) u$; and it is not difficult to show that $s_0(\varepsilon) > \varepsilon$ will still hold. Since $u_0(\omega_0) = \varepsilon$, $v(S) - v(S \setminus \{0\}) \geqslant \varepsilon$ must always hold for S which contains this agent of type 0. But for any S which contains exactly one agent of type 0, is missing at least one other type completely, and contains more than one agent, must satisfy $v(S) - v(S \setminus \{0\}) > \varepsilon$. For example, if S consists of one agent of type 0 and one agent of type 1, then for this agent of type 0, $v(S \setminus \{0\}) = 0$ and $v(S) = ((1-\varepsilon)/l) + \varepsilon$. Since the Shapley value for an agent of type 0 is just a weighted average

---

1. A simple modification of this example can be used to show that cardinal value allocations need not be symmetric, i.e., individuals with identical utility functions and identical initial endowment vectors need not obtain utility equivalent value allocations. Take the case $\varepsilon = 0$, and add a fourth person (agent 3) to the economy identical to agent 0. Then use the weights $\lambda_0 = \lambda_1 = \lambda_2 = 1$, $\lambda_3 = 0$. The new player will be a "dummy" in the side payment game, and thus has no effect on the value allocation and will receive $(\bar{x}_3, \bar{y}_3) = (0, 0)$. This still leaves open the more interesting possibility, however, of a nonsymmetric value allocation in which each agent receives a positive weight λ_a.

of the $v(S) - v(S \setminus \{0\})$ over S to which this agent belongs, it must be that $s_0(\varepsilon) > \varepsilon$.

The determination of a value allocation depends on each agent's preferences as well as his initial endowment vector. This in itself is appropriate, but the manner in which agents' preferences influence the outcome accounts for the peculiar nature of the above examples. The value of an agent a is a weighted average of terms of the form $v_u(S) - v_u(S \setminus \{a\})$, and the magnitude of such a term can be great even if the agent has no initial endowment, since his utility function enters the determination of $v_u(S)$ as if it were a production function. This is perfectly acceptable if utility is transferable, but unreasonable otherwise. In the above examples this is demonstrated clearly in the case $\varepsilon = 0$: Agent 0 has nothing to offer the other agents, and so his marginal contribution to a coalition should be zero. But by allowing him to "produce" utility with his superior "technology" the value allocation procedure assigns him a positive marginal contribution to some coalitions, and thus a positive value allocation. The "principle of irrelevant alternatives" is supposed to be justification for allowing utility transfers in the determination of v_u, since the solution concept permits no actual utility transfers to be made. Actually, the value allocation procedure does not require utility transfers to achieve $v_u(A)$, but it allows utility transfers in the determination of each $v_u(S)$, $S \subset\subset A$. This latter fact would be unimportant if v_u represented correctly the relative worth of the various coalitions, which in the above examples it clearly does not if $\varepsilon = 0$. One possible escape from this problem may be to exclude agents with no initial endowment from the economy before calculating the value allocation. Note, however, that $s_0(\varepsilon)$ in the above examples depends continuously on ε; the solution for small $\varepsilon > 0$ is not qualitatively different from that of $\varepsilon = 0$. If agent 0 was excluded at the outset, this would no longer be the case. It has been suggested that the bargaining game is qualitatively different for small $\varepsilon > 0$ than for $\varepsilon = 0$.[2] This, however, does not seem to correct the problem: the "wrong" outcome in the above examples for $\varepsilon = 0$ arises because the agents' utility functions are treated as production functions in assigning worth to coalitions, and this latter characteristic of the value allocation is present whether $\varepsilon = 0$ or $\varepsilon > 0$. Ultimately the issue is not with the Shapley value but whether or not the characteristic function v_u correctly reflects the relative worth of the various coalitions. The above examples suggest that it may not.[3]

---

2. This idea, as well as that of treating the agent with no initial endowment as a dummy, is from helpful conversations with Mordecai Kurz and Sergiu Hart. This brief discussion, of course, does not do justice to their position.

3. See Roth [7] for similar examples.

One must, therefore, accept the idea of a value allocation for precisely what its definition says that it is: an estimation of bargaining strength, under the assumption utility is transferable, stated in terms of commodity allocations such that no utility transfers are required in the final outcome. The solution concept is, I believe, of interest in itself; these examples, I hope, will warn against the danger of giving the value allocation too broad of an interpretation.

5 Proofs of the Lemma and Theorem

Proof of the Lemma First, part (i) of the Lemma is shown. Choose an $(\succsim, p, x) \in \mathscr{P}^* \times (\text{int } \mathbb{R}^l_+) \times \mathbb{R}^l_+$ such that $\succsim$ is locally nonsatiated and a sequence of points $(\succsim_n, p_n, x_n) \in \mathscr{P}^* \times (\text{int } \mathbb{R}^l_+) \times \mathbb{R}^l_+$ such that $\lim_{n \to \infty} (\succsim_n, p_n, x_n) = (\succsim, p, x)$. It must be shown that $\lim_{n \to \infty} M(\succsim_n, p_n, x_n) = (\succsim, p, x)$. Let α be any limit point of the sequence $\{M(\succsim_n, p_n, x_n)\}_{n=1}^{\infty}$,[4] and for each n choose $\tilde{x}_n \in \mathbb{R}^l_+$ such that $\tilde{x}_n \succsim_n x_n$ and $M(\succsim_n, p_n, x_n) = p_n \tilde{x}_n$. It may be assumed, by selecting subsequences if necessary, that $\lim_{n \to \infty} M(\succsim_n, p_n, x_n) = \alpha$ and that $\lim_{n \to \infty} \tilde{x}_n = \tilde{x}$ for some $\tilde{x} \in \mathbb{R}^l_+$. Let $x^* \in \mathbb{R}^l_+$ be such that $x^* \succsim x$ and $M(\succsim, p, x) = px^*$. Note that $\tilde{x}_n \succsim_n x_n$ for all n implies $\tilde{x} \succsim x$, so $M(\succsim, p, x) \leq p\tilde{x} = \alpha$. Thus it only remains to show $\alpha \leq M(\succsim, p, x)$. Let $N(x^*)$ denote a neighborhood of x^* in $\mathbb{R}^l_+$. Then it is asserted that there exists an $\bar{n}$ such that $N(x^*) \cap \{x' \in \mathbb{R}^l_+ : x' \succsim_n x_n\} \neq \emptyset$ for all $n \geq \bar{n}$. If this is false, then there exists a subsequence $\{x_{n_k}\}_{k=1}^{\infty}$ of $\{x_n\}_{n=1}^{\infty}$ such that $N(x^*) \cap \{x' \in \mathbb{R}^l_+ : x' \succsim_{n_k} x_{n_k}\} = \emptyset$ for all k. By local nonsatiation of $\succsim$, there exists a $y \in N(x^*)$ such that $y \succ x^*$ and also it must be that $x_{n_k} \succsim_{n_k} y$ for all k. Thus, in the limit, $x \succsim y \succ x^*$, a contradiction. Therefore one can choose a sequence $\{x'_n\}_{n=1}^{\infty}$ such that $x'_n \succ_n x_n$ and $\lim_{n \to \infty} x'_n = x^*$. Thus one has $M(\succsim_n, p_n, x_n) \leq p_n x'_n$ for all $n \geq 1$, and $\lim_{n \to \infty} p_n x'_n = px' = M(\succsim, p, x)$, so $\alpha \leq M(\succsim, p, x)$. Therefore $\alpha = M(\succsim, p, x)$, and the proof of (i) is complete.

Now part (ii) of the Lemma is proven. If $x \succsim y$, then clearly $M(\succsim, p, x) \geq M(\succsim, p, y)$ since $\{x' : x' \succsim x\} \subset \{x' : x' \succsim y\}$. Thus one needs to show that $x \succ y \Rightarrow M(\succsim, p, x) > M(\succsim, p, y)$. Let $\tilde{x}$ be such that $\tilde{x} \succsim x$ and $M(\succsim, p, x) = p\tilde{x}$. Then, since $\succsim$ has closed graph, $t\tilde{x} \succ y$ for $t < 1$ and t close to 1, so $M(\succsim, p, y) \leq tp\tilde{x}$ for some $t < 1$. $p\tilde{x} > 0$, since $\tilde{x} \succ y$ implies $\tilde{x} \neq 0$ by the hypothesis of part (ii). Thus $M(\succsim, p, y) < M(\succsim, p, x)$. This completes the proof of the Lemma.

---

4. The existence of finite α is guaranteed by the fact that $0 \leq M(\succsim_n, p_n, x_n) \leq p_n x_n$.

Proof of the Theorem First some preliminary observations. It is clear that $M(\gtrsim_a, p, x_a) \leq p \cdot x_a$ with equality if and only if p is normal to a plane of support of $\{x' : x' \gtrsim_a x_a\}$ at x_a. Thus, for each $S \subset A$, $v(p, S) \leq p(\sum_{a \in S} \omega_a)$, with equality if and only if there exists $\{x_a\}_{a \in S}$ such that $\sum_{a \in S}(x_a - \omega_a) = 0$ and $M(\gtrsim_a, p, x_a) = p \cdot x_a$ for each a. Note also, that since each $M(\gtrsim_a, \cdot, \cdot)$ is a continuous mapping of $(\text{int } \mathbb{R}_+^l) \times \mathbb{R}_+^l$ into $\mathbb{R}$, each $v(\cdot, S) : \text{int } \mathbb{R}_+^l \to \mathbb{R}$ is continuous and thus each $Sh_a : \text{int } \mathbb{R}_+^l \to \mathbb{R}$ is continuous. From the basic properties of the Shapley value, $\sum_{a \in A} Sh_a(p) = v(p, A) \leq p\omega$ and $Sh_a(p) \geq M(\gtrsim_a, p, \omega_a) \geq 0$ for each a. Since $\omega \gg 0$, $\omega_a \neq 0$ for at least one a, so $\sum_{a \in A} Sh_a(p) > 0$ for all p. For each $a \in A$, define $\phi_a : \text{int } \mathbb{R}_+^l \to R$ by $\phi_a(p) = Sh_a(p) / \sum_{a' \in A} Sh_{a'}(p)$. Clearly each ϕ_a is continuous and satisfies $\sum_{a \in A} \phi_a(p) = 1$ and $\phi_a(p) \geq 0$ for each $p \in \text{int } \mathbb{R}_+^l$ and $a \in A$. To prove the Theorem, it suffices to show that (3) has a solution, and this is accomplished simply by showing that the "economy" in which agent a has preference ordering $\gtrsim_a$ and income, at price vector p, of $\phi_a(p)p\omega$, has an equilibrium. For each $a \in A$ let $Z_a(p) = \{x' \in \mathbb{R}_+^l : x' \text{ is maximal for } \gtrsim_a \text{ in the set } \{x'' : px'' \leq \phi_a(p)p\omega\}\}$. Clearly, Z_a is upper-hemi-continuous at each $p \in \text{int } \mathbb{R}_+^l$ and is nonempty, compact, convex valued, and also satisfies $[x \in Z_a(p) \Rightarrow px = \phi_a(p)p\omega]$. Define $Z(p) = \sum_{a \in A} Z_a(p) - \omega$. Then Z, as a correspondence from $\text{int } \mathbb{R}_+^l$ to $\mathbb{R}^l$ is upper-hemi-continuous, convex compact valued, and satisfies "Walras law" $[x \in Z(p) \Rightarrow p \cdot x = 0]$. Thus a standard existence theorem can be applied (for example, Hildenbrand [3, Lemma 1, Chapter 2.2]), provided it can be shown that Z satisfies a suitable boundary condition, when restricted to the set $S_+^{l-1} = \{p \in \text{int } \mathbb{R}_+^l : \sum p_i = 1\}$. It is shown that if $\{p_n\}_{n=1}^\infty$ is a sequence in S_+^{l-1} which converges to a point $p \in BdS_+^{l-1}$, then any sequence $\{x_n\}_{n=1}^\infty$ such that $x_n \in Z(p_n)$ must satisfy $\lim_{n \to \infty} \inf u \cdot x_n = +\infty$, where $u \in \mathbb{R}^l$ is the vector all of whose coordinates are 1. Let $\alpha = \lim_{n \to \infty} \inf u \cdot x_n$, and by selecting a subsequence if necessary, assume $\alpha = \lim_{n \to \infty} u \cdot x_n$. Since $\sum_{a \in A} \phi_a(p) = 1$, for every n there exists an $a_n \in A$ such that $\phi_{a_n}(p_n) \geq 1/|A|$. Since A is finite, there is an $\bar{a} \in A$ and a subsequence $\{n_k\}_{k=1}^\infty$ of $\{n\}_{n=1}^\infty$ such that $\phi_{\bar{a}}(p_{n_k}) \geq 1/|A|$ for all k. Thus agent $\bar{a}$ has income $\phi_{\bar{a}}(p_{n_k})p_{n_k}\omega \geq \min_1(\omega_i)/|A| > 0$. Since $\gtrsim_{\bar{a}}$ is monotone, any sequence of points $x'_{n_k} \in Z_{\bar{a}}(p_{n_k})$ must satisfy $\lim_{n \to \infty} \inf u \cdot x'_{n_k} = +\infty$. Since each Z_a is bounded below, it must be that $\alpha = \lim_{n \to \infty} u \cdot x_n = \lim_{k \to \infty} u \cdot x_{n_k} = +\infty$. Thus Lemma 2, Chapter 2.2 of Hildenbrand [3] applies, and thus there exists a $\bar{p} \in S_+^{l-1}$ such that $0 \in Z(\bar{p})$, i.e. there exists $\bar{x}_a \in Z_a(\bar{p})$ such that $\sum_{a \in A}(\bar{x}_a - \omega_a) = 0$. $\bar{x}_a \in Z_a(\bar{p})$ implies that $M(\gtrsim_a, \bar{p}, \bar{x}_a) = \bar{p}\bar{x}_a = \phi_a(\bar{p})\bar{p}\omega$. In particular, $M(\gtrsim_a, \bar{p}, \bar{x}_a) = \bar{p}\bar{x}_a$ for each $a \in A$ implies $\sum_{a \in A} Sh_a(\bar{p}) = v(\bar{p}, A) = \bar{p}\omega$. Thus $M(\gtrsim_a, \bar{p}, \bar{x}_a) = \phi_a(\bar{p})\bar{p}\omega = Sh_a(\bar{p})$ for each a. This completes the proof of the Theorem.

Acknowledgments

A preliminary version of this paper was completed at Universität Bonn. Financial support of the Sonderforschungsbereich 21 is gratefully acknowledged. The paper has benefited from several discussions with Andreu Mas-Colell.

References

1. Aumann, R.: "Values of Markets with a Continuum of Traders," *Econometrica*, 43 (1975), 611–646 [Chapter 51].
2. Champsaur, P.: "Cooperation versus Competition," *Journal of Economic Theory*, 11 (1975), 394–417.
3. Hildenbrand, W.: *Core and Equilibria of a Large Economy*. Princeton, New Jersey: Princeton University Press, 1974.
4. Hurwicz, L., and H. Uzawa: "On the Integrability of Demand Functions," in *Preference, Utility and Demand*, ed. by J. S. Chipman, L. Hurwicz, M. K. Richter, and H. Sonnenschein. New York: Harcourt Brace Jovanovich, 1971.
5. Kannai, Y., and R. Mantel: "Nonconvexifiable Pareto Sets," *Econometrica*, 46 (1978), 571–576.
6. Mas-Colell, A.: "Competitive and Value Allocations of Large Exchange Economies," *Journal of Economic Theory*, 14 (1977), 419–438.
7. Roth, A.: "Values for Games without Side Payments: Some Difficulties with Current Concepts," *Econometrica*, 48 (1980), 457–465 [Chapter 61a].
8. Shapley, L.: "A Value for *n*-Person Games," in *Contributions to the Theory of Games, Vol. II*, ed. by A. W. Kuhn and A. W. Tucker. Princeton, N.J.: Princeton University Press, 1953.
9. ———: "Utility Comparison and the Theory of Games," in *La Decision*, ed. by G. Th. Guilbaud. Paris: Editions du CNRS, 1969.
10. Shapley, L., and M. Shubik: "Pure Competition, Coalition Power, and Fair Division," *International Economic Review*, 3 (1969), 337–362.

61c On the Non-Transferable Utility Value: A Comment on the Roth–Shafer Examples

1 Introduction

In game theory, the term "solution concept" denotes a correspondence between games and outcomes (or sets of outcomes). Two familiar examples of solution concepts are the Nash Equilibrium Point and the Core. Both are successful tools of economic analysis: applied to a variety of contexts, they yield important and interesting results. The Core is particularly successful in classical[1] market contexts.

Another solution concept is Shapley's value [47]; more generally,[2] his Non-Transferable-Utility (NTU) value[3] [49]. While not as well-known as the core, it is in some ways even more "successful": it has been applied to a broader variety of contexts, often yielding interesting results.[4]

Several years ago, A. Roth [43] constructed a class of examples in which, he argued, the NTU value looks strange and counterintuitive; specifically, in which there are very strong, compelling reasons leading to an alternative outcome, not consistent with the NTU value. He concluded that "at the very least, some modifications are required in the existing theory." While far from universally accepted in the profession, Roth's work has had a considerable echo.

The main purpose of this paper is to rebut Roth's. We make two points:

i. Roth's reasoning is unsound; specifically, the arguments for the alternative outcome are not nearly as compelling as they appear at first.

This paper originally appeared in *Econometrica* 53 (1985): 667–677. Reprinted with permission.

1. I.e., without political elements, public goods, taxation, increasing returns, fixed prices, and so on.

2. As defined in [47], the value applies to transferable utility (TU) games only. NTU games generalize TU games; every TU game is an instance of an NTU game, but not conversely. The NTU value, when applied to coalitional[24] TU games, coincides with the TU value.

3. Sometimes called the λ-*transfer* value, an unfortunate practice, as it involves no transfers. Shapley [49] used the term "λ-transfer value" in a different sense, closer to the plain meaning of the words.

4. Until the last decade, most of the applications of the value concept were in a TU environment. The literature on applications of the TU value is far too voluminous to be catalogued here. The last decade has seen more and more applications of the NTU value in the "strict" sense, i.e., to environments that are not TU. These include classical markets [1, 10, 11, 12, 20, 24, 28, 46, 50], production with increasing average returns [29], taxation [4, 5, 14, 15, 26, 27, 36, 37], public goods [6, 7, 40, 41], monopoly [16], rationing and fixed prices [3, 15], incomplete information [30], and general theories of justice [8, 9, 49, 51] (to which the core is inapplicable because, inter alia, it is often empty). Note also that the early works of Nash [31, 33] on bargaining and threatening in a two-person context are in fact applications of the "strict" NTU value. (For axiomatizations of the NTU value and additional treatments of a general nature, see [2, 21, 22, 23, 24, 25, 35, 38, 39, 42].)

ii. Even if the arguments were sound, the examples would by no means justify abandoning the NTU value as an analytic tool, or even modifying it. A solution concept is not a theorem, and one counterintuitive example is not sufficient to make us abandon an otherwise successful tool. Most popular solution concepts are beset by counterintuitive examples; we will adduce just two, one for the Nash Equilibrium Point, and one for the Core.

Point (ii) is presented in Section 2. Most of the remaining sections are devoted to our main thesis, Point (i); readers in a hurry may confine themselves to Section 3, which presents the gist of the argument informally.

Back-to-back with [43], W. Shafer [46] published a different set of examples meant to show that the NTU value may yield counterintuitive results. While similar in principle to Roth's, these examples are set in a special context that makes them in some ways more compelling. Nevertheless, they fit well into the general framework of the NTU value; this will be discussed in Section 8.

This paper focuses on the reasonableness or unreasonableness of various *outcomes* of the Roth and Shafer games; it is not concerned with the internal workings of the NTU value. Therefore we do not find it necessary to quote the definition of the NTU value, which may be found in almost any paper on the subject. Conceptual discussion of the definition as such may be found, e.g., in [1, 49].

2 Counterintuitive Examples for Other Solution Concepts

Consider first the Nash Equilibrium Point (EP), the game theoretic concept that is perhaps best-known and most frequently applied in economics. There are very simple, natural non-zero-sum two-person games that have a unique EP $\sigma = (\sigma_1, \sigma_2)$, which yields each player only his security level (i.e., his maxmin value, the amount he can guarantee for himself), but such that σ_i does not, in fact, *guarantee* the security level. For example, the game[5] in Figure 1 has a unique EP, consisting of (1/2,1/2) for each player, and the expected outcome is (3, 3). But in using those strategies, each player runs the risk of receiving less than 3 if the other should play his second strategy. This risk is quite unnecessary, since

---

5. Examples of this kind have been in the folklore of game theory for a long time. For a discussion, see, e.g., [18, p. 125].

2,6	4,2
6,0	0,4

Figure 1

player 1 has the maxmin strategy (3/4, 1/4) available, which assures him of 3 regardless; similarly player 2 has strategy (1/4, 3/4). Under these circumstances, it is hard to see why the players would use their equilibrium strategies.[6]

Next, we turn to the Core, also widely applied in economics. Consider a market in totally complementary goods, e.g., right and left gloves. There are four agents. Initially 1 and 2 hold one and two left gloves respectively, 3 and 4 hold one right glove each. (In coalitional form, $v(1234) = v(234) = 2$, $v(ij) = v(12j) = v(134) = 1$, $v(S) = 0$ otherwise, where $i = 1, 2$ and $j = 3, 4$.) The Core has a unique point, namely (0, 0, 1, 1); that is, the owners of the left gloves must simply give their merchandise, for nothing, to the owners of the right gloves. This in itself seems strange enough. It becomes even stranger when one realizes that Agent 2, simply by throwing away one glove—an action that he can perform by himself, without consulting anybody—can make the situation completely symmetric (as between 1, 2 and 3, 4). Appeals to "competition" ring hollow. With such small numbers—two traders on each side—the market can hardly be deemed competitive; certainly not here, where a single agent can, by his own actions, improve the situation so dramatically for himself.[7]

Do these examples imply that we should abandon or modify the EP or the Core? We think not. At some point, we should ask ourselves how such counterintuitive examples fit into the conceptual framework of game theory, and of theory in the social sciences in general. But not in this article. Here we wished only to show that at worst, the Roth example puts the NTU value into a class with the EP and the Core; and in the sequel, we mean to show that it doesn't even do that.

---

6. The equilibrium and maxmin strategies are mixed, but that is not an issue; if one excludes mixed strategies, one can still construct an example in which these phenomena occur, simply by explicitly adding rows and columns to the original game that contain the payoffs of the appropriate mixed strategies.

7. The archetype of this genre of examples is the market with one seller and two buyers [34, p. 610 ff.], in which the unique core point is (1, 0, 0). Cf. also [13, 48].

3 The Roth Example

Let p be a parameter with[8] $0 < p < 1/2$. There are three players, who must share 1. By himself, each player can get 0. If Players 1 and 3, or 2 and 3, form a coalition, then 3 gets a utility of $1 - p$ (the larger amount), and the other player gets p. If 1 and 2 form a coalition, they get 1/2 each. If all three form a coalition, they may use a random device of their choosing to pick a 2-person coalition, which must then divide as above. No other outcomes are possible.

The unique NTU value of this game is (1/3, 1/3, 1/3). But Roth argues that 3 is weak, because he can only offer 1 and 2 a payoff of p, which is $<1/2$. Players 1 and 2 would therefore spurn 3's offers, and gravitate toward each other. Roth concludes that the outcome *must* be (1/2, 1/2, 0); it is the "unique outcome ... consistent with the hypothesis that the players are rational utility maximizers ... the outcome (1/2, 1/2, 0) is *strictly* preferred by *both* players 1 and 2 to *every* other feasible outcome ... So ... there is really no conflict between players 1 and 2: their interests coincide in the choice of the outcome (1/2, 1/2, 0), and the rules permit them to achieve this outcome" [43, pp. 468–9; his emphasis].

At first, this reasoning sounds compelling. But let's look a little closer. Suppose the players and the rules have just been announced on television. The amount 1 to be shared may be fairly large, so the players are rather excited. Suddenly the phone rings in 1's home; 3 is on the line with an offer. At first 1 is tempted to dismiss it. But then he realizes that if he does so, and if 3 manages to get in touch with 2 before he (1) does, then he won't get anything at all out of the game, unless 2 also rejects 3's offer. "But wait a minute," 1 now says to himself; "2 will only reject 3's offer if he thinks that I will reject it. When he gets 3's phone call, he will go through the agonizing that I am going through now, and will realize that in his situation I would also agonize. We seem to be caught in a web of circular reasoning. It is rational for me to reject 3's offer only if it is rational for 2 to reject it, and this in turn depends on its being rational for me to reject it. In short, I should reject 3's offer only if it is pretty clear to start with that I should reject it. I'm beginning not to like this one bit."

At this point, 1 breaks into a cold sweat. "Are you still there?", he says anxiously into the receiver. "Yes," says 3, "but I'm getting a little impatient." 1 sighs with relief. "You have a deal," he says.

---

8. The end-points have special properties requiring separate treatment. When $p = 1/2$, the game is symmetric, and all agree that (1/3, 1/3, 1/3) is the appropriate "value." For a discussion of $p = 0$, see note 20.

To illustrate the force of this reasoning, suppose the amount to be divided is $100,000, and that $p = \$49,000$. When 1 gets 3's phone call, he must choose between (i) getting $49,000 with certainty, on the spot; and (ii) getting $50,000 if he is convinced that 2 is convinced that he (1) will reject 3's offer (or if he can get in touch with 2 before 3 does), and getting 0 otherwise. In my opinion there is little doubt that 1 would accept the $49,000, even though the $1,000 he foregoes is by no means a negligible sum.

In short, Roth's statements are simply incorrect. (1/2, 1/2, 0) is *not* the "unique outcome consistent with the hypothesis that the players are rational utility maximizers." Another outcome that may well be[9] consistent with this hypothesis is that resulting if any two players who meet immediately close a deal. Indeed, if each player thinks that the others will do this, then *to maximize his own utility*, he must do so as well. We are of course not asserting that rational utility maximization implies this as the unique outcome. But it is certainly *consistent* with rational utility maximization.

Where Roth went wrong is in ignoring the crucial distinction between a single rational decision maker, and several of them. If 1 and 2 had had a single "manager," his arguments would have been airtight. But one of the central questions of cooperative game theory has always been, which coalition will form? And to this question, Roth's arguments do not speak convincingly. It is true that each of 1 and 2 would have liked {1, 2} to form. It is also true that *acting together*, they can bring this about. But acting separately, neither one of them can bring it about. And it will not come about without a kind of mutual reliance that has little to do with ordinary individual utility maximization, and that, because of its riskiness, may be totally unreasonable.

4 A Fifty-Person Game

To underscore the distinction between Roth's "rationality" and the ordinary kind, consider the following 50-person game: Three million dollars are to be divided. Each of Players 1 through 49 can form a two-person coalition with Player 50, which *must* split 59:1 (in favor of 50), yielding the "small" partner $50,000. The only other coalition that can get anything consists of all the players 1 through 49, which *must* split evenly, yielding each partner about $61,000. As before, the all-player coalition has the option of choosing a smaller coalition by a random device.

---

9. Formally, this may depend on p and on the bargaining procedure. Cf. Section 5.

The full force of Roth's reasoning applies to this game; there is no reason that his kind of "rationality" should apply any the less to 49 people than to 2. So presumably, he would predict with certainty that Players 1 through 49 will form a coalition and split evenly; in the role of Player 1, he would reject any overtures from 50 with dignity but firmness. Perhaps we are irrational, but for his sake, we hope we are not Player 2; for he can be assured that we would accept any offers from 50 with alacrity, while he is out there trying to round up the other fellows.

5 Some Formal Bargaining Models

Let us return to the game of Section 3. In commenting on Roth's paper, J. Harsanyi [19] suggests that all cooperative solution notions be abandoned. To deal with cooperative games (such as the one before us), he suggests constructing formal bargaining models, and analyzing them as non-cooperative games. This program, which goes back to Nash [32], has some serious pitfalls, as we will see in Section 7. Nevertheless, it is useful as a touchstone for the informal kind of reasoning that characterizes both Roth's paper and the previous sections of this one; it clarifies and sharpens our thinking. Also, it enables us to apply the familiar formal concepts of non-cooperative game theory.[10]

One simple bargaining model is the following: A player i is picked at random and given the "initiative." That is, i chooses another player j, and makes him an offer. If j rejects the offer, i makes an offer to the remaining player k; but k does not know of the previous offer to j. If k also rejects i's offer, the coalition $\{j,k\}$ forms.

The interesting case is that in which 3 gets the initiative. This results in a subgame whose extensive and strategic forms are depicted in Figure 2. Its EP's include (r, r, μ) and (a, a, μ), where μ is any mixed strategy of 3; denote these EP's by R and A respectively.

The phrase used by Roth—"outcome of the game consistent with the hypothesis that the players are rational utility maximizers"—is an excellent description of an EP. Indeed, an EP is *defined* as an outcome at which each player maximizes his utility, given that the others are at this outcome. Roth claims that such an outcome *must* lead to the coalition $\{1, 2\}$. Since this is not the case for the equilibrium point A, we conclude that at least in this bargaining model, Roth's assertion is incorrect.

While EP's are always consistent with rationality, we certainly do not claim that they are always consistent with reasonableness; witness the

---

10. EP's and their variants.

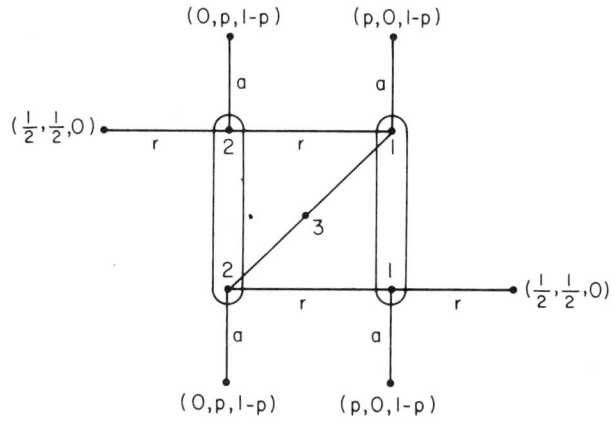

Figure 2

r and a denote, respectively, "reject 3's offer" and "accept 3's offer."

example in Section 2. But A does happen to be a rather reasonable specimen. First, it is "trembling hand" perfect in a very strong sense;[11] each players' action remains rational even if he is not entirely certain that all will play according to A. Second, for both 1 and 2 it is "strict," i.e., prescribes the *unique* best response to itself;[12] thus rationality not only allows 1 and 2 to accept an offer from 3, it *requires* them to do so (if each believes the other will accept). Of course R enjoys similar "reasonableness" properties.

When we compare A to R, we find that from the point of view of payoff, both 1 and 2 prefer R. But payoff is not the only consideration when

---

11. The definition [45] of a perfect EP requires only that there *exist* perturbations of the game with EP's close to it. In our case, this is so for *all* perturbations.

12. It is the lack of strictness that enables the pathology of the example in Section 2. Unfortunately, there are many important games that do not possess strict EP's. In our case, neither A nor R are strict for 3, but only because 3's choice never affects his payoff. Thus his choice is a matter of total indifference to him, so it is reasonable to assume that 1 and 2 perceive 3's strategy as mixed. (Harsanyi [18, p. 104] uses "strong" for what is here called "strict;" but "strong" has a different meaning in most of the literature.)

choosing an EP. There is also the problem of coordination.[13] How sure can each player be that the others will play according to it? And what is the cost if they do not?

From this point of view, A is distinctly preferable. Indeed, A requires no coordination at all; at the moment that 1 or 2 accepts 3's offer, he is *assured* a payoff of p, no matter what strategy the other one uses. On the other hand, R requires very much coordination: If 1 or 2 rejects 3's offer, he will get 0 unless the other one also rejects.

Taking all this into account, which EP seems more likely? This depends on p. If p is close to 1/2, the players will probably perceive the coordination and safety issues as paramount, and play a (i.e., accept 3's offer); if p is close to 0, they will probably forego safety for the lure of a higher payoff, i.e. play r. One can't set a precise boundary; moreover, considerations other than the size of p may enter, as we will see in Section 8. In any case, a blanket assertion that $(1/2, 1/2, 0)$ *must* be the outcome, no matter what p is, is totally unjustified.

Of course, if 1 or 2 gets the initiative, then all perfect EP's do result in $(1/2, 1/2, 0)$, though ordinary EP's need not.

Another bargaining model is the following: The three pairs of players are ordered at random, and in this order, are given the opportunity to agree; the first pair that does so forms a coalition. If no agreement is reached, all players get 0. Each player remembers which proposals he has rejected, but is not informed of proposals not involving him.

In one EP of this game, all players reject all proposals. This sounds rather unreasonable, and indeed this EP is neither perfect nor strict (for any player). It seems reasonable to consider only strategies in which 1 and 2 agree if they meet, and 3 makes an offer to the first player he meets. Indeed this is so at all perfect EP's. The only remaining issue is whether 1 and 2 should reject (r) or accept (a) 3's offer if they meet him. Figure 3 depicts the strategic form.

The analysis of this model is qualitatively similar to that of the first model, though there are differences. When $p > 1/4$, both (r,r) (henceforth R) and (a,a) (henceforth A) are perfect and strict EP's, and the comparison of A to R is much like in the first model. Unlike in the first model, A is here no longer an EP when $p < 1/4$. On the other hand, here A always yields to 3 more than his NTU value of 1/3, whereas in the first model it always yields him less.[14]

---

13. We use the term "coordination" for the problem of how players choose one EP from among a multiplicity of EP's. Harsanyi [18, p. 133] uses the same term in the related but narrower sense of choosing an EP from a multiplicity of EP's with the same payoff.

14. One must remember to average into the payoff the possibility that 1 or 2 gets the initiative.

	r	a
r	$\frac{1}{2},\frac{1}{2},0$	$\frac{1}{4},\frac{1}{4}+\frac{1}{2}p,\frac{1}{2}(1-p)$
a	$\frac{1}{4}+\frac{1}{2}p,\frac{1}{4},\frac{1}{2}(1-p)$	$\frac{1}{6}+\frac{1}{3}p,\frac{1}{6}+\frac{1}{3}p,\frac{2}{3}(1-p)$

Figure 3

6 Pre-Play Communication Doesn't Help

The analyses in the previous section *remain valid if pre-play communication is permitted*.[15] Suppose that before formal bargaining begins—i.e. before binding agreements can actually be made—there is a period during which players may converse and make tentative, non-binding agreements; moreover, all pairs of players actually get an opportunity to talk to each other. Suppose now that 1's thinking is such that in the absence of pre-play communication, he would play *a*; this is because his suspicion that 2 might also play *a* outweighs the prospect of the additional payoff otherwise. A pre-play conversation between 1 and 2 could alter 1's decision only if it could somehow allay his fear that 2 will play *a*. But if 2 really does intend to play *a*, he will want 1 to play *r*, since that improves 2's chances of getting an offer. To this end, 2 will be delighted to enter into a non-binding agreement with 1 to play *R*. Since 2 is motivated to make such an agreement no matter what he actually intends to play, such an agreement can give 1 no information. Similarly it can give 2 no information; the agreement is a dead letter as soon as it is made. The conversation between the players is therefore useless; in these games, it cannot help to resolve the coordination problem, and so cannot affect the outcome.

7 Conclusions from the Bargaining Models

One should neither overestimate nor underestimate the importance of the kind of formal bargaining model considered above. The quantitative results cannot be considered particularly significant. There are too many different possibilities for constructing a bargaining model, and the numerical results are too sensitive to its specific form. Moreover, "real" bargaining is too unstructured to be faithfully represented by such a model; it seems impossible adequately to model all the subtleties of

---

15. I.e., if the game is *vocal* [18, p. 112].

communication, timing, information, etc., that are inherent in real multi-person bargaining. *Any* formal model will have serious artificialities, which will distort the numerical results.[16]

In principle, though, Nash's program sounds very attractive. Indeed, if one accepts individual utility maximization as *the* driving force of game theory and economics, it is almost inescapable. At least, this is so if one views cooperative games simply as games in which binding agreements can be made, without reading other connotations into the word "cooperative."

Fortunately, Nash's program can be used qualitatively, without committing oneself to a specific bargaining model. Roth's solution is a case in point. Roth asserts that the logic of the situation implies that the outcome must *always* be (1/2, 1/2, 0). To refute this, it is in principle sufficient to point to one bargaining model in which it is not so, which we have done. Neither can (1/2, 1/2, 0) be considered as an amalgam arising from different bargaining models; its extreme nature—obviously 3 can't get *less* than 0—implies that if (1/2, 1/2, 0) is an average of outcomes, then it is *always* the outcome, and we are back to the previous argument.

There is another important qualitative use of Nash's program. Starting from analyses like those in Section 5, one asks which part of the reasoning depends critically on the specific bargaining model, and which is more generally valid. In our case, the numerical results depend critically on the specific bargaining model. But the reasoning according to which the coalitions {1, 3} or {2, 3} *can* form in equilibrium, and for large p are even likely to form, is of quite general validity.[17] Already in Section 3, where the setting was completely amorphous, we argued that these are likely

---

16. Another difficulty is that of multiple EP's; I have not stressed it because it is so well known. Non-uniqueness is of course ubiquitous in game theory as well as economics; but bargaining models are especially prone to having great swarms of EP's, which often render the analysis almost useless. Harsanyi and Selten [17] have developed several related methods, of considerable depth, for assigning unique EP's to games; but these methods are not nearly as persuasive as the more fundamental concepts of the noncooperative theory (EP's, perfect EP's, etc.).

17. One must be careful not to overstate the case. When we say that the conclusion is "of quite general validity," we do not mean that it holds in every conceivable bargaining model. We do mean that it holds in a wide range of fairly natural contexts; it does not depend on specific structures. It may well be possible to construct bargaining models for which the conclusion is false. If, for example, we demand that all bargaining take place in public, and that the extensive form be finite, then all *perfect* EP's do lead to (1/2, 1/2, 0), though ordinary EP's still need not. But this is a rather artificial context; the special circumstances enable one to work backwards from the end to obtain an antiintuitive result, much like in the hangman's paradox or the finitely repeated prisoner's dilemma. One man's meat is another man's poison, of course; Roth may say that a bargaining procedure cannot be considered "natural" *unless* all bargaining takes place in public. In any case, we take no dogmatic stance, and make no assertion that any particular outcome is the only "rational" one.

outcomes; what we are able to add now is that in fact, they correspond to perfect EP's of a bargaining game, though the setting remains amorphous. The reasoning according to which pre-play communication does not significantly change matters is also of quite general validity.

Let us state some conclusions. On the negative side, I think we have shown fairly conclusively that (1/2, 1/2, 0) need *not* be the outcome. It is more difficult to reach definitive conclusions on the positive side. Given a sufficiently abstract,[18] symmetric situation, it does appear that the coalitions {1, 3} and {2, 3} are less likely to form than {1, 2}; this is because {1, 2} will form as soon as its members have the opportunity to close a deal, which is not necessarily so for {1, 3} and {2, 3}. Moreover, the smaller p is, the less likely it is that {1, 3} or {2, 3} will form.[19] But whereas 3's *chance* of getting into a coalition is smaller than that of the others, his *payoff* $1 - p$ if he does get in is larger, and these two effects vary in opposite directions as p varies. All in all, perhaps (1/3, 1/3, 1/3)—the NTU value—does reflect the qualitative features of the situation quite well.

8 Some Special Contexts

We start with a political story. In many parliamentary democracies, cabinet posts in coalition governments are allotted to the parties in proportion to their seats in parliament, with the "leading" party (if there is one) getting the more important posts. If Parties 1, 2, and 3 elect 26 per cent, 26 per cent, and 48 per cent of parliament respectively, this yields Roth's game; the parameter p is[20] at most 35, and may be much smaller (depending on the importance of the "important" posts). Roth suggests that the smaller parties will *necessarily* form the government, that the large party is not only weaker than the small parties, but is actually completely powerless. But many people would say the opposite, that the large party has *more* power than the smaller parties.

For a market story, we turn to Shafer's example [46]. There are two goods: Agents 1, 2, 3 have endowments $(1 - \varepsilon, 0)$, $(0, 1 - \varepsilon)$, $(\varepsilon, \varepsilon)$, and utility functions $\sqrt{xy}$, $\sqrt{xy}$, $(x+y)/2$, respectively, where[21] $0 < \varepsilon < 1$. The NTU value gives 3 at least 1/6, even when ε is small.

18. This may not be the case in more concrete situations. See the next section.
19. If $p = 0$, then it appears that in most natural bargaining models, all perfect EP's do lead to (1/2, 1/2, 0) (though again, ordinary EP's need not). But in that case, (1/2, 1/2, 0) is also an NTU value (set $\lambda_1 = \lambda_2 = 1$, $\lambda_3 = 0$).
20. We are abstracting away from political ideology, i.e., assuming that all that matters is influence in the government.
21. Shafer also considers $\varepsilon = 0$, but we do not; see note 20.

The example is basically similar to Roth's, though less clear-cut.[22] What lends it credibility is the market context, in which endowments are more "visible" than utilities. In the previous story, 3 appeared as a large powerful political party; here he appears as a miserable peddler, with hardly any goods.

Shafer may well have a point. Game theory, as well as economics, typically provides multiple solutions. But game theoretic concepts apply to a "purified" or "processed" version of the original situation, such as the coalitional[23] or strategic[24] form. The processing removes some of the "glue" that gives the situation coherence; to choose among the multiple solutions, it may be necessary to restore some of this glue, to go back and look at the "raw," original situation.

T.C. Schelling [44] used the term "focal point" for an EP suggested by the particular context of a game. Suppose that two people arrange to meet, but neglect to specify where. If in the past they have frequently met at a certain bar, it will be natural for them to seek each other there; though there is nothing in the mathematical structure of the game that distinguishes this EP from any other, their mutual expectations reinforce each other and make it a likely outcome. When nations negotiate boundaries, rivers and watersheds are focal points. In deciding the level of poison gas or nuclear weapons that international convention tolerates in war, the zero level suggests itself as a focal point, even though one or both sides might prefer a different level.

In the above political context, history, custom, public opinion, and the sheer size of 3 generate a perception that it will probably lead the government; once there, this perception reinforces itself[25] and becomes a focal point. In Shafer's context, 3's puniness generates the opposite perception, that he will be excluded from the trading; it, too, reinforces itself and becomes a focal point.

The NTU value is "context-neutral." Based on the coalitional worth function only, it cannot take into account the peculiar features of each realization of this function. In discussing the situation, one can stay away from special contexts, as Roth did, and as we did in Sections 3–7. But if one does wish to tell stories, then more than one can be told; and again it

---

22. In Roth's example, 1 and 2 can achieve an outcome that is preferred by each of them to any other feasible outcome. Therefore if they meet before either one meets 3, it is a foregone conclusion that they will form a coalition, because there is no room for any argument between them. This is not true in Shafer's example; if they meet, 1 and 2 may argue, perhaps even disagree. This weakens them in itself, and also because the uncertain outcome if they meet diminishes whatever resolve they may have had to refuse offers from 3.

23. "Characteristic function."

24. "Normal."

25. As described in sections 3, 5, and 6.

turns out that on the whole, the outcome (1/3, 1/3, 1/3) is not a bad reflection of the qualitative situation.

9 The Conceptual Background of Roth's Solution

Underlying Roth's solution is the idea of domination, the same idea as that which underlies the core and the NM (von Neumann–Morgenstern) solution [33]. Recall that an outcome x dominates an outcome y iff there is a coalition that can achieve x, each of whose members prefers x to y. In Roth's game, (1/2, 1/2, 0) dominates all other outcomes; it is the only point in the core, and constitutes the only NM solution.

But domination, though unquestionably of fundamental importance, does not have the elemental persuasiveness of simple rationality (i.e., individual utility maximization). It is based on the principle of cooperation—that people should always act jointly to further common interests; and this goes substantially beyond rationality, which says only that an individual should always act in his own interests.[26] *Domination involves relying on others to cooperate*, and as we have seen above, that is not always the way to maximize utility.

Value theory, on the other hand, is not based on domination. The value of an individual is a kind of index or average, based on the strength of the coalitions of which he is a member, and of those of which he is not a member. No attempt is made to predict which coalitions will form; all coalitions are considered. This fits in well with the Nash program, which in Roth's game also leads to all coalitions.

In the context of cooperative games, the principle of cooperation may sound quite reasonable, and solution concepts based on domination certainly merit study. But the term "cooperative" is usually taken to mean only that players *can* make binding agreements, without any implication about what they *should* or are expected to do. The NTU value fits in well with this broader view, and certainly it, too, merits study.

We conclude by doffing our hat to Roth and Shafer. Their examples are ingenious, thought provoking, and far from transparent; and there is no doubt that they have led to a deeper understanding of the NTU value and of cooperative games in general.

---

26. It might be argued that rationality implies the principle of cooperation, since cooperation increases the utility of each individual in the group. But this argument is circular. The principle of cooperation will not be effective unless it is adopted by all involved; in general, it will not be rational for a single individual to adopt it. Moreover, it is by no means clear that it would be good for Society as a whole to adopt it. Thieves and murderers can also cooperate; in Roth's game, cooperation between 1 and 2 would be great for them, but not so good for 3.

Acknowledgments

This work was supported by National Science Foundation Grant SES 80-06654 at the Institute for Mathematical Studies in the Social Sciences, Stanford University, and by the Institute for Advanced Studies at the Hebrew University of Jerusalem. We gratefully acknowledge conversations and correspondence with K. Arrow, K. Binmore, M. Kurz, M. Maschler, A. Roth, and L. Shapley. Of course they are not responsible for the views expressed here. Roth, indeed, will speak for himself: He plans to publish a response.

References

1. Aumann, R.: "Values of Markets with a Continuum of Traders," *Econometrica*, 43 (1975), 611–646 [Chapter 51].
2. ———: "An Axiomatization of the Non-transferable Utility Value," *Econometrica*, 53 (1985), 599–612 [Chapter 60].
3. Aumann, R., and J. H. Drèze: "Values of Markets with Satiation or Fixed Prices," *Econometrica*, 54 (1986), 1271–1318 [Chapter 57].
4. Aumann, R., and M. Kurz: "Power and Taxes," *Econometrica*, 45 (1977), 1137–1161 [Chapter 52].
5. ———: "Power and Taxes in a Multicommodity Economy," *Israel Journal of Mathematics*, 27 (1977), 185–234 [Chapter 53].
6. Aumann, R., M. Kurz, and A. Neyman: "Power and Public Goods," *Journal of Economic Theory*, 42 (1987), 108–127 [Chapter 55].
7. ———: "Voting for Public Goods," *Review of Economic Studies*, 50 (1983), 677–694 [Chapter 56].
8. Brock, H. W.: "A New Theory of Social Justice Based on the Mathematical Theory of Games," in *Game Theory and Political Science*, ed. by P. C. Ordeshook. New York: NYU Press, 1978, pp. 563–626.
9. ———: "A Critical Discussion of the Work of John C. Harsanyi," *Theory and Decision*, 9(1978), 349–367.
10. Brown, D., and P. Loeb: "The Value of a Nonstandard Competitive Economy," *Israel Journal of Mathematics*, 25(1976), 71–86.
11. Champsaur, P.: "Cooperation vs. Competition," *Journal of Economic Theory*, 11 (1975), 394–417.
12. Cheng, H.: "On Dual Regularity and Value Convergence Theorems," *Journal of Mathematical Economics*, 8 (1981), 37–57.
13. Drèze, J. H., J. Jaskold-Gabszewicz, and A. Postlewaite: "Disadvantageous Monopolies and Disadvantageous Endowments," *Journal of Economic Theory*, 16 (1977), 116–121.
14. Gardner, R.: "Wealth and Power in a Collegial Polity," *Journal of Economic Theory*, 25 (1981), 353–366.
15. ———: "λ-transfer Value and Fixed-Price Equilibrium in Two-Sided Markets," in *Social Choice and Welfare*, ed. by P. K. Pattanaik and M. Salles. Amsterdam: North-Holland, 1983, pp. 301–323.
16. Guesnerie, R.: "Monopoly, Syndicate and Shapley Value: About Some Conjectures," *Journal of Economic Theory*, 15 (1977), 235–252.
17. Harsanyi, J. C.: "The Tracing Procedure," *International Journal of Game Theory*, 4 (1975), 61–94.
18. ———: *Rational Behavior and Bargaining Equilibrium in Games and Social Situations*. Cambridge: Cambridge University Press, 1977.
19. ———: "Comments on Roth's Paper, 'Values for Games without Side Payments'," *Econometrica*, 48 (1980), 477.

20. Hart, S.: "Values of Non-Differentiable Markets with a Continuum of Traders," *Journal of Mathematical Economics*, 4 (1977), 103–116.

21. ———: "Axiomatic Approaches to Coalitional Bargaining," in *Game Theoretic Models of Bargaining*, ed. by A. E. Roth. Cambridge: Cambridge University Press, 1985, pp. 305–319.

22. ———: "Non-Transferable Utility Games and Markets: Some Examples and the Harsanyi Solution," *Econometrica*, 53 (1985), 1445–1450.

23. ———: "An Axiomatization of Harsanyi's Non-Transferable Utility Solution," *Econometrica*, 53 (1985), 1295–1313.

24. Imai, H.: "On Harsanyi's Solution," *International Journal of Game Theory*, 12 (1983), 161–179.

25. Kern, R.: "The Shapley Transfer Value without Zero Weights," *International Journal of Game Theory*, 14 (1985), 73–92.

26. Kurz, M.: "Distortion of Preferences, Income Distribution and the Case for a Linear Income Tax," *Journal of Economic Theory*, 14 (1977), 291–298.

27. ———: "Income Distribution and Distortion of Preferences: the ℓ-Commodity Case," *Journal of Economic Theory*, 22 (1980), 99–107.

28. Mas-Colell, A.: "Competitive and Value Allocations of Large Exchange Economies," *Journal of Economic Theory*, 14 (1977), 419–438.

29. ———: "Remarks on the Game-Theoretic Analysis of a Simple Distribution of Surplus Problem," *International Journal of Game Theory*, 9 (1980), 125–140.

30. Myerson, R.: "Cooperative Games with Incomplete Information," *International Journal of Game Theory*, 13 (1984), 69–96.

31. Nash, J. F.: "The Bargaining Problem," *Econometrica*, 18 (1950), 155–162.

32. ———: "Non-Cooperative Games," *Annals of Mathematics*, 54 (1951), 286–295.

33. ———: "Two-Person Cooperative Games," *Econometrica*, 21 (1953), 128–140.

34. von Neumann, J., and O. Morgenstern: *Theory of Games and Economic Behavior*. Princeton: Princeton University Press, 1944.

35. Neyman, A.: "Values for Non-Transferable Utility Games with a Continuum of Players," Technical Report No. 351, School of Operations Research, Cornell University, 1977.

36. Osborne, M. J.: "An Analysis of Power in Exchange Economies," Technical Report No. 291, Economics Series, Institute for Mathematical Studies in the Social Sciences, Stanford University, 1979.

37. ———: "Why Do Some Goods Bear Higher Taxes Than Others?" *Journal of Economic Theory*, 32 (1984), 301–316.

38. Pečerskiy, S.: "The Shapley Value for Games without Side Payments," *Mat. Methody v. Social. Nauk.-Trudy Sem. Processy Optimal. Upavlenija II Sekcija*, 10 (1978), 43–62, 121 (in Russian) (Math. Reviews 80b: 90159).

39. Rosenmuller, J.: "Remark on the Transfer Operator and the Value-Equilibrium Equivalence Hypothesis," in *Mathematical Economics and Game Theory, Essays in Honor of Oskar Morgenstern*, ed. by R. Henn and O. Moeschlin, Lecture Notes in Economics and Mathematical Systems, Vol. 142. New York: Springer, 1977, pp. 108–127.

40. ———: "Values of Non-Sidepayment Games and Their Application in the Theory of Public Goods," in *Essays in Game Theory and Mathematical Economics in Honor of Oskar Morgenstern*, Vol. 4 of *Gesellschaft, Recht, Wirtschaft*. Mannheim/Wien/Zurich: Bibliographisches Institut, 1981, pp. 111–129.

41. ———: "On Values, Location Conflicts, and Public Goods," in *Proceedings of the Oskar Morgenstern Symposium at Vienna*, ed. by M. Deistler, E. Furst, and G. Schwodiauer. Wien: Physica-Verlag, 1982, pp. 74–100.

42. ———: "Selection of Values for Non-Sidepayment Games," *Methods of Operations Research* 48 (1984), 339–398.

43. Roth, A.: "Values for Games without Side Payments: Some Difficulties with Current Concepts," *Econometrica*, 48 (1980), 457–465 [Chapter 61a].

44. Schelling, T. C.: *The Strategy of Conflict*. Cambridge: Harvard University Press, 1960.

45. Selten, R.: "Re-examination of the Perfectness Concept for Equilibrium Points in Extensive Games," *International Journal of Game Theory*, 4 (1975), 25–55.

46. Shafer, W.: "On the Existence and Interpretation of Value Allocations," *Econometrica*, 48 (1980), 467–477 [Chapter 61b].

47. Shapley, L. S.: "A Value for n-Person Games," in *Contributions to the Theory of Games II*, ed. by H. W. Kuhn and A. W. Tucker. Princeton: Princeton University Press, 1953, pp. 307–317.

48. ———: "The Solutions of a Symmetric Market Game," in *Contributions to the Theory of Games IV*, ed. by A. W. Tucker and R. D. Luce. Princeton: Princeton University Press, 1959, pp. 145–162.

49. ———: "Utility Comparison and the Theory of Games," in *La Decision: Aggregation et Dynamique des Ordres de Preference*. Paris: Editions du Centre National de la Recherche Scientifique, 1969, pp. 251–263.

50. Shapley, L. S., and M. Shubik: "Pure Competition, Coalitional Power, and Fair Division," *International Economic Review*, 10 (1969), 337–362.

51. Yaari, M.: "Rawls, Edgeworth, Shapley, Nash: Theories of Distributive Justice Re-examined," *Journal of Economic Theory*, 24 (1981), 1–39.

61d On the Non-Transferable Utility Value: A Reply to Aumann

Alvin E. Roth

Aumann (1985a) seeks to dispel the questions raised about the NTU value in Roth (1980), by presenting an alternative analysis of the single-parameter class of games examined there. Here I explain why I find his response largely unpersuasive. Aumann's analysis does not apply to the full range of parameter values for the class of games in question, and, in any event, the noncooperative model he considers presents little support for the NTU value. The NTU value does not correspond to any equilibrium outcome of the noncooperative game he considers, or to any average of equilibrium outcomes of the game, or in any precise way to an average of equilibrium outcomes over different possible noncooperative realizations of the cooperative game. I argue that no consistent interpretation of the NTU value along these lines is generally possible. An additional example is presented to illuminate the discussion.

In a recent paper, Aumann (1985a) replies to criticisms of the NTU (nontransferable utility) value (Shapley, 1969) appearing in Roth (1980),[1] which were couched in terms of the following single parameter family of 3-person games. For any real number p, let $G(p)$ be the game in which the players can each assure themselves of achieving a payoff of at least 0; players 1 and 2 acting together can achieve (1/2, 1/2, 0), i.e., they can each get a payoff of 1/2, leaving 0 for player 3; players 1 and 3 acting together can achieve $(p, 0, 1 - p)$; and players 2 and 3 acting together can achieve $(0, p, 1 - p)$. All three players acting together can achieve any convex combination of (1/2, 1/2, 0), $(p, 0, 1 - p)$, and $(0, p, 1 - p)$. No sidepayments of any sort are feasible.

Readers unfamiliar with the definition of the NTU value can refer to my 1980 paper to see how it takes the value for this (and for any) NTU game to equal the value of an associated transferable utility game. The NTU value predicts the outcome (1/3, 1/3, 1/3) for $G(p)$, *regardless* of the value of p, since, if utility *were* transferable, the game would be symmetric, since the sum of the payoffs to every 2-person coalition is 1. In Roth (1980) I argued that this is unreasonable when p is strictly less than 1/2. When $p < 1/2$, players 1 and 2 have a common interest in the outcome (1/2, 1/2, 0), which yields both of them their highest feasible payoff,

---

This paper originally appeared in *Econometrica* 54 (1986): 981–984. Reprinted with permission.

1. Roth (1980) criticized not only the NTU value, but also a closely related concept due to Harsanyi (1963, 1977), which Shapley (1969, p. 260) cites as the motivation for the NTU value. However, since Aumann directs his comments at the NTU value, and since Harsanyi (1980) has indicated substantial agreement with the conclusions of Roth (1980), the comments in this paper will be directed at the NTU value alone.

and is the only outcome which does so. Since the rules of the game allow them to realize this common interest, I argued that, at least in the idealized case when it is common knowledge that the players are both rational, we should expect the outcome (1/2, 1/2, 0) to be the result of the game. While this is so for any value of p smaller than 1/2, it is perhaps most vivid when $p \leq 0$, in which case player 3 has nothing whatsoever to offer the other players. When $p < 0$, the game is perfectly playable, but it would be individually irrational for either player 1 or 2 to form a coalition with player 3, as would be necessary to justify the prediction of (1/3, 1/3, 1/3). The case for (1/2, 1/2, 0) in this situation seems completely compelling. Nevertheless, even in this case, the NTU value continues to predict the outcome (1/3, 1/3, 1/3).[2] Aumann considers the games $G(p)$ only for $p > 0$ in his reply. He constructs an extensive form game whose characteristic function equals $G(p)$, in which one of the three players is randomly "given the initiative" to form a coalition. If player 1 or 2 is chosen as the initiator, then (1/2, 1/2, 0) is the unique perfect equilibrium outcome. In the subgame which occurs, with probability 1/3, when player 3 is given the initiative, the outcome (1/2, 1/2, 0) is again a perfect equilibrium outcome, which results when players 1 and 2 each adopt a strategy of rejecting player 3's offer to form a coalition. But, in this subgame, it is also a perfect equilibrium for players 1 and 2 to each adopt a strategy of accepting player 3's offer, and for player 3 to randomize his choice of which player to approach first.

A detailed critique of Aumann's analysis of this game is contained in Roth (1984). Here I will simply note that the perfect equilibrium points Aumann discusses do not, under any interpretation, provide a noncooperative model that predicts the NTU value. The only perfect equilibria with payoffs other than (1/2, 1/2, 0) he finds are in the subgame (which occurs with probability 1/3) in which player 3 is the "initiator." Even if we assign these alternative equilibria probability *one* in the subgame in which they occur, the game's expected payoff to player 3 is still

---

2. When p is exactly equal to 0 (and only then), Shapley's original definition of the NTU value allows the TU game that determines the NTU value to be obtained by multiplying by 0 the utility that player 3 obtains at every outcome. In this degenerate case, (1/2, 1/2, 0) is selected by the NTU value as an outcome, along with (1/3, 1/3, 1/3). Since multiplying a player's utility by 0 does not yield an equivalent utility function (as does multiplying it by a positive number), Shapley and subsequent authors have taken pains to stress that including this case is a technical flaw in the definition, needed for the existence proof (e.g., Shapley (1969, p. 260): "Zero weights can occur, and must be allowed if the existence theorem is to hold in general"). In using the NTU value, authors almost always stress that the NTU values they use involve only positive weights. Aumann (1985b), in a recent axiomatic treatment, *defines* the NTU value to have positive weights. It is therefore quite a substantial although unremarked innovation on Aumann's part to dismiss the case $p = 0$ by saying simply "But in that case, (1/2, 1/2, 0) is also an NTU value."

strictly smaller than the 1/3 predicted by the NTU value. To see this, note that player 3 is chosen as initiator only with probability 1/3, so his expected payoff is no more than $(2/3)(0) + (1/3)(1-p)$ even under this extreme assumption.[3] So this game, at least, does not support the argument that the NTU value reflects some sort of average over the payoffs players might expect. Next we consider an example that shows that *no* interpretation of the NTU value as an average of what players might expect over different possible extensive-form realizations of a given cooperative game will be sustainable in general.[4]

Let q be a nonnegative parameter: I will concentrate on the cases $q = 0$ and $q = 1$. The case $q = 0$ has earlier been considered by Owen (1972) and Maschler (personal communication). There are three players; each can assure himself of getting 0; players 1 and 2 acting together can achieve the payoff vector $(1 + q, -q, 0)$; players 1 and 3 or players 2 and 3 acting together can achieve only the payoff vector $(0, 0, 0)$, and players 1, 2, and 3 acting all together can achieve any payoff (x_1, x_2, x_3) whose components are not less than -1 and which sum to 1. No sidepayments of any sort are feasible.

Players 1 and 2 together can engage in an activity with payoff $1 + q$ to player 1, and $-q$ to player 2. Since player 2 can assure himself of at least 0, no assumption other than his individual rationality is needed to conclude that players 1 and 2 will not form a coalition to carry out this activity when $q = 1$. When $q = 0$, this coalition offers no benefits to 2. No other one or two-player coalition has any productive activities available to it. Only the three-player coalition has any individually rational productive activities available to it, and its potential activities are symmetric in the payoffs they give to the players. A case could thus be made that the outcome $(1/3, 1/3, 1/3)$ is the most representative of the players' prospects in this game. At the least, to argue that player 3 should have a pos-

---

3. Of course, it might be argued that the game $G(p)$ could be represented by an extensive form game in which one of the three players was always the initiator. But there are three equally likely such games, corresponding to the three players who might be initiators, and once again, even if we assign the outcome $(1/2, 1/2, 0)$ favored by players 1 and 2 probability *zero* whenever *any* other perfect equilibrium outcomes exist, player 3's expected payoff over these representations is still less than 1/3.

4. In this regard Aumann criticizes the outcome $(1/2, 1/2, 0)$ in the game $G(p)$ by saying

"Neither can $(1/2, 1/2, 0)$ be considered as an amalgam arising from different bargaining models; its extreme nature—obviously 3 can't get *less* than 0—implies that if $(1/2, 1/2, 0)$ is an average of outcomes, then it is *always* the outcome, and we are back to the previous argument."

He defends $(1/3, 1/3, 1/3)$—the NTU value in the game $G(p)$—saying:

"But whereas 3's *chance* of getting into a coalition is smaller than that of the others, his *payoff* $1 - p$ if he does get in is larger, and these two effects vary in opposite directions as p varies. All in all, perhaps $(1/3, 1/3, 1/3)$—the NTU value—does reflect the qualitative features of the game quite well."

itive payoff it is sufficient to note that *no* individually rational, mutually profitable agreement can be reached without his cooperation. Nevertheless, the NTU value predicts (1/2, 1/2, 0), for any q.

The technical reason for this is clear: if utility *were* transferable, player 3 would be a dummy in the corresponding TU game. In that game, the coalition of players 1 and 2 would have a worth of 1, player 3 would add nothing, and so he would get nothing at the TU value, which is (1/2, 1/2, 0). Because this is feasible in the NTU game and the TU value in the TU game, it is *by definition* also the NTU value in the NTU game. But in the NTU game, player 3 is no dummy.

In the NTU game, players 1 and 2 can't get together on their own since a sidepayment from 1 to 2 would be required before any joint activity would be rational or profitable for player 2, and such sidepayments are impossible. But player 3 can act as an intermediary: when he is a member of the coalition along with 1 and 2, any distribution of the gains from cooperation is possible. Of course, if it were suddenly possible (as in the TU game) for player 1 to make sidepayments to player 2, then the services of player 3 would have no value. But to conclude *for this reason* that player 3's services are of absolutely no value even when such sidepayments *cannot* be made strikes me as absurd. Yet this is what the NTU value does.

The similarity between this example and the games $G(p)$ is that both were chosen to emphasize the great difference that can exist between an NTU game and the associated TU game. This difference is ignored by the NTU value. The difference between the two examples is that, loosely speaking, the roles of the payoff vectors (1/3, 1/3, 1/3) and (1/2, 1/2, 0) are reversed. The "averaging" arguments which Aumann made against (1/2, 1/2, 0) (and for (1/3, 1/3, 1/3)) in the games $G(p)$ (see footnote 5) would have to be reversed in the case of this example, where it is the NTU value which assigns 0 to the third player.

My conjecture is that the NTU value is an unreliable indicator for NTU games that are "very" different from the associated TU game. The behavior of the NTU value with respect to the individually irrational payoffs noted here seems particularly telling, since the individual irrationality would disappear if utility were transferable. That is, NTU games exhibit properties completely absent from the associated TU games. Thus an entire class of phenomena that can occur only in NTU games is ignored by the NTU value. This is inevitable in a solution concept that treats every NTU game as a TU game. Since situations modelled as NTU games often cannot be adequately modelled as TU games, this is a disturbing "hidden" feature of the NTU value, which gives special cause for caution in interpreting it. Just this problem with the NTU value accounts for some of its strangest predictions. Aside from the examples already noted here, recall that Shafer (1980) produced examples of an exchange economy in which the prediction of the NTU value was that one of the traders would receive a final endowment larger in every component than his initial endowment.

Let me close by emphasizing that I do not disagree with Aumann's conclusion that the NTU value "merits study." Our disagreement seems to concern what directions of study are likely to be most illuminating, and where they are likely to lead.[5]

Acknowledgments

I have benefited from lengthy correspondence and conversation on these matters with Bob Aumann, although we have failed to reach agreement. His willingness to explore these intellectual differences in a spirit of friendly inquiry has, from the beginning, been an inspiration without which it would have been difficult to pursue these matters. I am also indebted to many more of my colleagues than I can name here for their advice and encouragement in pursuing this matter. This work has been supported by grants from the National Science Foundation and the Office of Naval Research, and by Fellowships from the John Simon Guggenheim Memorial Foundation and the Alfred P. Sloan Foundation.

References

Aumann, Robert J. (1985a): "On the Non-transferable Utility Value: A Comment on the Roth–Shafer Examples," *Econometrica*, 53, 667–677 [Chapter 61c].

———(1985b): "An Axiomatization of the Non-transferable Utility Value," *Econometrica*, 53, 599–612 [Chapter 60].

Gardner, Roy (1983): "λ-Transfer Value and Fixed-Price Equilibrium in Two-Sided Markets," in *Social Choice and Welfare*, ed. by P. K. Pattanaik and M. Salles. Amsterdam: North-Holland.

Harsanyi, John C. (1963): "A Simplified Bargaining Model for the n-Person Cooperative Game," *International Economic Review*, 4, 194–200.

———(1977): *Rational Behavior and Bargaining Equilibrium in Games and Social Situations*. Cambridge: Cambridge University Press.

———(1980): "Comments on Roth's Paper: 'Values for Games Without Side Payments'," *Econometrica*, 48, 477.

Owen, Guillermo (1972): "Values of Games Without Sidepayments," *International Journal of Game Theory*, 1, 94–109.

Roth, Alvin E. (1980): "Values for Games Without Sidepayments: Some Difficulties With Current Concepts," *Econometrica*, 48, 457–465 [Chapter 61a].

Shafer, Wayne J. (1980): "On the Existence and Interpretation of Value Allocations," *Econometrica*, 48, 467–476 [Chapter 61b].

Shapley, Lloyd S. (1969): "Utility Comparison and the Theory of Games," in *La Decision: Aggregation et Dynamique des Ordres de Preference*. Paris: Editions du Centre National de la Recherche Scientifique, pp. 251–263.

---

5. Worth noting in this regard is a paper by Gardner (1983) who analyzes a two-sided market and establishes conditions under which the NTU value exhibits what he calls "Roth's paradox."

61e Rejoinder

1. In his *Reply*, Roth (1986) adheres to the position he took in (1980), in which he said that for $0 < p < 1/2$, "(1/2, 1/2, 0) is the unique outcome ... consistent with the hypothesis that the players are rational utility maximizers." We find this position untenable. The case for (1/2, 1/2, 0) depends on a kind of mutual reliance between Players 1 and 2 that goes far beyond ordinary individual utility maximization, and that, because of its riskiness, may be totally unreasonable. This was explained in our *Comment* (Aumann (1985)), using informal as well as formal arguments; we will not repeat them here.

2. Roth complains that no one of our formal bargaining models predicts the value. We never claimed that they do; indeed, we emphasized that "the quantitative results (of the formal bargaining models) cannot be considered particularly significant ..." (Aumann (1985, p. 673)). These models were meant primarily to rebut Roth's position that (1/2, 1/2, 0) is the *only* reasonable outcome of these games; taken together, they also lend qualitative support to the value. Roth has no response to this, nor to any of the less formal reasoning that forms the bulk of our *Comment*.

3. Our own position is not dogmatic. For $0 < p < 1/2$, the unique value is (1/3, 1/3, 1/3), the unique core point (1/2, 1/2, 0). For small p the core looks reasonable, the value strange. As p grows, the value becomes more and more reasonable, the core stranger and stranger. For large p it is the value that is reasonable, the core strange. The Harsanyi solution (Hart (1985)) which yields $(1/2 - p/3, 1/2 - p/3, 2p/3)$ seems in these games more "sensible"—less "extreme"—than either the core or the value. Nevertheless, both the core and the value reflect important qualitative features of the games; one would not want to dispense with either one. The different outcomes that different solution concepts yield represent different approaches or viewpoints; they illuminate the problem from various angles.

4. When $p = 0$, we agree unequivocally that (1/3, 1/3, 1/3) is inappropriate. But just at this point, where Roth's case appears finally to become transparent, it vanishes. The outcome (1/3, 1/3, 1/3) is then no longer *the* value; it is only *a* value, another one being precisely the outcome (1/2, 1/2, 0) preferred by Roth. One can't base a compelling argument against the value on this kind of multivalent situation.

Roth objects that the value (1/2, 1/2, 0) is supported by weights (1, 1, 0); he says that zero weights are a "technical flaw," and implies that they should not be used in conceptual discussion.[1] Zero weights are indeed associated with degeneracies; but they are in the game, not in the value. When $p = 0$, Player 3 can benefit no one but himself by joining a coalition. If one views the weight of a player as an endogenous measure of his importance and influence, "the weight he pulls" in Society, then in such a game it is quite natural to assign him weight 0. (See also (8) below.)

In brief, zero weights are associated with degenerate games, which it is best to avoid in conceptual discussion. But when they do occur, they cannot be dismissed, and are indeed quite natural.

5. When $p < 0$, the game is not superadditive; again, one cannot base compelling counterintuitive examples on such games. Roth calls them "perfectly playable." But if one does play them, then each player can assure 0 to himself, without the help of anyone else. In practice, therefore, each of the coalitions $\{1,3\}$ and $\{2,3\}$ can obtain $(0, 0)$. With a coalitional form (characteristic function) that reflects this, $(1/2, 1/2, 0)$ reappears as a value, and Roth's argument disappears.

6. Next, Roth brings up the old Maschler–Owen example, which involves completely different issues. We welcome the opportunity to discuss this.

Two variants, which we call V_0 and V_1, are at issue. In V_0, there are three players; by himself each one can get 0. If Players 1 and 3, or 2 and 3, form a coalition, then both get 0; if 1 and 2 form a coalition, then 1 gets a payoff of 1, and 2 gets nothing. All three together can share 1 in any way they please. The unique NTU value of this game is $(1/2, 1/2, 0)$. This seems strange because Player 3 fulfills an important function in enabling 1 and 2 to share the amount 1 between them; while he need not get the same payoff as they, it certainly appears that he should get *something* for his services.

Before going on, we call attention to a four-person TU game, communicated to us by S. Zamir, in which the core behaves somewhat like the value of V_0, but seems even stranger. Define v on the player set $\{1, 2, 3, 4\}$ by

$$v(S) = \begin{cases} 3 & \text{if } |S| = 4, \\ 0 & \text{if } |S \cap \{1,2,3\}| = 0 \text{ or } 1, \\ 2 & \text{otherwise.} \end{cases}$$

Intuitively, v is obtained from the majority game on $\{1, 2, 3\}$ by adding Player 4, who brings with him resources enabling all players together to get 3 rather than 2. The core of v consists of the single point $(1, 1, 1, 0)$.

---

1. But when it suits his purposes, he himself uses zero weights with impunity; they appear explicitly in (Roth (1980)), as an important component of his attack on Harsanyi's solution.

This is even stranger than the value of V_0. There, one may feel that the additional player should be compensated for his services in enabling 1 to transfer payoff to 2. Here, the additional player brings tangible resources, which actually increase total revenue by a significant amount. Yet he gets no part of this revenue.

What is happening is that without Player 4, the game is coreless. The three players must either vie for a spot on a two-player coalition, knowing that one of them will be left in the cold, or agree to a compromise that yields any two of them less than what they can get for themselves. Player 4's contribution is just enough to make the core nonempty; it is totally gobbled up by 1, 2, and 3, on the pretext that any two of them thereby get no more than what they could have gotten previously. But only *two* of the three could previously have gotten 1; to conclude for this reason that now *each one* of them should get 1 strikes us as absurd.

We do not wish to disparage the core. The usefulness of a solution concept is measured not by its behavior in contrived examples, but by the insights it yields into social models of some generality. In this respect, both the core and the value have rather good track records.

But in fact, the value of V_0 is not all that strange. Suppose you and your brother are bequeathed a house located in your town. You wish to take possession, and to send your brother half its value; to this end, you ask your bank to make the transfer. What should the bank's fee be? Most people would suggest a relatively small fixed sum, or perhaps a few promil of the amount to be transferred. Very few would suggest anything substantial.

It may be objected that there is competition among the banks, so they cannot take too much; at worst, you *could* bring the money by hand. But even if there is only one bank, and you are somehow prevented from bringing the money by hand, many people would be appalled if the bank took any substantial proportion. This view is expressed by the value, which is a measure of a player's contribution to the social product (see (8) below). The transfer under discussion does not change the total social product; enabling it makes no substantial, measurable contribution, such as is made by Player 4 in Zamir's game v. While this is not the only possible view, it is not an unreasonable one. On the other hand, the core of v makes no sense at all to us.

Even if one insists that the bank should get a positive proportion, it is not clear how much. The game has no obvious symmetries; any positive amount seems possible. Thus at worst, the value appears as an extreme or limiting point of the "reasonable outcomes." That can't be considered a compelling counterintuitive example.

We come next to V_1, which differs from V_0 only in that the coalition $\{1,2\}$ gets $(2,-1)$ rather than $(1, 0)$. As in V_0, the value is $(1/2, 1/2, 0)$. Roth argues that the payoffs to $\{1,2\}$ are individually irrational, and should therefore be ignored. What remains is symmetric in all players, so the value "should" be $(1/3, 1/3, 1/3)$.

This argument applies not to the value, but to formulations of the coalitional form that involve individual irrationalities. If one insists that individual irrationalities cannot occur in practice and that their prospect can have no effect on the final outcome, one should exclude them to start with. This procedure is quite standard in Game Theory; applied to V_1, it indeed yields the value $(1/3, 1/3, 1/3)$. That presents no problem for the value.

There is, however, another view, in which individual irrationalities do play a significant role. The statement that "Player 2 can guarantee himself 0" can be interpreted to refer to the *beginning* of bargaining only. During the course of bargaining, he may make commitments which, while undertaken with the expectation of profit, *can* also lead to loss; and if they do, he cannot at that time renege and go back to 0. For example, consider a two-person bargaining game in which the individually rational levels are 0, and the two players together can get either $(2,-1)$ or $(-1,2)$. In the "standard" solution, each player has an expected payoff of $1/2$, based on a coin-toss between $(2,-1)$ and $(-1,2)$; the price of the positive *expectation* is a commitment to accept a negative, individually irrational payoff if the toss goes against you. Similarly, in the course of bargaining, players may wish to commit themselves to joining certain coalitions under certain circumstances; or they may find it advantageous to forego certain options that they have. "Strategic risks" of this kind are in principle quite similar to the coin-toss in the above example. They are risky because one cannot be sure how the other players will respond; and while made in the expectation of profit, they may lead to loss, and even to individually irrational outcomes.

In V_1, suppose 2 commits himself to joining a coalition with 1, on pain of paying a fine in the amount of 1. Then 2's individually rational level drops to -1, and V_1 is transformed into a game V_0^* that is strategically equivalent to V_0. Reasoning as in V_0, we conclude that in V_0^*, Player 3 "should" enable 1 to make transfers to 2, without expecting a substantial fee. Thus 1 and 2 may expect to share the amount of $2 + (-1) = 1$. Since each one can get 0 at the beginning of the game, it seems reasonable for them to share this amount $1/2 - 1/2$; i.e., for 2 to extract, in return for his commitment, a promise from 1 to transfer to him the amount $3/2$, if (or when) 3 will allow it. This yields precisely the value.

To sum up, in analyzing V_1 one must first decide whether to exclude individual irrationalities at the outset, or to recognize them as important elements of the dynamics of bargaining. In the former case, we agree that the outcome should be (1/3, 1/3, 1/3), and indeed this is the value of the appropriately adjusted form of V_1. In the latter case, Roth's symmetry argument vanishes, and the value (1/2, 1/2, 0) appears as reasonable as in V_0.

7. The last example in the *Reply* is Shafer's, which we have already discussed (Aumann (1985, Section 8)). Roth finds it strange that there are exchange economies in which the value allocation yields one of the traders more of every commodity than his endowment. This is certainly interesting, but on reflection, not so strange. The utility function of the trader in question is "flatter" than that of the others; he is less particular, less risk-averse. Merchants like that have parlayed shoestrings into fortunes; the very willingness to take risks confers an advantage, even when not actually taken.[2] It is a strength of the value, not a weakness, that it reflects subtleties such as the effects of the utility function on bargaining strength.

8. Unable to make a compelling case with examples, Roth turns to the *definition* of value as such. He argues that since it allows Transfers of Utility, the endogenous TU game appearing in the definition has no clear relevance to the original NTU game.

This sounds cogent at first, but it does not survive examination. A TU game v is a function that associates with each coalition S a real number $v(S)$. One interpretation of $v(S)$ is as a sum of money (or other transferable good) that S may divide among its members in any way it wishes; this gives rise to the appellation "TU." But it is not the only interpretation. Even when utility is not transferable, one can take $v(S)$ simply as an appropriate numerical measure of the worth of S. That is what the NTU value does.

Measures or indexes like this are quite common in economics, accounting, and indeed all walks of life. A person's (or firm's) net worth is a sum of worths of assets that are very different from each other, e.g. in their liquidity; the total dollar figure is operationally meaningless, yet conveys important information. Similar remarks hold for GNP, national debt, price indexes, the mean of a distribution, the Gini index of inequality, and so on. Economic theory (e.g. taxation or growth theory) often uses social welfare functions—simple sums of utilities—even though util-

---

2. Suppose two agents must divide 6 dollars, with conflict payoffs of 0. If their utility functions are x and $\sqrt{x}$, the Nash bargaining solution yields 4 dollars to the risk neutral agent, and 2 dollars to the risk averse one, even though no overt risks are taken by any one.

ity is not transferable; nobody bats an eyelash. In Game Theory, the most frequent application of the TU value is to voting, committees and so on; coalitions S are assigned worth 1 if they can win, 0 if they can't, even though utility is not transferable.

The value of a player is meant as a measure of his social productivity, his contribution to the total social product. The explicit formula involves expected contributions to coalitions S that he may join. None of this requires transferable utility to make sense.

For the measure of S's worth, the maximum of a weighted sum of utilities—a kind of social welfare function—seems eminently reasonable. The weights are chosen so that the resulting value is feasible; an infeasible result would indicate that some people are overrated (or underrated), much like an imbalance between supply and demand indicates that some goods are overpriced (or underpriced). Note that zero weights fit very naturally into this picture, much like zero prices.

We do not claim that this is the only possible formulation of the intuitive concept in question. But it is reasonable and natural enough, and certainly does not suffer from the ills that Roth attributes to it.[3]

9. To conclude, not one single "clean," convincing counterintuitive example to the value has been adduced. The one remaining oddity—for indeterminately small positive p—comes nowhere near in sharpness and force to existing counterintuitive examples to other solution concepts such as the core, and cannot be considered a serious challenge to the value. The attacks on the definition of the value as such also fail to hold up.

From all this, the NTU Shapley value emerges stronger than ever. In applications, it often has substantial intuitive content; not infrequently, it yields important, unexpected insights (see the references of our *Comment*). It has several quite different characterizations, all of them intuitively meaningful; and it enjoys many important relationships with other solution concepts of economics and game theory. Its domain is very broad—it is almost always nonempty, in political contexts as well as all kinds of economic ones. On the other hand, it is "sharp," almost always containing only finitely many points, often quite few. And as we have seen, it is associated with unusually few anomalies or conceptual difficulties.

In brief, the value emerges as an eminently robust and useful cooperative solution concept. Al Roth's thoughtful probing has contributed significantly to this development.

---

3. Myerson (1986) provides a very pretty alternative rationale, both for the general definition of the NTU value, and for its realization in the Maschler–Owen game V_0.

Acknowledgments

This work was supported by National Science Foundation Grant SES83-20453 at the Institute for Mathematical Studies in the Social Sciences (Economics) at Stanford University, by the Institute for Mathematics and its Applications at the University of Minnesota, and by the Mathematical Sciences Research Institute, Berkeley. Technical services (typing) on some earlier versions were graciously provided through the good offices of A. Roth, and funded by the Office of Naval Research under Contract N 00014-84-K-0263 at the University of Pittsburgh. We are very grateful to S. Zamir for permission to publish his counterintuitive example to the core (Section 6); and to D. Kreps, the co-editor in charge of this paper, for several helpful suggestions, and especially for pointing out an error in an earlier version. We alone are responsible for all remaining errors.

References

Aumann, Robert J. (1985): "On the Non-transferable Utility Value: A Comment on the Roth–Shafer Examples," *Econometrica*, 53, 667–677 [Chapter 61c].

Hart, S. (1985): "Non-transferable Utility Games and Markets: Some Examples and the Harsanyi Solution," *Econometrica*, 53, 1445–1450.

Myerson, R. (1986): "An Introduction to Game Theory," in *Studies in Mathematical Economics*, ed. by S. Reiter, Vol. 25 of MAA Studies in Mathematics. Washington, D.C.: Mathematical Association of America, 1–61.

Roth, A. (1980): "Values for Games Without Side Payments: Some Difficulties with Current Concepts," *Econometrica*, 48, 457–465 [Chapter 61a].

———(1986): "On the Non-transferable Utility Value: A Reply to Aumann," *Econometrica*, 54, 981–984 [Chapter 61d].

62a Non-Symmetric Cardinal Value Allocations

Allen J. Scafuri and Nicholas C. Yannelis

1 Introduction

In a recent paper, Wayne Shafer [9, p. 472] presented an example of a non-symmetric cardinal value allocation[1] in which one agent is assigned a weight of zero and treated as a "dummy." He then left as an open problem the question of whether there can exist cardinal value allocations which assign different utilities to identical agents when all players are given strictly positive weights. This problem has also been alluded to in a somewhat different context by Champsaur [4, p. 389]. The purpose of this paper is to present an example which provides an affirmative answer to the question. We shall use the model and notation of Shafer's paper, to which the reader is referred for further explanation.

2 Shafer's Model and Example

A game with side payments $\Gamma = (A, V)$ consists of a finite set of agents A and a superadditive real valued function V defined on the power set of A, and such that $V(\phi) = 0$. Each $S \subset A$ is a coalition and $V(S)$ is the "payoff" which that coalition can guarantee its members. The *Shapley Value* (see [10]) of a game Γ attempts to define a "fair division" of payoffs by assigning to each player the expected value of the incremental gain he brings to all possible coalitions, assuming they are all equally likely to form. Imputations are assigned to players according to the rule

$$s_a = \sum_{S \subset A} \frac{(|S| - 1)!(|A| - |S|)!}{|A|!} [V(S) - V(S \setminus \{a\})].$$

A *finite exchange economy* is a sequence of ordered triples $E = \{(X_a, u_a, \omega_a)\}_{a \in A}$ where A is a finite set of agents, and (X_a, u_a, ω_a) are the characteristics of the agent. Here, $X_a \subset R^l_+$ is the consumption set of agent a; $\omega_a \in R^l_+$ is his initial endowment and $u_a : X_a \to R$ is his utility function (assumed unique up to a linear affine transformation). An allocation for E is $\{x_a\}_{a \in A}$ such that $x_a \in X_a$ for each a and $\sum_{a \in A}(x_a - \omega_a) = 0$. To each finite exchange economy E and each vector of "weights" $\lambda \in R^{|A|}_{++}$ we may associate a game $\Gamma = (A, V_{\lambda u})$ according

---

This paper originally appeared in *Econometrica* 52 (1984): 1365–1368. Reprinted with permission.

1. See Aumann [1] for a distinction between ordinal and cardinal value allocations.

to the rule

$$V_{\lambda u}(S) = \max\left\{\sum_{a \in S} \lambda_a u_a(x_a) \;\middle|\; \sum_{a \in S}(x_a - \omega_a) = 0\right\}.$$

An allocation $\{x_a\}_{a \in A}$ is a *cardinal value allocation* for E if for some set of weights $\lambda_a u_a(x_a)$ is the Shapley value of each player in the associated side payment game. Two agents are *identical* if they have the same characteristics; and a cardinal value allocation is *symmetric* if identical agents are assigned the same utility.

Shafer's example (2) both illustrates the concept of cardinal value allocation and points out a "peculiarity" of value. There are three agents $\{0,1,2\} = A$ and two commodities x and y. Utility functions and endowments are

$$u_0(x_0, y_0) = \left[\tfrac{1}{2}x_0^\beta + \tfrac{1}{2}y_0^\beta\right]^{1/\beta}, \quad \omega_0 = (0,0),$$

$$u_i(x_i, y_i) = \left[\tfrac{1}{2}x_i^\rho + \tfrac{1}{2}y_i^\rho\right]^{1/\rho}, \quad i = 1, 2,$$

$$\omega_1 = (1,0), \quad \omega_2 = (0,1)$$

for $0 \leq \rho < \beta \leq 1$. For weights $\lambda_0 = \lambda_1 = \lambda_2 = 1$, one may easily compute characteristic functions for possible coalitions upon noting that for all $x, y [1/2 x^\beta + 1/2 y^\beta]^{1/\beta} \geq [1/2 x^\rho + 1/2 y^\rho]^{1/\rho}$ with equality only if $x = y$. The Shapley values are

$$s_0 = \tfrac{1}{3}\left[(\tfrac{1}{2})^{1/\beta} - (\tfrac{1}{2})^{1/\rho}\right], \quad s_1 = s_2 = \frac{1 - s_0}{2},$$

and the value allocation is $(x_i, y_i) = (s_i, s_i)$ for $i = 0, 1, 2$. The value allocation thus assigns positive consumption to an agent with no endowment and is therefore not in the core, since clearly the coalition $\{1,2\}$ can block the above value allocation.

A further example constructed from this by Shafer shows that by adding a "dummy" player we can obtain a non-symmetric cardinal value allocation. Add an additional person (agent 3) with characteristics identical to agent 0 and assign weights $\lambda_3 = 0$ and $\lambda_0 = \lambda_1 = \lambda_2 = 1$. The Shapley values of agents 0, 1, 2 remain unchanged but that of agent 3 is $s_3 = 0$. Hence, the value allocation is $(x_i, y_i) = (s_i, s_i)$ for $i = 0, 1, 2, 3$ and agents 0 and 3 are not treated symmetrically. However, in Shafer's words, "this still leaves open the more interesting possibility of a non-symmetric value allocation in which each agent receives a positive weight."

3 A Non-Symmetric Cardinal Value Allocation with Positive Player Weights

Upon further reflection, it will become clear that Shafer almost provided the answer to his own question. The critical feature of his examples is that agent 0 always has a higher elasticity of substitution than agents 1 and 2 *and* that all utility functions are normalized on $u_i(1,1) = 1$. Thus, whenever weights assigned to the three agents are equal, the maximum value which defines a coalition's characteristic function is achieved by allowing agent 3 to consume all goods in any coalition of which he is a member and for which the endowments of x and y are not equal. The condition that $[1/2x^\beta + 1/2y^\beta]^{1/\beta} > [1/2x^\rho + 1/2y^\rho]^{1/\rho}$ implies agent 0 has a superior "utility producing technology." This is not true, however, for any choice of weights. Clearly, the choice of zero weight removes any advantage but it is not the only one which does. Given any choice of $\lambda < (1/2)^{((1/\rho)-(1/\beta))}$, we find that $\lambda[1/2x^\beta + 1/2y^\beta]^{1/\beta} < [1/2x^\rho + 1/2y^\rho]^{1/\rho}$ for any values except $x = y = 0$.

We may now construct an example of a non-symmetric cardinal value allocation with positive player weights by making use of the above observation. To Shafer's example (2), add another agent (3) identical to agent zero and assign weights $\lambda_0 = \lambda_1 = \lambda_2 = 1$ and $0 < \lambda_3 < (1/2)^{((1/\rho)-(1/\beta))}$. The agents' Shapley values are then

$$s_0 = \tfrac{1}{3}\left[(\tfrac{1}{2})^{1/\beta} - (\tfrac{1}{2})^{1/\rho}\right], \quad s_1 = s_2 = \frac{1 - s_0}{2}, \quad s_3 = 0,$$

and the value allocation is $(x_i, y_i) = (s_i, s_i)$, $i = 0, 1, 2, 3$, which clearly treats agents 0 and 3 in a non-symmetric manner.

4 Conclusions

The above example casts doubt on any interpretation of the weights as a meaningful "endogenous utility comparison" as has been suggested in Shapley [11]. This certainly reinforces the negative results given by [7, 8, 9, 12, and 13].[2]

Recently, Aumann [3] has axiomatized the nontransferable utility value (NTU). It is important to note that the axioms refer to values as payoff vectors only and the positive weights associated with the value do not appear explicitly in the axioms. However, no symmetry axiom is posed by Aumann. In fact, our counter-intuitive example shows that if the weights are to be endogenously determined, no symmetry axiom can be imposed.

We may conclude that there exist value allocations which treat identical agents quite differently. For purposes of comparison it should be noted that all competitive equilibrium allocations possess the equal treat-

---

2. See Aumann [2] for a critical discussion of the counter-intuitive examples of Roth [7] and Shafer [9].

ment property as do other game solution concepts such as the Shapley value when applied to games with transferable utility and the Nash solution. Finally, we have not examined the question of whether the unequal treatment of cardinal value allocations disappears as the economy becomes large. Champsaur [4], Mas-Colell [6], and Cheng [5] have shown that the set of cardinal value allocations converges to the set of competitive equilibrium allocations as the number of agents goes to infinity if a symmetry requirement is satisfied. It remains an open question whether the value convergence theorem remains valid for allocations without the equal treatment property.

Acknowledgments

We would like to thank William Thomson and an anonymous referee for several helpful comments.

References

1. Aumann, R. J.: "Values of Markets with a Continuum of Traders," *Econometrica*, 43 (1975), 611–646 [Chapter 51].
2. ———: "On the Non-Transferable Utility Value," *Econometrica*, 53 (1985), 667–677 [Chapter 61c].
3. ———: "An Axiomatization of the Non-Transferable Utility Value," *Econometrica*, 53 (1985), 559–612 [Chapter 60].
4. Champsaur, P.: "Cooperation vs. Competition," *Journal of Economic Theory*, 11 (1975), 394–417.
5. Cheng, H.: "On Dual Regularity and Value Convergence Theorems," *Journal of Mathematical Economics*, 8 (1981), 37–57.
6. Mas-Colell, A.: "Competitive and Value Allocations of Large Exchange Economies," *Journal of Economic Theory*, 14 (1977), 419–438.
7. Roth, A.: "Values for Games Without Side Payments: Some Difficulties with Current Concepts," *Econometrica*, 48 (1980), 457–465 [Chapter 61a].
8. Scafuri, A., and N. Yannelis: "Some Observations on Value in Public Goods Economies," Mimeo, University of Minnesota, 1984.
9. Shafer, W. J.: "On the Existence and Interpretation of Value Allocation," *Econometrica*, 48 (1980), 467–474 [Chapter 61b].
10. Shapley, L. S.: "A Value for an n-Person Game," in *Contributions to the Theory of Games* II, ed. by H. W. Kuhn and A. W. Tucker. Princeton, New Jersey: Princeton University Press, 1953.
11. ———: "Utility Comparison and the Theory of Games," in *La Decision*, ed. by Guilbaud. Paris, France: Editions du CNRS, 1969.
12. Thomson, W.: "On the Manipulability of Shapley Value," *International Journal of Game Theory*, 17 (1980), 101–127.
13. Yannelis, N. C.: "Existence and Fairness of Value Allocation Without Convex Preferences," *Journal of Economic Theory*, 31 (1983), 283–292.

62b Value, Symmetry, and Equal Treatment: A Comment on Scafuri and Yannelis[1]

Two participants i and j in a game or economic model are called *substitutes* if they enter the model in the same way; i.e., if interchanging them, while keeping all other elements of the model fixed, constitutes a symmetry of the model. In a market, this means that they have the same endowments and utility functions. In a TU (transferable utility) coalitional game v, it means that $v(S \cup i) = v(S \cup j)$ whenever S is a coalition containing neither i nor j. In an NTU (nontransferable utility) game V, it means that $V(S \cup i)$ transforms into $V(S \cup j)$ if we interchange the x^i and x^j axes,[2] whenever S contains either both i or j or neither one (Wooders, 1983).

Scafuri and Yannelis (1984) construct an example of a market in which the value does not provide *equal treatment*: i.e., there is a value that assigns different utilities to substitutes. They call their example "counterintuitive" and say that it "reinforces" allegedly negative results on the NTU value obtained by others.

It is difficult to understand this view. Equal treatment is provided by almost no multi-valued game-theoretic solution concept, including the most well known and widely applied. Neither the core, nor the solution of von Neumann and Morgenstern (1944), nor the bargaining set[3] provide equal treatment. If indeed this is a valid criticism, why pick just the NTU value as its target?

Markets with nonequal treatment cores abound; they are the rule, not the exception. An explicit example is a 3-agent market in complementary goods, e.g., perfectly divisible right and left "gloves." Agent 1 initially holds two right gloves, whereas 2 and 3 hold one left glove each; all agents have linear utilities in pairs of gloves. This corresponds to the TU game $v(123) = 2$, $v(12) = v(13) = 1$, $v(S) = 0$ otherwise. The core contains the point $(1, 0, 1)$, and so certainly does not provide equal treatment.

There is even a 3-person NTU game whose core, while nonempty, contains *no* equal treatment outcomes.[4] All three players acting together

---

This paper originally appeared in *Econometrica* 55 (1987): 1461–1464. Reprinted with permission.

1. Research supported by the Institute for Mathematical Studies in the Social Sciences, Economics, at Stanford University, under National Science Foundation Grant SES 83-20464, by the Mathematical Sciences Research Institute, Berkeley, and by a Ford Visiting Research Professorship of Economics at the University of California at Berkeley. We are grateful to Myrna H. Wooders for pointing out an error in our previous definition of equal treatment in NTU games, and referring us to the correct definition.

2. More precisely, that $\pi^{ij} V(S \cup i) = V(S \cup j)$, where π^{ij} is the reflection of payoff space in the hyperplane $x^i = x^j$.

3. Davis and Maschler (1967), Peleg (1967).

4. We are grateful to B. Peleg for bringing this example to our attention.

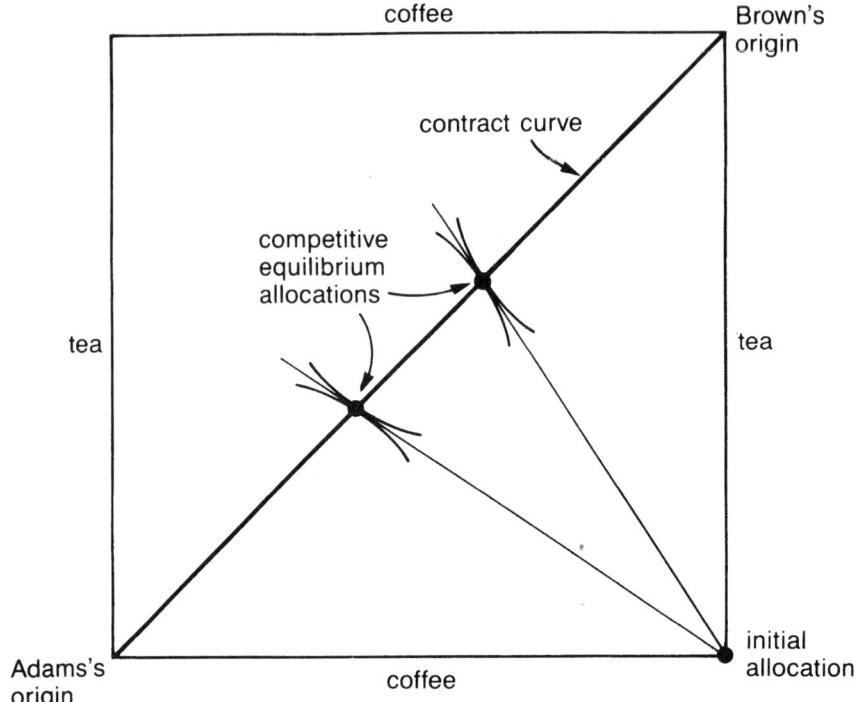

Figure 1

may divide a dollar in any way they please; any two may divide it fifty-fifty, but in no other way; acting alone, each player gets nothing. The game is totally symmetric—all players are substitutes for each other—but the only outcomes in the core are the permutations of (1/2, 1/2, 0), none of which is equal treatment.[5]

Technically, the competitive equilibrium does provide equal treatment; substitutes do always get the same utility. But even there, symmetric[6] agents need not. The Edgeworth box in Figure 1 illustrates a two-commodity, two-agent market with multiple equilibria, that is symmetric under a simultaneous interchange both of the agents and of the commodities. Adams relates to coffee and tea in the same way that Brown relates to tea and coffee, respectively. Adams is endowed with a kilo of tea; symmetrically, Brown is endowed with a kilo of coffee. The agents

---

5. This cannot happen with the NTU value; when the feasible set $V(N)$ is convex, the methods of Shapley (1969) always yield at least one equal treatment value.

6. We call Agents i and j *symmetric* if there is a symmetry of the game or economy that takes i to j (and hence also one that takes j to i; in Figure 1, the same symmetry does both).

are formally indistinguishable; yet there is a competitive equilibrium that assigns to Adams more tea *and* more coffee than it assigns to Brown. If one likes drama, one can fix things so that Adams gets 999 grams of tea and 999 grams of coffee, and Brown gets only one gram of each.

One must distinguish between symmetry and equal treatment. Symmetry of a multi-valued solution concept refers to the set of *all* outcomes assigned by it to a given game or economy; equal treatment refers separately to *each single* outcome. That the core is symmetric means that as a whole, it is invariant under any symmetry of the game or economy. Thus interchanging substitutes i and j does not change the core as a whole; in utility space, the core is its own reflection in the hyperplane $x^i = x^j$. But equal treatment would mean that substitutes get the same utility at each point of the core, that the hyperplane $x^i = x^j$ actually includes the core.

Symmetry is an eminently reasonable requirement. Indeed, almost all solution concepts in the literature, including the NTU value and the core, do satisfy it. So does the competitive equilibrium: In the coffee-tea example, while one equilibrium gives almost all the goods to Adams, another one does so for Brown.

But equal treatment is a horse of a different color. We have already noted that almost no multi-valued game-theoretic solution satisfies it.[7] Indeed, it seems much too strong for a reasonable general requirement. In the above "divide the dollar" game, the core calls for equal division between two of the three players, but does not say *which* two. This seems perfectly reasonable, but it is not equal treatment. Equal treatment would imply that all three players share the dollar equally. While that also makes sense, it cannot be considered the *only* reasonable outcome.[8]

The reader might object that this is all very well for concepts like the core, which represent some notion of stability; but that the value, which represents an index of power or an arbitrated outcome, should provide equal treatment. But closer examination reveals that this argument is unfounded. Any multi-valued solution concept, no matter what its intuitive content, associates a set of outcomes with each game or economy to which it applies. Choosing a single outcome from this set necessarily calls for more information about the underlying situation, information that is not provided by the game or economy as described. It is this additional information that may well distinguish between substitutes.

---

7. The one notable exception is the kernel (Davis and Maschler, 1965). But even it may contain outcomes at which symmetric players get unequal utility (Maschler and Peleg, 1967, 598–599).

8. For single-valued solution concepts, of course, symmetry implies equal treatment; this accounts for the fact, noted by Scafuri and Yannelis, that the TU value and the Nash bargaining solution, both of which are by definition single valued, do provide equal treatment.

In an NTU game, two players can show up as formal substitutes even though in fact they are vastly different. For example, this could happen if the units of payoff are dollars and cents respectively, and the utilities of all players are linear in money. Equal treatment would require that one player gets exactly 100 times the payoff of the other; and while this may be quite reasonable in some situations, one certainly would not want to insist on it. The point is that the description of an NTU game does not enable any exogenous comparison between the utilities of different players; from the mere fact that players are substitutes in a given model, one cannot conclude that they are truly "identical." They may be, and then again they may not; the value allows for both possibilities.

Scafuri and Yannelis write that their example "casts doubt on any interpretation of the weights as a meaningful 'endogenous utility comparison' as has been suggested in Shapley (1969)." But "endogenous" does not mean "unique." Scafuri and Yannelis will agree, I hope, that prices are endogenously determined by market forces; yet commodities that appear entirely symmetrically in a market can easily have different prices (as in the economy of Figure 1). The unknowns x and y appear symmetrically in the system $xy = 2$, $x + y = 3$, yet $x \neq y$ in both solutions. Endogeneity has nothing to do with equal treatment.

Turning to the example itself, we find it rather pathological. The allocation in question is indeed associated with nonzero weights; but as the authors themselves point out, the same allocation is also associated with the weight vector (1, 1, 1, 0). Zero weights symptomize degeneracy in the game itself, not in the value; they imply that there are some players who cannot, under any circumstances, contribute to the others. In our case, agents 0 and 3 are endowed with no goods whatsoever, which implies that they cannot contribute anything to other individuals or coalitions. We have not stressed these points because they are secondary; for the reasons stated above, it would not be at all surprising or disconcerting to find a perfectly "healthy" NTU game with a nonequal treatment value. But this particular example must be considered weak.

For the record, we note that with a continuum of agents, the value equivalence principle (e.g., Hart (1977)) shows that value allocations are competitive; this assures equal treatment.

References

Davis, M., and M. Maschler (1965): "The Kernel of a Cooperative Game," *Naval Research Logistics Quarterly*, 12, 223–259.

——— (1967): "Existence of Stable Payoff Configurations for Cooperative Games," in *Essays in Mathematical Economics in Honor of Oskar Morgenstern*, ed. by M. Shubik. Princeton: Princeton University Press, pp. 39–62; also, *Bulletin of the American Mathematical Society*, 69 (1963), 106–108.

Hart, S. (1977): "Values of Non-Differentiable Markets with a Continuum of Traders," *Journal of Mathematical Economics*, 4, 103–116.

Maschler, M., and B. Peleg (1967): "The Structure of the Kernel of a Cooperative Game," *SIAM Journal of Applied Mathematics*, 15, 569–604.

Peleg, B. (1967): "Existence Theorem for the Bargaining set $M_1^{(i)}$," in *Essays in Mathematical Economics in Honor of Oskar Morgenstern*, ed. by M. Shubik. Princeton: Princeton University Press, pp. 53–56; also, *Bulletin of the American Mathematical Society*, 69 (1963), 109–110.

Scafuri, A. J., and N. Yannelis (1984): "Non-Symmetric Cardinal Value Allocations," *Econometrica*, 52, 1365–1368 [Chapter 62a].

Shapley, L. S. (1969): "Utility Comparison and the Theory of Games," in *La Décision: Agrégation et dynamique des ordres de préférences*, ed. by G. Th. Guilbaud. Paris: Editions du Centre National de la Recherche Scientifique, pp. 251–263.

Von Neumann, J., and O. Morgenstern (1944): *Theory of Games and Economic Behavior*. Princeton: Princeton University Press (Third Edition, 1953).

Wooders, M. H. (1983): "The ε-core of a Large Replica Game," *Journal of Mathematical Economics*, 11, 277–300.

XII COALITIONAL GAMES: SURVEYS OF VALUE THEORY

Chapter 63 is the text of a lecture given at the International Congress of Mathematicians that took place in Helsinki in 1978; it surveyed the state of the art in the theory of the Shapley value at that time. This was four years after the publication of *Values of Non-Atomic Games*,[1] and there was a lot of activity in the area. Thus, much of the survey describes the developments in non-atomic value theory since the publication of the book, including the following breakthroughs on the abstract side:

• The complete solution of the "diagonal problem," which had been left open in the book: not all (non-atomic) values need be diagonal,[2] but *continuous* values *are*.[3]

• Neyman's[4] proof that voting games have asymptotic values, another difficult problem that had been left open in the book.

• Mertens's[5] development of general tools for analyzing games that are "kinked" (nondifferentiable) on the diagonal.

• Hart's[6] discovery of the close relationship between the value and the center of symmetry of the core in nondifferentiable market games.

• Hart's development of a value for non-atomic games that takes into account the size of a coalition, and not only its economic or political "clout."[7]

Also surveyed here are applications to various economic and political models, applications to cost sharing, and many other important advances. But many of the most significant developments in value theory are not included in this survey, for the simple reason that they came later.

Chapter 64 (like chapter 58) was presented orally as part of the NATO Advanced Study Institute that took place in July 1991 at the State University of New York at Stony Brook; the excellent written version is due to J.-F. Mertens. Unlike chapter 63, the purpose of this lecture was less to give a comprehensive survey of the most important recent developments, and more to inform the audience about the fundamentals of the theory and to touch on some of the later developments. The paper con-

---

1. By R. Aumann and L. Shapley, Princeton University Press, 1974.
2. A. Neyman and Y. Tauman, "The existence of nondiagonal axiomatic values," *Math. Oper. Res.* 1 (1976), 246–250; also Y. Tauman, "A nondiagonal value on a reproducing space," *Math. Oper. Res.* 2 (1977), 331–337.
3. A. Neyman, "Continuous values are diagonal," *Math. Oper. Res.* 2 (1977), 338–342.
4. "Singular games have asymptotic values," *Math. Oper. Res.* 6 (1981), 205–212.
5. "Values and derivatives," *Math. Oper. Res.* 5 (1980), 523–552.
6. "Asymptotic value of games with a continuum of players," *J. Math. Econ.* 4 (1977), 57–80, and "Values of nondifferentiable markets with a continuum of traders," ibid., 103–116.
7. "Measure-based values of market games," *Math. Oper. Res.* 5 (1980), 197–228.

tains (i) the basic definition, (ii) a brief proof of existence and uniqueness of a value for finite games satisfying Shapley's axioms, (iii) some elementary economic and political examples that, inter alia, compare the value with the core, (iv) some alternative characterizations of the value, including the potential approach of Hart and Mas-Colell, Young's axiomatization (which substitutes a monotonicity axiom for Shapley's additivity axiom), Neyman's observation that to derive the value of a game v, one may restrict the domain of Shapley's axioms to the additive group generated by the subgames of v, and (v) a succinct overview of the axiomatization of the NTU value (see chapter 60).

63 Recent Developments in the Theory of the Shapley Value

1 Introduction

The Shapley value is an a priori measure of a game's utility to its players; it measures what each player can expect to obtain, "on the average," by playing the game. Other concepts of cooperative game theory, such as the Core, Bargaining Set [6], and N–M Solution [26] predict outcomes (or sets of outcomes) that are in themselves stable, that cannot be successfully challenged or upset in some appropriate sense. Almost invariably, they fail to define a unique result; and in a significant proportion of the cases, they do not define any result at all.[1] The Shapley value, although it is not in any formal sense defined as an average of such "stable" outcomes, nevertheless can be considered a mean, which takes into account the various power relationships and possible outcomes.

It follows from this that the Shapley value may also be thought of as a reasonable compromise, the outcome of an arbitration procedure. A player should be willing to settle for a compromise that yields with certainty what he otherwise would only have expected in the mean. For example, the symmetric N–M solution of the 3-person majority game predicts one of the three payoff vectors (1/2, 1/2, 0), (1/2, 0, 1/2), and (0, 1/2, 1/2), corresponding to the three possible 2-person majorities. Before the beginning of bargaining, each player may figure that his chances of getting into a ruling coalition are 2/3, and conditional on this, his payoff is 1/2; the "expected outcome" would then be (1/3, 1/3, 1/3), and this is also the Shapley value. It would, therefore, also be a reasonable compromise; but it is not in itself stable, since it can be easily improved upon by any two-person coalition.

Mathematically, the Shapley value is perhaps the most tractable of all the concepts of cooperative game theory. This has led to the growth of a considerable theory, which in turn has enabled a wide range of applications to Economics and Political Science. Here we survey some of the more recent of these developments.

2 General Definition in the Transferable Utility Case

We begin by recalling that a *coalitional game*, or simply *game* for short, is a real-valued function v on the σ-field $\mathscr{C}$ of a measurable space $(I, \mathscr{C})$,

---

This chapter originally appeared in *Proceedings of the International Congress of Mathematicians, Helsinki, 1978*, pp. 995–1003, Academia Scientiarum Fennica, 1980. Reprinted with permission.

1. The Bargaining Set is the only one of these three covered by a general existence theorem.

with $v(\emptyset) = 0$. Here I is the *player space*, the members of $\mathscr{C}$ are *coalitions*, and $v(S)$ is the *worth* of a coalition S. A game is called *monotonic* if $S \supset T$ implies $v(S) \geq v(T)$.

Fix $(I, \mathscr{C})$. An *outcome* (or *payoff vector*) is a finitely additive game.[2] For each game v and automorphism (one-one bimeasurable function) Θ of $(I, \mathscr{C})$, define the game $\Theta_* v$ by $(\Theta_* v)(S) = v(\Theta S)$ for all S.

Let us be given a linear space Q of games, which is *symmetric* in the sense that $\Theta_* Q = Q$ for all Θ. An operator φ from Q to outcomes is called *symmetric* if $\varphi(\Theta_* v) = \Theta_*(\varphi v)$ for all v in Q and all automorphisms Θ; *monotonic* if φv is monotonic whenever v is; and *efficient* if $(\varphi v)(I) = v(I)$ for all v in Q. A *value* on Q is an operator from Q to outcomes that is linear, monotonic, symmetric, and efficient.

3 Finite Games

A game v is called *finite* if there is a finite subset N of I (a *support* of v) such that $v(S) = v(S \cap N)$ for all S. The finite games form a linear space on which there is a unique value; it is given by

$$(\psi v)(\{i\}) = E(v(S_i \cup \{i\}) - v(S_i)), \tag{3.1}$$

where S_i is the set of players (members of N) preceding i in a random order on N, and E is the expectation operator when each order on N has probability $1/|N|!$ [36]. It is easy to check that (3.1) does indeed define a value; as for uniqueness, perhaps the simplest proof is that of Dubey [7], who uses an induction on $|N|$ to show that every finite game is a linear combination of unanimity games (games for which $v(S) = 1$ or 0 according as $S \supset N$ or $S \not\supset N$).

4 Nonatomic Games, Partition Values, and the Diagonal Property

Diametrically opposed to the finite games are the *nonatomic games*, which model situations in which no individual player has any significance [2]. Examples are games of the form $f \circ \mu$, where μ is a nonatomic vector measure, and f is a real-valued function on the range of μ vanishing at 0. One approach to defining a value for a nonatomic game v is via approximations by finite games. Specifically, if Π is a measurable partition of I—i.e. a finite subfield of $\mathscr{C}$—we may define a finite game v_Π, whose support consists of the atoms of Π, by $v_\Pi = v|\Pi$; then v_Π is a kind of

---

2. Intuitively, the sharing of proceeds in an additive game involves no difficulties, so that by associating an additive game to a non-additive game, we have essentially specified an outcome.

finite approximant to v. Given a coalition S in $\mathscr{C}$, an increasing sequence $\{\Pi_1, \Pi_2, \ldots\}$ of such partitions is called *S-admissible* if $S \in \Pi_1$ and $\bigcup_i \Pi_i$ generates $\mathscr{C}$. A value φ on a space Q is called a *partition value* [30] if for each game v in Q and each coalition S, there is an S-admissible sequence $\{\Pi_1, \Pi_2, \ldots\}$ such that

$$\lim_{n \to \infty} (\psi v_{\Pi_n})(S) \to (\varphi v)(S), \tag{4.1}$$

where ψ is the value for finite games. If for a specific game v and outcome φv, (4.1) holds for all S and all S-admissible sequences, then we write $v \in \text{ASYMP}$ and call φv the *asymptotic value* [16] of v. Whereas the partition value is defined in terms of the imbedding space Q, the definition of asymptotic value is independent of any imbedding space; its existence depends on the game v only.

A partition value of a non-atomic game is a limit of values of large finite approximants. The asymptotic value is the strongest possible partition value; if it exists, then no matter how the player space is cut up,[3] in the limit the result is the same.

Are there values that are not partition values? This leads us to the *diagonal property* of values. Let v be a nonatomic nonnegative measure on $\mathscr{C}$ with $v(I) = 1$ ($v \in \text{NA}^1$ for short); Π a partition of I into many— say n—"small" sets; and Q_h the union of the first h atoms of Π in a random order on the atoms. For a fixed h, we will have $v(Q_h) \approx h/n$ with high probability; moreover, for fixed ε, if Π is sufficiently far out in some S-admissible sequence, then the probability is $> 1 - \varepsilon$ that $|v(Q_h) - (h/n)| < \varepsilon$ *simultaneously* for all h. Thus if $\mu \in (\text{NA}^1)^m$ (i.e. μ is an m-tuple of NA^1 measures), almost all the coalitions occurring in Formula (3.1) as applied to v_Π will have μ-measures very near the "diagonal" $D^m = \{(t, \ldots, t): t \in [0, 1]\}$ of the m-cube. In particular, let φ be a partition value; then

if φ is defined for two games v_1 and v_2 that agree on all coalitions S with $\mu(S)$ in some ε-neighborhood of D^m, then $\varphi v_1 = \varphi v_2$. (4.2)

Any value φ satisfying (4.2) for all vectors μ of NA^1 measures is called a *diagonal value*.

All the values treated in [2] were diagonal, and for a long time it was not known whether *all* values are diagonal. Finally, Neyman and Tauman [29] and Tauman [40][4] found examples of nondiagonal values. In particular, not all values are partition values.

---

3. E.g. into n intervals of "length" $1/n$, or into n of length $1/2n$ and n^2 of length $1/2n^2$.
4. [40] avoids a certain undesirable pathology in [29].

What, then, accounts for the diagonality of all previously considered values? In [27], Neyman answered this question by showing that all *continuous* values are diagonal; here continuity is w.r.t. (with respect to) the *variation* norm, defined by

$$\|v\| = \sup\left\{\sum_{i=1}^{k} |v(S_i) - v(S_{i-1})| : \emptyset = S_0 \subset S_1 \subset \cdots \subset S_k = I\right\}.$$

This norm plays a crucial role in the theory, and all previously considered values had been continuous w.r.t. it.

Closely related to the diagonal property is the *diagonal formula* for values. Let pNA denote the smallest variation-closed linear space containing all games $f \circ v$, where $v \in \mathrm{NA}^1$ and f is absolutely continuous. There is a unique value on pNA, and pNA $\subset$ ASYMP [16]. Suppose now that $\mu \in (\mathrm{NA}^1)^m$ and $f \in C^1(R^m)$. Then $f \circ \mu \in$ pNA, and

$$\varphi(f \circ \mu) = \left\langle \mu, \int_0^1 \nabla f(t, \ldots, t) dt \right\rangle \tag{4.3}$$

[2, Theorem B]. To understand (4.3), note that it follows from Lyapunov's theorem that for each t in $[0, 1]$ there is a coalition tI with $\mu(tI) = (t, \ldots, t)$: the tI are called *diagonal coalitions*, and may be considered "perfect samples" of I as far as $f \circ \mu$ is concerned. Let us now think of a "player" in a non-atomic game as an infinitesimal coalition ds; the marginal contribution of ds when added to tI is

$$(f \circ \mu)(tI \cup ds) - (f \circ \mu)(tI) = \langle \mu(ds), \nabla f(t, \ldots, t) \rangle.$$

Thus (4.3) says that *the value of a player is his average contribution to a diagonal coalition.*

This principle, which is of fundamental importance in the theory of nonatomic games and its applications, has been extended far beyond the space pNA for which it was originally established. The deepest and furthest-reaching work on this subject is due to J.-F. Mertens [20], who has established the existence of a value obeying a suitable analogue of (4.3) on a very large space of games, which even contains games not in ASYMP.

5 Political Applications

A *weighted majority* (WM) game is one of the form $f_q \circ v$, where v is a non-negative measure with $v(I) = 1$ (the *vote* measure), $0 < q < 1$ and

$f_q(x) = 0$ or 1 according as $x \leqslant q$ or $x > q$. Finite WM games appear already in [26]. Values of finite WM games were first studied by Shapley and Shubik [38], who interpreted them as measures of political power. They have since been applied to many voting situations, such as the UN security council, the US electoral college, state legislatures, multi-party parliaments, etc.; [18] is a good survey. Shapiro and Shapley [35], Milnor and Shapley [21], and Hart [11] studied values of *oceanic* games, i.e. WM games in which v contains a nonatomic part (the "ocean" of small voters) as well as some atoms (large voters); [21] contains an application to corporations with several large stockholders. An interesting qualitative conclusion is that when $q = 1/2$, a *single* atom has value larger than his vote, as might be expected; but this is often reversed when there are several atoms. For example, when v has 2 atoms and an ocean of measure $1/3$ each, then the atoms get only $1/4$ of the value each.

The above are asymptotic results on the values of the atoms when the largest "small" vote tends to 0. Calculating the values of the small voters themselves, even approximately, is much more difficult, and even when there are no atoms, the problem was open for many years. Only recently did A. Neyman [28] prove, in a remarkable tour-de-force of combinatorial reasoning, that $f_q \circ v \in \text{ASYMP}$ when $v \in \text{NA}^1$. Intuitively, his result says that the value of a coalition depends only on its total vote, not on the relative sizes of the voters. It can be used to prove that oceanic games are in ASYMP, and also that $f \circ v \in \text{ASYMP}$ when f is monotonic and continuous, and $v \in \text{NA}^1$. Also, there are close connections to renewal theory.

More complex political structures can also often be described by using WM games. A bicameral legislature is the product of 2 WM games, and the electoral college when the players are the individual citizens is a polynomial in WM games. Such games need not be in ASYMP; thus if $\mu, v \in \text{NA}^1$ and $\mu \neq v$, then $(f_{2/3} \circ \mu)(f_{2/3} \circ v) \notin \text{ASYMP}$; however, it is a member of a space with a partition value [30]. Whether there is a partition value on the algebra generated by all nonatomic WM games is an open question.

See [31] for an application using a non-symmetric variant of the value.

A variant of the Shapley value called the *Banzhaf value* has achieved some prominence in connection with political models. For finite games it is defined by (3.1), with the sole difference that now S_i varies over the set of all subsets of $N \setminus \{i\}$, each such coalition receiving probability $1/2^{|N|-1}$. In general, it is not efficient. An account of the theory and a very extensive bibliography may be found in [8].

6 Economic Applications

Games arising in economics often have a property called "homogeneity of degree 1;" roughly, this means that two coalitions differing from each other in their size only, but not in their composition, have worths proportional to their sizes. Examples are games $f \circ \mu$, where $\mu \in (\mathrm{NA}^1)^m$ and f is a function of m variables that is homogeneous of degree 1. Suppose now that φ is a partition value. A principle that is basic to many of the economic applications asserts that

if φ is defined for a superadditive[5] game v that is homogeneous of degree 1, then φv is in the core of v. (6.1)

(Recall that the *core* of a game v is the set of outcomes v such that $v(I) = v(I)$ and $v(S) \geq v(S)$ for all S.)

Let's demonstrate this in the particular case in which $v = f \circ \mu$, where $\mu \in (\mathrm{NA}^1)^m$ and f is a superadditive[6] function defined and homogeneous of degree 1 on the nonnegative orthant of R^m, and C^1 in its interior. Although $f \notin C^1(R^m)$, it can be shown that nevertheless $v \in \mathrm{pNA}$ and the diagonal formula (4.3) holds. Moreover the homogeneity of degree 1 and the superadditivity together yield the concavity of f. Since f is homogeneous of degree 1, $\nabla f(t, \ldots, t)$ is a constant, so (4.3) yields $\varphi v = \langle \mu, \nabla f(1, \ldots, 1) \rangle$. This means that φv is a function h of $\mu(S)$, i.e. $(\varphi v)(S) = h(\mu(S))$; and in fact h is the linear function with coefficients $\nabla f(1, \ldots, 1)$. By the efficiency of the value, $h(1, \ldots, 1) = (\varphi v)(I) = v(I) = f(1, \ldots, 1)$, and hence it follows that the graph of h is tangent to that of f at $(1, \ldots, 1)$. Since f is concave and h is linear, it follows that the graph of h always lies above that of f; but this implies that $(\varphi v)(S) \geq v(S)$ for all S, which together with the efficiency $(\varphi v)(I) = v(I)$ means that φv is in the core.

In this case a small additional argument, which depends on the actual tangency (i.e. the differentiability of f), yields that v is the *only* member of the core. This is true whenever $v \in \mathrm{pNA}$; pNA expresses a kind of differentiability property of a game. In general, though, the core will contain more than just the value. For example, when v is the minimum of two NA^1 measures, then the core consists of a non-degenerate interval (i.e. the set of all convex combinations of two different outcomes); in this case the asymptotic value exists and is the midpoint of the core. More generally, Hart [12] has proved that if a superadditive game v that is

---

5. $v(S \cup T) \geq v(S) + v(T)$ whenever $S \cap T = \emptyset$.
6. $f(x + y) \geq f(x) + f(y)$.

homogeneous of degree 1 has an asymptotic value φv, then φv is the center of symmetry of the core of v.

If the core has no center of symmetry,[7] there will be no asymptotic value; but not all is lost. If v is an NA^1 measure, an outcome φv is called a *v-value* if for all S, (4.1) holds for all S-admissible sequences of partitions whose atoms have equal (or in an appropriate sense almost equal) v-measures. Suppose now that μ in $(\text{NA}^1)^m$ is absolutely continuous w.r.t. v, with Radon–Nikodym derivative $d\mu/dv$ in $(L^2(v))^m$. Let f be superadditive and homogeneous of degree 1; then Hart [14] has shown that $v = f \circ \mu$ has a v-value, which has an interesting expression in terms of the core of v and the m-dimensional normal distribution whose covariance matrix is the same as that of $d\mu/dv$.

We come now to the applications. An important model in economic theory is that of the *exchange economy*. Like many economic models, it cannot be expressed as a transferable utility (TU) game as in §2; a more general concept—the *nontransferable utility* (NTU) game—is required. The most commonly used adaptation of the value to NTU games is that introduced[8] in [37], which culminated a long development to which many contributed; see in particular [24], [9]. We will not define the NTU value here; a brief treatment is in [1, §4]. It is enough for our purposes to note that the analysis involves the values of certain TU games auxiliary to the given NTU game.

In an exchange economy, the law of supply and demand defines *competitive prices* and, correspondingly, *competitive allocations* of goods and services. The TU games to which we are led from exchange economies are precisely the superadditive homogeneous games, and their cores are closely related to the cores of the "parent" NTU economies. The relationship between the value and the core expressed by (6.1), and the subsequent discussion, thus imply a close relationship between values and competitive allocations. More precisely, it can be proved that all allocations associated with an NTU value of a non-atomic exchange economy—i.e. all *value allocations*—are competitive. When the utility functions of the agents in the economy are sufficiently differentiable, we can assert the converse as well; in that case, therefore, the value allocations are the same as the competitive allocations.

Again, many people contributed to this development; see in particular [39], [5], [2], [4], [12], [13], [19], [14]. An excellent survey up to 1976 is in [13].

---

7. For example, the core of the minimum of 3 linearly independent measures is a triangle.
8. For an alternative approach, see Owen [32].

Models containing both political and economic elements, including in particular problems of taxation and redistribution, have been considered recently [1]. The TU games to which these models lead are products of pNA games with nonatomic WM games; the methods of Neyman [28] show that they have asymptotic values, and they are also amenable to the diagonal methods of Mertens [20].

Conceptually, these models differ from exchange economies in that threats play an important role. Games of this kind were treated by Nash [25], and much more generally by Harsanyi [9]. The worth $v(S)$ of a coalition S in an auxiliary TU game is now based as much on the harm that *S could* do to the players outside it as the good that it could do for itself. The value is of course efficient, so that it assumes that destructive threats are not actually carried out; this fits well our interpretation of the value as a reasonable compromise.[9] None of the pie gets thrown out, but how it gets cut up may depend on threats.

7 Cost Sharing

An interesting practical application of the Shapley value is to problems of cost sharing. For example, Littlechild and Owen [17] have considered the problem of airport landing fees. Runways (and other airport components) must be built large enough to accommodate the largest aircraft that will use them; but obviously it makes no sense to share the cost equally among all users, i.e. to charge the same landing fees to a jumbo jet and a private 4-seater. Here one defines a game v by considering the players to be individual aircraft landings, with $v(S)$ the hypothetical cost of building a facility that will accommodate the set S of landings. Each landing is then charged a fee precisely equal to its Shapley value. The efficiency condition assures that the fees will exactly cover the cost, the symmetry condition assures that similar users are charged the same fee, and the linearity condition assures that the cost of using two different and independent facilities is the sum of the costs of using each one separately. Monotonicity, of course, only says that you don't get paid for landing at an airport.

A spectacular recent application of this type is to telephone billing at large institutions. See Billera, Heath, and Raanan [3]; the system proposed by them has been adopted for internal telephone billing at Cornell University.

---

9. Value models in which threats do sometimes get carried out involve incomplete information; see [10], [23].

8 Other Contributions

A complete review of recent developments in the theory of the Shapley value is impossible in the space allotted to this paper. The quantifier "some", not "all", should be understood in the title; there have been many important contributions not covered here. We close by mentioning two conceptually innovative recent works: In [34], A. Roth formalized the idea that the value measures a game's utility to its players; and in [22], R. Myerson characterized the value in terms of communication networks connecting the players.

9 Conclusion

Much of the analysis in political and economic science has traditionally proceeded on an ad hoc basis, often using different methods and principles for each model under consideration. A unified approach to these disciplines is provided by game theory. Among the tools it provides, the Shapley value is particularly broadly and systematically applicable, and appears able to account for theoretical principles in widely diverse areas.

Acknowledgments

This work was supported by National Science Foundation Grant SOC75-21820-AO1 at the Institute for Mathematical Studies in the Social Sciences, Stanford University.

References

1. R. J. Aumann and M. Kurz., *Power and taxes in a multi-commodity economy*, Israel J. Math. **27** (1977), 185–234 [Chapter 53].

2. R. J. Aumann and L. S. Shapley, *Values of non-atomic games*, Princeton Univ. Press, Princeton, N. J. 1974.

3. L. J. Billera, D. C. Heath and J. Raanan, *Internal telephone billing rates—a novel application of non-atomic game theory*, Operations Res. **26** (1978), 956–965.

4. D. Brown and P. Loeb, *The values of non-standard exchange economies*, Israel J. Math. **25** (1977), 71–86.

5. P. Champsaur, *Cooperation versus competition*, J. Econom. Theory **11** (1975), 393–417.

6. M. Davis and M. Maschler, *Existence of stable payoff configurations for cooperative games*, Bull. Amer. Math. Soc. 69 (1963), 106–108.

7. P. Dubey, *On the uniqueness of the Shapley value*, Internat. J. Game Theory **4** (1975), 131–139.

8. P. Dubey and L. S. Shapley, *Mathematical properties of the Banzhaf power index*, Math. Operations Res. **4** (1979), 99–131.

9. J. C. Harsanyi, *A bargaining model for the cooperative n-person game*, Contributions to the Theory of Games, vol. IV. Ann. of Math. Studies, No. 40, A. W. Tucker and R. D. Luce, eds., Princeton Univ. Press, Princeton N. J. 1959, pp. 325–355.

10. J. C. Harsanyi and R. Selten, *A generalized Nash solution for two-person bargaining games with incomplete information*, Management Sci. **18** (1972), 80–106.

11. S. Hart, *Values of mixed games*, Internat. J. Game Theory **2** (1973), 69–85.

12. ———, *Asymptotic value of games with a continuum of players*, J. Math. Econom. **4** (1977), 57–80.

13. ———, *Values of non-differentiable markets with a continuum of traders*, J. Math. Econom. **4** (1977), 103–116.

14. ———, *Measure-based values of market games*, Math. Operations Res. **5** (1980), 197–228.

15. W. Hildenbrand, *Core and equilibria in a large economy*, Princeton Univ. Press, Princeton, N. J. 1974.

16. Y. Kannai, *Values of games with a continuum of players*, Israel J. Math. **4** (1966), 54–58.

17. S. C. Littlechild and G. Owen, *A simple expression for the Shapley value in a special case*, Management Sci. **20** (1973), 370–372.

18. W. F. Lucas, *Measuring power in weighted voting systems*, Case Studies in Applied Mathematics, C.U.P.M., Math. Assoc. of America, 1976, pp. 42–106.

19. A. Mas-Colell, *Competitive and value allocations of large exchange economies*, J. Econom. Theory **14** (1977), 419–438.

20. J.-F. Mertens, *Values and derivatives*, Math. Operations Res. **5** (1980), 523–552.

21. J. W. Milnor and L. S. Shapley, *Values of large games II: Oceanic games*, Math. Operations Res. **3** (1978), 290–307.

22. R. B. Myerson, *Graphs and cooperation in games*, Math. Operations Res. **2** (1977), 225–229.

23. ———, *Incentive compatibility and the bargaining problem*, Econometrica **47** (1979), 61–73.

24. J. F. Nash, *The bargaining problem*, Econometrica **18** (1950), 155–162.

25. ———, *Two-person cooperative games*, Econometrica **21** (1953), 128–140.

26. J. von Neumann and O. Morgenstern, *Theory of games and economic behavior*, Princeton Univ. Press, Princeton, N. J., 1944.

27. A. Neyman, *Continuous values are diagonal*, Math. Operations Res. **2** (1977), 338–342.

28. ———, *Singular games have asymptotic values*, Math. Operations Res. **6** (1981), 205–212.

29. A. Neyman and Y. Tauman, *The existence of non-diagonal axiomatic values*, Math. Operations Res. **1** (1976), 246–250.

30. ———, *The partition value*, Math. Operations Res. **4** (1979), 236–264.

31. G. Owen, *Political games*, Naval Res. Logist. Quart. **18** (1971), 345–355.

32. ———, *Values of games without side payments*, Internat. J. Game Theory **1** (1972), 95–109.

33. W. Riker and L. S. Shapley, *Weighted voting: A mathematical analysis for instrumental judgments*, Representation: Nomos X, J. R. Pennock and J. W. Chapman, eds., Atherton, New York, 1968, pp. 199–216.

34. A. E. Roth, *The Shapley value as a von Neumann–Morgenstern utility*, Econometrica **45** (1977), 657–664.

35. N. Z. Shapiro and L. S. Shapley, *Values of large games I: A limit theorem*, Math. Operations, Res. **3** (1978), 1–9.

36. L. S. Shapley, *A value for n-person games*, Contributions to the Theory of Games, Vol. II. Ann. of Math. Studies, No. 28, H. W. Kuhn and A. W. Tucker, eds. Princeton Univ. Press, Princeton, N. J., 1953, pp. 307–317.

37. ———, *Utility comparison and the theory of games*, La Decision-Colloques Internationaux du C.N.R.S., Paris, Editions du C.N.R.S., 1969, pp. 251–263.

38. L. S. Shapley and M. Shubik, *A method for evaluating the distribution of power in a committee system*, Amer. Pol. Sci. Rev. **48** (1954), 787–792.

39. ———, *Pure competition, coalitional power, and fair division*, Internat. Econom. Rev. **10** (1969), 337–362.

40. Y. Tauman, *A non-diagonal value on a reproducing space*, Math. Operations Res. **2** (1977), 331–337.

64 The Shapley Value

1 Introduction

The purpose of this chapter is to present an important solution concept for cooperative games, due to Lloyd S. Shapley (Shapley (1953)). In the first part, we will be looking at the transferable utility (TU) case, for which we will state the main theorem and study several examples. Afterwards, we will extend the axiomatic construction to the non-transferable utility (NTU) case.

2 The Shapley Value in the TU Case

2.1 A First Approach

Let N be a finite set of players and $n = |N|$. A game is a mapping $v: 2^N \to \mathbb{R}$ such that $v(\emptyset) = 0$. For S in 2^N (i.e., $S \subset N$), $v(S)$ may be interpreted as the *worth* of coalition S, i.e. what the players belonging to S can get together by coordinating their efforts. This models a game with transferable utility (or with side payments), i.e., where coalised players may reallocate the total utility within the coalition: it is sufficient to map every coalition to a single number, the coalition's total utility.

The unanimity game U_T associated with the coalition $T \subset N$ is defined by:

$$U_T(S) = \begin{cases} 1, & \text{if } S \supset T; \\ 0, & \text{otherwise.} \end{cases}$$

Given a set of players N, denote by $G(N)$ the set of all possible games with players in N. Let $E = \mathbb{R}^N$ be the space of payoff vectors and for $x \in E$ denote by $x(S)$ the sum $\sum_{n \in S} x_n$. We may then define a value as a mapping $\varphi: G(N) \to E$ such that:

$$\forall v \in G(N), (\varphi v)(N) = v(N) \qquad (a)$$

$$\forall v, w \in G(N), \varphi(v + w) = \varphi v + \varphi w \qquad (b)$$

$$\forall T \subseteq N, \forall \alpha \in \mathbb{R}, \varphi(\alpha U_T)_i = \begin{cases} \alpha/|T|, & \text{if } i \in T; \\ 0, & \text{otherwise.} \end{cases} \qquad (c)$$

Axioms (a) and (b) are the standard ones of efficiency and additivity, whereas axiom (c) is equivalent to the axioms: (i) neutral to permutation

---

This chapter originally appeared in *Game Theoretic Methods in General Equilibrium Analysis*, edited by J.-F. Mertens and S. Sorin, Kluwer Academic Publishers, Dordrecht, 1994, pp. 61–66. Reprinted with permission.

and (ii) null player. It is remarkable that no further conditions are required to determine the value uniquely as in the following (Shapley (1953)):

THEOREM 1 *For each N, there exists a unique value function; this value is given by*

$$(\varphi v)_i = \frac{1}{n!} \sum_{<} [v\{j | j \leqq i\} - v\{j | j \prec i\}],$$

where the sum extends over all total orders on the player set.

Before giving the proof, let us observe that the intuitive interpretation of the formula is the following: when a player joins a coalition, it may modify the worth of the coalition; the Shapley value gives to each player his average marginal contribution to the worth of all possible coalitions.

Proof It is easily seen that the function defined above is a value. To prove its uniqueness, it suffices to show that the above games $U_T(T \neq \emptyset, T \subseteq N)$ form a basis. Since their number equals the dimension of $G(N)$, it suffices to show they are linearly independent. Suppose they are not:

$$\exists (\alpha_i) \text{ such that } \sum \alpha_i U_{T_i} = 0 \text{ and } \alpha_j \neq 0 \text{ for some } j. \quad (1)$$

Among the subsets T_i such that $\alpha_i \neq 0$, there exists at least one coalition, say T_1, with a minimum number of players. Then rearranging (1):

$$U_{T_1} = -(1/\alpha_1) \sum_{j>1} \alpha_j U_{T_j};$$

yet $U_{T_1}(T_1) = 1$ and $U_{T_j}(T_1) = 0$ for $j > 1$ because in this case $T_j \not\subseteq T_1$.

Hence any game may be written as a linear combination of the unanimity games, and by axiom c), a value is uniquely determined on these games. QED

2.2 Examples

We shall now examine some examples to underline the differences between the Shapley value and another solution concept, the core.

Example 1 Majority game of 3 players.

$N = \{1, 2, 3\}$, $v(S) = 1$ if $|S| \geq 2$ and 0 otherwise. The game is symmetric; player i changes the worth of the coalition that precedes him if he is in position 2, which happens for two different orders. By either argument, $\varphi v = (1/3, 1/3, 1/3)$. On the other hand the core is empty, since there are always two players who can form a coalition and share what the third player gets.

Example 2 Market with one seller and two buyers.

$N = \{1, 2, 3\}$, $v(\{1, 2, 3\}) = v(\{1, 2\}) = v(\{1, 3\}) = 1$ and $v(S) = 0$ otherwise (1 is the seller). Player 2 (or player 3) changes the worth of the coalition that precedes him if player 1 is first and he is second, while player 1 contributes to the coalition as soon as he is not in first: $\varphi v = (2/3, 1/6, 1/6)$. Obviously $(1, 0, 0)$ is in the core, and nothing else is, because any other outcome could be blocked by a coalition of player 1 with one of the other players.

Example 3 A weighted voting game.

$N = \{1, 2, 3, 4\}$, with weights $(2, 1, 1, 1)$; the total weight is 5 and a majority of 3 wins. Player 1 is pivotal in position 2 or 3 (1 chance out of 2), while players 2, 3, and 4 are in symmetric positions; therefore $\varphi v = (1/2, 1/6, 1/6, 1/6)$. Once again the core is empty, since any outcome can be improved upon by the three players who get the least. Note that whereas the large player (player 1) has only 40% of the vote, he gets half the value.

Example 4 Another weighted voting game.

$N = \{1, 2, 3, 4, 5\}$, with weights $(3, 3, 1, 1, 1)$; the core is empty as before, and we have $\varphi v = (3/10, 3/10, 2/15, 2/15, 2/15)$. In this case the large players' value is *less* than their proportion of the vote; thus players 3, 4 and 5 would get less (1/3 instead of 2/5) if they were to unite into a single player with weight 3.

Example 5 Market game with 1,000,000 left gloves and 1,000,001 right gloves—one glove per player.

2,000,001 players, $v(N) = 1{,}000{,}000$. In this case the core has a single element, where the left glove owners get 1 (pair), and the right glove owners get 0. The Shapley value, on the other hand, assigns a total of 500,428 to the left glove owners and a total of 499,572 to the right glove owners.

2.3 Other Characterizations

2.3.1 The Potential

By theorem 1 it is possible to define the Shapley value through the marginal contributions of players: namely the value of a game may be seen as the vector of the players' "expected payoffs" as their expected marginal contributions to coalitions (with the appropriate interpretation, axiom (b) may be seen as an expected utility property). This idea has led to another approach based upon the "potential" of a game. Let us define in a general way the marginal contributions: map every game (N, v) to a real number $P(N, v)$, called the *potential* of the game, and let player i's marginal contribution be: $P(N, v) - P(N \setminus \{i\}, "v|N \setminus \{i\}")$. Then one could reasonably require that these marginal contributions satisfy an efficiency condition, i.e. add up to $v(N)$ for all players in N. It is clear inductively that this condition (with an appropriate definition of "$v|N \setminus \{i\}$") determines a unique potential function; moreover it has been shown that it leads precisely to the Shapley value (Hart and Mas-Colell (1989)).

2.3.2 The Monotonicity Principle

Alternative axiomatizations have been put forward. For instance (Young (1985)), it is possible to replace the additivity axiom and the null player axiom by some requirement related to the monotonicity of the value. More precisely, define φ to be an *allocation procedure* if it maps every game to a point in $\mathbb{R}^N$ and is efficient. The procedure φ is *symmetric* (anonymous) if for all permutations π of N, $\varphi_{\pi i}(\pi v) = \varphi_i(v)$, where $\pi v(S) = v(\pi^{-1}(S))$ for all S. The procedure φ satisfies *strong monotonicity* if:

$\forall$ games v, w $\forall i \in N, (\forall S \subset N, v^i(S) \geqslant w^i(S)) \Rightarrow (\varphi_i(v) \geqslant \varphi_i(w))$,

where $v^i(S) = v(S \cup \{i\}) - v(S)$.

In words, strong monotonicity means that the payoff to a player depends only on his marginal contributions—and monotonically. The result is then as follows:

THEOREM 2 *The Shapley value is the unique symmetric allocation procedure that is strongly monotonic.*

2.3.3 A Smaller Class of Games

The axiomatization of the Shapley value requires the application of the value axioms to all games. Yet, it is possible (Neyman (1989)) to derive the Shapley value of any given game v by applying the axioms to a smaller class of games, namely the additive group generated by the subgames of v, which yields a stronger characterization of the Shapley value. Given a game (N, v), and a coalition $S \subset N$, define the subgame v_S as the mapping: $2^N \to \mathbb{R}$ such that $v_S(T) = v(S \cap T)$; $v(S)$ may be viewed as the restriction of v to the subsets of S. Denote by $G(v)$ the additive group generated by the subgames of v:

$$G(v) = \{w \in G(N) | w = \sum k_i v_{S_i} \text{ with } k_i \text{ integers and } S_i \text{ coalitions}\}$$

If Q is a subset of $G(N)$, we say that a map $\Psi: Q \to \mathbb{R}^N$ obeys the *null player axiom* if:

$$\forall v \in Q, \forall i \in N, (v(S \cup \{i\}) = v(S), \forall S \subset N) \Rightarrow (\Psi_i(v) = 0)$$

The extension of the other axioms to a subset of $G(N)$ is straightforward. Then:

THEOREM 3 *Let $v \in G$. If a map Ψ from $G(v)$ into $\mathbb{R}^N$ is efficient, additive, and symmetric, and obeys the null player axiom, then it is the Shapley value.*

Note that, in this case too, it is possible to replace the additivity and null player axioms by strong monotonicity.

3 The Shapley Value in the NTU Case

If $x, y \in \mathbb{R}^N$, then we write $x \geqslant y$ if $x_i \geqslant y_i$ for all i. A set A in $\mathbb{R}^N$ is said to be *comprehensive* if $x \in A$ and $x \geqslant y$ implies $y \in A$. A convex set C in $\mathbb{R}^N$ is said to be *smooth* if it has a unique supporting hyperplane at each point of its frontier ∂C. An *NTU game* is a function V that assigns to each coalition S a convex comprehensive non-empty proper subset $V(S)$ of $\mathbb{R}^S$, such that:

1. $V(N)$ is smooth,
2. If $x, y \in \partial V(N)$ and $x \geqslant y$, then $x = y$,
3. $\forall S \subset N, \exists x \in \mathbb{R}^N$ s.t. $V(S) \times \{0^{N \setminus S}\} \subset V(N) + x$,

where $0^{N \setminus S}$ is the 0-vector in $\mathbb{R}^{N \setminus S}$. The interpretation of the NTU case is that, since no side payments are allowed, there is no possible reallocation. Thus to evaluate the worth of a coalition one has to take into account the payoffs of all players belonging to the coalition; maximizing the worth of a coalition is now a "multi-criterion" problem, in the currently popular jargon. Condition 2 says that the frontier of the grand coalition payoff-set contains only strict Pareto-optima and Condition 3 can be thought of as an extremely weak kind of monotonicity. If v is a TU game, then the NTU game V *corresponding* to v is defined by:

$$V(S) = \left\{ x \in \mathbb{R}^S \mid \sum_{i \in S} x^i \leqslant v(S) \right\};$$

thus we can speak of an NTU *unanimity game* as one corresponding to a TU unanimity game.

We still need to define the Shapley correspondence; the idea (Shapley 1969) is to associate a TU game with every NTU game and comparison vector, with the worth of a coalition S being the best it can get in terms of the comparison vector. More precisely, let V be an NTU game and $\lambda \in \mathbb{R}^N$ a comparison vector (i.e. $\lambda^i > 0, \forall i$). Define an auxiliary TU game v_λ as follows:

$$v_\lambda(S) = \sup\{\langle \lambda^S, x \rangle | x \in V(S)\};$$

the game v_λ is well defined if the supremum is finite for all S. A *Shapley value* of V is a point x in the closure $cl(V(N))$ of $V(N)$ such that for some λ, v_λ is well defined, and the vector $(\lambda^i x^i)$ is the Shapley value of v_λ. Now if Γ is the set of all NTU games with at least one Shapley value, the correspondence from Γ to $\mathbb{R}^N$ that assigns to every game V in Γ the set $\Lambda(V)$ of its Shapley values is the *Shapley correspondence*.

Define a *value correspondence* as a correspondence $\Phi: \Gamma \to \mathbb{R}^N$ satisfying the following axioms:

a. $\forall V \in \Gamma, \Phi(V) \neq \emptyset$;
b. $\forall V \in \Gamma, \Phi(V) \subset \partial V(N)$;
c. $\forall U, V \in \Gamma, \Phi(U + V) \supset (\Phi(U) + \Phi(V)) \cap \partial(U + V)(N)$;
d. For all unanimity games $U_T, \Phi(U_T) = \{\Pi_T/|T|\}$;
e. $\Phi(clV) = \Phi(V)$;
f. $\forall \lambda \in \mathbb{R}^N, \lambda > 0, \Phi(\lambda V) = \lambda \Phi(V)$;
g. $\forall V, W \in \Gamma$, if $V(N) \subset W(N)$ and $V(S) = W(S)$ for $S \neq N$, then $\Phi(V) \supset F(W) \cap V(N)$.

Axiom (a) is non-emptiness; axiom (b), efficiency, says that all values are Pareto optimal; axiom (c) says that if y and z are values of V and W and if $y + z$ is Pareto optimal in $V + W$, then it is a value of $V + W$; axiom (d) determines the values of the unanimity games (the values are unique); axiom (e) is closure invariance; axiom (f) is scale covariance; axiom (g) is the well-known independence of irrelevant alternatives (I.I.A.), and it says that a value y of a game W remains a value when one removes outcomes other than y from the set $W(N)$ of all feasible outcomes, without changing the $W(S)$ for $S \neq N$. We now have (Aumann (1985)):

THEOREM 4 *There is a unique value correspondence, and it is the Shapley value.*

It is noteworthy that removing I.I.A. (Axiom (g)) is not too damaging, as the following holds (Aumann (1985)):

THEOREM 5 *The Shapley correspondence is the maximal correspondence among those satisfying axioms* (a) *through* (f) (*i.e. if* Φ *satisfies axioms* (a) *through* (f), *then* $\Phi(V) \subset \Lambda(V)$ *for all games V in* Γ).

Acknowledgments

The author is very grateful to J.-F. Mertens, who prepared this chapter on the basis of the author's oral presentation.

References

Aumann, R. J. (1978) *Recent Developments in the Theory of the Shapley Value*, Proceedings of the International Congress of Mathematicians, Academia Scientiarum Fennica, Helsinki, 1980, 995–1003 [Chapter 63].

Aumann, R. J. (1985) *An Axiomatization of the Non-Transferable Utility Value*, Econometrica, 53, 599–612 [Chapter 60].

Hart S. (1990) *Advances in Value Theory*, in: "Game Theory and Applications," Ichiishi, T., A. Neyman and Y. Tauman eds., Academic Press, 166–175.

Hart S. and A. Mas-Colell (1989) *Potential, Value and Consistency*, Econometrica, 57, 589–614.

Neyman, A. (1989) *Uniqueness of the Shapley Value*, Games and Economic Behavior, 1, 116–118.

Shapley, L. S. (1953) *A Value for n-person Games*, in: "Contributions to the Theory of Games, II," H. Kuhn and A. W. Tucker, eds., Princeton: Princeton University Press.

Shapley, L. S. (1969) *Utility Comparison and the Theory of Games*, in: "La Décision: Agrégation et dynamique des ordres de préférence" (Paris: Editions du Centre National de la Recherche Scientifique, 251–263.

Young, H. P. (1985) *Monotonic Solutions of Cooperative Games*, International Journal of Game Theory, 14, 65–72.

XIII MATHEMATICAL METHODS

While most of my work has some mathematical component, the papers in this group are largely or exclusively mathematical; the decision-theoretic, game-theoretic, or economic motivation is either not mentioned at all in the body of the paper, or mentioned only very briefly. Needless to say, these papers *were* motivated by other, more applied work.

As it happens, all the chapters in this group are in the general area of measure, measurability, Borel structures, and so on, though with two distinct backgrounds: Probability (chapters 65–67 and 73) and non-atomic measure spaces of players or traders (chapters 68–72).

Chapters 65, 66, and 67 deal with the same topic—choosing a function at random. The motivation comes from chapter 28, "Mixed and Behavior Strategies in Infinite Extensive Games." For example, consider an extensive game with just two moves: Ann chooses an element x of a set X; Bob, being informed of x, chooses an element[1] of a set Y, where both X and Y are, say, copies of the unit interval [0,1]. A mixed strategy μ of Ann is a probability distribution on X, when X has its Borel structure—that is, all Borel subsets of X are assigned a probability by μ. A pure strategy of Bob is a function f from X to Y; in order that μ and f induce a probability distribution on Y, we must take f *measurable* (when Y, too, has its Borel structure). Thus a pure strategy of Bob is a member of the space Y^X of all measurable functions from X to Y; so a mixed strategy ν of Bob is a distribution on Y^X. To make sense of this, Y^X must have a measurable structure as well—that is, we must determine a σ-field of *measurable* subsets of Y^X, which are assigned probabilities by ν. In order that μ and ν then induce a probability distribution on Y, the function $\varphi\colon Y^X \times X \to Y$ defined by $\varphi(f, x) := f(x)$ must be measurable. These considerations led me to try to identify those measurable structures (σ-fields) on Y^X for which φ is measurable.

To my dismay, it turned out that there is *no* measurable structure on Y^X for which φ is measurable. In order that φ be measurable, one must restrict Bob's choice of a pure strategy to certain subsets F of Y^X, called *admissible*; it is, so to speak, impossible to mix over all of Y^X. These developments are the subject of chapters 65 ("Spaces of Measurable Transformations") and 66 ("Borel Structures for Function Spaces"); chapter 65 is a research announcement that sets forth the results, while chapter 66 contains the full proofs.

Chapter 67 ("On Choosing a Function at Random") approaches the problem slightly differently. It views mixed strategies not as distributions

---

1. This particular game is of perfect information, so mixed strategies have little interest here. However, it might well be that Bob is not informed of x itself, but only of the value $g(x)$ of some function g of x; then mixed strategies *are* of interest, and the considerations below still apply.

over pure strategies, but as random variables whose values are pure strategies. Thus a mixed strategy is a function θ from some sample space Ω to the space Y^X of pure strategies. Now there is a natural one–one correspondence between such functions θ and functions $\vartheta\colon \Omega \times X \to Y$, given for each ω in Ω by $\theta(\omega)(x) = \vartheta(\omega, x)$. Thus one may view mixed strategies as measurable functions from $\Omega \times X$ into Y. Chapter 65 shows that this approach yields results that are basically similar to those obtained with the "distribution" definition of mixed strategies; in particular, the set of pure strategies that a particular mixed strategy ϑ can actually pick is admissible in the sense defined above. It is this second, "random variable" approach that actually turned out most fruitful for the application in chapter 28.

Chapter 68, "Integrals of Set-Valued Functions," was originally motivated by the existence proof for competitive equilibrium when the set T of traders is a continuum (chapter 47). This proof depends crucially on set-valued aggregates (like the "aggregate demand set"). When T is finite, such aggregates are defined as sums; thus if $\mathbf{F}(t)$ is a subset of a Euclidean space for each $t \in T$, then the aggregate $\sum \mathbf{F} = \sum_{t \in T} \mathbf{F}(t)$ is the set of all sums $\sum_{t \in T} x_t$, where for each $t \in T$, the point x_t is selected from $\mathbf{F}(t)$. When T is a continuum, aggregates must be defined as integrals; thus we define $\int \mathbf{F}$ as the set of all integrals $\int \mathbf{f}(t)dt$, where the integral is over T, and $\mathbf{f}$ ranges over all integrable selections of a point valued function from the set-valued function $\mathbf{F}$. Chapter 68 establishes several basic properties of such integrals, many of which are needed for the existence proof in chapter 47. Inter alia, under appropriate conditions it is proved that $\int \mathbf{F}$ is non-empty, and that integration preserves both upper- and lowersemicontinuity, and so also continuity (i.e., if $\mathbf{F}_x(t)$ is continuous[2] in x for each fixed t, then also $\int \mathbf{F}_x$ is continuous in x).

The existence of competitive equilibrium in markets with a continuum of traders depends strongly on the preservation of uppersemicontinuity under integration, whose proof in chapter 68 makes heavy use of comparatively deep tools from functional analysis. Several years later I discovered an "elementary" proof, using only the basic facts about Lebesgue integrals, and Lyapunov's theorem on the range of a vector measure. This is chapter 69.

Chapters 70 through 72 were motivated by my book with Shapley, *Values of Non-Atomic Games*[3] (henceforth *VNAG*). Chapter 70, "A Variational Problem Arising in Economics," (joint with Micha Perles) arose

---

2. In the Hausdorff metric.
3. Princeton University Press, Princeton, 1974.

from the study of TU (transferable utility) market games with a non-atomic continuum of traders. In such a game, the worth $v(S)$ of a coalition S is defined as the maximum total utility that S can get by trading among its own members only; when S is a continuum, this total utility is expressed as an integral, whose maximum is not always attained. Chapter 70 identifies conditions under which the maximum *is* attained. In the process, heavy use is made of the theory of integrals of set-valued functions developed in chapter 68.

Chapter 71, "Random Measure Preserving Transformations," demonstrates that in a certain natural sense, it is *impossible* to define the notion of a random measure preserving transformation in a natural way. More (but still not quite) precisely, let $\mathscr{G}$ be the group of automorphisms (invertible Lebesgue measure preserving transformations) of the unit interval I. Then it is impossible to define a translation invariant probability measure μ on $\mathscr{G}$—that is, one for which $\mu(\mathscr{H}T) = \mu(\mathscr{H})$ for all measurable subsets $\mathscr{H}$ of $\mathscr{G}$ and all T in $\mathscr{G}$.

Though it is of interest on its own, the original motivation of chapter 71 came from an attempt to define values for non-atomic games by random order methods analogous to those used for finite games. To be more explicit, we must review some definitions. Define a *non-atomic game* as a real-valued function v on the Borel subsets S of the unit interval I; here I represents the player set, S a coalition, $v(S)$ the "worth" of S. *Payoff vectors* are identified with measures μ on the family $\mathscr{B}$ of all coalitions, where $\mu(S)$ represents the total payoff to S. If $\mathscr{R}$ is a total order on I, and $s \in I$, then $\{t: s\mathscr{R}t\}$ is the coalition of all players before s in the order $\mathscr{R}$; it is called an *initial $\mathscr{R}$-segment*. We call $\mathscr{R}$ *measurable* if the initial $\mathscr{R}$-segments are all in $\mathscr{B}$, and together they generate $\mathscr{B}$. Note that if $\mathscr{R}$ is measurable, then there is at most one measure $\varphi^{\mathscr{R}}v$ that coincides with v on the initial $\mathscr{R}$-segments; $(\varphi^{\mathscr{R}}v)(S)$ represents the total incremental worth contributed by the coalition S when the players are ordered in accordance with $\mathscr{R}$.

To define a value by random order methods, one needs (i) to define the notion of a "random order" $\mathscr{R}$, and (ii) to identify conditions under which the measure $\varphi^{\mathscr{R}}v$ exists. One may then define the value of a coalition S as the expectation of $(\varphi^{\mathscr{R}}v)(S)$ when $\mathscr{R}$ is chosen at random.

To address task (i), note that if T is an automorphism of I and $\mathscr{R}$ is an order on I, one may define an order $T\mathscr{R}$ on I by $Ts\ T\mathscr{R}\ Tt \Leftrightarrow s\mathscr{R}t$. If T were a "random" automorphism, and $\mathscr{R}$ is a fixed measurable order (say the natural order on I), then one could think of $T\mathscr{R}$ as a "random" order. Thus a notion of "random automorphism" might have led to a notion of "random order." That is what motivated chapter 71; as we have seen, the result was negative.

Sections 12 and 13 of *VNAG* show that other approaches to defining a random order on I also fail; the notion is in a certain basic sense undefinable. Task (ii) nevertheless remains of interest, since the measures $\varphi^{\mathscr{R}} v$ may be used in other ways to define a value; for example, they play a fundamental role in constructing the "mixing value" (*VNAG*, chapter II).

Call a game v *orderable* if $\varphi^{\mathscr{R}} v$ exists for all measurable $\mathscr{R}$—i.e., if for every measurable $\mathscr{R}$, every coalition S has an "incremental worth." Not all games are orderable; a sufficient condition for orderability is that v be *absolutely continuous*, in a sense closely related to that applying to measures (*VNAG*, sections 5 and 12, in particular proposition 12.8). The question arises whether this sufficient condition is also necessary—whether orderability implies absolute continuity. Chapter 72, written jointly with U. Rothblum, answers this question in the negative by providing a counterexample.

The last item in this collection is chapter 73, "Biconvexity and Bimartingales," which is joint with S. Hart. It was motivated by the theory of repeated games of incomplete information, in which I first became interested in the late sixties (see chapter 23 and the material on it in the introduction to group IV). One of the problems left open by that early work was to characterize the equilibrium payoffs in such games, in the spirit of the "Folk Theorem" for repeated games of complete information. It turns out that such a characterization lies much deeper than the Folk Theorem, even in the restricted case of two-person games in which there is incomplete information on one side only. The problem of finding it remained open for many years, and was finally solved only in 1985, by S. Hart.[4]

Hart's characterization is based on the idea of a *bimartingale*—a kind of generalized random walk in which a particle alternates between two types of move (horizontal and vertical moves in the plane, say). In the game context, this reflects a process in which the informed player reveals information in bits and pieces, the revelations alternating with sessions in which the players reach tentative agreements about future play. Geometrically, bimartingales are related to *biconvex* sets, like planar sets whose vertical and horizontal sections are all convex (e.g., crosses). Chapter 73 studies the geometry of biconvexity and bimartingales; inter alia, it provides characterizations in terms of support functions, in a sense analogous to that applying to ordinary convexity.

---

4. "Nonzero-sum two-person repeated games with incomplete information," *Math. Oper. Res.* 10, 117–153.

65 Spaces of Measurable Transformations[1]

By a *space* we shall mean a measurable space, i.e. an abstract set together with a σ-ring of subsets, called *measurable* sets, whose union is the whole space. The *structure* of a space will be the σ-ring of its measurable subsets. A *measurable transformation* from one space to another is a mapping such that the inverse image of every measurable set is measurable.

Let X and Y be spaces, F a set of measurable transformations from X into Y, and $\phi_F: F \times X \to Y$ the natural mapping defined by $\phi_F(f, x) = f(x)$. A structure R on F will be called *admissible* if ϕ_F, considered as a mapping from the product space $(F, R) \times X$ into Y, is a measurable transformation.[2] It may not be possible to define an admissible structure on F; if it is, F itself will also be called *admissible*. We are concerned with the problem of characterizing, for given X and Y, the admissible sets F and the admissible structures R on the admissible sets.

The following three theorems may be established fairly easily:

THEOREM A. *A set consisting of a single measurable transformation is admissible.*

THEOREM B. *A subset of an admissible set is admissible. Indeed, if $G \subset F$, R is an admissible structure on F, and R_G is the subspace structure on G induced[3] by R, then R_G is admissible on G.*

THEOREM C. *The union of denumerably many admissible sets is admissible. Indeed, if $F = \bigcup_{i=1}^{\infty} F_i$ and $R_1, R_2, \cdots$ are admissible structures on $F_1, F_2, \cdots$ respectively, then the structure R on F generated by the members of all the R_i is admissible on G.*

Much more can be said if X and Y are assumed to be *separable*, i.e. to have countably generated structures.[4] To state our theorems in this case we first define the concept of Banach class, closely related to that of Baire class. Let $\mathfrak{A}$ be an arbitrary class of meas-

---

[1] The author is much indebted to Professor P. R. Halmos, who suggested a number of significant improvements in the complete version of this note.

[2] (F, R) is the space whose underlying abstract set is F and whose structure is R.

[3] R_G consists of all intersections of G with members of R.

[4] The term is used by analogy with its topological use. We will also use the term "separable structure," meaning a countably generated structure.

This chapter originally appeared in *Bulletin of the American Mathematical Society* 66 (1960): 301–304. Reprinted with permission.

urable subsets of X. For each denumerable ordinal number $\alpha \geq 1$, we define classes $P_\alpha(\mathfrak{A})$ and $Q_\alpha(\mathfrak{A})$ inductively as follows: $Q_1(\mathfrak{A})$ consists of all denumerable unions of members of $\mathfrak{A}$, and $P_1(\mathfrak{A})$ consists of all complements of members of $Q_1(\mathfrak{A})$; supposing $Q_\beta(\mathfrak{A})$ and $P_\beta(\mathfrak{A})$ to have been defined for all $\beta < \alpha$, we define $Q_\alpha(\mathfrak{A}) = Q_1(\cup_{\beta<\alpha} P_\beta(\mathfrak{A}))$ and $P_\alpha(\mathfrak{A}) = P_1(\cup_{\beta<\alpha} P_\beta(\mathfrak{A}))$. $Q_\alpha(\mathfrak{A}) \cup P_\alpha(\mathfrak{A})$ is the set of all subsets of X which can be "reached from $\mathfrak{A}$" by performing at most α operations, where each operation consists of forming a denumerable union and a complement. If $\mathfrak{A}$ generates the structure of X, then the union (over α) of all the $Q_\alpha(\mathfrak{A})$ (or of the $P_\alpha(\mathfrak{A})$) is the set of all measurable subsets of X. If $\mathfrak{B}$ is a class of measurable subsets of Y and $\alpha \geq 0$ is a denumerable ordinal number, then we define $L_\alpha(\mathfrak{A}, \mathfrak{B})$ to be the set of all functions $f: X \to Y$ such that for all $B \in Q_1(\mathfrak{B})$, $f^{-1}(B) \in Q_{\alpha+1}(\mathfrak{A})$. If X and Y are separable and $\mathfrak{A}$ and $\mathfrak{B}$ are denumerable generating sets for their respective structures, then the union (over α) of all the $L_\alpha(\mathfrak{A}, \mathfrak{B})$ is the set of all measurable transformations from X into Y. It will be denoted Y^X. In this case $L_\alpha(\mathfrak{A}, \mathfrak{B})$ is called the *Banach class*[5] of order α for $(\mathfrak{A}, \mathfrak{B})$. A subset F of Y^X is said to be of *bounded* Banach class if there is an α and denumerable generating sets $\mathfrak{A}$, $\mathfrak{B}$ such that $F \subset L_\alpha(\mathfrak{A}, \mathfrak{B})$. It is important to note that the definition of bounded Banach class is independent of the choice of $\mathfrak{A}$ and $\mathfrak{B}$, i.e. that if $F \subset L_\alpha(\mathfrak{A}, \mathfrak{B})$, then for any other generating pair $\mathfrak{A}'$, $\mathfrak{B}'$, there is an α' such that $F \subset L_{\alpha'}(\mathfrak{A}', \mathfrak{B}')$. If X and Y are separable metric spaces and Y is pathwise connected, then the Banach classes coincide with the Baire classes (for appropriate choice of $\mathfrak{A}$ and $\mathfrak{B}$).

THEOREM D. *If X and Y are separable, then F is admissible if and only if it is of bounded Banach class.*

THEOREM E. *If X and Y are separable, then every admissible subset of Y^X has a separable admissible structure.*

A space Z and its structure are called *regular* if for all $x, y \in Z$, there is a measurable set in Z containing x but not y. It is known (cf. [2]) that a space is separable and regular if and only if it is isomorphic[6] to a subspace of I, where I denotes the unit interval $[0, 1]$ with the usual Borel structure.

THEOREM F. *If X and Y are separable and regular, then every admissible subset of Y^X has a separable and regular admissible structure.*

---

[5] Because of the work that Banach [1] did in characterizing these classes.

[6] Two spaces are said to be *isomorphic* if there is a $1-1$ correspondence between them that preserves measurability (in both directions).

The *natural* admissible structure on a given admissible set F is defined to be the smallest admissible structure on F, if it exists. Alternatively, it may be defined to be the intersection of all the admissible structures on F, in case this is admissible. Not every admissible set need have a natural admissible structure; the counter-example is due to P. R. Halmos.

If $a \in X$ and $B \subset Y$, define $F(a, B) = \{f : f \in F, f(a) \in B\}$. It is not hard to prove that if B is measurable and a is arbitrary, then every admissible structure on F must contain $F(a, B)$. A "converse" would be that the structure generated by the $F(a, B)$ is admissible, and it would follow that it is also natural.

THEOREM G. *If X and Y are separable metric spaces and F contains continuous functions only, then F has a natural admissible structure, which is generated by the set of all $F(a, B)$, where B is measurable and a is arbitrary.*

We now give some applications. A space is said to have the *discrete* structure if every subset is measurable. Let J be the space consisting of 0 and 1 only, and K the space of all positive integers, both with the discrete structure. If X is an arbitrary space, then X^J and X^K are both admissible, and possess natural admissible structures which make them isomorphic to $X \times X$ and $\times_{i=1}^{\infty} X_i$ respectively, where the X_i are copies of X. In particular, J^K is admissible and has a natural admissible structure which makes it isomorphic to I. These results are relatively trivial or at least easily derivable from known results.

The situation changes when we pass to exponent spaces with non-discrete structures. For example, J^I may be considered the set of all measurable subsets of I. It is not itself admissible. The set of all open subsets of I is admissible, as is the set of all closed subsets, the set of all G_δ, etc. In general, a subset F of J^I is admissible if and only if all members of F can be constructed from the open subsets of I by taking denumerable unions and intersections at most α times, where α is an arbitrary denumerable ordinal number (which is fixed for given F, but may differ for different F). I do not know whether or not every admissible subset of J^I has a natural admissible structure, but if F is admissible, then we may endow it with an admissible structure in such a way so that it will be isomorphic to a subset of I.

I^I is not admissible. The set of all continuous functions from I into I is admissible; more generally, a necessary and sufficient condition that a subset F of I^I be admissible is that there exist a denumerable ordinal number α such that all members of F are of Baire class α

at most. The set H of all continuous functions from I into I has a natural admissible structure; it is the Borel structure of H when considered as a metric space (in the uniform convergence topology). Again, I do not know whether or not every admissible subset of I^I has a natural admissible structure, but if F is admissible, we may endow it with an admissible structure in such a way so that it will be isomorphic to a subset of I.

The above theory may be applied to give a generalization of Kuhn's theorem [3] about optimal behavior strategies in games of perfect recall, to games in which there may be a continuum of alternatives at some of the moves.

A fuller account of the theory outlined above, together with proofs, will be published elsewhere.[7]

References

1. S. Banach, *Über analytisch darstellbare Operationen in abstrakten Räumen*, Fund. Math. vol. 17 (1931) pp. 283–295.

2. G. W. Mackey, *Borel structures in groups and their duals*, Trans. Amer. Math. Soc. vol. 85 (1957) pp. 134–165.

3. H. W. Kuhn, *Extensive games and the problem of information*, Contributions to the Theory of Games II, Princeton University Press, 1953, pp. 245–266.

---

[7] R. J. Aumann, *Borel Structures for Function Spaces*, Illinois J. Math. vol. 5 (1961) pp. 614–630 [Chapter 66].

66 Borel Structures for Function Spaces[1]

If X and Y are topological spaces, then Y^X denotes the set of all continuous mappings from X into Y. For a given topology on Y^X, we may ask whether the natural mapping $\varphi: Y^X \times X \to Y$ defined by $\varphi(f, x) = f(x)$ is continuous; if it is, then the topology on Y^X is said to be *admissible* [1]. It is always possible to find an admissible topology; for instance, the discrete topology on Y^X is always admissible. Moreover, when X is locally compact, Y^X has a unique smallest[2] admissible topology; this is the familiar "compact-open" topology. These and related questions concerning topologies for function spaces have been investigated in considerable detail by several authors [1, 4].

We are interested in the analogous situation when X and Y are Borel spaces[3] rather than topological spaces; in this case we define Y^X as the set of all Borel mappings[4] from X into Y. Unfortunately, it turns out that even for some of the simplest Borel spaces, it is impossible to define a Borel structure on Y^X so that φ is a Borel mapping; even if we impose the discrete structure on Y^X, φ will in general not be Borel. As a substitute, we may ask ourselves the following questions: For which *subsets* F of Y^X is it possible to impose a Borel structure on F so that $\varphi | F \times X$ will be Borel? If it is possible for a given F, what can we say about the appropriate structures? In particular, is there always a smallest such structure (corresponding to the compact-open topology)?

Let us introduce some terminology. We will write "space" instead of "Borel space", "structure" instead of "Borel structure", and φ_F instead of $\varphi | F \times X$. A structure R on F for which φ_F is Borel will be called *admissible*; a subset F of Y^X on which it is possible to impose an admissible structure is

---

Received November 14, 1960.

[1] The results proved here were announced in [2]. The author is grateful to Prof. P. R. Halmos for many helpful suggestions, and in particular for the counterexample in Section 7; also to Dr. M. Rabin for a number of helpful discussions.

[2] I.e., weakest, with fewest open sets. We remark that the local compactness condition on X may be replaced by certain other conditions on X and Y; see [1, 4].

[3] A *Borel space* is a set X together with a σ-ring of subsets of X called *Borel sets* whose union is all of X. The σ-ring of Borel sets is called the *Borel structure* of X, or simply its *structure*. The structure of the cartesian product of two Borel spaces X and Y is taken to be that generated by the *Borel rectangles*—the products of a Borel set in X and a Borel set in Y. Our definition of Borel space is slightly more general than Mackey's definition [7], in which it is demanded that the structure be a σ-field rather than a σ-ring; as it turns out, most of our theorems and examples refer to the more restricted kind of space anyway. In [5] and [2], the word "measurable" is used in the sense that "Borel" is used here.

[4] A *Borel mapping* is a mapping such that the inverse image of every Borel set is Borel. It is called "Borel function" in [7] and "measurable transformation" in [2, 5].

This chapter originally appeared in *Illinois Journal of Mathematics* 5 (1961): 614–630. Reprinted with permission.

also called *admissible*. We will be chiefly concerned with characterizing, for given X and Y, the admissible sets F and the admissible structures on them.

We first state our theorems, then give some illustrations and applications. The following three general theorems may be established fairly easily:

THEOREM A. *A set consisting of a single Borel mapping is admissible.*

THEOREM B. *A subset of an admissible set is admissible. Indeed, if $G \subset F$, R is an admissible structure on F, and R_G is the subspace structure on G induced*[5] *by R, then R_G is admissible on G.*

THEOREM C. *The union of denumerably many admissible sets is admissible. Indeed, if $F = \cup_{i=1}^{\infty} F_i$ and $R_1, R_2, \cdots$ are admissible structures on $F_1, F_2, \cdots$ respectively, then the structure R on F generated by the members of all the R_i is admissible on G.*

Much more can be said if X and Y are assumed to be *separable*, i.e., to have structures with countable generating families.[6] To state our theorems in this case, we need the concept of Banach class, closely related to that of Baire class. Let X be a space, $\mathfrak{A}$ a countable generating family for its structure. For each denumerable ordinal number $\alpha \geq 1$, we will define a family $Q_\alpha(\mathfrak{A})$ of Borel subsets of X; roughly $Q_\alpha(\mathfrak{A})$ consists of all those sets that can be constructed from $\mathfrak{A}$ by means of at most α operations, where each operation consists of forming a denumerable union and a complement. Thus the union of all the $Q_\alpha(\mathfrak{A})$ is precisely the structure of X. Now let $\mathfrak{B}$ be a countable generating family for Y, and for each denumerable ordinal $\alpha \geq 0$, define the *Banach class*[7] $L_\alpha(\mathfrak{A}, \mathfrak{B})$ to be the family of all functions $f: X \to Y$ such that for all $B \in Q_1(\mathfrak{B})$, $f^{-1}(B) \in Q_{\alpha+1}(\mathfrak{A})$. The union of all the Banach classes is precisely Y^X. If X and Y are separable metric spaces and Y is pathwise connected, and if $\mathfrak{A}$ and $\mathfrak{B}$ are appropriately chosen, then the Banach classes coincide with the Baire classes.[8]

Let $F \subset Y^X$. If we fix α but do not specify $\mathfrak{A}$ and $\mathfrak{B}$, then of course we can not say whether or not F is included in the αth Banach class $L_\alpha(\mathfrak{A}, \mathfrak{B})$. However, we will be interested not so much in the question of whether or not F is in a Banach class of a specified order, but rather of whether there exists a Banach class of *any* order which includes F. The answer to *this* question is independent of the choice of $\mathfrak{A}$ and $\mathfrak{B}$. In other words, if $\mathfrak{A}$ and $\mathfrak{A}'$ are countable generating families for the structure of X, and $\mathfrak{B}$ and $\mathfrak{B}'$ for that of Y,

---

[5] R_G consists of all intersections of G with members of R.

[6] As in [7], a generating family $\mathfrak{A}$ for the structure of a space X is a set of Borel subsets of X with the property that every σ-ring containing $\mathfrak{A}$ is the structure of X. The term "separable" is used by analogy with its use in topology; we will apply it indiscriminately to the space and to the structure.

[7] After the work that Banach [3] did in characterizing these families.

[8] In this case the structures of X and Y are taken to be those generated by the closed sets, and $\mathfrak{A}$ and $\mathfrak{B}$ are taken to consist of the open spheres with rational radius.

and if α is a denumerable ordinal such that $F \subset L_\alpha(\mathfrak{A}, \mathfrak{B})$, then there is a denumerable ordinal α' such that $F \subset L_{\alpha'}(\mathfrak{A}', \mathfrak{B}')$. In this case we will say that F is of *bounded* Banach class; this concept depends on X and Y only, not on any particular choice of countable generating families for their structures.

THEOREM D. *Assume that X and Y are separable. Then a necessary and sufficient condition for a subset of Y^X to be admissible is that it be of bounded Banach class.*

THEOREM E. *If X and Y are separable, then every admissible subset of Y^X has a separable admissible structure.*

A space Z and its structure are called *regular*[9] if for all $x, y \in Z$, there is a Borel set in Z containing x but not y. It is known [7] that a space is separable and regular if and only if it is isomorphic[10] to a subspace of I, where I denotes the unit interval [0, 1] with the usual Borel structure.[11]

THEOREM F. *If X and Y are separable and regular, then every admissible subset of Y^X has a separable and regular admissible structure.*

The *natural* admissible structure on a given admissible set F is defined to be the smallest admissible structure on F, if it exists. Alternatively, it may be defined to be the intersection of all the admissible structures on F, in case this is admissible. Not every admissible set need have a natural admissible structure; the counterexample, which is due to P. R. Halmos, is given in Section 7.

If $a \in X$ and $B \subset Y$, define $F(a, B) = \{f : f \in F, f(a) \in B\}$. It is not hard to prove that if B is Borel and a is arbitrary, then every admissible structure on F must contain $F(a, B)$. A "converse" would be that the structure generated by the $F(a, B)$ is admissible, and it would follow that it is also natural. This "converse" is not in general true; the best we have been able to establish is the following:

THEOREM G. *If X and Y are separable metric spaces and F contains continuous functions only, then F has a natural admissible structure, which is generated by the set of all $F(a, B)$, where B is Borel and a is arbitrary.*

We now give some applications. A space is said to have the *discrete* structure if every subset is Borel. Let J be the space consisting of 0 and 1 only, and K the space of all positive integers, both with the discrete structure. If X is an arbitrary space, then X^J and X^K are both admissible, and possess

---

[9] Mackey [7] calls this a "separated" space. We do not use this term because we wish to avoid confusion with "separable".

[10] Two spaces are *isomorphic* if there is a one-one correspondence between them that sends Borel sets into Borel sets (in both directions).

[11] Mackey [7] uses the term "countably generated" for what we call "separable and regular" spaces.

natural admissible structures which make them isomorphic to $X \times X$ and $\times_{i=1}^{\infty} X_i$ respectively, where the X_i are copies of X. In particular, J^K is admissible and has a natural admissible structure which makes it isomorphic to I. These results are relatively trivial or at least easily derivable from known results.

The situation changes when we pass to exponent spaces with nondiscrete structures. For example, J^I may be considered the set of all Borel subsets of I. It is not itself admissible. The set of all open subsets of I is admissible, as is the set of all closed subsets, the set of all G_δ, etc. In general, a subset F of J^I is admissible if and only if all members of F can be constructed from the open subsets of I by taking denumerable unions and intersections at most α times, where α is an arbitrary denumerable ordinal number (which is fixed for given F, but may differ for different F). Whether or not every admissible subset of J^I has a natural admissible structure remains an open question; but if F is admissible, then we may endow it with an admissible structure in such a way that it will be isomorphic to a subset of I.

I^I is not admissible. The set of all continuous functions from I into I is admissible; more generally, a necessary and sufficient condition that a subset F of I^I be admissible is that there exist a denumerable ordinal number α such that all members of F are of Baire class α at most. The set H of all continuous functions from I into I has a natural admissible structure; it is the Borel structure of H when considered as a metric space (in the uniform convergence topology). Again, whether or not every admissible subset of I^I has a natural admissible structure remains an open question; but if F is admissible, we may endow it with an admissible structure in such a way that it will be isomorphic to a subset of I.

Section 1 is devoted to a brief summary of terminology and to proving Theorems A, B, and C. In Section 2 we give the precise definition of Banach class and justify the remarks about these classes made above. Sections 3 and 4 are devoted to a proof of Theorem D when it is assumed that X and Y are regular as well as separable; in Section 3 we also establish the inadmissibility of J^I and I^I. In Sections 5 and 6 we prove Theorems F and G respectively. Section 7 is devoted to Halmos's counterexample. Finally, in Section 8 we prove Theorem E and remove the regularity restriction on the previous proof of Theorem D.

1. Theorems A, B, and C

We first lay down a number of conventions to which we will adhere throughout Sections 1 through 8. "Countable" and "denumerable" will mean "of cardinality at most $\aleph_0$". α, β, and γ will denote denumerable ordinal numbers, also when decorated with subscripts, primes, etc.; Ω will denote the first nondenumerable ordinal number. X, Y, and Z will denote spaces. The structures of X and Y will be denoted S and T respectively. F will be a set of Borel mappings from X into Y. In unquantified statements, the universal quantifier is to be understood. The symbol ∎ will signal the end of a proof.

In addition to the conventions laid down above, we will occasionally make use of conventions which will be valid only throughout a section or a part of it. The rule is that a convention stated within the statement or proof of a lemma is valid only until the proof is completed, and all other conventions are valid for the remainder of the section.

Let U and V be two σ-rings, not necessarily on the same abstract space. A function ψ from U into V is called a *homomorphism* if for all $A_1, A_2, \cdots$ in U we have

$$\psi(\cup_{i=1}^{\infty} A_i) = \cup_{i=1}^{\infty} \psi(A_i) \quad \text{and} \quad \psi(A_1 - A_2) = \psi(A_1) - \psi(A_2).$$

If ψ is also one-one and its inverse is a homomorphism, then it is called an *isomorphism*.[12]

LEMMA 1.1   *If $\psi : U \to V$ is a homomorphism onto, and if Γ generates U, then $\psi(\Gamma)$ generates V.*

The lemma is easily verified.

Proof of Theorem A.   Let $f \in Y^X$, and let B be a Borel subset of Y. Then $\varphi_{\{f\}}^{-1}(B) = \{f\} \times f^{-1}(B)$. Since f and B are both Borel, so is $f^{-1}(B)$. ∎

Proof of Theorem B.   The set U of all Borel subsets of $F \times X$ and the set V of all subsets of $G \times X$ of the form $C \cap (G \times X)$, where C is a Borel subset of $F \times X$, are both σ-rings. The function $\psi : U \to V$ defined by $\psi(C) = C \cap (G \times X)$ is a homomorphism onto. The set Γ of all rectangles of the form $D \times A$, where D and A are Borel subsets of F and X respectively, generates U. Hence by Lemma 1.1, $\psi(\Gamma)$ generates V. Now $\psi(\Gamma)$ is the set of all rectangles of the form $D' \times A$, where D' and A are Borel subsets of G and X respectively. In other words, $\psi(\Gamma)$ generates the structure of $G \times X$. This structure is therefore identical with V. Thus for a set to be Borel in $G \times X$, it is necessary and sufficient that it be Borel in V.

Let B be a Borel subset of Y. Then

$$\varphi_G^{-1}(B) = \{(f, x) : f \in G, x \in X, f(x) \in B\}$$
$$= \{(f, x) : f \in F, x \in X, f(x) \in B\} \cap \{(f, x) : f \in G\}$$
$$= \varphi_F^{-1}(B) \cap (G \times X).$$

Hence $\varphi_G^{-1}(B) \in V$, and hence it is a Borel subset of $G \times X$. Hence the structure R_G on G makes φ_G measurable; therefore it is admissible.

Proof of Theorem C.   If we take R to be the structure of F, then the structure on $F \times X$ is generated by all sets of the form $G \times A$, where G is a Borel subset of some F_i, and A is Borel in X. Hence every set that is Borel in $F_i \times X$ is also Borel in $F \times X$. Now let B be a Borel subset of Y. Then

---

[12] This is a σ-ring isomorphism, not to be confused with a space isomorphism.

$$\varphi_F^{-1}(B) = \{(f, x): f \in F, f(x) \in B\}$$
$$= \bigcup_{i=1}^{\infty} \{(f, x): f \in F_i, f(x) \in B\}$$
$$= \bigcup_{i=1}^{\infty} \varphi_{F_i}^{-1}(B);$$

but since by hypothesis each of the $\varphi_{F_i}^{-1}(B)$ is Borel in $F_i \times X$, it follows that $\varphi_F^{-1}(B)$ is Borel in $F \times X$. Hence the structure R makes φ_F a Borel mapping, and therefore it is admissible.

COROLLARY 1.2. *Every denumerable subset of Y^X is admissible.*

2. Banach classes

The definition of Banach class was sketched in the introduction. In this section we give the precise definition, and establish the less obvious properties of Banach classes. It will be assumed that X and Y are separable.

Let $\mathfrak{A}$ be an arbitrary family of Borel subsets of X. For each denumerable ordinal $\alpha \geq 1$, we define $P_\alpha(\mathfrak{A})$ and $Q_\alpha(\mathfrak{A})$ inductively as follows: $Q_1(\mathfrak{A})$ consists of all denumerable unions of members of $\mathfrak{A}$, and $P_1(\mathfrak{A})$ consists of all complements of members of $Q_1(\mathfrak{A})$; supposing $Q_\beta(\mathfrak{A})$ and $P_\beta(\mathfrak{A})$ to have been defined for all $\beta < \alpha$, we define

$$Q_\alpha(\mathfrak{A}) = Q_1(\bigcup_{\beta<\alpha} P_\beta(\mathfrak{A})) \quad \text{and} \quad P_\alpha(\mathfrak{A}) = P_1(\bigcup_{\beta<\alpha} P_\beta(\mathfrak{A})).$$

$Q_\alpha(\mathfrak{A}) \cup P_\alpha(\mathfrak{A})$ is the set of all subsets of X which can be "reached from $\mathfrak{A}$" by performing at most α operations, where each operation consists of forming a denumerable union and a complement; if $\mathfrak{A}$ generates the structure of X, then the union (over α) of all the $Q_\alpha(\mathfrak{A})$ (or of the $P_\alpha(\mathfrak{A})$) is the set of all Borel subsets of X. If $\mathfrak{B}$ is a family of Borel subsets of Y, we may define $P_\alpha(\mathfrak{B})$ and $Q_\alpha(\mathfrak{B})$ in a similar manner.

For the remainder of this section, let $\mathfrak{A}$ and $\mathfrak{B}$ denote countable generating families for the structures S of X and T of Y respectively. For each denumerable ordinal $\alpha \geq 0$, we define $L_\alpha(\mathfrak{A}, \mathfrak{B})$ to be the set of all functions $f: X \to Y$ such that for all $B \in Q_1(\mathfrak{B})$, we have $f^{-1}(B) \in Q_{\alpha+1}(\mathfrak{A})$.

LEMMA 2.1. $Y^X = \bigcup_{\alpha<\Omega} L_\alpha(\mathfrak{A}, \mathfrak{B})$.

This lemma follows without difficulty from the following lemma, by setting $Z = X$.

LEMMA 2.2. *A necessary and sufficient condition that a mapping $f: Z \to Y$ be Borel is that for every $B \in \mathfrak{B}, f^{-1}(B)$ is Borel in Z.*

Proof. Necessity is obvious. To prove sufficiency, let U be the set of all subsets B of Y such that $f^{-1}(B)$ is Borel. U includes $\mathfrak{B}$ and is a σ-ring; hence $U \supset T$. ∎

Let $F \subset Y^X$. We shall say that F is of *bounded Banach class w.r.t.* $(\mathfrak{A}, \mathfrak{B})$ if there is an α such that $F \subset L_\alpha(\mathfrak{A}, \mathfrak{B})$.

LEMMA 2.3. *The concept of bounded Banach class is independent of the choice of countable generating families. In other words, if $\mathfrak{A}'$ and $\mathfrak{B}'$ are any other countable generating families for S and T respectively, and if F is of bounded Banach class w.r.t. $(\mathfrak{A}, \mathfrak{B})$, then it is also of bounded Banach class w.r.t. $(\mathfrak{A}', \mathfrak{B}')$.*

Because of Lemma 2.3, we can speak of F being of *bounded Banach class* without referring to the generating families.

We supply only the idea of the proof; the details may be filled in by the reader. Suppose $f \in L_\alpha(\mathfrak{A}, \mathfrak{B})$. Each member of $\mathfrak{B}'$ is Borel, and so can be "reached" in denumerably many steps from $\mathfrak{B}$; since $\mathfrak{B}'$ is denumerable, there is a denumerable upper bound on the number of steps needed. It follows that there is also such an upper bound, say γ, if we start out with $B' \in Q_1(\mathfrak{B}')$; that is, B' can then be reached in at most γ steps from $\mathfrak{B}$ (independent of the choice of a particular B'). Hence $f^{-1}(B')$ can be reached in γ steps from sets of the form $f^{-1}(B)$, where $B \in \mathfrak{B}$. But all such $f^{-1}(B)$ can be reached in $\alpha + 1$ steps from $\mathfrak{A}$; so $f^{-1}(B')$ can be reached in $\alpha + 1 + \gamma$ steps from $\mathfrak{A}$. Now each member of $\mathfrak{A}$ can be reached in denumerably many steps from $\mathfrak{A}'$, and again $\mathfrak{A}$ has only denumerably many members; so there is an upper bound on the number of steps necessary. Adding this upper bound to $\alpha + 1 + \gamma$, we obtain a denumerable α' such that

$$f^{-1}(B') \in Q_{\alpha'+1}(\mathfrak{A}').$$

Since α' is independent of the choice of B' and f, it follows that $f \in L_{\alpha'}(\mathfrak{A}', \mathfrak{B}')$ and $L_\alpha(\mathfrak{A}, \mathfrak{B}) \subset L_{\alpha'}(\mathfrak{A}', \mathfrak{B}')$. ∎

We end this section with the following lemma, whose proof may be supplied by the reader:

LEMMA 2.4. *If $\alpha + 1 < \beta$, then $Q_\alpha(\mathfrak{A}) \subset Q_\beta(\mathfrak{A})$.*

3. Theorem D: Necessity

We assume throughout this section that X and Y are separable, and fix denumerable generating families $\mathfrak{A}$ and $\mathfrak{B}$ for their respective structures. Z will be an arbitrary Borel space. We will prove that every admissible subset of Y^X is of bounded Banach class.

The principal tool in the proof is Lemma 3.2, which says that if B is a Borel subset of the cartesian product $X \times Z$, then there is an α (depending on B) such that every Z-section[13] of B may be constructed in at most α steps from $\mathfrak{A}$. The idea of the proof is that since B is Borel, there must be an α such that B may be constructed in at most α steps from rectangles in $X \times Z$ whose X-factors are in $\mathfrak{A}$. If we copy this construction step by step, but within a given Z-section of $X \times Z$, then we obtain the desired construction of the corresponding Z-section of B.

To illustrate the general necessity proof, we use Lemma 3.2 to show that

---

[13] A section parallel to the X-axis.

J^I is not admissible. If it were, then $\varphi^{-1}(1)$ would be Borel in $J^I \times I$, and so by Lemma 3.2, all the J^I-sections of $\varphi^{-1}(1)$ would be constructible in at most α steps from $\mathfrak{A}$, where α is fixed. Such a section consists of the set of all $x \in I$ such that $f(x) = 1$, where f is an arbitrary but fixed characteristic function; in other words, *any* Borel set may be represented as such a section, and so would be constructible in α steps from $\mathfrak{A}$, where α is fixed. Now this is known to be impossible if, for example, we take $\mathfrak{A}$ to be a denumerable basis for the open sets of I (cf. [6], p. 207). ∎

J^I may be considered a subset of I^I; so by Theorem B, I^I is not admissible either.

We now give the formal proofs. Write $Q_\alpha(\mathfrak{A}) = Q_\alpha$, $P_\alpha(\mathfrak{A}) = P_\alpha$, $L_\alpha(\mathfrak{A}, \mathfrak{B}) = L_\alpha$. If $B \subset X \times Z$, let B^z denote the Z-section $\{x \in X : (x, z) \in B\}$ of B. For each denumerable α, let α'' denote the largest limit ordinal no larger than α, i.e., the smallest $\beta \leq \alpha$ such that $-\beta + \alpha$ is finite.[14] Let α' be the "finite tail" of α, i.e., $\alpha' = -\alpha'' + \alpha$.

LEMMA 3.1. *$\beta < \gamma$ implies $\beta + \beta' + 1 < \gamma + \gamma'$.*

Proof. If $\beta < \gamma''$, then since γ'' is a limit ordinal and β' is finite, it follows that $\beta + \beta' + 1 < \gamma'' \leq \gamma \leq \gamma + \gamma'$. If $\beta \geq \gamma''$, then $\beta' < \gamma'$, and hence $2\beta' + 1 < 2\gamma'$ (β' and γ' being finite). Hence

$$\beta + \beta' + 1 = (\beta - \beta') + 2\beta' + 1$$
$$= (\gamma - \gamma') + 2\beta' + 1 < \gamma - \gamma' + 2\gamma' = \gamma + \gamma'. \quad \blacksquare$$

LEMMA 3.2. *Let B be a Borel subset of $X \times Z$. Then there is an α such that every Z-section of B is in Q_α.*

Proof. We define N_α and M_α for each α as follows: N_1 is the set of all sets of the form $A \times C$, where $A \in Q_1$ and C is a Borel subset of Z; $M_1 = \{D_1 - D_2 : D_1, D_2 \in N_1\}$. Suppose N_β and M_β have been defined for all $\beta < \alpha$; define $N_\alpha = Q_1(\bigcup_{\beta < \alpha} M_\beta)$ and $M_\alpha = \{D_1 - D_2 : D_1, D_2 \in N_\alpha\}$.

We prove by transfinite induction on γ that $D \in N_\gamma$ implies that every Z-section of D is in $Q_{\gamma+\gamma'}$. For $\gamma = 1$ this follows from Lemma 3.1. Suppose we have shown it for all $\beta < \gamma$. Then if $\beta < \gamma$ and $D \in M_\beta$, we have $D = D_1 - D_2$, where $D_1, D_2 \in N_\beta$. Hence for arbitrary $z \in Z$, we have by induction hypothesis that $D_1^z, D_2^z \in Q_{\beta+\beta'}$. Hence $X - D_1^z \in P_{\beta+\beta'}$. But D_2^z, as a member of $Q_{\beta+\beta'}$, is of the form $\bigcup_{i=1}^{\infty} A_i$, where

$$A_i \in \bigcup_{\alpha < \beta+\beta'} P_\alpha \subset \bigcup_{\alpha < \beta+\beta'+1} P_\alpha \qquad \text{(for } i = 1, 2, \cdots\text{)}.$$

Setting $A_0 = X - D_1^z$, we obtain $(X - D_1^z) \cup D_2^z = \bigcup_{i=0}^{\infty} A_i$, where again $A_i \in \bigcup_{\alpha < \beta+\beta'+1} P_\alpha$ (for $i = 0, 1, 2, \cdots$). Hence $(X - D_1^z) \cup D_2^z \in Q_{\beta+\beta'+1}$, and hence

$$D^z = (D_1 - D_2)^z = D_1^z - D_2^z = X - ((X - D_1^z) \cup D_2^z) \in P_{\beta+\beta'+1}.$$

---

[14] $\alpha'' = 0$ if α is finite.

Now suppose $D \in N_\gamma$; then $D = \bigcup_{i=1}^\infty D_i$, where $D_i \in \bigcup_{\beta<\gamma} M_\beta$. Hence $D_i^z \in \bigcup_{\beta<\gamma} P_{\beta+\beta'+1} \subset \bigcup_{\alpha<\gamma+\gamma'} P_\alpha$, the last inclusion being a consequence of Lemma 3.1. Hence $D^z = \bigcup_{i=1}^\infty D_i^z \in Q_{\gamma+\gamma'}$. This completes the induction.

N_1 generates the Borel structure of $X \times Z$, and hence the Borel structure of $X \times Z$ is $\bigcup_{\alpha<\Omega} N_\alpha$; hence for some γ, $B \in N_\gamma$. Hence every Z-section of B is in $Q_{\gamma+\gamma'}$. ∎

To avoid confusion in the sequel, we now make the following remarks concerning notation. We are concerned with *elements* (e.g. members x of X); with *sets* (e.g. subsets A of X); and with *families* of sets (e.g. $Q_1(\mathfrak{A})$). Functions $f: X \to Y$ are defined in the first instance on elements only; but the definition can be extended to sets in the usual way, by writing $f(A) = \{f(x): x \in A\} =$ image of A under f. We will go one step further, and extend the definition of functions that were originally defined on elements to families as well. This is done in the natural way, by writing $f(\mathfrak{C}) = \{f(A): A \in \mathfrak{C}\}$ (for an arbitrary family $\mathfrak{C}$). Thus $f(x)$ is an element, $f(A)$ a set, and $f(\mathfrak{A})$ a family. Similar remarks hold for the inverse function; we will write $f^{-1}(\mathfrak{D})$ for $\{f^{-1}(B): B \in \mathfrak{D}\}$ (where $\mathfrak{D}$ is an arbitrary family). In this notation, for example, L_α can be defined as $\{f \in Y^X: f^{-1}(Q_1(\mathfrak{B})) \subset Q_{\alpha+1}\}$. In which sense f or f^{-1} is meant in a particular case will always be clear from the context.

Note that if $f: X \to Y$ is a Borel mapping, then $f^{-1}: T \to S$ is a homomorphism (see §1). Hence

(3.3) $\qquad f^{-1}(Q_1(\mathfrak{B})) \subset Q_1(f^{-1}(\mathfrak{B}))$.

LEMMA 3.4. *If F is an admissible subset of Y^X, then there is a γ such that $F \subset L_\gamma$.*

Proof. Impose an admissible structure U on F. Let $\mathfrak{B} = \{B_1, B_2, \cdots\}$. Then since B_j is Borel, $\varphi_F^{-1}(B_j)$ must be a Borel subset of $F \times X$. Hence by Lemma 3.2, there are α_j such that all F-sections of $\varphi_F^{-1}(B_j)$ are in Q_{α_j}. An F-section of $\varphi_F^{-1}(B_j)$ has the form $\{x: x \in X, f(x) \in B_j\}$, where $f \in F$; i.e., it has the form $f^{-1}(B_j)$. It follows that for all $f \in F$, $f^{-1}(B_j) \in Q_{\alpha_j}$. Let $\alpha = \sup \alpha_j + 1$; then by Lemma 2.4, $f^{-1}(B_j) \in Q_\alpha$ for all j, i.e., $f^{-1}(\mathfrak{B}) \subset Q_\alpha$. Hence

$$f^{-1}(Q_1(\mathfrak{B})) \subset Q_1(f^{-1}(\mathfrak{B})) \subset Q_1(Q_\alpha) \subset Q_1(P_{\alpha+1}) \subset Q_{\alpha+2},$$

where the first inclusion follows from (3.3) and the two last from the definitions of P_α and Q_α. Comparing the first and last members of this chain of inclusions, we deduce $f \in L_{\alpha+1}$. ∎

4. Theorem D: Sufficiency in the regular case

In this section we will assume that X and Y are regular as well as separable, and will prove that every subset of Y^X which is of bounded Banach class is admissible. We will retain the notation of the previous section, but choose $\mathfrak{A}$ and $\mathfrak{B}$ in a particular way, which we describe in the next paragraph (by Lemma 2.3, no loss of generality is involved). It is of course sufficient to prove that every L_γ is admissible.

Since X and Y are separable and regular, we may assume without loss of generality that they are subspaces of the unit interval I. Now as well as being considered a Borel space with the usual structure, I may also be considered a topological space with the usual topology.[15] This topology induces relative subset topologies on X and on Y, which we will call the natural topologies on X and on Y. We choose $\mathfrak{A}$ and $\mathfrak{B}$ to be denumerable bases for the natural topologies of X and Y respectively; then $\mathfrak{A}$ and $\mathfrak{B}$ also generate the Borel structures of X and Y. Note that the choice of particular denumerable bases for the natural topologies of X and Y does not affect the values of the Q_α and the P_α. In the sequel, references to topological concepts like continuity and the related Baire functions are to be understood as referring to the natural topologies. Note that Q_1 is the set of open sets of X, $Q_1(\mathfrak{B})$ is the set of open sets of Y, and L_0 is the set of continuous functions from X into Y.

LEMMA 4.1. *If $Y = I$ and $\gamma \geq 1$, then for every $f \in L_\gamma$, there is a sequence $f_1, f_2, \cdots$ of members of $\cup_{\beta<\gamma} L_\beta$ such that[16] $f = \lim_{i\to\infty} f_i$.*

Proof. It is sufficient to prove that when $Y = I$, the L_γ coincide with the Baire classes of order γ. This is known to be the case (cf. [3], p. 284, also [6], p. 294). ∎

If $Y = I$, then with every $f \in Y^X$, we may associate an infinite sequence $f_1, f_2, \cdots$ of members of Y^X as follows: By Lemma 2.1, for every $f \in Y^X$ there is a unique β such that $f \in L_\beta - \cup_{\alpha<\beta} L_\alpha$; we construct the f_i in such a way that $f = \lim_{i\to\infty} f_i$, and $f_i \in \cup_{\alpha<\beta} L_\alpha$. When $\beta = 0$, define $f_i = f$ for all i. The construction is possible by Lemma 4.1; however, it is not unique. Throughout the remainder of this section, we will consider the f_i as fixed; in other words, with each $f \in Y^X$ we associate a unique sequence $f_1, f_2, \cdots$.

Let $\lambda = \{\lambda_1, \cdots, \lambda_k\}$ be a finite sequence of positive integers. We will denote the sequence $\{\lambda_1, \cdots, \lambda_k, n\}$ by (λ, n). We will also use the notation f_λ instead of $f_{\lambda_1\cdots\lambda_k}$; thus $f_{\lambda n} = f_{(\lambda,n)}$. The empty sequence of integers will be denoted by $\varnothing$; we define $f_\varnothing = f$.

Let $\mathfrak{A} = \{A_1, A_2, \cdots\}$ and $\mathfrak{B} = \{B_1, B_2, \cdots\}$. If i and j are positive integers, define

$$E_\gamma(\lambda, i, j) = \{f : f_\lambda(A_i) \subset B_j, f_\lambda \in L_0, f \in L_\gamma\}.$$

Define R_γ to be the σ-ring generated by all sets of the form $E_\gamma(\lambda, i, j)$, where λ ranges over all finite sequences of positive integers, and i and j range over all positive integers.

LEMMA 4.2. *If $Y = I$, then R_γ is an admissible structure on L_γ.*

Proof. Impose the structure R_γ on L_γ; we wish to prove that it is ad-

---

[15] That generated by the open intervals.
[16] Pointwise.

missible. Define
$$C_\gamma(j, \lambda, \beta) = \{(f, x) : f_\lambda(x) \in B_j, f \in L_\gamma, f_\lambda \in L_\beta\}.$$
We will prove

(4.3) *For each j, each[17] λ, and each $\beta \leq \gamma$, $C_\gamma(j, \lambda, \beta)$ is Borel in $L_\gamma \times X$.*

The proof is by induction on β. If $\beta = 0$, we have
$$C_\gamma(j, \lambda, 0) = \{(f, x) : f_\lambda(x) \in B_j, f \in L_\gamma, f_\lambda \text{ is continuous}\}$$
$$= \{(f, x) : (\exists i)(x \in A_i, f_\lambda(A_i) \subset B_j, f \in L_\gamma, f_\lambda \in L_0)\}$$
$$= \bigcup_{i=1}^\infty \{(f, x) : f_\lambda(A_i) \subset B_j, f \in L_\gamma, f_\lambda \in L_0, x \in A_i\}$$
$$= \bigcup_{i=1}^\infty E_\gamma(\lambda, i, j) \times A_i$$

(the second equality follows from the openness of B_j). The last expression is a denumerable union of Borel rectangles in $L_\gamma \times X$, and is therefore a Borel set. Now suppose that (4.3) has been proved for all $\alpha < \beta$. For each j, let $B_{k(j,1)}$, $B_{k(j,2)}$, $\cdots$ be a sequence of basic open sets, each of whose closures is contained in B_j and for which $\bigcup_{i=1}^\infty B_{k(j,i)} = B_j$. Suppose $b_1, b_2, \cdots$ is a convergent sequence of points in Y. If $\lim_{n \to \infty} b_n \in B_j$, then there must be an i such that $\lim_{n \to \infty} b_n \in B_{k(j,i)}$, i.e.,

(4.4) $(\exists i)(\exists N)(\text{for all } n \geq N)(b_n \in B_{k(j,i)}).$

Conversely, if (4.4) holds, then $\lim_{n \to \infty} b_n$ must be in the closure of $B_{k(j,i)}$, and hence in B_j. Therefore
$$C_\gamma(j, \lambda, \beta) = \{(f, x) : f_\lambda(x) \in B_j, f \in L_\gamma, f_\lambda \in L_\beta\}$$
$$= \{(f, x) : \lim_{n \to \infty} f_{\lambda n}(x) \in B_j, f \in L_\gamma, f_\lambda \in L_\beta\}$$
$$= \{(f, x) : (\exists i)(\exists N)(\text{for all } n \geq N)(f_{\lambda n}(x) \in B_{k(j,i)}, f \in L_\gamma, f_\lambda \in L_\beta)\}$$
$$= \bigcup_{i=1}^\infty \bigcup_{N=1}^\infty \{(f, x) : (\text{for all } n \geq N)(f_{\lambda n}(x) \in B_{k(j,i)}, f \in L_\gamma), f_\lambda \in L_\beta\}$$
$$= \bigcup_{i=1}^\infty \bigcup_{N=1}^\infty \{(f, x) : (\text{for all } n \geq N)(f_{\lambda n}(x) \in B_{k(j,i)}, f \in L_\gamma), \text{ and}$$
$$(\text{for all } n \geq N)(\exists \alpha)(\alpha < \beta, f_{\lambda n} \in L_\alpha)\}$$
$$= \bigcup_{i=1}^\infty \bigcup_{N=1}^\infty \{(f, x) : (\text{for all } n \geq N)(\exists \alpha)$$
$$(\alpha < \beta, f_{\lambda n}(x) \in B_{k(j,i)}, f \in L_\gamma, f_{\lambda n} \in L_\alpha)\}$$
$$= \bigcup_{i=1}^\infty \bigcup_{N=1}^\infty \bigcap_{n=N}^\infty \bigcup_{\alpha<\beta} \{(f, x) : f_{\lambda n}(x) \in B_{k(j,i)}, f \in L_\gamma, f_{\lambda n} \in L_\alpha\}$$
$$= \bigcup_{i=1}^\infty \bigcup_{N=1}^\infty \bigcap_{n=N}^\infty \bigcup_{\alpha<\beta} C_\gamma(k(j, i), (\lambda, n), \alpha).$$

The set $C(k(j, i), (\lambda, n), \alpha)$ is a Borel subset of $L_\gamma \times X$ by induction hypothesis, and $\bigcup_{\alpha<\beta}$ is a denumerable union; hence the last expression obtained is Borel in $L_\gamma \times X$. This completes the inductive proof of (4.3).

---

[17] Including the empty one.

We now have

$$\varphi_{L_\gamma}^{-1}(B_j) = \{(f, x) : \varphi_{L_\gamma}(f, x) \in B_j, f \in L_\gamma\}$$
$$= \{(f, x) : f(x) \in B_j, f \in L_\gamma\}$$
$$= \{(f, x) : f_\varnothing(x) \in B_j, f \in L_\gamma, f_\varnothing \in L_\gamma\}$$
$$= C_\gamma(j, \varnothing, \gamma).$$

Hence by (4.3) and Lemma 2.2, φ_{L_γ} is Borel. ∎

Lemma 4.2 establishes the admissibility of L_γ when $Y = I$. In the general case, when Y is merely a subset of I, we may consider Y^X as a subset of I^X. Let $\mathfrak{B}'$ be a basis for the open sets of I, such that $\mathfrak{B}$ consists precisely of the intersections of Y with members of $\mathfrak{B}'$. Set $L_\gamma^I = L_\gamma(\mathfrak{A}, \mathfrak{B}')$; L_γ^I is admissible by Lemma 4.2. Now it may be established that $L_\gamma = L_\gamma^I \cap Y^X$; hence L_γ is also admissible. (Note that the admissibility of a set F is a property of F alone, and not of the function space in which F happens to be imbedded. Here $L_\gamma \subset Y^X \subset I^X$, and we have established the admissibility of L_γ in I^X; but this is no different from its admissibility in the intermediate space Y^X.)

We have established

LEMMA 4.5. *L_γ is admissible.*

5. Theorem F

We retain the conventions of Sections 3 and 4; in particular it is assumed that X and Y are separable and regular. We wish to prove that every admissible subset of Y^X has a separable and regular admissible structure. For this it suffices to show that L_γ has such a structure.

As in the previous section, we first assume $Y = I$.

LEMMA 5.1. *If $Y = I$, then the structure R_γ on L_γ is separable and regular.*

Proof. Separability is immediate. To prove regularity let f and g be distinct members of L_γ. There must be an i_1 such that $f_{i_1} \neq g_{i_1}$, for otherwise $f = \lim_i f_i = \lim_i g_i = g$. Suppose we have defined $i_1, \cdots, i_k$ so that $f_{i_1 \cdots i_k} \neq g_{i_1 \cdots i_k}$; then we can define i_{k+1} so that $f_{i_1 \cdots i_{k+1}} \neq g_{i_1 \cdots i_{k+1}}$. We thus obtain an infinite sequence $\{i_1, i_2, \cdots\}$. For each k, let α_k and β_k be such that

$$f_{i_1 \cdots i_k} \in L_{\alpha_k} - \bigcup_{\alpha < \alpha_k} L_\alpha \quad \text{and} \quad g_{i_1 \cdots i_k} \in L_{\beta_k} - \bigcup_{\beta < \beta_k} L_\beta.$$

$\{\alpha_1, \alpha_2, \cdots\}$ and $\{\beta_1, \beta_2, \cdots\}$ are both strictly decreasing sequences of ordinal numbers, and therefore they both terminate; that is, there are j and k such that $\alpha_i = 0$ for all $i \geq j$ and $\beta_i = 0$ for all $i \geq k$. Let $m = \max(j, k)$, and let $\lambda = \{i_1, \cdots, i_m\}$. Then $g_\lambda \in L_0$, $f_\lambda \in L_0$, and $g_\lambda \neq f_\lambda$; in other words, both f_λ and g_λ are continuous, and there is an $x \in X$ such that $g_\lambda(x) \neq f_\lambda(x)$. Hence there are disjoint basic open sets B_j and B_k such that $f_\lambda(x) \in B_j$ and

$g_\lambda(x) \in B_k$. From the continuity of f and g it now follows that there is an $A_i \in \mathfrak{A}$ such that $x \in A_i$, $f_\lambda(A_i) \subset B_j$, and $g_\lambda(A_i) \subset B_k$. Hence

$$f \in E_\gamma(\lambda, i, j), \quad \text{and} \quad g \in E_\gamma(\lambda, i, k);$$

but since B_j and B_k are disjoint, so are $E_\gamma(\lambda, i, j)$ and $E_\gamma(\lambda, i, k)$. ∎

We now drop the assumption $Y = I$ and assume only $Y \subset I$. We proceed as in the previous section, and construct R'_γ from $\mathfrak{B}'$ in the same way that R_γ was constructed from $\mathfrak{B}$ in the case $Y = I$. We have just seen that R'_γ is a separable and regular admissible structure for L^I_γ. Since

$$L_\gamma = L^I_\gamma \cap Y^X \subset L^I,$$

we may apply Theorem B, and deduce that $(R'_\gamma)_{L_\gamma}$ is a separable and regular admissible structure on L_γ. This completes the proof of Theorem F.

6. Theorem G

Let $\mathfrak{A}$ and $\mathfrak{B}$ be as in footnote 8; set $\mathfrak{A} = \{A_1, A_2, \cdots\}$, $\mathfrak{B} = \{B_1, B_2, \cdots\}$. For each j, let $B_{k(j,1)}$, $B_{k(j,2)}$, $\cdots$ be a sequence of members of $\mathfrak{B}$ each of whose closures is contained in B_j and for which $\bigcup_{m=1}^\infty B_{k(j,m)} = B_j$. For each i, let $\{a_{i1}, a_{i2}, \cdots\}$ be a denumerable dense set in A_i. Suppose

$$f: X \to Y$$

is continuous. If $f(x) \in B_j$, then there is an m for which $f(x) \in B_{k(j,m)}$. By the continuity of f, there is an i such that $x \in A_i$ and $f(A_i) \subset B_{k(j,m)}$; hence

(6.1) $\quad (\exists m)(\exists i)(\forall n)(x \in A_i, f(a_{in}) \in B_{k(j,m)})$.

Conversely, assume (6.1); since the a_{in} are dense in A_i and f is continuous, it follows that $f(A_i)$ is included in the closure of $B_{k(j,m)}$, which in turn is a subset of B_j. Hence $f(x) \in f(A_i) \subset B_j$. Thus we have shown that for continuous f, $f(x) \in B_j$ is equivalent to (6.1). Hence if F contains continuous functions only, then

$$\varphi_F^{-1}(B_j) = \{(f, x) : f(x) \in B_j, f \in F\}$$
$$= \{(f, x) : (\exists m)(\exists i)(\forall n)(x \in A_i, f(a_{in}) \in B_{k(j,m)}, f \in F)\}$$
$$= \bigcup_{m=1}^\infty \bigcup_{i=1}^\infty \bigcap_{n=1}^\infty F(a_{in}, B_{k(j,m)}) \times A_i.$$

If we impose on F the structure R generated by all the $F(a, B)$, then the last expression obtained is Borel in $F \times X$. Hence by Lemma 2.2, R is admissible. But by the remarks preceding the statement of Theorem G in the introduction, all admissible structures on F must contain R; hence R is a natural admissible structure. This completes the proof of Theorem G.

The statement that the natural structure on the set of all continuous members of I^I is the same as that induced by the uniform convergence topology is a consequence of Theorem G.

7. An admissible set without a natural admissible structure[18]

LEMMA 7.1. *Under the continuum hypothesis, there exists a one-one mapping θ of I onto itself such that $\theta = \theta^{-1}$ and such that if A is an uncountable Borel subset of I with an uncountable complement, then $\theta(A)$ is not Borel.*

Proof. Let $\{A_\alpha : \alpha < \Omega\}$ be a well-ordering of the Borel sets referred to above. Begin the definition of θ by interchanging some point of A_0 with some point of $I - A_0$. At stage α the mapping θ is already defined at a countable set. For each $\beta \leq \alpha$ find an unused point in A_α, and let θ interchange it with an unused point in $I - A_\beta$. When the induction is finished, let θ be the identity at all points not yet used. Then $\theta(A_\alpha) \neq A_\beta$ whenever $\beta \leq \alpha$, and hence $\theta(A_\alpha)$ is never measurable in I. ∎

Let[19] U be the set of all Borel subsets of I, $V = \theta(U)$, W the unit interval $[0, 1]$, S the structure on W generated by $U \cup V$, and D the *diagonal* of $W \times W$, i.e., the set $\{(x, x) : x \in W\}$.

LEMMA 7.2. *The σ-ring $(U \cap V) \times S$ on $W \times W$ does not contain D.*

Proof. Set $T = U \cap V$. By Lemma 7.1, for each set A in T, either A or $W - A$ is countable; let $q(A)$ be the countable one. If $C \in T \times S$, then every "horizontal" section C^x of C belongs to T. Write C^* for the union of all the sets $q(C^x)$, and let M be the class of all those sets C in $T \times S$ for which C^* is countable. M is a σ-ring included in $T \times S$ and containing all rectangles in $T \times S$; hence $M = T \times S$. Since M does not contain the diagonal, the proof of the lemma is complete.

LEMMA 7.3. *Both $U \times S$ and $V \times S$ contain D.*

Proof. $D \in U \times S$ follows from $D \in U \times U$. Next, $\theta(S)$ is a σ-ring and includes $\theta(U)$ and $\theta(V)$, i.e., V and U; hence $\theta(S) \supset S$. Define $\theta \times \theta : W \times W \to W \times W$ by $(\theta \times \theta)(x, y) = (\theta(x), \theta(y))$. Then

$$(\theta \times \theta)(U \times S) = \theta(U) \times \theta(S) \subset V \times S.$$

But since θ is onto, $(\theta \times \theta)(D) = D$. Hence from $D \in U \times S$ we obtain $D = (\theta \times \theta)(D) \in (\theta \times \theta)(U \times S) \subset V \times S$. ∎

Let X be the space whose underlying abstract space is W and whose structure is S. For each $t \in W$, define $f_t \in J^X$ by $f_t(s) = 0$ for $t \neq s$, $f_t(t) = 1$. Define $F \subset J^X$ by $F = \{f_t : t \in W\}$. Let $\lambda : W \to F$ be the natural map, defined by $\lambda(t) = f_t$, and let e be the identity on W. Define $D^* \subset F \times X$ by $D^* = (\lambda \times e)(D)$. Since $\varphi_F^{-1}(J) = \{D^*, F \times X - D^*\}$, it follows that a necessary and sufficient condition for a structure R on F to be admissible is that $R \times S$ contains D^*. Hence from Lemma 7.3 it follows that $\lambda(U)$ and $\lambda(V)$ are admissible, whereas it follows from Lemma 7.2 that $\lambda(U) \cap \lambda(V)$ (which is the same as $\lambda(U \cap V)$) is not admissible. Hence the intersection

---

[18] I am indebted to Prof. P. R. Halmos, who supplied the substance of this section.
[19] See the remarks preceding Lemma 3.4.

of all admissible sets on F is not admissible, and therefore F has no natural admissible structure.

X is separable and regular, and may therefore be considered a subset of I; however, it seems unlikely that X is isomorphic to I itself (note that the example depends on the continuum hypothesis). The question remains open as to whether the admissible subsets of such sets as J^I or I^I have natural admissible structures.

8. Dropping the regularity assumption: Theorems D and E

Our principal tool in this section is the idea of structure-preserving mapping. Suppose X and X_* are spaces with structures S and S_* respectively. A mapping $\pi: X \to X_*$ is called *structure-preserving* if the mapping it induces[19] on S is an isomorphism onto S_*. Given spaces X and Y which need not be regular, our procedure will be to construct spaces X_* and Y_* which *are* regular, and are connected to X and Y by means of structure-preserving (onto) mappings $\pi_X: X \to X_*$ and $\pi_Y: Y \to Y_*$. We will then deduce the desired theorems for X and Y from the corresponding theorems for X_* and Y_*.

Any space X may be divided into equivalence classes by means of the following relation: $x \equiv y$ if and only if every Borel set that contains x also contains y. X is regular if and only if each of these equivalence classes contains exactly one point. In the general case, X_* is defined to be the space of the equivalence classes in X, with the identification structure; and π_X is the identification mapping. That X_* is regular and that π_X is structure-preserving is not difficult to verify. Separability carries over from X to X_*.

Let F be a set of Borel mappings from X into Y. Any member f of F induces a unique Borel mapping $f_*: X_* \to Y_*$ for which the diagram

(8.1)
$$\begin{array}{ccc} X & \xrightarrow{f} & Y \\ \pi_X \downarrow & & \downarrow \pi_Y \\ X_* & \xrightarrow{f_*} & Y_* \end{array}$$

is commutative. Let us denote by F_* the set of all f_* induced in this way by members f of F, and by π_F the function from F to F_* defined by

$$\pi_F(f) = f_*.$$

The crucial step is the proof of

LEMMA 8.2. *F is admissible if and only if F_* is admissible.*

Proof. The "if" half is the easier one. Let R_* be an admissible structure on F_*. R_* induces a unique structure R on F for which π_F is structure-preserving. If we impose R_* on F_* and R on F, then in the commutative diagram

$$\begin{array}{ccc} F \times X & \xrightarrow{\varphi_F} & Y \\ \pi_F \times \pi_X \downarrow & & \downarrow \pi_Y \\ F_* \times X_* & \xrightarrow{\varphi_{F_*}} & Y_* \end{array}$$

$\pi_F \times \pi_X$ will be structure-preserving. The fact that φ_F is a Borel mapping then follows from the corresponding fact for φ_{F_*}, and from the fact that the vertical mappings are structure-preserving.

The "only if" half is slightly trickier, because if we start out with an admissible structure R on F, then there may be no structure R_* on F_* for which π_F is structure-preserving. We get around this difficulty by defining a subset G of F so that $\pi_F(G) = F_*$ and $\pi_F \mid G$ (which is the same as π_G) is one-one. The structure R_G is admissible on G by Theorem B, and induces a unique structure R_* on F_* for which π_G is structure-preserving. The remainder of the proof follows as before from the commutativity of the diagram

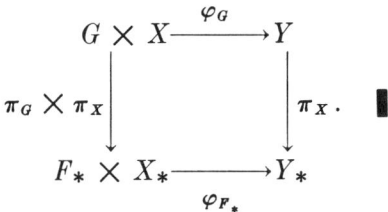

To prove Theorem E, suppose F to be an admissible subset of Y^X, where X and Y are separable but need not be regular. Then F_* is an admissible subset of $Y_*^{X_*}$, and hence by Theorem F has a separable and regular admissible structure, which we will call R_*. But then the structure R induced[20] by R_* on F will be separable. ∎

It remains only to establish sufficiency in Theorem D. Fix denumerable generating families $\mathfrak{A}$ and $\mathfrak{B}$ for the structures of X and Y respectively, and set $\mathfrak{A}_* = \pi_X(\mathfrak{A})$, $\mathfrak{B}_* = \pi_X(\mathfrak{B})$; $\mathfrak{A}_*$ and $\mathfrak{B}_*$ are denumerable generating families for the separable and regular spaces X_* and Y_*. From the commutativity of diagram (8.1) and the fact that π_X and π_Y are structure-preserving, it follows that

(8.3) $$L_\gamma(\mathfrak{A}, \mathfrak{B})_* = L_\gamma(\mathfrak{A}_*, \mathfrak{B}_*).$$

Since X_* and Y_* are separable and regular, we may apply Theorem D (cf. Section 4), and deduce that the right side of (8.3) is admissible. But then

---

[20] See the proof of Lemma 8.2.

the left side is admissible, and therefore by Lemma 8.2, $L_\gamma(\mathfrak{A}, \mathfrak{B})$ is also admissible.

References

1. R. Arens and J. Dugundji, *Topologies for function spaces*, Pacific J. Math., vol. 1 (1951), pp. 5-31.
2. R. J. Aumann, *Spaces of measurable transformations*, Bull. Amer. Math. Soc., vol. 66 (1960), pp. 301-304 [Chapter 65].
3. S. Banach, *Über analytisch darstellbare Operationen in abstrakten Räumen*, Fund. Math., vol. 17 (1931), pp. 283-295.
4. R. H. Fox, *On topologies for function spaces*, Bull. Amer. Math. Soc., vol. 51 (1945), pp. 429-432.
5. P. R. Halmos, *Measure theory*, Princeton, Van Nostrand, 1950.
6. F. Hausdorff, *Set theory*, New York, Chelsea, 1957.
7. G. W. Mackey, *Borel structures in groups and their duals*, Trans. Amer. Math. Soc., vol. 85 (1957), pp. 134-165.

The Hebrew University
   Jerusalem, Israel
Princeton University
   Princeton, New Jersey

67 On Choosing a Function at Random

1 Introduction

We use the prefix "*m-*" to abbreviate "measurable." Let X and Y be *m*-spaces[1] and let Y^X denote the set of all *m*-transformations from X into Y. We are interested in random procedures for choosing a member of Y^X. An example of such a random procedure is a stochastic process; here X is the time axis. Another example is a mixed strategy in a game, in which the player has to pick a member of Y on the basis of information which may vary over X (here Y^X is the set of pure strategies).

The relation of the concepts we use here to standard concepts from the theory of stochastic processes will be briefly explored in section 7.

2 Distributions over Function Space

There are two approaches to the problem of formalizing the intuitive notion of "random procedure for choosing a member of Y^X." First, we may define a distribution—i.e. probability measure—on Y^X. However, since Y^X is not endowed with any *m*-structure to begin with, we must impose an *m*-structure on it as part of the definition of "distribution." Thus a distribution on Y^X should be defined as a pair $(\mathscr{C}, \mu)$, where $\mathscr{C}$ is a σ-ring on Y^X and μ is a measure on $\mathscr{C}$.

Let us now choose a function f at random from Y^X, and a point x at random from X, and inquire as to the distribution of a point on a (not necessarily continuous) path, when both path and time are chosen at random. More precisely, let us fix distributions $(\mathscr{C}, \mu)$ on Y^X and ν on[2] X, and let $\varphi: Y^X \times X \to Y$ be the mapping defined by $\varphi(f, x) = f(x)$; we wish to regard φ as a random variable, and seek its distribution. Now φ has a distribution only if it is an *m*-transformation; indeed it is easily seen that for $B \subset Y$, we have $\text{Prob}\{f(x) \in B\} = (\mu \times \nu)\{\varphi^{-1}(B)\}$, so that $\varphi^{-1}(B)$ must be measurable in $Y^X \times X$ whenever B is measurable in Y. This places a restriction on the choice of the *m*-structure $\mathscr{C}$ on Y^X; it should be chosen so that φ is an *m*-transformation.

Rather surprisingly, it is in general *impossible* to define an *m*-structure on Y^X that will satisfy this condition. For example, it is impossible even

---

This chapter originally appeared in *Ergodic Theory*, edited by F. B. Wright, pp. 1–20. © 1963 Academic Press, Inc. Reprinted with permission.

1. Spaces on which there is defined a σ-ring of *m*-sets but not necessarily a measure. The σ-ring of *m*-sets is called the *m-structure* of the space.

2. Note that there is no need to specify an *m*-structure on X as part of the definition of distribution, as X is provided with an *m*-structure to begin with.

when X and Y are both copies of the unit interval with the usual Borel structure [2] (henceforth this m-space will be denoted I). Put in other words, we cannot define a probability distribution over *all* of Y^X that will enjoy reasonable properties. The question now arises whether we could not achieve our aim by restricting attention to a subset F of Y^X (such as the set of all continuous functions when X and Y are copies of the unit interval). More precisely, for $F \subset Y^X$ define $\varphi_F : F \times X \to Y$ by $\varphi_F(f, x) = f(x)$. Is it possible to impose an m-structure on F so that φ_F will be an m-transformation?

Obviously the answer depends on F; for $F = Y^X$ we have seen that it is negative, whereas when F has only one element it is trivially positive. We will say that an F for which the answer to the above question is positive is *admissible*; the appropriate m-structures will also be called *admissible*. Thus the admissible subsets of Y^X are precisely those over which a probability distribution can reasonably be defined. The problem of characterizing the admissible subsets of Y^X has been solved under fairly wide conditions on Y and X [1,2]. In particular, when Y and X are copies of I, then F is admissible if and only if it is a subset of some Baire class (of arbitrary finite or denumerable order). Thus for example the set of all continuous functions is admissible, as is the set of all functions with discontinuities of the first kind only, and so on.

Returning to our original problem, we see that the correct definition of a distribution over Y^X is a *triple* $(F, \mathscr{F}, \mu)$, where $\mathscr{F}$ is an admissible m-structure on the admissible set F, and μ is a probability measure on $\mathscr{F}$. Intuitively, this is a random procedure for picking a member of Y^X, under which the members of Y^X that "can actually occur" are precisely the members of F.

3 Random Variables over Function Space

In contrast to the first approach, which is based on the idea of a "distribution" over Y^X, the second approach is based on the idea of a "random variable" with values in Y^X. Specifically, let $\Omega = (\Omega, \beta, \lambda)$ be an arbitrary probability space which will serve as our sample space. Intuitively, our "random variable" is a function Θ from Ω to Y^X. Here again we are interested in the distribution of $f(x)$, when both f and x are chosen at random; we may expect that some condition must be placed on Θ to ensure that $f(x)$ *has* a distribution. Fortunately, the appropriate condition is *not* that Θ be an m-transformation, because this would again involve defining an m-structure on Y^X. To state the correct condition, we recall that to every function from Ω to Y^X there is a corresponding function from $\Omega \times X$ to Y; to $\Theta: \Omega \to Y^X$ there corresponds the function $\vartheta: \Omega \times X \to Y$ defined by $\vartheta(\omega, x) = \Theta(\omega)(x)$. It is easily seen that if $x \in X$ is chosen according to a distribution v, then for $B \subset Y$, we have $\text{Prob}\{f(x) \in B\} = (\lambda \times v)\{\vartheta^{-1}(B)\}$. Thus the correct condition is that the function ϑ be an m-transformation. Clearly the simplest procedure—which we adopt—is to define a *random variable* varying over Y^X with sample space Ω to be an m-transformation ϑ from $\Omega \times X$ into Y.

4 Ranges and Admissible Sets

The purpose of this note is to investigate the relation between the concepts of "random variable" and "distribution" defined in the previous sections. Let us define the *range* of a random variable ϑ to be the set of all functions in Y^X of the form $\vartheta(\omega, \cdot)$, where $\omega \in \Omega$. The range is the set of points in Y^X that "can actually occur" under ϑ; thus it corresponds to the notion of an admissible set under the "distribution" definition. Our first question is to what extent this correspondence can be made precise.

THEOREM 1 *Every range is admissible. Conversely, for every admissible set F there is a sample space Ω and a random variable ϑ such that the range of ϑ is F.*

Proof The converse is trivial, because we may take $\Omega = (F, \mathscr{F})$ and $\vartheta = \varphi_F$, where $\mathscr{F}$ is an admissible m-structure on F. For the first part, let ϑ be a random variable with range R. As in the previous section, denote $\vartheta(\omega, \cdot)$ by $\Theta(\omega)$; thus $\Theta: \Omega \to R \subset Y^X$. For every $f \in R$ choose one member ω of Ω, such that $\Theta(\omega) = f$; let Ω' be the subspace of Ω obtained in this way, with the subspace structure (a set is measurable in Ω' if and only if it is the intersection of Ω' with an m-set in Ω). Let ϑ' be the restriction of ϑ to $\Omega' \times X$. Now the restriction of an m-transformation to a subspace is still an m-transformation; hence if we give $\Omega' \times X$ the subspace structure (i.e. as a subspace of $\Omega \times X$), then ϑ' will be an m-transformation. But it is easily verified that the subspace structure on $\Omega' \times X$ is the same as the product structure; hence ϑ' is an m-transformation also when $\Omega' \times X$ has the product structure.

Ω' and R are in one–one correspondence under the correspondence $\omega \leftrightarrow \Theta(\omega)$. Let us impose on R the structure corresponding to that of Ω'; then Ω' and R are isomorphic. Hence $\Omega' \times X$ and $R \times X$ are also isomorphic. Let us denote the isomorphism by $\zeta: R \times X \to \Omega' \times X$; we have $\zeta(\Theta(\omega), x) = (\omega, x)$, where on the right side ω is uniquely defined because of the definition of Ω'. Now $\varphi_R(\Theta(\omega), x) = \Theta(\omega)(x) = \vartheta(\omega, x) = \vartheta'(\omega, x) = \vartheta'\zeta(\Theta(\omega), x)$; thus $\varphi_R = \vartheta'\zeta$. But both ϑ' and ζ are

m-transformations, and therefore φ_R is also an m-transformation. Therefore R is admissible, and the proof of Theorem 1 is complete.

5 Ranges and Admissible Sets when the Sample Space Is Standard

Up to now we have allowed the sample space Ω to be arbitrary; in proving that every admissible set F is the range of some random variable, we even allowed Ω to vary with F. In some applications, though, Ω is a copy of the unit interval I with Lebesgue measure, or can be taken as such without loss of generality. This restriction does not change the fact that every range is admissible, but it does cast doubt on the converse.

Let us further restrict our consideration to the case $X = Y = I$. (This restriction is not as severe as it may at first seem; according to a theorem of Mackey [6], every Borel subset of every separable metric topological space, when endowed with the subspace structure, is isomorphic[3] to I.) As we remarked above, there is in this case an elegant characterization of admissible sets, namely as arbitrary subsets of Baire classes of arbitrary order. We seek now a characterization of ranges.

THEOREM 2 *If X, Y, and Ω are copies[4] of I, then every range is a subset of some Baire class, and every Baire class is a subset of some range.*[5]

This theorem does not give a complete characterization of ranges, similar to the complete characterization of admissible sets mentioned above. For example, I do not even know whether every Baire class is a range; on the other hand, it is highly likely that there exist subsets of Baire classes that are not ranges.[6] What Theorem 2 does do is give an "order of magnitude" characterization for ranges; a range can be as large as a Baire class of arbitrarily high (denumerable) order, but no larger.

Proof That every range is a subset of some Baire class is a trivial consequence of Theorem 1 and the fact that every admissible set is a subset of some Baire class (proved in [2]). To prove the second part of Theorem 2, we define a transfinite sequence $\{F_\alpha\}$, where α ranges over all denu-

---

3. Two m-spaces are *isomorphic* if there is a one–one correspondence between them, which carries m-sets onto m-sets in both directions.

4. Strictly speaking, Ω is a copy of (I, λ) rather than of I. However we are dealing with measurability properties rather than with measure properties, and will henceforth (for the remainder of the section) ignore the measure on Ω.

5. "Baire class" is a topological concept, so its appearance in a theorem that deals with m-structures should be explained. Since X and Y are in one–one correspondence with I, we can impose on them topologies corresponding to the standard topology on I; the theorem pertains to these topologies. For an intrinsic characterization of Baire classes in terms of m-structures, see [1, 2].

6. The entire discussion is under the assumption $\Omega = X = Y = I$.

merable ordinals, inductively as follows: F_0 is the set of all continuous functions from X to Y; F_α is the set of all functions that are pointwise upper limits of sequences of functions in $\bigcup_{\beta<\alpha} F_\beta$. We will prove by induction on α that F_α is a range; since the α'th Baire class is clearly a subset of F_α, this will complete our proof. To start the induction, let $F_0(a, B) = \{f \in F_0 : f(a) \in B\}$, where $a \in X$ and $B \subset Y$. Let $\mathscr{F}_0$ be the m-structure on F_0 that is generated by all sets of the form $\mathscr{F}_0(r, U)$, where r is rational and U is open. (It happens that $\mathscr{F}_0$ is also generated by the uniform convergence topology on F_0, but this does not concern us here.)

LEMMA 1 $(F_0, \mathscr{F}_0)$ *is an admissible pair.*

Proof   We first show that if a is arbitrary and B is measurable, then $F_0(a, B) \in \mathscr{F}_0$. Indeed, let $r_1, r_2, \ldots$ be a sequence of rationals converging to a. Then for open U, $f(a) \in U$ if and only if $(\exists N)(\forall n \geqslant N)(f(r_n) \in U)$; thus $F_0(a, U) = \bigcup_{N=1}^\infty \bigcap_{n=N}^\infty F_0(r_n, U)$, and hence $F_0(a, U) \in \mathscr{F}_0$. Since the m-structure of Y is generated by the open sets, and the mapping $B \to F_0(a, B)$ carries unions into unions and complements into complements, it follows that $F_0(a, B) \in \mathscr{F}_0$ for all a and all measurable B. The lemma now follows from theorem G of [2].

LEMMA 2 $(F_0, \mathscr{F}_0)$ *is isomorphic to* I.

Proof   Denote the infinite product $I \times I \times \ldots$ by I^∞; the members of I^∞ are sequences $t = \{t_1, t_2, \ldots\}$. The m-structure of I^∞ is generated by cylinder sets of the form $I \times \cdots \times I \times U \times I \times \ldots$, i.e. by the sets $\{t : t_i \in U\}$, where U is open; such a set will be denoted $I^\infty(i, U)$.

Let $\{r_1, r_2, \ldots\}$ be the set of all rationals in I. Define a mapping $\xi : F_0 \to I^\infty$ by $\xi(f) = \{f(r_1), f(r_2), \ldots\}$. We first show

i. ξ *is an isomorphism between* F_0 *and its image* $\xi(F_0)$. Since a continuous function is completely determined by its values on the rationals, $\xi(f_1) = \xi(f_2)$ implies $f_1 = f_2$; in other words, F_0 and its image $\xi(F_0)$ are in one–one correspondence under ξ. Moreover, if $\xi(F_0)$ is considered a subspace of I^∞ with the subspace structure, then ξ maps the generator $F_0(r_i, U)$ of $\mathscr{F}_0$ onto the generator $\xi(F_0) \cap I^\infty(i, U)$ of the structure of $\xi(F_0)$. Hence the generators are also in one–one correspondence under ξ, and a fortiori ξ is an isomorphism between F_0 and $\xi(F_0)$.

Next, we show

ii. $\xi(F_0)$ *is an m-set in* I^∞. This follows from the remark that a function f on the rationals between 0 and 1 can be extended to a continuous function on I if and only if it is uniformly continuous on the rationals. Thus we have $t \in \xi(F_0)$ if and only if

$$(\forall k)(\exists j)(|t_m - t_n| < 1/k \quad \text{whenever} \quad |r_m - r_n| < 1/j).$$

In other words,

$$\xi(F_0) = \bigcap_{k=1}^{\infty} \bigcup_{j=1}^{\infty} \bigcap_{(m,n) \in A(j)} \{t: |t_m - t_n| < 1/k\},$$

where $A(j) = \{(p,q): |r_p - r_q| < 1/j\}$. Since it is well known (and easily proved) that the set in curly brackets is an m-set in I^∞, this demonstrates (ii).

I^∞ is known to be isomorphic to I; from (ii) it follows that $\xi(F_0)$ passes under this isomorphism to a Borel subset of I. According to a theorem of Mackey [6], every non-countable Borel subspace B of I is isomorphic to I. Hence $\xi(F_0)$ is isomorphic to I. The lemma now follows from (i).[7]

Lemma 2 says that Ω and F_0 are isomorphic; we may therefore assume without loss of generality that they are identical, and define $\vartheta = \varphi_{F_0}$. Then by Lemma 1, ϑ is an m-transformation, and by Lemma 2, its range is F_0. This starts our induction.

For the inductive step, let α be a finite or denumerable ordinal, and suppose it has been shown that F_β is a range for all $\beta < \alpha$. Let $\Omega_1, \Omega_2, \ldots$ be a sequence of copies of Ω, and let $\vartheta_1: \Omega_1 \times X \to Y$, $\vartheta_2: \Omega_2 \times X \to Y, \ldots$ be a sequence of random variables such that each F_α with $\alpha < \beta$ is the range of infinitely many of the ϑ_i. The infinite product $\Omega_1 \times \Omega_2 \times \ldots$ is isomorphic to Ω, and as before we suppose without loss of generality that it equals Ω; thus for $\omega \in \Omega$, we may write $\omega = \{\omega_1, \omega_2, \ldots\}$. Define $\vartheta: \Omega \times X \to Y$ by $\vartheta(\omega, x) = \limsup_{i \to \infty} \vartheta_i(\omega_i, x)$. From the fact that the ϑ_i are m-transformations, it follows that ϑ is an m-transformation; furthermore it may be seen that the range of ϑ is exactly F_α. This completes the proof of Theorem 2.

Note that we started our induction by showing that F_0 is a range, but it would have been sufficient to show that F_0 is a *subset* of some range; this is in fact easier. We chose to show that F_0 is a range, because this lemma is of interest in itself.

6 The Distribution of a Random Variable

Let $\vartheta: \Omega \times X \to Y$ be a random variable; we wish to define the concept of "the distribution of ϑ". According to section 2, this must be of the

---

7. I am grateful to B. Peleg for pointing out an error in the original proof of this lemma.

form $(F, \mathscr{F}, \mu)$, where F and $\mathscr{F}$ are admissible and μ is a measure on $\mathscr{F}$. The natural definitions would be as follows: F is the range of ϑ. The m-structure is the *identification* structure; that is, a subset G of F is in $\mathscr{F}$ if and only if $\Theta^{-1}(G)$ is an m-set in Ω. The measure μ is defined by $\mu(G) = \lambda\Theta^{-1}(G)$, where λ is the measure on Ω. These definitions are "natural" in the sense that the induced distribution on Y is the same if ϑ or if $(F, \mathscr{F}, \mu)$ is used.

The only trouble with this "natural" definition is that I do not know whether as defined, the structure $\mathscr{F}$ is admissible.[8] Indeed, let $\iota: X \to X$ be the identity, and define $\Theta \times \iota: \Omega \times X \to F \times X$ by $(\Theta \times \iota)(\omega, x) = (\Theta(\omega), x)$. Then $\vartheta = \varphi_F \circ (\Theta \times \iota)$, and hence $\vartheta^{-1} = (\Theta \times \iota)^{-1}\varphi_F^{-1}$. Now let B be an m-subset of Y. Then $\vartheta^{-1}(B)$ is an m-subset of $\Omega \times X$, and hence $(\Theta \times \iota)^{-1}\varphi_F^{-1}(B)$ also is. We know that $(F, \mathscr{F})$ is an identification space of Ω under the identification map Θ. If we only knew that $(F, \mathscr{F}) \times X$ is an identification space of $\Omega \times X$ under $\Theta \times \iota$, then the measurability of $\varphi_F^{-1}(B)$ would follow from that of $(\Theta \times \iota)^{-1}\varphi_F^{-1}(B)$, and we could deduce that $\mathscr{F}$ is admissible. The proposition that "if Θ is an identification map and ι an identity map, then $\Theta \times \iota$ is also an identification map" is intuitively very compelling, but unfortunately I have not succeeded in proving it.[9] Let us call this proposition the "identification space hypothesis"; only the following special cases are known to me:

MACKEY'S RESULT[10]   *The identification space hypothesis holds if the domains and images of both Θ and ι are analytic*[11] *m-spaces.*

ERNEST'S RESULT[12]   *The identification space hypothesis holds if Θ carries m-sets onto m-sets.*

Though the hypotheses of both these theorems are quite general, I do not know whether they hold in the situation under consideration, even for some of the simplest cases, (e.g. when $X = Y = \Omega = I$, and ϑ is the random variable with range F_1 that we defined in the previous section).

Our conclusion is that for all we know at the present, a given random variable may have no distribution.

---

8. By Theorem 1, the set F must be admissible. But the choice of the structure in the proof of that theorem is not unique, and so cannot be used for the current purpose. Even if we arbitrarily pick one of the structures that fit that proof, the resulting distribution may violate the "naturalness" condition of the previous paragraph.

9. The topological analogue is false; the counter-example is an adaptation of an example in Kelley's book [4, p. 132, example G].

10. Private correspondence with Professor G. W. Mackey.

11. I.e., isomorphic with analytic subspaces of I; cf. [6].

12. Private correspondence with Professor J. Ernest.

7 Relation to Other Concepts

What we call a "random variable with values in Y^X" is called a "measurable random function" by Loève [5, p. 502]. A somewhat similar object is called a "measurable stochastic process" by Doob [3, p. 60]; however, Doob fixes the m-structure of X to be the class of all Lebesgue measurable sets. Our "distributions" are closely related to what Doob [3, p. 67] calls a "process of function space type" (i.e. a process in which the sample space is the function space itself). The differences are that Doob considers the set of *all* functions from X into Y, whereas we consider only m-transformations; and Doob imposes a fixed m-structure on function space, namely that generated by all sets of the form $F(a, B) = \{f : f(a) \in B\}$ (of course without regard to admissibility).

What distinguishes the problem discussed here from those of much of the theory of stochastic processes is that we allow x as well as f to vary at random, and inquire as to the distribution of $f(x)$ as a function of both f and x. This makes simultaneous measurability in both variables essential. In stochastic processes one is also interested in the distribution of $f(x)$; but usually only f varies at random, and attention is fixed on some finite x-set. Simultaneous measurability in both variables is then often useful, but not essential.

8 Open Questions

i. Characterize ranges of random variables when $X = Y = \Omega = I$.

ii. In particular, is every Baire class a range in this case?

iii. Prove or disprove the identification space hypothesis.

iv. Does every random variable have a distribution (in the sense of section 6)?

References

1. R. J. Aumann, Spaces of measurable transformations, *Bull. Amer. Math. Soc.* 66 (1960), pp. 301–304 [Chapter 65].

2. R. J. Aumann, Borel structures for function spaces, *Ill. Jour. of Math.* 5 (1961), pp. 614–630 [Chapter 66].

3. J. L. Doob, *Stochastic Processes*, New York, John Wiley (1953).

4. J. L. Kelley, *General Topology*, Princeton, Van Nostrand (1955).

5. M. Loève, *Probability Theory* (second edition), Princeton, Van Nostrand (1960).

6. G. W. Mackey, Borel structures in groups and their duals, *Trans. Amer. Math. Soc.* 85 (1957), pp. 134–165.

68 Integrals of Set-Valued Functions

Introduction

Set-valued functions have been of interest for some time. Fixed-point theorems for such functions were proved by Kakutani [1], Eilenberg and Montgomery [2], and others; furthermore, set-valued functions have been used repeatedly in Economics (see for example Arrow and Debreu [3], McKenzie [4] and Vind [5]). Integrals of set-valued functions have been studied in connection with statistical problems; see Kudo [6] and Richter [7]. Lately integrals of set-valued functions have arisen in connection with economic problems [8, 9], and we here extend the basic theory of such integrals.

Let T be the unit interval $[0, 1]$. For each t in T, let $F(t)$ be a nonempty subset of euclidean n-space E^n. Let $\mathscr{F}$ be the set of all point-valued functions f from T to E^n such that f is integrable over T and $f(t) \in F(t)$ for all t in T. Define

$$\int_T F(t)\, dt = \left\{ \int_T f(t)\, dt : f \in \mathscr{F} \right\}$$

i.e., the set of all integrals of members of $\mathscr{F}$. This notion is a natural generalization of the integral of point-valued functions on the one hand, and of the sum of a finite number of sets on the other hand. It is closely connected with Vind's "set valued measures" [5]; if for measurable subsets S of T, we set $\nu(S) = \int_S F(t)\, dt$, then ν is a set-valued measure.

The following conventions will be used: Instead of $\int_T F(t)\, dt$, $\int_T f(t)\, dt$, etc., we will write $\int F$, $\int f$, etc.; when it is necessary to integrate over a subset S of T we will write $\int_S F$, etc., but otherwise the range of integration will not be specified and will be understood to be all of T. When we refer to "all" or "each" t in T we will mean "almost all," and when we say that "there is a t in T" with some property, then we shall mean that the set of

---

Research partially supported by the Office of Naval Research under contract number N62558-3586.

points t in T with the property in question is of positive Lebesgue measure. "Measurable" will mean "Lebesgue measurable". Coordinates of points in E^n will be denoted by superscripts. If x and y are in E^n, $x \geqslant y$ will mean[1] $x^i \geqslant y^i$ for all i. The scalar product is denoted $x \cdot y$, and $|x|$ means $(|x^1|, \cdots, |x^n|)$. Set-theoretic subtraction will be denoted \.

The following basic theorem is due to Richter [7]:

THEOREM 1. *$\int F$ is convex.*

It is natural to ask under what conditions $\int F$ (or equivalently, $\mathscr{F}$) is nonempty. The function F will be called *Borel-measurable* if its graph $\{(t, x) : x \in F(t)\}$ is a Borel subset[1a] of $T \times E^n$. It will be called *integrably bounded* if there is a point-valued integrable function h from T to E^n such that $|x| \leqslant h(t)$ for all x and t such that $x \in F(t)$.

THEOREM 2. *If F is Borel-measurable and integrably bounded, then $\int F$ is nonempty.*

Neither hypothesis in this theorem can be omitted, as we shall show by counter-example. This theorem follows immediately from a basic "measurable choice theorem" due to von Neumann [11].

The function F will be called *nonnegative* if $x \geqslant 0$ for all x and t such that $x \in F(t)$. For each t in T, $F^*(t)$ will denote the convex hull of $F(t)$.

THEOREM 3. *If F is nonnegative and Borel-measurable, then $\int F = \int F^*$.*

This theorem looks like a trivial consequence of Theorem 1, but it isn't. Neither hypothesis can be removed. Note that the theorem remains true if F is bounded from below (or from above) by an integrable point-valued function h, so that in effect the nonnegativity condition is a weakening of the condition of integrable boundedness.

The function F will be called *closed* if $F(t)$ is closed for each t.

THEOREM 4. *If F is closed and integrably bounded, then $\int F$ is compact.*

Note that the hypothesis of Borel-measurability is not needed here. Theorem 4 was proved by Kudo [6] for F that are convex-valued and Borel measurable (in addition to the conditions given here). Richter [7] proved a version slightly different from Kudo's, which does not require convexity, but which still requires measurability.

---

[1] Note that this differs from the standard usage, in which this relation is denoted $\geq$.

[1a] This definition is justified by the fact that a point-valued function is Borel-measurable if and only if its graph is a Borel set (see [10], p. 365, Proposition 4, and p. 398, Proposition 2). For compact convex valued functions, this definition coincides with that of Kudo [6] and Richter [7], who use the measurability of the support function.

We now turn to a generalization of Lebesgue's dominated convergence theorem; this may be considered our chief result. If A_1, A_2, $\cdots$ are subsets of E^n, then by definition [10, p. 241 ff.], $x \in \liminf A_k$ if and only if every neighborhood of x intersects all the A_k with sufficiently high k, and $x \in \limsup A_k$ if and only if every neighborhood of x intersects infinitely many A_k. If $\liminf A_k = \limsup A_k = A$, then we write $A = \lim A_k$, or $A_k \to A$.

THEOREM 5. *If $F_k(t) \to F(t)$ for all t, and all the F_k are Borel-measurable and bounded by the same integrable point-valued function, then $\int F_k \to \int F$.*

Theorem 5 may be restated in terms of continuous set-valued functions. This will be done in Section 5. Neither of the conditions can be omitted in this theorem. The proof of Theorem 5 makes use of two analogues of Fatou's lemma (Propositions 4.1 and 5.1), which are of some interest in their own right.

In the final section, we consider an application to extreme points of sets of vector functions, as treated by Karlin in [12].

Nowhere in this paper do we use the fact that the measure space T is totally finite. In particular, the theorems remain true when T is $(0, \infty)$ or $(-\infty, \infty)$. What *is* essential (to at least some of the theorems) is that T be nonatomic.

PROOF OF THEOREM 2

It will be convenient in this section to replace the euclidean space E^n by an arbitrary separable and complete metric space X. The function F will still be defined on $T = [0, 1]$, but its values will now be subsets of X. A point-valued function f from T to X will be called *Lebesgue-measurable* if $f^{-1}(U)$ is a Lebesgue-measurable subset of T for every open (or equivalently, Borel) subset of X. Recall that an *analytic* subset of X is the continuous image of a Borel subset of X [10, p. 360]; the set-valued function F will be called *analytic* if its graph is an analytic subset of $T \times X$. The following proposition follows from a lemma of von Neumann [11, p. 448, lemma 5]:

PROPOSITION 2.1. *If F is an analytic set-valued function from T to X, then there is a Lebesgue-measurable point-valued function $f : T \to X$ such that $f(t) \in F(t)$ for all t.*

Since every Borel-measurable F is analytic, it follows by setting $X = E^n$ that if F is a Borel-measurable function from T to E^n, then there is an f satisfying the conclusions of Proposition 2.1. If, moreover, F is integrably bounded, then it follows that f is integrable. Then $\int f \in \int F$, and Theorem 2 is proved.

Proof of Theorem 3

The proof is by induction on the dimension n of the space. For $n = 0$ the theorem is immediate, since $F^*(t) = F(t) = \{0\}$ for all t. Suppose the theorem true for dimensions less than n. If the theorem is false in dimension n, then for appropriate F we have $\int F^* \setminus \int F \neq \phi$; let $x \subset \int F^* \setminus \int F$. By Theorem 1, $\int F$ is convex, so it has a supporting hyperplane passing through x; that is, there is a vector $a \in E^n$ such that

$$a \cdot y \leqslant a \cdot x \qquad (3.1)$$

for all $y \in \int F$.

Since $x \in \int F^*$, it follows that $x = \int f^*$, where $f^*(t) \in F^*(t)$ for all t. Furthermore, f^* may be chosen Borel-measurable; for, every Lebesgue measurable function is equivalent to (i.e., differs on a set of measure 0 from) a Borel measurable function,[2] and according to our convention, "all" means "almost all." Now recall Caratheodory's theorem, which states that if D and D^* are subsets of E^n such that D^* is the convex hull of D, then every point of D^* is a convex combination of $n+1$ points of D (see Eggleston [13], p. 34 ff.). According to this, for each t there are positive real numbers $\varphi_0(t), \cdots, \varphi_n(t)$ summing to 1, and members $g_0(t), \cdots, g_n(t)$ of $F(t)$, such that

$$f^*(t) = \sum_{j=0}^{n} \varphi_j(t) g_j(t). \qquad (3.2)$$

We wish to show that the φ_j and the g_j obeying (3.2) can be chosen measurable, and with g_0 integrable. To this end, note that (3.2) implies that for at least one of the g_j, $\sum_{i=1}^{n} g_j^i(t) \leqslant \sum_{i=1}^{n} f^{*i}(t)$. Since the indexing of the g_j is of no significance, this means that the subset $G(t)$ of $E^{(n+1)+n(n+1)}$ defined by

$$G(t) = \Big\{(\xi_0, \cdots, \xi_n, x_0, \cdots, x_n) : 0 < \xi_j \leqslant 1 \text{ and } x_j \in F(t) \text{ for all } j,$$

$$\sum_{j=0}^{n} \xi_j = 1, \sum_{i=1}^{n} x_0^i \leqslant \sum_{i=1}^{n} f^{*i}(t), \text{ and } f^*(t) = \sum_{j=0}^{n} \xi_j x_j \Big\}$$

is nonempty for all t. Furthermore, the graph of G is a Borel subset of $E^{(n+1)+n(n+1)+1}$; this follows from the Borel measurability of F and f^*. So by Proposition 2.1, we may choose measurable φ_j and g_j so that

$$(\varphi_0(t), \cdots, \varphi_n(t), g_0(t), \cdots, g_n(t)) \in G(t) \qquad \text{for all} \quad t.$$

---

[2] This is immediate for simple functions, and every measurable function is the limit of a sequence of simple functions.

Then the φ_j and the g_j are measurable and obey (3.2). Furthermore, from $g_0(t) \in F(t)$ and the nonnegativity of F it follows that $g_0(t) \geqslant 0$. Hence

$$0 \leqslant g_0^1(t) \leqslant \sum_{i=1}^{n} g_0^i(t) \leqslant \sum_{i=1}^{n} f^{*i}(t),$$

and therefore the integrability of g_0^i follows from that of f^*. Similarly all the g_0^i are integrable, i.e., g_0 is integrable.

We now show that

$$a \cdot g_j(t) \leqslant a \cdot f^*(t) \tag{3.3}$$

for all t and j. Indeed, suppose that

$$a \cdot g_k(t) > a \cdot f^*(t) \tag{3.4}$$

for some k and t, say for $t \in S$, where S has positive measure. For each t, there is a j obeying

$$a \cdot g_j(t) \geqslant a \cdot f^*(t); \tag{3.5}$$

for otherwise, since $\varphi_j(t) > 0$, we have

$$a \cdot f^*(t) = \sum_{j=1}^{n+1} \varphi_j(t) a \cdot g_j(t) < \sum_{j=1}^{n+1} \varphi_j(t) a \cdot f^*(t) = a \cdot f^*(t),$$

an absurdity. Let us denote the first j that fulfills (3.5) by $j(t)$. Define a function f by

$$f(t) = \begin{cases} g_k(t) & \text{when} & t \in S \\ g_{j(t)}(t) & \text{when} & t \notin S. \end{cases}$$

Clearly f is measurable, but possibly it is not integrable. For each positive integer m, let $U(m) = \{t : f(t) \leqslant (m, \cdots, m)\}$, and define a sequence of integrable functions f_m by

$$f_m(t) = \begin{cases} f(t) & \text{when} & t \in U(m) \\ g_0(t) & \text{when} & t \notin U(m). \end{cases}$$

Then $f_m(t) \in F(t)$ for all t, and so (3.1) yields

$$a \cdot \int f_m \leqslant a \cdot \int f^* \tag{3.6}$$

for all f. Now $\bigcup_{m=1}^{\infty} U(m) = T$; therefore

$$\int_{T \setminus U(m)} a \cdot (g_0 - f^*) \to 0. \tag{3.7}$$

Furthermore, for sufficiently large m, $U(m) \cap S$ has positive measure. For such m, then, it follows from (3.4) that

$$\int_{U(m) \cap S} a \cdot f_m = \int_{U(m) \cap S} a \cdot g_k > \int_{U(m) \cap S} a \cdot f^*.$$

In other words, if

$$\epsilon(m) = \int_{S \cap U(m)} a \cdot (f_m - f^*),$$

then $\epsilon(m)$ is monotone increasing in m, and $\epsilon(m) > 0$ for sufficiently large m, say for $m \geqslant m_0$. Now from (3.7) it follows that for m sufficiently large, say $m \geqslant m_1$, we have

$$\int a \cdot (f_m - f^*) = \int_{T \setminus U(m)} a \cdot (g_0 - f^*) \geqslant -\frac{\epsilon(m_0)}{2}.$$

Furthermore, by (3.5) and the definition of f,

$$\int_{U(m) \setminus S} a \cdot f_m = \int_{U(m) \setminus S} a \cdot f \geqslant \int_{U(m) \setminus S} a \cdot f^*.$$

So if $m \geqslant \max(m_1, m_0)$, then

$$a \cdot \int f_m - a \cdot \int f^* = \int a \cdot (f_m - f^*) = \int_{U(m) \cap S} + \int_{U(m) \setminus S} + \int_{T \setminus U(m)}$$

$$\geqslant \epsilon(m_0) + 0 - \frac{\epsilon(m_0)}{2} > 0,$$

contradicting (3.6). This demonstrates (3.3).

Next, we demonstrate

$$a \cdot g_j(t) = a \cdot f^*(t) \tag{3.8}$$

for all t and j. Indeed, suppose that there is a j—say $j = k$—such that for some t, strong inequality holds in (3.3), i.e., $a \cdot g_j(t) < a \cdot f^*(t)$. Applying (3.3) and $\varphi_k(t) > 0$, we deduce

$$a \cdot f^*(t) = \sum_{j=0}^{n} \varphi_j(t) a \cdot g_j(t) < \sum_{j=0}^{n} \varphi_j(t) a \cdot f^*(t) = a \cdot f^*(t),$$

an absurdity. This proves (3.8).

Let H be the hyperplane $\{y : a \cdot y = 0\}$. Define $E(t) = [F(t) - f^*(t)] \cap H$, and let $E^*(t)$ be the convex hull of $E(t)$. From (3.8) it follows that $g_j(t) - f^*(t) \in E(t)$, and so by (3.2), $0 = f^*(t) - f^*(t) \in E^*(t)$. Since H is of dimension $n - 1$, we may apply the induction hypothesis to E, and deduce

$0 \in \int E$. Let e be such that $e(t) \in E(t)$ for all t and $\int e = 0$. Then for each t, $e(t) + f^*(t) \in F(t)$, and $\int [e + f^*] = \int f^* = x$. Hence $x \in \int F$ after all, contradicting our assumption. This completes the proof of Theorem 3.

To see that Theorem 3 is false if it is not assumed that F is Borel-measurable, let $n = 1$, and let g be the characteristic function of a subset of T with inner measure 0 and outer measure 1. Let $F(t)$ contain the two points $g(t)$ and 2 only. Then $\int F = \{2\}$ but $\int F^* = [1, 2]$.

To see that Theorem 3 is false without the nonnegativity assumption, let $n = 1$ and let $F(t) = \{1/t, -1/t\}$. Then $\int F = \phi$ and $\int F^* = (-\infty, \infty)$.

Proof of Theorem 4

The proof of Theorem 4 is a consequence of the following analogue of Fatou's lemma.

PROPOSITION 4.1. *If F_1, F_2, $\cdots$ is a sequence of set-valued functions that are all bounded by the same integrable point-valued function h, then*

$$\int \limsup F_k \supset \limsup \int F_k .$$

PROOF. Suppose $x \in \limsup \int F_k$. Then x is a limit point of a sequence $\int f_k$, where $f_k(t) \in F_k(t)$ for each k and t; that is, there is a subsequence of $\int f_k$ converging to x. We wish to show that $x \in \int \limsup F_k$; for this purpose we may assume without loss of generality that x is actually the limit of the $\int f_k$, i.e. that the subsequence converging to x is the whole original sequence.

The f_k can be considered real-valued functions on $\{1, \cdots, n\} \times T$; since they are integrable, it follows that they are members of the Banach space $L^1 = L^1(\{1, \cdots, n\} \times T)$. Because the f_k are all bounded by the integrable function h, it follows that there is a subsequence with a weak limit, which we call f (see Dunford-Schwartz [14], Theorem IV.8.9, p. 292). Again, we may assume without loss of generality that f_k actually converges to f weakly.

Let $\{\epsilon_1, \epsilon_2, \cdots\}$ be a sequence of positive real numbers tending to 0. Since $\{f_1, f_2, \cdots\}$ approaches f weakly, it follows that also $\{f_m, f_{m+1}, \cdots\}$ approaches f weakly, for all m. Hence from a known theorem, it follows that there is a sequence of convex combinations of $f_m, f_{m+1}, \cdots$ that approaches f in the norm of L^1 ([14] Corollary V. 3.14, p. 422). For each m, let g_m be such a convex combination with $\|g_m - f\| \leq \epsilon_m$, where $\| \|$ denotes the L^1 norm. Then $g_m \to f$ in the norm of L^1. Hence by a known theorem, there is a subsequence of the g_m that converges to f almost everywhere (see [14], Theorem III, 3.6(i), p. 122, for a proof that norm convergence implies convergence in measure, and Corollary III, 6.13(a), p. 150, for a proof that convergence in measure implies convergence a.e. of a subsequence); without loss of generality[3] let it be the whole original sequence. So for all t, $g_m(t) \to f(t)$, and $g_m(t)$ is a convex combination of $\{f_m(t), f_{m+1}(t), \cdots\}$. Since the latter are points of E^n, it follows from Caratheodory's theorem [13, pp. 34 ff.] that

$$g_m(t) = \sum_{j=0}^n \theta_{jm}(t) \, e_{jm}(t),$$

where the $\theta_j(t)$ are nonnegative and sum to 1, and $e_{0m}(t), \cdots, e_{nm}(t)$ are chosen from among $f_m(t), f_{m+1}(t), \cdots$. Now for each t, we can choose a subsequence of the $g_m(t)$ such that all the corresponding subsequences of $\{\theta_{0m}(t)\}, \cdots, \{\theta_{nm}(t)\}, \{e_{0m}(t)\}, \cdots,$ and $\{e_{nm}(t)\}$ converge. The limits of the θ's in these subsequences must be nonnegative numbers summing to 1, and the limits of the e's must be limit points of the $f_m(t)$. Hence for each t,

$$f(t) = \lim g_m(t) = \sum_{j=0}^n \theta_j e_j(t),$$

where the θ_j are nonnegative and sum to 1, and the $e_j(t)$ are limit points of $\{f_1(t), f_2(t), \cdots\}$. So if we let $G(t)$ be the set of limit points of $\{f_k(t)\}$, and $G^*(t)$ the convex hull of $G(t)$, then we have shown that $f(t) \in G^*(t)$ for each t. Hence $\int f \in \int G^*$.

As we have noted above, every Lebesgue-measurable function is equivalent to a Borel-measurable function; so we may assume that the f_k are Borel-measurable. Then it follows easily that G is Borel-measurable. So we may apply Theorem 3 and deduce that $\int G^* = \int G$. Hence $\int f \in \int G$. But since $f_k(t) \in F_k(t)$, it follows that every limit point of the $F_k(t)$ will be a member of $\limsup F_k(t)$, i.e., $G(t) \subset \limsup F_k(t)$. Hence $\int f \in \int \limsup F_k$. But from the weak convergence of f_k to f it follows that $\int f = \lim \int f_k = x$, so that $x \in \int \limsup F_k$. This completes the proof of Proposition 4.1.

Suppose now that F is closed and integrably bounded; set $F_1 = F_2 = \cdots = F$. Then $\limsup F_k = \mathrm{cl}\, F = F$ and $\limsup \int F_k = \mathrm{cl} \int F$, where cl denotes "closure" (see [10], p. 243, IV.6). So by Proposition 4.1,

$$\int F = \int \limsup F_k \supset \limsup \int F_k = \mathrm{cl} \int F,$$

and hence $\int F$ is closed. Since it is bounded by the integral of the function h that bounds F, it follows that it is compact, proving Theorem 4.

It is possible to prove Theorem 4 somewhat more simply by a direct application of [14, V.3.14]; but Proposition 4.1 is interesting for its own sake, and is needed in the proof of Theorem 5.

---

[3] This involves cutting down the original sequence of f_k's and using the corresponding subsequence of the g_k's.

If we replace the integrable boundedness condition in Theorem 4 by the assumption that F is nonnegative, then $\int F$ need not even be closed. For a counterexample, let $n = 2$, let $g(t) = ((1-t)/t, t/(1-t))$, and let $F(t) = \{0, g(t)\}$. Then F is the union of the open positive quadrant with the origin $\{0\}$.

Proof of Theorem 5

We first prove another analogue of Fatou's lemma, as follows:

PROPOSITION 5.1. *If all the F_k are Borel-measurable and bounded by the same integrable point-valued function, then*

$$\int \liminf F_k \subset \liminf \int F_k .$$

PROOF. Let $x \in \int \liminf F_k$. Then $x = \int f$, where $f(t) \in \liminf F_k(t)$ for each t. Since f is equivalent to a Borel-measurable function, it may be assumed to be Borel-measurable. Now the space $E^n \times E^n \times \cdots$ may be metrized so that it is complete and so that its topology is the usual product topology [10, p. 313]. For each t, define a subset $G(t)$ of $E^n \times E^n \times \cdots$ by

$$G(t) = \{(x_1, x_2, \cdots) : x_1 \in F_1(t), x_2 \in F_2(t), \cdots, \text{ and } \lim x_k = f(t)\}.$$

Then the statement $f(t) \in \liminf F_k(t)$ is precisely equivalent to the statement $G(t) \neq \phi$ [10, p. 242], and hence $G(t) \neq \phi$ for each t. Furthermore, G is easily seen to be Borel-measurable and hence analytic. So by Proposition 2.1, there is a measurable function g from T to $E^n \times E^n \times \cdots$ such that $g(t) \in G(t)$ for each t; that is, a sequence $f_1, f_2, \cdots$ of measurable functions from T to E^n, such that $f_k(t) \in F_k(t)$ for each t, and $\lim f_k(t) = f(t)$. Now since $f_k(t) \in F_k(t)$, all the f_k are bounded by the same integrable point valued function. Hence from Lebesgue's bounded convergence theorem, it follows that $\int f_k \to \int f = x$. But $\int f_k \in \int F_k$, and so $x \in \liminf \int F_k$ [10, p. 242]. This completes the proof of Proposition 5.1.

If $F(t) = \liminf F_k(t) = \limsup F_k(t)$ for all t, then by Propositions 4.1 and 5.1,

$$\int F = \int \liminf F_k \subset \liminf \int F_k \subset \limsup \int F_k \subset \int \limsup F_k = \int F;$$

hence equality holds throughout, and so $\lim \int F_k$ exists and equals $\int \lim F_k$. This proves Theorem 5.

Theorem 5 is false without the Borel-measurability assumption. Indeed, let $n = 1$, let g be the characteristic function of a subset of T with inner measure

0 and outer measure 1, and let $F_k(t) = \{g(t)/k\}$. Then $\lim \int F_k = \phi$ but $\int \lim F_k = \{0\}$.

Let A be an arbitrary subset of a metric space X, and let G_x be a set-valued function defined for x in A, whose values are subsets of E^n. G is called *upper-semicontinuous* if $x_n \to x$ implies $G_x \supset \lim \sup G_{x_n}$ for all x_n and x in A; *lower-semicontinuous* if $x_n \to x$ implies $G_x \subset \lim \inf G_{x_n}$ for all x_n and x in A; and *continuous* if $x_n \to x$ implies $G_x = \lim G_{x_n}$ for all x_n and x in A. This is the same as the standard definition of these terms (see, for example, Karlin [15], p. 409).

COROLLARY 5.2. *Let $F_x(t)$ be a set-valued function defined for $t \in T$ and $x \in A$, all of whose values are bounded by the same integrable point-valued function, and such that F_x is Borel-measurable for each fixed $x \in A$. Then if $F_x(t)$ is upper-semicontinuous in x for each fixed t, then $\int F_x$ is upper-semicontinuous; if $F_x(t)$ is lower-semicontinuous in x for each fixed t, then $\int F_x$ is lower semicontinuous; and if $F_x(t)$ is continuous for each fixed t, then $\int F_x$ is continuous.*

PROOF. Follows from Propositions 4.1 and 5.1.

APPLICATIONS TO EXTREME POINTS OF SETS OF VECTOR FUNCTIONS

In this section, we make use of Proposition 2.1 only; the other results will not be used. Let A be a compact convex subset of E^n, B the set of its extreme points, cl (B) the closure of B. If $n \geqslant 3$, it is not necessarily true that cl $(B) = B$. Let $\mathscr{M}_A$, $\mathscr{M}_B$, and $\mathscr{M}_{\text{cl}(B)}$ be the sets of all measurable functions from T to A, B, and cl (B) respectively. Since the space of all measurable functions from T to E^n has a linear structure, we may discuss the extreme points of $\mathscr{M}_A$. In [12], Karlin proved that the set of extreme points of $\mathscr{M}_A$ includes $\mathscr{M}_B$ and is included in $\mathscr{M}_{\text{cl}(B)}$.

PROPOSITION 6.1. *The set of extreme points of $\mathscr{M}_A$ is precisely $\mathscr{M}_B$.*

PROOF. Clearly, every point of $\mathscr{M}_B$ is an extreme point of $\mathscr{M}_A$. Conversely, let f be an extreme point of $\mathscr{M}_A$. If f is not in $\mathscr{M}_B$, then for some t, $f(t)$ is not an extreme point of A. Hence for each t we may choose $g(t)$ and $h(t)$ in A so that $f(t) = \frac{1}{2}g(t) + \frac{1}{2}h(t)$, and $g(t)$ and $h(t)$ differ from $f(t)$ for at least some t. Because of Proposition 2.1, g and h may be chosen so as to be measurable as well. Then g and h are in $\mathscr{M}_A$ and $f = \frac{1}{2}g + \frac{1}{2}h$, contradicting the extreme point property of f. This proves the proposition.

Another situation treated by Karlin in [12] is the following: Let $\mu_1, \cdots, \mu_m$ be a set of nonatomic totally finite measures on T, and let $a_1, \cdots, a_m$, $b_1, \cdots, b_m$ be in E^n. Let A and B be as above. Let $\mathscr{G}$ be the subset of $\mathscr{M}_A$

consisting of functions f such that $b_i \leqslant \int f d\mu_i \leqslant a_i$ for $i = 1, \cdots, m$. Karlin proved that the extreme points of $\mathscr{G}$ are contained in $\mathscr{M}_{\text{cl}(B)}$.

PROPOSITION 6.2. *The extreme points of $\mathscr{G}$ are contained in $\mathscr{M}_B$.*

PROOF. The proof follows Karlin's ideas, but use of Proposition 2.1 makes it simpler and yields the stronger result. Let f be an extreme point of $\mathscr{G}$, and suppose it is not in $\mathscr{M}_B$. Construct g and h as in the previous proof, and let $e = \frac{1}{2}g - \frac{1}{2}h$, so that $f + e = g \in \mathscr{M}_A$ and $f - e = h \in \mathscr{M}_A$. For $S \subset T$, set $\mu(S) = \{\int_S e d\mu_1, \cdots, \int_S e d\mu_m\}$. Then μ is a vector measure of dimension nm. Applying Lyapunov's theorem on vector measures [16], we obtain a subset S of T such that $\mu(S) = \frac{1}{2}\mu(T)$. Define e' on T by $e'(t) = e(t)$ for $t \in S$, and $e'(t) = -e(t)$ for $t \notin S$. Now define f_1 and f_2 by $f_1 = f + e'$, $f_2 = f - e'$. Then $f = \frac{1}{2}f_1 + \frac{1}{2}f_2$, f_1 and f_2 are in $\mathscr{M}_A$, and

$$\int f_1 d\mu_i = \int f d\mu_i + \int_S e d\mu_i - \int_{T \setminus S} e d\mu_i$$

$$= \int f d\mu_i + \tfrac{1}{2} \int e d\mu_i - \tfrac{1}{2} \int e d\mu_i = \int f d\mu_i.$$

Hence $f_1 \in \mathscr{G}$. Similarly $f_2 \in \mathscr{G}$, and the proof is complete.

If F is the set-valued function defined by $F(t) = A$ for all t, then $\mathscr{F} = \mathscr{M}_A$. Propositions analogous to 6.1 and 6.2 can be proved in the general case, when F need not be constant, but is assumed to have compact convex values. In this connection, we note that Theorems 3 and 4 of this paper are generalizations of Theorem 3 and Corollary 1 of [12].

ACKNOWLEDGMENTS

We gratefully acknowledge some very helpful conversations with Lester Dubins, Yakar Kannai, Bezalel Peleg, and Micha Perles. Also, we wish to thank Gerard Debreu for calling our attention to some of the references.

REFERENCES

1. S. KAKUTANI, A generalization of Brouwer's fixed point theorem. *Duke Math. J.* 8 (1941), 457-459.
2. S. EILENBERG AND D. MONTGOMERY, Fixed point theorems for multi-valued transformations. *Am. J. Math.* 8 (1946), 214-222.
3. K. J. ARROW AND G. DEBREU, Existence of an equilibrium for a competitive economy. *Econometrica* 22 (1954), 265-290.
4. L. W. MCKENZIE, On the existence of general equilibrium for a competitive market. *Econometrica* 27 (1959), 54-71.
5. K. VIND, Edgeworth allocations in an exchange economy with many traders. *Econ. Rev.* 5 (1964), 165-177.

6. H. KUDO, Dependent experiments and sufficient statistics. *Nat. Sci. Rept. Ochanomizu Univ., Tokyo* **4** (1954), 151-163.
7. H. RICHTER, Verallgemeinerung eines in der Statistik benötigten Satzes der Masstheorie. *Math. Annalen* **150** (1963), 85-90 (for correction to this article see same volume, pp. 440-441).
8. R. J. AUMANN, Existence of a competitive equilibrium in markets with a continuum of traders. *Econometrica* **34** (1966), 1-17 [Chapter 47].
9. R. J. AUMANN AND M. A. PERLES, A variational problem arising in economics. *J. Math. Anal. Appl.* **11** (1965), 465-503 [Chapter 70].
10. C. KURATOWSKI, "Topologie I." Monografie Matematyczne, Warsaw, 1948.
11. J. VON NEUMANN, On rings of operators. Reduction theory. *Ann. Math.* **50** (1949), 401-485.
12. S. KARLIN, Extreme points of vector functions. *Proc. Am. Math. Soc.* **4** (1953), 603-610.
13. H. G. EGGLESTON, "Convexity." Cambridge Univ. Press, 1958.
14. N. DUNFORD AND J. T. SCHWARTZ, "Linear Operators, Part I." Interscience, New York, 1958.
15. S. KARLIN, "Mathematical Methods and Theory in Games, Programming, and Economics," Vol. I. Addison-Wesley, Reading, Mass., 1959.
16. A. LYAPUNOV, Sur les fonctions-vecteurs complètement additives. *Bull. Acad. Sci. URSS, Sér. Math.* **4** (1940), 465-478.

69 An Elementary Proof that Integration Preserves Uppersemicontinuity

The purpose of this 'classroom note' is to prove, using comparatively elementary tools, that under appropriate boundedness conditions, the upper-semicontinuity of a relation[1] $F_p(t)$ in a parameter p is preserved by integration over t. Most of the previous proofs of this result [Aumann (1964), Schmeidler (1970), Hildenbrand (1974)] are far deeper, and all are more involved. In addition to the elementary facts of Lebesgue integration – such as Fatou's lemma and the Dominated Convergence Theorem – we will use only the Convexity Theorem of Lyapunov (1940), and a measurable selection theorem for measurable compact-valued correspondences.[2] It should be noted that Lyapunov's theorem has been proved in an entirely elementary fashion by Halmos (1948). As for the selection theorem, this depends on the measurability of analytic sets, and so is perhaps our most advanced tool. Nevertheless, it too can be proved from 'scratch' in a couple of pages, and certainly it is significantly simpler than the general selection theorem of von Neumann (1949).

The result proved here is the deepest of the lemmas needed for the existence of competitive equilibria in markets with a non-atomic continuum of traders [see e.g. Schmeidler (1969) or Hildenbrand (1974)]. Thus the Existence Theorem is brought within the reach of relatively unsophisticated audiences.

Let $(T, \mathscr{C}, \mu)$ be a complete non-atomic measure space, and let P be a metric space. The integral $\int x(t)\mu(dt)$ of a function x on T will be denoted $\int x$. A *selection* from a correspondence F on T is a function x on T such that $x(t) \in F(t)$ for all t. The *integral* of such a correspondence F is the set of all integrals of its selections; it is denoted $\int F$. A relation F on T to a Euclidean space E^n is *bounded by* a non-negative real-valued function h if $\|x\| \leq h(t)$ whenever $x \in F(t)$. A function x on T is *bounded by* h if $\|x(t)\| \leq h(t)$ for all t. A relation on P to E^n is *upper-semicontinuous* if its graph is closed.

[1] We use the term 'relation' to mean 'set-valued function'.
[2] Relations with non-empty values.

For each p in P and t in T, let $F_p(t) \subset E^n$. We wish to prove that if all the relations F_p are bounded by the same integrable function h, and if for each fixed t, $F_p(t)$ is uppersemicontinuous in p, then also $\int F_p$ is uppersemicontinuous in p. This may be restated as follows:

Lemma. Let F be a correspondence from T to E^n. Let $\{x_k\}$ be a sequence of measurable functions from T to E^n, all of which are bounded by the same integrable function h. Assume that for each t, each limit point of $\{x_k(t)\}$ belongs to $F(t)$. Then each limit point of $\int x_k$ belongs to $\int F$.

Before proving the lemma, we quote the selection theorem that we shall need. A relation on T to E^n is *measurable* if its graph is measurable in the product σ-field $\mathscr{C} \times \mathscr{B}$, where $\mathscr{B}$ denotes the Borel sets of E^n.

Compact-valued selection theorem. Every compact-valued measurable correspondence on T to E^n has a measurable selection.

This can be proved in a few lines from the 'projection theorem', which in turn is essentially equivalent to the absolute measurability of analytic sets.[3]

The proof of the lemma is by induction on n. For $n = 0$ there is nothing to prove. Suppose the lemma has been proved up to (and including) $n-1$; we will prove it for n. Let x be a limit point of $\int x_k$; we must show $x \in \int F$. W.l.o.g. (without loss of generality) we may assume $x = \lim \int x_k$; otherwise we can restrict attention to a subsequence of the originally given sequence. Let $D(t)$ be the set of all limit points of $x_k(t)$. Then $D(t) \subset F(t)$, and so it is sufficient to prove that $x \in \int D$. Contrariwise, suppose $x \notin \int D$. A well-known theorem of Richter (1963) states that the integral of a correspondence on a non-atomic measure space is convex,[4] and hence $\int D$ is convex. Since $x \notin \int D$, there is a hyperplane through x that supports $\int D$. W.l.o.g. we may assume that the vector $(1, 0, \ldots, 0)$ is orthogonal to this hyperplane; i.e., that

$$x^1 \leq \inf \{y^1 : y \in \int D\}. \qquad (1)$$

Define $\delta(t) = \liminf x_k^1(t)$; then δ is measurable. Since the x_k are bounded by h, so is D; in particular $D(t)$ is compact, and hence for each t there is an $x(t)$ in $D(t)$ with $x^1(t) = \delta(t)$. Hence if we define

$$B(t) = D(t) \cap \{y \in E^n : y^1 = \delta(t)\}, \qquad (2)$$

then $B(t)$ is compact and non-empty, and B is measurable. Hence from the

---

[3] See Hildenbrand (1974, Prop. 3 of D.II.3, p. 60; and D.I.11, p. 40).
[4] This theorem can be proved in a few lines from Lyapunov's Convexity Theorem [see Hildenbrand (1974, p. 62)].

Compact-Valued Selection Theorem it follows that there is a measurable function y such that for all t, $y^1(t) = \delta(t)$ and $y(t) \in D(t)$. Hence $\int y \in \int D$, and so from (1) it follows that

$$x^1 \leq \int \delta. \tag{3}$$

Hence from Fatou's lemma and the definition of δ we deduce

$$0 \leq \limsup \int |x_k^1 - \delta| = \limsup [\int (x_k^1 - \delta) + 2 \int \max(0, \delta - x_k^1)]$$

$$\leq \limsup \int (x_k^1 - \delta) + 2 \limsup \int \max(0, \delta - x_k^1)$$

$$\leq \int \lim x_k^1 - \int \delta + 2 \int \max(0, \delta - \liminf x_k^1) = x^1 - \int \delta \leq 0.$$

Hence $\int |x_k^1 - \delta| \to 0$ as $k \to \infty$, and hence there is a subsequence of $\{x_k^1\}$ that tends a.e. to δ. W.l.o.g. we may assume that it is the entire sequence; that is,

$$x_k^1(t) \to \delta(t) \text{ a.e.} \tag{4}$$

Hence it follows that a.e. $D(t) \subset \{y \in E^n : y^1 = \delta(t)\}$. Now set $H = \{y \in E^n : y^1 = 0\}$, and define

$$D^*(t) = D(t) - (\delta(t), 0, \ldots, 0),$$

$$x_k^*(t) = x_k(t) - (x_k^1(t), 0, \ldots, 0).$$

Then $D^*(t) \subset H$ and $x_k^*(t) \in H$. From (4) it follows that every limit point of $\{x_k^*(t)\}$ belongs to $D^*(t)$. Hence from the induction hypothesis it follows that every limit point of $\int x_k^*$ belongs to $\int D^*$. Hence from Lebesgue's Dominated Convergence Theorem, and from (4), we obtain

$$x = \lim \int x_k = \lim \int x_k^* + \int (\lim x_k^1, 0, \ldots, 0)$$

$$= \lim \int x_k^* + \int (\delta, 0, \ldots, 0) \in \int D^* + \int (\delta, 0, \ldots, 0) = \int D.$$

This completes the proof of the lemma.

References

Aumann, R. J., 1964, Integrals of set-valued functions, J. Math. Anal. Appl. 12, 1–12 [Chapter 68].
Halmos, P.R., 1948, The range of a vector measure, Bull. Amer. Math. Soc. 54, 416–421.
Hildenbrand, W., 1974, Core and equilibria of a large economy (Princeton University Press, Princeton).

Lyapunov, A., 1940, Sur les fonctions-vecteur completement additives, Bull. Acad. Sci. URSS Ser. Math. 4, 465–478.

Richter, H., 1963, Verallgemeinerung eines in der Statistik benötigten Satzes der Masstheorie, Math. Annalen 150, 85–90.

Schmeidler, D., 1969, Competitive equilibria in markets with a continuum of traders and incomplete preferences, Econometrica 37, 635–645.

Schmeidler, D., 1970, Fatou's lemma in several dimensions, Proc. Amer. Math. Soc. 24, 300–306.

von Neumann, J., 1949, On rings of operators, reduction theory, Ann. of Math. 50, 401–485.

70 A Variational Problem Arising in Economics
with M. Perles

INTRODUCTION

Let $u(x, t)$ be a real function of the two real variables x and t, where $x \geq 0$ and $0 \leq t \leq 1$. Suppose that u is continuous and monotone increasing in x for each fixed t. Consider the following problem:

Find an integrable function $\mathbf{x}(t)$ that maximizes $\int_0^1 u(\mathbf{x}(t), t)\, dt$, subject to the conditions $\mathbf{x}(t) \geq 0$ for each t, and $\int_0^1 \mathbf{x}(t)\, dt = 1$.

Unfortunately, the maximum in this problem need not be attained, even when the supremum of the integral is finite, and even when u is very "regular." In fact, $u(x, t) = xt$ provides a counterexample; we have $\int_0^1 t\mathbf{x}(t)\, dt < 1$ for any feasible choice of $\mathbf{x}$, but the supremum is 1.

Intuitively, the maximum is not attained in this problem because it is "worthwhile" to concentrate all the area $\int_0^1 \mathbf{x}(t)\, dt$ at our disposal on a t-interval that is close to 1; i.e., to choose $\mathbf{x}(t)$ large for t close to 1, and 0 elsewhere. If we would assume $u(x, t) = o(x)$ as $x \to \infty$, this might no longer be worthwhile. This assumption, when made for each t separately, is still not sufficient to ensure that the maximum is attained; some kind of uniformity condition is needed. Uniform convergence of $u(x, t)/x$ to 0 is sufficient, but not necessary; it turns out that the proper condition is that of *integrable* convergence, which we shall now define.

DEFINITION 1. Let $f(x, t)$ be a real valued function for $x \geq 0$ and $0 \leq t \leq 1$. Then $f(x, t) = o(x)$ as $x \to \infty$, *integrably* in t, if for each $\epsilon > 0$ there is an integrable function $\eta(t)$, such that $|f(x, t)| \leq \epsilon x$ whenever $x \geq \eta(t)$.

Integrable convergence reduces to uniform convergence when $\eta(t)$ is a constant, or equivalently, when it is bounded. The two concepts are not equivalent; $x^{1/2}/t^{1/4} = o(x)$ integrably, but not uniformly. The relation

---

The research described in this paper was partially supported by the U.S. Office of Naval Research, Logistics and Mathematical Statistics Branch, under contract No. N62558-3586.

This chapter originally appeared in *Journal of Mathematical Analysis and Applications* 11 (1965): 488–503. © Academic Press, Inc. Reprinted with permission.

between uniform and integrable convergence is roughly similar to that between a uniformly bounded and an integrably bounded sequence.

The main result of this paper holds when x may have values in an arbitrary euclidean space E^n (rather than E^1). Let us recall that a vector x in E^n is called nonnegative ($x \geqslant 0$) if all its coordinates are nonnegative,[1] and that a real function $g(x)$ defined for $x \geqslant 0$ is called *nondecreasing (increasing)* if it is nondecreasing (increasing) in each variable separately. Also, g is called *upper-semicontinuous* if $g(x) = \limsup_{y \to x} g(y)$ for all x in the domain of definition of g. The unit interval $[0, 1]$ will be denoted by T.

MAIN THEOREM. *Let $u(x, t)$ be a Borel-measurable[2] nonnegative real-valued function defined for $x \geqslant 0$ in E^n and t in T, which is nondecreasing and upper-semicontinuous in x for each fixed t. Assume further that*

(A) $u(\xi, \cdots, \xi, t) = o(\xi)$ *as* $\xi \to \infty$, *integrably in t.*

Let $a \geqslant 0$ be in E^n, and let $\mathscr{P}(u, a)$ be the problem:

Maximize $\int_T u(\mathbf{x}(t), t)\, dt$ subject to $\mathbf{x}(t) \geqslant 0$ for all $t \in T$ and $\int_T \mathbf{x}(t)\, dt = a$. Then $\mathscr{P}(u, a)$ has a solution.

The asymptotic condition (A) may be replaced by a similar condition along any *positive* ray. Because u is nondecreasing, any one of these conditions is equivalent to the condition that $u(x, t) = o(\|x\|)$ as $\|x\| \to \infty$, integrably in t, where $\|x\|$ is any one of the usual norms on E^n.

The condition that u be nondecreasing can be dispensed with, if the condition $\int_T \mathbf{x}(t)\, dt = a$ is replaced by $\int_T \mathbf{x}(t)\, dt \leqslant a$, and certain other slight changes are made (cf. Section 6).

The proof will be in two stages. First we will prove (Section 2) that the main theorem holds when u is concave. This proof depends on arguments involving weak compactness. For nonconcave u, we define (Section 3) the concavified function u^* in a manner similar to that of Shapley and Shubik [1], and show (Section 4) that the problems $\mathscr{P}(u, a)$ and $\mathscr{P}(u^*, a)$ have the same value and have solutions in common (where the *value* of $\mathscr{P}(u, a)$ is the supremum of $\int_T u(\mathbf{x}(t), t)\, dt$ under the restrictions on $\mathbf{x}$). In particular, we construct a solution of $\mathscr{P}(u, a)$ from a solution of $\mathscr{P}(u^*, a)$. In this part of the proof we make use of the integral of a set-valued function, as studied in [2].

The proof is an existence proof, and does not yield a characterization of the solution. A characterization is given in Section 5; it is relatively easily obtained—much more easily than the existence. The problem is really an infinite dimensional analogue of a nonlinear programming problem in the sense of Kuhn and Tucker [3], and the characterization is derived from reasoning similar to that leading to their characterizations.

---

[1] Note that this *differs* from the standard usage, in which this relation is denoted $\geqq$.
[2] In all variables simultaneously.

Suppose that u is actually increasing. If we assume $u(\xi, \cdots, \xi, t) = o(\xi)$ uniformly, then we may conclude that $\mathscr{P}(u, a)$ has a bounded solution. Indeed, for any p with $1 \leqslant p \leqslant \infty$ we may define the notion $f(\xi, t) = o(\xi)$ *p-integrably* by adding the condition $\eta(t) \in L^p$ to Definition 1. Then if it is assumed that $u(\xi, \cdots, \xi, t) = o(\xi)$ *p*-integrably, it may be concluded (Section 6) that $\mathscr{P}(u, a)$ has a solution in L^p, and indeed all solutions are in L^p. This conclusion does not hold if u is merely nondecreasing (Section 6). Various other counterexamples are presented in Section 6, and also the generalization of the main theorem to u that are not even nondecreasing that we discussed above.

In Section 1 we explain the notations, terminology, and conventions used in the paper.

Throughout the paper, the unit interval T may be replaced by the real line $(-\infty, \infty)$, the half line $(0, \infty)$, or indeed any Borel set on the real line. The proofs are not affected.

Following are indications of two of the economic applications. The problem treated here arose in connection with an investigation of markets with a continuum of traders and transferable utilities, being conducted by L. S. Shapley and one of the authors. There x and t stand for a commodity bundle and a trader respectively; $\int_T u(\mathbf{x}(t), t) \, dt$ represents the aggregate utility of the coalition T under the commodity-assignment $\mathbf{x}$. If a is the aggregate (initial) commodity bundle held by the coalition, then the value of $\mathscr{P}(u, a)$—if it is attained—is the maximum aggregate utility that the coalition T can assure itself by trading among its own members.

In other economic applications, t stands for "time" rather than "trader." One of the interpretations possible in this direction is that x stands for a vector of resources, $u(x, t)$ is the (discounted) return from using the vector x of resources at time t, and a represents the total amount of resources available. Then $\int_T u(\mathbf{x}(t), t) \, dt$ is the total value of a program $\mathbf{x}$ of resource use, and the problem $\mathscr{P}(u, a)$ is that of finding a program that will maximize this value.[3]

1. Preliminaries

The Borel-measurability of u will be assumed throughout the paper. This is needed mainly to assure the Lebesgue-measurability of $u(\mathbf{x}(t), t)$ for each Lebesgue-measurable $\mathbf{x}$. Throughout Sections 1-5 it will be assumed that u is nonnegative.

---

[3] In a recent publication [7], M. Yaari has treated such a problem for the case in which x is one-dimensional and $u(x, t) = \alpha(t) g(x)$, where α is bounded and continuous and g is concave. Such problems have also been discussed by Arrow, Chakravarty, Karlin, Koopmans, Strotz, and others.

A number of conventions: $x \cdot y$ is the scalar product of x and y. The symbol 0 denotes the origin of E^n as well as the number zero. If $\mathbf{y}$ is a function on T, we write $\int \mathbf{y}$ for $\int_T \mathbf{y}(t)\, dt$. Abusing our notation, we write $u(\mathbf{x})$ for the function on T whose value at t is $u(\mathbf{x}(t), t)$; in particular, therefore, $\int u(\mathbf{x})$ means $\int_T u(\mathbf{x}(t), t)\, dt$. The nonnegative orthant $\{x \in E^n : x \geqslant 0\}$ is denoted P. Superscripts will be used exclusively to denote coordinates. $x \geqslant y$ means $x^i \geqslant y^i$ for[4] all i. We will use the phrase "all t in T" to mean the same as "almost all t in T"; the two phrases will be used interchangeably. The closure of a set B is denoted cl (B). The vector $(1, \cdots, 1)$ in E^n will be denoted e, and the vector $(0, \cdots, 0, 1, 0, \cdots, 0)$, where 1 is in the ith place, will be denoted e_i. For $x \in P$, we will write Σx instead of $\sum_{i=1}^n x^i$. When we say that u is "increasing," "continuous," etc., we mean "increasing in x for each fixed t," "continuous in x for each fixed t," etc. The symbol $\setminus$ denotes set-theoretic subtraction, and μ denotes Lebesgue measure on T. val $\mathscr{P}(u, a)$ denotes the value of $\mathscr{P}(u, a)$.

It is convenient to view the space of all integrable functions from T to E^n as a Banach space, with norm $\int \Sigma |\mathbf{x}|$. If we write $x^i(t) = x(t; i)$, then we see that this space is precisely $L^1(T \times \{1, \cdots, n\})$. All references to weak convergence, strong convergence, etc. will refer to this space.

A real function f on P will be called *concave* if

$$f(\theta x + (1 - \theta) y) \geqslant \theta f(x) + (1 - \theta) f(y)$$

for all x, y in P and θ in $[0, 1]$. A concave function must be continuous in the interior of P, but may have jumps on the boundary. If it is upper-semicontinuous, then it is a fortiori continuous everywhere.

2. The Concave Case

Lemma 2.1. *Let a be given, and suppose that u is continuous and nondecreasing, and satisfies the asymptotic condition* (A). *Let* $\mathbf{x}, \mathbf{y}_1, \mathbf{y}_2, \cdots$ *be integrable functions from T to P such that $\int \mathbf{y}_j \leqslant a$ and $\mathbf{y}_j \to \mathbf{x}$ almost everywhere (a.e.). Then*

$$\int u(\mathbf{y}_j) \to \int u(\mathbf{x}).$$

Proof. Without loss of generality (w.l.o.g.) let $\Sigma a = 1$. Since $\mathbf{y}_j \to \mathbf{x}$ a.e., it follows that $u(\mathbf{y}_j)$ converges a.e. to $u(\mathbf{x})$. Let $\epsilon > 0$ be given, and choose an integrable η such that $\mathbf{x}(t) < \eta(t)e$ for all t, and $u(\xi e, t) < \epsilon \xi$ whenever $\xi \geqslant \eta(t)$. Let $U = U_k = \{t : \mathbf{y}_k(t) \leqslant \eta(t)e\}$. Since convergence

---
[4] Note that this *differs* from the standard usage, in which this relation is denoted $\geqq$.

a.e. implies convergence in measure,[5] it follows that $\mu(T \setminus U_k) \to 0$ as $k \to \infty$. Let χ_U denote the characteristic function of U. Then

$$\int |u(\mathbf{y}_k) - u(\mathbf{x})| = \int_U + \int_{T \setminus U}$$

$$\leq \int \chi_U |u(\mathbf{y}_k) - u(\mathbf{x})| + \int_{T \setminus U} u(\mathbf{x}) + \int_{T \setminus U} u(\mathbf{y}_k).$$

The integrand in the first term of the last line is bounded by $u(\eta e)$, which in turn is $\leq \epsilon \eta$; so we may apply Lebesgue's dominated convergence theorem and deduce that the first term tends to 0. The second term tends to 0 because $\mu(T \setminus U) \to 0$ as $k \to \infty$. The third term is

$$\leq \epsilon \int_{T \setminus U} \max_i y_k^i \leq \epsilon \int \sum y_k \leq \epsilon \sum a = \epsilon.$$

Hence

$$\limsup_{k \to \infty} \int |u(\mathbf{y}_k) - u(\mathbf{x})| \leq \epsilon,$$

and hence it vanishes. This proves the lemma.

PROPOSITION 2.2. *Suppose that u is continuous, concave and nondecreasing, and satisfies the asymptotic condition (A). Then val $\mathscr{P}(u, a)$ is attained.*

PROOF. Assume w.l.o.g. that $u(0, t) = 0$ for all t, and that $\Sigma a = 1$. Let $\alpha = \text{val } \mathscr{P}(u, a)$. Clearly val $\mathscr{P}(u, y)$ is a nondecreasing function of y; let $D = \{y \in P : y \leq a \text{ and val } P(u, y) = \alpha\}$. Let $\lambda = \inf \{\Sigma y : y \in D\}$, $\{b_k\}$ a sequence of points in D such that $\Sigma b_k \to \lambda$, and let $\{c_k\}$ be a convergent subsequence of $\{b_k\}$; set $c = \lim_k c_k$. For each k, we have val $\mathscr{P}(u, c_k) = \alpha$; let $\{\mathbf{x}_k\}$ be a sequence of integrable functions such that $\int \mathbf{x}_k = c_k$ for all k, and $\int u(\mathbf{x}_k)$ is a nondecreasing sequence that approaches α as $k \to \infty$. We wish to show that $\{\mathbf{x}_k\}$ has a weakly convergent subsequence.

To show this, it is sufficient to show that for each i and each decreasing sequence $\{S_m\}$ of subsets of T with void intersection, we have

$$\int_{S_m} \mathbf{x}_k^i \to 0 \qquad \text{as} \qquad m \to \infty, \qquad \text{uniformly in } k \qquad (2.3)$$

---

[5] In case T has infinite Lebesgue measure, we may use the measure $\nu(S) = \int_S u(\mathbf{x})$ rather than Lebesgue measure μ. Since $\mathbf{x}$ is integrable we always have $\nu(T) < \infty$, and can therefore deduce that convergence a.e. implies convergence in the measure ν, which is what is needed below.

(Dunford-Schwartz [4], p. 292, Theorem IV.8.9). Fix i. If $c^i = 0$, (2.3) follows easily from $\int x_k^i = c_k^i \to c^i = 0$. Assume therefore that $c^i > 0$. Let $\epsilon > 0$ be given, and assume w.l.o.g. that $\frac{1}{2}\epsilon < c^i$. Let

$$b = c - \tfrac{1}{2}\epsilon e_i + \sum_{j=1}^{n} \min\left(\frac{\epsilon}{4n}, c^j - a^j\right) e_j \quad \text{and} \quad \beta = \operatorname{val} \mathscr{P}(u, b).$$

Since $\sum b < \lambda$, it follows that $b \notin D$, and hence $\beta < \alpha$. Let $\gamma = \frac{1}{2}(\alpha - \beta)$. Since $\int u(\mathbf{x}_k) \to \alpha$, it follows that $\int u(\mathbf{x}_k) > \beta + \gamma$ for $k > k_0 = k_0(\epsilon)$. Choose an integrable η so that $u(\xi e, t) \leq \gamma \xi$ whenever $\xi \geq \eta(t)$; a fortiori, $u(x, t) \leq \gamma \sum x$ whenever $x^i \geq \eta(t)$. For each k, let

$$U = U_k = \{t : \mathbf{x}_k^i(t) \leq \eta(t)\}.$$

Choose k_1 so that $k_1 \geq k_0$ and $c_k^j \leq \min(a^j, c^j + (\epsilon/4n)) = b^j$ for all $k \geq k_1$ and all j. Then

$$\int_{T\setminus U} \mathbf{x}_k^i < \tfrac{1}{2}\epsilon \quad \text{for} \quad k \geq k_1. \tag{2.4}$$

Indeed, if $\int_{T\setminus U} \mathbf{x}_k^i \geq \tfrac{1}{2}\epsilon$, then

$$\int_U \mathbf{x}_k^i \leq c_k^i - \tfrac{1}{2}\epsilon \leq b^i, \quad \text{and} \quad \int_U \mathbf{x}_k^j \leq c_k^j \leq b^j \quad \text{for all } j \neq i;$$

then $\int_U \mathbf{x}_k \leq b$, and hence $\int_U u(\mathbf{x}_k) \leq \beta$. Hence

$$\int u(\mathbf{x}_k) = \int_U + \int_{T\setminus U} \leq \beta + \int_{T\setminus U} \gamma \sum \mathbf{x}_k$$

$$\leq \beta + \gamma \sum \int \mathbf{x}_k = \beta + \gamma \sum c_k \leq \beta + \gamma \sum a = \beta + \gamma.$$

This contradicts the definitions of k_1 and k_0, and so proves (2.4).

Since η and $\mathbf{x}_1, \cdots, \mathbf{x}_{k_1}$ are integrable, we may choose m_0 so that whenever $m > m_0$ we have $\int_{S_m} \eta < \tfrac{1}{2}\epsilon$ and $\int_{S_m} \mathbf{x}_k^i < \epsilon$ for all $k \leq k_1$. Then if $m > m_0$ and $k \leq k_1$, then $\int_{S_m} \mathbf{x}_k^i < \epsilon$; if $k > k_1$, then by (2.4), we again have

$$\int_{S_m} \mathbf{x}_k^i = \int_{S_m \cap U} + \int_{S_m \setminus U} \leq \int_{S_m} \eta + \int_{T\setminus U} \mathbf{x}_k^i < \tfrac{1}{2}\epsilon + \tfrac{1}{2}\epsilon = \epsilon.$$

This proves (2.3).

We conclude that $\{\mathbf{x}_k\}$ has a subsequence converging weakly to some $\mathbf{x}$. Then there is a sequence of functions converging strongly (i.e., in norm) to $\mathbf{x}$, each one of which is a (finite) convex combination of $\mathbf{x}_1, \mathbf{x}_2, \cdots$ [4, p. 422, corollary V.3.14]. Now every strongly convergent sequence in L^1 has a subsequence that converges a.e. to the same limit; so there is a sequence $\{y_j\}$ of convex combinations of $x_1, x_2, \cdots$ that converges a.e. to x.

From the concavity of u and the fact that the sequence $\int u(x_k)$ is increasing, it follows that $\int u(y_j) \geqslant \int u(x_1)$; combining this with Lemma 2.1, we deduce that $\int u(x) \geqslant \int u(x_1)$. But for each k, the sequence $\{x_k, x_{k+1}, \cdots\}$ converges weakly to x, so we may conclude in the same way that $\int u(x) \geqslant \int u(x_k)$ for each k. Hence $\int u(x) \geqslant \alpha$. But $\{z : \int z \leqslant a\}$ is weakly closed, so $\int x \leqslant a$. Let $d = \int x$; then $\int [x + a - d] = a$. So

$$\alpha \leqslant \int u(x) \leqslant \int u(x + a - d) \leqslant \text{val } \mathscr{P}(u, a) = \alpha.$$

Hence equality holds throughout, and the proposition is proved.

3. Concavification

Let f be a nonnegative real-valued function on P and let

$$F = \{(\nu, x) \in E^{n+1} : x \geqslant 0, 0 \leqslant \nu \leqslant f(x)\}.$$

Let F^* be the convex hull of F. If there is a function f^* on P such that

$$F^* = \{(\nu, x) \in E^{n+1} : x \geqslant 0, 0 \leqslant \nu \leqslant f^*(x)\},$$

then f^* is called the *concavification* of f. The definite article is justified by the fact that there can be at most one concavification. There may be none, as has been shown by Shapley and Shubik [1].

When it exists, the concavification is always concave; this follows from the definition. If f is concave, then f^* exists and equals f.

PROPOSITION 3.1. *If f is nondecreasing and upper-semicontinuous, and $f(\xi e) = o(\xi)$ as $\xi \to \infty$, then f has a nondecreasing and continuous concavification.*

PROOF. A slightly weaker form of this lemma has been proved by Shapley and Shubik[6] [1, Theorem 3]. We first show that F^* is closed.

Let $(\nu, x) \in \text{cl } (F^*)$. Then there is a sequence $\{(\nu_k, x_k)\}$ of members of F^* that tends to (ν, x). Now recall Caratheodory's theorem, which states that if F and F^* are subsets of E^{n+1} such that F^* is the convex hull of F, then every point of F^* is a convex combination of $n + 2$ points of F [5, p. 34 ff.]. Hence we have

$$(\nu_k, x_k) = \sum_{j=1}^{n+2} \alpha_{kj}(\lambda_{kj}, y_{kj}),$$

---

[6] Their theorem contains a monotonicity assumption that is slightly stronger than ours.

where $\alpha_{kj} \geq 0$, $\sum_{j=1}^{n+2} \alpha_{kj} = 1$, and $(\lambda_{kj}, y_{kj}) \in F$. The sequence of points $(\alpha_{k1}, \cdots, \alpha_{k,n+2})$ in E^{n+2} has a limit point, which we call $(\alpha_1, \cdots, \alpha_{n+2})$; w.l.o.g. assume it is the limit. We have $\alpha_j \geq 0$ and $\sum_{j=1}^{n+2} \alpha_j = 1$. Some of the α_j may vanish, but not all of them can vanish; for convenience, assume that α_j vanishes for $j \leq m$, and does not vanish for $j > m$.

Let $\beta = \min(\alpha_{m+1}, \cdots, \alpha_{n+2}) > 0$. For sufficiently large k, we have $\alpha_{kj} > \beta/2$ for all $j > m$, and $(\nu_k, x_k) < (\nu, x) + (1, e)$. Hence

$$(\nu, x) + (1, e) > (\nu_k, x_k) \geq \sum_{j=m+1}^{n+2} \alpha_{kj}(\lambda_{kj}, y_{kj}) \geq \frac{\beta}{2}(\lambda_{kj}, y_{kj})$$

for k sufficiently large and all $j > m$. It follows that the sequences $\{(\lambda_{kj}, y_{kj})\}$ are bounded when $j > m$. They therefore have limit points (λ_j, y_j) and we may assume w.l.o.g. that $(\lambda_{kj}, y_{kj}) \to (\lambda_j, y_j)$ as $k \to \infty$. Hence

$$(\nu, x) = \sum_{j=m+1}^{n+2} \alpha_j(\lambda_j, y_j) + \lim_{k \to \infty} \sum_{j=1}^{m} \alpha_{kj}(\lambda_{kj}, y_{kj}). \tag{3.2}$$

We conclude from this that the limit on the right-hand side of (3.2) exists; denote it by (λ, y).

Define real numbers ζ and ζ_{kj} by $\zeta = \Sigma y$, $\zeta_{kj} = \Sigma y_{kj}$. Then

$$\alpha_{kj}\zeta_{kj} \leq \sum_{j=1}^{m} \alpha_{kj}\zeta_{kj} < \zeta + 1$$

for sufficiently large k and all $j \leq m$. For given $\epsilon > 0$, let η be such that $f(\xi e) \leq \epsilon\xi$ for all $\xi \geq \eta$. Then because f is nondecreasing and $(\lambda_{kj}, y_{kj}) \in F$, we have

$$\lambda_{kj} \leq f(y_{kj}) \leq f(\zeta_{kj}e) \leq \begin{cases} \epsilon\zeta_{kj} & \text{if } \zeta_{kj} \geq \eta \\ f(\eta e) & \text{if } \zeta_{kj} \leq \eta. \end{cases}$$

In any case
$$\lambda_{kj} \leq \max(\epsilon\zeta_{kj}, f(\eta e)).$$
Hence
$$\alpha_{kj}\lambda_{kj} \leq \max(\epsilon\alpha_{kj}\zeta_{kj}, \alpha_{kj}f(\eta e)) \leq \max(\epsilon(\zeta + 1), \alpha_{kj}f(\eta e))$$

for k sufficiently large and all $j \leq m$. Since $\alpha_{kj} \to 0$ as $k \to \infty$ and $f(\eta e)$ is fixed, it follows that

$$\limsup_k \alpha_{kj}\lambda_{kj} \leq \epsilon(\zeta + 1).$$

But since ϵ was chosen arbitrarily, it follows that the lim sup vanishes and so the limit exists and vanishes. Hence $\lambda = 0$.

Now (λ_j, y_j) is the limit of (λ_{kj}, y_{kj}) for $j > m$; because f is upper-semicontinuous, F is closed, so from $(\lambda_{kj}, y_{kj}) \in F$ it follows that $(\lambda_j, y_j) \in F$. Now this means that $\lambda_j \leq f(y_j)$. Because f is nondecreasing and $y \geq 0$, it follows that $\lambda_j \leq f(y_j + y)$; hence $(\lambda_j, y_j + y) \in F$. But since $\alpha_j = 0$ for $j \leq m$, it follows that $\sum_{j=m+1}^{n+2} \alpha_j = 1$. Hence from (3.2) and $\lambda = 0$ we conclude

$$(\nu, x) = \sum_{j=m+1}^{n+2} \alpha_j(\lambda_j, y_j) + (0, y) = \sum_{j=m+1}^{n+2} \alpha_j(\lambda_j, y_j + y).$$

Since $(\lambda_j, y_j + y) \in F$, it follows that $(\nu, x) \in F^*$, and we have proved that F^* is closed.

If η is chosen so that $f(\xi e) \leq \xi$ whenever $\xi \geq \eta$, then for all x,

$$f(x) \leq f\left(e \sum x\right) \leq \max\left(\sum x, f(\eta e)\right) \leq \sum x + f(\eta e).$$

It follows that

$$F \subset \left\{(\nu, x) : 0 \leq x, 0 \leq \nu \leq \sum x + f(\eta e)\right\}.$$

The right side of this inclusion is convex, so F^* is also included in it. Hence if $(\nu, x) \in F^*$, then $\nu \leq \sum x + f(\eta e)$. Hence for each x, the set $\{\nu : (\nu, x) \in F^*\}$ is bounded, and because F^* is closed, it is compact. Hence the maximum of this set is attained, and this maximum is precisely $f^*(x)$. The monotonicity of f^* follows from that of f. To show that f^* is continuous, suppose that x is a point of discontinuity. Let $x_k \to x$ and $f^*(x_k) \to \psi \neq f^*(x)$. Since F^* is closed, f^* is upper-semicontinuous, so $\psi > f^*(x)$ is impossible. Hence $\psi < f^*(x)$; let $\theta = f^*(x) - \psi > 0$. For $z > 0$ sufficiently small, it follows from the upper-semicontinuity of f^* that $f^*(x + z) < f^*(x) + \frac{1}{2}\theta$. Now for k sufficiently large, $x - x_k < z$, and hence $\frac{1}{2}x_k + \frac{1}{2}(x + z) > x$. Hence

$$f^*(x) \leq f^*(\tfrac{1}{2}x_k + \tfrac{1}{2}(x+z)) \leq \tfrac{1}{2}f^*(x_k) + \tfrac{1}{2}f^*(x+z)$$
$$\leq \tfrac{1}{2}f^*(x_k) + \tfrac{1}{2}f^*(x) + \tfrac{1}{4}\theta.$$

Letting $k \to \infty$, we deduce

$$\tfrac{1}{2}f^*(x) \leq \tfrac{1}{2}\psi + \tfrac{1}{4}\theta < \tfrac{1}{2}(\psi + \theta) = \tfrac{1}{2}f^*(x),$$

a contradiction. This completes the proof of the proposition.

LEMMA 3.3. *Suppose u is upper-semicontinuous and nondecreasing, and satisfies the asymptotic condition (A). For fixed t, let $u^*(x, t)$ be the concavification of $u(x, t)$. Then u^* also satisfies the asymptotic condition (A).*

PROOF. Let $\epsilon > 0$ be given. Set

$$\mathbf{F}(t) = \{(v, x) : 0 \leqslant x, 0 \leqslant v \leqslant u(x, t)\}. \tag{3.4}$$

Choose η in accordance with Definition 1 to correspond to $\epsilon/(n+1)$. Let

$$\mathbf{H}(t) = \left\{(v, x) : 0 \leqslant x, 0 \leqslant v \leqslant \frac{\epsilon}{n+1} \sum x + u(\eta(t)e, t)\right\}.$$

Then $\mathbf{F}(t) \subset \mathbf{H}(t)$, and since $\mathbf{H}(t)$ is convex it follows that also $\mathbf{F}^*(t) \subset \mathbf{H}(t)$. Hence for each ξ and t,

$$u^*(\xi e, t) \leqslant \epsilon \frac{n}{n+1} \xi + u(\eta(t) e, t).$$

Hence if $\xi \geqslant (n+1) u(\eta(t) e, t)/\epsilon$, then $u^*(\xi e, t) \leqslant \epsilon \xi$. From condition (A) and the integrability of η it follows that $u(\eta(t)e, t)$ is integrable. This proves the lemma.

LEMMA 3.5. *Under the conditions of Lemma 3.3, the Borel-measurability of u^* follows from that of u.*

PROOF. From the definition of u^*, it follows that

$$u^*(x, t) = \max \left(\sum_{i=1}^{k} \alpha_i u(x_i, t) : k > 0, \alpha_i \geqslant 0 \quad \text{and} \right.$$

$$\left. x_i \geqslant 0 \text{ for } i = 1, \cdots, k, \sum_{i=1}^{k} \alpha_i = 1, \sum_{i=1}^{k} \alpha_i x_i = x \right). \tag{3.6}$$

We claim that $u^*(x, t) < \gamma$ if and only if there is a positive integer m such that if k is a positive integer, $\beta_1, \cdots, \beta_k$ are nonnegative *rational* numbers summing to 1, and $y_1, \cdots, y_k$ are rational points in P such that

$$\sum_{i=1}^{k} \beta_i y_i < x + \frac{e}{m},$$

then

$$\sum_{i=1}^{k} \beta_i u(y_i, t) < \gamma - \frac{1}{m} \tag{3.7}$$

Indeed, the "only if" part of the previous sentence follows from the monotonicity of u^* and the fact that for sufficiently large m,

$$u^*\left(x + \frac{e}{m}, t\right) < \gamma - \frac{1}{m}$$

(because of the continuity of u^*). To demonstrate the "if" part, assume that $u^*(x, t) \geq \gamma$, and let the max in (3.6) be assumed at $k, \alpha_1, \cdots, \alpha_k, x_1, \cdots, x_k$. Choose nonnegative rational numbers $\beta_1, \cdots, \beta_k$ summing to 1 and rational points y_i such that $y_i \geq x_i$ for all i, with the β_i sufficiently close to the α_i and the y_i sufficiently close to the x_i so that

$$\sum_{i=1}^{k} \beta_i u(x_i, t) \geq \sum_{i=1}^{k} \alpha_i u(x_i, t) - \frac{1}{m} = u^*(x, t) - \frac{1}{m} \geq \gamma - \frac{1}{m} \qquad (3.8)$$

and

$$\sum_{i=1}^{k} \beta_i y_i < \sum_{i=1}^{k} \alpha_i x_i + \frac{e}{m} = x + \frac{e}{m}. \qquad (3.9)$$

Then from (3.8) and $y_i \geq x_i$ we deduce

$$\sum_{i=1}^{k} \beta_i u(y_i, t) \geq \gamma - \frac{1}{m}.$$

and from (3.9) and (3.7) we deduce

$$\sum_{i=1}^{k} \beta_i u(y_i, t) < \gamma - \frac{1}{m}.$$

This contradiction establishes our claim, and the lemma follows without difficulty.

4. Proof of the Main Theorem

PROPOSITION 4.1. *Suppose that for each fixed t, u has a Borel-measurable[¹] concavification u^*. Then $\mathscr{P}(u, a)$ and $\mathscr{P}(u^*, a)$ have the same value, and $\mathscr{P}(u, a)$ is solvable if and only if $\mathscr{P}(u^*, a)$ is solvable. Furthermore, every solution of $\mathscr{P}(u, a)$ solves $\mathscr{P}(u^*, a)$.*

PROOF. We make use of the theory of integrals of set-valued functions [2]. Let $\mathbf{F}$ be a function defined on T, whose values are subsets of E^{n+1}. Then $\int_T \mathbf{F}(t)\, dt$, or $\int \mathbf{F}$ for short, is defined to be the set of all vectors of the form $\int \mathbf{f}$, where $\mathbf{f}$ is a point-valued function such that $\mathbf{f}(t) \in \mathbf{F}(t)$ for all t. The function $\mathbf{F}$ is said to be *Borel-measurable* if $\{(x, t) : x \in \mathbf{F}(t)\}$ is a Borel subset of $E^{n+1} \times T$. The fact that we need in the proof of our main theorem is that if $\mathbf{F}^*(t)$ denotes the convex hull of $\mathbf{F}(t)$, and $\mathbf{F}$ is Borel-measurable and takes only values that are subsets of P, then $\int \mathbf{F}^* = \int \mathbf{F}$ [2, Theorem 3].

[¹] The Borel-measurability of u^* is needed only to assure that the problem $\mathscr{P}(u^*, a)$ is well defined (cf. Section 1).

For a given u, define $\mathbf{F}(t)$ by (3.4). Since u is Borel-measurable, so is $\mathbf{F}$. Then it may be verified that

(4.2) $\mathscr{P}(u, a)$ has a solution $\mathbf{x}_0$ if and only if $V = \{v : (v, a) \in \int \mathbf{F}\}$ has a maximum, and then the value of $\mathscr{P}(u, a)$ is max V.

Now let $\mathbf{F}^*(t)$ be the convex hull of $\mathbf{F}(t)$, and let $V^* = \{v : (v, a) \in \int \mathbf{F}^*\}$. Since $\mathbf{F}$ is Borel-measurable and takes only values that are subsets of P, it follows that $\int \mathbf{F} = \int \mathbf{F}^*$, and therefore $V = V^*$. Proposition 4.1 now follows from (4.2).

The main theorem follows from Propositions 2.2, 3.1, and 4.1, and Lemmas 3.3 and 3.5.

5. A Characterization of the Solution

THEOREM 5.1. *Let u be nondecreasing, and let $a > 0$. Then a necessary and sufficient condition for a nonnegative $\mathbf{x}$ to solve $\mathscr{P}(u, a)$ is that $\int \mathbf{x} = a$ and there is a c in P such that*

$$u(x, t) - u(\mathbf{x}(t), t) \leqslant c \cdot (x - \mathbf{x}(t)) \tag{5.2}$$

for all t in T and $x \in P$. If u is increasing then $c \geqslant 0$ may be replaced by $c > 0$.

In the case in which u is differentiable in x, (5.2) implies

$$[\partial u/\partial x^i]_{x=\mathbf{x}(t)} \leqslant c^i$$

for all t and i, with equality holding whenever $\mathbf{x}^i(t) > 0$. That is, the partial derivatives, when evaluated at $\mathbf{x}(t)$, are constant for $\mathbf{x}^i(t) > 0$, and are at most equal to this constant for $\mathbf{x}^i(t) = 0$.

During the course of the proof we shall make use of Proposition 2.1 of [2], which states that for each Borel-measurable set valued function $\mathbf{F}(t)$, there is a point-valued Lebesgue-measurable function $\mathbf{f}(t)$ such that $\mathbf{f}(t) \in \mathbf{F}(t)$ for each t.

The proof of Theorem 5.1 is similar to the proof in Künzi and Krelle [6] of the Kuhn-Tucker theorem; we merely sketch it. Sufficiency is trivial. To prove necessity, we first show that there is a c in P such that

$$\int [u(\mathbf{y}) - u(\mathbf{x})] \leqslant c \cdot \int [\mathbf{y} - \mathbf{x}] \tag{5.3}$$

for all nonnegative and integrable y. Define K_1, $K_2 \subset E^{n+1}$ by

$$K_1 = \left\{(y^0, y) \in E^{n+1} : \text{there is a nonnegative integrable } \mathbf{y} \text{ such that } y^0 \leqslant \int u(\mathbf{y}) \text{ and } y \leqslant a - \int \mathbf{y}\right\}$$

$$K_2 = \left\{(y^0, y) \in E^{n+1} : y^0 > \int u(\mathbf{x}) \text{ and } y \geqslant 0\right\}.$$

Then K_1 is convex; for, we may define

$$\mathbf{K}_1(t) = \{(y^0, y) \in E^{n+1} : \text{there is a } z \in P \text{ such that } y^0 \leqslant u(z, t) \text{ and } y \leqslant a - z\},$$

and then by using Proposition 2.1 of [2] it may be established that $K_1 = \int \mathbf{K}_1$. Since every integral of a set-valued function is convex (Theorem 1 of [2]), it follows that K_1 is convex. K_2 is clearly convex and is disjoint from K_1, so there is a hyperplane separating K_1 and $\text{cl}(K_2)$, i.e., there are d^0 and $d = (d^1, \cdots, d^n)$ such that

$$d^0 y_1^0 + d \cdot y_1 \leqslant d^0 y_2^0 + d \cdot y_2$$

whenever $(y_1^0, y_1) \in K_1$ and $(y_2^0, y_2) \in \text{cl}(K_2)$. Then $d^0 > 0$ and $d \geqslant 0$, and we obtain

$$d^0 \int_0 [u(\mathbf{y}) - u(\mathbf{x})] \leqslant d \cdot \int [\mathbf{y} - \mathbf{x}];$$

dividing by d^0, we obtain (5.3). If u is increasing, it is easily established that $c > 0$.

To deduce (5.2) from (5.3), suppose that for all t in a set S of positive measure, there is an $x \in P$ such that

$$u(x, t) - u(\mathbf{x}(t), t) > c \cdot (x - \mathbf{x}(t)).$$

Define $\mathbf{y}(t)$ to be such an x when it exists, and $\mathbf{y}(t) = \mathbf{x}(t)$ otherwise. Integrating, we obtain a contradiction to (5.3). The possibility of choosing an appropriate $\mathbf{y}$ that is measurable follows from proposition 2.1 of [2].

6. COUNTEREXAMPLES AND GENERALIZATIONS

If $u(x, t)$ is not upper-semicontinuous for each fixed t, then it need not satisfy the main theorem, even if it is concave. Let $n = 1$, and let

$$u(x, t) = \begin{cases} x & \text{when} \quad x \leqslant 2 \\ 2 & \text{when} \quad x \geqslant 2 \end{cases}$$

when $0 \leqslant t \leqslant \frac{1}{2}$, and

$$u(x, t) = \begin{cases} 0 & \text{when} \quad x = 0 \\ 2 & \text{when} \quad x > 0 \end{cases}$$

when $\frac{1}{2} < t \leqslant 1$. Then the value of $\mathscr{P}(u, 1)$ is 2, but it is not achieved. It is possible to adjust this example so that u is increasing in x for each fixed t. It is also possible to construct an example of a u that satisfies all the conditions of the main theorem except that it may fail to be upper-semicontinuous at a point in the interior of P, and that does not satisfy the main theorem.

In the main theorem, the assumption that u is nonnegative may be replaced by the assumption that $u(0, t)$ is integrable.

The assumption that $u(x, t)$ is nondecreasing for each fixed t cannot be removed, as may be seen from the example $n = 1$, $u(x, t) = e^{-x}$, $a = 1$, in which the sup is 1 but is not attained. However, if we change the condition $\int x = a$ to read $\int x \leqslant a$, then the monotonicity assumption can be replaced by a far weaker assumption, which, roughly speaking, says that u is bounded on compact subsets of P. For $x \in E^n$, let[8]

$$\| x \| = \max(|x^1|, \cdots, |x^n|).$$

THEOREM 6.1. *Let $a \geqslant 0$. Suppose that u is continuous,[9] that*

$$u(x, t) = o(\| x \|)$$

as $\| x \| \to \infty$, integrably in t, and that for every integrable real function η there is an integrable real function ζ such that $\| x \| \leqslant \eta(t)$ implies $|u(x, t)| \leqslant \zeta(t)$ for all x and t. Let $\mathscr{P}$ be the problem:

$$\text{Maximize } \int u(\mathbf{x}) \text{ subject to } \mathbf{x}(t) \geqslant 0 \text{ for all } t \text{ and } \int \mathbf{x} \leqslant a.$$

Then $\mathscr{P}$ has a solution.

REMARK. All nondecreasing u satisfying the asymptotic condition (A) for which $u(0, t)$ is integrable satisfy the boundedness condition of this theorem.

PROOF. We define a "nondecreasification" u' of u as follows:

$$u'(x, t) = \max\{u(y, t) : 0 \leqslant y \leqslant x\}.$$

The max is attained because u is continuous. Clearly u' is nondecreasing. It may be verified that u' is Borel-measurable in both variables, continuous in x, and satisfies the asymptotic condition (A). Hence $\mathscr{P}(u', a)$ has a solution $\mathbf{x}_0$. Then for each t, there is a $y \in P$ such that $y \leqslant \mathbf{x}_0(t)$ and $u'(\mathbf{x}_0(t), t) = u(y, t)$. The function $u'(\mathbf{x}_0(t), t)$ is a Borel-measurable function of t. Hence

$$\{(y, t) : 0 \leqslant y \leqslant \mathbf{x}_0(t) \text{ and } u'(\mathbf{x}_0(t), t) = u(y, t)\}$$

is Borel-measurable. According to Proposition 2.1 of [2], there is a nonnegative measurable function $\mathbf{y}_0$ such that $\int \mathbf{y}_0 \leqslant \int \mathbf{x}_0 = a$ and $u(\mathbf{y}_0(t), t) = u'(\mathbf{x}_0(t), t)$

---

[8] The L^∞ norm we use here may be replaced by any of the usual norms on E^n, without affecting Theorem 6.1.

[9] The theorem can also be proved when u is only assumed to be upper-semicontinuous. However, the notion of analytic set must then be used instead of Borel set, and we do not wish to get involved in those complications here.

for all t. Hence

$$\int u(\mathbf{x}) \leqslant \int u'(\mathbf{x}) \leqslant \int u'(\mathbf{x}_0) \leqslant \int u(\mathbf{y}_0)$$

for all nonnegative integrable $\mathbf{x}$ such that $\int \mathbf{x} \leqslant a$, and so $\mathbf{y}_0$ solves $\mathscr{2}$. The proof of the theorem is complete.

Our final result deals with an extension of the main theorem to L^p.

THEOREM 6.2. *Let* $1 \leqslant p \leqslant \infty$ *and* $a \geqslant 0$. *Suppose* u *is nonnegative, increasing, and upper-semicontinuous,*[10] *and that*

(A_p) $u(\xi e, t) = o(\xi)$ *as* $\xi \to \infty$, *p-integrably in* t.

Then $\mathscr{P}(u, a)$ *has a solution in* L^p, *and indeed all solutions are in* L^p.

PROOF. Assume without loss of generality that $a > 0$.
By the main theorem, there is a solution $\mathbf{x}$ in L^1. By Theorem 5.1, every solution $\mathbf{x}$ satisfies (5.2) with $c > 0$. Let $\epsilon = \min(c^1, \cdots, c^n)$, and let η in L^p be such that $u(\xi e, t) < \epsilon \xi$ whenever $\xi \geqslant \eta(t)$. For each t, let

$$\xi(t) = \max(\mathbf{x}^1(t), \cdots, \mathbf{x}^n(t)).$$

By setting $x = 0$ in (5.2), we obtain

$$u(\xi(t)e, t) \geqslant u(\mathbf{x}(t), t) \geqslant c \cdot \mathbf{x}(t) \geqslant \epsilon \xi(t).$$

Hence $\xi(t) < \eta(t)$, and the theorem follows.

Theorem 6.2 cannot be extended to nondecreasing u when $p > 1$. For example, let $n = 1$, and

$$u(x, t) = \begin{cases} xt^{1/2} & \text{for} & x \leqslant t^{-1/2} \\ 1 & \text{for} & x \geqslant t^{-1/2} \end{cases}$$

Then u satisfies (A_p) for all p, and $\mathscr{P}(u, 2)$ has the solution $t^{-1/2}$, but no solution in L^2.

ACKNOWLEDGMENTS

It gives us pleasure to acknowledge the help of S. Agmon; a suggestion of Professor Agmon led to a very considerable shortening of the proof of Proposition 2.2 (the main theorem in the concave case). We also gratefully acknowledge a number of helpful conversations with Y. Kannai and with D. Leviatan.

REFERENCES

1. L. S. SHAPLEY AND M. SHUBIK. Quasi-Cores in a Monetary Economy with Nonconvex Preferences. *Econometrica* 32 (1966), 805–827.
2. R. J. AUMANN. Integrals of set-valued functions. *J. Math. Anal. Appl.* 12 (1965), 1–12 [Chapter 68].

---

[10] Cf. Section 1.

3. H. W. KUHN AND A. W. TUCKER. Non-linear programming. *Proc. Second Berkeley Symp. Math. Statist. Probab.*, pp. 481-492.
4. N. DUNFORD AND J. T. SCHWARTZ. "Linear Operators." Part I. Interscience, New York, 1958.
5. H. EGGLESTON. "Convexity." Cambridge Univ. Press, 1958.
6. H. P. KÜNZI AND W. KRELLE. "Nichtlineare Programmierung." Springer, Berlin, 1962.
7. M. E. YAARI. On the existence of an optimal plan in a continuous-time allocation process. *Econometrica* **32** (1964), 576-590.

71 Random Measure-Preserving Transformations

1. Introduction

It is the purpose of this note to show that it is impossible to define a probability measure on the group $\mathcal{G}$ of invertible measure-preserving transformations from the unit interval onto itself, if it is demanded that the measure on $\mathcal{G}$ obey two fairly "natural" conditions. One of these is an invariance condition on the measure, and the other asserts that certain distinguished subsets of $\mathcal{G}$ are measurable.

One reason for trying to construct such a probability measure is the following: the group $\mathcal{G}$ has been topologized in at least two different ways (see Halmos [3]); in one of those topologies (the "weak" topology) it has been proved that the set $\mathcal{E}$ of ergodic transformations (and in fact, the set $\mathcal{W}$ of weakly mixing transformations) is of the second category, and the set $\mathcal{S}$ of strongly mixing transformations is of the first category (see [3], p. 77 ff.). Corresponding to this information about the "topological size" of $\mathcal{E}$, $\mathcal{W}$, and $\mathcal{S}$, it would have been natural to seek information about the measures of these (and possibly other) subsets of $\mathcal{G}$. One could have hoped, for example, that "almost every transformation is ergodic." However, one needs first to have an appropriate measure on $\mathcal{G}$.

Another motivation comes from game theory. One of the characterizations of the Shapley value [4] of a cooperative n-person game involves a random ordering of the players. Recently games in which the player set may be a (possibly atomless) measure space have attracted attention, in part because of their applications to economics and politics. (For a comprehensive list of references, see Debreu [2].) One approach to defining the Shapley value for such games would involve the notion of a "random ordering" of the measure space of players. Replacing "ordering" with "measure-preserving transformation," leads to the question that we have answered (negatively) in this note.

The theorem of this paper provides additional evidence of the comparative intractability of function spaces when viewed from the measure-theoretic rather than from the topological viewpoint (compare with [1]).

A precise statement of the theorem is given in section 2, and it is proved in

This research was supported by the Office of Naval Research under task NR 047-006, and by the Army, Navy, Air Force and NASA under a contract administered by the Office of Naval Research, Task NR 042-242. Reproduction in whole or in part is permitted for any purpose of the United States Government.

section 3. The "naturalness" of the invariance condition is discussed in the last section.

2. Statement of the theorem

Let **I** be the measure algebra of Lebesgue measurable subsets of the unit interval I, modulo the sets of measure 0 (that is, the algebra in which sets differing by sets of measure 0 are not distinguished). Let $\mathcal{G}$ be the group of Lebesgue measure preserving automorphisms of **I**; the members of this group may be thought of as invertible measure-preserving transformations from I onto itself, where two transformations are identified if they differ on a set of measure 0 only. We will treat members of **I** as if they were subsets of I, speaking of unions, intersections, inclusions, and so on. Lebesgue measure will be denoted by λ throughout. No confusion will result.

THEOREM. *There is no pair* (Γ, μ), *where* Γ *is a σ-field of subsets of* $\mathcal{G}$, *and μ is a probability measure on* Γ, *for which*

(2.1) *for each* $\mathcal{H} \in \Gamma$ *and* $T \in \mathcal{G}$, *we have* $\mathcal{H}T \in \Gamma$, *and* $\mu(\mathcal{H}T) = \mu(\mathcal{H})$;

(2.2) *for all E and F in* **I**, *the function from* $\mathcal{G}$ *to the reals defined by* $f(T) = \lambda(E \cap TF)$ *is Γ-measurable.*

A few words of explanation are in order. "Probability measure," of course, means that $\mu(\mathcal{G}) = 1$. Condition (2.1) is the right-invariance condition; it says that if a Γ-measurable set of transformations is multiplied on the right by a single transformation, then it remains Γ-measurable, and its μ-measure (probability) remains unchanged. Without some such condition it would be trivially possible to construct a probability measure on $\mathcal{G}$, for example by concentrating all the probability on one transformation T. Condition (2.2) is a measurability assumption which seems very reasonable.

The theorem remains true if $\mathcal{H}T$ is replaced by $T\mathcal{H}$ in condition (2.1), that is, if right invariance is replaced by left invariance. Condition (2.2) remains unchanged.

3. Proof of the theorem

It will be assumed throughout that there is given a pair (Γ, μ) obeying the specifications of the theorem, and this will lead eventually to a contradiction.

Often it will be convenient to use the language of probability, that is, to replace μ by "Prob," $\int_\mathcal{G} \mu(dT)$ by "Exp" (for "Expectation"), and so on. "Variance" will be abbreviated by "Var," and "Covariance" by "Cov"; like "Exp," these two operators will be applied exclusively to random variables defined on the probability space $(\mathcal{G}, \Gamma, \mu)$.

LEMMA 1. *Let* $D, F_1, F_2 \in$ **I**, *and* $\lambda(F_1) = \lambda(F_2)$. *Then*

(1) $$\operatorname{Exp} \lambda(D \cap TF_1) = \operatorname{Exp} \lambda(D \cap TF_2).$$

PROOF. Let S be a member of $\mathcal{G}$ such that $SF_1 = F_2$. Define measures η_1 and η_2 on the closed unit interval $[0, 1]$ by

(2) $$\eta_i[0, \alpha] = \mu\{T: \lambda(D \cap TF_i) \leq \alpha\}$$

for $i = 1, 2$. Then η_i is the distribution of the random variable $\lambda(D \cap TF_i)$, and

(3) $$\eta_2[0, \alpha] = \mu\{T: \lambda(D \cap TSF_1) \leq \alpha\}$$
$$= \mu\{TS: \lambda(D \cap TSF_1) \leq \alpha\}$$
$$= \mu\{U: \lambda(D \cap UF_1) \leq \alpha\}$$
$$= \eta_1[0, \alpha],$$

where the second equality follows from (2.1) and the third by setting $U = TS$ and noting that as T runs over $\mathcal{G}$, so does U. From this it follows that

(4) $$\text{Exp } \lambda(D \cap TF_2) = \int_0^1 \alpha \eta_2(d\alpha) = \int_0^1 \alpha \eta_1(d\alpha) = \text{Exp } \lambda(D \cap TF_1),$$

which is the assertion of the lemma.

LEMMA 2. *For all* $D, F, \in \mathbf{I}$,

(5) $$\text{Exp } \lambda(D \cap TF) = \lambda(D)\lambda(F).$$

PROOF. For an arbitrary but fixed positive integer m, let $F_1, \cdots, F_m$ be disjoint members of $\mathbf{I}$ with equal measure, whose union is I. Then $\lambda(F_i) = 1/m$ for all i. From lemma 1 it follows that $\text{Exp } \lambda(D \cap TF_i)$ does not depend on i; let us denote it by γ. Now

(6) $$\lambda(D) = \text{Exp } \lambda(D) = \text{Exp } \lambda(D \cap TI)$$
$$= \text{Exp } \lambda\left(D \cap T \bigcup_{i=1}^m F_i\right) = \text{Exp } \sum_{i=1}^m \lambda(D \cap TF_i)$$
$$= \sum_{i=1}^m \text{Exp } \lambda(D \cap TF_i) = m\gamma.$$

Hence, $\gamma = \lambda(D)(1/m) = \lambda(D)\lambda(F_i)$ for $i = 1, \cdots, m$.

Now whenever $\lambda(F) = 1/m$ for some m, it is possible to set $F_1 = F$ and to find $m - 1$ sets $F_2, \cdots, F_m$ satisfying the above conditions; hence, whenever $\lambda(F)$ is the reciprocal of an integer, the assertion of the lemma is established. But each measurable set $F \in \mathbf{I}$ is a countable union of sets whose measures are reciprocals of integers; and since Exp is countably additive for nonnegative random variables, the assertion of the lemma follows in the general case as well.

Before stating the next lemma, we introduce the following notation: for $D, E, F \in \mathbf{I}$ and $E \cap F = \emptyset$, we write

(7) $$g(E, F) = g_D(E, F) = \text{Exp } [\lambda(D \cap TE)\lambda(D \cap TF)].$$

LEMMA 3. *Let* $D, E, F_1, F_2 \in \mathbf{I}$, *and* $F_1 \cap E = F_2 \cap E = \emptyset$, $\lambda(F_1) = \lambda(F_2)$. *Then* $g_D(E, F_1) = g_D(E, F_2)$.

PROOF. The proof is similar to that of lemma 1. This time, let S be a member of $\mathcal{G}$ such that both $SF_1 = F_2$ and $SE = E$. Define measures η_1 and η_2 on $[0, 1]$ by $\eta_i[0, \alpha] = \mu\{T: \lambda(D \cap TE)\lambda(D \cap TF_i) \leq \alpha\}$; then because of (2.1), $\eta_2[0, \alpha] = \eta_1[0, \alpha]$ for all α, and hence

(8) $$\int_0^1 \alpha \eta_2(d\alpha) = \int_0^1 \alpha \eta_1(d\alpha);$$

but that is precisely what the lemma asserts.

LEMMA 4. *Let $D, E, F \in \mathbf{I}$, and $E \cap F = \emptyset$. Then*

(9) $$g_D(E, F) \leq \lambda^2(D)\lambda(E)\lambda(F).$$

PROOF. If $\lambda(E)$ or $\lambda(F)$ vanish, there is nothing to prove; assume, therefore, that $\lambda(E) > 0$, $\lambda(F) > 0$, and so $\lambda(E) < 1$, $\lambda(F) < 1$. For an arbitrary but fixed positive integer m, let $F_1, \cdots, F_m$ be disjoint members of $\mathbf{I}$ with equal measure, whose union is $I - E$. Then $\lambda(F_i) = (1 - \lambda(E))/m$ for all i. From lemma 3 it follows that $g(E, F_i)$ does not depend on i; denote it by γ. Now

(10) $$g(E, I\backslash E) = g\left(E, \bigcup_{i=1}^m F_i\right) = \sum_{i=1}^m g(E, F_i) = m\gamma.$$

Hence,

(11) $$g(E, F_i) = \gamma = g(E, I\backslash E)/m = \lambda(F_i)g(E, I\backslash E)/(1 - \lambda(E)).$$

Whenever $F \subset I\backslash E$ and $\lambda(F) = (1 - \lambda(E))/m$ for some m, it is possible to set $F_1 = F$ and to find $m - 1$ sets $F_2, \cdots, F_m$ satisfying the above conditions; hence, for such F, we have

(12) $$g(E, F) = \lambda(F) \frac{g(E, I\backslash E)}{(1 - \lambda(E))}.$$

But each set $F \subset I\backslash E$ is a countable union of such F; and so (12) follows for all F with $E \cap F = \emptyset$. Now

(13) $$\begin{aligned} g(E, I\backslash E) &= \mathrm{Exp}\left[\lambda(D \cap TE)(\lambda(D) - \lambda(D \cap TE))\right] \\ &\leq \mathrm{Exp}\left[\max_{0 \leq \beta \leq \lambda(D)} \beta(\lambda(D) - \beta)\right] \\ &= \mathrm{Exp}\left[\lambda^2(D)/4\right] = \lambda^2(D)/4. \end{aligned}$$

Then choosing E_0 so that $\lambda(E_0) = \frac{1}{2}$, and applying (12), we find

(14) $$g(E_0, F) \leq \lambda(F) \frac{\lambda^2(D)/4}{\frac{1}{2}} = \lambda(F)\lambda(E_0)\lambda^2(D)$$

whenever $E_0 \cap F = \emptyset$. Now by using (12) and the symmetry of g in its two variables, we obtain

(15) $$g(E, F) = \lambda(E) \frac{g(F, I\backslash F)}{(1 - \lambda(F))}$$

whenever $E \cap F = \emptyset$. Setting $E = E_0$ in (15) and combining with (14), we deduce

(16) $$g(F, I\backslash F)/(1 - \lambda(F)) \leq \lambda(F)\lambda^2(D)$$

whenever $E_0 \cap F = \emptyset$. Combining this with (15), we obtain

(17) $$g(E, F) \leq \lambda(E)\lambda(F)\lambda^2(D)$$

whenever $E_0 \cap F = \emptyset$ and $E \cap F = \emptyset$. Now whenever $\lambda(F) \leq \frac{1}{2}$, it is possible to choose E_0 so that $E_0 \cap F = \emptyset$; since $E \cap F = \emptyset$ by the hypothesis of the lemma, the lemma is proved in those cases. When $\lambda(F) > \frac{1}{2}$, we may express F as the union of two disjoint subsets each of measure $\leq \frac{1}{2}$: the lemma then follows from the additivity in F both of $g(E, F)$ and of $\lambda(F)$.

LEMMA 5. *For all* $D, F \in \mathbf{I}$,

$$\operatorname{Var} \lambda(D \cap TF) = 0. \tag{18}$$

PROOF. If $D = I$ there is nothing to prove. Therefore, assume that $\lambda(D) < 1$. Let $F_1, \cdots, F_n$ be disjoint members of $\mathbf{I}$, with equal measures, such that $\bigcup_{i=1}^{n} F_i = F$; then $\lambda(F_i) = \lambda(F)/n$ for all i. Assume $n > 1$. Define random variables $\mathbf{x}_1, \cdots, \mathbf{x}_n$ by $\mathbf{x}_i = \lambda(D \cap TF_i)$. Then

$$\operatorname{Var} \lambda(D \cap TF) = \sum_{i=1}^{n} \operatorname{Var} \mathbf{x}_i + 2 \sum_{i>j} \operatorname{Cov}(\mathbf{x}_i, \mathbf{x}_j). \tag{19}$$

Now by lemmas 2 and 4,

$$\begin{aligned}
\operatorname{Cov}(\mathbf{x}_i, \mathbf{x}_j) &= g(F_i, F_j) - (\operatorname{Exp} \lambda(D \cap TF_i))(\operatorname{Exp} \lambda(D \cap TF_j)) \\
&\leq \lambda(F_i)\lambda(F_j)\lambda^2(D) - (\lambda(D)\lambda(F_i))(\lambda(D)\lambda(F_j)) \\
&= 0.
\end{aligned} \tag{20}$$

On the other hand, $\mathbf{x}_j$ is clearly bounded by $\lambda(F_i) = \lambda(F)/n \leq 1/n$, so $\operatorname{Var} \mathbf{x}_i \leq 1/n^2$. Hence

$$\operatorname{Var} \lambda(D \cap TF) \leq n/n^2 = 1/n. \tag{21}$$

Letting $n \to \infty$, we deduce the conclusion of the lemma.

Suppose now that $F = [0, \frac{1}{2}]$. Then it follows from lemmas 2 and 5 that with probability one, TF intersects every rational interval D in a set of measure $\frac{1}{2}\lambda(D)$. But then with probability one, TF is a set of density $\frac{1}{2}$ at each point; whereas, it is known that there are no such Lebesgue measurable sets. This is the contradiction that establishes our theorem.

The corresponding theorem when right invariance is replaced by left invariance can be proved in a similar manner. Alternatively, if (Δ, ν) is a pair satisfying (2.2) and the left-invariant analogue of (2.1), define

$$\Gamma = \{\mathcal{K} \subset \mathcal{G} : \mathcal{K}^{-1} \in \Delta\}, \tag{22}$$

where

$$\mathcal{K}^{-1} = \{T \in \mathcal{G} : T^{-1} \in \mathcal{K}\}; \tag{23}$$

and define μ on Γ by $\mu(\mathcal{K}) = \nu(\mathcal{K}^{-1})$. Then it may be verified that (μ, Γ) satisfies (2.1) and (2.2), and so contradicts the main theorem; this establishes the left-invariant version.

4. Discussion of the invariance condition

The invariance condition (2.1) seems rather strong. One may ask whether weaker conditions might not be devised, under which it *would* be possible to define a probability measure on $\mathcal{G}$, while still retaining the intuitive notion that the measure is distributed "uniformly" over $\mathcal{G}$. Certainly, the current result does not entirely exclude such a possibility, and we will not pretend that the last word on the subject has been said.

However, it should be pointed out that we have proved more than appears at first sight. The invariance condition (2.1) is used only twice in the proof, namely in the proofs of lemmas 1 and 3. Thus one could substitute these lemmas for condition (2.1) and still obtain the same result. Although the statements of these lemmas are more involved and less concise than (2.1), their direct intuitive appeal is perhaps greater than that of (2.1): both lemmas assert that the measure on $\mathcal{G}$ does not "discriminate" between sets F_1 and F_2 in $\mathbf{I}$ of equal Lebesgue measure. It is really only these "non-discrimination" conditions, in addition to condition (2.2), that are needed to prove the nonexistence of a measure on $\mathcal{G}$.

Note added in proof. Following is an extremely short proof of the theorem of this paper, for which I am indebted to Professor Harry Furstenberg. Let S in $\mathcal{G}$ be strongly mixing. Fix D and F in $\mathbf{I}$, and define random variables $\mathbf{y}_n = \mathbf{y}_n(T)$ by $\mathbf{y}_n = \lambda(D \cap TS^n F)$. Then for each T, $\mathbf{y}_n(T) = \lambda(T^{-1}D \cap S^n F) \to \lambda(D)\lambda(F)$ as $n \to \infty$. Because of the invariance condition (2.1), all the $\mathbf{y}_n$ have the same distribution; since they tend to the constant $\lambda(D)\lambda(F)$ pointwise, it follows that $\mathbf{y}_n = \lambda(D)\lambda(F)$ for each n with probability 1. By setting $n = 0$ we complete the proof.

REFERENCES

[1] R. J. AUMANN, "Borel structures for function spaces," *Illinois J. Math.*, Vol. 5 (1961), pp. 614–630 [Chapter 66].
[2] G. DEBREU, "Integration of correspondences," *Proceedings of the Fifth Berkeley Symposium on Mathematical Statistics and Probability*, Berkeley and Los Angeles, University of California Press, 1966, Vol. II, Part I, pp. 351–372.
[3] P. R. HALMOS, *Lectures on Ergodic Theory*, New York, Chelsea, 1956.
[4] L. S. SHAPLEY, "A Value for n-person games," *Contributions to the Theory of Games II*, Ann. of Math. Study 28, Princeton, Princeton University Press (1953), pp. 309–317.

72 Orderable Set Functions and Continuity III: Orderability and Absolute Continuity
with U. Rothblum

1. Introduction. This paper is one of a series of studies (cf. [1], [5], [6]) in which orderability and various continuity notions for set functions are investigated and related to each other. Throughout we assume familiarity with the concepts summarized in § 2 of [6]. Our main result (§ 5) concerns the *absolute continuity* of set functions (see [1, § 5] or § 2 of this paper). In [1, Prop. 12.8] it was shown that every absolutely continuous set function is orderable; here (§ 5) we construct an example to show that the converse is false. The example is a function of two nonatomic measures, and is in a sense "simplest possible": In § 4 we show that for functions of a single nonatomic measure, orderability and absolute continuity are equivalent.

2. Notations and definitions. We refer the reader to § 2 of [6] for a summary of some notations and definitions from [1] and [5] that will be used in this paper. Familiarity with the above section will be assumed throughout our discussions.

For x in the Euclidean space E^n, $\|x\|$ will always mean the summing norm, i.e., $\|x\| = \sum_{i=1}^{n} |x_i|$. If $x, y \in E^n$, write $x \leq y$ if $x_i \leq y_i$ for all i. If μ is a vector measure $(\mu_1, \cdots, \mu_n)$, then $\sum \mu$ will denote $\sum_{i=1}^{n} \mu_i$.

We next summarize some definitions and conventions from [1] which were not used in [6] and will be needed in this paper. The norm on BV is the *variation norm*, defined by

$$\|v\| = \inf\{u(I) + w(I) \mid u - w = v, \text{ where } u \text{ and } w \text{ are monotonic}\}.$$

A *chain* is a nondecreasing sequence of sets of the form $\emptyset = S_0 \subset S_1 \subset \cdots \subset S_n = I$. A *link* of this chain is a pair of successive elements. A *subchain* is a set of links. A chain will be identified with the subchain consisting of all links. If v is a set function and Λ is a subchain of a chain, then the *variation of v over Λ* is defined by $\|V\|_\Lambda = \sum |v(S_i) - v(S_{i-1})|$, where the sum ranges over $\{i \mid \{S_{i-1}, S_i\} \in \Lambda\}$. For a fixed Λ, $\|\cdot\|_\Lambda$ is a pseudonorm on BV, i.e., it enjoys all the properties of a norm except $\|v\|_\Lambda = 0 \Rightarrow v = 0$. It is known (see [1, Prop. 4.1]) that for every $v \in BV$, $\|v\| = \sup \|v\|_\Lambda$, where the supremum is taken over all subchains Λ. It is also known that

---

Received by the editors August 14, 1975. This work was supported by the National Science Foundation under Grant GS-3269 at the Institute for Mathematical Studies in the Social Sciences, Stanford University, Stanford, California.

This chapter originally appeared in *SIAM Journal on Control and Optimization* 15(1): 156–162. © 1977 by The Society for Industrial and Applied Mathematics. All rights reserved. Reprinted with permission.

the linear subspaces M, NA, WC and ORD are closed subspaces of BV [5, Prop. 4.2 and 4.3].

A set function v is said to be *absolutely continuous with respect to* a set function w (written $v \ll w$) [1, p. 35] if for every $\varepsilon > 0$ there is a $\delta > 0$ such that for every chain Ω and every subchain Λ of Ω, $\|w\|_\Lambda \leq \delta$ implies $\|v\|_\Lambda \leq \varepsilon$. Note that this relation is transitive, and that if v and w are measures, it coincides with the usual notion of absolute continuity. A set function is *absolutely continuous* if there is a measure $\mu \in \mathrm{NA}^+$ such that $v \ll \mu$. The set of all absolutely continuous set functions forms a closed linear subspace of BV [1, Prop. 5.2], denoted AC. Finally, pNA denotes the closed subspace of BV spanned by all powers of nonatomic measures.

3. Weak continuity and absolute continuity.
A real-valued function on a subset of E^n is said to be *monotonically absolutely continuous* if for every $\varepsilon > 0$ there is a $\delta > 0$ such that if $x_1 \leq y_1 \leq x_2 \leq \cdots \leq x_n \leq y_n$, then

$$\sum_{i=1}^{n} \|y_i - x_i\| \leq \delta \Rightarrow \sum_{i=1}^{n} |f(y_i) - f(x_i)| \leq \varepsilon.$$

If the domain of f is one-dimensional, then monotonic absolute continuity coincides with the usual absolute continuity.

PROPOSITION 1. *Let μ be an n-dimensional σ-additive measure whose components are in NA^+ and are mutually singular. Let f be a real-valued function on the range of μ with $f(0) = 0$. Let $v = f \circ \mu$. Then $v \ll \sum \mu \Leftrightarrow f$ is monotonically absolutely continuous.*

Proof. The direction $\Leftarrow$ is obvious. To prove the direction $\Rightarrow$, recall Lyapunov's theorem [4], according to which the range of a nonatomic σ-additive vector measure is convex and compact. From this and the mutual singularity it follows that if $x_1 \leq y_1 \leq x_2 \leq \cdots \leq x_n \leq y_n$, then there exist $S_1, T_1, \cdots, S_n, T_n$ in $\mathscr{C}$ such that $\mu(S_i) = x_i$, $\mu(T_i) = y_i$, and $S_1 \subseteq T_1 \subseteq \cdots \subseteq S_n \subseteq T_n$, completing the proof of Proposition 1.

PROPOSITION 2. *Let $v \in \mathrm{BV}$ and $\mu, \xi \in M^+$. If $v \ll \xi$, then $v \underset{w}{\ll} \mu$ if and only if $v \ll \mu$.*

Proof. Sufficiency of the condition is obvious. To see the necessity, let $\xi = \xi^{ac} + \xi^\perp$ be the Lebesgue decomposition of ξ with respect to μ, i.e., $\xi^\perp$ and ξ^{ac} are nonnegative measures such that $\xi^{ac} \leq \mu$ and $\xi^\perp \perp \mu$ [3, Thm. C, p. 134]. Let $A \in \mathscr{C}$ be such that $\xi^\perp(A) = 0$ and $\mu(I \setminus A) = 0$.

We shall show that $v \ll \xi^{ac}$, and since $\xi^{ac} \ll \mu$ it will follow that $v \ll \mu$. Let $\delta > 0$ correspond to a given ε in accordance with the absolute continuity $v \ll \xi$; i.e.,

(3.1) $\qquad$ for any subchain Λ, $\qquad \|\xi\|_\Lambda \leq \delta \Rightarrow \|v\|_\Lambda \leq \varepsilon.$

We shall prove that $v \ll \xi^{ac}$ by showing that

(3.2) $\qquad$ for any subchain Λ, $\qquad \|\xi^{ac}\|_\Lambda \leq \delta \Rightarrow \|v\|_\Lambda \leq \varepsilon.$

If we intersect each set in each link of Λ with A then we get a subchain Λ^* such that $\|\xi\|_{\Lambda^*} = \|\xi^{ac}\|_\Lambda \leq \delta$, and therefore by (3.1), $\|v\|_{\Lambda^*} \leq \varepsilon$. But because $v \underset{w}{\ll} \mu$ and $\mu(I \setminus A) = 0$, it follows that $\|v\|_\Lambda = \|v\|_{\Lambda^*} \leq \varepsilon$. This proves (3.2).

COROLLARY 1. *Let $\mu = (\mu_1, \mu_2, \cdots, \mu_n)$ be an n-dimensional vector of measures in NA^+. Let f be a real-valued function on the range of μ, such that $v = f \circ \mu \in BV$. Then $v \in AC$ if and only if $v \ll \sum \mu$.*

Proof. Sufficiency of the condition is obvious. To verify the necessity note that $f \circ \mu \underset{w}{\leq} \sum \mu$ and use Proposition 2.

COROLLARY 2.[1] *Let $v = f \circ \mu$, where $\mu \in NA^+$; then $v \in pNA$ if and only if $v \in AC$.*

Proof. The fact that $pNA \subseteq AC$ has been proved in [1, Cor. 5.3]. Now let $f \circ \mu \in AC$. Then by Corollary 1, $f \circ \mu \ll \mu$, and hence by Proposition 1 and Theorem C in [1], $f \circ \mu \in pNA$.

COROLLARY 3. *The inclusions $BV \supseteq WC \supseteq AC$ are strict.*

Proof. The unanimity game v defined by

$$v(S) = \begin{cases} 1, & S = I, \\ 0, & \text{otherwise}, \end{cases}$$

shows that $BV \neq WC$. Next, let λ be Lebesgue measure, and let g be the Cantor function, which is not absolutely continuous; then $g \circ \lambda \in WC$, and by Propositions 1 and 2, $g \circ \lambda \notin AC$.

4. Ordered absolute continuity. Let $\mathcal{R}$ be a measurable order. A chain $\emptyset = S_0 \subseteq S_1 \subseteq \cdots \subseteq S_m = I$ is called an $\mathcal{R}$-*chain* if all the S_i are $\mathcal{R}$-initial segments. Note that an $\mathcal{R}$-chain is defined by a finite sequence of elements in I, $\infty \overset{\mathcal{R}}{\geq} s_m \overset{\mathcal{R}}{\geq} \cdots \overset{\mathcal{R}}{\geq} s_1 \overset{\mathcal{R}}{\geq} s_0 = -\infty$, such that $I(s_i, \mathcal{R}) = S_i$.

If v and w are in BV, then v is said to be *ordered absolutely continuous with respect to w* (written $v \underset{o}{\ll} w$), if for every measurable order $\mathcal{R}$ and $\varepsilon > 0$ there exists a $\delta > 0$ such that for every $\mathcal{R}$-chain Ω and every subchain Λ of Ω, $\|w\|_\Lambda \leq \delta$ implies $\|v\|_\Lambda \leq \varepsilon$. Note that the relation is transitive.

PROPOSITION 3. *Let $v \in BV$, $\mu \in M^+$.[2] Then v is ordered absolutely continuous with respect to μ if and only if $v \in ORD$ and $v \underset{w}{\leq} \mu$.*

Proof. First assume that v is ordered absolutely continuous. It is easily verified that this implies $v \underset{w}{\leq} \mu$. Using the argument of the proof of Proposition 12.8 of [1] we obtain that[3] $v \in ORD$. This completes the proof of one direction.

To prove the second direction, let us assume $v \underset{w}{\leq} \mu$ and $v \in ORD$. By [5, Thm. 3.2], we know that $v \underset{w}{\leq} \mu$ implies that $\varphi^\mathcal{R} v \underset{w}{\leq} \mu$ for all measurable orders $\mathcal{R}$. Recall that weak continuity and absolute continuity between members of M coincide [2, § III. 4.3, p. 131]; hence $\varphi^\mathcal{R} v \ll \mu$ for all measurable orders $\mathcal{R}$. But then it follows that $v \underset{o}{\ll} \mu$.

A set is said to be *ordered absolutely continuous* if there is a measure $\mu \in NA^+$ such that v is ordered absolutely continuous with respect to μ. The set of all ordered absolutely continuous functions in BV is denoted OAC.

---

[1] Cf. [1, Thm. C].

[2] One may extend this theorem and require only $\mu \in M$, and not $\mu \in M^+$. This would slightly complicate the proof.

[3] In Proposition 12.8 of [1] one assumes $v \ll \mu$ and obtains in addition to $v \in ORD$, also that $\varphi^\mathcal{R} v \ll \mu$ uniformly in $\mathcal{R}$. Here we assume only $v \underset{o}{\ll} \mu$, and can also obtain $\varphi^\mathcal{R} v \ll \mu$, but not uniformly.

COROLLARY 4. $\text{ORD} \cap \text{WC} = \text{OAC}$

COROLLARY 5. OAC *is a closed linear subspace of* BV.

Remark. One may conjecture that if $v \in \text{ORD}$ and every point in I is v-null then there exists a measure $\mu \in \text{NA}^+$ such that $v \underset{w}{\lesssim} \mu$. If this is true then clearly it should yield that OAC equals the set of all set functions in ORD for which every point is null.

PROPOSITION 4. *Let* $v = f \circ \mu$, *where* $\mu \in \text{NA}^+$; *then* $v \ll \mu$ *if and only if* $v \underset{o}{\lesssim} \mu$.

Proof. If $v \ll \mu$, then trivially $v \underset{o}{\lesssim} \mu$. Assume now that $v \underset{o}{\lesssim} \mu$. By Proposition 3, $v \in \text{ORD}$ and $v \underset{w}{\lesssim} \mu$. Let $\mathcal{R}$ be an arbitrary fixed measurable order, then by [5, Thm. 3.2], $\varphi^{\mathcal{R}} v \underset{w}{\lesssim} \mu$. Since weak continuity and absolute continuity between totally finite measures coincide, it follows that $\varphi^{\mathcal{R}} v \ll \mu$. For a given ε, let δ be given in accordance with the absolute continuity $\varphi^{\mathcal{R}} v \ll \mu$; i.e., for every subchain Λ,

$$(4.1) \qquad \|\mu\|_\Lambda \leq \delta \Rightarrow \|\varphi^{\mathcal{R}} v\|_\Lambda \leq \varepsilon.$$

We shall show that $v \ll \mu$ by showing that for every subchain Λ,

$$(4.2) \qquad \|\mu\|_\Lambda \leq \delta \Rightarrow \|v\|_\Lambda \leq \varepsilon.$$

Let Λ be a subchain satisfying $\|\mu\|_\Lambda \leq \delta$ whose links are $\{S_j, T_j | 1 \leq j \leq m\}$, where $\varnothing \subseteq S_1 \subseteq T_1 \subseteq S_2 \subseteq \cdots \subseteq S_m \subseteq T_m \subseteq I$. Let

$$\bar{S}_j = \cap \{I(s, \mathcal{R}) | s \in I, \mu(I(s, \mathcal{R})) > \mu(S_j)\},$$
$$\bar{T}_j = \cap \{I(s, \mathcal{R}) | s \in I, \mu(I(s, \mathcal{R})) > \mu(T_j)\}.$$

By [1, Lem. 12.15] it follows that for $1 \leq j \leq m$, $\bar{S}_j, \bar{T}_j$ are measurable and that

$$(4.3) \qquad \mu(\bar{S}_j) = \mu(S_j) \quad \text{and} \quad \mu(\bar{T}_j) = \mu(T_j).$$

Note also that $\bar{S}_j$ and $\bar{T}_j$ are $\mathcal{R}$-initial sets; hence, by [6, Lem. 2], it follows that for $1 \leq j \leq m$,

$$(4.4) \qquad (\varphi^{\mathcal{R}} v)(\bar{T}_j) = v(\bar{T}_j) \quad \text{and} \quad (\varphi^{\mathcal{R}} v)(\bar{S}_j) = v(\bar{S}_j).$$

Let $\bar{\Omega}$ be the chain $\varnothing \subseteq \bar{S}_1 \subseteq \bar{T}_1 \subseteq \bar{S}_2 \subseteq \cdots \subseteq \bar{S}_m \subseteq \bar{T}_m \subseteq I$ and let $\bar{\Lambda}$ be a subchain of $\bar{\Omega}$ whose links are $\{\bar{S}_j, \bar{T}_j\}$, $1 \leq j \leq m$. Note that (4.3) implies that $\|\mu\|_{\bar{\Lambda}} = \|\mu\|_\Lambda \leq \delta$. Hence, by (4.1), $\|\varphi^{\mathcal{R}} v\|_{\bar{\Lambda}} \leq \varepsilon$, and therefore (4.4) and (4.3) imply that

$$\varepsilon \geq \|\varphi^{\mathcal{R}} v\|_{\bar{\Lambda}} = \|v\|_{\bar{\Lambda}} = \sum_{j=1}^{m} |f(\mu(\bar{T}_j)) - f(\mu(\bar{S}_j))|$$
$$= \sum_{j=1}^{m} |f(\mu(T_j)) - f(\mu(S_j))| = \|v\|_\Lambda.$$

We have established (4.2), thus completing the proof of Proposition 4.

COROLLARY 6. *Let* $v = f \circ \mu$ *where* $\mu \in \text{NA}^+$; *then*

$$v \in \text{AC} \Leftrightarrow v \ll \mu \Leftrightarrow v \in \text{OAC} \Leftrightarrow v \underset{o}{\lesssim} \mu \Leftrightarrow v \in \text{ORD} \Leftrightarrow v \in \text{pNA}.$$

Proof. The above follows from Proposition 2, Corollary 2, Proposition 3 and Proposition 4.

Remark. It clearly follows from Corollary 6 that if we wish to construct an example of the form $v = f \circ \mu$ that is in ORD\AC, then μ has to be at least two-dimensional.

5. ORD includes AC strictly. It was proved in [1, Prop. 12.8] that ORD $\supseteq$ AC. We are now going to construct an example of a set function in ORD that is not in AC.[4] The example that we are going to describe appears, in a different context, at the beginning of § 9 in [1]. For each $k \geq 2$ let $A_k \subset [0, 1]^2$ be the parallelogram whose vertices are: $(2^{-k}, 0)$, $(2^{-k} + 4^{-k}, 0)$, $(2^{-k+1} + 4^{-k}, 1)$ and $(2^{-k+1}, 1)$ (see Fig. 1). Define a nondecreasing continuous function f on the square such that for $x \in A_k$,

$$f(x) = f(x_1, x_2) = 2^k x_1 + 2^{-k+1} - 1;$$

for x between A_k and A_{k-1},

$$f(x) = 2^{-k+1} + x_2 + \frac{(x_1 - 2 \cdot 4^{-k})}{(1 - 2^{-k} + x_2)};$$

for x to the right of A_2 let f be defined by the same formula that defines f on A_2, i.e., $f(x) = 4x_1 - 1/2$; and finally for $x_1 = 0$ let $f(x) = x_2$. Let μ be any 2-dimensional vector measure on $(I, \mathscr{C})$, whose range if $[0, 1]^2$. We shall show that $v = f \circ \mu \in \text{ORD} \setminus \text{AC}$.

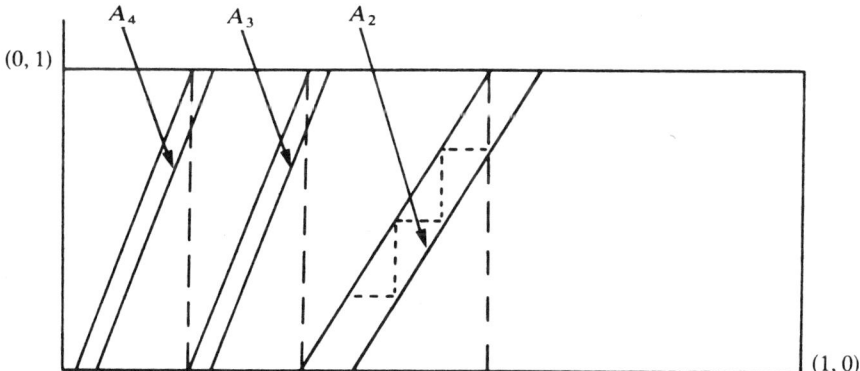

To show that $v \notin \text{AC}$, let

$$(2^{-k}, 0) = x_1^k \leq x_2^k \leq \cdots \leq x_n^k = (2^{-k+1}, 1)$$

be a "staircase" sequence of points in A_n, i.e., each point differs from the preceding one in one coordinate only (see Fig. 1). On the vertical segments of this sequence, f does not change; all the change is concentrated on the horizontal segments. But the total length of the horizontal segments goes to 0, whereas the total change in f is 1. Therefore f is not monotonically absolutely continuous,

---

[4] One can easily see that by "smoothing" our example one can get a set function in MIX [1, § 13] that is not in AC.

therefore $f \circ \mu$ is not absolutely continuous with respect to $\Sigma \mu$ (Proposition 1), and therefore $f \circ \mu \notin AC$ (Corollary 1).

Let us now prove that $v \in ORD$. Set $\mu = \mu_1 + \mu_2$. We shall show that v is ordered absolutely continuous with respect to μ, and then use Proposition 3. Let $\mathcal{R}$ be a fixed measurable order. For a given $\varepsilon > 0$ we may choose a $1 > \delta_1 > 0$ such that

(5.1) $$\|x - y\| \leq \delta_1 \Rightarrow |f(x) - f(y)| \leq \varepsilon/2.$$

This is possible because of the uniform continuity of f in $[0, 1]^2$.

Let J_1 denote the intersection of all $\mathcal{R}$-initial segments of μ_1-measure >0. By [1, Lem. 12.15] it follows that J_1 is measurable and $\mu_1(J_1) = 0$. Let J denote the intersection of all $\mathcal{R}$-initial segments of μ-measure $> \mu(J_1) + \delta_1$. By the same lemma[5] we mentioned before, it follows that J is measurable and $\mu(J) = \mu(J_1) + \delta_1$, therefore $J \supseteq J_1$. Finally, observe that $\|\mu(J) - \mu(J_1)\| = \delta_1$; hence by (5.1) it follows that $|v(J) - v(J_1)| \leq \varepsilon/2$.

Now let p be an integer ≥ 2 such that $2^{p-1} \geq 1/\mu_1(J)$. Note that p depends only on $\mathcal{R}$ and ε. One can easily verify that f fulfills a Lipschitz condition on $\{x \in [0, 1]^2 | x_1 \geq \mu_1(J)\}$ with constant 2^p, i.e., $\|f(y) - f(x)\| \leq 2^p \|x - y\|$; this implies that if $S, T \in \mathcal{C}$ and $J \subseteq S \subseteq T$, then $\|v(T) - v(S)\| \leq 2^p \{\mu(T) - \mu(S)\}$. Define $\delta = \min\{\delta_1, (2^p + 1)^{-1} \varepsilon/2\}$ and note that δ depends only on $\mathcal{R}$ and ε.

Let Λ be a subchain of an $\mathcal{R}$-chain Ω, with links $\{S_i, T_i\}$ ($1 \leq i \leq n$), where $\emptyset \subseteq S_1 \subseteq T_1 \subseteq S_2 \subseteq \cdots \subseteq S_n \subseteq T_n \subseteq I$. By definition of $\mathcal{R}$-chain, S_i and T_i are $\mathcal{R}$-initial segments ($1 \leq i \leq n$). We shall show that $\|\mu\|_\Lambda \leq \delta$ implies $\|v\|_\Lambda \leq \varepsilon$, which implies that v is ordered absolutely continuous with respect to μ, and hence by Proposition 4 that $v \in ORD$.

Let $\|\mu\|_\Lambda \leq \delta$, i.e., $\|\mu\|_\Lambda = \sum_{i=1}^n \{\mu(T_i) - \mu(S_i)\} \leq \delta$. Without loss of generality we may assume that if $T_i \supseteq J$, then $S_i \supseteq J$; otherwise split $\{S_i, T_i\}$ into two links $\{S_i, J\}$ and $\{J, T_i\}$. Similarly we may assume that if $S_i \subseteq J_1$, then $T_i \subseteq J_i$. Note that since μ and v are monotonic, $\|\mu\|_\Lambda$ and $\|v\|_\Lambda$ remain unchanged. Let

$$I_1 = \{1 \leq i \leq n | T_i \subseteq J_1\},$$
$$I_2 = \{1 \leq i \leq n | J \subseteq S_i\},$$
$$I_3 = \{1 \leq i \leq n | J_1 \subseteq S_i \subseteq T_i \subseteq J\}.$$

I_1, I_2 and I_3 are disjoint, and by our previous assumption $I_1 \cup I_2 \cup I_3 = \{1, 2, \cdots, n\}$. Now

$$\|v\|_\Lambda = \sum_{i=1}^n |v(T_i) - v(S_i)| = \sum_{l=1}^3 \sum_{i \in I_l} \{v(T_i) - v(S_i)\}$$

$$\leq \sum_{i \in I_1} \{\mu_2(T_i) - \mu_2(S_i)\} + \sum_{i \in I_2} \{2^p (\mu(T_i) - \mu(S_i))\} + v(J) - v(J_1)$$

$$\leq \delta + 2^p \cdot \delta + \frac{\varepsilon}{2} \leq \frac{\varepsilon}{2} + \frac{\varepsilon}{2} = \varepsilon.$$

---

[5] The lemma must be modified to apply to measures μ in NA^+ for which $\mu(I) \neq 1$. Note that $\mu(J_1) + \delta_1 < \mu(I)$.

This completes the proof that $v \in \text{ORD} \setminus \text{AC}$. Hence we have shown

(5.2) $\qquad\qquad\qquad$ ORD includes AC strictly.

REFERENCES

[1] R. J. AUMANN AND L. S. SHAPLEY, *Values of Non-Atomic Games*, Princeton University Press, Princeton, NJ, 1974.
[2] N. DUNFORD AND J. T. SCHWARTZ, *Linear Operators, Part I*, Interscience, New York, 1958.
[3] P. R HALMOS, *Measure Theory*, Van Nostrand, Princeton, 1950.
[4] A. LYAPUNOV, *Sur les fonctions-vecteurs complèment additives*, Bull. Acad. Sci. U.S.S.R. Ser. Math., 4 (1940), pp. 465–478.
[5] U. G. ROTHBLUM, *On orderable set functions and continuity. I*, Israel J. Math., 16 (1973), pp. 375–397.
[6] ———, *Orderable set functions and continuity. II: Set functions with infinitely many null points*, SIAM J. Control and Optimization, 15 (1977), pp. 144–155.

73 Bi-Convexity and Bi-Martingales
with S. Hart

0. Introduction

Let $\mathcal{X}$ and $\mathcal{Y}$ be compact convex subsets of Euclidean spaces (usually of different dimensions), with generic elements x and y. A subset of $\mathcal{X} \times \mathcal{Y}$ is *bi-convex* if each of its x- and y-sections is convex. The *bi-convex hull* of a set is the smallest bi-convex set containing it. A real function $f(x, y)$ on a bi-convex subset of $\mathcal{X} \times \mathcal{Y}$ is *bi-convex* if it is convex in each variable x and y separately. A *bi-martingale* is a martingale with values in $\mathcal{X} \times \mathcal{Y}$ whose x- and y-coordinates change only one at a time. (For detailed definitions and illustrative examples, see Sections 1, 2 and 3.)

These concepts arise in the analysis of repeated games of incomplete information [3]. In this paper we explore the relationships between them.

A martingale can be viewed as a splitting process. A particle (mass point) in space splits into several new particles, whose centroid is the starting point. Each of the new particles then splits, and the process is repeated again and again.

Research partially supported by NSF grants at the Institute for Mathematical Studies in the Social Sciences, Stanford University. The second author has also been partially supported by the Deutsche Forschungsgemeinschaft. We thank Andreu Mas-Colell, Jean-Francois Mertens, Abraham Neyman and Lloyd S. Shapley for many useful discussions.

This chapter originally appeared in *Israel Journal of Mathematics* 54 (1986): 159–180. Reprinted with permission.

Eventually a cloud forms; if we confine ourselves to a bounded subset of space, then by the martingale convergence theorem, the cloud converges to a limit cloud. At each stage, the starting point is the centroid of the cloud, and therefore lies in its convex hull. It also lies in the convex hull of the limit cloud. (Here, mass corresponds to probability, and centroid to expectation.)

If the martingale is a bi-martingale, at each stage either all particles split "horizontally", or all particles split "vertically". Therefore, at each stage the starting point is in the bi-convex hull of the cloud. Rather surprisingly, though, it need not be in the bi-convex hull of the limit cloud (see Example 2.5). This is so even in the special case in which the bi-martingale is *almost finite* (i.e. the mass that continues to split after n stages tends to 0 as $n \to \infty$).

Given the limit cloud, what *can* we say about the starting point? To answer this question, we must examine more carefully the notion of convex hull and its generalizations to bi-convexity. The convex hull co(A) of a set A can be defined as the smallest convex set containing A; this is the definition that corresponds to the definition of bi-convex hull given above. But co(A) can also be defined by a process of separation, as follows: First, one removes all the points z that can be strictly separated from A by a convex function f (i.e., $f(z) > \sup f(A)$). This yields the closed convex hull B_1 of A; obviously $B_1 \supset A$. Define B_2 by removing from B_1 all points that can be strictly separated from A by a convex function defined on B_1 only. The reader may convince himself that iterated finitely often, this process leads to co(A), where it ends.

One may also apply this process of separation to bi-convexity, substituting bi-convex functions for convex functions. The process may then require transfinitely many iterations; but it, too, must eventually end. We call the result bi-co$^\#(A)$; it always contains bi-co(A), but, unlike in the case of convexity, it is in general different (Example 2.5).

Suppose that in the iterative process that leads to co(A), we limit ourselves to separating functions that, in addition to being convex, are continuous. Then it may be seen that we will never get beyond the closed convex hull — the first iteration will also be the last. But if we demand that the separating functions be continuous only on A, then again, a finite number of iterations lead to co(A).

Similarly, in the case of bi-convexity we may separate by bi-convex functions that are continuous on A. Again, the process must converge (after a possibly transfinite number of stages). The result, which we call bi-co*(A), may be different both from bi-co(A) and from bi-co$^\#(A)$ (see Section 5); of course, bi-co$(A) \subset$ bi-co$^\#(A) \subset$ bi-co*(A).

Our main results (Section 4) may now be stated as follows:

(1) If A is the limit cloud of a bi-martingale, then the set of all possible starting points is bi-co*(A) (see Theorem 4.7).

(2) If we restrict ourselves to bi-martingales that are almost finite (see definition above), then the set of all possible starting points is bi-co$^\#(A)$ (Theorem 4.3).

To complete the picture, we note that

(3) If we restrict ourselves to *finite* bi-martingales (i.e. those that actually remain fixed after a bounded number of stages), then the set of all possible starting points is bi-co(A).

1. Bi-martingales

Let $\mathcal{X}$ and $\mathcal{Y}$ be compact convex subsets of some Euclidean spaces (of different dimensions, in general). Let $(\Omega, \mathcal{F}, P)$ be an atomless probability space. A sequence $\{Z_n\}_{n=1}^{\infty} \equiv \{(X_n, Y_n)\}_{n=1}^{\infty}$ of $(\mathcal{X} \times \mathcal{Y})$-valued random variables is a *bi-martingale* if:

(1.1) There exists a non-decreasing sequence $\{\mathcal{F}_n\}_{n=1}^{\infty}$ of finite subfields[1] of $\mathcal{F}$, such that $\{Z_n\}_n$ is a martingale with respect to $\{\mathcal{F}_n\}_n$.

(1.2) For each $n = 1, 2, \ldots$, either $X_n = X_{n+1}$ or $Y_n = Y_{n+1}$ (a.s.).

(1.3) Z_1 is constant (a.s.).

The martingale condition (1.1) means, first, that Z_n is $\mathcal{F}_n$-measurable, and second, that $E(Z_{n+1} | \mathcal{F}_n) = Z_n$ (a.s.), for all $n = 1, 2, \ldots$. By (1.3), we thus have $E(Z_n) = Z_1$ for all n. Since $\mathcal{X}$ and $\mathcal{Y}$ are compact, the sequence $\{Z_n\}$ forms a bounded martingale, hence it has an almost everywhere limit $Z_\infty \equiv (X_\infty, Y_\infty)$.

Let A now be a measurable subset of $\mathcal{X} \times \mathcal{Y}$. We will consider the following set:

(1.4) $A^* = \{z \in \mathcal{X} \times \mathcal{Y} \mid \text{there exists a bi-martingale } \{Z_n\}_{n=1}^{\infty} \text{ converging to } Z_\infty, \text{ such that } Z_\infty \in A \text{ and } Z_1 = z \text{ (a.s.)}\}.$

Without condition (1.2), A^* becomes just co(A), the convex hull[2] of A; the

---

[1] A field is finite if it contains finitely many elements; this finiteness condition will turn out to be inessential — see Remark 4.11.

[2] Indeed, every point in co(A) can be obtained (by Caratheodory's theorem). Conversely, we have $z = E(Z_\infty)$ where $P(Z_\infty \in A) = 1$, which implies $z \in \overline{\text{co}}(A)$ ($=$ the closed convex hull of A). If $z \notin \text{co}(A)$, then there exists a supporting hyperplane, i.e., $\lambda \neq 0$ such that $\lambda \cdot z = \sup\{\lambda \cdot a \mid a \in A\}$. But this implies $P(Z_\infty \in A') = 1$, where $A' = \{a \in A \mid \lambda \cdot a = \lambda \cdot z\}$, and A' is a set of lower dimension than A. The proof is now completed by induction.

same will happen if we drop (1.3) (and replace $Z_1 = z$ by $E(Z_1) = z$ in the definition of A^*). However, the set A^* as given by (1.4) is in general strictly included in co(A). For example, as we will see later, if we take $\mathscr{X} = \mathscr{Y} = [0, 1]$ and $A = \{(0,0), (1,0), (0,1)\}$, then A^* is the L-shaped set $\{(x, y) \in [0,1] \times [0,1] \mid x = 0 \text{ or } y = 0\}$.

REMARK 1.5. One can represent a bi-martingale $\{Z_n\}_{n=1}^{\infty}$ as a rooted tree, with the values of Z_n attached to its nodes at level n, and the probabilities $P(Z_{n+1} \mid \mathscr{F}_n)$ attached to its branches from there. For example, see Fig. 1.1, where

$$1 = \alpha_1 + \beta_1 = \alpha_2^1 + \beta_2^1 = \alpha_2^2 + \beta_2^2 = \ldots; \qquad 0 \leq \alpha_1, \beta_1, \alpha_2^1, \beta_2^1, \alpha_2^2, \beta_2^2, \ldots;$$

$$z_1 = \alpha_1 z_2^1 + \beta_1 z_2^2, \qquad z_2^1 = \alpha_2^1 z_3^1 + \beta_2^1 z_3^2, \ldots;$$

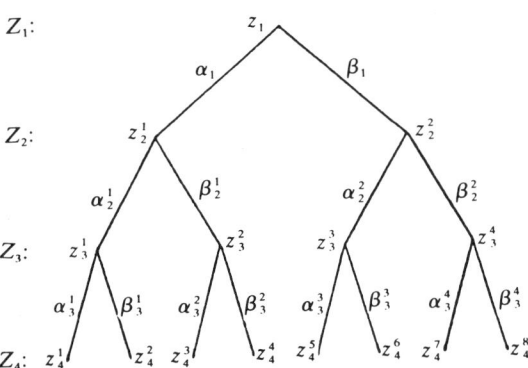

Fig. 1.1.

and (writing $z_i^j \equiv (x_i^j, y_i^j)$)

$$x_1 = x_2^1 = x_2^2, \quad y_2^1 = y_3^1 = y_3^2, \quad y_2^2 = y_3^3 = y_3^4, \ldots.$$

Note that the total probability of each node is the product of the probabilities along the unique path connecting the node to the root.

Conversely, each such tree structure gives rise to a bi-martingale; this is the (only) reason we required P to be an atomless measure. It follows that the specific choice of the probability space is of no consequence, as long as it is atomless. Thus, A^* is determined by *distributions* of bi-martingales, and not by the bi-martingales themselves.

REMARK 1.6. If $\{Z_n\}_{n=1}^{\infty}$ is a bi-martingale, $Z_n \to Z_\infty$ and $Z_n \in A$ a.s., then $Z_n \in A^*$ for all n. But $A^* \subset \operatorname{co}(A)$, therefore A^* does not change if we replace $\mathscr{X}$ and $\mathscr{Y}$ by other compact convex sets whose product contains A.

REMARK 1.7. Call a bi-martingale *binary* if all nodes in the associated tree (cf. Remark 1.5) have at most two immediate successors. Note that A^* does not change if we consider only binary bi-martingales. The interest in these is due to the following: Let $\{Z_n\}_{n=1}^{\infty}$ be a sequence of $(\mathscr{X} \times \mathscr{Y})$-valued random variables, where $Z_n \equiv (X_n, Y_n)$, and denote by $X_n^{(i)}$ and $Y_n^{(i)}$ the coordinates of X_n and Y_n, respectively. Then $\{Z_n\}$ is a binary bi-martingale if and only if it satisfies (1.1), (1.3), and for each i and j, the sequence $\{X_n^{(i)} \cdot Y_n^{(j)}\}_{n=1}^{\infty}$ is a real-valued martingale (with respect to the same $\{\mathscr{F}_n\}_{n=1}^{\infty}$ as in (1.1)). This follows from the easily checked fact that for real numbers, if $x = \alpha x' + (1 - \alpha)x''$, $y = \alpha y' + (1 - \alpha)y''$, $xy = \alpha x'y' + (1 - \alpha)x''y''$ and $0 < \alpha < 1$, then either $x = x' = x''$ or $y = y' = y''$. It is however no longer true for convex combinations of more than two points; e.g.,

$$(\tfrac{3}{2}, \tfrac{3}{2}; \tfrac{3}{2} \cdot \tfrac{3}{2}) = \tfrac{1}{4}(0,0;0\cdot 0) + \tfrac{3}{8}(3,1;3\cdot 1) + \tfrac{3}{8}(1,3;1\cdot 3).$$

2. Bi-convex sets

A convex combination $(x, y) = \sum_{i=1}^{m} \alpha_i (x_i, y_i)$ (with $\alpha_i \geq 0$, $\sum_{i=1}^{m} \alpha_i = 1$) will be called *bi-convex* if either $x_1 = x_2 = \cdots = x_m = x$ or $y_1 = y_2 = \cdots = y_m = y$. A set B is a *bi-convex set* if it contains all the bi-convex combinations of its elements. Thus, B is bi-convex if for all $x \in \mathscr{X}$ and $y \in \mathscr{Y}$, its sections $B_x. \equiv \{y \in \mathscr{Y} \mid (x, y) \in B\}$ and $B_{.y} \equiv \{x \in \mathscr{X} \mid (x, y) \in B\}$ are convex sets. An example of a bi-convex set that is not convex is again $B = \{(x, y) \in [0, 1] \times [0, 1] \mid x = 0$ or $y = 0\}$. Another example is the graph of the subdifferential mapping of a convex function (cf. [5], Theorem 23.5).

Next, we want to define the bi-convex hull of a given subset A of $\mathscr{X} \times \mathscr{Y}$. There are two ways to proceed.

First, define inductively the sequence of sets $\{A_n\}_{n=1}^{\infty}$ as follows: $A_1 = A$ and A_{n+1} is the set of all bi-convex combinations of elements of A_n (for $n = 1, 2, \ldots$). Let $B = \bigcup_{n=1}^{\infty} A_n$ be the limit of this sequence. Second, let B' be the intersection of all bi-convex sets that contain A.

PROPOSITION 2.1. $B = B' =$ *the smallest*[3] *bi-convex set containing* A.

The proof is straightforward; we will call the set obtained the *bi-convex hull* of A, and will denote it bi-co(A).

[3] Relative to set inclusion.

An interesting question is: does there exist an n such that, analogous to Caratheodory's theorem, bi-co$(A) = A_n$? The answer is (in general) no.

EXAMPLE 2.2. Let $\mathscr{X} = \mathscr{Y} = [0,1]$, and for all $m = 1, 2, \ldots$ define

$$z_{2m} = \left(1 - \frac{1}{2^{m-1}}, 1 - \frac{3}{2^{m+2}}\right), \quad z_{2m+1} = \left(1 - \frac{3}{2^{m+2}}, 1 - \frac{1}{2^m}\right),$$

$$w_{2m} = \left(1 - \frac{1}{2^{m-1}}, 1 - \frac{1}{2^m}\right), \quad w_{2m+1} = \left(1 - \frac{1}{2^m}, 1 - \frac{1}{2^m}\right),$$

and put $z_1 = w_1 = (0,0)$. Then w_n is a bi-convex combination of z_n and w_{n-1} (for $n = 2, 3, \ldots$), namely $w_n = \frac{4}{5} z_n + \frac{1}{5} w_{n-1}$. Now let $A = \{z_n\}_{n=1}^\infty$; then it can be checked that $w_n \in A_n$ but $w_n \notin A_{n-1}$, for each $n = 2, 3, \ldots$ (see Fig. 2.1). ∎

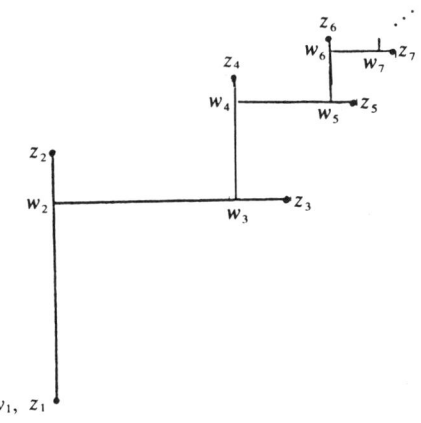

Fig. 2.1.

Note that by adding the point $(1,1)$ to the set A in Example 2.2, one obtains a *closed* (hence compact) set A with bi-co$(A) \supsetneq A_n$ for all n.

How are bi-convex sets related to bi-martingales?

PROPOSITION 2.3. *For any set A, A^* is a bi-convex set containing* bi-co(A).

PROOF. To see that A^* is a bi-convex set, recall the tree structure in Remark 1.5. Given a collection of m such trees, with roots $z_1, \ldots, z_m$, where, say, $x_1 = \cdots = x_m = x$, we construct for every non-negative $\alpha_1, \cdots, \alpha_m$ with $\sum_{i=1}^m \alpha_i = 1$ a new tree as follows. The root is $z = (x, y)$, where $y = \sum_{i=1}^m \alpha_i y_i$; it has m branches to nodes $z_1, \ldots, z_m$, with probabilities $\alpha_1, \ldots, \alpha_m$ (respectively); from each such node z_i, we follow the corresponding given tree. This shows that if $z_1, \ldots, z_m$ belong to A^*, then $z \in A^*$ too.

The inclusion $A^* \supset A$ is obtained by considering constant bi-martingales; it implies that $A^* \supset \text{bi-co}(A)$. ∎

REMARK 2.4. The set A_n corresponds precisely to those bi-martingales $\{Z_m\}_{m=1}^\infty$ for which the limit Z_∞ is attained at most in n steps (i.e., $Z_n = Z_\infty$).

Are the two sets A^* and bi-co(A) actually equal? The following example shows that this is not the case in general.

EXAMPLE 2.5. Again, let $\mathcal{H} = \mathcal{Y} = [0,1]$. Let $z_1 = (\frac{1}{3}, 0)$, $z_2 = (0, \frac{2}{3})$, $z_3 = (\frac{2}{3}, 1)$ and $z_4 = (1, \frac{1}{3})$, then $A = \{z_1, z_2, z_3, z_4\}$ is clearly a bi-convex set, i.e., $A = \text{bi-co}(A)$. Let $w_1 = (\frac{1}{3}, \frac{1}{3})$, $w_2 = (\frac{1}{3}, \frac{2}{3})$, $w_3 = (\frac{2}{3}, \frac{2}{3})$ and $w_4 = (\frac{2}{3}, \frac{1}{3})$; we will show that all these points belong to A^* (as we will see in Section 4, A^* is precisely the bi-convex hull of all the points z_i and w_i, $1 \leq i \leq 4$; it consists of the square whose vertices are the w_i's, together with the four line segments $[w_i, z_i]$; see Fig. 2.2).

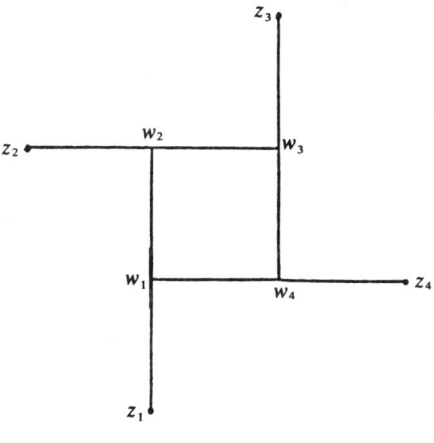

Fig. 2.2.

Indeed, consider the following tree (see Fig. 2.3): the root is w_1; every node w_i has two sons, z_i and w_{i+1} (where $i+1$ is taken modulo 4), with probability $\frac{1}{2}$ each; every node z_i has one son z_i only. It is easily seen that this tree defines a bi-martingale $\{Z_n\}_{n=1}^\infty$ with $Z_1 = w_1$; the probability that A is never reached is zero (this happens only along the rightmost path in the tree, whose probability is $\lim_{n \to \infty} (\frac{1}{2})^n = 0$), thus $w_1 \in A^*$. A similar construction proves that w_2, w_3 and w_4 belong to A^* too. ∎

This example points out the difference between "finite" bi-martingales (which generate only $\bigcup_{n=1}^\infty A_n = \text{bi-co}(A)$; see Remark 2.4), and "infinite" ones. In Section 4 we will make this distinction (and another one) more precise.

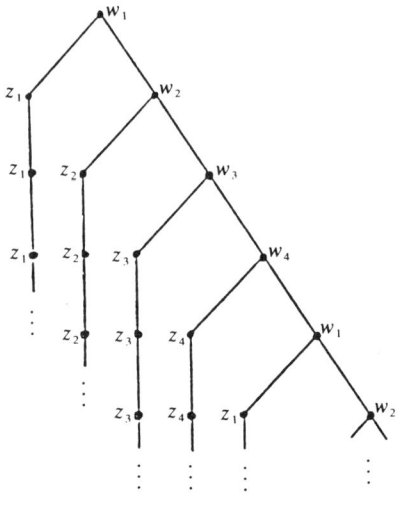

Fig. 2.3.

3. Bi-convex functions

In the previous section we saw that bi-convex sets are not sufficient to characterize A^*. We thus approach the problem in a dual way — by separation. In the case of convexity, it is enough to consider affine[4] functions: any point outside a convex set can be separated from it by such a function. However, this does not generalize to the bi-convex case: the corresponding bi-affine functions separate strictly less than the larger class of bi-convex functions.

Let $B \subset \mathcal{X} \times \mathcal{Y}$ be a bi-convex set, and let $f: B \to \mathbf{R}$, where $\mathbf{R}$ denotes the real line. The function f is *bi-convex* (*bi-affine*) if $f(x, \cdot)$ is a convex (affine) function on $B_x. = \{y \in \mathcal{Y} \mid (x, y) \in B\}$ for all $x \in \mathcal{X}$, and $f(\cdot, y)$ is a convex (affine) function on $B_{.y} = \{x \in \mathcal{X} \mid (x, y) \in B\}$ for all $y \in \mathcal{Y}$; i.e.,

$$f(\lambda'x' + \lambda''x'', y) \leq \lambda'f(x', y) + \lambda''f(x'', y)$$

and

$$f(x, \lambda'y' + \lambda''y'') \leq \lambda'f(x, y') + \lambda''f(x, y'')$$

for all $\lambda', \lambda'' \geq 0$, $\lambda' + \lambda'' = 1$, and $(x', y), (x'', y), (x, y'), (x, y'') \in B$. Note that f is bi-affine if we have equalities above; it has to be of the form

$$f(x, y) = \sum_{i,j} \alpha_{ij} x^{(i)} y^{(j)} + \sum_i \beta_i x^{(i)} + \sum_j \gamma_j y^{(j)} + \delta,$$

---

[4] We use *affine* for a function that is both convex and concave; it is sometimes called *linear*.

where i and j denote the coordinates of x and y, respectively, and α_{ij}, β_i, γ_j and δ are real constants.

The following is immediate.

PROPOSITION 3.1. *Let $f: B \to \mathbf{R}$ be a bi-convex function.[5] Then, for all real α, the set $\{(x, y) \in B \mid f(x, y) \leq \alpha\}$ is a bi-convex set.*

As in the standard convex case, the converse is of course not true in general.

We can now define the notion of separation. Let B be a bi-convex set, $B \supset A$ (the set A is assumed fixed throughout). Then a point $z \in B$ is (*strongly bi-*) *separated from A with respect to*[6] B if there exists a bounded bi-convex function f on B such that $f(z) > \sup f(A) \equiv \sup\{f(a) \mid a \in A\}$. Let us denote by $\text{ns}(B) (\equiv \text{ns}_A(B))$ the set of all points $z \in B$ that cannot be separated from A; thus, $z \in \text{ns}(B)$ if and only if $z \in B$ and, for all bi-convex functions f defined on B, we have $f(z) \leq \sup f(A)$.

From Proposition 3.1 one readily obtains

PROPOSITION 3.2. *Let B be a bi-convex set, $B \supset A$. Then the set $\text{ns}(B)$ is bi-convex, and $\text{ns}(B) \supset \text{bi-co}(A)$.*

In general, we cannot expect the opposite inclusion to hold, since even for ordinary convexity, the analogous assertion cannot be made: if B is a convex set and $A \subset B$, then the set of points in B that cannot be separated from A by a convex function on B need *not* be included in $\text{co}(A)$. This set *is*, however, included in $\overline{\text{co}}(A)$, the closed convex hull of A, and so the question arises whether, similarly, we can assert that $\text{ns}(B)$ is included in $\overline{\text{bi-co}}(A)$. The answer is no; this is further evidence for the non-finite dimensional character of bi-convexity (see Example 2.2).

EXAMPLE 3.3. Consider again Example 2.5, and let $B = \mathcal{X} \times \mathcal{Y}$. Let f be a bi-convex function on B, and assume that it separates at least one of the points w_1, w_2, w_3, w_4 from A. Let i be such that $f(w_i) \geq f(w_j)$ for all $1 \leq j \leq 4$, then f separates w_i from A. Now f is bi-convex, thus

$$f(w_i) \leq \tfrac{1}{2} f(z_i) + \tfrac{1}{2} f(w_{i-1})$$

(where we define $w_0 \equiv w_4$). But $z_i \in A$, thus $f(w_i) > f(z_i)$, which implies $f(w_{i-1}) > f(w_i)$, contradicting the choice of i. This shows that $w_i \in \text{ns}(B)$ for $1 \leq i \leq 4$. On the other hand, $w_i \notin A$; since A is itself closed and bi-convex, it

---

[5] We always assume that the domain of definition B of a bi-convex function is a bi-convex set.
[6] The domain does indeed matter — see Example 3.5.

follows that ns(B) is *not* included in $\overline{\text{bi-co}(A)}$. From $w_i \in \text{bi-co}(A)$ it follows that ns$(B) \supset C \equiv \text{bi-co}\{z_i, w_i \mid 1 \le i \le 4\}$ (by Proposition 3.2). At the end of this section we will show that actually ns$(B) = C$. ∎

We claimed that separation by bi-affine functions is not sufficient; the following example shows that bi-convex functions may indeed separate more.

EXAMPLE 3.4. Let $\mathscr{X} = \mathscr{Y} = [0, 1]$, $A = \{(0,0), (\frac{1}{2}, 0), (0, \frac{1}{2}), (1, 1)\}$, $B = \mathscr{X} \times \mathscr{Y}$ (see Fig. 3.1). It is easy to see that bi-co(A) consists of A together with the two line segments $[(0, 0), (\frac{1}{2}, 0)]$ and $[(0, 0), (0, \frac{1}{2})]$. Consider now the following function f on B:

$$f(x, y) = \begin{cases} xy, & 0 \le x, y \le 1/2, \\ -3xy + 2x + 2y - 1, & \text{otherwise.} \end{cases}$$

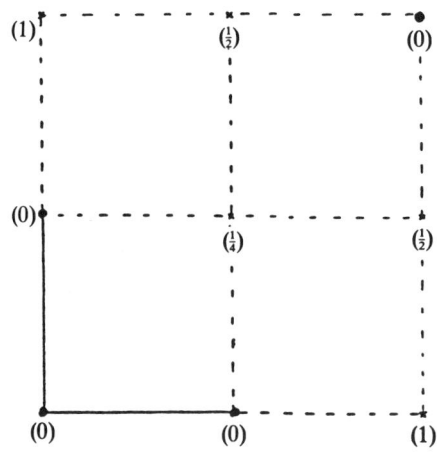

Fig. 3.1.

It can be checked that $f(x, y) \ge 0$ for all $(x, y) \in [0, 1] \times [0, 1]$, $f(x, y) = 0$ if and only if $(x, y) \in \text{bi-co}(A)$, and f is bi-convex (actually, it is *piecewise bi-affine*; it is obtained by putting $f(0, 0) = f(0, \frac{1}{2}) = f(\frac{1}{2}, 0) = f(1, 1) = 0$, $f(\frac{1}{2}, \frac{1}{2}) = \frac{1}{4}$, $f(0, 1) = f(1, 0) = 1$ and $f(\frac{1}{2}, 1) = f(1, \frac{1}{2}) = \frac{1}{2}$, and then extending it bi-affinely in each of the four small squares). Therefore, f separates every point not in bi-co(A) from A; thus, ns$(B) = \text{bi-co}(A)$.

Now let g be a bi-affine function on B; we will show that it cannot separate the point $(\frac{1}{4}, \frac{1}{4})$ from A. Indeed, let α, β, γ, δ be the values of g at the points of $A : (0, 0), (\frac{1}{2}, 0), (0, \frac{1}{2}), (1, 1)$ (respectively). Without loss of generality, assume

$g(\frac{1}{4},\frac{1}{4}) = 0$; thus, $\alpha, \beta, \gamma, \delta$ are all negative. Using repeatedly the fact that g is bi-affine, we obtain:

$$g(\tfrac{1}{4}, 0) = \tfrac{1}{2}g(0,0) + \tfrac{1}{2}g(\tfrac{1}{2}, 0) = \tfrac{1}{2}\alpha + \tfrac{1}{2}\beta,$$

$$g(\tfrac{1}{4}, \tfrac{1}{2}) = 2g(\tfrac{1}{4}, \tfrac{1}{4}) - g(\tfrac{1}{4}, 0) = -\tfrac{1}{2}\alpha - \tfrac{1}{2}\beta,$$

$$g(\tfrac{1}{2}, \tfrac{1}{2}) = 2g(\tfrac{1}{4}, \tfrac{1}{2}) - g(0, \tfrac{1}{2}) = -\alpha - \beta - \gamma,$$

$$g(\tfrac{1}{2}, 1) = 2g(\tfrac{1}{2}, \tfrac{1}{2}) - g(\tfrac{1}{2}, 0) = -2\alpha - 3\beta - 2\gamma,$$

$$g(0, 1) = 2g(0, \tfrac{1}{2}) - g(0, 0) = 2\gamma - \alpha,$$

$$g(1, 1) = 2g(\tfrac{1}{2}, 1) - g(0, 1) = -3\alpha - 6\beta - 6\gamma.$$

But $g(1,1) = \delta$, thus $-3\alpha - 6\beta - 6\gamma = \delta$, which is impossible since $\alpha, \beta, \gamma, \delta < 0$. ∎

Finally, we show that the separation does depend on the domain of definition B (in the regular convex case, all the separation is obtained by affine functions, which can always be extended to the whole space; this is so for neither convex nor bi-convex functions).

EXAMPLE 3.5. Let $\mathcal{X} = \mathcal{Y} = [0, 2]$, $A = \{(x, y) \mid 1 < x, y < 2 \text{ or } x = y = 1\}$ (i.e., A is an open square together with one of its corners). Let $B = \mathcal{X} \times \mathcal{Y}$; then we claim that the points $(x, 1)$ and $(1, y)$, for $1 < x, y < 2$, belong to ns(B). Indeed, let f be a bi-convex function on B, then

$$f(x, 1) \leq \frac{\varepsilon}{1+\varepsilon} f(x, 0) + \frac{1}{1+\varepsilon} f(x, 1+\varepsilon)$$

for every $0 < \varepsilon < 1$. Since $(x, 1+\varepsilon) \in A$ for $1 < x < 2$ and $0 < \varepsilon < 1$, we obtain when $\varepsilon \to 0$ that $f(x, 1) \leq \sup f(A)$, thus $(x, 1) \in$ ns(B). Similarly for $(1, y)$.

Now let $B = [1, 2) \times [1, 2)$; then the following bi-convex function separates all points $(x, 1)$ and $(1, y)$ (for $1 < x, y < 2$:

$$f(x, y) = \begin{cases} x - 1, & y = 1, \\ y - 1, & x = 1, \\ 0, & \text{otherwise.} \end{cases}$$

∎

What are the continuity properties of bi-convex functions? As we shall now see, they parallel those of convex functions (cf. [5]). A real function f defined on a set B is *lower-semi-continuous* at a point $\bar{z} \in B$ if

$$\liminf_{z \to \bar{z}} f(z) = f(\bar{z})$$

(or, equivalently, if $\liminf_{n\to\infty} f(z_n) \geq f(\bar{z})$ for every sequence $\{z_n\}_{n=1}^{\infty} \subset B$ such that $z_n \to \bar{z}$). It is *upper-semi-continuous* at $\bar{z}$ if

$$\limsup_{z\to\bar{z}} f(z) = f(\bar{z}),$$

and it is *continuous* at $\bar{z}$ if it is both lower- and upper-semi-continuous there. The following results should be compared with Theorems 7.4 and 10.2 in [5].

Let $B \subset \mathcal{X} \times \mathcal{Y}$ and let $z = (x, y) \in B$. The point z is *bi-relatively interior* to B if z is interior to B relative to $\text{aff}(\text{proj}_\mathcal{X} B) \times \text{aff}(\text{proj}_\mathcal{Y} B)$, where the affine space generated by a set C is denoted $\text{aff}(C)$. For example, let $\mathcal{X} = \mathcal{Y} = [0, 1]$ and let $B = \{(t, t) \mid 0 < t < 1\}$, then every point of B is a relatively interior point, but none is bi-relatively interior. Note also that on this set B, any function f is bi-convex!

PROPOSITION 3.6. *Let f be a bi-convex function on a bi-convex set B, and let $\bar{z} = (\bar{x}, \bar{y})$ be a bi-relatively interior point of B. Then f is lower-semi-continuous at $\bar{z}$.*

PROOF. Without loss of generality, assume $\bar{z}$ is actually interior to B. Let U be a closed cube around $\bar{x}$, and V a closed cube around $\bar{y}$, such that $U \times V \subset B$. Let $z = (x, y) \in U \times V$; express it as a bi-convex combination of the vertices of $U \times V$, say $z = \sum_{i=1}^{I} \alpha_i z_i$; then

$$f(z) \leq \sum_{i=1}^{I} \alpha_i f(z_i),$$

which implies that f is bounded from above on $U \times V$ (by $\max\{f(z) \mid z \text{ vertex of } U \times V\}$).

Now let $(x, y) \in U \times V$; continue the straight line (in U) through x and $\bar{x}$, past $\bar{x}$, until it intersects the boundary of U at a point x'; define y' similarly. Then $\bar{x} = \lambda x + \lambda' x'$ and $\bar{y} = \mu y + \mu' y'$, where $\lambda, \lambda', \mu, \mu' \geq 0$, $\lambda + \lambda' = \mu + \mu' = 1$. Since f is a bi-convex function,

$$f(\bar{x}, \bar{y}) \leq \lambda f(x, \bar{y}) + \lambda' f(x', \bar{y}) \leq \lambda \mu f(x, y) + \lambda \mu' f(x, y') + \lambda' f(x', \bar{y}).$$

As $(x, y) \to (\bar{x}, \bar{y})$, we have $\lambda' \to 0$ and $\mu' \to 0$ (the boundaries of U and V are at a positive distance from $\bar{x}$ and $\bar{y}$, respectively). Together with the boundedness from above of f on $U \times V$, this implies that only the first term matters, thus

$$f(\bar{x}, \bar{y}) \leq \liminf_{(x,y)\to(\bar{x},\bar{y})} f(x, y). \blacksquare$$

Again, let $B \subset \mathscr{X} \times \mathscr{Y}$ and $z = (x, y) \in B$. We say that B is *locally bi-simplicial at z* if there exist a neighborhood U of x in $\mathscr{X}$, a neighborhood V of y in $\mathscr{Y}$, a collection of simplices $S_1, S_2, \ldots, S_n$ in $\mathscr{X}$ and a collection of simplices $T_1, T_2, \ldots, T_m$ in $\mathscr{Y}$, such that (putting $S = \bigcup_{i=1}^{n} S_i$ and $T = \bigcup_{j=1}^{m} T_j$), $S \times T \subset B$ and $(U \times V) \cap B = (U \times V) \cap (S \times T)$ (compare with [5, p. 84]). Examples of sets that are locally bi-simplicial at all their points are sets $B = C \times D$, where $C \subset \mathscr{X}$ and $D \subset \mathscr{Y}$ are (relatively) open convex sets, or polyhedral sets. If we consider again $\mathscr{X} = \mathscr{Y} = [0, 1]$ and $B = \{(t, t) \mid 0 < t < 1\}$, then B is locally bi-simplicial at none of its points (although it is locally simplicial at all of them).

PROPOSITION 3.7. *Let f be a bi-convex function on a bi-convex set B, and let $\bar{z} = (\bar{x}, \bar{y}) \in B$. If B is locally bi-simplicial at $\bar{z}$, then f is upper-semi-continuous at $\bar{z}$.*

PROOF. Without loss of generality, assume each S_i has $\bar{x}$ as one of its vertices, and each T_j has $\bar{y}$ as one of its vertices (if this is not so, partition the corresponding simplex into smaller ones with this property). It suffices to show that f is upper-semi-continuous on each $S_i \times T_j$. Let $x_0 = \bar{x}, x_1, \ldots, x_p$ be the vertices of S_i, and $y_0 = \bar{y}, y_1, \ldots, y_q$ the vertices of T_j; then each $x \in S_i$ and each $y \in T_j$ can be expressed as $x = \sum_{r=0}^{p} \lambda_r x_r$ and $y = \sum_{s=0}^{q} \mu_s y_s$, with $\lambda_r, \mu_s \geq 0$ and $\sum_{r=0}^{p} \lambda_r = \sum_{s=0}^{q} \mu_s = 1$. Hence

$$f(x, y) \leq \sum_{r=0}^{p} \lambda_r f(x_r, y) \leq \sum_{r=0}^{p} \sum_{s=0}^{q} \lambda_r \mu_s f(x_r, y_s).$$

As $(x, y) \to (\bar{x}, \bar{y})$, we have $\lambda_0 \to 1$ and $\mu_0 \to 1$, thus $\lambda_r \to 0$ and $\mu_s \to 0$ for all $r \neq 0$ and $s \neq 0$, implying that $\limsup f(x, y) \leq f(\bar{x}, \bar{y})$. ∎

COROLLARY 3.8. *Let f be a bi-convex function on a bi-convex set B. Then f is continuous at all its bi-relatively interior points.*

PROOF. If z is a bi-relatively interior point of B, then B is locally bi-simplicial at z. ∎

We now complete the analysis of Examples 3.3 and 3.5. Consider first Example 3.3; we wish to show that $\text{ns}(B)$ does not contain any points outside C. Indeed, the function

$$f(x, y) = [x - \tfrac{2}{3}]_+ \cdot [y - \tfrac{1}{3}]_+$$

(where $[\lambda]_+ \equiv \text{Max}\{\lambda, 0\}$ for real λ), separates from A all points in the positive orthant with origin at w_4. In a similar way the other three orthants (with origins w_1, w_2 and w_3) are also separated.

Functions of this type, i.e.

(3.9) $$f(x, y) = [g(x)]_+ [h(y)]_+,$$

where g and h are affine functions on $\mathscr{X}$ and $\mathscr{Y}$, respectively, are often useful (for applications, see Section 7 in [3]). In Example 3.5, they suffice to show that

$$\text{ns}(\mathscr{X} \times \mathscr{Y}) = [1, 2) \times [1, 2),$$
$$\text{ns}([1, 2) \times [1, 2)) = A$$

(note that $A = A^*$, since A is a convex set). However, functions of this type do not separate everything that bi-convex functions do (see Example 3.4).

4. Main results

In this section we will obtain a characterization of A^* by separation properties. The main result is Theorem 4.7; see also Theorem 4.3 (these correspond to (1) and (2) at the end of the Introduction.)

In Section 3 we have defined, for every bi-convex set B that contains A, the set $\text{ns}(B)$ of all points of B that cannot be separated from A by any bi-convex function. As we saw in Example 3.5, one may have to apply the operator "ns" repeatedly in order to obtain the desired set A^* (see also Example 5.5 and Remark 5.7).

Formally, one defines inductively $B_0 = \mathscr{X} \times \mathscr{Y}$, $B_{\alpha+1} = \text{ns}(B_\alpha)$ for every ordinal α, and $B_\alpha = \bigcap_{\beta < \alpha} B_\beta$ for every limit ordinal[7] α. Since $\text{ns}(B) \subset B$ for every[8] B, and $B \subset B'$ implies $\text{ns}(B) \subset \text{ns}(B')$, one obtains a non-increasing sequence of sets $\{B_\alpha\}_\alpha$, with limit $C \equiv B_\gamma$ for some ordinal γ. (In the introduction, the limit set C was denoted bi-co$^*(A)$.)

PROPOSITION 4.1. *The limit set C satisfies $C = \text{ns}(C)$. Moreover, it is the largest such set, i.e., if $B = \text{ns}(B)$ then $B \subset C$.*

PROOF. Since $C = B_\gamma$ is the limit of the above sequence, we have $B_{\gamma+1} = B_\gamma$, or $\text{ns}(C) = C$. If $B = \text{ns}(B)$, then $B \subset B_0 = \mathscr{X} \times \mathscr{Y}$, and $B \subset B_\beta$ for all $\beta < \alpha$ implies $B = \text{ns}(B) \subset \text{ns}(B_\beta)$, thus $B \subset B_\alpha$; transfinite induction then gives $B \subset B_\gamma = C$. ∎

Does this set C coincide with A^*? Example 5.1 will show that this is not the case in general. What then is this set C? Consider Example 2.5: the points w_i

---

[7] Equivalently, define $B_\alpha = \bigcap_{\beta < \alpha} \text{ns}(B_\beta)$ for *every* ordinal α. Note that one may take $B_0 = \text{co}(A)$.

[8] We will always assume throughout this section that B is a bi-convex set containing A.

(for $1 \leq i \leq 4$) belong to A^* but not to bi-co(A). Actually, w_i can be obtained by a bi-martingale which a.s. reaches A in finite time (i.e., one need not go to the limit Z_∞, since on almost every path $Z_n \in A$ for all n large enough; how large depends on the path).

Formally,[9] let $\{Z_n\}_{n=1}^\infty$ be a bi-martingale, $Z_n \to Z_\infty$ a.s., and let $\{\mathcal{F}_n\}_{n=1}^\infty$ be the corresponding sequence of (finite) fields; that is, $\mathcal{F}_n$ is the field generated by $(Z_1, Z_2, \ldots, Z_n)$. Put $\mathcal{F}_\infty = \lim_{n \to \infty} \mathcal{F}_n$ (a σ-field), and denote by $\mathbf{N}$ the set of positive integers $\{1, 2, \ldots\}$. A *stopping time* N is a random variable with values in $\mathbf{N}_\infty \equiv \mathbf{N} \cup \{\infty\}$, such that the event $\{N = n\}$ belongs to $\mathcal{F}_n$ for every $n \in \mathbf{N}_\infty$. Intuitively, this means that N depends only on the "past" — i.e., $Z_1, Z_2, \ldots, Z_N$, but not on the "future". A stopping time N is *a.s. finite* if $P(N < \infty) = 1$; it is *a.s. bounded* if there exists $n_0 < \infty$ such that $P(N \leq n_0) = 1$. Note that if we only consider values of Z_n that have positive probability, the finiteness of the fields $\mathcal{F}_n$ implies that "a.s. bounded" is the same as "everywhere bounded", which is the same as "everywhere finite" (by König's Lemma); this however differs from "a.s. finite" (see Fig. 2.3 for an example).

We now define $A^\#$ as the set of all $z \in \mathcal{X} \times \mathcal{Y}$ such that there exists a bi-martingale $\{Z_n\}_{n=1}^\infty$ with $Z_1 = z$, together with an a.s. finite stopping time N, such that $Z_N \in A$ (a.s.). Note that if we require the stopping time to be bounded, then bi-co(A) is obtained (see Remark 2.4), whereas A^* corresponds to the case that the stopping time need not be a.s. finite. In a similar way to Proposition 2.3, we have

PROPOSITION 4.2. *$A^\#$ is a bi-convex set, satisfying*

$$\text{bi-co}(A) \subset A^\# \subset A^*.$$

Example 2.5 shows that bi-co(A) may be a proper subset of $A^\#$; Example 5.1 will show that $A^\#$ may be a proper subset of A^*.

THEOREM 4.3. *The largest set C satisfying $C = \text{ns}(C)$ is precisely $A^\#$.*

Thus, $A^\#$ is the largest set that contains A and such that no bounded bi-convex function defined on $A^\#$ can separate any of its points from A. We divide the proof into two parts.

PROPOSITION 4.4. $\text{ns}(A^\#) = A^\#$.

PROOF. Let $z \in A^\#$, $\{Z_n\}_{n=1}^\infty$ a bi-martingale with $Z_1 = z$, and let N be an a.s. finite stopping time with $Z_N \in A$ (a.s.). For every bounded bi-convex function f

---

[9] References for the following are, e.g., [1, Ch. 9], [4, Ch. IV–V].

defined on $A^\#$, the sequence $\{f(Z_n)\}_{n=1}^\infty$ is a real bounded sub-martingale (i.e., $E[f(Z_{n+1})|\mathcal{F}_n] \geq f(Z_n)$ for every n; this follows from the fact that f is bi-convex and $\{Z_n\}$ is a bi-martingale[10]). Since N is an a.s. finite stopping time, we obtain $f(z) = f(Z_1) \leq E(f(Z_N))$. But $Z_N \in A$ a.s., thus $f(Z_N) \leq \sup f(A)$, hence $f(z) \leq \sup f(A)$, which proves that $z \in \operatorname{ns}(A^\#)$. ∎

PROPOSITION 4.5. *Let B satisfy $B = \operatorname{ns}(B)$. Then $B \subset A^\#$.*

PROOF. Define a real function $\varphi \equiv \varphi_B$ on B by $\varphi(z) = \inf P(Z_n \notin A$ for all $n \geq 1)$ for every $z \in B$, where the infimum is taken over all bi-martingales $\{Z_n\}_{n=1}^\infty$ with $Z_1 = z$ and $Z_n \in B$ for all $n \geq 1$. If is straightforward to check that φ is a non-negative bi-convex function on B, and moreover that $\varphi(a) = 0$ for all $a \in A$. Since $\operatorname{ns}(B) = B$ we cannot separate by φ, thus $\varphi(z) = 0$ for all $z \in B$.

Now $1 - \varphi(z) = \sup P(Z_n \in A$ for some $n \geq 1) = \sup P(Z_N \in A)$, the supremum being taken over all bi-martingales $\{Z_n\}_{n=1}^\infty$ as above and over all a.s. finite stopping times N. Therefore it remains to prove that this supremum is achieved for each $z \in B$. This is a standard argument[11]; we will briefly sketch it here.

Choose $0 < \rho < 1$; for every $z \in B$, $\varphi(z) = 0$; hence there exists a bi-martingale $\{Z_n\}_{n=1}^\infty$ together with a stopping time N, such that $Z_1 = z$, $Z_n \in B$ for all n, $P(N < \infty) = 1$, and $P(Z_N \in A) > \rho$; since once A is reached, the bi-martingale can remain constant, we may replace N by an integer $m \equiv m(z)$ large enough, such that $P(Z_m \in A) > \rho$. (We will say that $\{Z_n\}$ and $m(z)$ "correspond" to z.)

Consider the bi-martingale $\{Z_n\}$ corresponding to z, and follow it up to step $m = m(z)$; from each point $z' = Z_m$ that does not belong to A (but does however belong to B), continue with the bi-martingale corresponding to z', for $m(z')$ more steps, and so on. The total probability that A is reached in finite time is then at least

$$\rho + (1-\rho)\rho + (1-\rho)^2\rho + \cdots,$$

which converges to 1. This completes the proof. ∎

PROOF OF THEOREM 4.3. Propositions 4.4 and 4.5 give the two inclusions $A^\# \subset C$ and $C \subset A^\#$, respectively. ∎

We have thus seen that separating by *all* bi-convex functions leads to $A^\#$, which is included in A^*. Now we will show that suitably restricting the family of functions used for separation leads to A^*.

---

[10] Note that $Z_n \in A^\#$ for all n, thus $f(Z_n)$ is well defined.
[11] E.g., it follows from Corollary 3.8.1 in [2].

Let B be a bi-convex set containing A. Let $\mathscr{C}(B) \equiv \mathscr{C}_A(B)$ be the set of all real functions on B that are bi-convex, bounded, and continuous at each point of A (continuity is not required on all B, but just on A). Note that functions of the type (3.9) belong to $\mathscr{C}(B)$ for any B. Let $\mathrm{nsc}(B)$ be the set of all $z \in B$ that are not separated from A by any $f \in \mathscr{C}(B)$; that is, such that $f(z) \leq \sup f(A)$ for all $f \in \mathscr{C}(B)$. One immediately obtains

PROPOSITION 4.6. *For every B, the set $\mathrm{nsc}(B)$ is bi-convex, and $\mathrm{ns}(B) \subset \mathrm{nsc}(B) \subset B$.*

We now define the set D as the largest set such that $\mathrm{nsc}(D) = D$. As was the case for the set C, we obtain D as the limit of the sequence $\{B_\alpha\}_\alpha$, where $B_0 = \mathscr{X} \times \mathscr{Y}$ and $B_\alpha = \bigcap_{\beta < \alpha} \mathrm{nsc}(B_\beta)$ for all ordinals α. (In the introduction, the limit set D was denoted bi-co*(A).)

THEOREM 4.7. *Assume A is a closed set. Then the largest set D satisfying $D = \mathrm{nsc}(D)$ is precisely A^*.*

Thus, A^* is the largest set that contains A, and such that no bi-convex function defined on A^* and continuous on A, can separate any point in A^* from A.

PROOF. It will follow from Propositions 4.8 and 4.9.

PROPOSITION 4.8. $\mathrm{nsc}(A^*) = A^*$.

PROOF. Let $z \in A^*$, $\{Z_n\}_{n=1}^\infty$ a bi-martingale with $Z_1 = z$, $Z_n \to Z_\infty$, $P(Z_\infty \in A) = 1$, and let $f \in \mathscr{C}(A^*)$. Since $Z_n \in A^*$ for all n, we obtain a bounded (real) sub-martingale $\{f(Z_n)\}_{n=1}^\infty$; moreover, $Z_\infty \in A$ implies $f(Z_n) \to f(Z_\infty)$, therefore $f(z) = f(Z_1) \leq E(f(Z_\infty)) \leq \sup f(A)$. ∎

PROPOSITION 4.9. *Assume A is a closed set, and let B satisfy $B = \mathrm{nsc}(B)$. Then $B \subset A^*$.*

PROOF. For every z, let $d(z, A)$ denote the distance of z from the set A; A being a closed set, $d(z, A) = 0$ if and only if $z \in A$. Define a real function $\psi \equiv \psi_B$ on B by $\psi(z) = \inf E[d(Z_\infty, A)]$ for every $z \in B$, where the infimum is taken over all bi-martingales $\{Z_n\}_{n=1}^\infty$ satisfying $Z_1 = z$, $Z_n \in B$ for all n, and $Z_n \to Z_\infty$ (a.s.). It is again easy to see that ψ is a bounded bi-convex function. One possible bi-martingale for z is the constant one (namely, $Z_n = z$ for all n); therefore $\psi(z) \leq d(z, A)$, which shows that ψ vanishes and is continuous at every point of A. But $B = \mathrm{nsc}(B)$, hence ψ does not separate any point of B from A — thus ψ is identically zero on all B.

To complete the proof we will now show that the infimum in the definition of $\psi(z)$ is indeed achieved[12] for all $z \in B$. Since $Z_n \to Z_\infty$ implies $d(Z_n, A) \to d(Z_\infty, A)$, we have $E[d(Z_n, A)] \to E[d(Z_\infty, A)]$; therefore, for every $z \in B$ and every $\rho > 0$ there exists a bi-martingale $\{Z_n\}_{n=1}^\infty$ (with $Z_1 = z$ and $Z_n \in B$ for all n), and an integer m such that $E[d(Z_m, A)] < \rho$. After stage m, continue with bi-martingales corresponding to each $z' = Z_m$ and $\rho/2$, for $m' \equiv m'(z')$ more steps; follow then bi-martingales corresponding to $z'' = Z_{m+m'}$ and $\rho/3$, and so on. This construction yields a new bi-martingale $\{\bar{Z}_n\}_{n=1}^\infty$ with $\bar{Z}_1 = z$ and $\bar{Z}_n \in B$ for all n; let $\bar{Z}_\infty$ be its a.s. limit. We also obtain an increasing sequence $\{N_k\}_{k=1}^\infty$ of finite stopping times ($N_1 = m, N_2 = m + m', \ldots$) such that $E[d(\bar{Z}_{N_k}, A)] < \rho/k$ for all $k \geq 1$. Therefore $E[d(\bar{Z}_\infty, A)] = 0$. ∎

REMARK 4.10. It can be easily checked in the proof of Proposition 4.8 that it suffices for f to be just upper-semi-continuous rather than continuous at each point of A. Therefore Theorem 4.7 remains true if one allows separation by this type of bounded bi-convex functions too. In what regards checking upper-semi-continuity, recall Proposition 3.7.

REMARK 4.11. The finiteness of the fields $\mathscr{F}_n$ does not play any role in the proofs of Propositions 4.4 and 4.8. Together with Theorems 4.3 and 4.7, this implies that neither $A^\#$ nor A^* will change if this finiteness condition is dropped from the definition of a bi-martingale.

5. Examples

This section is devoted to three examples, settling (in the negative) some questions regarding A^*:
1. Is $A^\#$ equal to A^*?
2. Is $(A^*)^*$ equal to A^*?
3. If A is a closed set, is A^* closed too?

Since the motivation for the study of A^* came from game theory (see [3]), we will take in all three examples the set A to be compact and piecewise algebraic (i.e., a finite union of sets defined by algebraic functions). It is thus conjectured that the same phenomena appear in the game theoretic context as well.

All three examples use an idea similar to Example 2.2; however, by making each of the two spaces $\mathscr{X}$ and $\mathscr{Y}$ two-dimensional, one can obtain a kind of a "rotating staircase", which eliminates unwanted interaction between the various "steps". To get a geometric picture, imagine in Fig. 2.1 that $\mathscr{X}$ becomes

---

[12] Again, one may apply Corollary 3.8.1 in [2].

two-dimensional — a plane perpendicular to the page — whereas $\mathcal{Y}$ remains one-dimensional; "rotate" slightly each of the "steps" $w_2z_3, w_4z_5, \ldots$ in the $\mathcal{X}$-plane, around $w_3, w_5, \ldots$ (respectively).

EXAMPLE 5.1. Let $\mathcal{X} = \mathcal{Y} = [0,1]^2 \subset \mathbf{R}^2$. Let $T = [0, 0.2]$; for every $t \in T$, let[13]

$$b_t = (1, 3t - 2t^2; 2t, 4t^2), \qquad c_t = (t, t^2; 1, 3t - 2t^2),$$
$$d_t = (2t, 4t^2; 2t, 4t^2), \qquad e_t = (t, t^2; 2t, 4t^2).$$

Let $B = \{b_t\}_{t \in T}$, $C = \{c_t\}_{t \in T}$, $D = \{d_t\}_{t \in T \setminus \{0\}}$, $E = \{e_t\}_{t \in T \setminus \{0\}}$, and define $A = B \cup C \cup \{O\}$, where $O = (0, 0; 0, 0)$.

PROPOSITION 5.2. (1) $D \cup E \subset A^*$.
(2) $(D \cup E) \cap A^\# = \emptyset$.

PROOF. (1) For every $t \in T$, $t \neq 0$, we have

$$d_t = \frac{t}{1-t} b_t + \frac{1-2t}{1-t} e_t \qquad (y \text{ constant}),$$

$$e_t = \frac{t}{1-t} c_t + \frac{1-2t}{1-t} d_{t/2} \qquad (x \text{ constant}).$$

We thus obtain a bi-martingale, represented in tree form in Fig. 5.1. As $t \to 0$, both d_t and e_t converge to $O \in A$; therefore the above bi-martingale converges

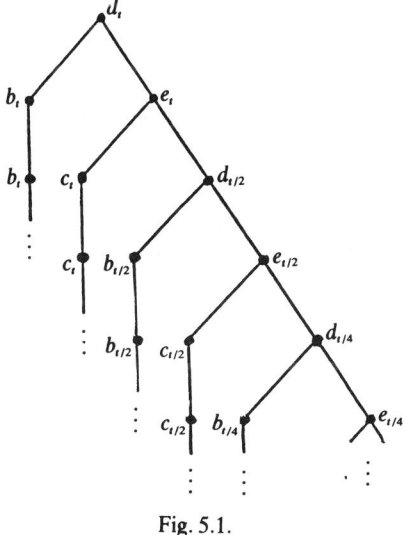

Fig. 5.1.

---

[13] We write a point z as $z = (x^{(1)}, x^{(2)}; y^{(1)}, y^{(2)})$, where $x = (x^{(1)}, x^{(2)}) \in \mathcal{X}$ and $y = (y^{(1)}, y^{(2)}) \in \mathcal{Y}$.

to A with probability one (recall that all b_t and c_t belong to A). Similarly for e_t, showing that d_t and e_t belong to A^* for all $t \in T \setminus \{0\}$.

(2) Let $\{Z_n\}_{n=1}^{\infty}$ be a bi-martingale in $\mathscr{X} \times \mathscr{Y}$ and let N be an a.s. finite stopping time, with $Z_1 = d_t$ (for some $t \in T$, $t \neq 0$) and $Z_N \in A$ (a.s.). Define now the sets

$$F_+ = \{z \subset \mathscr{X} \times \mathscr{Y} \mid x^{(2)} > 0 \text{ and } y^{(2)} > 0\}, \qquad F_0 = \{z \in \mathscr{X} \times \mathscr{Y} \mid x^{(2)} = y^{(2)} = 0\},$$

and $F = F_+ \cup F_0$. Since F is a convex set and $F \supset A$, we obtain $F \supset A^*$. Therefore $Z_n \in F$ for all n. Moreover, $Z_n \in F_+$ implies $Z_{n+1} \in F_+$, since $x^{(2)}$ and $y^{(2)}$ cannot both change. But $Z_1 = d_t \in F_+$, therefore $Z_n \in F_+$ for all n, hence $Z_N \in F_+$, which implies that $d_t \in (A \cap F_+)^*$. Now $z \in A \cap F_+$ implies $z = b_t$ or $z = c_t$ for some $0 < t \leq 0.2$, hence $x^{(1)} + y^{(1)} \geq 1$. From this it follows that $x^{(1)} + y^{(1)} \geq 1$ for every $z \in \text{co}(A \cap F_+)$, hence for every $z \in (A \cap F_+)^*$, but d_t does not satisfy this inequality. Similarly for e_t. ∎

It is instructive to compute $\varphi \equiv \varphi_B$ for $B = \mathscr{X} \times \mathscr{Y}$ in this example (see the proof of Proposition 4.5). By considering the bi-martingale constructed in the proof of (1) above, we have (put $t_n \equiv t/2^n$):

$$\varphi(d_t) = \prod_{n=0}^{\infty} \left(\frac{1-2t_n}{1-t_n}\right)^2 = (1-2t)^2,$$

and

$$\varphi(e_t) = \frac{1-2t}{1-t} \prod_{n=1}^{\infty} \left(\frac{1-2t_n}{1-t_n}\right)^2 = (1-2t)(1-t).$$

As $t \to 0$, both $d_t \to O$ and $e_t \to O$, but $\varphi(d_t) \to 1$ and $\varphi(e_t) \to 1$; since $\varphi(O) = 0$, φ is indeed *not* continuous at O (which belongs to A).

The next example shows that the $*$ operator is not idempotent; namely, in general $(A^*)^* \supsetneq A^*$. Actually, we will even show that $(A_2)^* \supsetneq A^*$, where A_2 is the set of all bi-convex combinations of the elements of A (see Section 2).

Note, however, that $(A^*)^* = A^*$ (if N_1 and $N_2 \equiv N_2(\omega_{N_1})$ are a.s. finite stopping times, then so is $N_1 + N_2$). Thus, Example 5.3 provides a further instance of the "$*$" operator being different from the "$\#$" one (indeed: we must have either $A^* \neq A^\#$, or $A^* = A^\# \equiv B$ and then $B^* \neq B^\#$).

EXAMPLE 5.3. Let $\mathscr{X} = \mathscr{Y} = [-1,1]^2 \subset \mathbf{R}^2$. Let $T = [0, 0.2]$, and define the sets B, C, D and E as in Example 5.1. Let $g = (-1, 0; 0, 0)$, and put $A = B \cup C \cup \{g\}$.

PROPOSITION 5.4. (1) $O \in A_2$ (where $O = (0, 0; 0, 0)$).
(2) $D \cup E \subset (A_2)^*$.
(3) $(D \cup E) \cap A^* = \emptyset$.

PROOF. (1) $O = \frac{1}{2}g + \frac{1}{2}b_0$ (y is constant).

(2) Follows immediately from Proposition 5.2(1) and (1) above.

(3) Define the sets F_+, F_0 and F as in the proof of Proposition 5.2(2). Since $A \subset F$ and F is a convex set, we obtain $A^* \subset F$.

Let $z \in A^* \cap F_+$; we will show that $z \notin D \cup E$. Let $\{Z_n\}_{n=1}^\infty$ be a bi-martingale, $Z_1 = z$, $Z_n \to Z_\infty$, $Z_\infty \in A$ (a.s.). As in the previous Proposition, we again obtain $Z_n \in A^* \cap F_+$ for all n.

Consider the function

$$f(z) = f(x^{(1)}, x^{(2)}; y^{(1)}, y^{(2)}) = [-x^{(1)}]_+ [y^{(2)}]_+$$

(see (3.9)). It is a bi-convex, bounded and continuous function. It vanishes on A (since $x^{(1)} < 0$ only at g, where $y^{(2)} = 0$), thus it must vanish on A^* (by Proposition 4.8). Therefore $Z_n \in A^* \cap F_+$ implies $0 \le X_n^{(1)}$ (= the first $\mathcal{X}$ coordinate of Z_n) hence $0 \le X_\infty^{(1)}$. Hence Z_∞ cannot equal g, and we have $Z_\infty \in B \cup C$ (a.s.) and $z \in (B \cup C)^*$.

Finally, $x^{(1)} + y^{(1)} \ge 1$ on $B \cup C$, thus on $(B \cup C)^*$; but this is not so on $D \cup E$, completing the proof that $(D \cup E) \cap A^* = \emptyset$. ∎

The last example is concerned with topological properties of A^* and the other sets we dealt with: bi-co(A) and $A^\#$. If A is a closed set, so will be each of the sets A_n for $n \ge 2$ (see Section 2). However, it may well be the case that none of bi-co(A), $A^\#$ and A^* are closed.

EXAMPLE 5.5. Let $\mathcal{X} = \mathcal{Y} = [-1, 1]^2 \subset \mathbf{R}^2$. Let $T = [0, 0.1]$, $T' = [0.1, 0.2]$, and define for every $t \in T \cup T' = [0, 0.2]$

$$b_t = (-1, -3t - 2t^2; t, t^2), \quad c_t = (2t, 4t^2; -1, -3t - 2t^2),$$

$$d_t = (t, t^2; t, t^2), \quad e_t = (2t, 4t^2; t, t^2).$$

Let $B = \{b_t\}_{t \in T}$, $C = \{c_t\}_{t \in T}$, $D = \{d_t\}_{t \in T\setminus\{0\}}$, $D' = \{d_t\}_{t \in T'}$, $E = \{e_t\}_{t \in T\setminus\{0\}}$, and put $A = B \cup C \cup D'$.

PROPOSITION 5.6. (1) $D \cup E \subset \text{bi-co}(A)$.
(2) $O = (0, 0; 0, 0) \notin A^*$.

Since $d_t, e_t \to O$ as $t \to 0$, the point O belongs to the *closure* of each one of the sets bi-co(A), $A^\#$ and A^*, but does not belong to any one of these sets.

PROOF. (1) For every $t \in T \setminus \{0\}$, d_t is a bi-convex combination of b_t and e_t (with y constant), and e_t is a bi-convex combination of c_t and d_{2t} (with x constant). Therefore bi-co(A) contains e_t for all $0.1/2 \le t \le 0.2/2$, hence d_t for all those t, hence e_t for all $0.1/4 \leq t \leq 0.2/4$, and so on. Thus $d_t, e_t \in \text{bi-co}(A)$ for all $0 < t \leq 0.1$.

(2) Let $F = \{z \in \mathscr{X} \times \mathscr{Y} \mid x^{(2)} \leq 0 \text{ and } y^{(2)} \leq 0\}$ and $F_0 = \{z \in \mathscr{X} \times \mathscr{Y} \mid x^{(2)} = y^{(2)} = 0\}$. We claim that $A^* \cap F = A^* \cap F_0$. Indeed, for every $u > 0$ consider the function $[-3u - 2u^2 - x^{(2)}]_+ [u^2 - y^{(2)}]_+$. It vanishes on A (since only b_t, for $t > u$, satisfies $x^{(2)} < -3u - 2u^2$; but then $u^2 - y^{(2)} = u^2 - t^2 < 0$). Therefore it vanishes on A^*. Let $z \in A^*$ with $x^{(2)} < 0$; then $x^{(2)} < -3u - 2u^2$ for some $u > 0$ (small enough), hence $y^{(2)} \geq u^2 > 0$. In a similar way, $y^{(2)} < 0$ implies $x^{(2)} > 0$, thus $z \in A^* \cap F$ only when $x^{(2)} = y^{(2)} = 0$, or $z \in A^* \cap F_0$.

Consider now a bi-martingale $\{Z_n\}_{n=1}^{\infty}$ with $Z_1 = O$, $Z_n \to Z_\infty$, $Z_\infty \in A$ (a.s.). We claim that $Z_n \in A^* \cap F_0$ implies $Z_{n+1} \in A^* \cap F_0$. Indeed, assume without loss of generality that $X_{n+1} = X_n$. Then $X_{n+1}^{(2)} = X_n^{(2)} = 0$ and $E(Y_{n+1}^{(2)} \mid \mathscr{F}_n) = Y_n^{(2)} = 0$. If $Y_{n+1}^{(2)} \leq 0$, then $Z_{n+1} \in A^* \cap F = A^* \cap F_0$, or $Y_{n+1}^{(2)} = 0$; thus $Y_{n+1}^{(2)} = 0$ throughout. Now $Z_1 = O \in A^* \cap F_0$, therefore $Z_n \in A^* \cap F_0$ for all n, implying that $Z_\infty \in F_0$. But $A \cap F_0 = \{b_0, c_0\}$, thus $O \in \{b_0, c_0\}^*$, which is clearly impossible (both b_0 and c_0 satisfy $x^{(1)} + y^{(1)} = 1$). ■

REMARK 5.7. In Example 5.5, the point O is a bi-relatively interior point of $\mathscr{X} \times \mathscr{Y}$, therefore any bi-convex function is continuous there (recall Corollary 3.8). Therefore O belongs to both $\text{ns}(\mathscr{X} \times \mathscr{Y})$ and $\text{nsc}(\mathscr{X} \times \mathscr{Y})$. This shows that even if A is a closed set — as in Example 5.5 — one may have to apply the operators ns and nsc more than once in order to obtain $A^\#$ and A^*, respectively (in Example 3.5, the set A was not closed).

REFERENCES

1. K. L. Chung, *A Course in Probability Theory*, Academic Press, New York, 1974.
2. L. E. Dubins and L. J. Savage, *Inequalities for Stochastic Processes: How to Gamble if You Must*, Dover, New York, 1976.
3. S. Hart, *Nonzero-sum two-person repeated games with incomplete information*, Mathematics of Operations Research **10** (1985), 117–153.
4. P. A. Meyer, *Probability and Potentials*, Blaisdell Publishing Co., 1966.
5. R. T. Rockafellar, *Convex Analysis*, Princeton University Press, 1970.

Author Index for Both Volumes

Anscombe, F. J., 17
Brandenburger, A., 37
Bruno, M., 7
Dreze, J., 44, 57
Gardner, R. J., 54
Hahn, F., 7
Hart, S., 73
Katznelson, Y., 30
Kruskal, J. B., 11, 12
Kurz, M., 52, 53, 55, 56
Maschler, M., 29, 42, 45
Myerson, R., 59
Neyman, A., 55, 56
Peleg, B., 38, 43, 49
Perles, M., 70
Rabinowitz, P., 43
Radner, R., 30
Rosenthal, R. W., 30, 54
Roth, A., 61a, 61d
Rothblum, U., 72
Savage, L. J., 18
Scafuri, A. J., 62a
Sen, A., 7
Shafer, W., 61b
Shapley, L. S., 22
Sorin, S., 24
Weiss, B., 30
Yannelis, N., 62a

The chapter(s) that each individual authored or coauthored is listed, after his name, by its number(s) in this collection. Volume I comprises chapters 1–37 and volume II chapters 38–73.

Journal Index for Both Volumes

Annals of Mathematical Statistics, 17
Annals of Mathematics, 10
Annals of Statistics, 32
Bulletin of the American Mathematical Society, 38, 65
Bulletin of the Research Council of Israel, 26
Econometrica, 7, 14, 33, 37, 46, 47, 51, 52, 57, 60, 61, 62
Games and Economic Behavior, 24, 36
Illinois Journal of Mathematics, 66
International Economic Review, 50
International Journal of Game Theory, 44
Israel Journal of Mathematics, 53, 73
Journal of Economic Theory, 19, 45, 48, 54, 55
Journal of Mathematical Analysis and Applications, 68, 70
Journal of Mathematical Economics, 31, 49, 69
Journal of the Society for Industrial and Applied Mathematics, 27
Management Science, 29
Mathematics of Computation, 43
Mathematics of Operations Research, 30
Naval Research Logistics Quarterly, 11, 12, 16
Pacific Journal of Mathematics, 21
Review of Economic Studies, 56
SIAM Journal on Control and Optimization, 72
Transactions of the American Mathematical Society, 39
World Politics, 3

The chapters appearing in each journal are listed after the journal's name, by their number(s) in this collection. Volume I comprises chapters 1–37 and volume II chapters 38–73.

Citation Index

1 Introduction

Just as the references in each of the papers in this collection provide an overview of the work leading to that paper, so the items in the following citation index provide an overview of the subsequent developments. Together, the references and the citations yield a rough picture of the scientific context in which the work is embedded.

The citations to each chapter are listed separately. Within each chapter's list, the citations are listed alphabetically by author; this should make it easier to follow various strands of work, as represented by various authors. It should also make it easier to find a particular citation, if the author is known.

The citations were culled from three categories of sources: journal articles, collections (such as anniversary or memorial volumes, published conference proceedings, and so on), and monographs or texts. The breadth of coverage varies widely among these three categories. The coverage of the journal literature is the most complete. The main sources here are the three major citation indexes, covering the Sciences (SCI), the Social Sciences (SSCI), and the Arts & Humanities (AHCI). Each of these is covered from its inception (1960 for the SCI, 1969 for the SSCI, and 1977 for the AHCI) through 1998. These are supplemented by the early volumes of the *International Journal of Game Theory* (1971–1986), the *Journal of Mathematical Economics* (1974–1979), *Mathematics of Operations Research* (1976–1979), *Games and Economic Behavior* (1989–1990), and *Economic Theory* (1991–1995), covering the period from the inception of each until it was included in the major indexes listed above. We have also added a few early citations on an ad hoc basis, particularly to Chapter 10.

The coverage in the second and third categories is much less complete. These two categories cover only books found in my own personal library, and those of some of my colleagues in Jerusalem. While this is somewhat haphazard, one suspects that there may be a correlation between these books and citations of works in this collection. In any case, we would be grateful to hear of additional items, both for their intrinsic interest and for possible future editions of this collection. Section 2 below lists the collections that the index covers, together with the abbreviations used for these books in the index; Section 3 does the same for the texts and monographs. Section 4 lists abbreviations for journals occurring frequently in the index (we have not provided a complete list of all journals appearing in the index). Section 5 lists abbreviations for frequently occurring generic words (like "Journal" or "Economic"). Even within these confines, errors and omissions have doubtless crept in. We would be grateful to be apprised of them.

Citations of preprints have been replaced by citations of the corresponding subsequent publication. Doctoral dissertations, and other unpublished material, are not included in the index. When the same item appears in several sources, like a collection *and* a journal, only the more accessible source (usually the journal) is used.

Each of the two volumes in this collection contains the citations of the chapters in that volume. For convenience, the list of abbreviations is printed in both volumes.

2 Collections and their Abbreviations

Adv Game Th
M. Dresher, L. S. Shapley & A. W. Tucker, Editors, *Advances in Game Theory*, Princeton University Press, 1964.

Aliprantis, Border & Luxemburg
C. D. Aliprantis, K. C. Border & W. A. J. Luxemburg, Editors, *Positive Operators, Riesz Spaces, and Economics, Proceedings of a Conference at Caltech, Pasadena, California, April 16–20, 1990*, Springer, 1991.

Arrow and Econ Policy
G. R. Feiwel, Editor, *Arrow and the Foundations of the Theory of Economic Policy*, Macmillan, 1987.

Arrow and Econ Th
G. R. Feiwel, Editor, *Arrow and the Ascent of Modern Economic Theory*, Macmillan, 1987.

Ashenfelter & Rees
O. Ashenfelter & A. Rees, Editors, *Discrimination in Labor Markets*, Princeton University Press, 1973.

Barg Inc Inf
P. B. Linhart, R. Radner & M. A. Satterthwaite, Editors, *Bargaining with Incomplete Information*, Academic Press, 1992.

Bicchieri & Dalla Chiara
C. Bicchieri & M. L. Dalla Chiara, Editors, *Knowledge, Belief, and Strategic Interaction*, Cambridge University Press, 1992.

Binmore & Dasgupta 1986
K. Binmore & P. Dasgupta, Editors, *Economic Organizations as Games*, Basil Blackwell, 1986.

Binmore & Dasgupta 1987
K. Binmore & P. Dasgupta, Editors, *The Economics of Bargaining*, Basil Blackwell, 1987.

Brittanica
Encyclopedia Brittanica, 15th Edition, 1990.

Coal Behav
S. Groennings, E. W. Kelley & M. Leiserson, Editors, *The Study of Coalition Behavior*, Holt, Reinhart & Winston, 1970.

CORE 20th Ann
B. Cornet & H. Tulkens, Editors, *Contributions to Operations Research and Economics: The Twentieth Anniversary of CORE*, MIT Press, 1989.

Cost Allocation
H. P. Young, Editor, *Cost Allocation: Methods, Principles, Applications*, Elsevier Science Publishers B.V., 1985.

Debreu Festschr
W. Hildenbrand & A. Mas-Colell, Editors, *Contributions to Mathematical Economics in Honor of Gerard Debreu*, Elsevier Science Publishers B.V., 1986.

Dreze Essays
J. H. Dreze, *Essays on Economic Decisions under Uncertainty*, Cambridge University Press, 1987.

Dreze Festschr
J. J. Gabszewicz, J. F. Richard & L. A. Wolsey, Editors, *Economic Decision-Making: Games, Econometrics and Optimisation—Contributions in Honour of Jacques H. Dreze*, Elsevier Science Publishers B.V., 1990.

Frontiers of Economics
K. J. Arrow & S. Honkapohja, Editors, *Frontiers of Economics*, Basil Blackwell, 1985.

Frontiers Game Th
K. Binmore, A. Kirman & P. Tani, Editors, *Frontiers of Game Theory*, MIT Press, 1993.

Hagen Conf I
O. Moeschlin & D. Pallaschke, Editors, *Game Theory and Related Topics, Proceedings of the Seminar on Game Theory and Related Topics, Bonn/Hagen, September 1978*, North-Holland, 1979.

Hagen Conf II
O. Moeschlin & D. Pallaschke, Editors, *Game Theory and Mathematical Economics, Proceedings of the Seminar on Game Theory and Mathematical Economics, Bonn/Hagen, October 1980*, North-Holland, 1981.

Hahn
F. Hahn, Editor, *The Economics of Missing Markets, Information and Games*, Clarendon Press, 1989.

Hahn Festschr
P. Dasgupta, D. Gale, O. Hart & E. Maskin, Editors, *Economic Analysis of Markets and Games: Essays in Honor of Frank Hahn*, MIT Press, 1992.

Handb Exp Econ
J. H. Kagel & A. Roth, Editors, *Handbook of Experimental Economics*, Princeton University Press, 1995.

Handb Game Th
R. J. Aumann & S. Hart, Editors, *Handbook of Game Theory with Economic Applications*, Elsevier Science Publishers B.V., Vols 1 (1992), 2 (1994).

Handb Math Econ
Handbook of Mathematical Economics, Elsevier Science Publishers B.V., K. J. Arrow & M. D. Intriligator, Editors, Vols 1–3 (1981); W. Hildenbrand & H. Sonnenschein, Editors, Vol 4 (1991).

Handb Math Psych
R. D. Luce, R. R. Bush & E. Galanter, Editors, *Handbook of Mathematical Psychology*, Wiley, 1965.

Harsanyi Essays
J. C. Harsanyi, *Essays on Ethics, Social Behavior, and Scientific Explanation*, D. Reidel, 1976.

Harsanyi Festschr
R. Selten, Editor, *Rational Interaction, Essays in Honor of John C. Harsanyi*, Springer, 1992.

Harsanyi Papers Game Th
J. Harsanyi, *Papers in Game Theory*, D. Reidel, 1982.

Haveman & Margolis
R. H. Haveman & J. Margolis, Editors, *Public Expenditures and Policy Analysis*, Rand McNally, 1970.

Heertje I, II, III
A. Heertje, Editor, *The Makers of Modern Economics*, Harvester Wheatsheaf, Vols 1 (1993), 2 (1995), 3 (1997).

Holler, Coalitions
M. J. Holler, Editor, *Coalitions and Collective Action*, Physica, 1984.

Int Encycl Soc Sci
D. L. Sills, Editor, *International Encyclopedia of the Social Sciences*, Macmillan, 1968.

La Decision
Y. Aharoni et al, *La Decision*, Centre National de la Recherche Scientifique, Paris, 1969.

Lect Notes
Lecture Notes in Economics and Mathematical Systems, Springer.

Lect Notes 102
J.-P. Aubin, Editor, *Analyse Convexe et Ses Applications*, Lect Notes 102, 1974.

Lect Notes 141
R. Henn & O. Moeschlin, Editors, *Mathematical Economics and Game Theory: Essays in Honor of Oskar Morgenstern*, Lect Notes 141, 1977.

Lect Notes 157
R. Henn, B. Korte & W. Oettli, Editors, *Optimization and Operations Research*, Lect Notes 157, 1978.

Lect Notes 177
G. Fandel & T. Gal, Editors, *Multiple Criteria Decision Making: Theory and Application*, Lect Notes 177, 1980.

Lect Notes 213
R. Tietz, Editor, *Aspiration Levels in Bargaining and Economic Decision Making*, Lect Notes 213, 1983.

Lect Notes 226
G. Hammer & D. Pallaschke, Editors, *Selected Topics in Operations Research and Mathematical Economics*, Lect Notes 226, 1984.

Lect Notes 244
C. D. Aliprantis, O. Burkinshaw & N. J. Rothman, Editors, *Advances in Equilibrium Theory*, Lect Notes 244, 1985.

Lect Notes 259
E. J. Anderson & A. B. Philpott, Editors, *Infinite Programming*, Lect Notes 259, 1985.

Lect Notes 264
H. F. Sonnenschein, Editor, *Models of Economic Dynamics*, Lect Notes 264, 1986.

Lect Notes 353
G. Ricci, Editor, *Decision Processes in Economics*, Lect Notes 353, 1991.

Lect Notes 389
B. Dutta, D. Mookherjee, T. Parthasarathy, T. Raghavan, D. Ray & S. Tijs, Editors, *Game Theory and Economic Applications*, Lect Notes 389, 1992.

Lect Notes 453
A. Tangian & J. Gruber, Editors, *Constructing Scalar-Valued Objective Functions*, Lect Notes 453, 1997.

Leondes
Leondes, C. T., Editor, *Control and Dynamic Systems, Vol 17: Advances in Theory and Applications*, Academic Press, 1981.

Mitiagin
C. Mitiagin, Editor, *Mathematical Economics and Functional Analysis* (in Russian), Nauka, 1974.

Morgenstern Festschr
M. Shubik, Editor, *Essays in Mathematical Economics in Honor of Oskar Morgenstern*, Princeton University Press, 1967.

Morgenstern Memorial
R. J. Aumann et al, *Essays in Game Theory and Mathematical Economics in Honor of Oskar Morgenstern*, Wissenschaftsverlag, Bibliographisches Institut, 1981.

Nakamura Papers
M. Suzuki, Editor, *Game Theory and Social Choice, Selected Papers of Kenjiro Nakamura*, Keiso Shuppan Service Centre: Tokyo, 1981.

The New Palgrave
J. Eatwell, M. Milgate & P. Newman, Editors, *The New Palgrave, A Dictionary of Economics*, Vols 1–4, Macmillan, London & Basingstoke, 1987.

The New Palgrave Game Theory
J. Eatwell, M. Milgate & P. Newman, Editors, *The New Palgrave Game Theory*, Macmillan, London & Basingstoke, 1989.

Ohio Conf
T. Ichiishi, A. Neyman & Y. Tauman, Editors, *Game Theory and Applications, Proceedings of the International Conference on Game Theory and Applications held at the Ohio State University, June 1987*, Academic Press, 1990.

Proc Berkeley Symp 1965
L. M. Le Cam & J. Neyman, Editors, *Proceedings of the Fifth Berkeley Symposium on Mathematical Statistics and Probability*, University of California Press, 1967.

Proc Berkeley Symp 1970
L. M. Le Cam, J. Neyman & E. L. Scott, Editors, *Proceedings of the Sixth Berkeley Symposium on Mathematical Statistics and Probability*, University of California Press, 1972.

Proc Econometr Soc Congr 1980
W. Hildenbrand, Editor, *Advances in Economic Theory, Invited Papers for the Fourth World Congress of the Econometric Society at Aix-En-Provence*, Cambridge University Press, 1980.

Proc Econometr Soc Congr 1985
T. F. Bewley, Editor, *Advances in Economic Theory: Fifth World Congress*, Cambridge University Press, 1987.

Proc Econometr Soc Congr 1990
J. J. Laffont, Editor, *Advances in Economic Theory: Sixth World Congress*, Vols 1–2, Cambridge University Press, 1992.

Proc Econometr Soc Congr 1995
D. M. Kreps & K. F. Wallis, Editors, *Advances in Economics and Econometrics: Theory and Applications, Seventh World Congress*, Vol 1, Cambridge University Press, 1997.

Proc Int Congr Math 1974
R. D. James, Editor, *Proceedings of the International Congress of Mathematicians, 1974*, Canadian Mathematical Congress, 1975.

Proc Int Congr Math 1986
A. M. Gleason, Editor, *Proceedings of the International Congress of Mathematicians, 1986*, American Mathematical Society, 1987.

Proc Int Congr Phil Sci 1975
R. Butts & J. Hintikka, Editors, *Proceedings of the Fifth International Congress of Logic, Methodology and Philosophy of Science*, D. Reidel, 1977.

Proc Int Symp Econ Th
W. A. Barnett, Editor, *Equilibrium Theory and Applications: Proceedings of the Sixth International Symposium in Economic Theory and Econometrics*, Cambridge University Press, 1991.

Proc Int Symp Soc Ch
W. A. Barnett, H. Moulin, M. Salles & N. J. Schofield, Editors, *Social Choice, Welfare, and Ethics: Proceedings of the Eighth International Symposium in Economic Theory and Econometrics*, Cambridge University Press, 1995.

Proc Symp Appl Math AMS
W. F. Lucas, Editor, *Game Theory and its Applications, Proceedings of Symposia in Applied Mathematics*, Vol 24, American Mathematical Society, 1981.

Rapoport
A. Rapoport, Editor, *Game Theory as a Theory of Conflict Resolution*, D. Reidel, 1974.

Reiter
S. Reiter, Editor, *Studies in Mathematical Economics*, Studies in Mathematics, Vol 25, The Mathematical Association of America, 1986.

Roth: Bargaining
A. E. Roth, Editor, *Game-Theoretic Models of Bargaining*, Cambridge University Press, 1985.

Roth: Shapley Value
A. E. Roth, Editor, *The Shapley Value: Essays in Honor of Lloyd S. Shapley*, Cambridge University Press, 1988.

Sauermann
H. Sauermann, Editor, *Beiträge zur Experimentellen Wirtschaftsforschung*, Vol 2, J. C. B. Mohr, 1970.

Selten
R. Selten, Editor, *Game Equilibrium Models*, Vols 1–4, Springer, 1991.

Selten Festschr
W. Albers, W. Güth, P. Hammerstein, B. Moldovanu & E. Van Damme, Editors, *Understanding Strategic Interaction: Essays in Honor of Reinhard Selten*, Springer, 1997.

Shapira
Z. Shapira, Editor, *Organizational Decision Making*, Cambridge University Press, 1997.

Shapley Festschr
T. E. S. Raghavan, T. S. Ferguson, T. Parthasarathy & O. J. Vrieze, Editors, *Stochastic Games and Related Topics, Essays in Honor of Professor L. S. Shapley*, Kluwer Academic Publishers, 1991.

Shubik
M. Shubik, Editor, *Game Theory and Related Approaches to Social Behavior*, John Wiley & Sons, 1964.

Stony Brook ASI 1991
J.-F. Mertens & S. Sorin, Editors, *Game-Theoretic Methods in General Equilibrium Analysis, Proceedings of the NATO Advanced Study Institute on Game-Theoretic Methods in General Equilibrium Analysis, held at Long Island, New York, July 1–12, 1991*, Springer, 1994.

Stony Brook ASI 1994
S. Hart & A. Mas-Colell, Editors, *Cooperation: Game-Theoretic Approaches, Proceedings of the NATO Advanced Study Institute on Cooperation: Game-Theoretic Approaches, held at SUNY, Stony Brook, New York, July 18–29, 1994*, Springer, 1997.

Summer Tel Aviv 1990
E. Karni et al, *Summer in Tel Aviv* 1990, *Workshop in Economic Theory*, Faculty of Social Sciences, Tel Aviv University, 1990.

Summer Tel Aviv 1991
D. M. Kreps et al, *Summer in Tel Aviv* 1991, *Workshop in Economic Theory*, Faculty of Social Sciences, Tel Aviv University, 1991.

TARK I
J. Y. Halpern, Editor, *Theoretical Aspects of Reasoning about Knowledge, Proceedings of the 1986 Conference*, Morgan Kaufmann, 1986.

TARK II
M. Y. Vardi, Editor, *Theoretical Aspects of Reasoning about Knowledge, Proceedings of the Second Conference*, Morgan Kaufmann, 1988.

TARK III
R. Parikh, Editor, *Theoretical Aspects of Reasoning about Knowledge, Proceedings of the Third Conference* (TARK 1990), Morgan Kaufmann, 1990.

TARK IV
Y. Moses, Editor, *Theoretical Aspects of Reasoning about Knowledge, Proceedings of the Fourth Conference* (TARK 1992), Morgan Kaufmann, 1992.

TARK V
R. Fagin, Editor, *Theoretical Aspects of Reasoning about Knowledge, Proceedings of the Fifth Conference* (TARK 1994), Morgan Kaufmann, 1994.

TARK VI
Y. Shoham, Editor, *Theoretical Aspects of Reasoning about Knowledge, Proceedings of the Sixth Conference* (TARK 1996), Morgan Kaufmann, 1996.

TARK VII
I. Gilboa, Editor, *Theoretical Aspects of Rationality and Knowledge, Proceedings of the Seventh Conference* (TARK 1998), Morgan Kaufmann, 1998.

Toulon Conf
A. Mensch, Editor, *Theory of Games: Techniques and Applications, Proceedings of a Conference under the Aegis of the NATO Scientific Affairs Committee, Toulon, June 29–July 3, 1964*, The English University Press, 1966.

Turin Conf
K. J. Arrow, E. Colombatto, M. Perlman & C. Schmidt, Editors, *The Rational Foundations of Economic Behaviour, Proceedings of the IEA Conference held in Turin, Italy, 1993*, Macmillan, 1996.

Vienna Conf
S. J. Brams, A. Schotter & G. Schwodiauer, Editors, *Applied Game Theory, Proceedings of a Conference at the Institute for Advanced Studies, Vienna, June 1978*, Physica, 1979.

Weil
J. W. Weil, Editor, *Control Theory and Topics in Functional Analysis, Vols 1–3, Proceedings of an International Seminar Course, International Centre for Theoretical Physics, Trieste, Italy, September 11–November 29, 1974*, International Atomic Energy Agency, Vienna, 1976.

Weintraub
E. R. Weintraub, Editor, *Toward a History of Game Theory*, Duke University Press, 1992.

3 Texts and Monographs and their Abbreviations

Note

1. In the Index, Texts and Monographs are abbreviated by the family name(s) of their author(s).

2. Different books by the same author(s) are distinguished either (i) by year of publication, or (ii) the item appearing first in the list below is abbreviated by the name(s) of the author(s)

only; subsequent items are identified in some additional way. (Thus, "Kreps" stands for Kreps's text, *Microeconomic Theory*; his monograph, *Notes on the Theory of Choice*, is abbreviated "Kreps—*Notes Th. Choice*.")

3. For texts and monographs where the references are organized by chapter, the chapter number of the citation is indicated in the index.

Aubin, J.-P., *Mathematical Methods of Game and Economic Theory*, North-Holland, 1979.

———, *Dynamic Economic Theory*, Springer, 1997.

Aoki, M., *The Cooperative Game Theory of the Firm*, Clarendon Press, 1984.

Arrow, K. J. & F. H. Hahn, *General Competitive Analysis*, Holden-Day, 1971.

Aumann, R. J., *Lectures on Game Theory*, Westview Press, 1989.

Aumann, R. J. & M. B. Maschler, *Repeated Games with Incomplete Information*, MIT Press, 1995.

Aumann, R. J. & L. S. Shapley, *Values of Non-Atomic Games*, Princeton University Press, 1974.

Balasko, Y., *Foundations of the Theory of General Equilibrium*, Academic Press, 1988.

Binmore, K., *Essays on the Foundations of Game Theory*, Basil Blackwell, 1990.

———, *Game Theory and the Social Contract I: Playing Fair*, MIT Press, 1994.

———, *Game Theory and the Social Contract II: Just Playing*, MIT Press, 1998.

Border, K. C., *Fixed Point Theorems with Applications to Economics and Game Theory*, Cambridge University Press, 1985.

Brams, S. J., *Negotiation Games: Applying Game Theory to Bargaining and Arbitration*, Routledge, 1990.

———, *Theory of Moves*, Cambridge University Press, 1994.

———, *Game Theory and Politics*, Macmillan, 1975.

Burde, G. & H. Zieschang, *Knots*, Walter de Gruyter, 1985.

Charreton, R., *Économie Politique*, Editions Technip, 1988.

Cornwall, R., *Introduction to the Use of General Equilibrium Analysis*, North-Holland, 1984.

Crowell, R. H. & R. H. Fox, *Introduction to Knot Theory*, Ginn and Company, 1963.

Davis, M., *Game Theory: A Nontechnical Introduction*, Basic Books, 1970.

Driessen, T., *Cooperative Games: Solutions and Applications*, Kluwer Academic Publishers, 1988.

Dubois, D. & H. Prade, *Fuzzy Sets and Systems: Theory and Applications*, Academic Press, 1980.

Duffie, D., *Security Markets*, Academic Press, 1988.

Ellickson, B., *Competitive Equilibrium: Theory and Applications*, Cambridge University Press, 1993.

Epstein, R. A., *The Theory of Gambling and Statistical Logic*, Academic Press, 1967.

Fagin, R., J. Y. Halpern, Y. Moses & M. Y. Vardi, *Reasoning about Knowledge*, MIT Press, 1995.

Ferguson, T. S., *Mathematical Statistics: A Decision Theoretic Approach*, Academic Press, 1967.

Fishburn, P. C., *Decision and Value Theory*, John Wiley & Sons, 1964.

———, *Utility Theory for Decision Making*, John Wiley & Sons, 1970.

———, *Mathematics of Decision Theory*, Mouton, 1972.

———, *The Theory of Social Choice*, Princeton University Press, 1973.

———, *The Foundations of Expected Utility*, D. Reidel, 1982.

———, *Nonlinear Preference and Utility Theory*, Johns Hopkins University Press, 1988.

Fouraker, L. E. & S. Siegel, *Bargaining Behavior*, McGraw-Hill, 1963.

Frank, S. A., *Foundations of Social Evolution*, Princeton University Press, 1998.

Friedman, J. W., *Oligopoly and the Theory of Games*, North-Holland, 1977.

———, *Game Theory with Applications to Economics*, Oxford University Press, 1986.

Fudenberg, D. & D. M. Kreps, *Lectures on Learning and Equilibrium in Strategic Form Games*, CORE Lecture Series, Louvain-la-Neuve, Université Catholique de Louvain, 1992.

Fudenberg, D. & J. Tirole, *Game Theory*, MIT Press, 1991.

———, *Dynamic Models of Oligopoly*, Harwood Academic Publishers, 1986.

Germeier, Y. B., *Non-Antagonistic Games*, D. Reidel, 1986.

Gibbons, R., *A Primer in Game Theory*, Harvester Wheatsheaf, 1992.

Greenberg, J., *The Theory of Social Situations*, Cambridge University Press, 1990.

Guerrien, B., *La Theorie des Jeux*, Economica, 1993.

Guiasu, S. & M. Malitza, *Coalition and Connection in Games*, Pergamon Press, 1980.

Güth, W., *Spieltheorie und Ökonomische (Bei)Spiele*, Springer, 1999.

Harsanyi, J. C., *Rational Behavior and Bargaining Equilibrium in Games and Social Situations*, Cambridge University Press, 1977.

Harsanyi, J. C. & R. Selten, *A General Theory of Equilibrium Selection in Games*, MIT Press, 1988.

Hart, O., *Firms, Contracts and Financial Structure*, Clarendon Press, 1996.

Hermes, H. & J. P. LaSalle, *Functional Analysis and Optimal Control*, Academic Press, 1969.

Hildenbrand, W., *Core and Equilibria of a Large Economy*, Princeton University Press, 1974.

Hildenbrand, W. & A. P. Kirman, *Introduction to Equilibrium Analysis*, North-Holland, 1976; *Equilibrium Analysis*, North-Holland, 1988.

Howard, N., *Paradoxes of Rationality: Theory of Metagames and Political Behavior*, MIT Press, 1971.

Intriligator, M. D., *Mathematical Optimization and Economic Theory*, Prentice-Hall, 1971.

Klein, E., *Mathematical Methods in Theoretical Economics*, Academic Press, 1973.

Klein, E. & A. C. Thompson, *Theory of Correspondences*, John Wiley & Sons, 1984.

Krantz, D. H., R. D. Luce, P. Suppes & A. Tversky, *Foundations of Measurement*, Vol 1, Academic Press, 1971.

Kreps, D. M., *A Course in Microeconomic Theory*, Princeton University Press, 1990.

———, *Game Theory and Economic Modelling*, Clarendon Press, 1990.

———, *Notes on the Theory of Choice*, Westview Press, 1988.

Laffont, J.-J., *Fundamentals of Public Economics*, MIT Press, 1988.

Magill, M. & M. Quinzii, *Theory of Incomplete Markets*, MIT Press, 1996.

Makarov, V. L. & A. M. Rubinov, *Mathematical Theory of Economic Dynamics and Equilibria*, Springer, 1977.

Mas-Colell, A., *The Theory of General Economic Equilibrium: A Differentiable Approach*, Cambridge University Press, 1989.

Mas-Colell, A., M. D. Whinston & J. R. Green, *Microeconomic Theory*, Oxford University Press, 1995.

McMillan, J., *Game Theory in International Economics*, Harwood Academic Publishers, 1986.

Moulin, H., *Game Theory for the Social Sciences*, Hermann, 1981.

———, *Cooperative Microeconomics: A Game-Theoretic Introduction*, Princeton University Press, 1995.

Myerson, R. B., *Game Theory: Analysis of Conflict*, Harvard University Press, 1991.

Nicholson, W., *Microeconomic Theory: Basic Principles and Extensions*, Harcourt Brace Jovanovich, 1992.

Nikaido, H., *Convex Structures and Economic Theory*, Academic Press, 1968.

Ordeshook, P. C., *Game Theory and Political Theory*, Cambridge University Press, 1989.

Osborne, M. J. & A. Rubinstein, *Bargaining and Markets*, Academic Press, 1990.

———, *A Course in Game Theory*, MIT Press, 1994.

Owen, G., *Game Theory*, Academic Press, 1968, 1982, 1995.

Parthasarathy, T. & T. E. S. Raghavan, "Some Topics in Two-Person Games," *Modern Analytic and Computational Methods in Science and Mathematics* 22, Elsevier, 1971.

Peleg, B., *Game Theoretic Analysis of Voting in Committees*, Cambridge University Press, 1984.

Rader, T., *Theory of Microeconomics*, Academic Press, 1972.

Rapoport, Amnon, *Experimental Studies of Interactive Decisions*, Kluwer Academic Publishers, 1990.

Rapoport, Anatol, *N-Person Game Theory*, University of Michigan Press, 1970.

Rasmusen, E., *An Introduction to the Theory of Games*, Basil Blackwell, 1989.

Rauhut, B., N. Schmitz & E.-W. Zachow, *Spieltheorie*, Teubner, 1979.

Roth, A. E. & M. A. O. Sotomayor, *Two-Sided Matching*, Cambridge University Press, 1990.

Scarf, H. E., *The Computation of Economic Equilibria*, Yale University Press, 1973.

Selten, R., *Preispolitik der Mehrproduktenunternehmung in der statischen Theorie*, Springer, 1970.

Sen, A., *On Ethics and Economics*, Blackwell Publishers, 1992.

Shubik, M., *Game Theory in the Social Sciences*, MIT Press, 1982.

———, *A Game-Theoretic Approach to Political Economy: Game Theory in the Social Sciences*, Vol 2, MIT Press, 1984.

Straffin, P. D., *Game Theory and Strategy*, The Mathematical Association of America, 1993.

Szép, J. & F. Forgó, *Introduction to the Theory of Games*, D. Reidel, 1985.

Takayama, A., *Mathematical Economics*, Dryden Press, 1974.

Thomson, W. & T. Lensberg, *Axiomatic Theory of Bargaining with a Variable Number of Agents*, Cambridge University Press, 1989.

Tirole, J., *The Theory of Industrial Organization*, MIT Press, 1988.

Van Damme, E., *Refinements of the Nash Equilibria Concept*, Springer, 1983.

———, *Stability and Perfection of Nash Equilibrium*, Springer, 1987.

Vilkas, E. I. & E. Z. Maiminas, *Solutions: Theory, Information, Modelisation* (in Russian), Radio i Svyaz', 1981.

Vorob'ev, N. N., *Game Theory: Lectures for Economists and Systems Scientists*, Springer, 1977.

Weibull, J. W., *Evolutionary Game Theory*, MIT Press, 1995.

Weintraub, E. R., *General Equilibrium Analysis*, Cambridge University Press, 1985.

Worobjow, N. N., *Entwicklung der Spieltheorie*, VEB Deutscher Verlag der Wissenschaften, 1975.

4 Journal Abbreviations

Acc Rev	*Accounting Review*
AER	*American Economic Review*

Am Economist	American Economist
Am J Pol Sci	American Journal of Political Science
Am Pol Sci Rev	American Political Science Review
Ann Math	Annals of Mathematics
Ann Math Stat	Annals of Mathematical Statistics
Ann Prob	Annals of Probability
Ann Stat	Annals of Statistics
Annual Rev Psych	Annual Review of Psychology
Arch Math	Archiv der Mathematik
Atti Acad Lincei	Atti dell'Accademia Nazionale dei Lincei, Rendiconti, Classe di Scienze Fisiche, Matematiche e Naturali
Aut Rem Contr	Automation and Remote Control, USSR
Behav Res	Behavior Research Methods, Instruments & Computers
Behav Sci	Behavioral Science
Bell J	Bell Journal of Economics
Brit J Pol Sci	British Journal of Political Science
Bull Acad Polon	Bulletin de l'Academie Polonaise des Sciences, Serie des Sciences Mathematiques, Astronomiques et Physiques
Bull AMS	Bulletin of the American Mathematical Society
Bull SMF	Bulletin de la Societe Mathematique de France
Can J Econ	Canadian Journal of Economics—Revue Canadienne d'Économique
Coll Math	Colloquium Mathematicum
Confl Manag	Conflict Management and Peace Science
CR Acad Sci	Comptes Rendus Hebdomadaires des Seances de l'Academie des Sciences, Serie A: Sciences Mathematiques
Cybernet & Syst	Cybernetics and Systems Analysis
Doklady Bolg	Doklady Bolgarskoi Akademii Nauk
Doklady SSSR	Doklady Akademii Nauk SSSR
Econ & Phil	Economics and Philosophy
Econ J	Economic Journal
Econ Let	Economics Letters
Econ Rec	Economic Record
Econ Th	Economic Theory
Économie Appl	Économie Appliquée
Eng Cybernetics	Engineering Cybernetics, USSR
Eur Econ Rev	European Economic Review
Eur J OR	European Journal of Operational Research
Eur J Pop	European Journal of Population—Revue Européenne de Demographie
Eur J Soc Psych	European Journal of Social Psychology
Ev Ecol	Evolutionary Ecology
Fuzzy	Fuzzy Sets and Systems
GEB	Games and Economic Behavior
Gen Systems	General Systems
Hist Pol Econ	History of Political Economy

Hitotsubashi J Econ	Hitotsubashi Journal of Economics
IEEE Trans Aut Contr	IEEE Transactions on Automatic Control
IEEE Trans Man & Cyber	IEEE Transactions on Systems, Man and Cybernetics
IEEE Trans Sci & Cyber	IEEE Transactions on Systems Science and Cybernetics
IER	International Economic Review
IJGT	International Journal of Game Theory
IJM	Israel Journal of Mathematics
Indiana U Math J	Indiana University Mathematics Journal
Inf Sci	Information Sciences
Int J Contr	International Journal of Control
Int J Systems Sci	International Journal of Systems Science
Int Org	International Organization
Int Phil Q	International Philosophical Quarterly
Int Stud Q	International Studies Quarterly
J Acc Res	Journal of Accounting Research
J Appr Th	Journal of Approximation Theory
J Assoc Comp Mach	Journal of the Association for Computing Machinery
J Bus	Journal of Business
J Comp Econ	Journal of Comparative Economics
J Confl Res	Journal of Conflict Resolution
J Dev Econ	Journal of Development Economics
J Diff Eq	Journal of Differential Equations
J Econ Behav	Journal of Economic Behavior & Organization
J Econ Dyn	Journal of Economic Dynamics & Control
J Econ Issues	Journal of Economic Issues
J Exp Soc Psych	Journal of Experimental Social Psychology
J Financ Res	Journal of Financial Research
J Finance	Journal of Finance
J Int Econ	Journal of International Economics
J Law & Econ	Journal of Law & Economics
J Legal St	Journal of Legal Studies
J London Math Soc	Journal of the London Mathematical Society
J Math Anal Appl	Journal of Mathematical Analysis and Applications
J Math Mech	Journal of Mathematics & Mechanics
J Math Psych	Journal of Mathematical Psychology
J Math Pures et Appl	Journal de Mathematiques Pures et Appliquees
J Math Sociol	Journal of Mathematical Sociology
J Multiv Anal	Journal of Multivariate Analysis
J Opt Th Appl	Journal of Optimization Theory and Applications
J Pers Soc Psych	Journal of Personality and Social Psychology
J Phil	Journal of Philosophy
J Publ Econ	Journal of Public Economics
J Reine & Ang Math	Journal für die Reine und Angewandte Mathematik
J Risk Unc	Journal of Risk and Uncertainty
J Royal Stat Soc	Journal of the Royal Statistical Society Series B: Methodological

Jahrb Nat-ok Stat	Jahrbücher für Nationalökonomie und Statistik
JASA	Journal of the American Statistical Association
JEL	Journal of Economic Literature
JEP	Journal of Economic Perspectives
JET	Journal of Economic Theory
JME	Journal of Mathematical Economics
JPE	Journal of Political Economy
Large Scale Systems/Th Appl	Large Scale Systems—Theory and Applications
Lect Notes	Lecture Notes in Economics and Mathematical Systems, Springer
Lect Notes Math	Lecture Notes in Mathematics, Springer
Ling & Phil	Linguistics and Philosophy
Manag Dec Econ	Managerial and Decision Economics
Manag Sci	Management Science
Manag Sci Appl	Management Science Series B: Applications
Manag Sci Th	Management Science Series A: Theory
Manus Math	Manuscripta Mathematica
Math Ann	Mathematische Annalen
Math Notes	Mathematical Notes
Math Scand	Mathematica Scandinavica
Math Soc Sci	Mathematical Social Sciences
Mich Law Rev	Michigan Law Review
MOR	Mathematics of Operations Research
NRLQ	Naval Research Logistics Quarterly
Omega	Omega—International Journal of Management Science
OR	Operations Research
OR Letters	Operations Research Letters
Ox Econ Papers	Oxford Economic Papers—New Series
P Peace Sci	Papers, Peace Science Society International
Pac J Math	Pacific Journal of Mathematics
Phil Sci	Philosophy of Science
Phil Stud	Philosophical Studies
PNAS	Proceedings of the National Academy of Sciences (USA)
Proc AMS	Proceedings of the American Mathematical Society
Proc IEEE	Proceedings of the IEEE
Proc Jap Acad	Proceedings of the Japan Academy Series A: Physical and Biological Sciences
Publ Choice	Public Choice
QJE	Quarterly Journal of Economics
Qual & Quant	Quality & Quantity
Rand J	Rand Journal of Economics
Reg Sci	Regional Science and Urban Economics
RES	Review of Economic Studies
Rev d'Écon Politique	Revue d'Économie Politique
Rev Econ	Revue Economique

Rev Fr Aut	Revue Francaise d'Automatique, Informatique et Recherche Operationelle
Rev Fr RO	Revue Francaise de Recherche Operationelle
Rivista	Rivista Internazionale di Scienze Economiche e Commerciali
Sbornik	Mathematics of the USSR—Sbornik
Scand J Econ	Scandinavian Journal of Economics
SIAM J Appl Math	SIAM Journal on Applied Mathematics
SIAM J Contr	SIAM Journal on Control and Optimization
SIAM J Discr Math	SIAM Journal on Discrete Mathematics
SIAM Rev	SIAM Review
Sib Math J	Siberian Mathematical Journal
Simul & Gaming	Simulation & Gaming
Soc Choice Welf	Social Choice and Welfare
Soc Psych Q	Social Psychology Quarterly
Socio-Econ Plan Sci	Socio-Economic Planning Sciences
South Econ J	Southern Economic Journal
Strat Manag J	Strategic Management Journal
Th & Dec	Theory and Decision
Th Prob Appl	Theory of Probability and its Applications
Theoria	Theoria (Sweden)
Trans AMS	Transactions of the American Mathematical Society
U Chic Law Rev	University of Chicago Law Review
Vestnik	Vestnik Leningradskogo Universiteta Seriya Matematiki, Mekhaniki, Astronomii
Virg Law Rev	Virginia Law Review
Water Res	Water Resources Research
Wisc Law Rev	Wisconsin Law Review
Z Ges Staatsw	Zeitschrift für die Gesamte Staatswissenschaft—Journal of Institutional and Theoretical Economics
Z Nat-ok	Zeitschrift für Nationalokonomie
Z Soz-psych	Zeitschrift für Sozialpsychologie
Z Wahrscheinl	Zeitschrift für die Wahrscheinlichkeitstheorie und Verwandte Gebiete

5 Generic Abbreviations

Acad	Academy, Accademia
Acc	Accounting
Adv	Advances
Agr	Agriculture, Agricultural
Am	American
Anal	Analysis
Ann	Annals (of), Annalen
Annl	Annual
Appl	Applications, Applied
Arch	Archives

Behav	Behavior, Behavioral
Brit	British
Bull	Bulletin (of)
Comp	Computing, Computation(al)
Contr	Control
Cyb	Cybernetic(s)
Econ	Economic(s), Economist, Économie
Eng	Engineering
Env	Environment(al)
Eur	European, Européennes
Fin	Finance
Fr	Français
Gen	General
IEEE	Institute of Electrical and Electronics Engineers
Ind	Industrial
Inf	Information
Int	International, Internazionale
J	Journal (of)
Let	Letters
Manag	Management
Math	Mathematics, Mathematical
Meth	Methods
Opt	Optimization
OR	Operations (Operational) Research
Pap	Papers
Phil	Philosophy, Philosophical
Pol	Politics, Political
Polon	Polonais (Polish)
Prob	Probability
Proc	Proceedings
Psych	Psychology, Psychological
Publ	Public
Q	Quarterly
Reg	Regional
Res	Research
Rev	Review(s) (of), Revue
RO	Recherche Operationelle
Scand	Scandinavian
Sci	Science(s)
SIAM	Society for Industrial and Applied Mathematics
Soc	Society, Social
Sociol	Sociology
Stat	Statistical, Statistics, Statistician
Strat	Strategic, Strategy
Stud	Studies (in)

Sys	Systems
Th	Theory, Theoretical
Trans	Transactions
U	University (of)
Z	Zeitschrift (für)

Chapter 38   Von Neumann–Morgenstern Solutions to Cooperative Games without Side Payments

Bloomfield, S. D., 1976, Social Choice Interpretation of the von Neumann–Morgenstern Game, *Econometrica* 44, 105–114.

Bloomfield, S. & R. Wilson, 1972, The Postulates of Game Theory, *J Math Sociol* 2, 221–234.

Chien, T. Q., 1976, Generalized Cooperative Games and Markets, *Kybernetika* 12, 328–354.

Chwe, M. S. Y., 1994, Farsighted Coalitional Stability, *JET* 63, 299–325.

Debreu, G., 1976, Least Concave Utility Functions, *JME* 3, 121–129.

Dreze, J. H., S. Gepts & J. J. Gabszewicz, 1969, On Cores and Competitive Equilibria, in *La Decision*, 91–114.

Gelovani, V. A. & E. R. Smoljakov, 1986, On the Existence of Stable States in Game Problems, *Th & Dec* 20, 189–203.

Harsanyi, J. C., 1974, Equilibrium-Point Interpretation of Stable Sets and a Proposed Alternative Definition, *Manag Sci Th* 20, 1472–1495.

Hart, S., 1974, Formation of Cartels in Large Markets, *JET* 7, 453–466.

Henss, R., 1985, Negotiation Results and Coalition Formation in Cooperative Normal-Form Games: Theoretical Solution Concepts, *Z Soz-psych* 16, 91–100.

Holler, M. J., 1991, Three Characteristic Functions and Tentative Remarks on Credible Threats, *Qual & Quant* 25, 29–35.

Ichiishi, T., 1981, A Social Coalitional Equilibrium Existence Lemma, *Econometrica* 49, 369–377.

Ishikawa, S. & K. Nakamura, 1979, Existence of the Core of a Characteristic Function Game with Ordinal Preferences, *J OR Soc Japan* 22, 225–232.

Ishikawa, S. & K. Nakamura, 1980, Representation of Characteristic Function Games by Social Choice Functions, *IJGT* 9, 191–199.

Maschler, M., 1964, Stable Payoff Configurations for Quota Games, in *Adv Game Th*, 477–499.

Michener, H. A., Y. C. Choi & D. C. Dettman, 1986, Stability by Deterrence in Cooperative Non-Sidepayment N-Person Games, *J Econ Behav* 7, 375–402.

Michener, H. A., D. C. Dettman, J. M. Ekman & Y. C. Choi, 1985, A Comparison of the Alpha-Characteristic and Beta-Characteristic Functions in Cooperative Non-Sidepayment n-Person Games, *J Math Sociol* 11, 307–330.

Michener, H. A., J. M. Ekman & D. C. Dettman, 1986, Predictive Superiority of the Beta-Characteristic Function in Cooperative Non-Sidepayment n-Person Games, *Th & Dec* 21, 99–128.

Michener, H. A., K. Potter, C. G. Depies & G. B. Macheel, 1984, A Test of the Core Solution in Finite Strategy Non-Sidepayment Games, *Math Soc Sci* 8, 141–168.

Michener, H. A., K. Potter, G. B. Macheel & C. G. Depies, 1984, A Test of the von Neumann–Morgenstern Stable Set Solution in Cooperative Non-Sidepayment n-Person Games, *Behav Sci* 29, 13–27.

Michener, H. A. & M. S. Salzer, 1989, Comparative Accuracy of Value Solutions in Non-Sidepayment Games with Empty Core, *Th & Dec* 26, 205–233.

Myerson, R. B., 1980, Conference Structures and Fair Allocation Rules, *IJGT* 9, 169–182.

Myerson, R. B., 1992, Fictitious Transfer Solutions in Cooperative Game Theory, in *Harsanyi Festschr*, 13–33.

Nakamura, K., 1973, Psi-Stability of a Cooperative Game without Side Payments, *IJGT* 2, 129–140.

Nakamura, K. & M. Suzuki, 1981, Social Decision and Coalition Power, in *Nakamura Papers*, 3 14.

Otten, G. J., P. Borm, B. Peleg & S. Tijs, 1998, The MC-Value for Monotonic NTU-Games, *IJGT* 27, 37–47.

Packel, E. W., 1981, A Stochastic Solution Concept for n-Person Games, *MOR* 6, 349–362.

Peleg, B., 1963, Bargaining Sets of Cooperative Games without Side Payments, *IJM* 1, 197–200.

Peleg, B., 1963, Solutions to Cooperative Games without Side Payments, *Trans AMS* 106, 280–292.

Peleg, B., 1965, Utility Functions of Money for Clear Games, *NRLQ* 12, 57–63.

Peleg, B., 1966, The Independence of Game Theory of Utility Theory, *Bull AMS* 72, 995–999.

Peleg, B., 1984, Core Stability and Duality of Effectivity Functions, in *Lect Notes* 226, 272–287.

Poza, H. B. & R. L. Kashyap, 1973, A Solution for a Class of Stochastic Cooperative Games, *Int J Contr* 17, 705–720.

Reichardt, R., 1967, Three-Person Games with Imperfect Coalitions: A Sociologically Relevant Concept in Game Theory, *Gen Systems* 13, 189.

Roth, A. E., 1980, Values for Games without Sidepayments: Some Difficulties with Current Concepts, *Econometrica* 48, 457–465.

Rozen, V. V., 1988, Cooperative Games with Quasiordered Outcomes, *Cybernetics* 24, 774–781.

Sakurai, M. M. & J. M. Brennan, 1988, Computing the von Neumann–Morgenstern Characteristic Function $v(S)$ for Cooperative n-Person Transferable Utility Normal-Form Games: LP and Saddlepoint Solutions, *Behav Res* 20, 367–371.

Scarf, H. E., 1967, The Core of an n-Person Game, *Econometrica* 35, 50–69.

Schotter, A. & G. Schwodiauer, 1980, Economics and the Theory of Games: A Survey, *JEL* 18, 479–527.

Sen, M., 1983, Implementable Social Choice Rules: Characterization and Correspondence Theorems under Strong Nash Equilibrium, *JME* 11, 1–24.

Shapley, L. S. & H. E. Scarf, 1974, On Cores and Indivisibility, *JME* 1, 23–37.

Shapley, L. S. & M. Shubik, 1966, Quasi-Cores in a Monetary Economy with Nonconvex Preferences, *Econometrica* 34, 805–827.

Shapley, L. S. & M. Shubik, 1969, The Core of an Economic System with Externalities, *AER* 59, 678–684.

Shapley, L. S. & M. Shubik, 1977, Trade Using One Commodity as a Means of Payment, *JPE* 85, 937–968.

Shehory, O. & S. Kraus, 1998, Methods for Task Allocation via Agent Coalition Formation, *Artificial Intelligence* 101, 165–200.

Shubik, M., 1981, Game Theory Models and Methods in Political Economy, in *Handb Math Econ* 1, 285–330.

Stearns, R. E., 1964, Three-Person Cooperative Games without Side Payments, in *Adv Game Th*, 377–406.

Stearns, R. E., 1964, On the Axioms for a Cooperative Game without Side Payments, *Proc AMS* 15, 82–86.

Tack, W. H., 1991, Rational Action in Social Situations, *Z Soz-psych* 22, 151–165.

Van der Laan, G., D. Talman & Z. F. Yang, 1998, Cooperative Games in Permutational Structure, *Econ Th* 11, 427–442.

Vilkov, V. B., 1977, Convex Games without Side Payments, *Vestnik*, 21–24.

Vilkov, V. B. & T. E. Kulakovskiya, 1975, The Core Solution in Games without Side Payments, *Vestnik*, 14–18.

Vilkov, V. B. & O. A. Malafeev, 1988, A Note on an N-M-Solution Property, *Vestnik*, 24–26.

Wilson, R., 1971, Stable Coalition Proposals in Majority Rule Voting, *JET* 3, 254–271.

Yasuda, Y., 1970, A Note on the Core of a Cooperative Game without Side Payments, *NRLQ* 17, 143.

Texts and Monographs: Border; Cornwall; Davis; Friedman 1977; Howard; Myerson; Peleg; Shubik 1982; Takayama.

In this collection: Chapters 2, 39, 41, 42, 46, and 51.

Chapter 39 The Core of a Cooperative Game without Side Payments

Abdou, J., 1991, Maxmin and Minmax for Coalitional Game Forms, *GEB* 3, 267–277.

Asdrubali, P., 1996, Coalitional Instability of the Distributive Lindahl Equilibrium, *Econ Th* 8, 565–575.

Aubin, J. P., 1981, Cooperative Fuzzy Games, *MOR* 6, 1–13.

Berka, M., 1982, Stable Outcomes and Alpha-Cores in Continuous Games, *Kybernetika* 18, 201–214.

Bush, W. C. & L. S. Mayer, 1974, Some Implications of Anarchy for the Distribution of Property, *JET* 8, 401–412.

Calvo, E. & J. J. Lasaga, 1997, Probabilistic Graphs and Power Indexes: An Application to the Spanish Parliament, *J Th Pol* 9, 477–501.

Chien, T. Q., 1976, Generalized Cooperative Games and Markets, *Kybernetika* 12, 328–354.

Chwe, M. S. Y., 1994, Farsighted Coalitional Stability, *JET* 63, 299–325.

Dreze, J. H., S. Gepts & J. J. Gabszewicz, 1969, On Cores and Competitive Equilibria, in *La Decision*, 91–114.

Holly, C., 1994, An Exchange Economy can have an Empty α-Core, *Econ Th* 4, 453–461.

Howard, N., 1974, "General" Metagames: An Extension of the Metagame Concept, in *Rapoport*, 539–552.

Howard, N., 1975, The First Geneva Conference: Examples of a Dynamic Theory of Games, *P Peace Sci* 24, 1–28.

Kajii, A., 1992, A Generalization of Scarf's Theorem: An Alpha-Core Existence Theorem without Transitivity or Completeness, *JET* 56, 194–205.

Kalai, E., A. Postlewaite & J. Roberts, 1979, A Group Incentive Compatible Mechanism Yielding Core Allocations, *JET* 20, 13–22.

Kats, A., 1974, Non-Cooperative Monopolistic Games and Monopolistic Market Games, *IJGT* 3, 251–260.

Kolpin, V., 1993, Shared Facility Games with Variable Utilization, *IER* 34, 387–400.

Mertens, J. F., 1980, A Note on the Characteristic Function of Supergames, *IJGT* 9, 189–190.

Mertens, J. F., 1987, Supergames, in *The New Palgrave* 4, 551–552.

Mertens, J. F., 1987, Repeated Games, in *Proc Int Congr Math* 1986, Vol 2, 1528–1577.

Michener, H. A., Y. C. Choi & D. C. Dettman, 1986, Stability by Deterrence in Cooperative Non-Sidepayment n-Person Games, *J Econ Behav* 7, 375–402.

Michener, H. A., P. D. Clancy & K. Yuen, 1984, Do Outcomes of n-Person Sidepayment Games Fall in the Core?, in *Holler, Coalitions*, 269–282.

Michener, H. A., D. C. Dettman, J. M. Ekman & Y. C. Choi, 1985, A Comparison of the Alpha-Characteristic and Beta-Characteristic Functions in Cooperative Non-Sidepayment n-Person Games, *J Math Sociol* 11, 307–330.

Michener, H. A., J. M. Ekman & D. C. Dettman, 1986, Predictive Superiority of the Beta-Characteristic Function in Cooperative Non-Sidepayment n-Person Games, *Th & Dec* 21, 99–128.

Michener, H. A., K. Potter, C. G. Depies & G. B. Macheel, 1984, A Test of the Core Solution in Finite Strategy Non-Sidepayment Games, *Math Soc Sci* 8, 141–168.

Michener, H. A., K. Potter, G. B. Macheel & C. G. Depies, 1984, A Test of the von Neumann–Morgenstern Stable Set Solution in Cooperative Non-Sidepayment n-Person Games, *Behav Sci* 29, 13–27.

Michener, H. A., K. Potter & M. M. Sakurai, 1983, On the Predictive Efficiency of the Core Solution in Side-Payment Games, *Th & Dec* 15, 11–28.

Michener, H. A., G. D. Richardson & M. S. Salzer, 1989, Extensions of Value Solutions in Constant-Sum Non-Sidepayment Games, *J Confl Res* 33, 530–553.

Michener, H. A. & M. S. Salzer, 1989, Comparative Accuracy of Value Solutions in Non-Sidepayment Games with Empty Core, *Th & Dec* 26, 205–233.

Michener, H. A., M. S. Salzer, M. S. Reimer & J. Lee, 1991, A Comparison of Lambda-Transfer and Core Solutions in Constant-Sum Non-Sidepayment Games, *Behav Sci* 36, 115–132.

Michener, H. A. & K. Yuen, 1982, A Competitive Test of the Core Solution in Side-Payment Games, *Behav Sci* 27, 57–68.

Moulin, H. & B. Peleg, 1982, Cores of Effectivity Functions and Implementation Theory, *JME* 10, 115–145.

Nakamura, K., 1973, Psi-Stability of a Cooperative Game without Side Payments, *IJGT* 2, 129–140.

Nakamura, K. & M. Suzuki, 1981, Social Decision and Coalition Power, in *Nakamura Papers*, 3–14.

Otten, G. J., P. Borm, T. Storcken & S. Tijs, 1995, Effectivity Functions and Associated Claim Game Correspondences, *GEB* 9, 172–190.

Pattanaik, P. K., 1976, Counter-Threats and Strategic Manipulation under Voting Schemes, *RES* 43, 11–18.

Peleg, B., 1963, Solutions to Cooperative Games without Side Payments, *Trans AMS* 106, 280–292.

Peleg, B., 1965, Utility Functions of Money for Clear Games, *NRLQ* 12, 57–63.

Peleg, B., 1980, A Theory of Coalition Formation in Committees, *JME* 7, 115–134.

Qin, C. Z., 1994, The Inner Core of an n-Person Game, *GEB* 6, 431–444.

Rajan, R., 1989, Endogenous Coalition Formation in Cooperative Oligopolies, *IER* 30, 863–876.

Ray, D. & R. Vohra, 1997, Equilibrium Binding Agreements, *JET* 73, 30–78.

Reichardt, R., 1967, 3-Person Games with Imperfect Coalitions: A Sociologically Relevant Concept in Game Theory, *Gen Systems* 13, 189.

Rubinstein, A., 1979, Equilibrium in Supergames with the Overtaking Criterion, *JET* 21, 1–9.

Sakurai, M. M. & J. M. Brennan, 1988, Computing the von Neumann–Morgenstern Characteristic Function $v(S)$ for Cooperative n-Person Transferable Utility Normal Form Games: LP and Saddlepoint Solutions, *Behav Res* 20, 367–371.

Scarf, H. E., 1971, On the Existence of a Cooperative Solution for a General Class of n-Person Games, *JET* 3, 169–181.

Schleicher, H., 1976, Strategic Analysis of a Category of Election Games, *Économie Appl* 29, 127–153.

Schotter, A., 1973, Core Allocations and Competitive Equilibrium: A Survey, *Z Nat-ok* 33, 281–313.

Schotter, A. & G. Schwodiauer, 1980, Economics and the Theory of Games: A Survey, *JEL* 18, 479–527.

Selten, R., 1997, Descriptive Approaches to Cooperation, in *Stony Brook ASI* 1994, 289–328.

Shapley, L. S. & M. Shubik, 1966, Quasi-Cores in a Monetary Economy with Nonconvex Preferences, *Econometrica* 34, 805–827.

Shapley, L. S. & M. Shubik, 1967, Concepts and Theories of Pure Competition, in *Morgenstern Festschr*, 63–79.

Shehory, O. & S. Kraus, 1998, Methods for Task Allocation via Agent Coalition Formation, *Artificial Intelligence* 101, 165–200.

Sherali, H. D. & R. Rajen, 1986, A Game Theoretic Mathematical Programming Analysis of Cooperative Phenomena in Oligopolistic Markets, *OR* 34, 683–697.

Shubik, M., 1970, A Curmudgeon's Guide to Microeconomics, *JEL* 8, 405–434.

Shubik, M., 1981, Game Theory Models and Methods in Political Economy, in *Handb Math Econ* 1, 285–330.

Sorin, S., 1992, Repeated Games with Complete Information, in *Handb Game Th* 1, 71–107.

Stearns, R. E., 1964, Three-Person Cooperative Games without Side Payments, in *Adv Game Th*, 377–406.

Stearns, R. E., 1964, On the Axioms for a Cooperative Game without Side Payments, *Proc AMS* 15, 82–86.

Tran, Q. C., 1983, A General Exchange Economy, *Kybernetika* 19, 299–308.

Van Damme, E. E. C., 1981, History-Dependent Equilibrium Points in Dynamic Games, in *Hagen Conf II*, 27–38.

Van der Laan, G., D. Talman & Z. F. Yang, 1998, Cooperative Games in Permutational Structure, *Econ Th* 11, 427–442.

Vasin, A. A., 1978, Strong Equilibrium Points in Some Supergames, *Vestnik Moskovskogo Universiteta Seriya Fiziki i Astronomii*, 30–39.

Weber, R. J., 1994, Games in Coalitional Form, in *Handb Game Th* 2, 1285–1303.

Wilson, R., 1971, Stable Coalition Proposals in Majority Rule Voting, *JET* 3, 254–271.

Yano, M., 1990, A Local Theory of Cooperative Games, *IJGT* 19, 301–324.

Yasuda, Y., 1970, A Note on the Core of a Cooperative Game without Side Payments, *NRLQ* 17, 143.

Zhao, J., 1991, The Equilibria of a Multiple Objective Game, *IJGT* 20, 171–182.

Zhao, J. G., 1992, The Hybrid Solutions of an n-Person Game, *GEB* 4, 145–160.

Zhao, J., 1996, The Hybrid Equilibria and Core Selection in Exchange Economies with Externalities, *JME* 26, 387–407.

Zhao, J. G., 1997, A Cooperative Analysis of Covert Collusion in Oligopolistic Industries, *IJGT* 26, 249–266.

Zhou, L., 1994, A New Bargaining Set of an n-Person Game and Endogenous Coalition-Formation, *GEB* 6, 512–526.

Texts and Monographs: Aubin 1979; Aumann; Aumann & Maschler; Aumann & Shapley; Cornwall; Friedman 1977, 1986; Howard; Myerson; Peleg; Scarf; Shubik 1982; Takayama.

In this collection: Chapters 2, 23 (p. 414), 28, 41, and 46.

Chapter 41 A Survey of Cooperative Games without Side Payments

Au, W. T. & H. A. Michener, 1994, A Probabilistic Theory of Coalition Formation in n-Person Side-Payment Games, *J Math Sociol* 19, 165–188.

Baudier, E., 1973, Competitive Equilibrium in a Game, *Econometrica* 41, 1049–1068.

Berl, J. E., R. D. McKelvey, P. C. Ordeshook & M. D. Winer, 1976, Experimental Test of a Core in a Simple *n*-Person Cooperative Non-Sidepayment Game, *J Confl Res* 20, 453–479.

Billera, L. J., 1970, Existence of General Bargaining Sets for Cooperative Games without Side Payments, *Bull AMS* 76, 375–379.

Billera, L. J., 1970, Some Theorems on the Core of an *n*-Person Game without Side Payments, *SIAM J Appl Math* 18, 567–579.

Bloch, F., 1996, Sequential Formation of Coalitions in Games with Externalities and Fixed Payoff Division, *GEB* 14, 90–123.

Bloomfield, S. & R. Wilson, 1972, The Postulates of Game Theory, *J Math Sociol* 2, 221–234.

Boehm, V., 1974, The Core of an Economy with Production, *RES* 41, 429–436.

Borglin, A., 1973, Price Characterization of Stable Allocations in Exchange Economies with Externalities, *JET* 6, 483–494.

Borm, P. E. M. & S. H. Tijs, 1992, Strategic Claim Games corresponding to an NTU Game, *GEB* 4, 58–71.

Brams, S. J. & D. Wittman, 1981, Nonmyopic Equilibria in 2×2 Games, *Confl Manag* 6, 39–62.

Breyer, F. & R. J. Gardner, 1980, Libera, Paradox, Game Equilibrium, and the Gibbard Optimum, *Publ Choice* 35, 469–481.

Chwe, M. S. Y., 1994, Farsighted Coalitional Stability, *JET* 63, 299–325.

Dreze, J. H., 1987, Investment under Private Ownership: Optimality, Equilibrium and Stability, in *Dreze Essays*, 261–297.

Dreze, J. H. & D. de la Vallee Poussin, 1971, A Tatonnement Process for Public Goods, *RES* 38, 133–150.

Einy, E. & B. Peleg, 1995, Coalition-Proof Communication Equilibria, in *Proc Int Symp Soc Ch*, 289–300.

Fiorina, M. P. & C. R. Plott, 1978, Committee Decisions under Majority Rule: An Experimental Study, *Am Pol Sci Rev* 72, 575–598.

Fudenberg, D. & J. Tirole, 1985, Preemption and Rent Equalization in the Adoption of New Technology, *RES* 52, 383–401.

Fumas, V. S. & A. B. Whinston, 1982, Subsidy-Free Welfare Games, *South Econ J* 49, 389–405.

Gardner, R. J., 1977, The Borda Game, *Publ Choice* 30, 43–50.

Gonedes, N. J., 1975, Information Production and Capital Market Equilibrium, *J Finance* 30, 841–864.

Greenberg, J., 1994, Coalition Structures, in *Handb Game Th* 2, 1305–1337.

Hammond, P. J., M. Kaneko & M. H. Wooders, 1989, Continuum Economies with Finite Coalitions: Core, Equilibria, and Widespread Externalities, *JET* 49, 113–134.

Hart, S., 1985, Axiomatic Approaches to Coalitional Bargaining, in *Roth: Bargaining*, 305–319.

Hart, S. & M. Kurz, 1983, Endogenous Formation of Coalitions, *Econometrica* 51, 1047–1064.

Hart, S. & M. Kurz, 1984, Stable Coalition Structures, in *Holler, Coalitions*, 235–258.

Haurie, A., 1973, Optimality in Multicriterion Perturbed Systems with Application to Linear Control Systems with Quadratic Costs, *Rev Fr Aut* 7, 91.

Haurie, A., 1973, Pareto Optimal Decisions for a Coalition of a Subset of Players, *IEEE Trans Aut Contr* 18, 144–149.

Haurie, A., 1975, Some Properties of the Characteristic Function and Core of a Multistage Game of Coalitions, *IEEE Trans Aut Contr* 20, 238–241.

Henss, R., 1985, Negotiation Results and Coalition Formation in Cooperative Normal Form Games: Theoretical Solution Concepts, *Z Soz-psych* 16, 91–100.

Holler, M. J., 1991, 3 Characteristic Functions and Tentative Remarks on Credible Threats, *Qual & Quant* 25, 29–35.

Ichiishi, T., 1988, Core-Like Solutions for Games with Probabilistic Choice of Strategies, *Math Soc Sci* 15, 51–60.

Ichiishi, T. & S. Webber, 1978, Some Theorems on the Core of a Non-Sidepayment Game with a Measure Space of Players, *IJGT* 7, 95–112.

Imai, H., 1983, On Harsanyi's Solution, *IJGT* 12, 161–179.

Kalai, E., 1975, Excess Functions for Cooperative Games without Sidepayments, *SIAM J Appl Math* 29, 60–71.

Kalai, E., 1975, On Game-Type Subsets, *IJGT* 4, 141–150.

Kalai, E., 1977, Proportional Solutions to Bargaining Situations: Interpersonal Utility Comparisons, *Econometrica* 45, 1623–1630.

Kalai, E., A. Postlewaite & J. Roberts, 1979, A Group Incentive Compatible Mechanism Yielding Core Allocations, *JET* 20, 13–22.

Kalai, E., A. Postlewaite & J. Roberts, 1979, Arbitration of Exchange Situations with Public Goods, in *Vienna Conf*, 198–203.

Kalai, E. & D. Schmeidler, 1977, An Admissible Set Occurring in Various Bargaining Situations, *JET* 14, 402–411.

Lucas, W. F., 1977, The Existence Problem for Solutions, in *Lect Notes* 141, 64–75.

Lucas, W. F., 1992, Von Neumann–Morgenstern Stable Sets, in *Handb Game Th* 1, 543–590.

Maschler, M., 1992, The Bargaining Set, Kernel, and Nucleolus, in *Handb Game Th* 1, 591–667.

McKelvey, R. D., P. C. Ordeshook & M. D. Winer, 1978, The Competitive Solution for n-Person Games without Transferable Utility, with an Application to Committee Games, *Am Pol Sci Rev* 72, 599–615.

Mertens, J. F., 1980, A Note on the Characteristic Function of Supergames, *IJGT* 9, 189–190.

Mertens, J. F., 1987, Repeated Games, in *Proc Int Congr Math* 1986, Vol 2, 1528–1577.

Michener, H. A., Y. C. Choi & D. C. Dettman, 1986, Stability by Deterrence in Cooperative Non-Sidepayment n-Person Games, *J Econ Behav* 7, 375–402.

Michener, H. A., P. D. Clancy & K. Yuen, 1984, Do Outcomes of n-Person Sidepayment Games Fall in the Core?, in *Holler, Coalitions*, 269–282.

Michener, H. A., D. C. Dettman, J. M. Ekman & Y. C. Choi, 1985, A Comparison of the Alpha-Characteristic and Beta-Characteristic Functions in Cooperative Non-Sidepayment n-Person Games, *J Math Sociol* 11, 307–330.

Michener, H. A., J. M. Ekman & D. C. Dettman, 1986, Predictive Superiority of the Beta-Characteristic Function in Cooperative Non-Sidepayment n-Person Games, *Th & Dec* 21, 99–128.

Michener, H. A., K. Potter, C. G. Depies & G. B. Macheel, 1984, A Test of the Core Solution in Finite Strategy Non-Sidepayment Games, *Math Soc Sci* 8, 141–168.

Michener, H. A., K. Potter, G. B. Macheel & C. G. Depies, 1984, A Test of the von Neumann–Morgenstern Stable Set Solution in Cooperative Non-Sidepayment n-Person Games, *Behav Sci* 29, 13–27.

Michener, H. A., M. S. Salzer, M. S. Reimer & J. Lee, 1991, A Comparison of Lambda-Transfer and Core Solutions in Constant-Sum Non-Sidepayment Games, *Behav Sci* 36, 115–132.

Michener, H. A. & K. Yuen, 1983, A Test of $M_1^{(i)}$ Bargaining Sets in Sidepayment Games, *J Confl Res* 27, 109–135.

Milgrom, P. & J. Roberts, 1996, Coalition-Proofness and Correlation with Arbitrary Communication Possibilities, *GEB* 17, 113–128.

Myerson, R. B., 1992, Fictitious Transfer Solutions in Cooperative Game Theory, in *Harsanyi Festschr*, 13–33.

Nakamura, K., 1973, Psi-Stability of a Cooperative Game without Side Payments, *IJGT* 2, 129–140.

Nakamura, K. & M. Suzuki, 1981, Social Decision and Coalition Power, in *Nakamura Papers*, 3–14.

Otten, G. J., P. Borm, T. Storcken & S. Tijs, 1995, Effectivity Functions and Associated Claim Game Correspondences, *GEB* 9, 172–190.

Owen, G., 1971, Values of Games without Side Payments, *IJGT* 1, 95–109.

Owen, G., 1990, Stable Outcomes in Spatial Voting Games, *Math Soc Sci* 19, 269–279.

Packel, E. W., 1981, A Stochastic Solution Concept for *n*-Person Games, *MOR* 6, 349–362.

Plott, C. R., 1976, Axiomatic Social Choice Theory: Overview and Interpretation, *Am J Pol Sci* 20, 511–596.

Radner, R., 1986, Repeated Partnership Games with Imperfect Monitoring and No Discounting, *RES* 53, 43–57.

Rajan, R., 1989, Endogenous Coalition Formation in Cooperative Oligopolies, *IER* 30, 863–876.

Rapoport, A. & J. P. Kahan, 1982, The Power of a Coalition and Payoff Disbursement in 3-Person Negotiable Conflicts, *J Math Sociol* 8, 193–224.

Ray, D., 1989, Credible Coalitions and the Core, *IJGT* 18, 185–187.

Rosenthal, R. W., 1971, External Economies and Cores, *JET* 3, 182–188.

Rosenthal, R. W., 1972, Cooperative Games in Effectiveness Form, *JET* 5, 88–101.

Roth, A. E., 1976, Subsolutions and the Supercore of Cooperative Games, *MOR* 1, 43–49.

Rubinstein, A., 1979, Equilibrium in Supergames with the Overtaking Criterion, *JET* 21, 1–9.

Sakurai, M. M. & J. M. Brennan, 1988, Computing the von Neumann–Morgenstern Characteristic Function $v(S)$ for Cooperative *n*-Person Transferable Utility Normal-Form Games: LP and Saddlepoint Solutions, *Behav Res* 20, 367–371.

Schofield, N., 1975, Game Theoretic Analysis of Olson's Game of Collective Action, *J Confl Res* 19, 441–461.

Schwodiauer, G., 1969, Structured Propensity to Monopolize in Oligopolistic Markets, *Jahrb Nat-ok Stat* 183, 465–486.

Sharkey, W. W., 1981, Convex Games without Side Payments, *IJGT* 10, 101–106.

Shepsle, K. A., 1974, On the Size of Winning Coalitions, *Am Pol Sci Rev* 68, 505–518.

Sorenson, J. R., J. T. Tschirhart & A. B. Whinston, 1978, Private Good Clubs and the Core, *J Publ Econ* 10, 77–95.

Sorin, S., 1992, Repeated Games with Complete Information, in *Handb Game Th* 1, 71–107.

Taylor, M., 1971, Mathematical Political Theory, *Brit J Pol Sci* 1, 339–382.

Weber, S., 1979, On ε-Cores of Balanced Games, *IJGT* 8, 241–250.

Weber, S., 1981, Some Results on the Weak Core of a Non-Side-Payment Game with Infinitely Many Players, *JME* 8, 101–111.

Weber, S., 1981, On the Core and Weak Core of Quasi-Balanced Games, in *Hagen Conf II*, 141–152.

Welty, G. A., 1971, Mill's Principle of Government as a Basis of Democracy, *Monist* 55, 51.

Wilson, R., 1971, Stable Coalition Proposals in Majority Rule Voting, *JET* 3, 254–271.

Wilson, R., 1972, The Game-Theoretic Structure of Arrow's General Possibility Theorem, *JET* 5, 14–20.

Yasuda, Y., 1970, A Note on the Core of a Cooperative Game without Side Payments, *NRLQ* 17, 143.

Texts and Monographs: Aubin 1979; Aumann; Aumann & Maschler; Aumann & Shapley; Friedman 1977; Intriligator Ch. 6; Laffont Ch. 1; Myerson; Ordeshook Ch. 7; Shubik 1982; Straffin; Takayama; Worobjow.

In this collection: Chapter 51.

Chapter 42 The Bargaining Set for Cooperative Games

Aivazian, V. A., I. Lipnowski & J. L. Callen, 1987, The Coase Theorem and Coalitional Stability, *Economica* 54, 517–520.

Albers, W., 1997, A Model of Boundedly Rational Experienced Bargaining in Characteristic Function Games, in *Selten Festschr*, 365–385.

Albers, W. & G. Albers, 1983, Prominence and Aspiration Adjustment in Location Games, in *Lect Notes* 213, 243–258.

Albers, W., H. Crott & J. K. Murnighan, 1985, The Formation of Blocs in an Experimental Study of Coalition Formation, *J Occupational Behaviour* 6, 33–48.

Allen, B. & S. Sorin, 1994, General Equilibrium and Cooperative Games: Basic Results, in *Stony Brook ASI* 1991, 17–33.

Anderson, R. M., W. Trockel & L. Zhou, 1997, Nonconvergence of the Mas-Colell and Zhou Bargaining Sets, *Econometrica* 65, 1227–1239.

Antimonov, S. G. & Y. P. Ivanilov, 1973, Problems of Cooperation for Multisector Production Models, *Eng Cybernetics* 10, 753.

Au, W. T. & Michener, H. A., 1994, A Probabilistic Theory of Coalition Formation in n-Person Side-Payment Games, *J Math Sociol* 19, 165–188.

Baron, D. P., 1991, A Spatial Bargaining Theory of Government Formation in Parliamentary Systems, *Am Pol Sci Rev* 85, 137–164.

Baron, D. P., 1993, Government Formation and Endogenous Parties, *Am Pol Sci Rev* 87, 34–47.

Bennett, E., 1983, The Aspiration Approach to Predicting Coalition Formation and Payoff Distribution in Sidepayment Games, *IJGT* 12, 1–28.

Bennett, E., 1983, Characterization Results for Aspirations in Games with Sidepayments, *Math Soc Sci* 4, 229–241.

Bennett, E., 1984, A New Approach to Predicting Coalition Formation and Payoff Distribution in Characteristic Function Games, in *Holler, Coalitions*, 60–80.

Bennett, E., 1997, Multilateral Bargaining Problems, *GEB* 19, 151–179.

Bennett, E. & E. Van Damme, 1991, Demand Commitment Bargaining: The Case of Apex Games, in *Selten* 3, 118–140.

Bennett, E. & W. R. Zame, 1988, Bargaining in Cooperative Games, *IJGT* 17, 279–300.

Bergstresser, K. & P. L. Yu, 1977, Domination Structures and Multicriteria Problems in n-Person Games, *Th & Dec* 8, 5–48.

Bettenhausen, K. & J. K. Murnighan, 1985, The Emergence of Norms in Competitive Decision-Making Groups, *Administrative Sci Q* 30, 350–372.

Billera, L. J., 1970, Existence of General Bargaining Sets for Cooperative Games without Side Payments, *Bull AMS* 76, 375–379.

Bird, C., 1975, A Class of Convex Nuclei Solution Concepts from Differences in Coalition Excesses, *SIAM J Appl Math* 29, 503–510.

Bird, C., 1975, Inessential Games and Non-Imposed Solutions to Allocation Problems, *Publ Choice* 22, 91–102.

Bird, C. G. & K. O. Kortanek, 1974, Game Theoretic Approaches to Some Air Pollution Regulation Problems, *Socio-Econ Plan Sci* 8, 141–147.

Bloch, F., 1996, Sequential Formation of Coalitions in Games with Externalities and Fixed Payoff Division, *GEB* 14, 90–123.

Bonacich, P., 1979, A Single Measure for Point and Interval Predictions of Coalition Theorems, *Behav Sci* 24, 85–93.

Borch, K. H., 1987, Johansen, Leif, in *The New Palgrave* 2, 1020–1022.

Bradley, G. H. & M. Shubik, 1974, A Note on the Shape of a Pareto-Optimal Surface, *JET* 8, 530–538.

Brams, S. J. & D. Wittman, 1981, Nonmyopic Equilibria in 2×2 Games, *Confl Manag* 6, 39–62.

Buckley, J. J. & T. E. Westen, 1976, Bargaining Set Theory and Majority Rule, *J Confl Res* 20, 481–496.

Calvo, E., J. J. Lasaga & E. Winter, 1996, The Principle of Balanced Contributions and Hierarchies of Cooperation, *Math Soc Sci* 31, 171–182.

Chang, C., 1991, A Bisection Property of the Kernel, *IJGT* 20, 1–11.

Chang, C. & Y. J. Lee, 1993, A Non-Weakly Balanced Game with Nonempty Bargaining Set, *JME* 22, 195–198.

Charnes, A. & S. C. Littlechild, 1975, Formation of Unions in n-Person Games, *JET* 10, 386–402.

Charnes, A., S. Littlechild & S. Sorensen, 1973, Core-Stem Solutions of n-Person Essential Games, *Socio-Econ Plan Sci* 7, 649–660.

Chwe, M. S. Y., 1994, Farsighted Coalition Stability, *JET* 63, 299–325.

Cochinard, S., 1995, The Coalition Concept in Game Theory, *Rev d'Écon Politique* 105, 633–655.

Cross, J. G., 1967, Some Theoretic Characteristics of Economic and Political Coalitions, *J Confl Res* 11, 184–195.

Crott, H. W. & W. Albers, 1981, The Equal Division Kernel: An Equity Approach to Coalition Formation and Payoff Distribution in n-Person Games, *Eur J Soc Psych* 11, 285–305.

Davis, M., 1990, Game Theory, in *Brittanica*, Vol 19, 643–649.

Davis, M. & M. Maschler, 1963, Existence of Stable Payoff Configurations for Cooperative Games, *Bull AMS* 69, 106–108.

Davis, M. & M. Maschler, 1965, The Kernel of a Cooperative Game, *NRLQ* 12, 223–259.

Davis, M. & M. Maschler, 1967, Existence of Stable Payoff Configurations for Cooperative Games, in *Morgenstern Festschr*, 39–52.

Driessen, T. S. H., 1986, Solution Concepts of k-Convex n-Person Games, *IJGT* 15, 201–229.

Driessen, T. S. H., 1998, A Note on the Inclusion of the Kernel in the Core of the Bilateral Assignment Game, *IJGT* 27, 301–303.

Driessen, T. S. H. & S. H. Tijs, 1992, The Core and the Tau-Value for Cooperative Games with Coalition Structures, in *Lect Notes* 389, 146–169.

Dubey, P. & A. Neyman, 1984, Payoffs in Nonatomic Economies: An Axiomatic Approach, *Econometrica* 52, 1129–1150.

Dubey, P. & A. Neyman, 1997, An Equivalence Principle for Perfectly Competitive Economies, *JET* 75, 314–344.

Dutta, B., D. Ray, R. Vohra & K. Sengupta, 1989, A Consistent Bargaining Set, *JET* 49, 93–112.

Einy, E., R. Holzman, D. Monderer & B. Shitovitz, 1997, Core Equivalence Theorems for Infinite Convex Games, *JET* 76, 1–12.

Einy, E., D. Monderer & D. Moreno, 1998, The Least Core, Kernel and Bargaining Sets of Large Games, *Econ Th* 11, 585–601.

Einy, E. & D. Wettstein, 1996, Equivalence between Bargaining Sets and the Core in Simple Games, *IJGT* 25, 65–71.

Felsenthal, D. S., 1979, Aspects of Coalition Payoffs: The Case of Israel, *Comparative Pol Stud* 12, 151–168.

Forman, R. & J. D. Laing, 1982, Metastability and Solid Solutions of Collective Decisions, *Math Soc Sci* 2, 397–420.

Fouraker, L. E. & S. Siegel, *Bargaining Behavior*, McGraw-Hill, 1963.

Friend, K. E., J. D. Laing & R. J. Morrison, 1977, Game-Theoretic Analysis of Coalition Behavior, *Th & Dec* 8, 127–157.

Funk, S. G., A. Rapoport & J. P. Kahan, 1980, Quota vs. Positional Power in a 4-Person Apex Game, *J Exp Soc Psych* 16, 77–93.

Gafgen, G., 1970, Trends in Mathematical Economics of Prior Times, *Kyklos* 23, 597–619.

Grandmont, J. M., 1977, The Logic of the Fixed-Price Method, *Scand J Econ* 79, 169–186.

Grandmont, J. M., G. Laroque & Y. Younes, 1978, Equilibrium with Quantity Rationing and Recontracting, *JET* 19, 84–102.

Granot, D., 1977, Cooperative Games in Stochastic Characteristic Function Form, *Manag Sci* 23, 621–630.

Greenberg, J., 1992, On the Sensitivity of von Neumann and Morgenstern Abstract Stable Sets: The Stable and the Individually Stable Bargaining Sets, *IJGT* 21, 41–55.

Groennings, S., E. W. Kelley & M. Leiserson, 1970, Introduction, in *Coal Behav*, 1–7.

Grofman, B., 1982, A Dynamic Model of Protocoalition Formation in Ideological n-Space, *Behav Sci* 27, 77–90.

Hardin, R., 1976, Hollow Victory: The Minimum Winning Coalition, *Am Pol Sci Rev* 70, 1202–1214.

Harsanyi, J. C., 1979, A New General Solution Concept for Both Cooperative and Non-cooperative Games, *Pap Rhineland-Westphalian Acad Sci* 287, 7–28; also in *Harsanyi Papers Game Th*, 211–232.

Hart, O. D., 1977, Takeover Bids and Stock Market Equilibrium, *JET* 16, 53–83.

Hart, S., 1997, Classical Cooperative Theory I: Core-Like Concepts, in *Stony Brook ASI 1994*, 35–41.

Hildebrand, D. K., J. D. Laing & H. Rosenthal, 1974, Prediction Logic and Quasi-Independence in Empirical Evaluation of Formal Theory, *J Math Sociol* 3, 197.

Hoffman, E. & E. W. Packel, 1982, A Stochastic Model of Committee Voting with Exogenous Costs: Theory and Experiments, *Behav Sci* 27, 43–56.

Horowitz, A. D., 1973, The Competitive Bargaining Set for Cooperative n-Person Games, *J Math Psych* 10, 265–289.

Horowitz, A. D., 1974, Test of the Kernel and Two Bargaining Set Models in Four- and Five-Person Games, in *Rapoport*, 161–192.

Horowitz, A. D., 1977, A Test of the Core, Bargaining Set, Kernel and Shapley Models in n-Person Quota Games with One Weak Player, *Th & Dec* 8, 49–65.

Kahan, J. P., 1983, On Choosing between Bayes and von Neumann, *Manag Sci* 29, 1334–1336.

Kahan, J. P. & A. Rapoport, 1974, Test of the Bargaining Set and Kernel Models in Three-Person Games, in *Rapoport*, 119–160.

Kahan, J. P. & A. Rapoport, 1977, When You Don't Need to Join: Effects of Guaranteed Payoffs on Bargaining in 3-Person Cooperative Games, *Th & Dec* 8, 97–126.

Kahan, J. P. & A. Rapoport, 1979, Influence of Structural Relationships on Coalition Formation in 4-Person Apex Games, *Eur J Soc Psych* 9, 339–361.

Kahan, J. P. & A. Rapoport, 1980, Coalition Formation in the Triad when Two are Weak and One is Strong, *Math Soc Sci* 1, 11–37.

Kalai, E., E. A. Pazner & D. Schmeidler, 1976, Collective Choice Correspondences as Admissible Outcomes of Social Bargaining Processes, *Econometrica* 44, 233–240.

Kalai, G., M. Maschler & G. Owen, 1975, Asymptotic Stability and Other Properties of Trajectories and Transfer Sequences Leading to the Bargaining Sets, *IJGT* 4, 193–213.

Kaneko, M., 1978, Price Oligopoly as a Cooperative Game, *IJGT* 7, 137–150.

Kaufmann, M. & W. H. Tach, 1975, Coalition Formation and Profit Sharing in Strategically Equivalent 3-Person Coalitions, *Z Soz-psych* 6, 227–245.

Kelley, E. W., 1970, Bargaining in Coalition Situations, in *Coal Behav*, 273–296.

Kikuta, K., 1997, The Kernel for Reasonable Outcomes in a Cooperative Game, *IJGT* 26, 51–59.

Komorita, S. S., 1979, An Equal Excess Model of Coalition Formation, *Behav Sci* 24, 369–381.

Komorita, S. S., 1984, Coalition Bargaining, *Adv Experimental Soc Psych* 18, 183–245.

Komorita, S. S. & A. L. Ellis, 1988, The Level of Aspiration in Coalition Bargaining, *J Pers Soc Psych* 54, 421–431.

Komorita, S. S., A. L. Ellis & K. F. Aquino, 1989, Coalition Bargaining: A Comparison of Theories based on Allocation Norms and Theories based on Bargaining Strength, *Soc Psych Q* 52, 183–196.

Komorita, S. S., T. P. Hamilton & D. A. Kravitz, 1984, Effects of Alternatives in Coalition Bargaining, *J Exp Soc Psych* 20, 116–136.

Komorita, S. S. & D. A. Kravitz, 1979, Effects of Alternatives in Bargaining, *J Exp Soc Psych* 15, 147–157.

Komorita, S. S. & D. A. Kravitz, 1981, Effects of Prior Experience on Coalition Bargaining, *J Pers Soc Psych* 40, 675–686.

Komorita, S. S. & C. E. Miller, 1986, Bargaining Strength as a Function of Coalition Alternatives, *J Pers Soc Psych* 51, 325–332.

Kurz, M., 1988, Coalitional Value, in *Roth: Shapley Value*, 155–173.

Laing, J. D. & R. J. Morrison, 1974, Sequential Games of Status, *Behav Sci* 19, 177–196.

Laing, J. D. & B. Slotznick, 1991, When Anyone Can Veto: A Laboratory Study of Committees Governed by Unanimous Rule, *Behav Sci* 36, 179–195.

Lee, M., R. D. McKelvey & H. Rosenthal, 1979, Game Theory and the French Apparentements of 1951, *IJGT* 8, 27–53.

Leiserson, M., 1970, Game Theory and the Study of Coalition Behavior, in *Coal Behav*, 255–272.

Leopoldwildburger, U., 1992, Payoff Divisions on Coalition Formation in a 3-Person Characteristic Function Experiment, *J Econ Behav* 17, 183–193.

Lucas, W. F., 1966, Solution Theory for n-Person Games in Partition Function Form, in *Toulon Conf*, 131–134.

Lucas, W. F. & M. Rabie, 1982, Games with No Solutions and Empty Cores, *MOR* 7, 491–500.

Mannix, E. A., 1994, Will We Meet Again? Effects of Power, Distribution Norms, and Scope of Future Interaction in Small Group Negotiation, *Int J Conflict Manag* 5, 343–368.

Maor, M., 1995, Intraparty Determinants of Coalition Bargaining, *J Th Pol* 7, 65–91.

Mares, M., 1976, Bargaining in Games, *Kybernetika* 12, 3–20.

Mares, M., 1976, Suggestion of a Cooperative Market Model, *Kybernetika* 12, 439–450.

Mares, M., 1976, Cooperative Games Connected with Markets, *Kybernetika* 12, 451–461.

Mares, M., 1978, General Coalition Games, *Kybernetika* 14, 245–260.

Maschler, M., 1963, n-Person Games with Only 1, $n-1$, and n-Person Permissible Coalitions, *J Math Anal Appl* 6, 230–256.

Maschler, M., 1963, The Power of a Coalition, *Manag Sci* 10, 8–29.

Maschler, M., 1966, The Inequalities that Determine the Bargaining Set $M_1^{(i)}$, *IJM* 4, 127–134.

Maschler, M., 1992, The Bargaining Set, Kernel, and Nucleolus, in *Handb Game Th* 1, 591–667.

Maschler, M. & B. Peleg, 1966, A Characterization, Existence Proof, and Dimension Bounds for the Kernel of a Game, *Pac J Math* 18, 289–328.

Maschler, M. & B. Peleg, 1967, The Structure of the Kernel of a Cooperative Game, *SIAM J Appl Math* 15, 569–604.

Maschler, M., B. Peleg & L. S. Shapley, 1971, The Kernel and Bargaining Set for Convex Games, *IJGT* 1, 73–93.

Mas-Colell, A., 1987, Cooperative Equilibrium, in *The New Palgrave* 1, 659–662.

Mas-Colell, A., 1989, An Equivalence Theorem for a Bargaining Set, *JME* 18, 129–139.

McKelvey, R. D. & P. C. Ordeshook, 1979, An Experimental Test of Several Theories of Committee Decision Making under Majority Rule, in *Vienna Conf*, 152–167.

McKelvey, R. D., P. C. Ordeshook & M. D. Winer, 1978, Competitive Solutions for n-Person Games without Transferable Utility, with an Application to Committee Games, *Am Pol Sci Rev* 72, 599–615.

Medlin, S. M., 1976, Effects of Grand Coalition Payoffs on Coalition Formation in 3-Person Games, *Behav Sci* 21, 48–61.

Megiddo, N., 1974, On the Nonmonotonicity of the Bargaining Set, the Kernel and the Nucleolus of a Game, *SIAM J Appl Math* 27, 355–358.

Michener, H. A., Y. C. Choi & D. C. Dettman, 1986, Stability by Deterrence in Cooperative Non-Sidepayment n-Person Games, *J Econ Behav* 7, 375–402.

Michener, H. A. & D. J. Myers, 1998, Probabilistic Coalition Structure Theories: An Empirical Comparison in Four-Person Superadditive Sidepayment Games, *J Confl Res* 42, 830–860.

Michener, H. A. & D. J. Myers, 1998, An Empirical Comparison of Probabilistic Coalition Structure Theories in Three-Person Sidepayment Games, *Theory & Decision* 45, 37–82.

Michener, H. A. & K. Potter, 1981, Generalizability of Tests in n-Person Sidepayment Games, *J Confl Res* 25, 733–749.

Michener, H. A., M. M. Sakurai, K. Yuen & T. J. Kasen, 1979, Competitive Test of the $M_1^{(i)}$ and the $M_1^{(im)}$ Bargaining Sets, *J Confl Res* 23, 102–119.

Michener, H. A. & K. Yuen, 1983, A Test of the $M_1^{(im)}$ Bargaining Sets in Sidepayment Games, *J Confl Res* 27, 109–135.

Miller, C. E., 1979, Coalition Formation in Triads with Single-Peaked Payoff Curves, *Behav Sci* 24, 75–84.

Miller, C. E., 1980, Coalition Formation in Characteristic Function Games: A Competitive Test of 3 Theories, *J Exp Soc Psych* 16, 61–76.

Miller, C. E., 1980, Effects of Payoffs and Resources on Coalition Formation: A Test of 3 Theories, *Soc Psych Q* 43, 154–164.

Miller, C. E., 1980, A Test of 4 Theories of Coalition Formation: Effects of Payoffs and Resources, *J Pers Soc Psych* 38, 153–164.

Miller, C. E. & R. Crandall, 1980, Experimental Research on the Social Psychology of Bargaining and Coalition Formation, in Paulus, P. B. (Ed.), *Psychology of Group Influence*, Lawrence Erlbaum Assoc., Hillsdale, N.J., 333–374.

Miller, G. J. & J. A. Oppenheimer, 1982, Universalism in Experimental Committees, *Am Pol Sci Rev* 76, 561–574.

Moulin, H., 1995, An Appraisal of Cooperative Game Theory, *Rev d'Écon Politique* 105, 617–632.

Murnighan, J. K., 1978, Models of Coalition Behavior: Game Theoretic, Social Psychological, and Political Perspectives, *Psychological Bulletin* 85, 1130–1153.

Murnighan, J. K., 1981, Defectors, Vulnerability, and Relative Power: Some Causes and Effects of Leaving a Stable Coalition, *Human Relations* 34, 589–609.

Murnighan, J. K., 1982, Evaluating Theoretical Predictions in the Social Sciences: Coalition Theories and Other Models, *Behav Sci* 27, 125–130.

Murnighan, J. K., 1985, Coalitions in Decision-Making Groups, *Organizational Analogs, Organizational Behavior and Human Decision Processes* 35, 1–26.

Murnighan, J. K. & A. E. Roth, 1977, Effects of Communication and Information Availability in an Experimental Study of a 3-Person Game, *Manag Sci* 23, 1336–1348.

Murnighan, J. K. & A. E. Roth, 1978, Large Group Bargaining in a Characteristic Function Game, *J Confl Res* 22, 299–317.

Murnighan, J. K. & E. Szwajkowski, 1979, Coalition Bargaining in 4 Games that Include a Veto Player, *J Pers Soc Psych* 37, 1933–1946.

Muto, S., 1990, Resale-Proofness and Coalition-Proof Nash Equilibria, *GEB* 2, 337–361.

Muto, S. & M. Nakayama, 1992, Stable Outcomes in Sequential Information Trading: An Application of the Bargaining Set, in *Lect Notes* 389, 409–441.

Myerson, R. B., 1986, An Introduction to Game Theory, in *Reiter*, 1–61.

Naumova, N. I., 1976, Existence of Some Bargaining Sets for Games with a Discrete Set of Players, *Vestnik*, 47–54.

Naumova, N. I., 1978, M-Systems of Relations and their Application in Cooperative Games, *Vestnik*, 60–66.

Niou, E. M. S. & P. C. Ordeshook, 1986, A Theory of the Balance of Power in International Systems, *J Confl Res* 30, 685–715.

Niou, E. M. S. & P. C. Ordeshook, 1987, Preventive War and the Balance of Power: A Game-Theoretic Approach, *J Confl Res* 31, 387–419.

Niou, E. M. S. & P. C. Ordeshook, 1988, A Theory of the Balance of Power: An Experimental Test, *Simul & Gaming* 19, 415–439.

Nurmi, H., 1980, Game Theory and Power Indexes, *Z Nat-ok* 40, 35–38.

Nurmi, H., 1980, Majority Rule: Second Thoughts and Refutations, *Qual & Quant* 14, 743–765.

Oppenheimer, J., 1979, Outcomes of Logrolling in the Bargaining Set and Democratic Theory: Some Conjectures, *Publ Choice* 34, 419–434.

Ordeshook, P. C. & M. Winer, 1980, Coalitions and Spatial Policy Outcomes in Parliamentary Systems: Some Experimental Results, *Am J Pol Sci* 24, 730–752.

Peleg, B., 1963, Quota Games with a Continuum of Players, *IJM* 1, 48–53.

Peleg, B., 1963, Bargaining Sets of Cooperative Games without Side Payments, *IJM* 1, 197–200.

Peleg, B., 1964, On the Bargaining Set M_0 of m-Quota Games, in *Adv Game Th*, 501–512.

Peleg, B., 1966, The Independence of Game Theory of Utility Theory, *Bull AMS* 72, 995–999.

Peleg, B., 1967, Existence Theorem for the Bargaining Set $M_1^{(i)}$, in *Morgenstern Festschr*, 53–56.

Philips, L. & R. M. Harstad, 1991, Interaction between Resource Extraction and Futures Markets: A Game-Theoretic Analysis, in *Selten* 2, 289–307.

Polzer, J. T., E. A. Mannix & M. A. Neale, 1998, Interest Alignment and Coalitions in Multiparty Negotiation, *Acad Manag J* 41, 42–54.

Potters, J., R. Poos, S. Tijs & S. Muto, 1989, Clan Games, *GEB* 1, 275–293.

Potters, J. & H. Reijnierse, 1995, Gamma-Component Additive Games, *IJGT* 24, 49–56.

Rajan, R., 1989, Endogenous Coalition Formation in Cooperative Oligopolies, *IER* 30, 863–876.

Rapoport, A., 1974, Introduction to Game Theory as a Theory of Conflict Resolution, in *Rapoport*, 1–14.

Rapoport, A., 1984, Variability in Payoff Disbursement in Coalition Formation Experiments, *Eur J Soc Psych* 14, 265–280.

Rapoport, A., 1985, A Note on the Equal Division Kernel and the Alpha-Power Model, *J Math Sociol* 11, 65–76.

Rapoport, A., 1987, Comparison of Theories for Payoff Disbursement of Coalition Values, *Th & Dec* 22, 13–47.

Rapoport, A. & J. P. Kahan, 1974, Computer Controlled Research on Bargaining and Coalition Formation, *Behav Res Meth & Instrumentation* 6, 87–93.

Rapoport, A. & J. P. Kahan, 1976, When Three is not Always Two against One: Coalitions in Experimental Three-Person Cooperative Games, *J Exp Soc Psych* 12, 253–273.

Rapoport, A. & J. P. Kahan, 1979, Standards of Fairness in 4-Person Monopolistic Cooperative Games, in *Vienna Conf*, 74–95.

Rapoport, A. & J. P. Kahan, 1982, The Power of a Coalition and Payoff Disbursement in 3-Person Negotiable Conflicts, *J Math Sociol* 8, 193–224.

Rapoport, A. & J. P. Kahan, 1984, Coalition Formation in a 5-Person Market Game, *Manag Sci* 30, 326–343.

Reijnierse, H., M. Maschler, J. Potters & S. Tijs, 1996, Simple Flow Games, *GEB* 16, 238–260.

Riker, W. H., 1967, Bargaining in a Three-Person Game, *Am Pol Sci Rev* 61, 642–656.

Riker, W. H., 1977, Minimum Winning Coalitions, *Am Pol Sci Rev* 71, 1056–1059.

Riker, W. H., 1992, The Entry of Game Theory into Political Science, *Hist Pol Econ* 24, 207–223.

Rinaldi, S., R. Soncini-Sessa & A. B. Whinston, 1979, Stable Taxation Schemes in Regional Environmental Management, *J Env Econ & Manag* 6, 29–50.

Rosenthal, R. W., 1972, Cooperative Games in Effectiveness Form, *JET* 5, 88–101.

Roth, A. E., 1976, Subsolutions and the Supercore of Cooperative Games, *MOR* 1, 43–49.

Roth, B. M., 1979, Competing Norms of Distribution in Coalitional Games, *J Confl Res* 23, 513–537.

Schjodt, U. & B. Sloth, 1994, Bargaining Sets with Small Coalitions, *IJGT* 23, 49–55.

Schleicher, H., 1976, Strategic Analysis of a Category of Election Games, *Économie Appl* 29, 127–153.

Schmeidler, D., 1969, The Nucleolus of a Characteristic Function Game, *SIAM J Appl Math* 17, 1163–1170.

Schofield, N., 1975, Game Theoretic Analysis of Olson's Game of Collective Action, *J Confl Res* 19, 441–461.

Schofield, N., 1978, Generalized Bargaining Sets for Cooperative Games, *IJGT* 7, 183–199.

Schofield, N., 1980, The Bargaining Set in Voting Games, *Behav Sci* 25, 120–129.

Schotter, A. & G. Schwodiauer, 1980, Economics and the Theory of Games: A Survey, *JEL* 18, 479–527.

Schwodiauer, G., 1969, Structured Propensity to Monopolize in Oligopolistic Markets, *Jahrb Nat-ok Stat* 183, 465–486.

Seibold, D. R. & T. M. Steinfatt, 1979, Creative Alternative Games: Exploring Interpersonal Influence Processes, *Simul & Gaming* 10, 429–457.

Seidl, C., 1975, Liberal Values, *Z Nat-ok* 35, 257–292.

Selten, R., 1970, Psychologiche Faktoren bei Koalitionsverhandlungen, in *Sauermann*, 100.

Selten, R., 1981, A Noncooperative Model of Characteristic Function Bargaining, in *Morgenstern Memorial*, 131–151.

Selten, R., 1983, Equal Division Payoff Bounds for 3-Person Characteristic Function Experiments, in *Lect Notes* 213, 265–275.

Selten, R., 1987, Equity and Coalition Bargaining in Experimental Three-Person Games, in Roth, A. E. (Ed.), *Laboratory Experimentation in Economics*, Cambridge University Press, 42–98.

Selten, R., 1991, Balance of Power in a Parlor Game, in *Selten* 4, 150–209.

Selten, R., 1997, Descriptive Approaches to Cooperation, in *Stony Brook ASI* 1994, 289–328.

Selten, R., 1998, Features of Experimentally Observed Bounded Rationality, *Eur Econ Rev* 42, 413–436.

Selten, R. & W. Krischker, 1983, Comparison of Two Theories for Characteristic Function Experiments, in *Lect Notes* 213, 259–264.

Sengupta, A. & K. Sengupta, 1994, Viable Proposals, *IER* 35, 347–359.

Serrano, R., 1997, A Comment on the Nash Program and the Theory of Implementation, *Econ Let* 55, 203–208.

Serrano, R., 1997, Reinterpreting the Kernel, *JET* 77, 58–80.

Shapley, L. S. & M. Shubik, 1966, Quasi-Cores in a Monetary Economy with Nonconvex Preferences, *Econometrica* 34, 805–827.

Shenoy, P. P., 1979, On Coalition Formation: A Game-Theoretical Approach, *IJGT* 8, 133–164.

Shenoy, P. P., 1980, On Committee Decision-Making: A Game-Theoretical Approach, *Manag Sci* 26, 387–400.

Shepsle, K. A., 1974, On the Size of Winning Coalitions, *Am Pol Sci Rev* 68, 505–518.

Sherali, H. D. & R. Rajen, 1986, A Game Theoretic Mathematical Programming Analysis of Cooperative Phenomena in Oligopolistic Markets, *OR* 34, 683–697.

Shimomura, K. I., 1997, Quasi-Cores in Bargaining Sets, *IJGT* 26, 283–302.

Shitovitz, B., 1989, The Bargaining Set and the Core in Mixed Markets with Atoms and an Atomless Sector, *JME* 18, 377–383.

Shubik, M., 1964, Game Theory and the Study of Social Behavior: An Introductory Exposition, in *Shubik*, 3–77.

Shubik, M., 1972, Gaming and Game Theory, *Manag Sci Appl* 18, 37–53.

Shubik, M., 1981, Game Theory Models and Methods in Political Economy, in *Handb Math Econ* 1, 285–330.

Shubik, M., 1997, On the Trail of a White Whale: The Rationalizations of a Mathematical Institutional Economist, in *Heertje III*, 96–121.

Shubik, M. & L. Van der Heyden, 1978, Logrolling and Budget Allocation Games, *IJGT* 7, 151–162.

Spinetto, R., 1974, The Geometry of Solution Concepts for n-Person Cooperative Games, *Manag Sci Th* 20, 1292–1299.

Stark, O., 1984, Bargaining, Altruism, and Demographic Phenomena, *Population and Development Rev* 10, 679–692.

Stearns, R. E., 1968, Convergent Transfer Schemes for n-Person Games, *Trans AMS* 134, 449–459.

Stefanescu, A., 1994, Solutions of Transferable Utility Cooperative Games, *Rairo-RO/OR* 28, 369–387.

Stefanescu, A., 1996, Coalitional Stability and Rationality in Cooperative Games, *Kybernetika* 32, 483–490.

Sussangkarn, C., 1978, Equilibrium Payoff Configurations for Cooperative Games with Transferability, *J Confl Res* 22, 121–141.

Sussangkarn, C., 1983, A Unique Bargaining Solution based on Competitive Commitments, *MOR* 8, 205–214.

Tack, W. H., 1988, New Approaches in Research on Cooperative Conflict Resolution by Group Decisions, *Acta Psychologica* 68, 113–136.

Tavernier, E. M. & M. P. Hartley, 1997, Agriculture and the Environment: The Demand for Conflict Resolution Mechanisms, *J Sustainable Agr* 10, 11–23.

Taylor, M., 1971, Mathematical Political Theory, *Brit J Pol Sci* 1, 339–382.

Vind, K., 1992, Two Characterizations of Bargaining Sets, *JME* 21, 89–97.

Vohra, R., 1991, An Existence Theorem for a Bargaining Set, *JME* 20, 19–34.

Vohra, R., 1994, Bargaining Sets, in *Stony Brook ASI* 1991, 51–58.

Wang, Q. & M. Parlar, 1989, Static Game Theory Models and their Applications in Management Science, *Eur J OR* 42, 1–21.

Wilson, R., 1971, Stable Coalition Proposals in Majority Rule Voting, *JET* 3, 254–271.

Wolf, G. & M. Shubik, 1977, Beliefs about Coalition Formation in Multiple Resource 3-Person Situations, *Behav Sci* 22, 99–106.

Wu, L. S. Y., 1977, A Dynamic Theory for a Class of Games with Nonempty Cores, *SIAM J Appl Math* 32, 328–338.

Zhou, L., 1994, A New Bargaining Set of an *n*-Person Game and Endogenous Coalition Formation, *GEB* 6, 512–526.

Texts and Monographs: Aubin 1979; Binmore 1998; Davis; Driessen; Germeier; Greenberg; Guiasu & Malitza; Güth; Harsanyi; Harsanyi & Selten; Howard; Moulin Ch. 6; Myerson; Ordeshook Ch. 9; Owen 1968 Ch. 9, 1982 Ch. 11, 1995 Ch. 13; Amnon Rapoport Ch. 8, 9, 10, 11, 13, 15; Anatol Rapoport; Selten; Shubik 1982, 1984; Straffin; Szép & Forgó; Vilkas & Maiminas; Worobjow.

In this collection: Chapters 43, 44, and 59.

Chapter 43   A Method of Computing the Kernel of *n*-Person Games

Bitter, D., 1982, The Kernel for the Grand Coalition of the Four-Person Game, *IJGT* 11, 215–239.

Granot, D., F. Granot & W. P. R. Zhu, 1997, The Reactive Bargaining Set of Some Flow Games and of Superadditive Simple Games, *IJGT* 26, 207–214.

Horowitz, A. D., 1973, The Competitive Bargaining Set for Cooperative *n*-Person Games, *J Math Psych* 10, 265–289.

Maschler, M., 1966, The Inequalities that Determine the Bargaining Set $M_1^{(i)}$, *IJM* 4, 127–134.

Maschler, M., 1992, The Bargaining Set, Kernel, and Nucleolus, in *Handb Game Th* 1, 591–667.

Maschler, M. & B. Peleg, 1966, A Characterization, Existence Proof, and Dimension Bounds for the Kernel of a Game, *Pac J Math* 18, 289–328.

Maschler, M. & B. Peleg, 1967, The Structure of the Kernel of a Cooperative Game, *SIAM J Appl Math* 15, 569–604.

Maschler, M., B. Peleg & L. S. Shapley, 1979, Geometric Properties of the Kernel, Nucleolus, and Related Solution Concepts, *MOR* 4, 303–338.

Megiddo, N., 1974, Kernels of Compound Games with Simple Components, *Pac J Math* 50, 531–555.

Michener, H. A. & K. Potter, 1981, Generalizability of Tests in *n*-Person Sidepayment Games, *J Confl Res* 25, 733–749.

Michener, H. A. & M. M. Sakurai, 1976, A Research Note on the Predictive Adequacy of the Kernel, *J Confl Res* 20, 129–141.

Peleg, B., 1965, The Kernel of the Composition of Characteristic Function Games, *IJM* 3, 127–138.

Peleg, B., 1981, Coalition Formation in Simple Games with Dominant Players, *IJGT* 10, 11–33.

Schofield, N., 1976, Kernel and Payoffs in European Government Coalitions, *Publ Choice* 26, 24–49.

Schofield, N., 1982, Bargaining Set Theory and Stability in Coalition Governments, *Math Soc Sci* 3, 9–31.

Wolsey, L. A., 1976, The Nucleolus and Kernel for Simple Games of Special Valid Inequalities for 0–1 Linear Integer Programs, *IJGT* 5, 227–238.

Chapter 44   Cooperative Games with Coalition Structures

Amer, R. & F. Carreras, 1995, Games and Cooperation Indexes, *IJGT* 24, 239–258.

Amer, R. & F. Carreras, 1997, Cooperation Indexes and Weighted Shapley Values, *MOR* 22, 955–968.

Au, W. T. & H. A. Michener, 1994, A Probabilistic Theory of Coalition Formation in *n*-Person Side-Payment Games, *J Math Sociol* 19, 165–188.

Azoff, E. & C. Bird, 1979, On the Formation of Coalitions in Infinite Player Games, *JME* 6, 203–213.

Ball, M. A., 1995, Bargaining in *n*-Person Cooperative Games with Linearly Distributed, Transferable Utility, *Proc Royal Soc London A* 451, 349–365.

Ball, M. A., 1998, Coalition Structure and the Equilibrium Concept in *n*-Person Games with Transferable Utility, *Proc Royal Soc London A* 454, 1509–1522.

Bennett, E., 1983, The Aspiration Approach to Predicting Coalition Formation and Payoff Distribution in Sidepayment Games, *IJGT* 12, 1–28.

Bennett, E. & E. Van Damme, 1991, Demand Commitment Bargaining: The Case of Apex Games, in *Selten* 3, 118–140.

Binmore, K., A. Kirman & P. Tani, 1993, Introduction: Famous Gamesters, in *Frontiers Game Th*, 1–25.

Bloch, F., 1996, Sequential Formation of Coalitions in Games with Externalities and Fixed Payoff Division, *GEB* 14, 90–123.

Brandenburger, A. M. & H. Stuart, 1996, Value-Based Business Strategy, *J Econ & Manag Strat* 5, 5–24.

Carreras, F. & A. Magaña, 1997, The Multilinear Extension of the Quotient Game, *GEB* 18, 22–31.

Chang, C., 1988, A Note on the von Neumann–Morgenstern Solution, *IJGT* 17, 311–314.

Chang, C., 1988, An Absorbing Set for Cooperative Games, *OR Letters* 7, 149–153.

Chang, C., 1991, The Bisection Property of the Kernel, *IJGT* 20, 1–11.

Chang, C. & C. Y. Kan, 1992, The Kernel, the Bargaining Set and the Reduced Game, *IJGT* 21, 75–83.

Chang, C. & C. Y. Kan, 1994, A Study on Decomposable Convex Games, *GEB* 7, 35–38.

Charnes, A. & S. C. Littlechild, 1975, Formation of Unions in *n*-Person Games, *JET* 10, 386–402.

Cochinard, S., 1995, The Coalition Concept in Game Theory, *Rev d'Écon Politique* 105, 633–655.

Deb, R. & S. Weber, 1996, The Nakamura Theorem for Coalition Structures of Quota Games, *IJGT* 25, 189–198.

Demange, G., 1994, Intermediate Preferences and Stable Coalition Structures, *JME* 23, 45–58.

Derks, J. J. M. & R. P. Gilles, 1995, Hierarchical Organization Structures and Constraints on Coalition Formation, *IJGT* 24, 147–163.

Donsimoni, M. P., N. S. Economides & H. M. Polemarchakis, 1986, Stable Cartels, *IER* 27, 317–327.

Dreze, J. H. & J. Greenberg, 1980, Hedonic Coalitions: Optimality and Stability, *Econometrica* 48, 987–1003.

Driessen, T. S. H., 1991, A Survey of Consistency Properties in Cooperative Game Theory, *SIAM Rev* 33, 43–59.

Driessen, T. S. H. & S. H. Tijs, 1984, Extensions and Modifications of the Tau-Value for Cooperative Games, in *Lect Notes* 226, 252–261.

Driessen, T. S. H. & S. H. Tijs, 1992, The Core and the Tau-Value for Cooperative Games with Coalition Structure, in *Lect Notes* 389, 146–169.

Einy, E., 1985, On Connected Coalitions in Dominated Simple Games, *IJGT* 14, 103–125.

Esteban, J., 1991, The Social Viability of Money: Competitive Equilibria and the Core of Overlapping Generations Economies: Foreword, *Lect Notes* 372.

Funk, S. G., A. Rapoport & J. P. Kahan, 1980, Quota vs. Positional Power in the 4-Person Apex Game, *J Exp Soc Psych* 16, 77–93.

Gerard-Varet, L. A. & S. Zamir, 1987, Remarks on the Reasonable Set of Outcomes in a General Coalition Function Form Game, *IJGT* 16, 123–143.

Greenberg, J., 1979, Stability when Mobility is Restricted by the Existing Coalition Structure, *JET* 21, 213–221.

Greenberg, J., 1979, Efficiency and Altruism in Cooperative Games with Coalition Structures, in *Vienna Conf*, 204–213.

Greenberg, J., 1994, Coalition Structures, in *Handb Game Th* 2, 1305–1337.

Greenberg, J. & J. Kats, 1980, Unilateral Transfers and Pareto Optimality, *Econometrica* 48, 777–779.

Greenberg, J. & S. Weber, 1993, Stable Coalition Structures with a Unidimensional Set of Alternatives, *JET* 60, 62–82.

Greenberg, J. & S. Weber, 1993, Stable Coalition Structures in Consecutive Games, in *Frontiers Game Th*, 103–115.

Hart, S., 1987, Shapley Value, in *The New Palgrave* 4, 318–320.

Hart, S. & M. Kurz, 1983, Endogenous Formation of Coalitions, *Econometrica* 51, 1047–1064.

Hart, S. & M. Kurz, 1984, Stable Coalition Structures, in *Holler, Coalitions*, 235–258.

Henss, R., 1986, Cooperation, Stability and Trust in a Cooperative Normal Form Game, *Z Soz-psych* 17, 31–39.

Henss, R., 1986, Bargaining Strength in 3-Person Characteristic-Function Games with $v(I) > \emptyset$: A Reanalysis of Kahan and Rapoport, 1977, *Th & Dec* 21, 267–282.

Ichiishi, T., 1977, Coalition Structures in a Labor Managed Market Economy, *Econometrica* 45, 341–360.

Johansen, L., 1982, Cores, Aggressiveness and the Breakdown of Cooperation in Economic Games, *J Econ Behav* 3, 1–37.

Kahan, J. P. & A. Rapoport, 1980, Coalition Formation in the Triad when Two are Weak and One is Strong, *Math Soc Sci* 1, 11–37.

Kirman, A., C. Oddou & S. Weber, 1986, Stochastic Communication and Coalition Formation, *Econometrica* 54, 129–138.

Kurz, M., 1988, Coalitional Value, in *Roth: Shapley Value*, 155–173.

Le Breton, M. & S. Weber, 1995, Stability of Coalition Structures and the Principle of Optimal Partitioning, in *Proc Int Symp Soc Ch*, 301–319.

Levy, A. & R. P. McLean, 1989, Weighted Coalition Structure Values, *GEB* 1, 234–249.

Maschler, M., 1990, Consistency, in *Ohio Conf*, 183–186.

Maschler, M., 1992, The Bargaining Set, Kernel, and Nucleolus, in *Handb Game Th* 1, 591–667.

Maschler, M., B. Peleg & L. S. Shapley, 1979, Geometric Properties of the Kernel, Nucleolus, and Related Solution Concepts, *MOR* 4, 303–338.

Meca-Martínez, A., J. Sánchez-Soriano, I. García-Jurado & S. Tijs, 1998, Strong Equilibria in Claim Games corresponding to Convex Games, *IJGT* 27, 211–217.

Michener, H. A. & D. J. Myers, 1998, Probabilistic Coalition Structure Theories: An Empirical Comparison in Four-Person Superadditive Sidepayment Games, *J Confl Res* 42, 830–860.

Michener, H. A. & D. J. Myers, 1998, An Empirical Comparison of Probabilistic Coalition Structure Theories in Three-Person Sidepayment Games, *Theory & Decision* 45, 37–82.

Michener, H. A. & K. Potter, 1981, Generalizability of Tests in n-Person Sidepayment Games, *J Confl Res* 25, 733–749.

Moldovanu, B., 1989, A Note on a Theorem of Aumann and Dreze, *IJGT* 18, 471–476.

Moldovanu, B., 1990, Stable Bargained Equilibria for Assignment Games without Side Payments, *IJGT* 19, 171–190.

Myerson, R. B., 1977, Graphs and Cooperation in Games, *MOR* 2, 225–229.

Myerson, R. B., 1980, Conference Structures and Fair Allocation Rules, *IJGT* 9, 169–182.

Owen, G., 1977, Values of Games with A Priori Unions, in *Lect Notes* 141, 76–88.

Pederzoli, G., 1991, Communication Games, in *Lect Notes* 353, 170–189.

Peleg, B., 1981, Coalition Formation in Simple Games with Dominant Players, *IJGT* 10, 11–33.

Peleg, B., 1986, On the Reduced Game Property and its Converse, *IJGT* 15, 187–200.

Peleg, B., 1990, Axiomatizations of the Core, the Nucleolus, and the Prekernel, in *Ohio Conf*, 176–182.

Perezcastrillo, J. D., 1994, Cooperative Outcomes through Noncooperative Games, *GEB* 7, 428–440.

Potters, J. A. M. & S. H. Tijs, 1992, The Nucleolus of a Matrix Game and Other Nucleoli, *MOR* 17, 164–174.

Rapoport, A. & J. P. Kahan, 1982, The Power of a Coalition and Payoff Disbursement in 3-Person Negotiable Conflicts, *J Math Sociol* 8, 193–224.

Rapoport, A. & J. P. Kahan, 1983, Standards of Fairness in 3-Quota, 4-Person Games, in *Lect Notes* 213, 337–351.

Rapoport, A. & J. P. Kahan, 1984, Coalition Formation in a 5-Person Market Game, *Manag Sci* 30, 326–343.

Reny, P. & M. H. Wooders, 1996, The Partnered Core of a Game without Side Payments, *JET* 70, 298–311.

Reny, P. & M. H. Wooders, 1997, Credible Threats of Secession, Partnership, and Commonwealths, in *Selten Festschr*, 305–312.

Schotter, A. & G. Schwodiauer, 1980, Economics and the Theory of Games: A Survey, *JEL* 18, 479–527.

Sengupta, A. & K. Sengupta, 1994, Viable Proposals, *IER* 35, 347–359.

Shenoy, P. P., 1978, Coalition Formation in Simple Games: Mathematical Analysis of Caplow's and Gamson's Theories, *J Math Psych* 18, 177–194.

Shenoy, P. P., 1979, On Coalition Formation: A Game-Theoretical Approach, *IJGT* 8, 133–164.

Sherali, H. D. & R. Rajen, 1986, A Game Theoretic Mathematical Programming Analysis of Cooperative Phenomena in Oligopolistic Markets, *OR* 34, 683–697.

Shimomura, K. I., 1997, Quasi-Cores in Bargaining Sets, *IJGT* 26, 283–302.

Sussangkarn, C., 1978, Equilibrium Payoff Configurations for Cooperative Games with Transferability, *J Confl Res* 22, 121–141.

Tesfatsion, L., 1984, Games, Goals, and Bounded Rationality, *Th & Dec* 17, 149–175.

Thomson, W., 1990, The Consistency Principle, in *Ohio Conf*, 187–215.

Vandenbrink, R., 1997, An Axiomatization of the Disjunctive Permission Value for Games with a Permission Structure, *IJGT* 26, 27–43.

Vandenbrink, R. & R. P. Gilles, 1996, Axiomatizations of the Conjunctive Permission Value for Games with Permission Structures, *GEB* 12, 113–126.

Winter, E., 1991, On Nontransferable Utility Games with Coalition Structures, *IJGT* 20, 53–63.

Winter, E., 1992, Consistency and the Potential for Values of Games with Coalition Structures, *GEB* 4, 132–144.

Winter, E. & M. H. Wooders, 1994, An Axiomatization of the Core for Finite and Continuum Games, *Soc Choice Welf* 11, 165–175.

Wooders, M., 1994, Large Games and Economies with Effective Small Groups, in *Stony Brook ASI* 1991, 145–196.

Yi, S.-S., 1997, Stable Coalition Structures with Externalities, *GEB* 20, 201–237.

Zhou, L., 1994, A New Bargaining Set of an n-Person Game and Endogenous Coalition Formation, *GEB* 6, 512–526.

Texts and Monographs: Myerson; Amnon Rapoport Ch. 11, 13; Shubik 1982.
In this collection: Chapters 2 and 59.

Chapter 45 Game-Theoretic Analysis of a Bankruptcy Problem from the Talmud

Agastya, M., 1997, Adaptive Play in Multiplayer Bargaining Situations, *RES* 64, 411–426.

Arin, J. & E. Inarra, 1998, A Characterization of the Nucleolus for Convex Games, *GEB* 23, 12–24.

Armstrong, C. W., 1998, Sharing a Fish Resource: Bargaining-Theoretical Analysis of an Applied Allocation Rule, *Marine Policy* 22, 119–134.

Aumann, R. J., 1990, The Shapley Value, in *Ohio Conf*, 158–165.

Aumann, R. J., 1998, On the State of the Art in Game Theory, *GEB* 24, 181–210.

Balinsky, M. L. & G. Demange, 1989, An Axiomatic Approach to Proportionality between Matrices, *MOR* 14, 700–719.

Barberà, S., M. O. Jackson & A. Neme, 1997, Strategy-Proof Allotment Rules, *GEB* 18, 1–21.

Benoit, J. P., 1997, The Nucleolus is Contested Garment Consistent: A Direct Proof, *JET* 77, 192–196.

Bevia, C., 1996, Identical Preferences, Lower Bound Solutions and Consistency in Economies with Indivisible Goods, *Soc Choice Welf* 13, 113–126.

Chang, C. & T. Driessen, 1995, (Pre)Kernel Catchers for Cooperative Games, *OR Spektrum* 17, 23–26.

Chun, Y. S., 1988, The Proportional Solution for Rights Problems, *Math Soc Sci* 15, 231–246.

Chun, Y. S., 1988, The Equal-Loss Principle for Bargaining Problems, *Econ Let* 26, 103–106.

Chun, Y. S., 1989, A New Axiomatization of the Shapley Value, *GEB* 1, 119–130.

Chun, Y. S., 1989, A Noncooperative Justification for Egalitarian Surplus Sharing, *Math Soc Sci* 17, 245–261.

Chun, Y. S. & W. Thomson, 1992, Bargaining Problems with Claims, *Math Soc Sci* 24, 19–33.

Dagan, N., 1996, New Characterizations of Old Bankruptcy Rules, *Soc Choice Welf* 13, 51–59.

Dagan, N., R. Serrano & O. Volij, 1997, A Noncooperative View of Consistent Bankruptcy Rules, *GEB* 18, 55–72.

Dagan, N. & O. Volij, 1993, The Bankruptcy Problem: A Cooperative Bargaining Approach, *Math Soc Sci* 26, 287–297.

Dagan, N. & O. Volij, 1997, Bilateral Comparisons and Consistent Fair Division Rules in the Context of Bankruptcy Problems, *IJGT* 26, 11–25.

Derks, J. & J. Kuipers, 1997, On the Core of Routing Games, *IJGT* 26, 193–205.

Driessen, T. S. H., 1991, A Survey of Consistency Properties in Cooperative Game Theory, *SIAM Rev* 33, 43–59.

Driessen, T. S. H., 1997, Tree Enterprises and Bankruptcy Ventures: A Game-Theoretic Similarity due to a Graph-Theoretic Proof, *Discrete Appl Math* 79, 105–117.

Dutta, B., 1990, The Egalitarian Solution and Reduced Game Properties in Convex Games, *IJGT* 19, 153–169.

Einy, E., R. Holzman, D. Monderer & B. Shitovitz, 1997, Core Equivalence Theorems for Infinite Convex Games, *JET* 76, 1–12.

Einy, E. & B. Shitovitz, 1996, Convex Games and Stable Sets, *GEB* 16, 192–201.

Farmer, A. & J. Tiefenthaler, 1995, Fairness Concepts and the Intrahousehold Allocation of Resources, *J Dev Econ* 47, 179–189.

Fleurbaey, M., 1994, On Fair Compensation, *Th & Dec* 36, 277–307.

Fleurbaey, M., 1995, Three Solutions for the Compensation Problem, *JET* 65, 505–521.

Franks, J. R. & W. N. Torous, 1989, An Empirical Investigation of United States Firms in Reorganization, *J Finance* 44, 747–769.

Granot, D. & F. Granot, 1992, On Some Network Flow Games, *MOR* 17, 792–841.

Hamers, H., 1997, On the Concavity of Delivery Games, *Eur J OR* 99, 445–458.

Hart, S. & A. Mas-Colell, 1989, Potential, Value, and Consistency, *Econometrica* 57, 589–614.

Honda, T., 1997, An Optimum Yield and Allocation Policy of the Yield for Scallop Aquaculture at Oshima Substation in Hokkaido, *Nippon Suisan Gakkaishi* 63, 939–946.

Jackson, M. & H. Moulin, 1992, Implementing a Public Project and Distributing its Cost, *JET* 57, 125–140.

Kanda, H. & S. Levmore, 1994, Explaining Creditor Priorities, *Virg Law Rev* 80, 2103–2154.

Kennet, D. M. & D. J. Gabel, 1997, Fury Distributed Cost Pricing, Ramsey Pricing, and Shapley Value Pricing: A Simulated Welfare Analysis for the Telephone Exchange, *Rev Ind Organization* 12, 485–499.

Lensberg, T., 1987, Stability and Collective Rationality, *Econometrica* 55, 935–961.

Lensberg, T., 1988, Stability and the Nash Solution, *JET* 45, 330–341.

Maschler, M., 1990, Consistency, in *Ohio Conf*, 183–186.

Maschler, M., 1992, The Bargaining Set, Kernel, and Nucleolus, in *Handb Game Th* 1, 591–667.

Maschler, M. & G. Owen, 1989, The Consistent Shapley Value for Hyperplane Games, *IJGT* 18, 389–407.

Moldovanu, B., 1990, Stable Bargaining Equilibria for Assignment Games without Side Payments, *IJGT* 19, 171–190.

Moulin, H., 1987, The Pure Compensation Problem: Egalitarianism versus Laissez-Fairism, *QJE* 102, 769–783.

Moulin, H., 1987, Equal or Proportional Division of a Surplus, and Other Methods, *IJGT* 16, 161–186.

Moulin, H., 1990, Monotonic Surplus Sharing and the Utilization of Common Property Resources, in *Ohio Conf*, 282–299.

Moulin, H., 1992, All Sorry to Disagree: A General Principle for the Provision of Nonrival Goods, *Scand J Econ* 94, 37–51.

Otten, G. J., H. Peters & O. Volij, 1996, Two Characterizations of the Uniform Rule for Division Problems with Single-Peaked Preferences, *Econ Th* 7, 291–306.

Peleg, B., 1985, An Axiomatization of the Core of Cooperative Games without Side Payments, *JME* 14, 203–214.

Peleg, B., 1986, On the Reduced Game Property and its Converse, *IJGT* 15, 187–200.

Peleg, B., 1989, An Axiomatization of the Core of Market Games, *MOR* 14, 448–456.

Peleg, B., 1992, Axiomatizations of the Core, in *Handb Game Th* 1, 397–412.

Potters, J., R. Poos, S. Tijs & S. Muto, 1989, Clan Games, *GEB* 1, 275–293.

Schokkaert, E. & B. Overlaet, 1989, Moral Intuitions and Economic Models of Distributive Justice, *Soc Choice Welf* 6, 19–31.

Serrano, R., 1995, Strategic Bargaining, Surplus Sharing Problems and the Nucleolus, *JME* 24, 319–329.

Sertel, M. R., 1992, The Nash Bargaining Solution Manipulated by Pre-Donations is Talmudic, *Econ Let* 40, 45–55.

Tadenuma, K. & W. Thomson, 1991, No-Envy and Consistency in Economies with Indivisible Goods, *Econometrica* 59, 1755-1767.

Thomson, W., 1990, The Consistency Principle, in *Ohio Conf*, 187-215.

Thomson, W., 1994, Consistent Solutions to the Problem of Fair Division when Preferences are Single-Peaked, *JET* 63, 219-245.

Thomson, W., 1994, Resource-Monotonic Solutions to the Problem of Fair Division when Preferences are Single-Peaked, *Soc Choice Welf* 11, 205-223.

Thomson, W. & L. Zhou, 1993, Consistent Solutions in Atomless Economies, *Econometrica* 61, 575-587.

Young, H. P., 1985, Methods and Principles of Cost Allocation, in *Cost Allocation*, 3-29.

Young, H. P., 1987, On Dividing an Amount according to Individual Claims or Liabilities, *MOR* 12, 398-414.

Young, H. P., 1988, Distributive Justice in Taxation, *JET* 44, 321-335.

Young, H. P., 1994, Cost Allocation, in *Handb Game Th* 2, 1193-1235.

Texts and Monographs: Driessen; Thomson & Lensberg.

In this collection: Chapter 2.

Chapter 46 Markets with a Continuum of Traders

Allen, B., 1985, Expectations Equilibria with Dispersed Forecasts, *J Math Anal Appl* 109, 279-301.

Allen, B. & S. Sorin, 1994, General Equilibrium and Cooperative Games: Basic Results, in *Stony Brook ASI* 1991, 17-33.

Alo, R. A., A. Dekorvin & L. Kunes, 1980, Maps whose Values are Closed Convex Subsets of a Banach Space, *J Math Anal Appl* 76, 1-9.

Alo, R. A., A. Dekorvin & C. E. Roberts, 1979, An Optimal Sampling Theorem for Convex Set Valued Martingales, *J Reine & Ang Math* 310, 1-6.

Alo, R. A., A. Dekorvin & C. E. Roberts, 1980, On Some Properties of Continuous Multimeasures, *J Math Anal Appl* 75, 402-410.

Anderson, R. M., 1986, Notions of Core Convergence, in *Debreu Festschr*, 25-46.

Anderson, R. M., 1988, The Second Welfare Theorem with Nonconvex Preferences, *Econometrica* 56, 361-382.

Anderson, R. M., 1991, Non-Standard Analysis with Applications to Economics, in *Handb Math Econ*, 2145-2208.

Anderson, R. M., 1992, The Core in Perfectly Competitive Economies, in *Handb Game Th* 1, 413-457.

Anderson, R. M., 1994, Core Convergence in Perfectly Competitive Economies, in *Stony Brook ASI* 1991, 35-45.

Anderson, R. M. & W. R. Zame, 1997, Edgeworth's Conjecture with Infinitely Many Commodities: L^1, *Econometrica* 65, 225-273.

Anderson, R. M. & W. R. Zame, 1998, Edgeworth's Conjecture with Infinitely Many Commodities: Commodity Differentiation, *Econ Th* 11, 331-377.

Araujo, A. & A. Mas-Colell, 1978, Notes on the Smoothing of Aggregate Demand, *JME* 5, 113-127.

Arkin, V. I. & V. L. Levin, 1974, Variational Problems for Functions of Several Variables and a Resource Distribution Model, in *Mitiagin*, 7-34.

Armstrong, T. E., 1985, Remarks related to Finitely Additive Exchange Economies, in *Lect Notes* 244, 188-204.

Armstrong, T. E. & M. K. Richter, 1984, The Core-Walras Equivalence, *JET* 33, 116-151.

Armstrong, T. E. & M. K. Richter, 1986, Existence of Nonatomic Core-Walras Allocations, *JET* 38, 137–159.

Arrow, K. J., 1970, The Organization of Economic Activity: Issues pertinent to the Choice of a Market versus Nonmarket Allocation, in *Haveman & Margolis*, 59–73.

Arrow, K. J. & M. D. Intriligator, 1981, Historical Introduction, in *Handb Math Econ* 1, 1–14.

Arrow, K. J. & T. J. Kehoe, 1994, Distinguished Fellow: Scarf, Herbert, Contributions to Economics, *JEP* 8, 161–181.

Aubin, J. P., 1974, Core and Undesignated Game Equilibria without Side Payments, *CR Acad Sci* 279, 963–966.

Aubin, J. P., 1981, Cooperative Fuzzy Games, *MOR* 6, 1–13.

Aubin, J. P., 1981, Locally Lipschitz Cooperative Games, *JME* 8, 241–262.

Aumann, R. J., 1998, On the State of the Art in Game Theory, *GEB* 24, 181–210.

Balder, E. J. & A. Rustichini, 1994, An Equilibrium Result for Games with Private Information and Infinitely Many Players, *JET* 62, 385–393.

Basile, A., 1993, Finitely Additive Nonatomic Coalition Production Economies: The Core-Walras Equivalence, *IER* 34, 983–994.

Basu, K., 1994, Group Rationality, Utilitarianism, and the Escher Waterfall, *GEB* 7, 1–9.

Benassy, J. P., 1989, Market Size and Substitutability in Imperfect Competition: A Bertrand–Edgeworth–Chamberlin Model, *RES* 56, 217–234.

Berliant, M., Y. Y. Papageorgiou & P. Wang, 1990, On Welfare Theory and Urban Economics, *Reg Sci* 20, 245–261.

Berliant, M. & T. Tenraa, 1991, On the Continuum Approach of Spatial and Some Local Public Goods or Product Differentiation Models: Some Problems, *JET* 55, 95–120.

Bester, H., 1984, Core and Equilibrium in Incomplete Markets, *Z Nat-ok* 44, 255–266.

Bewley, T. F., 1973, Edgeworth's Conjecture, *Econometrica* 41, 425–454.

Boehm, V., 1973, On Cores and Equilibria of Productive Economies with a Measure Space of Consumers: An Example, *JET* 6, 409–412.

Boehm, V., 1974, The Core of an Economy with Production, *RES* 41, 429–436.

Bouzitat, M. J., 1966, Presentation Synthetique de la Theorie des Jeux, in *Toulon Conf*, 3–39.

Brock, W. A., 1972, On Models of Expectations that Arise from Maximizing Behavior of Economic Agents over Time, *JET* 5, 348–376.

Brown, D. J. & A. Robinson, 1972, A Limit Theorem on the Cores of Large Standard Exchange Economies, *PNAS* 69, 1258–1260.

Brown, D. J. & A. Robinson, 1974, Cores of Large Standard Exchange Economies, *JET* 9, 245–254.

Brown, D. J. & A. Robinson, 1975, Nonstandard Exchange Economies, *Econometrica* 43, 41–55.

Burkov, V. N., V. V. Kondratev, V. A. Malchano & A. V. Shchepki, 1977, Models and Mechanisms of Operation of Hierarchical Systems (Review), *Aut Rem Contr* 38, 1667–1687.

Butnariu, D., 1987, Values and Cores of Fuzzy Games with Infinitely Many Players, *IJGT* 16, 43–68.

Candeal, J. C., G. Chichilnisky & E. Induráin, 1997, Topological Aggregation of Preferences: The Case of a Continuum of Agents, *Soc Choice Welf* 14, 333–343.

Chae, S., 1987, Short-Run Core Equivalence in an Overlapping Generations Model, *JET* 43, 170–183.

Champsaur, P., 1976, Symmetry and Continuity Properties of Lindahl Equilibria, *JME* 3, 19–36.

Charreton, R., 1973, Decentralization of Economic Choices Seen through the Method of Linear Programming Resolution by Decomposition, *Rev Fr Aut* 7, 53–76.

Cheng, H. C., 1981, On Dual Regularity and Value Convergence Theorems, *JME* 8, 37–57.

Cheng, H. H. C., 1987, The Coalitional Approach to Core Theory, *JME* 16, 247–258.

Chipman, J. S., 1965, A Survey of the Theory of International Trade II: The Neo-Classical Theory, *Econometrica* 33, 685–760.

Chipman, J. S., 1970, External Economies of Scale and Competitive Equilibrium, *QJE* 84, 347–385.

Codognato, G. & J. Gabszewicz, 1991, Cournot–Walras Equilibria in an Exchange Economy, *Rev Econ* 42, 1013–1026.

Codognato, G. & J.-J. Gabszewicz, 1993, Cournot–Walras Equilibria in Markets with a Continuum of Traders, *Econ Th* 3, 453–464.

Cornwall, R. R., 1969, The Use of Prices to Characterize the Core of an Economy, *JET* 1, 353–373.

Cornwall, R. R., 1970, Convexity and Continuity Properties of Preference Functions, *Z Nat-ok* 30, 35–52.

Debreu, G., 1967, Integration of Correspondences, in *Proc Berkeley Symp* 1965, Vol 2, 351–372.

Debreu, G., 1967, Preference Functions on Measure Spaces of Economic Agents, *Econometrica* 35, 111–122.

Debreu, G., 1972, Smooth Preferences, *Econometrica* 40, 603–615.

Debreu, G., 1975, Four Aspects of the Mathematical Theory of Economic Equilibrium, in *Proc Int Congr Math* 1974, Vol 1, 65–77.

Debreu, G., 1981, Existence of Competitive Equilibrium, in *Handb Math Econ* 2, 697–743.

Debreu, G., 1984, Economic Theory in the Mathematical Mode, *AER* 74, 267–278; also in *Scand J Econ* 86, 393–410.

Debreu, G., 1986, Theoretic Models: Mathematical Form and Economic Content, *Econometrica* 54, 1259–1270.

Debreu, G., 1987, Mathematical Economics, in *The New Palgrave* 3, 399–403.

Debreu, G., 1991, The Mathematization of Economic Theory, *AER* 81, 1–7.

Debreu, G. & D. Schmeidler, 1972, The Radon–Nikodym Derivative of a Correspondence, in *Proc Berkeley Symp* 1970, Vol 2, 41–56.

Defalvard, H., 1998, Value and Contracts in light of Turgot [1769], *Rev Econ* 49, 1573–1599.

Destanne de Bernis, G., 1975, Limits of Reformulations of General Economic Equilibrium, *Rev Econ* 26, 884–930.

Diamond, P., 1989, Fixed Points of Iterates of Multivalued Mappings, *J Math Anal Appl* 143, 252–258.

Dierker, E., 1974, Topological Methods in Walrasian Economics, *Lect Notes* 92.

Dierker, H., 1975, Equilibria and Core of Large Economies, *JME* 2, 155–169.

Dreze, J., J. J. Gabszewicz, D. Schmeidler & K. Vind, 1972, Cores and Prices in an Exchange Economy with an Atomless Sector, *Econometrica* 40, 1091–1108.

Dreze, J. H., S. Gepts & J. J. Gabsewicz, 1969, On Cores and Competitive Equilibria, in *La Decision*, 91–114.

Dubey, P., A. Mas-Colell & M. Shubik, 1980, Efficiency Properties of Strategic Market Games: An Axiomatic Approach, *JET* 22, 339–362.

Dubey, P. & A. Neyman, 1988, Payoffs in Nonatomic Economies: An Axiomatic Approach, *Econometrica* 52, 1129–1150.

Dubey, P. & A. Neyman, 1994, An Axiomatic Approach to the Equivalence Phenomenon, in *Stony Brook ASI* 1991, 137–144.

Dubey, P. & A. Neyman, 1997, An Equivalence Principle for Perfectly Competitive Economies, *JET* 75, 314–344.

Dubey, P. & L. S. Shapley, 1994, Noncooperative General Exchange with a Continuum of Traders: Two Models, *JME* 23, 253–293.

Dubey, P. & M. Shubik, 1977, A Closed Economic System with Production and Exchange Modelled as a Game of Strategy, *JME* 4, 253–287.

Duffie, D. & H. Sonnenschein, 1989, Collected Papers of Arrow, Kenneth J., Vol 2: General Equilibrium, *JEL* 27, 565–598.

Eichberger, J., 1989, A Note on Bankruptcy Rules and Credit Constraints in Temporary Equilibrium, *Econometrica* 57, 707–715.

Einy, E., R. Holzman, D. Monderer & B. Shitovitz, 1996, Core and Stable Sets of Large Games Arising in Economics, *JET* 68, 200–211.

Einy, E., R. Holzman, D. Monderer & B. Shitovitz, 1997, Core Equivalence Theorems for Infinite Convex Games, *JET* 76, 1–12.

Einy, E. & B. Shitovitz, 1995, The Optimistic Stability of the Core Mapping in Public Goods Production Economies, *Econ Th* 6, 523–528.

Einy, E. & B. Shitovitz, 1997, Stability of the Core Mapping in Games with a Countable Set of Players, *IJGT* 26, 45–50.

Einy, E. & D. Wettstein, 1996, Equivalence between Bargaining Sets and the Core in Simple Games, *IJGT* 25, 65–71.

Ellickson, B., 1979, Competitive Equilibrium with Local Public Goods, *JET* 21, 46–61.

Emmons, D. W., 1984, Existence of Lindahl Equilibria in Measure Theoretic Economies without Ordered Preferences, *JET* 34, 342–359.

Emmons, D. W. & A. J. Scafuri, 1985, Value Allocations: An Exposition, in *Lect Notes* 244, 55–78.

Emmons, D. W. & N. C. Yannelis, 1985, On Perfectly Competitive Economies: Loeb Economies, in *Lect Notes* 244, 145–172.

Evans, A. W., 1972, The Pure Theory of City Size in an Industrial Economy, *Urban Studies* 9, 49–77.

Farrell, M. J., 1970, Edgeworth Bounds for Oligopoly Prices, *Economica* 37, 341–361.

Fisher, M. R., 1976, The Economic Contribution of Michael James Farrell, *RES* 43, 371–382.

Flåm, S. D., 1995, Learning Competitive Market Balance, *Econ Th* 6, 511–518.

Friedman, D., 1984, On the Efficiency of Experimental Double Auction Markets, *AER* 74, 60–72.

Friedman, J., 1981, Oligopoly Theory, in *Handb Math Econ* 2, 491–534.

Furth, D., 1998, The Core of the Inductive Limit of a Direct System of Economies with a Communication Structure, *JME* 30, 433–472.

Gabszewicz, J. J., 1975, Coalitional Fairness of Allocations in Pure Exchange Economies, *Econometrica* 43, 661–668.

Gabszewicz, J. J., 1977, Asymmetric Duopoly and the Core, *JET* 14, 172–179.

Gabszewicz, J. J., 1985, Imperfect Competition in General Equilibrium: An Overview of Recent Work (Comment), in *Frontiers of Economics*, 150–169.

Gabszewicz, J. J. & J. F. Mertens, 1971, An Equivalence Theorem for the Core of an Economy whose Atoms are not Too Big, *Econometrica* 39, 713–721.

Gabszewicz, J. J. & B. Shitovitz, 1992, The Core in Imperfectly Competitive Economies, in *Handb Game Th* 1, 459–483.

García-Cutrín, J. & C. Hervés-Beloso, 1993, A Discrete Approach to Continuum Economies, *Econ Th* 3, 577–583.

Garfield, E., 1985, The 1983 Nobel Prizes 3: The Economics and Literature Awards Go to Debreu, Gerard and Golding, William, *Current Contents* 8, 3–11.

Geanakoplos, J., 1987, Arrow–Debreu Model of General Equilibrium, in *The New Palgrave* 1, 116–124.

Georgescu, N., 1974, Dynamic Models and Economic Growth, *Économie Appl* 27, 529–563.

Georgescu, N., 1979, Methods in Economic Science, *J Econ Issues* 13, 317–328.

Glazyrina, I., 1997, Edgeworth's Conjecture in Atomless Economies with a Nonseparable Commodity Space, *JME* 27, 79–90.

Golan, A., 1991, The Discrete-Continuous Choice of Economic Modeling, or Quantum Economic Chaos, *Math Soc Sci* 21, 261–286.

Gowdy, J. M., 1979, Economic Growth Models and Regional Steady-State Planning, *Growth and Change* 10, 37–42.

Grandmont, J. M., G. Laroque & Y. Younes, 1978, Equilibrium with Quantity Rationing and Recontracting, *JET* 19, 84–102.

Green, E. J., 1984, Continuum and Finite-Player Noncooperative Models of Competition, *Econometrica* 52, 975–993.

Greenberg, J., 1975, Efficiency of Tax Systems Financing Public Goods in General Equilibrium Analysis, *JET* 11, 168–195.

Greenberg, J. & B. Shitovitz, 1986, A Simple Proof of the Equivalence Theorem for Oligopolistic Mixed Markets, *JME* 15, 79–83.

Greenberg, J. & B. Shitovitz, 1994, The Optimistic Stability of the Core of Mixed Markets, *JME* 23, 379–386.

Gretsky, N. E. & J. M. Ostroy, 1985, Thick and Thin Market Nonatomic Exchange Economies, in *Lect Notes* 244, 107–130.

Gretsky, N. E., J. M. Ostroy & W. R. Zame, 1992, The Nonatomic Assignment Model, *Econ Th* 2, 103–127.

Guesnerie, R., 1975, Pareto Optimality in Non-Convex Economies, *Econometrica* 43, 1–29.

Guesnerie, R., 1989, First-Best Allocations of Resources with Nonconvexities in Production, in *CORE 20th Ann*, 99–143.

Haller, H., 1990, Large Random Graphs in Pseudo-Metric Spaces, *Math Soc Sci* 20, 147–164.

Haller, H., 1991, Corporate Production and Shareholder Cooperation under Uncertainty, *IER* 32, 823–842.

Hammond, P. J., 1987, Markets as Constraints: Multilateral Incentive Compatibility in Continuum Economies, *RES* 54, 399–412.

Hammond, P. J., 1987, On Reconciling Arrow's Theory of Social Choice with Harsanyi's Fundamental Utilitarianism, in *Arrow and Econ Policy*, 179–221.

Hammond, P. J., M. Kaneko & M. H. Wooders, 1989, Continuum Economies with Finite Coalitions: Core, Equilibria, and Widespread Externalities, *JET* 49, 113–134.

Hart, S., 1974, Formation of Cartels in Large Markets, *JET* 7, 453–466.

Hart, S., 1977, Values of Non-Differentiable Markets with a Continuum of Traders, *JME* 4, 103–116.

Hart, S., 1979, Values of Large Market Games, in *Vienna Conf*, 187–197.

Hart, S., W. Hildenbrand & E. Kohlberg, 1974, On Equilibria: Correlations as Distributions on the Commodity Space, *JME* 1, 159–166.

Hart, S. & A. Mas-Colell, 1996, Harsanyi Values of Large Economies: Nonequivalence to Competitive Equilibria, *GEB* 13, 74–99.

Hausman, D., 1983, Are there Causal Relations among Dependent Variables, *Phil Sci* 50, 58–81.

Helpman, E. & J. J. Laffont, 1975, Moral Hazard in General Equilibrium Theory, *JET* 10, 8–23.

Hildenbrand, K., 1972, Continuity of the Equilibrium-Set Correspondence, *JET* 5, 152.

Hildenbrand, W., 1968, On the Core of an Economy with a Measure Space of Economic Agents, *RES* 35, 443–452.

Hildenbrand, W., 1970, On Economies with Many Agents, *JET* 2, 161–188.

Hildenbrand, W., 1970, Existence of Equilibria for Economies with Production and a Measure Space of Consumers, *Econometrica* 38, 608–623.

Hildenbrand, W., 1972, Metric Measure Spaces of Economic Agents, in *Proc Berkeley Symp* 1970, Vol 2, 81–95.

Hildenbrand, W., 1975, Distributions of Agents' Characteristics, *JME* 2, 129–138.

Hildenbrand, W., 1976, A Mathematical Model of Economic Equilibrium, *Z Angewandte Mathematik und Mechanik* 56, 16–21.

Hildenbrand, W., 1981, The Core of an Economy, in *Handb Math Econ* 2, 831–877.

Hildenbrand, W., 1987, Cores, in *The New Palgrave* 1, 666–670.

Hildenbrand, W., 1993, Edgeworth, Francis Ysidro: Perfect Competition and the Core, *Eur Econ Rev* 37, 477–490.

Hildenbrand, W. & A. P. Kirman, 1973, Size Removes Inequity, *RES* 40, 305–319.

Hildenbrand, W. & J. F. Mertens, 1972, Upper Hemi-Continuity of the Equilibrium-Set Correspondence for Pure Exchange Economies, *Econometrica* 40, 99–108.

Honkapohja, S., 1977, Money and Core in a Sequence Economy with Transaction Costs, *Eur Econ Rev* 10, 241–251.

Housman, D., 1988, Infinite-Player Noncooperative Games with Incomplete Information, *MOR* 13, 488–496.

Husseinov, F., 1994, Interpretation of Aubin Fuzzy Coalitions and their Extension: Relaxation of Finite Exchange Economies, *JME* 23, 499–516.

Ichiishi, T. & S. Weber, 1978, Some Theorems on the Core of a Non-Sidepayment Game with a Measure Space of Players, *IJGT* 7, 95–112.

Imai, H., 1983, Voting, Bargaining, and Factor Income Distribution, *JME* 11, 211–233.

Jaynes, G., M. Okuno & D. Schmeidler, 1978, Efficiency in an Atomless Economy with Fiat Money, *IER* 19, 149–156.

Kaneko, M., 1978, Price Oligopoly as a Cooperative Game, *IJGT* 7, 137–150.

Kaneko, M., 1981, The Nash Social Welfare Function for a Measure Space of Individuals, *JME* 8, 173–200.

Kaneko, M., 1982, The Optimal Progressive Income Tax: Existence and the Limit Tax Rates, *Math Soc Sci* 3, 193–222.

Kaneko, M. & M. H. Wooders, 1986, The Core of a Game with a Continuum of Players and Finite Coalitions: The Model and Some Results, *Math Soc Sci* 12, 105–137.

Kaneko, M. & M. H. Wooders, 1989, The Core of a Continuum Economy with Widespread Externalities and Finite Coalitions: From Finite to Continuum Economies, *JET* 49, 135–168.

Kannai, Y., 1969, Countably Additive Measures in Cores of Games, *J Math Anal Appl* 27, 227–240.

Kannai, Y., 1970, Continuity Properties of the Core of a Market, *Econometrica* 38, 791–815.

Keisler, H. J., 1992, A Law of Numbers for Fast Price Adjustment, *Trans AMS* 332, 1–51.

Keisler, H. J., 1996, Getting to a Competitive Equilibrium, *Econometrica* 64, 29–49.

Khan, M. A., 1976, Oligopoly in Markets with a Continuum of Traders: An Asymptotic Interpretation, *JET* 12, 237–293.

Khan, M. A., 1985, On Extensions of the Cournot–Nash Theorem, in *Lect Notes* 244, 79–106.

Khan, M. A., 1987, Perfect Competition, in *The New Palgrave* 3, 831–834.

Khan, M. A. & A. Yamazaki, 1981, On the Cores of Economies with Indivisible Commodities and a Continuum of Traders, *JET* 24, 218–225.

Kim, S. H., 1997, Continuous Nash Equilibria, *JME* 28, 69–84.

Kirman, A. P., 1981, Measure Theory with Applications to Economics, in *Handb Math Econ* 1, 159–209.

Kirman, A. P., 1987, Measure Theory, in *The New Palgrave* 3, 434–436.

Kirman, A. P. & D. Sondermann, 1972, Arrow's Theorem, Many Agents, and Invisible Dictators, *JET* 5, 267–277.

Klein, C., 1979, Atomless Economies with Countably Many Agents, in *Hagen Conf I*, 337–342.

Knoer, E. M., 1980, Economies as Distributions: Implications for Aggregation and Stability, *JET* 22, 439–450.

Kobayashi, T., 1980, Equilibrium Contracts for Syndicates with Differential Information, *Econometrica* 48, 1635–1665.

Koopmans, T. C., 1974, Is the Theory of Competitive Equilibrium With It?, *AER* 64, 325–329.

Laffont, J. J., 1975, Macroeconomic Constraints, Economic Efficiency and Ethics: Introduction to Kantian Economics, *Economica* 42, 430–437.

Landa, J., 1976, An Exchange Economy with Legally Binding Contracts: A Public Choice Approach, *J Econ Issues* 10, 905–922.

Lauwers, L., 1997, Topological Aggregation: The Case of an Infinite Population, *Soc Choice Welf* 14, 319–332.

Lauwers, L., 1998, Intertemporal Objective Functions: Strong Pareto versus Anonymity, *Math Soc Sci* 35, 37–55.

Lengwiler, Y., 1998, Endogenous Endowments and Equilibrium Starvation in a Walrasian Economy, *JME* 30, 37–58.

Leroy, S. F., 1980, Entry and Equilibrium under Adjustment Costs, *JET* 23, 348–360.

Levy, A. & L. S. Shapley, 1997, Individual and Collective Wage Bargaining, *IER* 38, 969–991.

Lewis, A. A., 1985, Loeb-Measurable Solutions to Star-Finite Games, *Math Soc Sci* 9, 197–247.

Lewis, A. A., 1990, On the Independence of Core Equivalence Results from Zermelo-Fraenkel Set Theory, *Math Soc Sci* 19, 55–95.

Lewis, A. A., 1990, A Game Theoretic Equivalence to the Hahn–Banach Theorem, *Math Soc Sci* 20, 199–214.

Lewis, A. A., 1991, On the Effective Content of Asymptotic Verifications of Edgeworth's Conjecture, *Math Soc Sci* 22, 275–324.

Loeb, P. & S. Rashid, 1987, Lyapunov's Theorem, in *The New Palgrave* 3, 259–261.

Makarov, V. L., V. M. Marakulin, A. N. Kozyrev & V. A. Vasilev, 1989, Equilibrium, Rationing, and Stability: A Summary of the Proceedings of the 2nd Novosibirsk School of Mathematical Economics, *Matekon* 25, 4–95.

Makowski, L., 1979, Value Theory with Personalized Trading, *JET* 20, 194–212.

Makowski, L. & J. M. Ostroy, 1998, Arbitrage and the Flattening Effect of Large Numbers, *JET* 78, 1–31.

Mandy, D. M., 1992, Nonuniform Bertrand Competition, *Econometrica* 60, 1293–1330.

Manelli, A. M., 1991, Core Convergence without Monotone Preferences and Free Disposal, *JET* 55, 400–415.

Manelli, A. M., 1991, Monotonic Preferences and Core Equivalence, *Econometrica* 59, 123–138.

Mas-Colell, A., 1975, A Model of Equilibrium with Differentiated Commodities, *JME* 2, 263–295.

Mas-Colell, A., 1977, Competitive and Value Allocations of Large Exchange Economies, *JET* 14, 419–438.

Mas-Colell, A., 1978, A Note on the Core Equivalence Theorem: How Many Blocking Allocations Are There?, *JME* 5, 207–215.

Mas-Colell, A., 1979, A Refinement of the Core Equivalence Theorem, *Econ Let* 3, 307–310.

Mas-Colell, A., 1980, The Cournotian Foundations of Walrasian Equilibrium Theory: An Exposition of Recent Theory, in *Proc Econometr Soc Congr* 1980, 183–224.

Mas-Colell, A., 1982, Perfect Competition and the Core, *RES* 49, 15–30.

Mas-Colell, A., 1986, An Introduction to the Differentiable Approach in the Theory of Economic Equilibrium, in *Reiter*, 160–198.

Mas-Colell, A., 1989, An Equivalence Theorem for a Bargaining Set, *JME* 18, 129–139.

Mas-Colell, A. & W. Neuefeind, 1977, Some Generic Properties of Aggregate Excess Demand and an Application, *Econometrica* 45, 591–599.

Mas-Colell, A. & X. Vives, 1993, Implementation in Economies with a Continuum of Agents, *RES* 60, 613–629.

Mauldin, R. D., 1986, Coalitional Convex Preference Orders are Almost Surely Convex, *J Math Anal Appl* 114, 548–551.

McLennan, A. & H. Sonnenschein, 1991, Sequential Bargaining as a Noncooperative Foundation for Walrasian Equilibrium, *Econometrica* 59, 1395–1424.

Milchtaich, I., 1998, Vector Measure Games based on Measures with Values in an Infinite Dimensional Vector Space, *GEB* 24, 25–46.

Moeseke, P. V., 1979, Value Cores for Finite Agents, *Econ Rec* 55, 76–81.

Mougeot, M., 1980, Des théorèmes d'existence aux procédures de planification, *Économies et Sociétés* 14, 1179–1231.

Muench, T. J., 1972, The Core and the Lindahl Equilibrium of an Economy with a Public Good: An Example, *JET* 4, 241–255.

Murnighan, J. K. & A. E. Roth, 1977, Effects of Communication and Information Availability in an Experimental Study of a 3-Person Game, *Manag Sci* 23, 1336–1348.

Murnighan, J. K. & E. Szwajkowski, 1979, Coalition Bargaining in Four Games that Include a Veto Player, *J Pers Soc Psych* 37, 1933–1946.

Nagahisa, R., 1992, Walrasian Social Choice in a Large Economy, *Math Soc Sci* 24, 73–78.

Narens, L., 1992, Cores of Dense Exchange Economies, *Math Soc Sci* 24, 277–292.

Newman, P., 1987, Optimality and Efficiency, in *The New Palgrave* 3, 727–729.

Nicola, P. C., 1970, Stability of General Economic Equilibrium with Distributors, *Rivista Int Scienze Sociali* 78, 433.

O'Brien, D. P., 1992, Economists and Data, *Brit J Ind Relations* 30, 253–285.

Oddou, C., 1976, Existence and Equivalence Theorems for Production Economies, *Econometrica* 44, 265–281.

Oddou, C., 1982, The Core of a Coalition Production Economy, *JME* 9, 1–21.

Okuda, H. & B. Shitovitz, 1985, Core Allocations and the Dimension of the Cone of Efficiency Price Vectors, *JET* 35, 166–171.

Ostroy, J. M., 1980, The No-Surplus Condition as a Characterization of Perfectly Competitive Equilibrium, *JET* 22, 183–207.

Ostroy, J. M., 1981, Differentiability as Convergence to Perfectly Competitive Equilibrium, *JME* 8, 59–73.

Ostroy, J. M., 1984, A Reformulation of the Marginal Productivity Theory of Distribution, *Econometrica* 52, 599–630.

Ostroy, J. M., 1984, On the Existence of Walrasian Equilibrium in Large-Square Economies, *JME* 13, 143–163.

Ostroy, J. M. & W. R. Zame, 1994, Nonatomic Economies and the Boundaries of Perfect Competition, *Econometrica* 62, 593–633.

Otani, Y. & J. Sicilian, 1990, Limit Properties of Equilibrium Allocations of Walrasian Strategic Games, *JET* 51, 295–312.

Paoli, J. M., 1966, L'Equilibre Concurrentiel Retrouve, *Rev Fr RO* 10, 213.

Papageorgiou, G. J., 1977, Fundamental Problems of Theoretical Planning, *Env & Planning A* 9, 1329–1356.

Papageorgiou, N. S., 1986, Efficiency and Optimality in Economies Described by Coalitions, *J Math Anal Appl* 116, 497–512.

Papageorgiou, N. S., 1986, On the Efficiency and Optimality of Allocations II, *SIAM J Contr* 24, 452–479.

Papageorgiou, N. S., 1997, Convex Integral Functionals, *Trans AMS* 349, 1421–1436.

Pascoa, M. R., 1993, Noncooperative Equilibrium and Chamberlinian Monopolistic Competition, *JET* 60, 335–353.

Pascoa, M. R., 1997, Monopolistic Competition and Non-Neighboring Goods, *Econ Th* 9, 129–142.

Peck, J. & K. Shell, 1990, Liquid Markets and Competition, *GEB* 2, 362–377.

Peleg, B., 1963, Quota Games with a Continuum of Players, *IJM* 1, 48–53.

Peleg, B., 1966, The Independence of Game Theory of Utility Theory, *Bull AMS* 72, 995–999.

Peleg, B., 1969, Equilibrium Points for Games with Infinitely Many Players, *J London Math Soc* 44, 292–294.

Perry, M. & P. J. Reny, 1994, A Noncooperative View of Coalition Formation and the Core, *Econometrica* 62, 795–817.

Pirrong, S. C., 1992, An Application of Core Theory to the Analysis of Ocean Shipping Markets, *J Law & Econ* 35, 89–131.

Platteau, J. P., 1994, Behind the Market Stage where Real Societies Exist 1: The Role of Public and Private Order Institutions, *J Development Stud* 30, 533–577.

Prescott, E. C. & J. V. Riosrull, 1992, Classical Competitive Analysis of Economies with Islands, *JET* 57, 73–98.

Qin, C.-Z., 1994, An Inner Core Equivalence Theorem, *Econ Th* 4, 311–317.

Rader, T., 1979, Nice Demand Functions II, *JME* 6, 253–262.

Rader, T., 1987, Production as Indirect Exchange, in *The New Palgrave* 3, 1000–1002.

Rajan, A. V., 1997, A Remark on the Equilibrium Set of Pure Exchange Economies, *Econ Th* 10, 373 379.

Rajan, A. V., 1997, Generic Properties of the Core and Equilibria of Pure Exchange Economies, *JME* 27, 471–486.

Rashid, S., 1978, Existence of Equilibrium in Infinite Economies with Production, *Econometrica* 46, 1155–1164.

Reif, N. & H. Wiesmeth, 1978, Pareto Ordering of Distributions, *JME* 5, 185–204.

Richter, M. K., 1971, Coalitions, Core, and Competition, *JET* 3, 323–334.

Roberts, D. J., 1973, Existence of Lindahl Equilibrium with a Measure Space of Consumers, *JET* 6, 355–381.

Roberts, J., 1987, Large Economies, in *The New Palgrave* 3, 132–133.

Roberts, J., 1987, Perfectly and Imperfectly Competitive Markets, in *The New Palgrave* 3, 837–841.

Rogerson, R., 1988, Indivisible Labor, Lotteries and Equilibrium, *J Monetary Econ* 21, 3–16.

Rosenmüller, J., 1977, Remark on the Transfer Operator and the Value-Equilibrium Equivalence Hypothesis, in *Lect Notes* 141, 108–127.

Rosenmüller, J., 1997, Finite Convergence of the Core in a Piecewise Linear Market Game, in *Selten Festschr*, 286–304.

Roth, A. E., 1988, An Introduction to the Shapley Value, in *Roth: Shapley Value*, 1–27.

Roumasset, J., 1979, Sharecropping, Production Externalities, and the Theory of Contracts, *Am J Agr Econ* 61, 640–647.

Rust, J., 1985, Stationary Equilibrium in a Market for Durable Assets, *Econometrica* 53, 783–805.

Rustichini, A. & N. C. Yannelis, 1991, Edgeworth's Conjecture in Economies with a Continuum of Agents and Commodities, *JME* 20, 307–326.

Sancho, F., 1986, The Core of a Large Economy with Personal Risks, *IER* 27, 407–414.

Sattinger, M., 1984, The Value of an Additional Firm in Monopolistic Competition, *RES* 51, 321–332.

Schjodt, U. & B. Sloth, 1994, Bargaining Sets with Small Coalitions, *IJGT* 23, 49–55.

Schmeidler, D., 1969, Competitive Equilibria in Markets with a Continuum of Traders and Incomplete Preferences, *Econometrica* 37, 578–585.

Schmeidler, D., 1972, A Remark on the Core of an Atomless Economy, *Econometrica* 40, 579–580.

Schotter, A., 1973, Core Allocations and Competitive Equilibrium: A Survey, *Z Nat-ok* 33, 281–313.

Schotter, A. & G. Schwodiauer, 1980, Economics and the Theory of Games: A Survey, *JEL* 18, 479–527.

Scotchmer, S. & J. F. Thisse, 1992, Space and Competition: A Puzzle, *Ann Reg Sci* 26, 269–286.

Scotchmer, S. & J. F. Thisse, 1992, The Implications of Space for Competition, *Rev Econ* 44, 653–669.

Segal, L. A., 1991, The Infinite and the Infinitesimal in Models for Natural Phenomena, *Rev Modern Physics* 63, 225–238.

Selten, R. & M. H. Wooders, 1991, A Game Equilibrium Model of Thin Markets, in *Selten* 3, 242–282.

Shaikh, A., 1982, Neo-Ricardian Economics: A Wealth of Algebra, a Poverty of Theory, *Rev Radical Political Econ* 14, 67–83.

Shapley, L. S. & M. Shubik, 1966, Quasi-Cores in a Monetary Economy with Nonconvex Preferences, *Econometrica* 34, 805–827.

Sharkey, W. W., 1989, Game Theoretic Modeling of Increasing Returns to Scale, *GEB* 1, 370–431.

Shell, K. & R. Wright, 1993, Indivisibilities, Lotteries, and Sunspot Equilibria, *Econ Th* 3, 1–17.

Shibusawa, H., 1997, General Equilibrium vs. Optimum, and the Allocation of Land for Transportation in a Closed Information-Oriented City with Traffic Congestion, *Pap Reg Sci* 76, 321–342.

Shitovitz, B., 1973, Oligopoly in Markets with a Continuum of Traders, *Econometrica* 41, 467–501.

Shitovitz, B., 1974, On Some Problems Arising in Markets with Some Large Traders and a Continuum of Small Traders, *JET* 8, 458–470.

Shitovitz, B., 1982, On Exploitation in a Class of Differentiable Mixed Markets, *Econ Let* 9, 301–304.

Shitovitz, B., 1982, Some Notes on the Core of a Production Economy with Some Large Traders and a Continuum of Small Traders, *JME* 9, 99–105.

Shitovitz, B., 1983, The Proportion of Blocking Coalitions in Atomless Economies, *JME* 12, 247–255.

Shitovitz, B., 1983, A Note on the Equivalence between Two Cones Generated by a Correspondence, *Econ Let* 12, 295–297.

Shitovitz, B., 1989, The Bargaining Set and the Core in Mixed Markets with Atoms and an Atomless Sector, *JME* 18, 377–383.

Shitovitz, B., 1992, Coalitionally Fair Allocations in Smooth Mixed Markets with an Atomless Sector, *Math Soc Sci* 25, 27–40.

Shitovitz, B., 1994, An Equivalence Theorem for the Core of a Duopolistic Market Game with Concave Utilities, *JME* 23, 101–106.

Shitovitz, B., 1997, A Comparison between the Core and the Monopoly Solutions in a Mixed Exchange Economy, *Econ Th* 10, 559–563.

Shitovitz, B. & M. Spiegel, 1998, Cournot–Nash and Lindahl Equilibria in Pure Public Good Economies, *JET* 83, 1–18.

Shubik, M., 1975, The General Equilibrium Model is Incomplete and not Adequate for the Reconciliation of Micro and Macroeconomic Theory, *Kyklos* 28, 545–573.

Shubik, M., 1975, The Role of Numbers and Information in Competition, *Rev Econ* 26, 605–621.

Shubik, M., 1976, Theory of Money and Financial Institutions 27: Beyond General Equilibrium, *Économie Appl* 29, 319–337.

Shubik, M., 1981, Game Theory Models and Methods in Political Economy, in *Handb Math Econ* 1, 285–330.

Shubik, M., 1987, Cournot, Antoine Augustin, in *The New Palgrave* 1, 708–712.

Shubik, M., 1990, Strategic Market Game Models of Exchange Economies, in *Ohio Conf*, 252–272.

Shubik, M., 1997, On the Trail of a White Whale: The Rationalizations of a Mathematical Institutional Economist, in *Heertje III*, 96–121.

Shubik, M. & J. G. Zhao, 1991, A Strategic Market Game of a Finite Exchange Economy with a Mutual Bank, *Math Soc Sci* 22, 257–274.

Smith, V. L., 1974, Optimal Costly Firm Entry in General Equilibrium, *JET* 9, 397–417.

Sondermann, D., 1974, Economies of Scale and Equilibria in Coalition Production Economies, *JET* 8, 259–291.

Sondermann, D., 1975, Smoothing Demand by Aggregation, *JME* 2, 201–223.

Sondermann, D., 1980, Uniqueness of Mean Maximizers and Continuity of Aggregate Demand, *JME* 7, 135–144.

Spivak, A., 1980, Efficient Allocations under a General Transaction Technology, *JET* 22, 465–476.

Sroka, J. J., 1993, The Value of Certain pNA Games through Infinite-Dimensional Banach Spaces, *IJGT* 22, 123–139.

Starr, R. M., 1969, Quasiequilibria in Markets with Non-Convex Preferences, *Econometrica* 37, 25–38.

Sun, Y. N., 1997, Integration of Correspondences on Loeb Spaces, *Trans AMS* 349, 129–153.

Suzuki, T., 1995, Nonconvex Production Economies, *JET* 66, 158–177.

Takekuma, S. I., 1982, Price Formation and Cooperative Behavior of Firms: A Limit Theorem on Competition among Firms, *Hitotsubashi J Econ* 22, 44–61.

Telser, L. G., 1994, The Usefulness of Core Theory in Economics, *JEP* 8, 151–164.

Telser, L. G., 1996, Competition and the Core, *JPE* 104, 85–107.

Thomson, W. & L. Zhou, 1993, Consistent Solutions in Atomless Economies, *Econometrica* 61, 575–587.

Tintner, G. & J. A. Licari, 1970, The Stochastic View of Economics, *Am Economist* 14, 4–10.

Tran, Q. C., 1983, The General Exchange Economy, *Kybernetika* 19, 299–308.

Trockel, W., 1976, A Limit Theorem on the Core, *JME* 3, 247–264.

Trockel, W., 1984, Market Demand, *Lect Notes* 223.

Urai, K., 1994, On the Existence of Equilibria in Economies with Infinitely Many Agents and Commodities: The Direct System of Economies, *JME* 23, 339–359.

Van Damme, E., 1994, Banking: A Survey of Recent Microeconomic Theory, *Oxford Rev Econ Policy* 10, 14–33.

Vanliedekerke, L. & L. Lauwers, 1997, Sacrificing the Patrol: Utilitarianism, Future Generations and Infinity, *Econ & Phil* 13, 159–174.

Varian, H. R., 1984, Debreu, Gerard: Contributions to Economics, *Scand J Econ* 86, 4–16.

Vasilev, V. A., 1997, An Equivalence Theorem for the Core and Coordinated Allocations in Mixed Economies, *Doklady SSSR* 352, 446–450.

Vega Redondo, F., 1997, The Evolution of Walrasian Behavior, *Econometrica* 65, 375–384.

Vickers, J., 1995, Concepts of Competition, *Ox Econ Papers* 47, 1–23.

Vind, K., 1964, Edgeworth-Allocations in an Exchange Economy with Many Traders, *IER* 5, 165–177.

Vind, K., 1995, Perfect Competition or the Core, *Eur Econ Rev* 39, 1733–1745.

Vives, X., 1988, Aggregation of Information in Large Cournot Markets, *Econometrica* 56, 851–876.

Vives, X., 1993, Information, Flexibility, and Competition, *J Japanese & Int Econ* 7, 219–237.

Wako, J., 1991, Strong Core and Competitive Equilibria of an Exchange Market with Indivisible Goods, *IER* 32, 843–852.

Walker, D. A., 1973, Edgeworth's Theory of Recontract, *Econ J* 83, 138–149.

Weintraub, E. R., 1977, Micro-Foundations of Macroeconomics: A Critical Survey, *JEL* 15, 1–23.

Weiss, E. A., 1981, Finitely Additive Exchange Economies, *JME* 8, 221–240.

Welty, G. A., 1971, Mill's Principle of Government as a Basis of Democracy, *Monist* 55, 51.

White, L. H., 1990, Restoring an Altered Menger, *Hist Pol Econ* 22, 349–358.

Wiesmeth, H., 1994, The Charges Approach to Environmental Policy in an Economy with Homogeneous Jurisdictions, *Ann OR* 54, 79–96.

Wilson, R. B., 1987, Exchange, in *The New Palgrave* 2, 202–207.

Winter, E. & M. H. Wooders, 1994, An Axiomatization of the Core for Finite and Continuum Games, *Soc Choice Welf* 11, 165–175.

Wooders, M. H., 1989, A Tiebout Theorem, *Math Soc Sci* 18, 33–55.

Wooders, M. H., 1992, Inessentiality of Large Groups and the Approximate Core Property: An Equivalence Theorem, *Econ Th* 2, 129–147.

Wooders, M., 1994, Large Games and Economies with Effective Small Groups, in *Stony Brook ASI* 1991, 145–196.

Wooders, M. H. & W. R. Zame, 1988, Values of Large Finite Games, in *Roth: Shapley Value*, 195–206.

Yamazaki, A., 1978, On Pseudo-Competitive Allocations and the Core of a Large Economy, *JME* 5, 217–228.

Yamazaki, A., 1981, Diversified Consumption Characteristics and Conditionally Dispersed Endowment Distributions: Regularizing Effect and Existence of Equilibria, *Econometrica* 49, 639–654.

Yamazaki, A., 1983, Competitive Firm Structures and Equilibria in a Coalition Production Economy, *Hitotsubashi J Econ* 24, 69–94.

Yamazaki, A., 1984, Walras Degrees and the Probability of a Blocking Coalition at Pareto Allocations, *JME* 13, 105–121.

Yannelis, N. C., 1985, Values and Fairness, in *Lect Notes* 244, 205–235.

Yannelis, N. C., 1987, Equilibria in Noncooperative Models of Competition, *JET* 41, 96–111.

Zhou, J. X., 1995, On the Existence of Equilibrium for Abstract Economies, *J Math Anal Appl* 193, 839–858.

Zhou, L., 1992, Strictly Fair Allocations in Large Exchange Economies, *JET* 57, 158–175.

Texts and Monographs: Aubin 1979, 1997; Arrow & Hahn; Aumann & Shapley; Balasko; Charreton; Ellickson; Greenberg; Hildenbrand; Hildenbrand & Kirman; Intriligator Ch. 6, 10; Klein; Klein & Thompson; Mas-Colell; Mas-Colell, Whinston & Green Ch. 18; Myerson; Nicholson Ch. 8; Nikaido; Owen 1968 Ch. 10, 1982 Ch. 12, 1995 Ch. 14; Rasmusen; Roth & Sotomayor; Shubik 1982, 1984; Szép & Forgó; Takayama; Weintraub; Worobjow.

In this collection: Chapters 1, 2, 15, 41, 47, 48, 51, 54, and 58.

Chapter 47 Existence of Competitive Equilibria in Markets with a Continuum of Traders

Allen, B., 1985, Expectations Equilibria with Dispersed Forecasts, *J Math Anal Appl* 109, 279–301.

Allen, B. & S. Sorin, 1994, General Equilibrium and Cooperative Games: Basic Results, in *Stony Brook ASI* 1991, 17–33.

Alo, R. A., R. Kleyle & A. Dekorvin, 1987, Emergence of a Dominant Course of Action in a General Feedback Loop when Goal Uncertainty is Present, *J Am Soc Inf Sci* 38, 111–117.

Anderson, R. M., 1978, An Elementary Core Equivalence Theorem, *Econometrica* 46, 1483–1487.

Anderson, R. M., 1982, A Market Value Approach to Approximate Equilibria, *Econometrica* 50, 127–136.

Arkin, V. I. & V. L. Levin, 1974, Variational Problems for Functions of Several Variables and a Resource Distribution Model, in *Mitiagin*, 7–34.

Arrow, K. J., 1968, Economic Equilibrium, in *Int Encycl Soc Sci*, Vol 4, 376–389.

Arrow, K. J., 1970, The Organization of Economic Activity: Issues Pertinent to the Choice of a Market versus Nonmarket Allocation, in *Haveman & Margolis*, 59–73.

Arrow, K. J., 1973, The Theory of Discrimination, in *Ashenfelter & Rees*, 3–33.

Arrow, K. J. & M. D. Intriligator, 1981, Historical Introduction, in *Handb Math Econ* 1, 1–14.

Artstein, Z., 1972, Set-Valued Measures, *Trans AMS* 165, 103–125.

Aubin, J. P., 1981, Cooperative Fuzzy Games, *MOR* 6, 1–13.

Balasko, Y. & K. Shell, 1980, The Overlapping Generations Model I: The Case of Pure Exchange without Money, *JET* 23, 281–306.

Balcer, Y., 1980, Equilibrium in an Exchange Economy with Non-Convex Preferences: A Simple Approach, *JET* 23, 236–242.

Basci, E., S. Ozyildirim & K. Aydogan, 1996, A Note on Price Volume Dynamics in an Emerging Stock Market, *J Banking & Fin* 20, 389–400.

Bewley, T. F., 1973, Edgeworth's Conjecture, *Econometrica* 41, 425–454.

Bidard, C., 1978, Is Equilibrium Profit Worthless?, *Rev Econ* 29, 565–573.

Bresson, Y., 1969, Abandoning Classical Supply and Demand Curves, *Rev Econ* 20, 84–116.

Brown, D. J., 1976, Existence of a Competitive Equilibrium in a Nonstandard Exchange Economy, *Econometrica* 44, 537–546.

Butnariu, D., 1987, Values and Cores of Fuzzy Games with Infinitely Many Players, *IJGT* 16, 43–68.

Codognato, G. & J.-J. Gabszewicz, 1993, Cournot–Walras Equilibria in Markets with a Continuum of Traders, *Econ Th* 3, 453–464.

Debreu, G., 1967, Integration of Correspondences, in *Proc Berkeley Symp* 1965, Vol 2, 351–372.

Debreu, G., 1967, Preference Functions on Measure Spaces of Economic Agents, *Econometrica* 35, 111–122.

Debreu, G., 1972, Smooth Preferences, *Econometrica* 40, 603–615.

Debreu, G., 1981, Existence of Competitive Equilibrium, in *Handb Math Econ* 2, 697–743.

Destanne de Bernis, G., 1975, Limits of Reformulations of General Economic Equilibrium, *Rev Econ* 26, 884–930.

Dierker, E., 1974, Topological Methods in Walrasian Economics, *Lect Notes* 92.

Drenick, R. F., 1981, Large Scale System Theory in the 1980's, *Large Scale Systems/Th Appl* 2, 29–43.

Dreze, J. H., S. Gepts & J. J. Gabszewicz, 1969, On Cores and Competitive Equilibria, in *La Decision*, 91–114.

Dubey, P. & A. Neyman, 1984, Payoffs in Nonatomic Economies: An Axiomatic Approach, *Econometrica* 52, 1129–1150.

Dubey, P. & A. Neyman, 1997, An Equivalence Principle for Perfectly Competitive Economies, *JET* 75, 314–344.

Dubey, P. & L. S. Shapley, 1994, Noncooperative General Exchange with a Continuum of Traders: Two Models, *JME* 23, 253–293.

Dubey, P. & M. Shubik, 1979, Bankruptcy and Optimality in a Closed Trading Mass Economy Modelled as a Non-Cooperative Game, *JME* 6, 115–134.

Duffie, D. & H. Sonnenschein, 1989, Collected Papers of Arrow, Kenneth J., Vol 2: General Equilibrium, by K. J. Arrow, *JEL* 27, 565–598.

Ellis, D. F., 1975, Non-Convexity and Optimal Allocation of Risk in an Exchange Economy, *Am Economist* 19, 10–18.

Emmons, D. W. & A. J. Scafuri, 1985, Value Allocations: An Exposition, in *Lect Notes* 244, 55–78.

Emmons, D. W. & N. C. Yannelis, 1985, On Perfectly Competitive Economies: Loeb Economies, in *Lect Notes* 244, 145–172.

Feiwel, G. R., 1987, The Potentials and Limits of Economic Analysis: The Contributions of Kenneth J. Arrow, in *Arrow and Econ Th*, 1–187.

Flåm, S. D., 1995, Learning Competitive Market Balance, *Econ Th* 6, 511–518.

García-Cutrín, J. & C. Hervés-Beloso, 1993, A Discrete Approach to Continuum Economies, *Econ Th* 3, 577–583.

Gerber, R. I., 1985, Existence and Description of a Housing Market Equilibrium, *Reg Sci* 15, 383–401.

Grandmont, J. M. & Y. Younes, 1972, On the Role of Money and the Existence of a Monetary Equilibrium, *RES* 39, 355–372.

Greenberg, J., B. Shitovitz & A. Wieczorek, 1979, Existence of Equilibria in Atomless Production Economies with Price Dependent Preferences, *JME* 6, 31–41.

Grimaud, A. & J. J. Laffont, 1989, Existence of a Spatial Equilibrium, *J Urban Econ* 25, 213–218.

Hammond, P. J., 1987, On Reconciling Arrow's Theory of Social Choice with Harsanyi's Fundamental Utilitarianism, in *Arrow and Econ Policy*, 179–221.

Hart, S., W. Hildenbrand & E. Kohlberg, 1974, On Equilibrium Allocations as Distributions on the Commodity Space, *JME* 1, 159–166.

Heller, W. P. & R. M. Starr, 1976, Equilibrium with Non-Convex Transactions Costs: Monetary and Non-Monetary Economies, *RES* 43, 195–215.

Hildenbrand, W., 1972, Metric Measure Spaces of Economic Agents, in *Proc Berkeley Symp* 1970, Vol 2, 81–95.

Hildenbrand, K., 1972, Continuity of the Equilibrium Set Correspondence, *JET* 5, 152–162.

Hildenbrand, W., 1970, On Economies with Many Agents, *JET* 2, 161–188.

Hildenbrand, W., 1970, Existence of Equilibria for Economies with Production and a Measure Space of Consumers, *Econometrica* 38, 608–623.

Hildenbrand, W. & J. F. Mertens, 1972, Upper Hemi-Continuity of the Equilibrium-Set Correspondence for Pure Exchange Economies, *Econometrica* 40, 99–108.

Ichiishi, T., 1977, Coalition Structures in a Labor-Managed Market Economy, *Econometrica* 45, 341–360.

Ichiishi, T. & S. Weber, 1978, Some Theorems on the Core of a Non-Sidepayment Game with a Measure Space of Players, *IJGT* 7, 95–112.

Jones, L. E., 1986, Special Problems Arising in the Study of Economies with Infinitely Many Commodities, in *Lect Notes* 264, 184–205.

Kannai, Y., 1969, Countably Additive Measures in Cores of Games, *J Math Anal Appl* 27, 227–240.

Kannai, Y., 1970, Continuity Properties of the Core of a Market, *Econometrica* 38, 791–815.

Katsenelinboigen, A., 1978, Studies in Soviet Economic Planning: Preface, *Int J Pol* 8, 11.

Khan, M. A., 1975, Some Approximate Equilibria, *JME* 2, 63–86.

Khan, M. A., 1985, On Extensions of the Cournot–Nash Theorem, in *Lect Notes* 244, 79–106.

Khan, M. A., 1987, Perfect Competition, in *The New Palgrave* 3, 831–834.

Khan, M. A. & Y. N. Sun, 1997, The Capital Asset Pricing Model and Arbitrage Pricing Theory: A Unification, *PNAS* 94, 4229–4232.

Khan, M. A. & R. Vohra, 1984, Equilibrium in Abstract Economies without Ordered Preferences and with a Measure Space of Agents, *JME* 13, 133–142.

Khan, M. A. & A. Yamazaki, 1981, On the Cores of Economies with Indivisible Commodities and a Continuum of Traders, *JET* 24, 218–225.

Kihlstrom, R. E. & J. J. Laffont, 1979, General Equilibrium: An Entrepreneurial Theory of Firm Formation Based on Risk Aversion, *JPE* 87, 719–748.

Kim, S. H., 1997, Continuous Nash Equilibria, *JME* 28, 69–84.

Kirman, A. P., 1981, Measure Theory with Applications to Economics, in *Handb Math Econ* 1, 159–209.

Kleyle, R. & A. Dekorvin, 1985, A 2-Phase Approach to Making Decisions Involving Goal Uncertainty, *J Inf Sci* 11, 161–171.

Knoer, E. M., 1980, Economies as Distributions: Implications for Aggregation and Stability, *JET* 22, 439–450.

Kolokoltsov, V. N., 1991, Introduction of a New Maslov-Type Currency (Coupons) as a Means of Solution of Market Games under Nonequilibrium Prices, *Doklady SSSR* 320, 1310–1314.

Konishi, H., 1996, Voting with Ballots and Feet: Existence of Equilibrium in a Local Public Goods Economy, *JET* 68, 480–509.

Koopmans, T. C., 1974, Is the Theory of Competitive Equilibrium With It?, *AER* 64, 325–329.

Lantner, R., 1976, Power Phenomena: Analysis of Relational Structures between Industrial Groups, *Économie Appl* 29, 297–317.

Ledyard, J. O., 1977, Incentive Compatible Behavior in Core-Selecting Organizations, *Econometrica* 45, 1607–1621.

Lengwiler, Y., 1998, Endogenous Endowments and Equilibrium Starvation in a Walrasian Economy, *JME* 30, 37–58.

Leroy, S. F., 1980, Entry and Equilibrium under Adjustment Costs, *JET* 23, 348–360.

Levy, A. & L. S. Shapley, 1997, Individual and Collective Wage Bargaining, *IER* 38, 969–991.

Lewis, A. A., 1990, On the Independence of Core Equivalence Results from Zermelo–Fraenkel Set Theory, *Math Soc Sci* 19, 55–95.

Lin, K. P., 1979, Some Generic Properties of Competitive Equilibrium with Externalities and without Convex Preferences, in Liu, P. T. & J. G. Sutinen (Eds.), *Control Theory in Mathematical Economics, Lecture Notes in Pure and Applied Mathematics 47, 3rd Kingston Conference on Differential Games and Control Theory, Univ. Rhode Isl., Kingston, R. I., June 5–8, 1978*, Marcel Dekker, New York, 205–220.

Loeb, P. & S. Rashid, 1987, Lyapunov's Theorem, in *The New Palgrave* 3, 259–261.

Lozada, G. A., 1996, Existence of Equilibria in Exhaustible Resource Industries: Nonconvexities and Discrete vs. Continuous Time, *J Econ Dyn* 20, 433–444.

Mas-Colell, A., 1975, A Model of Equilibrium with Differentiated Commodities, *JME* 2, 263–295.

Mas-Colell, A., 1977, Regular Nonconvex Economies, *Econometrica* 45, 1387–1407.

Mas-Colell, A., 1977, Indivisible Commodities and General Equilibrium Theory, *JET* 16, 443–456.

Nagahisa, R., 1992, Walrasian Social Choice in a Large Economy, *Math Soc Sci* 24, 73–78.

Neuberg, L. G., 1994, Essays on Philosophy and Economic Methodology, by D. M. Hausman, *Kyklos* 47, 464–467.

Noguchi, M., 1997, Economies with a Continuum of Consumers, a Continuum of Suppliers and an Infinite Dimensional Commodity Space, *JME* 27, 1–21.

Noguchi, M., 1997, Economies with a Continuum of Agents with the Commodity-Price Pairing (l_∞, l_1), *JME* 28, 265–287.

Oddou, C., 1976, Existence and Equivalence Theorems for Production Economies, *Econometrica* 44, 265–281.

Oddou, C., 1982, The Core of a Coalition Production Economy, *JME* 9, 1–21.

Okuno, M. & I. Zilcha, 1981, A Proof of the Existence of Competitive Equilibrium in an Overlapping Generations Exchange Economy with Money, *IER* 22, 239–252.

Ostroy, J. M. & W. R. Zame, 1994, Nonatomic Economies and the Boundaries of Perfect Competition, *Econometrica* 62, 593–633.

Podczeck, K., 1997, Markets with Infinitely Many Commodities and a Continuum of Agents with Nonconvex Preferences, *Econ Th* 9, 385–426.

Prescott, E. C. & J. V. Riosrull, 1992, Classical Competitive Analysis of Economies with Islands, *JET* 57, 73–98.

Rader, T., 1979, Nice Demand Functions II, *JME* 6, 253–262.

Rader, T., 1987, Production as Indirect Exchange, in *The New Palgrave* 3, 1000–1002.

Rajan, A. V., 1997, Generic Properties of the Core and Equilibria of Pure Exchange Economies, *JME* 27, 471–486.

Rashid, S., 1978, Existence of Equilibrium in Infinite Economies with Production, *Econometrica* 46, 1155–1164.

Rashid, S., 1979, The Relationship between Measure Theoretic and Non-Standard Exchange Economies, *JME* 6, 195–202.

Rupp, W., 1979, The Riesz Presentation of Additive and Sigma-Additive Set-Valued Measures, *Math Ann* 239, 111–118.

Rupp, W., 1979, The Riesz Presentation of Additive and Sigma-Additive Set-Valued Measures, *Pac J Math* 84, 445–453.

Rustichini, A. & N. C. Yannelis, 1991, Edgeworth's Conjecture in Economies with a Continuum of Agents and Commodities, *JME* 20, 307–326.

Schmeidler, D., 1969, Competitive Equilibria in Markets with a Continuum of Traders and Incomplete Preferences, *Econometrica* 37, 578–585.

Shell, K., 1971, Notes on the Economics of Infinity, *JPE* 79, 1002–1011.

Shell, K. & R. Wright, 1993, Indivisibilities, Lotteries, and Sunspot Equilibria, *Econ Th* 3, 1–17.

Shitovitz, B., 1973, Oligopoly in Markets with a Continuum of Traders, *Econometrica* 41, 467–501.

Shubik, M., 1976, Theory of Money and Financial Institutions 27: Beyond General Equilibrium, *Économie Appl* 29, 319–337.

Starr, R. M., 1969, Quasi-Equilibria in Markets with Non-Convex Preferences, *Econometrica* 37, 25–38.

Suzuki, T., 1995, Nonconvex Production Economies, *JET* 66, 158–177.

Telser, L. G., 1994, The Usefulness of Core Theory in Economics, *JEP* 8, 151–164.

Telser, L. G., 1996, Competition and the Core, *JPE* 104, 85–107.

Trockel, W., 1976, A Limit Theorem on the Core, *JME* 3, 247–264.

Varian, H. R., 1974, Equity, Envy, and Efficiency, *JET* 9, 63–91.

Varian, H. R., 1976, Two Problems in the Theory of Fairness, *J Publ Econ* 5, 249–260.

Wagner, D. H., 1977, A Survey of Measurable Selection Theorems, *SIAM J Contr* 15, 859–903.

Weiss, E. A., 1981, Finitely Additive Exchange Economies, *JME* 8, 221–240.

Wiley, J. S., 1987, Antitrust and Core Theory, *U Chic Law Rev* 54, 556–589.

Yamazaki, A., 1978, An Equilibrium Existence Theorem without Convexity Assumptions, *Econometrica* 46, 541–555.

Yamazaki, A., 1981, Diversified Consumption Characteristics and Conditionally Dispersed Endowment Distributions: Regularizing Effect and Existence of Equilibria, *Econometrica* 49, 639–654.

Yamazaki, A., 1983, Continuous Preference Relations which are Observable in Markets, *Hitotsubashi J Econ* 23, 40–47.

Yannelis, N. C., 1985, Values and Fairness, in *Lect Notes* 244, 205–235.

Yannelis, N. C., 1987, Equilibria in Noncooperative Models of Competition, *JET* 41, 96–111.

Yannelis, N. C., 1990, On the Upper and Lower Semicontinuity of the Aumann Integral, *JME* 19, 373–389.

Zhou, J. X., 1995, On the Existence of Equilibrium for Abstract Economies, *J Math Anal Appl* 193, 839–858.

Texts and Monographs: Aubin 1979, 1997; Arrow & Hahn; Balasko; Charreton; Duffie; Ellickson; Hildenbrand; Hildenbrand & Kirman; Intriligator Ch. 10; Klein; Klein & Thompson; Makarov & Rubinov; Nikaido; Shubik 1982, 1984; Szép & Forgó.

In this collection: Chapters 15, 41, and 68.

Chapter 48 Disadvantageous Monopolies

Bloch, F. & S. Ghosal, 1997, Stable Trading Structures in Bilateral Oligopolies, *JET* 74, 368–384.

Champsaur, P., 1975, Cooperation versus Competition, *JET* 11, 394–417.

Champsaur, P. & G. Laroque, 1976, A Note on the Core of Economies with Atoms or Syndicates, *JET* 13, 458–471.

Charnes, A. & S. C. Littlechild, 1975, Formation of Unions in *n*-Person Games, *JET* 10, 386–402.

d'Aspremont, C., A. Jacquemin, J. J. Gabszewicz & J. A. Weymark, 1983, On the Stability of Collusive Price Leadership, *Can J Econ* 16, 17–25.

Dreyer, J. S. & A. Schotter, 1980, Power Relationships in the International Monetary Fund: The Consequences of Quota Changes, *Rev Econ & Stat* 62, 97–106.

Dreze, J. H., J. J. Gabszewicz & A. Postlewaite, 1977, Disadvantageous Monopolies and Disadvantageous Endowments, *JET* 16, 16–121.

Gabszewicz, J. J., 1985, Imperfect Competition in General Equilibrium: An Overview of Recent Work (Comment), in *Frontiers of Economics*, 150–169.

Gabszewicz, J. J. & B. Shitovitz, 1992, The Core in Imperfectly Competitive Economies, in *Handb Game Th* 1, 459–483.

Gardner, R. J., 1977, Shapley Value and Disadvantageous Monopolies, *JET* 16, 513–517.

Granot, D. & M. Maschler, 1997, The Reactive Bargaining Set: Structure, Dynamics and Extension to NTU Games, *IJGT* 26, 75–95.

Greenberg, J. & B. Shitovitz, 1977, Advantageous Monopolies, *JET* 16, 394–402.

Guesnerie, R., 1977, Monopoly, Syndicate, and Shapley Value: About Some Conjectures, *JET* 15, 235–251.

Hildenbrand, W., 1981, The Core of an Economy, in *Handb Math Econ* 2, 831–877.

Kalai, E., A. Postlewaite & J. Roberts, 1978, Barriers to Trade and Disadvantageous Middlemen: Non-Monotonicity of the Core, *JET* 19, 200–209.

Kats, A., 1974, Non-Cooperative Monopolistic Games and Monopolistic Market Games, *IJGT* 3, 251–260.

Legros, P., 1987, Disadvantageous Syndicates and Stable Cartels: The Case of the Nucleolus, *JET* 42, 30–49.

Maschler, M., 1976, An Advantage of the Bargaining Set over the Core, *JET* 13, 184–192.

Maschler, M., 1992, The Bargaining Set, Kernel, and Nucleolus, in *Handb Game Th* 1, 591–667.

Maskin, E. & D. Newbery, 1990, Disadvantageous Oil Tariffs and Dynamic Consistency, *AER* 80, 143–156.

Morgenstern, O. & G. Schwodiauer, 1976, Competition and Collusion in Bilateral Markets, *Z Nat-ok* 36, 217–245.

Okuno, M., A. Postlewaite & J. Roberts, 1980, Oligopoly and Competition in Large Markets, *AER* 70, 22–29.

Postlewaite, A. & R. W. Rosenthal, 1974, Disadvantageous Syndicates, *JET* 9, 324–326.

Postlewaite, A. & M. Webb, 1984, The Possibility of Recipient-Harming, Donor-Benefiting Transfers with More Than Two Countries, *J Int Econ* 16, 357–364.

Salant, S. W., S. Switzer & R. J. Reynolds, 1983, Losses from Horizontal Mergers: The Effects of an Exogenous Change in Industry Structure on Cournot–Nash Equilibrium, *QJE* 98, 185–199.

Schjodt, U. & B. Sloth, 1994, Bargaining Sets with Small Coalitions, *IJGT* 23, 49–55.

Schotter, A., 1979, Disadvantageous Syndicates in Public Goods Economies, *AER* 69, 927–933.

Schotter, A., 1979, Voting Weights as Power Proxies: Some Theoretical and Empirical Results, in *Vienna Conf*, 58–73.

Schotter, A. & G. Schwodiauer, 1980, Economics and the Theory of Games: A Survey, *JEL* 18, 479–527.

Sherali, H. D. & R. Rajan, 1986, A Game Theoretic Mathematical Programming Analysis of Cooperative Phenomena in Oligopolistic Markets, *OR* 34, 683–697.

Shitovitz, B., 1974, On Some Problems Arising in Markets with Some Large Traders and a Continuum of Small Traders, *JET* 8, 458–470.

Shitovitz, B., 1982, Some Notes on the Core of a Production Economy with Some Large Traders and a Continuum of Small Traders, *JME* 9, 99–105.

Shubik, M., 1981, Game Theory Models and Methods in Political Economy, in *Handb Math Econ* 1, 285–330.

Shubik, M., 1985, The Many Approaches to the Study of Monopolistic Competition, *Eur Econ Rev* 27, 97–114.

Texts and Monographs: Aumann & Shapley; Cornwall; Ordeshook Ch. 8; Shubik 1982, 1984.

In this collection: Chapter 1.

Chapter 49 A Note on Gale's Example

Bhagwati, J. N., 1987, Immiserizing Growth, in *The New Palgrave* 2, 718–720.

Bhagwati, J. N., R. A. Brecher & T. Hatta, 1983, The Generalized Theory of Transfers and Welfare: Bilateral Transfers in a Multilateral World, *AER* 73, 606–618.

Bhagwati, J. N., R. A. Brecher & T. Hatta, 1984, The Paradoxes of Immiserizing Growth and Donor-Enriching Recipient-Immiserizing Transfers: A Tale of Two Literatures, *Weltwirtschaftliches Archiv (Rev World Econ)* 120, 228–243.

Chichilnisky, G., 1983, The Transfer Problem with 3 Agents Once Again: Characterization, Uniqueness and Stability, *J Dev Econ* 13, 237–248.

Chichilnisky, G. & W. Thomson, 1987, The Walrasian Mechanism from Equal Division is not Monotonic with respect to Variations in the Number of Consumers, *J Publ Econ* 32, 119–124.

Chun, Y. S., 1988, The Proportional Solution for Rights Problems, *Math Soc Sci* 15, 231–246.

Demeza, D., 1983, The Transfer Problem in a Many-Country World: Is It Better to Give Than Receive?, *Manchester School Econ Soc Stud* 51, 266–275.

Donsimoni, M. P. & H. M. Polemarchakis, 1989, Intertemporal Equilibrium and Disadvantageous Growth, *Eur Econ Rev* 33, 59–65.

Donsimoni, M. P. & H. M. Polemarchakis, 1994, Redistribution and Welfare, *JME* 23, 235–242.

Dreze, J. H., J. J. Gabszewicz & A. Postlewaite, 1977, Disadvantageous Monopolies and Disadvantageous Endowments, *JET* 16, 16–121.

Falkinger, J., 1990, Innovator-Imitator Trade and the Welfare Effects of Growth, *J Japanese & Int Econ* 4, 157–172.

Foster, J. E., M. K. Majumdar & T. Mitra, 1990, Inequality and Welfare in Market Economies, *J Publ Econ* 41, 351–367.

Galor, O. & H. M. Polemarchakis, 1987, Intertemporal Equilibrium and the Transfer Paradox, *RES* 54, 147–156.

Grinols, E. L., 1987, Transfers and the Generalized Theory of Distortions and Welfare, *Economica* 54, 477–491.

Guesnerie, R. & J. J. Laffont, 1978, Advantageous Reallocations of Initial Resources, *Econometrica* 46, 835–841.

Hands, D. W., 1986, Gross Substitutes and Immiserizing Growth, *Econ Let* 21, 353–356.

Hatta, T., 1984, Immiserizing Growth in a Many-Commodity Setting, *J Int Econ* 17, 335–345.

Jones, R. W., 1984, The Transfer Problem in a Three-Agent Setting, *Can J Econ* 17, 1–14.

Kemp, M. C. & S. Kojima, 1985, Tied Aid and the Paradoxes of Donor Enrichment and Recipient Impoverishment, *IER* 26, 721–729.

Leonard, D. & R. Manning, 1983, Advantageous Reallocations: A Constructive Example, *J Int Econ* 15, 291–295.

Majumdar, M. & T. Mitra, 1985, A Result on the Transfer Problem in International Trade Theory, *J Int Econ* 19, 161–170.

Mantel, R. R., 1984, Substitutability and the Welfare Effects of Endowment Increases, *J Int Econ* 17, 325–334.

Moulin, H., 1985, Egalitarianism and Utilitarianism in Quasi-Linear Bargaining, *Econometrica* 53, 49–67.

Moulin, H., 1990, Fair Division under Joint Ownership: Recent Results and Open Problems, *Soc Choice Welf* 7, 149–170.

Moulin, H. & W. Thomson, 1988, Can Everyone Benefit from Growth? Two Difficulties, *JME* 17, 339–345.

Polemarchakis, H. M., 1979, Equity, Efficiency, and Advantageous Randomness, *QJE* 93, 465–470.

Polemarchakis, H. M., 1983, On the Transfer Paradox, *IER* 24, 749–760.

Polterovich, V. M. & V. A. Spivak, 1980, The Budgetary Paradox in the Model of Economic Equilibrium, *Matekon* 16, 3–22.

Polterovich, V. M. & V. A. Spivak, 1983, Gross Substitutability of Point-to-Set Correspondences, *JME* 11, 117–140.

Rao, M., 1993, To Transfer or to Destroy, *Soc Choice Welf* 10, 177–184.

Ravallion, M., 1984, How Much is a Transfer Payment Worth to a Rural Worker?, *Ox Econ Papers* 36, 478–489.

Safra, Z., 1983, Manipulation by Reallocating Initial Endowments, *JME* 12, 1–17.

Safra, Z., 1987, Strategic Reallocation of Endowments, in *The New Palgrave* 4, 516–518.

Safra, Z., 1990, Connectedness of the Set of Manipulable Equilibria, *Math Soc Sci* 19, 45–53.

Sau, R., 1986, Economic Consequences of a Dominant Coalition, *Econ & Pol Weekly* 21, 1068–1073.

Yano, M., 1991, International Transfers: Strategic Losses and the Blocking of Mutually Advantageous Transfers, *IER* 32, 371–382.

Texts and Monographs: Cornwall.

Chapter 50   On the Rate of Convergence of the Core

Anderson, R. M., 1986, Notions of Core Convergence, in *Debreu Festschr*, 25–46.

Anderson, R. M., 1992, The Core in Perfectly Competitive Economies, in *Handb Game Th* 1, 413–457.

Anderson, R. M. & W. R. Zame, 1998, Edgeworth's Conjecture with Infinitely Many Commodities: Commodity Differentiation, *Econ Th* 11, 331–377.

Cheng, H. C., 1981, What is the Normal Rate of Convergence of the Core?, I, *Econometrica* 49, 73–83.

Cheng, H. C., 1982, Generic Examples on the Rate of Convergence of the Core, *IER* 23, 309–321.

Cheng, H. C., 1983, The Best Rate of Convergence of the Core, *IER* 24, 629–636.

Hildenbrand, W., 1981, The Core of an Economy, in *Handb Math Econ* 2, 831–877.

Tiplitz, C., 1984, Disaggregation in Economic Models, *NRLQ* 31, 213–228.

Texts and Monographs: Mas-Colell; Shubik 1984.

In this collection: Chapter 1.

Chapter 51   Values of Markets with a Continuum of Traders

Allen, B., 1985, Expectations Equilibria with Dispersed Forecasts, *J Math Anal Appl* 109, 279–301.

Beja, A. & I. Gilboa, 1990, Values for Two-Stage Games: Another View of the Shapley Axioms, *IJGT* 19, 17–31.

Brock, H. W., 1979, Game Theoretic Account of Social Justice, *Th & Dec* 11, 293–365.

Brown, D. J. & P. A. Loeb, 1976, Values of Nonstandard Exchange Economies, *IJM* 25, 71–86.

Cheng, H. C., 1981, On Dual Regularity and Value Convergence Theorems, *JME* 8, 37–57.

Dubey, P. & A. Neyman, 1984, Payoffs in Nonatomic Economies: An Axiomatic Approach, *Econometrica* 52, 1129–1150.

Dubey, P. & A. Neyman, 1994, An Axiomatic Approach to the Equivalence Phenomenon, in *Stony Brook ASI* 1991, 137–144.

Dubey, P. & A. Neyman, 1997, An Equivalence Principle for Perfectly Competitive Economies, *JET* 75, 314–344.

Emmons, D. W. & A. J. Scafuri, 1985, Value Allocations: An Exposition, in *Lect Notes* 244, 55–78.

Gardner, R. J., 1977, Shapley Value and Disadvantageous Monopolies, *JET* 16, 513–517.

Gardner, R., 1981, Wealth and Power in a Collegial Polity, *JET* 25, 353–366.

Geanakoplos, J., 1984, Utility Functions for Debreu Excess Demands, *JME* 13, 1–9.

Guesnerie, R., 1977, Monopoly, Syndicate, and Shapley Value: About Some Conjectures, *JET* 15, 235–251.

Hammond, P. J., M. Kaneko & M. H. Wooders, 1989, Continuum Economies with Finite Coalitions: Core, Equilibria, and Widespread Externalities, *JET* 49, 113–134.

Hart, S., 1977, Values of Non-Differentiable Markets with a Continuum of Traders, *JME* 4, 103–116.

Hart, S., 1979, Values of Large Market Games, in *Vienna Conf*, 187–197.

Hart, S., 1980, Measure-Based Values of Market Games, *MOR* 5, 197–228.

Hart, S., 1987, Shapley Value, in *The New Palgrave* 4, 318–320.

Hart, S., 1994, Value Equivalence Theorems: The TU and NTU Cases, in *Stony Brook ASI* 1991, 113–120.

Hart, S. & A. Mas-Colell, 1996, Harsanyi Values of Large Economies: Nonequivalence to Competitive Equilibria, *GEB* 13, 74–99.

Imai, H., 1983, On Harsanyi's Solution, *IJGT* 12, 161–179.

Imai, H., 1983, Voting, Bargaining, and Factor Income Distribution, *JME* 11, 211–233.

Kannai, Y., 1977, Concavifiability and Constructions of Concave Utility Functions, *JME* 4, 1–56.

Kreps, D. M. & E. L. Porteus, 1979, Temporal von Neumann–Morgenstern and Induced Preferences, *JET* 20, 81–109.

Laroque, G., 1978, Fixed Price Equilibrium: Some Results in Local Comparative Statics, *Econometrica* 46, 1127–1154.

Laroque, G., 1981, On the Local Uniqueness of the Fixed Price Equilibria, *RES* 48, 113–129.

Lewis, A. A., 1985, Loeb-Measurable Solutions to Star-Finite Games, *Math Soc Sci* 9, 197–247.

Magill, M. & M. Quinzii, 1992, Real Effects of Money in General Equilibrium, *JME* 21, 301–342.

Makowski, L. & J. M. Ostroy, 1992, The Existence of Perfectly Competitive Equilibrium a la Wicksteed, in *Hahn Festschr*, 370–403.

Mas-Colell, A., 1977, Recoverability of Consumers' Preferences from Market Demand Behavior, *Econometrica* 45, 1409–1430.

Mas-Colell, A., 1977, Competitive and Value Allocations of Large Exchange Economies, *JET* 14, 419–438.

Mas-Colell, A., 1978, A Note on the Core Equivalence Theorem: How Many Blocking Allocations Are There?, *JME* 5, 207–215.

Mas-Colell, A., 1980, Remarks on the Game-Theoretic Analysis of a Simple Distribution of Surplus Problem, *IJGT* 9, 125–140.

Mas-Colell, A., 1987, Cooperative Equilibrium, in *The New Palgrave* 1, 659–662.

Mertens, J. F., 1988, Nondifferentiable TU Markets: The Value, in *Roth: Shapley Value*, 235–264.

Mertens, J. F., 1994, The TU Value: The Non-Differentiable Case, in *Stony Brook ASI* 1991, 81–94.

Moeseke, P. V., 1979, Value Cores for Finite Agents, *Econ Rec* 55, 76–81.

Osborne, M. J., 1984, Why Do Some Goods Bear Higher Taxes Than Others?, *JET* 32, 301–316.

Roberts, J., 1987, Large Economies, in *The New Palgrave* 3, 132–133.

Roberts, J., 1987, Perfectly and Imperfectly Competitive Markets, in *The New Palgrave* 3, 837–841.

Rosenthal, R. W., 1976, Lindahl's Solution and Values for a Public-Goods Example, *JME* 3, 37–41.

Roth, A. E., 1988, Introduction to the Shapley Value, in *Roth: Shapley Value*, 1–27.

Schotter, A. & G. Schwodiauer, 1980, Economics and the Theory of Games: A Survey, *JEL* 18, 479–527.

Spulber, D. F., 1986, Value Allocations with Economies of Scale, *Econ Let* 21, 107–111.

Stadler, W., 1976, Sufficient Conditions for Preference Optimality, *J Opt Th Appl* 18, 119–140.

Vanpraag, B. M. & J. Linthorst, 1976, Municipal Welfare Functions, *Reg Sci* 6, 51–79.

Wooders, M. H., 1992, Inessentiality of Large Groups and the Approximate Core Property: An Equivalence Theorem, *Econ Th* 2, 129–147.

Wooders, M. H. & W. R. Zame, 1987, Large Games: Fair and Stable Outcomes, *JET* 42, 59–93.

Wooders, M. H. & W. R. Zame, 1988, Values of Large Finite Games, in *Roth: Shapley Value*, 195–206.

Zhou, L., 1992, Strictly Fair Allocations in Large Exchange Economies, *JET* 57, 158–175.

Texts and Monographs: Binmore 1998; Mas-Colell; Mas-Colell, Whinston & Green Ch. 18; Shubik 1982, 1984.

In this collection: Chapters 1, 2, 52, 53, 54, 55, 57, 58, 60, and 61c.

Chapter 52 Power and Taxes

Agmon, T., 1986, The World Debt Crisis: International Lending on Trial, by M. P. Claudon, *J Finance* 41, 525–527.

Agmon, T. & J. K. Deitrich, 1983, International Lending and Income Redistribution: An Alternative View of Country Risk, *J Banking & Fin* 7, 483–495.

Alt, J. E., 1983, The Evolution of Tax Structures, *Publ Choice* 41, 181–222.

Aoki, M., 1980, A Model of the Firm as a Stockholder-Employee Cooperative Game, *AER* 70, 600–610.

Austin, D. A., 1995, Coordinated Action in Local Public Goods Models: The Case of Secession without Exclusion, *J Publ Econ* 58, 235–256.

Becker, G. S., 1983, A Theory of Competition among Pressure Groups for Political Influence, *QJE* 98, 371–400.

Becker, G. S., 1985, Public Policies, Pressure Groups, and Dead Weight Costs, *J Publ Econ* 28, 329–347.

Benoit, J. P. & M. J. Osborne, 1995, Crime, Punishment, and Social Expenditure, *Z Ges Staatsw* 151, 326–347.

Binswanger, H. P. & K. Deininger, 1997, Explaining Agricultural and Agrarian Policies in Developing Countries, *JEL* 35, 1958–2005.

Brams, S. J., 1997, Game Theory and Emotions, *Rationality & Soc* 9, 91–124.

Brams, S. J. & M. P. Hessel, 1984, Threat Power in Sequential Games, *Int Stud Q* 28, 23–44.

Brito, D. L. & W. H. Oakland, 1977, Some Properties of an Optimal Income Tax, *IER* 18, 407–423.

Brooks, J. M., A. Dor & H. S. Wong, 1997, Hospital Insurer Bargaining: An Empirical Investigation, *J Health Econ* 16, 417–434.

Caillaud, B., P. Rey, R. Guesnerie & J. Tirole, 1988, Government Intervention in Production and Incentives Theory: A Review of Recent Contributions, *Rand J* 19, 1–26.

Cheng, L., 1986, The Impact of Changes in Relative Weights on the Optimal Solution of a Maximization Problem, *JME* 15, 143–150.

Crawford, V. P. & H. R. Varian, 1979, Distortion of Preferences and the Nash Theory of Bargaining, *Econ Let* 3, 203–206.

d'Aspremont, C. & L. A. Gerard-Varet, 1992, Non-Transferable Utility and Bayesian Incentives, in *Harsanyi Festschr*, 145–158.

Einy, E. & A. Neyman, 1990, On Non-Atomic Weighted Majority Games, *JME* 19, 391–403.

Eyckmans, J., S. Proost & E. Schokkaert, 1993, Efficiency and Distribution in Greenhouse Negotiations, *Kyklos* 46, 363–397.

Gardner, R., 1981, Wealth and Power in a Collegial Polity, *JET* 25, 353–366.

Gardner, R., 1984, Power and Taxes in a One-Party State: The USSR, 1925–1929, *IER* 25, 743–755.

Grafstein, R., 1990, Missing the Archimedean Point: Liberalism's Institutional Presuppositions, *Am Pol Sci Rev* 84, 177–193.

Hart, O. & J. Moore, 1990, Property Rights and the Nature of the Firm, *JPE* 98, 1119–1158.

Hart, S., 1987, Shapley Value, in *The New Palgrave* 4, 318–320.

Imai, H., 1983, On Harsanyi's Solution, *IJGT* 12, 161–179.

Imai, H., 1983, Voting, Bargaining, and Factor Income Distribution, *JME* 11, 211–233.

Intriligator, M. D., 1979, Income Redistribution: A Probabilistic Approach, *AER* 69, 97–105.

Johnson, W. R., 1987, Income Redistribution as Human Capital Insurance, *J Human Resources* 22, 269–280.

Joneslee, M. W., 1980, Maximum Acceptable Physical Risk and a New Measure of Financial Risk Aversion, *Econ J* 90, 550–568.

Kiander, J., 1991, Strike Threats and the Bargaining Power of Insiders, *Scand J Econ* 93, 349–362.

Konrad, K. A. & S. Skaperdas, 1993, Self-Insurance and Self-Protection: A Nonexpected Utility Analysis, *Geneva Papers on Risk & Insurance Theory* 18, 131–146.

Kurz, M., 1977, Distortion of Preferences, Income Distribution, and the Case for a Linear Income Tax, *JET* 14, 291–298.

Kurz, M., 1994, Game Theory and Public Economics, in *Handb Game Th* 2, 1153–1192.

Lindbeck, A., 1985, Redistribution Policy and the Expansion of the Public Sector, *J Publ Econ* 28, 309–328.

Livne, Z. A., 1989, Axiomatic Characterizations of the Raiffa and the Kalai–Smorodinsky Solutions to the Bargaining Problem, *OR* 37, 972–980.

Mertens, J. F., 1980, Values and Derivatives, *MOR* 5, 523–552.

Mirrlees, J. A., 1981, The Theory of Optimal Taxation, in *Handb Math Econ* 3, 1197–1249.

Neck, R., 1989, Sax, Emil: Contributions to Public Economics, *J Econ Stud* 16, 23–46.

Neyman, A., 1979, Asymptotic Values of Mixed Games, in *Hagen Conf I*, 71–81.

Neyman, A., 1985, Semivalues of Political Economic Games, *MOR* 10, 390–402.

Neyman, A., 1988, Weighted Majority Games Have Asymptotic Values, *MOR* 13, 556–580.

Osborne, M. J., 1984, Why Do Some Goods Bear Higher Taxes Than Others?, *JET* 32, 301–316.

Peck, R. M., 1986, Power and Linear Income Taxes: An Example, *Econometrica* 54, 87–94.

Peck, R. M., 1988, Power, Majority Voting, and Linear Income Tax Schedules, *J Publ Econ* 36, 53–67.

Peltzman, S., 1980, The Growth of Government, *J Law & Econ* 23, 209–287.

Potters, J. & F. Vanwinden, 1992, Lobbying and Asymmetric Information, *Publ Choice* 74, 269–292.

Przeworski, A., 1986, The Feasibility of Universal Grants under Democratic Capitalism, *Theory and Soc* 15, 695–707.

Przeworski, A., 1991, Could We Feed Everyone? The Irrationality of Capitalism and the Infeasibility of Socialism, *Pol & Soc* 19, 1–38.

Przeworski, A. & M. Wallerstein, 1986, Popular Sovereignty, State Autonomy, and Private Property, *Arch Eur Sociol* 27, 215–259.

Przeworski, A. & M. Wallerstein, 1988, Structural Dependence of the State on Capital, *Am Pol Sci Rev* 82, 11–29.

Ramseyer, J. M. & M. Nakazato, 1989, Tax Transitions and the Protection Racket: A Reply, *Virg Law Rev* 75, 1155–1175.

Richter, W. F., 1981, A Normative Justification of Progressive Taxation: How to Compromise on Nash and Kalai–Smorodinsky, in *Hagen Conf II*, 241–247.

Roseackerman, S., 1982, The Power to Tax: Analytical Foundations of a Fiscal Constitution, by G. Brennan & J. Buchanan, *Mich Law Rev* 80, 872–884.

Roth, A., 1979, Axiomatic Models of Bargaining, *Lect Notes* 170.

Rubinstein, A., Z. Safra & W. Thomson, 1992, On the Interpretation of the Nash Bargaining Solution and its Extension to Nonexpected Utility Preferences, *Econometrica* 60, 1171–1186.

Schotter, A. & G. Schwodiauer, 1980, Economics and the Theory of Games: A Survey, *JEL* 18, 479–527.

Shubik, M., 1981, Game Theory Models and Methods in Political Economy, in *Handb Math Econ* 1, 285–330.

Sjoblom, K., 1985, Voting for Social Security, *Publ Choice* 45, 225–240.

Skaperdas, S. & L. Gan, 1995, Risk Aversion in Contests, *Econ J* 105, 951–962.

Snyder, J. M. & G. H. Kramer, 1988, Fairness, Self-Interest, and the Politics of the Progressive Income Tax, *J Publ Econ* 36, 197–230.

Sobel, J., 1981, Distortion of Utilities and the Bargaining Problem, *Econometrica* 49, 597–619.

Svejnar, J., 1986, Bargaining Power, Fear of Disagreement, and Wage Settlements: Theory and Evidence from United States Industry, *Econometrica* 54, 1055–1078.

Tanaka, H., 1998, Redistribution Tax under Non-Benevolent Governments, *Publ Choice* 96, 325–343.

Telser, L. G., 1982, Voting and Paying for Public Goods: An Application of the Theory of the Core, *JET* 27, 376–409.

Thomson, W., 1988, The Manipulability of the Shapley Value, *IJGT* 17, 101–127.

Vanvelthoven, B. & F. Vanwinden, 1991, A Positive Model of Tax Reform, *Publ Choice* 72, 61–86.

Verbon, H. A. A., 1986, Altruism, Political Power and Public Pensions, *Kyklos* 39, 343–358.

Weingast, B. R., 1979, A Rational Choice Perspective on Congressional Norms, *Am J Pol Sci* 23, 245–262.

Wickstrom, B. A., 1984, Economic Justice and Economic Power: An Inquiry into Distributive Justice and Political Stability, *Publ Choice* 43, 225–249.

Yaari, M. E., 1981, Rawls, Edgeworth, Shapley, Nash: Theories of Distributive Justice Reexamined, *JET* 24, 1–39.

Texts and Monographs: Aoki; Brams 1990, 1994; Charreton; Sen; Shubik 1982, 1984.

In this collection: Chapters 1, 2, 53, 55, 56, 58, 60, and 61c.

Chapter 53 Power and Taxes in a Multi-Commodity Economy

Brock, H. W., 1978, A Critical Discussion of the Work of Harsanyi, *Th & Dec* 9, 349–367.

Brock, H. W., 1979, Game Theoretic Account of Social Justice, *Th & Dec* 11, 293–265.

Brock, H. W., 1992, Game Theory, Symmetry, and Scientific Discovery, in *Harsanyi Festschr*, 391–418.

Crawford, V. P. & H. R. Varian, 1979, Distortion of Preferences and the Nash Theory of Bargaining, *Econ Let* 3, 203–206.

Einy, E. & B. Peleg, 1991, Linear Measures of Inequality for Cooperative Games, *JET* 53, 328–344.

Gardner, R., 1981, Wealth and Power in a Collegial Polity, *JET* 25, 353–366.

Gardner, R., 1984, Power and Taxes in a One-Party State: The USSR, 1925–1929, *IER* 25, 743–755.

Hart, S., 1979, Values of Large Market Games, in *Vienna Conf*, 187–197.

Hart, S., 1980, Measure-Based Values of Market Games, *MOR* 5, 197–228.

Imai, H., 1983, Voting, Bargaining, and Factor Income Distribution, *JME* 11, 211–233.

Kleinberg, N. L. & J. H. Weiss, 1985, Equivalent *n*-Person Games and the Null Space of the Shapley Value, *MOR* 10, 233–243.

Kurz, M., 1980, Income Distribution and Distortion of Preferences: The 1-Commodity Case, *JET* 22, 99–106.

Kurz, M., 1994, Game Theory and Public Economics, in *Handb Game Th* 2, 1153–1192.

Monderer, D., 1986, Measure-Based Values of Nonatomic Games, *MOR* 11, 321–335.

Monderer, D., 1989, Asymptotic Measure-Based Values of Nonatomic Games, *MOR* 14, 737–744.

Monderer, D., 1989, Weighted Majority Games Have Many μ-Values, *IJGT* 18, 321–325.

Muto, S., 1989, Limit Properties of Power Indexes in a Class of Representative Systems, *IJGT* 18, 361–388.

Neyman, A., 1979, Asymptotic Values of Mixed Games, in *Hagen Conf I*, 71–81.

Neyman, A., 1985, Semivalues of Political Economic Games, *MOR* 10, 390–402.

Osborne, M. J., 1984, Why Do Some Goods Bear Higher Taxes Than Others?, *JET* 32, 301–316.

Peck, R. M., 1986, Power and Linear Income Taxes: An Example, *Econometrica* 54, 87–94.

Peck, R. M., 1988, Power, Majority Voting, and Linear Income Tax Schedules, *J Publ Econ* 36, 53–67.

Qin, C.-Z., 1994, An Inner Core Equivalence Theorem, *Econ Th* 4, 311–317.

Sobel, J., 1981, Distortion of Utilities and the Bargaining Problem, *Econometrica* 49, 597–619.

Tauman, Y., 1981, Values of Markets with a Majority Rule, in *Hagen Conf II*, 103–122.

Tauman, Y., 1982, A Characterization of Vector Measure Games in pNA, *IJM* 43, 75–96.

Telser, L. G., 1982, Voting and Paying for Public Goods: An Application of the Theory of the Core, *JET* 27, 376–409.

Thomson, W., 1988, The Manipulability of the Shapley Value, *IJGT* 17, 101–127.

Texts and Monographs: Shubik 1982, 1984.

In this collection: Chapters 52, 55, 56, 58, 61c, and 63.

Chapter 54   Core and Value for a Public Goods Economy: An Example

Greenberg, J. & B. Shitovitz, 1984, Aumann–Shapley Prices as a Scarf Social Equilibrium, *JET* 34, 380–382.

Kurz, M., 1994, Game Theory and Public Economics, in *Handb Game Th* 2, 1153–1192.

Wooders, M. H. & W. R. Zame, 1988, Values of Large Finite Games, in *Roth: Shapley Value*, 195–206.

Texts and Monographs: Shubik 1984.

Chapter 55   Power and Public Goods

Aumann, R. J., 1990, The Shapley Value, in *Ohio Conf*, 158–165.

Kleinberg, N. L. & J. H. Weiss, 1985, Equivalent *n*-Person Games and the Null Space of the Shapley Value, *MOR* 10, 233–243.

Kurz, M., 1994, Game Theory and Public Economics, in *Handb Game Th* 2, 1153–1192.

Osborne, M. J., 1984, Why Do Some Goods Bear Higher Taxes Than Others?, *JET* 32, 301–316.

Telser, L. G., 1982, Voting and Paying for Public Goods: An Application of the Theory of the Core, *JET* 27, 376–409.

Texts and Monographs: Shubik 1984.

In this collection: Chapters 1, 56, 58, and 61c.

Chapter 56   Voting for Public Goods

Kleinberg, N. L. & J. H. Weiss, 1985, Equivalent *n*-Person Games and the Null Space of the Shapley Value, *MOR* 10, 233–243.

Kurz, M., 1994, Game Theory and Public Economics, in *Handb Game Th* 2, 1153–1192.

Osborne, M. J., 1984, Why Do Some Goods Bear Higher Taxes Than Others?, *JET* 32, 301–316.

Peck, R. M., 1988, Power, Majority Voting, and Linear Income Tax Schedules, *J Publ Econ* 36, 53–67.

Telser, L. G., 1982, Voting and Paying for Public Goods: An Application of the Theory of the Core, *JET* 27, 376–409.

Texts and Monographs: Shubik 1984.

In this collection: Chapters 1, 55, 58, and 61c.

Chapter 57   Values of Markets with Satiation or Fixed Prices

Aumann, R. J., 1998, On the State of the Art in Game Theory, *GEB* 24, 181–210.

Barbera, S. & M. O. Jackson, 1995, Strategy-Proof Exchange, *Econometrica* 63, 51–87.

Dreze, J. H. & H. Muller, 1980, Optimality Properties of Rationing Schemes, *JET* 23, 131–149.

Thomson, W. & L. Zhou, 1993, Consistent Solutions in Atomless Economies, *Econometrica* 61, 575–587.

Texts and Monographs: Charreton.

In this collection: Chapters 58 and 61c.

Chapter 59 Endogenous Formation of Links between Players and of Coalitions: An Application of the Shapley Value

Ball, M. A., 1998, Coalition Structure and the Equilibrium Concept in n-Person Games with Transferable Utility, *Proc Royal Soc London A* 454, 1509–1522.

Bienenstock, E. J. & P. Bonacich, 1997, Network Exchange as a Cooperative Game, *Rationality & Soc* 9, 37–65.

Bloch, F., 1996, Sequential Formation of Coalitions in Games with Externalities and Fixed Payoff Division, *GEB* 14, 90–123.

Bonacich, P. & E. J. Bienenstock, 1997, Latent Classes in Exchange Networks: Sets of Positions with Common Interests, *J Math Sociol* 22, 1–28.

Borm, P., G. Owen & S. Tijs, 1992, On the Position Value for Communication Situations, *SIAM J Discr Math* 5, 305–320.

Cochinard, S., 1995, The Coalition Concept in Game Theory, *Rev d'Écon Politique* 105, 633–655.

Dutta, B. & S. Mutuswami, 1997, Stable Networks, *JET* 76, 322–344.

Dutta, B., A. van den Nouweland & S. Tijs, 1998, Link Formation in Cooperative Situations, *IJGT* 27, 245–256.

Feinberg, Y., 1998, An Incomplete Cooperation Structure for a Voting Game can be Strategically Stable, *GEB* 24, 2–9.

Gilles, R. P., G. Owen & R. Vandenbrink, 1992, Games with Permission Structures: The Conjunctive Approach, *IJGT* 20, 277–293.

Greenberg, J., 1994, Coalition Structures, in *Handb Game Th* 2, 1305–1337.

Horowitz, J. K. & R. E. Just, 1995, Political Coalition Breaking and Sustainability of Policy Reform, *J Dev Econ* 47, 271–286.

Jackson, M. O. & A. Wolinsky, 1996, A Strategic Model of Social and Economic Networks, *JET* 71, 44–74.

Meca-Martínez, A., J. Sánchez-Soriano, I. García-Jurando & S. Tijs, 1998, Strong Equilibria in Claim Games corresponding to Convex Games, *IJGT* 27, 211–217.

Michener, H. A. & D. J. Myers, 1998, Probabilistic Coalition Structure Theories: An Empirical Comparison in Four-Person Superadditive Sidepayment Games, *J Confl Res* 42, 830–860.

Michener, H. A. & D. J. Myers, 1998, An Empirical Comparison of Probabilistic Coalition Structure Theories in Three-Person Sidepayment Games, *Theory & Decision* 45, 37–82.

Nowak, A. S. & T. Radzik, 1994, The Shapley Value for n-Person Games in Generalized Characteristic Function Form, *GEB* 6, 150–161.

Qin, C. Z., 1996, Endogenous Formation of Cooperation Structures, *JET* 69, 218–226.

Rasch, B. E., 1993, Coalition Theory: Riker's Size Principle, *Tidsskrift for Samfunnsforskning* 34, 53–75.

Ray, D. & R. Vohra, 1997, Equilibrium Binding Agreements, *JET* 73, 30–78.

Van den Nouweland, A., P. Borm & S. Tijs, 1992, Allocation Rules for Hypergraph Communication Situations, *IJGT* 20, 255–268.

Weber, R. J., 1994, Games in Coalitional Form, in *Handb Game Th* 2, 1285–1303.

Yi, S.-S., 1997, Stable Coalition Structures with Externalities, *GEB* 20, 201–237.

Texts and Monographs: Myerson.

Chapter 60 An Axiomatization of the Non-Transferable Utility Value

Agastya, M., 1996, Multiplayer Bargaining Situations: A Decision Theoretic Approach, *GEB* 12, 1–20.

Aumann, R. J., 1990, The Shapley Value, in *Ohio Conf*, 158–165.

Beja, A. & I. Gilboa, 1990, Values for Two-Stage Games: Another View of the Shapley Axioms, *IJGT* 19, 17–31.

Bergstrom, T. C. & H. R. Varian, 1985, When Do Market Games Have Transferable Utility?, *JET* 35, 222–233.

Binmore, K. G., 1985, Bargaining and Coalitions, in *Roth: Bargaining*, 269–304.

Blackorby, C., W. Bossert & D. Donaldson, 1994, Generalized Ginis and Cooperative Bargaining Solutions, *Econometrica* 62, 1161–1178.

Borm, P., H. Keiding, R. P. McLean, S. Oortwijn & S. Tijs, 1992, The Compromise Value for NTU-Games, *IJGT* 21, 175–189.

Brock, H. W., 1992, Game Theory, Symmetry, and Scientific Discovery, in *Harsanyi Festschr*, 391–418.

Chun, Y., 1990, Minimal Cooperation in Bargaining, *Econ Let* 34, 311–316.

Chun, Y. & W. Thomson, 1990, The Nash Solution and Uncertain Disagreement Points, *GEB* 2, 213–223.

Chun, Y. S. & W. Thomson, 1992, Bargaining Problems with Claims, *Math Soc Sci* 24, 19–33.

d'Aspremont, C. & L. A. Gerard-Varet, 1992, Non-Transferable Utility and Bayesian Incentives, in *Harsanyi Festschr*, 145–158.

Dubey, P. & A. Neyman, 1994, An Axiomatic Approach to the Equivalence Phenomenon, in *Stony Brook ASI* 1991, 137–144.

Dubey, P. & A. Neyman, 1997, An Equivalence Principle for Perfectly Competitive Economies, *JET* 75, 314–344.

Emmons, D. W. & A. J. Scafuri, 1985, Value Allocations: An Exposition, in *Lect Notes* 244, 55–78.

Gardner, R., 1987, A Theory of the Spoils System, *Publ Choice* 54, 171–185.

Gul, F., 1989, Bargaining Foundations of the Shapley Value, *Econometrica* 57, 81–95.

Güth, W. & B. Kalkofen, 1989, Unique Solutions for Strategic Games, *Lect Notes* 328.

Hart, S., 1985, An Axiomatization of Harsanyi's Non-Transferable Utility Solution, *Econometrica* 53, 1295–1313.

Hart, S., 1985, Non-Transferable Utility Games and Markets: Some Examples and the Harsanyi Solution, *Econometrica* 53, 1445–1450.

Hart, S., 1985, Axiomatic Approaches to Coalitional Bargaining, in *Roth: Bargaining*, 305–319.

Hart, S., 1990, Advances in Value Theory, in *Ohio Conf*, 166–175.

Hart, S., 1994, The Harsanyi Value, in *Stony Brook ASI* 1991, 105–111.

Hart, S., 1997, Classical Cooperative Theory II: Value-Like Concepts, in *Stony Brook ASI* 1994, 43–49.

Kalai, E. & D. Samet, 1985, Monotonic Solutions to General Cooperative Games, *Econometrica* 53, 307–327.

Kern, R., 1985, The Shapley Transfer Value without Zero Weights, *IJGT* 14, 73–92.

Levy, A. & R. P. McLean, 1991, An Axiomatization of the Weighted NTU Value, *IJGT* 19, 339–351.

Mas-Colell, A., 1987, Cooperative Equilibrium, in *The New Palgrave* 1, 659–662.

Michener, H. A., G. D. Richardson & M. S. Salzer, 1989, Extensions of Value Solutions in Constant-Sum Non-Sidepayment Games, *J Confl Res* 33, 530–553.

Michener, H. A. & M. S. Salzer, 1989, Comparative Accuracy of Value Solutions in Non-Sidepayment Games with Empty Core, *Th & Dec* 26, 205–233.

Michener, H. A., M. S. Salzer, M. S. Reimer & J. Lee, 1991, A Comparison of Lambda-Transfer and Core Solutions in Constant-Sum Non-Sidepayment Games, *Behav Sci* 36, 115–132.

Myerson, R. B., 1986, An Introduction to Game Theory, in *Reiter*, 1–61.

Myerson, R. B., 1992, Fictitious Transfer Solutions in Cooperative Game Theory, in *Harsanyi Festschr*, 13–33.

Otten, G. J., P. Borm, B. Peleg & S. Tijs, 1998, The MC-Value for Monotonic NTU-Games, *IJGT* 27, 37–47.

Otten, G. J., P. Borm & S. Tijs, 1996, A Note on the Characterizations of the Compromise Value, *IJGT* 25, 427–435.

Peck, R. M., 1988, Power, Majority Voting, and Linear Income Tax Schedules, *J Publ Econ* 36, 53–67.

Peleg, B., 1985, An Axiomatization of the Core of a Cooperative Game without Side Payments, *JME* 14, 203–214.

Peleg, B., 1986, On the Reduced Game Property and its Converse, *IJGT* 15, 187–200.

Peleg, B., 1989, An Axiomatization of the Core of Market Games, *MOR* 14, 448–456.

Peleg, B., 1992, Axiomatizations of the Core, in *Handb Game Th* 1, 397–412.

Peleg, B., J. Potters & S. Tijs, 1996, Minimality of Consistent Solutions for Strategic Games, in particular for Potential Games, *Econ Th* 7, 81–93.

Peters, H., 1986, Simultaneity of Issues and Additivity in Bargaining, *Econometrica* 54, 153–169.

Peters, H. & K. Vrieze, 1994, Nash Refinement of Equilibria, *J Opt Th Appl* 83, 355–373.

Qin, C. Z., 1993, The Inner Core and the Strictly Inhibitive Set, *JET* 59, 96–106.

Rosenthal, E. C., 1990, Monotonicity of the Core and Value in Dynamic Cooperative Games, *IJGT* 19, 45–57.

Roth, A. E., 1988, Introduction to the Shapley Value, in *Roth: Shapley Value*, 1–27.

Samet, D., 1985, An Axiomatization of the Egalitarian Solutions, *Math Soc Sci* 9, 173–181.

Van Damme, E. E. C., 1986, The Nash Bargaining Solution is Optimal, *JET* 38, 78–100.

Winter, E., 1991, On Nontransferable Utility Games with Coalition Structures, *IJGT* 20, 53–63.

Winter, E., 1992, On Bargaining Position Descriptions in Nontransferable Utility Games: Symmetry versus Asymmetry, *IJGT* 21, 191–211.

Texts and Monographs: Güth; Myerson

In this collection: Chapters 61c and 64.

Chapter 61 The Non-Transferable Utility Value Controversy, Part 1

Agastya, M., 1996, Multiplayer Bargaining Situations: A Decision Theoretic Approach, *GEB* 12, 1–20.

Aumann, R. J., 1998, On the State of the Art in Game Theory, *GEB* 24, 181–210.

Beja, A. & I. Gilboa, 1990, Values for 2-Stage Games: Another View of the Shapley Axioms, *IJGT* 19, 17–31.

Bennett, E., 1983, Characterization Results for Aspirations in Games with Sidepayments, *Math Soc Sci* 4, 229–241.

Bergstrom, T. C. & H. R. Varian, 1985, When Do Market Games Have Transferable Utility?, *JET* 35, 222–233.

Binmore, K. G., 1985, Bargaining and Coalitions, in *Roth: Bargaining*, 269–304.

Borm, P., H. Keiding, R. P. McLean, S. Oortwijn & S. Tijs, 1992, The Compromise Value for NTU-Games, *IJGT* 21, 175–189.

Dutta, B. & D. Ray, 1991, Constrained Egalitarian Allocations, *GEB* 3, 403–422.

Gardner, R., 1987, A Theory of the Spoils System, *Publ Choice* 54, 171–185.

Hart, S., 1985, An Axiomatization of Harsanyi's Non-Transferable Utility Solution, *Econometrica* 53, 1295–1313.

Hart, S., 1985, Non-Transferable Utility Games and Markets: Some Examples and the Harsanyi Solution, *Econometrica* 53, 1445–1450.

Hart, S., 1985, Axiomatic Approaches to Coalitional Bargaining, in *Roth: Bargaining*, 305–319.

Hart, S., 1987, Shapley Value, in *The New Palgrave* 4, 318–320.

Hart, S., 1990, Advances in Value Theory, in *Ohio Conf*, 166–175.

Krasa, S. & N. C. Yannelis, 1994, The Value Allocation of an Economy with Differential Information, *Econometrica* 62, 881–900.

Mailath, G. J., 1998, Do People Play Nash Equilibrium? Lessons from Evolutionary Game Theory, *JEL* 36, 1347–1374.

Otten, G. J., P. Borm, B. Peleg & S. Tijs, 1998, The MC-Value for Monotonic NTU-Games, *IJGT* 27, 37–47.

Peleg, B., 1985, An Axiomatization of the Core of Cooperative Games without Side Payments, *JME* 14, 203–214.

Peleg, B., 1992, Axiomatizations of the Core, in *Handb Game Th* 1, 397–412.

Peters, H. & E. E. C. Van Damme, 1991, Characterizing the Nash and Raiffa Bargaining Solutions by Disagreement Point Axioms, *MOR* 16, 447–461.

Roth, A. E., 1988, Introduction to the Shapley Value, in *Roth: Shapley Value*, 1–27.

Serrano, R., 1997, Reinterpreting the Kernel, *JET* 77, 58–80.

Sprumont, Y., 1990, Population Monotonic Allocation Schemes for Cooperative Games with Transferable Utility, *GEB* 2, 378–394.

Winter, E., 1992, On Bargaining Position Descriptions in Nontransferable Utility Games: Symmetry versus Asymmetry, *IJGT* 21, 191–211.

Winter, E., 1994, The Demand Commitment Bargaining and Snowballing Cooperation, *Econ Th* 4, 255–273.

In this collection: Chapters 1, 2, 34, and 60.

Chapter 62 The Non-Transferable Utility Value Controversy, Part 2

Emmons, D. W. & A. J. Scafuri, 1985, Value Allocations: An Exposition, in *Lect Notes* 244, 55–78.

Krasa, S. & N. C. Yannelis, 1994, The Value Allocation of an Economy with Differential Information, *Econometrica* 62, 881–900.

Peleg, B., 1985, An Axiomatization of the Core of Cooperative Games without Side Payments, *JME* 14, 203–214.

Peleg, B., 1992, Axiomatizations of the Core, in *Handb Game Th* 1, 397–412.

Roth, A. E., 1988, Introduction to the Shapley Value, in *Roth: Shapley Value*, 1–27.

Winter, E., 1994, The Demand Commitment Bargaining and Snowballing Cooperation, *Econ Th* 4, 255–273.

Yannelis, N. C., 1985, Values and Fairness, in *Lect Notes* 244, 205–235.

Chapter 63 Recent Developments in the Theory of the Shapley Value

Hart, S., 1987, Shapley Value, in *The New Palgrave* 4, 318–320.

Hart, S., 1990, Advances in Value Theory, in *Ohio Conf*, 166–175.

In this collection: Chapter 64.

Chapter 64 The Shapley Value

In this collection: Chapter 58.

Chapter 65 Spaces of Measurable Transformations

Harsanyi, J. C., 1973, Games with Randomly Disturbed Payoffs: A New Rationale for Mixed Strategy Equilibrium Points, *IJGT* 2, 1–23.

In this collection: Chapter 67.

Chapter 66 Borel Structures for Function Spaces

Armbruster, W. & W. Böge, 1979, Bayesian Game Theory, in *Hagen Conf I*, 17–28.

Christensen, J. P. R., 1971, On Some Properties of Effros's Borel Structure on Spaces of Closed Subsets, *Math Ann* 195, 17–23.

Cressie, N., 1978, A Strong Limit Theorem for Random Sets, *Adv Appl Prob*, 36–46.

Dudley, R. M., 1978, A Central Limit Theorem for Empirical Measures, *Ann Prob* 6, 899–929.

Harsanyi, J. C., 1973, Games with Randomly Disturbed Payoffs: A New Rationale for Mixed Strategy Equilibrium Points, *IJGT* 2, 1–23.

Kirschner, H. P., 1975, Risk-Equivalence of 2 Methods of Randomization in Statistics, *J Multiv Anal* 6, 159–166.

Rao, B. V., 1970, On Borel Structures, *Coll Math* 21, 199–204.

Rao, B. V., 1971, Borel Structures for Function Spaces, *Coll Math* 23, 33–38.

Rustichini, A., 1993, Mixing on Function Spaces, *Econ Th* 3, 183–191.

Stinchcombe, M. B., 1992, Maximal Strategy Sets for Continuous Time Game Theory, *JET* 56, 235–265.

In this collection: Chapters 28, 65, 67, and 71.

Chapter 67 On Choosing a Function at Random

Harsanyi, J. C., 1967, Games with Incomplete Information Played by "Bayesian" Players I: The Basic Model, *Manag Sci* 14, 159–182.

Harsanyi, J. C., 1995, Games with Incomplete Information, *AER* 85, 291–303.

In this collection: Chapter 28.

Chapter 68 Integrals of Set-Valued Functions

Abreu, D., D. Pearce & E. Stacchetti, 1990, Toward a Theory of Discounted Repeated Games with Imperfect Monitoring, *Econometrica* 58, 1041–1063.

Allen, B. & S. Sorin, 1994, General Equilibrium and Cooperative Games: Basic Results, in *Stony Brook ASI* 1991, 17–33.

Alo, R. A., A. Dekorvin & L. Kunes, 1980, Maps whose Values are Closed Convex Subsets of a Banach Space, *J Math Anal Appl* 76, 1–9.

Alo, R. A., A. Dekorvin & C. E. Roberts, 1979, An Optimal Sampling Theorem for Convex Set-Valued Martingales, *J Reine & Ang Math* 310, 1–6.

Alo, R. A., A. Dekorvin & C. E. Roberts, 1980, On Some Properties of Continuous Multi-measures, *J Math Anal Appl* 75, 402–410.

Aronsson, G., 1973, A New Approach to Nonlinear Controllability, *J Math Anal Appl* 44, 763–772.

Artstein, Z., 1972, Set-Valued Measures, *Trans AMS* 165, 103–125.

Artstein, Z., 1974, On the Calculus of Closed Set-Valued Functions, *Indiana U Math J* 24, 433–441.

Artstein, Z., 1974, On a Variational Problem, *J Math Anal Appl* 45, 404–415.

Artstein, Z., 1975, Weak Convergence of Set-Valued Functions and Control, *SIAM J Contr* 13, 865–878.

Artstein, Z., 1980, Discrete and Continuous Bang-Bang and Facial Spaces, or Look for the Extreme Points, *SIAM Rev* 22, 172–185.

Artstein, Z., 1989, Parametrized Integration of Multifunctions with Applications to Control and Optimization, *SIAM J Contr* 27, 1369–1380.

Artstein, Z. & J. A. Burns, 1975, Integration of Compact Set-Valued Functions, *Pac J Math* 58, 297–307.

Artstein, Z. & R. A. Vitale, 1975, A Strong Law of Large Numbers for Random Compact Sets, *Ann Prob* 3, 879–880.

Artstein, Z. & R. J. B. Wets, 1988, Approximating the Integral of a Multifunction, *J Multiv Anal* 24, 285–308.

Aubin, J. P., H. Frankowska & C. Olech, 1986, Controllability of Convex Processes, *SIAM J Contr* 24, 1192–1211.

Baddeley, A. & I. Molchanov, 1998, Averaging of Random Sets based on their Distance Functions, *J Math Imaging & Vision* 8, 79–92.

Balder, E. J. & C. Hess, 1995, Fatou's Lemma for Multifunctions with Unbounded Values, *MOR* 20, 175–188.

Ban, J., 1990, Radon–Nikodym Theorem and Coalitional Expectation of Fuzzy-Valued Measures and Variables, *Fuzzy* 34, 383–392.

Ban, J., 1991, Ergodic Theorems for Random Compact Sets and Fuzzy Variables in Banach Spaces, *Fuzzy* 44, 71–82.

Ban, J., 1991, Sequences of Random Fuzzy Sets, *Int J Gen Sys* 20, 17–22.

Banks, H. T. & M. Q. Jacobs, 1970, A Differential Calculus for Multifunctions, *J Math Anal Appl* 29, 246–272.

Banks, H. T. & M. Q. Jacobs, 1970, The Optimization of Trajectories of Linear Functional Differential Equations, *SIAM J Contr* 8, 461–488.

Barany, I. & R. A. Vitale, 1993, Random Convex Hulls: Floating Bodies and Expectations, *J Appr Th* 75, 130–135.

Batukhtin, V. D. & A. I. Subbotin, 1972, Conditions for Terminating a Pursuit Game, *Eng Cybernetics* 10, 1–5.

Bewley, T. F., 1977, The Permanent Income Hypothesis: Theoretical Formulation, *JET* 16, 252–292.

Billera, L. J. & B. Sturmfels, 1992, Fiber Polytopes, *Ann Math* 135, 527–549.

Borges, R., 1967, Ecken des Wertebereiches von Vektorintegralen, *Math Ann* 173, 53–58.

Bradley, M. & R. Datko, 1977, Some Analytic and Measure Theoretic Properties of Set-Valued Mappings, *SIAM J Contr* 15, 625–635.

Bressan, A. & F. Flores, 1995, Multivariable Aumann Integrals and Controlled Wave Equations, *J Math Anal Appl* 189, 315–334.

Bridgland, T. F., 1970, Trajectory Integrals of Set-Valued Functions, *Pac J Math* 33, 43–68.

Bridgland, T. F., 1972, Extreme Limits of Compact-Valued Functions, *Trans AMS* 170, 149–163.

Brown, D. J., 1976, Existence of a Competitive Equilibrium in a Nonstandard Exchange Economy, *Econometrica* 44, 537–546.

Buckley, J. J., 1974, Graphs of Measurable Selections, *Proc AMS* 44, 78–80.

Buckley, J. J., 1992, Fuzzy Complex Analysis II: Integration, *Fuzzy* 49, 171–179.

Butnariu, D., 1989, Measurability Concepts for Fuzzy Mappings, *Fuzzy* 31, 77–82.

Byrne, C. L., 1978, Remarks on the Set-Valued Integrals of Debreu and Aumann, *J Math Anal Appl* 62, 243–246.

Castaing, C., 1967, Sur les Multi-Applications Mesurables, *Rev Fr d'Informatique & RO* 1, 91–126.

Cellina, A., 1970, Multivalued Differential Equations and Ordinary Differential Equations, *SIAM J Appl Math* 18, 533–538.

Chigir, S. A., 1977, Linear Problem of Convergence with Incomplete Information, *PMM J Appl Math & Mechanics* 41, 610–616.

Chow, S. N., L. A. Karlovitz & A. Lasota, 1972, The Integral Form of the Mean Value Theorem for Nondifferentiable Mappings, *J Math Anal Appl* 38, 214–222.

Chukwu, E. N., 1979, Controllability of Delay Systems with Restrained Controls, *J Opt Th Appl* 29, 301–320.

Clarke, F. H., 1981, A Variational Proof of the Aumann Theorem, *Appl Math & Opt* 7, 373–378.

Cornwall, R. R., 1970, Convexity and Continuity Properties of Preference Functions, *Z Nat-ok* 30, 35–52.

Cornwall, R. R., 1972, Conditions for the Graph and the Integral of a Correspondence to be Open, *J Math Anal Appl* 39, 771–792.

Cressie, N. & G. M. Laslett, 1987, Random Set Theory and Problems of Modeling, *SIAM Rev* 29, 557–574.

Datko, R., 1970, Measurability Properties of Set-Valued Mappings in a Banach Space, *SIAM J Contr* 8, 226–238.

Dauer, J. P., 1972, A Controllability Technique for Nonlinear Systems, *J Math Anal Appl* 37, 442–451.

Deblasi, F. S. & A. Lasota, 1968, Daniell's Method in the Theory of the Aumann–Hakuhara Integral of Set-Valued Functions, *Atti Acad Lincei* 45, 252.

Deblasi, F. S. & A. Lasota, 1969, A Characterization of the Integral of Set-Valued Functions, *Atti Acad Lincei* 46, 154.

Debreu, G., 1967, Integration of Correspondences, in *Proc Berkeley Symp* 1965, Vol 2, 351–372.

Debreu, G., 1981, Existence of Competitive Equilibrium, in *Handb Math Econ* 2, 697–743.

Delbaen, F. & W. Schachermayer, 1998, The Fundamental Theorem of Asset Pricing for Unbounded Stochastic Processes, *Math Ann* 312, 215–250.

Diamond, P., 1988, Interval-Valued Random Functions and the Kriging of Intervals, *Math Geology* 20, 145–165.

Diamond, P., 1989, Fuzzy Kriging, *Fuzzy* 33, 315–332.

Ding, Z. H., M. Ma & A. Kandel, 1997, Existence of the Solutions of Fuzzy Differential Equations with Parameters, *Inf Sci* 99, 205–217.

Doitchinov, B. D. & V. Veliov, 1993, Parametrizations of Integrals of Set-Valued Mappings and Applications, *J Math Anal Appl* 179, 483–499.

Dontchev, A. L. & V. M. Veliov, 1983, Singular Perturbations in the Mayer Problem for Linear Systems, *SIAM J Contr* 21, 566–581.

Dubois, D. & H. Prade, 1987, Fuzzy Numbers, *Fuzzy* 24, 259–262.

Dupacova, J. & R. Wets, 1988, Asymptotic Behavior of Statistical Estimators and of Optimal Solutions of Stochastic Optimization Problems, *Ann Stat* 16, 1517–1549.

Dynkin, E. B., 1974, Stochastic Dynamic Models of Economic Equilibrium, in *Proc Int Congr Math* 1974, 509–515.

Dynkin, E. B. & I. V. Evstigneev, 1976, Regular Conditional Expectations of Correspondences, *Th Prob Appl* 21, 325–338.

Dzhafarov, V. J., 1985, The Stability of Guaranteed Results in the Problem of Feedback Control, *Doklady SSSR* 285, 27–31.

Fryszkowski, A., 1990, Continuous Selections of Aumann Integrals, *J Math Anal Appl* 145, 431–446.

Fryszowski, A. & T. Rzezuchowski, 1991, A Continuous Version of the Filippov–Wazewski Relaxation Theorem, *J Diff Eq* 94, 254–265.

Gil, M. A. & P. Jain, 1992, A Comparison of Experiments in Statistical Decision Problems with Fuzzy Utilities, *IEEE Trans Man & Cyber* 22, 662–670.

Gil, M. A., M. Lopez-Diaz & H. Lopez-Garcia, 1998, The Fuzzy Hyperbolic Inequality Index associated with Fuzzy Random Variables, *Eur J OR* 110, 377–391.

Groves, T. & S. Hart, 1982, Efficiency of Resource Allocation by Uninformed Demand, *Econometrica* 50, 1453–1482.

Guo, C. M., D. L. Zhang & C. X. Wu, 1998, Generalized Fuzzy Integrals of Fuzzy-Valued Functions, *Fuzzy* 97, 123–128.

Guo, C. M., D. L. Zhang & C. X. Wu, 1998, Fuzzy-Valued Fuzzy Measures and Generalized Fuzzy Integrals, *Fuzzy* 97, 255–260.

Hajek, O., 1979, Discontinuous Differential Equations I, *J Diff Eq* 32, 149–170.

Halkin, H. & C. Hendricks, 1968, Subintegrals of Set-Valued Functions with Semianalytic Graphs, *PNAS* 59, 365.

Hart, S. & E. Kohlberg, 1974, Equally Distributed Correspondences, *JME* 1, 167–174.

Hermes, H., 1968, Calculus of Set-Valued Functions and Control, *J Math Mech* 18, 47–59.

Hermes, H., 1969, Existence and Properties of Solutions of $x \in R(t, x)$, *Stud Appl Math*, 188–193.

Hermes, H., 1970, The Generalized Differential Equation $x \in R(t, x)$, *Advances in Mathematics* 4, 149–169.

Hermes, H., 1971, On the Structure of Attainable Sets for Generalized Differential Equations and Control Systems, *J Diff Eq* 9, 141–154.

Hermes, H., 1972, The Geometry of Time-Optimal Control, *SIAM J Contr* 10, 221–229.

Hess, C., 1991, Convergence of Conditional Expectations for Unbounded Random Sets, Integrands, and Integral Functionals, *MOR* 16, 627–649.

Hiai, F., 1978, Radon–Nikodym Theorems for Set-Valued Measures, *J Multiv Anal* 82, 96–118.

Hiai, F., 1985, Convergence of Conditional Expectations and Strong Laws of Large Numbers for Multivalued Random Variables, *Trans AMS* 291, 613–627.

Hiai, F. & H. Umegaki, 1977, Integrals, Conditional Expectations, and Martingales of Multivalued Functions, *J Multiv Anal* 7, 149–182.

Hildenbrand, W., 1970, Existence of Equilibria for Economies with Production and a Measure Space of Consumers, *Econometrica* 37, 608–623.

Hildenbrand, W., 1972, Metric Measure Spaces of Economic Agents, in *Proc Berkeley Symp* 1970, Vol 2, 81–95.

Hildenbrand, W. & J. F. Mertens, 1971, Fatou's Lemma in Several Dimensions, *Z Wahrscheinl* 17, 151–155.

Himmelberg, C. J. & F. Van Vleck, 1971, Selection and Implicit Function Theorems for Multifunctions with a Souslin Graph, *Bull Acad Polon* 19, 911.

Hogan, W. W., 1973, Point-to-Set Maps in Mathematical Programming, *SIAM Rev* 15, 591–603.

Inoue, H., 1991, A Strong Law of Large Numbers for Fuzzy Random Sets, *Fuzzy* 41, 285–291.

Inoue, H., 1995, Randomly Weighted Sums for Exchangeable Fuzzy Random Variables, *Fuzzy* 69, 347–354.

Jacobs, M. Q., 1968, An Approximation of Integrals of Multivalued Functions, *SIAM Rev* 10, 476–477.

Jacobs, M. Q., 1968, Measurable Multivalued Mappings and Lusin's Theorem, *Trans AMS* 134, 471–481.

Jacobs, M. Q., 1969, On the Approximation of Integrals of Multivalued Functions, *SIAM J Contr* 7, 158–177.

Jang, L. C. & J. S. Kwon, 1998, Convergences of Sequences of Set-Valued and Fuzzy-Set-Valued Functions, *Fuzzy* 93, 241–246.

Jang, L. C. & J. S. Kwon, 1998, A Uniform Strong Law of Large Numbers for Partial Sum Processes of Banach Space-Valued Random Sets, *Stat & Prob Let* 38, 21–25.

Jang, L. C. & J. S. Kwon, 1998, A Uniform Strong Law of Large Numbers for Partial Sum Processes of Fuzzy Random Variables Indexed by Sets, *Fuzzy* 99, 97–103.

Johnson, J. A., 1974, Extreme Measurable Selections, *Proc AMS* 44, 107–112.

Jonasson, J., 1998, On Positive Random Objects, *J Th Prob* 11, 81–125.

Kaleva, O., 1987, Fuzzy Differential Equations, *Fuzzy* 24, 301–317.

Kannai, Y., 1970, Continuity Properties of the Core of a Market, *Econometrica* 38, 791–815.

Kannan, R., 1977, Monotonicity and Measurability, in Lakshmikantham, V. (Ed.), *Nonlinear Systems and Applications, Proceedings of an International Conference, Univ. of Texas, Arlington, TX, July 19–23, 1976*, Academic Press, New York, 189–198.

Kannan, R., 1977, Random Operator Equations, in Bednarek, A. R. & L. Cesari (Eds.), *Dynamical Systems, Proceedings of an International Symp., Univ. of Florida, Gainesville, FL, March 24–26, 1976*, Academic Press, New York, 113–137.

Kerimov, A. K., 1976, Nonlinear Differential Games, *Izvestiya Akademii Nauk Azerbaidzhanskoi SSR Seriya Fiziko-Tekhnicheskikh i Matematicheskikh Nauk*, 37–41.

Khan, M. A., 1985, Equilibrium Points of Nonatomic Games over a Nonreflexive Banach Space, *J Appr Th* 43, 370–376.

Khan, M. A., 1985, On Extensions of the Cournot–Nash Theorem, in *Lect Notes* 244, 79–106.

Khan, M. A., 1986, Equilibrium Points of Nonatomic Games over a Banach Space, *Trans AMS* 293, 737–749.

Khan, M. A., K. P. Rath & Y. Sun, 1997, On the Existence of Pure Strategy Equilibria in Games with a Continuum of Players, *JET* 76, 13–46.

Khan, M. A. & Y. N. Sun, 1995, Pure Strategies in Games with Private Information, *JME* 24, 633–653.

Khan, M. A. & Y. Sun, 1996, Integrals of Set-Valued Functions with a Countable Range, *MOR* 21, 946–954.

Khan, M. A. & Y. Sun, 1996, Nonatomic Games on Loeb Spaces, *PNAS* 93, 15518–15521.

Khan, M. A. & R. Vohra, 1984, Equilibrium in Abstract Economies without Ordered Preferences and with a Measure Space of Agents, *JME* 13, 133–142.

Kim, S. H., 1997, Continuous Nash Equilibria, *JME* 28, 69–84.

Kim, Y. K. & B. M. Ghil, 1997, Integrals of Fuzzy Number-Valued Functions, *Fuzzy* 86, 213–222.

Kirman, A. P., 1981, Measure Theory with Applications to Economics, in *Handb Math Econ* 1, 159–209.

Kisielewicz, M., 1981, On the Trajectories of Generalized Functional-Differential Systems of Neutral Type, *J Opt Th Appl* 33, 255–266.

Klement, E. P., 1988, A Radon–Nikodym Theorem for Fuzzy-Valued Measures, *Fuzzy* 27, 45–51.

Klement, E. P., M. L. Puri & D. A. Ralescu, 1986, Limit Theorems for Fuzzy Random Variables, *Proc Royal Soc London A* 407, 171–182.

Koutsougeras, L. C. & N. C. Yannelis, 1994, Convergence and Approximation Results for Non-Cooperative Bayesian Games: Learning Theorems, *Econ Th* 4, 843–857.

Krbec, P., 1976, Weak Stability of Multivalued Differential Equations, *Czechoslovak Math J* 26, 470–476.

Kruse, R., 1987, On the Variance of Random Sets, *J Math Anal Appl* 122, 469–473.

Li, L. S., 1995, Random Fuzzy Sets and Fuzzy Martingales, *Fuzzy* 69, 181–192.

Li, S. M. & Y. Ogura, 1998, Convergence of Set-Valued Sub- and Supermartingales in the Kuratowski–Mosco Sense, *Ann Prob* 26, 1384–1402.

Li, L. S. & Z. H. Sheng, 1998, The Fuzzy Set-Valued Measures Generated by Fuzzy Random Variables, *Fuzzy* 97, 203–209.

Lopez-Diaz, M. & M. A. Gil, 1998, The Lambda-Average Value and the Fuzzy Expectation of a Fuzzy Random Variable, *Fuzzy* 99, 347–352.

Lopez-Diaz, M. & M. A. Gil, 1998, Reversing the Order of Integration in Iterated Expectations of Fuzzy Random Variables and Statistical Applications, *J Stat Planning & Inference* 74, 11–29.

Luu, D. Q., 1986, Representation Theorems for Multi-Valued (Regular) L^1–amarts, *Math Scand* 58, 5–22.

Lyashenko, N. N., 1979, Limit Theorems for Sums of Independent Random Sets, *Th Prob Appl* 24, 438–440.

Markov, S. M., 1977, Differential Calculus for Interval-Valued Functions based on Extended Interval Arithmetic, *Doklady Bolg* 30, 1377–1380.

Markov, S. M., 1978, Calculus for Interval Functions of a Real Variable Using Extended Interval Arithmetic, *Doklady Bolg* 31, 373–376.

Markov, S. M., 1979, Calculus for Interval Functions of a Real Variable, *Computing* 22, 325–337.

Mas-Colell, A. & W. R. Zame, 1996, The Existence of Security Market Equilibrium with a Nonatomic State Space, *JME* 26, 63–84.

Meister, H. & O. Moeschlin, 1988, Unbiased Set-Valued Estimators with Minimal Risk, *J Math Anal Appl* 130, 426–438.

Menezes, F. M. & P. K. Monteiro, 1997, Sequential Asymmetric Auctions with Endogenous Participation, *Th & Dec* 43, 187–202.

Mertens, J. F., 1988, Nondifferentiable TU Markets: The Value, in *Roth: Shapley Value*, 235–264.

Molchano, I. S., E. Omey & E. Kozarovitzky, 1995, An Elementary Renewal Theorem for Random Compact Convex Sets, *Adv Appl Prob* 27, 931–942.

Mupasiri, D., 1995, Complex Extreme Measurable Selections, *J Australian Math Soc A* 58, 222–231.

Nanda, S., 1989, On Integration of Fuzzy Mappings, *Fuzzy* 32, 95–101.

Olech, C., 1976, Existence Theory in Optimal Control, in *Weil*, 291–328.

Ouyang, H. 1988, Topological Properties of the Spaces of Regular Fuzzy Sets, *J Math Anal Appl* 129, 346–361.

Papageorgiou, N. S., 1985, On the Efficiency and Optimality of Random Allocations, *J Math Anal Appl* 105, 113–135.

Papageorgiou, N. S., 1985, On Abstract Conditional Expectations, *J Math Anal Appl* 111, 35–48.

Papageorgiou, N. S., 1985, On the Theory of Banach Space-Valued Multifunctions I: Integration and Conditional Expectation, *J Multiv Anal* 17, 185–206.

Papageorgiou, N. S., 1985, Caratheodory Convex Integral Operations and Probability Theory, *Pac J Math* 116, 155–184.

Papageorgiou, N. S., 1985, Representation of Set-Valued Operators, *Trans AMS* 292, 557–572.

Papageorgiou, N. S., 1985, Stochastic Nonsmooth Analysis and Optimization in Banach Spaces, in *Lect Notes* 259, 226–242.

Papageorgiou, N. S., 1986, On the Efficiency and Optimality of Allocations II, *SIAM J Contr* 24, 452–479.

Papageorgiou, N. S., 1986, Random Differential Inclusions in Banach Spaces, *J Diff Eq* 65, 287–303.

Papageorgiou, N. S., 1986, The Integral Theory of Ioffe Fans, *J Math Anal Appl* 113, 544–561.

Papageorgiou, N. S., 1986, A Stability Result for Differential Inclusions in Banach Spaces, *J Math Anal Appl* 118, 232–246.

Papageorgiou, N. S., 1987, On Cesari's Property (Q), *J Opt Th Appl* 53, 259–268.

Papageorgiou, N. S., 1994, On Fatou's Lemma and Parametric Integrals for Set-Valued Functions, *J Math Anal Appl* 187, 809–825.

Papageorgiou, N. S., 1997, Convex Integral Functionals, *Trans AMS* 349, 1421–1436.

Papageorgiou, N. S. & D. A. Kandilakis, 1987, Convergence in Approximation and Nonsmooth Analysis, *J Appr Th* 49, 41–54.

Parthasarathy, T., 1972, Selection Theorems and their Applications, *Lect Notes Math* 263, Ch. 7.

Pascoa, M. R., 1993, Noncooperative Equilibrium and Chamberlinian Monopolistic Competition, *JET* 60, 335–353.

Perles, M. A. & M. Maschler, 1981, The Superadditive Solution for Nash Bargaining Games, *IJGT* 10, 163–193.

Pianigiani, G., 1977, On the Fundamental Theory of Multivalued Differential Equations, *J Diff Eq* 25, 30–38.

Polovinkin, E. S., 1983, The Integration of Multivalued Mappings, *Doklady SSSR* 271, 1069–1074.

Pucci, P. & G. Vitillaro, 1984, A Representation Theorem for Aumann Integrals, *J Math Anal Appl* 102, 86–101.

Puri, M. L. & D. A. Ralescu, 1983, A Strong Law of Large Numbers for Banach Space-Valued Random Sets, *Ann Prob* 11, 222–224.

Puri, M. L. & D. A. Ralescu, 1985, A Concept of Normality for Fuzzy Random Variables, *Ann Prob* 13, 1373–1379.

Puri, M. L. & D. A. Ralescu, 1985, Limit Theorems for Random Compact Sets in Banach Space, *Math Proc Cambridge Phil Soc* 97, 151–158.

Puri, M. L. & D. A. Ralescu, 1986, Fuzzy Random Variables, *J Math Anal Appl* 114, 409–422.

Puri, M. L. & D. A. Ralescu, 1991, Convergence Theorem for Fuzzy Martingales, *J Math Anal Appl* 160, 107–122.

Puri, M. L., D. A. Ralescu & S. S. Ralescu, 1987, Gaussian Random Sets in Banach Space, *Th Prob Appl* 31, 526–529.

Qi, J. Z., 1991, On the Solution Set of Differential Inclusions in Banach Space, *J Diff Eq* 93, 213–237.

Qin, C.-Z., 1994, An Inner Core Equivalence Theorem, *Econ Th* 4, 311–317.

Ralescu, A. L. & D. A. Ralescu, 1984, Probability and Fuzziness, *Inf Sci* 34, 85–92.

Ranguin, M., 1982, The Holder Property of the Minimal Time Function of an Autonomous Linear System, *Rairo-Automatique-Sys Anal Contr* 16, 329–340.

Rath, K. P., 1992, A Direct Proof of the Existence of Pure Strategy Equilibria in Games with a Continuum of Players, *Econ Th* 2, 427–433.

Ratschek, H. & G. Schroder, 1971, The Concept of Derivation of Interval Functions, *Computing* 7, 172–187.

Rechtschaffen, E. E. M., 1976, Equivalences between Differential Games and Optimal Controls, *J Opt Th Appl* 18, 73–79.

Roberts, D. J., 1973, Existence of Lindahl Equilibrium with a Measure Space of Consumers, *JET* 6, 355–381.

Robertson, A. P., 1974, Measurable Selections, *Proc Royal Soc Edinburgh A* 72, 1–7.

Rockafellar, R. T., 1969, Measurable Dependence of Convex Sets and Functions on Parameters, *J Math Anal Appl* 28, 4–25.

Romanflores, H. & R. C. Bassanezi, 1997, On Multivalued Fuzzy Entropies, *Fuzzy* 86, 169–177.

Rupp, W., 1977, Set-Valued Measures and Continuations, *Manus Math* 22, 137–150.

Rupp, W., 1979, The Riesz Presentation of Additive and Sigma-Additive Set-Valued Measures, *Math Ann* 239, 111–118.

Rupp, W., 1979, The Riesz Presentation of Additive and Sigma-Additive Set-Valued Measures, *Pac J Math* 84, 445–453.

Saintpierre, J. & S. Sajid, 1997, Integration of Multifunctions with respect to a Parametrized Measure, *CR Acad Sci* 324, 55–60.

Schmeidler, D., 1969, Competitive Equilibria in Markets with a Continuum of Traders and Incomplete Preferences, *Econometrica* 37, 578–585.

Schmeidler, D., 1970, Fatou's Lemma in Several Dimensions, *Proc AMS* 24, 300–306.

Shaw, W. H., 1977, Boundary-Value Problems and Periodic Solutions for Contingent Differential Equations, *J Math Anal Appl* 57, 610–624.

Shitovitz, B., 1973, Oligopoly in Markets with a Continuum of Traders, *Econometrica* 41, 467–501.

Shitovitz, B., 1992, Coalitionally Fair Allocations in Smooth Mixed Markets with an Atomless Sector, *Math Soc Sci* 25, 27–40.

Silin, D. B., 1998, Viscosity Solutions via Unbounded Set-Valued Integration, *Nonlinear Anal Th Meth Appl* 31, 55–90.

Smith, S. & D. Q. Mayne, 1988, An Exact Penalty Algorithm for Optimal Control Problems with Control and Terminal Constraints, *Int J Contr* 48, 257–271.

Stich, W. J. A., 1988, An Integral for Nonmeasurable Correspondences and the Shapley Interval, *Manus Math* 61, 215–221.

Stojakovic, M., 1992, Fuzzy Conditional Expectation, *Fuzzy* 52, 53–60.

Stojakovic, M., 1994, Fuzzy Random Variables, Expectation, and Martingales, *J Math Anal Appl* 184, 594–606.

Stojakovic, M., 1998, Representation of Fuzzy-Valued Mappings by the Sequence of Single-Valued Mappings, *Fuzzy* 98, 375–381.

Stojakovic, M. & Z. Stojakovic, 1996, Support Functions for Fuzzy Sets, *Proc Royal Soc London A* 452, 421–438.

Subrahmanyam, P. V. & S. K. Sudarsanam, 1996, A Note on Fuzzy Volterra Integral Equations, *Fuzzy* 81, 237–240.

Sun, Y., 1996, Distributional Properties of Correspondences on Loeb Spaces, *J Functional Anal* 139, 68–93.

Sun, Y. N., 1997, Integration of Correspondences on Loeb Spaces, *Trans AMS* 349, 129–153.

Suslov, S. I., 1991, A Nonlinear Bang-Bang Principle in R^n, *Math Notes* 49, 518–523.

Tolstonogov, A. A., 1975, Radon–Nikodym and Lyapunov Theorems for Multivalued Measures, *Doklady SSSR* 225, 1023–1026.

Tolstonogov, A. A., 1977, Support Functions of Convex Compacta, *Math Notes* 22, 604–609.

Tolstonogov, A. A., 1981, Differential Inclusions in Banach Space with a Nonconvex Right-Hand Side: The Existence of Solutions, *Sib Math J* 22, 625–637.

Tolstonogov, A. A., 1982, On the Structure of the Solution Set for Differential Inclusions in a Banach Space, *Sbornik* 118, 1–15.

Tolstonogov, A. A., 1995, Extreme Continuous Selectors of Multivaried Maps and their Applications, *J Diff Eq* 122, 161–180.

Valadier, M., 1971, Measurable Multiapplications with Compact Convex Values, *J Math Pures et Appl* 50, 265–297.

Veliov, V. M., 1987, On the Convexity of Integrals of Multivalued Mappings: Applications in Control Theory, *J Opt Th Appl* 54, 541–563.

Vitale, R. A., 1979, Approximation of Convex Set-Valued Functions, *J Appr Th* 26, 301–316.

Vitale, R. A., 1988, An Alternate Formulation of Mean Value for Random Geometric Figures, *J Microscopy* (Oxford) 151, 197–204.

Vitale, R. A., 1990, The Brunn–Minkowski Inequality for Random Sets, *J Multiv Anal* 33, 286–293.

Vitale, R. A., 1991, The Transitive Expectation of a Random Set, *J Math Anal Appl* 160, 556–562.

Wagner, D. H., 1975, The Integral of a Convex Hull-Valued Function, *J Math Anal Appl* 50, 548–559.

Wagner, D. H., 1976, The Integral of a Set-Valued Function with Semi-Closed Values, *J Math Anal Appl* 55, 616–633.

Wagner, D. H., 1977, A Survey of Measurable Selection Theorems, *SIAM J Contr* 15, 859–903.

Wagner, D. H. & L. D. Stone, 1974, Necessity and Existence Results on Constrained Optimization of Separable Functionals by a Multiplier Rule, *SIAM J Contr* 12, 356–372.

Wang, K. N. & A. N. Michel, 1996, Stability Analysis of Differential Inclusions in Banach Space with Applications to Nonlinear Systems with Time Delays, *IEEE Trans Circuits & Systems Appl* 43, 617–626.

Wang, R. M. & Z. P. Wang, 1997, Set-Valued Stationary Processes, *J Multiv Anal* 63, 180–198.

Watanabe, N. & T. Imaizumi, 1993, A Fuzzy Statistical Test of Fuzzy Hypotheses, *Fuzzy* 53, 167–178

Weil, W., 1982, An Application of the Central Limit Theorem for Banach Space-Valued Random Variables to the Theory of Random Sets, *Z Wahrscheinl* 60, 203–208.

Weil, W., 1995, The Estimation of Mean Shape and Mean Particle Number in Overlapping Particle Systems in the Plane, *Adv Appl Prob* 27, 102–119.

Wu, H. C., 1997, Fuzzy-Valued Integrals of Fuzzy-Valued Measurable Functions with respect to Fuzzy-Valued Measures based on Closed Intervals, *Fuzzy* 87, 65–78.

Wu, C. X. & S. J. Song, 1998, Existence Theorem to the Cauchy Problem of Fuzzy Differential Equations under Compactness-Type Conditions, *Inf Sci* 108, 123–134.

Wu, C. X., S. J. Song & E. S. Lee, 1996, Approximate Solutions, Existence, and Uniqueness of the Cauchy Problem of Fuzzy Differential Equations, *J Math Anal Appl* 202, 629–644.

Wu, C. X., D. L. Zhang, C. M. Guo & C. Wu, 1998, Fuzzy Number Fuzzy Measure and Fuzzy Integrals. (I). Fuzzy Integrals of Functions with respect to Fuzzy Number Fuzzy Measures, *Fuzzy* 98, 355–360.

Wu, C. X., D. L. Zhang, B. K. Zhang & C. M. Guo, 1998, Fuzzy Number Fuzzy Measures and Fuzzy Integrals. (II). Fuzzy Integrals of Fuzzy-Valued Functions with respect to Fuzzy Number Fuzzy Measures on Fuzzy Sets, *Fuzzy* 101, 137–141.

Xue, X. P., M. H. Ha & M. Ma, 1994, Random Fuzzy Number Integrals in Banach Spaces, *Fuzzy* 66, 97–111.

Yamazaki, A., 1978, An Equilibrium Existence Theorem without Convexity Assumptions, *Econometrica* 46, 541–555.

Yamazaki, A., 1981, Diversified Consumption Characteristics and Conditionally Dispersed Endowment Distributions: Regularizing Effect and Existence of Equilibria, *Econometrica* 49, 639–654.

Yannelis, N. C., 1989, Weak Sequential Convergence in $L^p(\mu, X)$, *J Math Anal Appl* 141, 72–83.

Yannelis, N. C., 1990, On the Upper and Lower Semicontinuity of the Aumann Integral, *JME* 19, 373–389.

Yannelis, N. C. & A. Rustichini, 1991, Equilibrium Points of Non-Cooperative Random and Bayesian Games, in *Aliprantis, Border & Luxemburg*, 23–48.

Zeephongsekul, P. & G. Xia, 1996, On Fuzzy Debugging of Software Programs, *Fuzzy* 83, 239–247.

Zhang, D. L. & C. M Guo, 1994, The Countable Additivity of Set-Valued Integrals and F-Valued Integrals, *Fuzzy* 66, 113–117.

Zhang, D. L. & C. M. Guo, 1994, The Fubini Theorem for F-Valued Integrals, *Fuzzy* 62, 355–358.

Zhang, D. & C. M. Guo, 1995, Fuzzy Integrals of Set-Valued Mappings and Fuzzy Mappings, *Fuzzy* 75, 103–109.

Zhang, D. & C. Guo, 1995, Generalized Fuzzy Integrals of Set-Valued Functions, *Fuzzy* 76, 365–373.

Zhang, D. L. & C. M. Guo, 1996, Integrals of Set-Valued Functions for Perpendicular-to-Decomposable Measures, *Fuzzy* 78, 341–346.

Zhang, D. L. & Z. X. Wang, 1993, Fuzzy Integrals of Fuzzy-Valued Functions, *Fuzzy* 54, 63–67.

Zhang, D. L. & Z. X. Wang, 1993, On Set-Valued Fuzzy Integrals, *Fuzzy* 56, 237–241.

Zhang, D. Z., O. Y. He, E. S. Lee & R. R. Yager, 1993, On Fuzzy Random Sets and their Mathematical Expectations, *Inf Sci* 72, 123–142.

Zhang, W. X., T. Li, J. F. Ma & A. J. Li, 1990, Set-Valued Measures and Fuzzy Set-Valued Measures, *Fuzzy* 36, 181–188.

Zhu, Q. J., N. Zhang & Y. He, 1992, An Algorithm for Determining the Reachability Set of a Linear Control System, *J Opt Th Appl* 72, 333–353.

Texts and Monographs: Aubin 1979, 1997; Aumann & Shapley; Dubois & Prade; Ellickson; Hermes & LaSalle; Hildenbrand; Klein & Thompson.

In this collection: Chapters 15, 47, 69, and 70.

Chapter 69 An Elementary Proof that Integration Preserves Uppersemicontinuity

Allen, B., 1986, General Equilibrium with Information Sales, *Th & Dec* 21, 1–33.

Allen, B. & S. Sorin, 1994, General Equilibrium and Cooperative Games: Basic Results, in *Stony Brook ASI* 1991, 17–33.

Debreu, G., 1981, Existence of Competitive Equilibrium, in *Handb Math Econ* 2, 697–743.

Khan, M. A., K. P. Rath & Y. Sun, 1997, On the Existence of Pure Strategy Equilibria in Games with a Continuum of Players, *JET* 76, 13–46.

Papageorgiou, N. S., 1985, On the Theory of Banach Space-Valued Multifunctions I: Integration and Conditional Expectation, *J Multiv Anal* 17, 185–206.

Papageorgiou, N. S., 1994, On Fatou's Lemma and Parametric Integrals for Set-Valued Functions, *J Math Anal & Appl* 187, 809–825.

Rashid, S., 1978, Existence of Equilibrium in Infinite Economies with Production, *Econometrica* 46, 1155–1164.

Rath, K. P., 1992, A Direct Proof of the Existence of Pure Strategy Equilibria in Games with a Continuum of Players, *Econ Th* 2, 427–433.

Rath, K. P., 1994, Some Refinements of Nash Equilibria of Large Games, *GEB* 7, 92–103.

Rath, K. P., 1998, Perfect and Proper Equilibria of Large Games, *GEB* 22, 331–342.

Yamazaki, A., 1978, An Equilibrium Existence Theorem without Convexity Assumptions, *Econometrica* 46, 541–555.

Yamazaki, A., 1981, Diversified Consumption Characteristics and Conditionally Dispersed Endowment Distributions: Regularizing Effect and Existence of Equilibria, *Econometrica* 49, 639–654.

Yannelis, N. C., 1990, On the Upper and Lower Semicontinuity of the Aumann Integral, *JME* 19, 373–389.

Texts and Monographs: Ellickson.

Chapter 70 A Variational Problem Arising in Economics

Artstein, Z., 1974, On a Variational Problem, *J Math Anal Appl* 45, 404–415.

Artstein, Z., 1980, Generalized Solutions to Continuous Time Allocation Processes, *Econometrica* 48, 899–922.

Artstein, Z., 1984, Convergence Rates for the Optimal Values of Allocation Processes, *MOR* 9, 348–355.

Artstein, Z. & J. Greenberg, 1980, Exhaustion Time in Continuous Allocation Processes, *J Math Anal Appl* 78, 378–399.

Artstein, Z. & S. Hart, 1981, A Law of Large Numbers for Random Sets and Allocation Processes, *MOR* 6, 485–492.

Artstein, Z. & R. J. B. Wets, 1988, Approximating the Integral of a Multifunction, *J Multiv Anal* 24, 285–308.

Artstein, Z. & R. J. B. Wets, 1989, Decentralized Allocation of Resources among Many Producers, *JME* 18, 303–324.

Aubin, J. P. & I. Ekeland, 1975, Minimization of Integral Criteria, *CR Acad Sci* 281, 285–288.

Aubin, J. P. & I. Ekeland, 1976, Estimates of the Duality Gap in Nonconvex Optimization, *MOR* 1, 225–245.

Balder, E. J., 1979, A Useful Compactification for Optimal Control Problems, *J Math Anal Appl* 72, 391–398.

Balder, E. J., 1984, A Unifying Note on Fatos's Lemma in Several Dimensions, *MOR* 9, 267–275.

Balder, E. J., 1984, A General Approach to Lower Semicontinuity and Lower Closure in Optimal Control Theory, *SIAM J Contr* 22, 570–598.

Balder, E. J., 1985, Elimination of Randomization in Statistical Decision Theory Reconsidered, *J Multiv Anal* 16, 260–264.

Balder, E. J., 1988, Fatou's Lemma in Infinite Dimensions, *J Math Anal Appl* 136, 450–465.

Balder, E. J., 1994, New Existence Results for Optimal Controls in the Absence of Convexity: The Importance of Extremality, *SIAM J Contr* 32, 890–916.

Berliocchi, H. & J. M. Lasry, 1973, Normal Integrands and Parametric Measurements in the Calculus of Variations, *Bull SMF* 101, 129–184.

Butnariu, D., 1987, Values and Cores of Fuzzy Games with Infinitely Many Players, *IJGT* 16, 43–68.

Chichilnisky, G., 1981, Existence and Characterization of Optimal Growth Paths Including Models with Non-Convexities in Utilities and Technologies, *RES* 48, 51–61.

Cox, J. C. & C. F. Huang, 1991, A Variational Problem Arising in Financial Economics, *JME* 20, 465–487.

Dubey, P. & A. Neyman, 1984, Payoffs in Nonatomic Economies: An Axiomatic Approach, *Econometrica* 52, 1129–1150.

Dubey, P. & A. Neyman, 1994, An Axiomatic Approach to the Equivalence Phenomenon, in *Stony Brook ASI* 1991, 137–144.

Dubey, P. & A. Neyman, 1997, An Equivalence Principle for Perfectly Competitive Economies, *JET* 75, 314–344.

Ekeland, I., 1974, Quelques applications de l'analyse convexe a la resolution des problemes d'optimisation non convexes, in *Lect Notes* 102, 102–114.

Groves, T. & S. Hart, 1982, Efficiency of Resource Allocation by Uninformed Demand, *Econometrica* 50, 1453–1482.

Hart, S., 1977, Values of Non-Differentiable Markets with a Continuum of Traders, *JME* 4, 103–116.

Hart, S., 1979, Values of Large Market Games, in *Vienna Conf*, 187–197.

Hart, S., 1980, Measure Based Values of Market Games, *MOR* 5, 197–228.

Hart, S., 1981, A Variational Problem Arising in Economics: Approximate Solutions and the Law of Large Numbers, in *Hagen Conf II*, 281–290.

He, H. & N. D. Pearson, 1991, Consumption and Portfolio Policies with Incomplete Markets and Short-Sale Constraints: The Infinite Dimensional Case, *JET* 54, 259–304.

Lasry, J. M. & H. Berliocchi, 1972, New Applications of Parametrized Measurements, *CR Acad Sci* 274, 1623–1626.

Lefevre, F., 1994, Addendum: The Shapley Value of a Perfectly Competitive Market May Not Exist, in *Stony Brook ASI* 1991, 95–104.

Magill, M. J. P., 1981, Infinite Horizon Programs, *Econometrica* 49, 679–711.

Maruyama, T., 1979, An Extension of the Aumann–Perles Variational Problem, *Proc Jap Acad* 55, 348–352.

Mertens, J. F., 1988, Nondifferentiable TU Markets: The Value, in *Roth: Shapley Value*, 235–264.

Penot, J. P., 1979, Non-Convex Analysis: Proceedings of the Pau Colloquium, May 23–25, 1977, *Bull SMF*, 7–24.

Petrakis, E., E. Rasmusen & S. Roy, 1997, The Learning Curve in a Competitive Economy, *Rand J* 28, 248–268.

Puri, M. L. & D. A. Ralescu, 1983, Differentials of Fuzzy Functions, *J Math Anal Appl* 91, 552–558.

Puri, M. L. & D. A. Ralescu, 1986, Fuzzy Random Variables, *J Math Anal Appl* 114, 409–422.

Qin, C.-Z., 1994, An Inner Core Equivalence Theorem, *Econ Th* 4, 311–317.

Rosenmüller, J., 1977, Remark on the Transfer Operator and the Value-Equilibrium Equivalence Hypothesis, in *Lect Notes* 141, 108–127.

Wagner, D. H., 1976, The Integral of a Set-Valued Function with Semi-Closed Values, *J Math Anal Appl* 55, 616–633.

Wagner, D. H., 1977, A Survey of Measurable Selection Theorems, *SIAM J Contr* 15, 859–903.

Wagner, D. H. & L. D. Stone, 1974, Necessity and Existence Results on Constrained Optimization of Separable Functionals by a Multiplier Rule, *SIAM J Contr* 12, 356–372.

Texts and Monographs: Aubin 1979; Aumann & Shapley.

In this collection: Chapters 51, 53, and 68.

Chapter 71 Random Measure-Preserving Transformations

Deneckere, R., 1986, On the Existence of Random Measure Preserving Bijections, *JME* 15, 267–274.

Kallman, R. R., 1985, Uniqueness Results for Groups of Measure Preserving Transformations, *Proc AMS* 95, 87–90.

Texts and Monographs: Aumann & Shapley.

Chapter 72 Orderable Set Functions and Continuity III: Orderability and Absolute Continuity

Rothblum, U. G., 1977, Orderable Set Functions and Continuity II: Set Functions with Infinitely Many Null Points, *SIAM J Contr* 15, 144–155.

Chapter 73 Bi-Convexity and Bi-Martingales

Bergin, J. & W. B. MacLeod, 1993, Efficiency and Renegotiation in Repeated Games, *JET* 61, 42–73.

Cave, J., 1987, Equilibrium and Perfection in Discounted Supergames, *IJGT* 16, 15–41.

Forges, F., 1984, A Note on Nash Equilibria in Infinitely Repeated Games with Incomplete Information, *IJGT* 13, 179–187.

Forges, F., 1988, Communication Equilibria in Repeated Games with Incomplete Information, *MOR* 13, 191–231.

Forges, F., 1990, Equilibria with Communication in a Job Market Example, *QJE* 105, 375–398.

Forges, F., 1990, Repeated Games with Incomplete Information, in *Ohio Conf*, 64–76.

Forges, F., 1992, Repeated Games of Incomplete Information: Non-Zero-Sum, in *Handb Game Th* 1, 155–177.

Forges, F. & E. Minelli, 1997, A Property of Nash Equilibria in Repeated Games with Incomplete Information, *GEB* 18, 159–175.

Hart, S., 1985, Non-Zero-Sum 2-Person Repeated Games with Incomplete Information, *MOR* 10, 117–153.

Ma, J. P., 1995, An Infinitely Repeated Rental Model with Incomplete Information, *Econ Let* 49, 261–266.

Matousek, J. & P. Plechac, 1998, On Functional Separately Convex Hulls, *Discrete & Comp Geometry* 19, 105–130.

Mertens, J. F., 1987, Repeated Games, in *The New Palgrave* 4, 151–152.

Mertens, J. F., 1987, Repeated Games, in *Proc Int Congr Math* 1986, Vol 2, 1528–1577.

Simon, R. S., S. Spiez & H. Torunczyk, 1995, The Existence of Equilibria in Certain Games, Separation for Families of Convex Functions and a Theorem of Borsuk–Ulam Type, *IJM* 92, 1–21.

Sorin, S., 1994, Implementation with Plain Conversation, in *Stony Brook ASI* 1991, 261–268.

Texts and Monographs: Aumann & Maschler.

Name Index

Abel, N. H., **I:** 249, 252–254, 412
Abner (biblical), **I:** 107
Abraham ben David, Rabbi, **II:** 143, 145
Abreu, D., **I:** 483, 485
Agmon, S., **II:** 637
Akerlof, G., **I:** 18, 39
Alfasi, Rabbi Isaac, **II:** 139, 143
Allais, M., **I:** 476, 485
Allen, B., **II:** 431, 445
Allen, S. G., **I:** 226, 229
Alt, F., **I:** 201, 208
Anderson, John Bayard, **I:** 138
Anderson, R. M., **I:** 24, 39, 84, 99. **II:** 213
Anscombe, F. J., **I:** 189, 190, 295, 306, 310, 567, 584, 591
Antonelli, G. A., **I:** 636, 648
Archimedes, **I:** 263–265, 301
Arens, R., **II:** 581, 597
Aristotle, **I:** 115
Arkoff, A., **I:** 65, 106, 110, 111
Armbruster, W., **I:** 650, 653, 666, 667
Arrow, K. J., **I:** 3–5, 18, 24, 39, 75, 98, 99, 127, 131–134, 135 ff., 422, 597, 649, 650, 667. **II:** 167, 169, 171, 187, 213, 251, 256–258, 270, 283, 430, 433, 445, 520, 607, 617, 625
Arrow, S., **I:** 132, 133
Artstein, Z., **I:** 87, 99
Ashenfelter, O., **I:** 39
Asscher, N., **I:** 82, 99
Aumann, S., **II:** 60, 135, 152
Aumann, Y., **II:** 143, 146, 152
Axelrod, R., **I:** 22, 39, 70, 71, 92, 100, 444, 472, 476, 484, 485

Bach, Johann Sebastian, **I:** 6, 17
Bacon, N. T., **I:** 40
Baire, R.-L., **II:** 577–579, 582, 584, 590, 600, 602, 606
Baker, C. W., **I:** 35
Balassa, B., **I:** 422
Balinski, M. L., **I:** 97, 100. **II:** 94, 110, 152
Banach, S., **I:** 256. **II:** 172, 185, 577, 578, 580, 582–584, 586, 587, 589, 590, 597, 613, 626
Banzhaf, J. F., **II:** 555, 559
Basu, K., **I:** 481, 485, 636, 647
Bator, F. M., **II:** 171
Bayes, T., **I:** 41, 42, 79, 100, 102, 431, 433–435, 532, 556, 591, 596–598, 602 ff., 634, 647, 648, 667, 668
Becker, G. S., **I:** 312
Bellman, R., **I:** 551, 552
Ben Porath, E., **I:** 470–472, 483–485, 636, 648
Berbee, H., **I:** 87, 100
Berge, C., **I:** 357, 358, 380, 393
Bergstrom, T., **I:** 119
Bernheim, B. D., **I:** 613
Bernoulli, J., **I:** 537. **II:** 28, 50, 420
Bertrand, J., **I:** 25, 40

Bewley, T., **I:** 24, 40, 68, 100, 119, 425
Bicchieri, C., **I:** 636, 648
Billera, L. J., **I:** 26, 37, 40, 62, 82, 88, 89, 98, 100. **II:** 558, 559
Binmore, K. G., **I:** 22, 40, 62, 72, 100, 480, 483–486, 635, 636, 648, 650, 664, 667. **II:** 520
Bixby, R., **I:** 26, 40, 62, 88, 89, 100
Blackwell, D., **I:** 68, 98, 100, 300–302, 321, 330, 334–336, 353, 424, 427, 551, 552. **II:** 28, 29, 52, 53
Boege, W., **I:** 650, 653, 666, 667
Bogardi, I., **I:** 36, 40, 97, 100
Böhm, V., **I:** 119
Bonanno, G., **I:** 636, 648
Bondareva, O. N., **I:** 25, 26, 40, 53, 62, 88, 89, 100. **II:** 39, 55
Bonini, C., **I:** 46
Borel, E., **I:** 52, 100, 103, 106, 188, 280, 281, 283, 286–288, 507, 509, 517, 521, 522, 534, 535, 539, 549, 550, 621. **II:** 178–181, 226, 261, 293, 326, 335, 342, 343, 361, 362, 573, 575, 578, 580, 581 ff., 600, 602–604, 606, 608–610, 613–616, 620, 624, 625, 632–634, 636, 644
Borns, M., **I:** xi. **II:** xi
Boskin, M. J., **II:** 281, 284
Bott, R., **I:** 31, 40
Brachfeld, M., **I:** 17
Braithwaite, R. B., **I:** 100, 104
Brams, S. J., **I:** 44, 100, 103
Brand, A., **I:** 265
Brandenburger, A., **I:** 557, 613, 634, 636, 643, 647, 648, 650, 651, 653, 664, 666, 667
Breiman, L., **I:** 534, 552
Brito, D. L., **I:** 191, 311, 312
Brock, H. W., **II:** 507, 520
Broida, T. R., **I:** 226, 229
Brouwer, L. E. J., **I:** 89, 101, 102. **II:** 26, 176, 617
Brown, D. J., **I:** 24, 40, 84, 100. **II:** 417, 429, 507, 520, 557, 559
Bruno, M., **I:** 4, 145, 152
Burger, E., **II:** 37, 38, 53
Burns, Robert, **I:** 597
Butters, G., **I:** 38, 40

Cain, G. G., **II:** 283
Cantor, G., **II:** 647
Capone, Al, **I:** 422
Carathéodory, C., **I:** 549. **II:** 610, 614, 655, 658
Carr, J. C., **I:** 35
Carter, Jimmy, **I:** 138
Case, J. H., **I:** 52, 100
Castaing, C., **I:** 287, 289
Cauchy, A. L., **I:** 540, 546, 547, 551
Cave, J., **I:** 421, 436, 443, 444
Cesaro, E., **I:** 68, 330, 412, 545. **II:** 422
Chakravarty, S., **II:** 625

Champsaur, P., **I**: 62, 101, 119. **II**: 199, 217, 222, 245, 256, 390, 392, 399, 419, 420, 429, 445, 485, 494, 497, 505, 507, 520, 537, 540, 557, 559
Chang, C., **II**: 133
Chapman, J. W., **I**: 44, 104
Charnes, A., **I**: 251, 256
Cheng, H. C., **I**: 24, 40. **II**: 507, 520, 540
Chernoff, H., **I**: 300, 302
Chipman, J. S., **I**: 208. **II**: 505
Cho, I.-K., **I**: 481, 486
Chung, K. L., **II**: 667, 674
Coase, R. H., **II**: 378–380, 382
Cobb, C. W., **II**: 304
Cohen, P., **I**: 288
Conner, J. R., **I**: 43
Coombs, C. H., **I**: 256, 274
Cooper, W. W., **I**: 251, 256
Cornet, B., **I**: 119
Cournot, A. A., **I**: 18, 40, 43
Crawford, V., **I**: 25, 40
Cromwell, P. R., **I**: 161
Crowell, R. H., **I**: 161, 181
Curio, E., **I**: 92, 104

Dalkey, N. C., **I**: 356, 383, 384, 393, 493, 494, 595, 596
Dalla Chiara, M. L., **I**: 648
Dantzig, G. B., **I**: 66, 101, 186, 193, 208
Dasgupta, P., **I**: 40, 485, 621, 647, 650, 667
d'Aspremont, C., **I**: 18, 40, 119, 120
David, P. A., **I**: 478, 486
Davidson, D., **I**: 300, 303
Davis, M., **I**: 27, 40, 52, 82, 97, 101. **II**: 41, 55, 59, 91, 93, 102, 107, 110, 113, 123, 132, 148, 152, 161, 169, 541, 543, 544, 551, 559
Davis, R. G., **I**: 237, 256, 266
Davis, R. L., **I**: 256, 274
de Finetti, B., **I**: 300, 303. **II**: 236, 239, 256
De Gaulle, Charles, **II**: 260
de la Vallèe Poussin, C. J., **I**: 120
Debreu, G., **I**: 16, 24, 39, 40, 84, 98, 99, 101, 127, 135, 140, 249, 252, 253, 256, 279, 280, 287, 289, 421, 634. **II**: 47, 53, 158, 160, 162, 164, 167–170, 173, 187, 205, 206, 212, 213, 217, 221, 225, 242, 245, 251, 256, 302, 320, 325, 333, 340, 358, 383, 385, 407, 429, 607, 617, 639, 644
de Groot, M. H., **I**: 595, 596
Dehn, M., **I**: 161, 181
Deistler, M., **I**: 44. **II**: 521
Dekel, E., **I**: 470–472, 613, 643, 648, 650, 651, 653, 666, 667
Delbaen, F., **I**: 119
Deschamps, R., **I**: 120
Diamond, P. A., **I**: 18, 40
Dierker, E., **I**: 24, 40. **II**: 212, 213
Dierker, H., **I**: 24, 40
Dolbear, F. T., Jr., **II**: 379, 382

Doob, J. L., **II**: 606
Doppler, C. A., **I**: 10
Douglas, P. C., **II**: 304
Drazen, A., **II**: 383, 429
Dresher, M. A., **I**: 101, 104, 503, 552, 591. **II**: 3, 27, 29, 55, 63, 110, 132, 358, 462
Dreze, J. H., **I**: 25, 41, 87, 97, 100, 101, 119–121, 125–127, 190, 305, 310, 615. **II**: 60, 113, 218, 383, 384, 387, 390, 405, 406, 429, 440, 445, 448, 462, 507, 509, 520
Dubey, P., **I**: 74, 85, 101. **II**: 552, 555, 559
Dubins, L., **I**: 25, 41, 306. **II**: 617, 668, 670, 674
Dugundji, J., **II**: 581, 597
Dunford, N., **II**: 176, 183, 187, 613, 614, 618, 628, 638, 647, 651
Dvoretzky, A., **I**: 551, 552

Eatwell, J., **I**: 47, 667, 668
Edgeworth, F. Y., **I**: 24, 40, 41, 46, 60, 67, 84, 101, 140. **II**: 54, 160, 168, 170, 191, 192, 206, 209, 213, 522, 542, 617
Eggleston, H. G., **I**: 272, 273. **II**: 187, 250, 256, 610, 614, 618, 629, 638
Eilenberg, S., **II**: 41, 55, 607, 617
Eisele, T., **I**: 653, 667
Ellsberg, D., **I**: 476, 486
Erdos, P., **I**: 9, 538, 552
Ernest, J., **II**: 605
Euclid, **I**: 76
Everett, H., **I**: 67, 101, 358, 393

Farquharson, R., **II**: 24, 26
Farrell, J., **I**: 620
Farrell, M. J., **II**: 171
Fatou, P., **II**: 183, 279, 326, 609, 613, 615, 619, 621, 622
Feiwel, G. R., **I**: 127 ff., 135 ff.
Feller, W., **I**: 332, 340, 353
Fellner, W., **I**: 415
Fenchel, W., **I**: 270, 273. **II**: 236, 238, 249, 256
Ferguson, T. S., **I**: 68, 100, 302
Fiacco, A. V., **I**: 46
Fishburn, P., **II**: 242, 256
Fogelman, F., **I**: 87, 101
Foley, D. K., **I**: 25, 41. **II**: 335, 337
Forges, F., **I**: 20, 41, 119, 482, 486
Foster, D., **I**: 479, 486
Fouraker, L. E., **I**: 16, 41
Fourier, J.-B. J., **I**: 533, 544, 545, 547
Fox, M., **I**: 506, 521
Fox, R. H., **II**: 581, 597
Fraenkel, A. H., **I**: 76
Freedman, D. A., **I**: 25, 41
Friedman, J. W., **I**: 22, 41
Fubini, G., **I**: 538, 540, 548
Fudenberg, D., **I**: 469–472, 481, 482, 486, 632, 634, 650, 668
Furst, E., **I**: 44. **II**: 521
Furstenberg, H., **II**: 644

Gabszewicz, J. J., **I:** 18, 25, 40, 41, 87, 101, 119, 120, 615. **II:** 509, 520
Gale, D., **I:** 25, 41, 50, 51, 85, 87, 88, 101, 157, 356, 357, 393, 485, 621, 647. **II:** 158, 171, 187, 201–203
Galil, Z., **I:** 87, 101
Gardner, R. J., **I:** 36, 41. **II:** 218, 335, 337, 507, 520, 527
Gately, D., **I:** 36, 41
Gauss, C. F., **II:** 205, 211, 212, 226, 237, 238, 240, 251
Geanakoplos, J., **I:** 649, 650, 658, 668
Gepts, S., **I:** 25, 41, 87, 101, 119, 120
Gerard-Varet, L.-A., **I:** 120
Gibbard, A., **I:** 138
Gilboa, Y., **I:** 471, 473, 484, 486
Gillette, D., **I:** 357, 358, 393
Gillies, D. B., **I:** 60, 66, 87, 101, 357, 393. **II:** 5, 8, 11, 13, 17, 26, 35, 55, 113, 114, 132, 167, 170
Gini, **II:** 533
Girshick, M. A., **I:** 300–302
Gödel, K., **I:** 288, 289
Gomory, R., **I:** 240, 256
Good, I. J., **I:** 300, 303
Gorman, W., **II:** 356
Greenberg, J., **I:** 119
Gray, J. H., **I:** 35
Griesmer, J. H., **I:** 18, 41. **II:** 108, 110
Grodal, B., **I:** 24, 41
Grossman, S. J., **I:** 593, 596
Groves, T., **II:** 494
Grünbaum, B., **I:** 277
Guesnerie, R., **I:** 36, 41. **II:** 507, 520
Guggenheim, J. S., **II:** 527
Guilbaud, G. T., **I:** 101, 105, 279, 280. **II:** 256, 430, 505, 540, 545
Gurk, H. M., **II:** 108, 110
Güth, W., **I:** 20, 45, 480, 486

Hadamard, J. S., **II:** 95
Hahn, F. H., **I:** 4, 24, 39, 145, 152, 485, 621, 634, 647, 648, 667. **II:** 213
Hai Gaon, Rabbi, **II:** 139
Hall, R. E., **II:** 281, 283
Halmos, P. R., **I:** 513, 521. **II:** 577, 579, 581, 583, 584, 594, 597, 619, 621, 639, 644, 646, 651
Hamilton, Alexander, **II:** 152
Hammond, P., **I:** 22, 41, 415, 422
Hansen, **I:** 120
Harary, F., **II:** 36, 55
Harrison, J. M., **I:** 609, 613
Harsanyi, J. C., **I:** 4, 11, 16, 19, 20, 23, 36, 41, 42, 54, 55, 72, 74, 76 ff., 153, 154, 415, 434, 435, 523, 524, 532, 563, 584, 585, 587, 588, 591, 593 ff., 597, 603, 611, 613–616, 620–623, 625, 634, 650, 651, 663, 668. **II:** 13, 26, 40, 53, 54, 152, 258, 263, 266, 271, 272, 284, 285, 291, 292, 316, 331, 333, 337, 341, 344–346, 352, 358–360, 363, 433, 436, 437, 445, 469, 483, 485, 487, 488, 492, 494, 508, 512 ff., 523, 527, 529, 530, 535, 557, 558, 560
Hart, O., **I:** 485, 621, 647
Hart, S., **I:** 22, 31–33, 36, 42, 53, 62, 74, 77, 81, 84, 85, 87, 97, 98, 102, 119, 472, 486. **II:** 356, 358, 381, 382, 384, 408, 417, 419, 430, 431, 445, 448, 462, 485, 494, 502, 507, 521, 529, 535, 544, 545, 549, 550, 555–557, 560, 566, 569, 576, 653, 666, 670, 674
Hausdorff, F., **I:** 418, 448, 468. **II:** 495, 574, 588, 590, 597
Hausner, M., **I:** 250, 252, 256, 257, 261, 263, 265, 269, 273
Heaney, J. P., **I:** 36, 46
Heath, D. C., **I:** 37, 40, 98, 100. **II:** 558, 559
Heller, J. E. I., **I:** 209
Henn, R., **II:** 521
Henry, C., **I:** 120
Hermann, J., **I:** 100
Herstein, I. N., **I:** 250, 252, 256, 257, 264, 273
Hicks, J., **I:** 101. **II:** 26, 54, 284
Hilbert, D., **I:** 509, 520
Hildenbrand, W., **I:** 24, 42, 99, 102, 103, 119, 421, 634. **II:** 221, 256, 261, 284, 299, 320, 325, 326, 333, 384, 417, 430, 495, 504, 505, 560, 619–621
Hintikka, J., **I:** 622
Hitler, Adolph, **I:** 642
Hofbauer, J., **I:** 479, 486
Honkapohja, S., **I:** 3, 5, 135
Hotelling, H., **I:** 18, 40, 42
Howson, J. T., **I:** 98, 103
Hu, T. C., **I:** 102, 105
Hurwicz, L., **I:** 18, 42, 208. **II:** 497, 505

Ibn Ezra, Rabbi Abraham, **II:** 143, 153
Ichiishi, T., **II:** 569
Imai, H., **I:** 36, 42. **II:** 507, 521
Intriligator, M. D., **II:** 430
Isaacs, R., **I:** 52, 102
Isbell, J. R., **I:** 511, 521. **II:** 16, 18, 26, 37, 40, 42, 52–54, 107, 110

Jacob (biblical), **I:** 422, 430
Jacquemin, A. E., **I:** 120
Jaedicke, R., **I:** 46
Jahnssonin, Y., **I:** 3
Jaskold-Gabszewicz. *See* Gabszewicz
Jensen, J. L., **I:** 424
Jentzsch, G., **II:** 3, 27–29, 48–50, 53, 54
Jervis, R., **I:** 619, 620
Joab (biblical), **I:** 107
Jordan, M.-E. C., **I:** 167
Joseph (biblical), **I:** 422, 430
Justman, M., **II:** 114, 118

Kadane, J. B., **I:** 613, 614, 625
Kahn, Herman, **I:** 67

Kahneman, D., **I:** 595, 596
Kakutani, S., **I:** 66, 102, 280, 414. **II:** 7, 23, 26, 607, 617
Kalai, E., **I:** 19, 42, 72, 76, 102, 442, 470, 473, 481, 483, 486, 597, 614. **II:** 150, 153
Kalisch, G. K., **II:** 36, 55, 169, 170
Kandori, M., **I:** 479, 486
Kaneko, M., **I:** 25, 42, 88, 102
Kannai, Y., **I:** 24, 42, 87, 101, 102, 253, 256, 279, 289. **II:** 140, 152, 224, 236, 256, 263, 284, 287, 309, 333, 344, 358, 362, 382, 408, 430, 495, 497, 498, 505, 553, 554, 560, 617, 637
Karlin, S., **II:** 256, 609, 616–618, 625
Katznelson, Y., **I:** 492, 533, 545, 547, 552
Kelley, J. L., **I:** 249, 256. **II:** 605, 606
Kepler, Johannes, **I:** 66, 111
Kern, R., **II:** 507, 521
Kerrich, J. E., **I:** 295, 303
Kestelman, H., **I:** 538, 552
Kimball, G. E., **I:** 208
Kissinger, Henry, **I:** 67
Klee, V. L., **I:** 270
Kleitman, D., **II:** 140, 152
Knoer, E. M., **I:** 25, 40
Knuth, D. E., **I:** 25, 42
Kohlberg, E., **I:** 19, 42, 68, 83, 100, 102, 119, 425–427, 436, 442, 470, 473, 481, 486, 597, 614. **II:** 118, 132, 151, 153
Kolmogorov, A. N., **I:** 308, 338, 622
König, J., **II:** 667
Koopman, B. O., **I:** 300, 303
Koopmans, T. C., **I:** 101, 103. **II:** 171, 625
Kortanek, K. O., **I:** 46
Kraft, C. H., **I:** 245, 256, 300, 303
Krawczyk, J., **I:** 408
Krelle, W., **II:** 634, 638
Kreps, D. M., **I:** 19, 22, 42, 319, 442, 444, 445, 471, 473, 481, 486, 597, 609, 613, 614, 617, 630, 633, 634, 650, 668. **II:** 535
Kripke, S., **I:** 622
Kruskal, J. B., **I:** 185, 193, 194, 202, 208, 209, 214–217, 219, 222, 228, 229, 236, 237, 256, 266
Kudo, H., **II:** 607, 608, 618
Kuhn, H. W., **I:** 40, 41, 45, 48, 50, 66, 67, 86, 101, 103, 105, 193, 208, 256, 329, 353, 356–359, 368, 386, 393, 479, 486, 491, 494, 495, 500, 501, 503, 505, 511–513, 518, 520–523, 532, 584, 591, 609, 614, 636, 637, 648. **II:** 52, 55, 133, 217, 256, 284, 334, 358, 430, 445, 462, 463, 483, 494, 505, 522, 540, 569, 580, 624, 634, 638
Künzi, H. P., **II:** 634, 638
Kuratowski, C., **I:** 285, 289. **II:** 608, 609, 615, 618
Kurz, M., **I:** 22, 28, 35, 36, 39, 42, 75, 100, 121, 129, 319, 415, 419, 421, 422, 430, 431, 443, 444, 472. **II:** 217, 218, 257, 258, 276, 283, 285, 300, 302, 333, 339 ff., 359, 360, 362, 363, 365, 366, 375–378, 380–382, 433, 435, 445, 448, 462, 483, 485, 494, 502, 507, 520, 521, 557–559

Lagrange, J. L., **II:** 250
Langenhop, C. E., **I:** 501
Laplace, P. S., **II:** 96
Larkey, P. D., **I:** 613, 614
Laroque, G., **II:** 199
Lebesgue, H., **I:** 280, 282, 283, 287, 288, 507, 508, 536–539, 545, 546, 548, 549, 552, 571. **II:** 162, 163, 165, 169, 172, 205, 322, 335, 353, 354, 376, 431, 574, 575, 606, 608–610, 615, 619, 621, 625–627, 634, 640, 643, 644, 646, 647
Lefevre, F., **I:** 119
Lehrer, E., **I:** 483, 486
Lemke, L. E., **I:** 98, 103
Lensberg, T., **I:** 97, 103. **II:** 152, 153
Leviatan, D., **II:** 637
Levine, D. K., **I:** 470–472, 481, 482, 486, 598, 614, 632, 634
Lewis, A., **I:** 483, 484, 486
Lewis, D. K., **I:** 20, 42, 81, 103, 555, 650, 668
Lewy, I., **II:** 136, 153
Lincoln, Abraham, **I:** 12
Lindahl, E., **I:** 25, 42. **II:** 335–337
Lindenstrauss, J., **I:** 280, 281, 283, 285
Linial, N., **I:** 484, 486
Linnaeus, **I:** 14
Lipschitz, **I:** 90. **II:** 650
Littlechild, S. C., **I:** 37, 43, 97, 103. **II:** 143, 153, 558, 560
Loeb, P., **I:** 84, 100. **II:** 507, 520, 557, 559
Loehman, E. T., **I:** 36, 43
Loève, M., **II:** 606
Lucas, W. F., **I:** 32, 43, 46, 60, 65, 79, 100, 103. **II:** 113, 129, 133, 555, 560
Luce, R. D., **I:** 15, 39, 41, 43, 46, 60, 67, 79, 81, 101, 103, 105, 107, 112, 257, 259, 261, 274, 291, 294, 297, 298, 303, 321, 328, 330, 338, 353, 498, 501, 530, 532. **II:** 5–8, 11, 13, 17, 24, 26, 27, 32, 43, 54–55, 63, 65, 84, 86, 89, 94, 110, 132, 170, 284, 333, 337, 358, 445, 483, 560
Lyapunov, A., **I:** 66, 103, 551, 573–575, 591. **II:** 185, 336, 554, 617–620, 622, 646, 651

Machiavelli, Niccolò, **I:** 41
Mackey, G. W., **I:** 281, 289, 509, 522. **II:** 578, 580–583, 597, 602, 604–606
Mailath, G., **I:** 479, 486
Maimonides, **I:** 430. **II:** 143
Malinvaud, E., **II:** 350, 406, 430
Malouf, M., **I:** 16, 43, 44
Manes, R. P., **I:** 36, 43
Mantel, R., **II:** 495, 497, 498, 505
Marschak, J., **I:** 257, 274
Martin, D. A., **I:** 52, 103

Marx, Karl, **I:** 117
Maschler, M., **I:** 27, 40, 43, 45, 72, 76, 82, 83, 97, 100, 101, 103, 105, 318, 395, 417, 426, 427, 429, 436, 437, 492, 560, 582, 583, 591. **II:** 41, 46, 55, 59–61, 63, 64, 68, 78, 88, 89, 91, 93, 102, 107, 110, 111, 113, 114, 122–124, 131–133, 135, 148, 149, 151–153, 170, 401, 430, 448, 462, 469, 520, 525, 530, 534, 541, 543–545, 551, 559
Mas-Colell, A., **I:** 24, 26, 36, 43, 62, 74, 84, 85, 89, 97, 99, 102, 103, 421, 634. **II:** 236, 256, 415, 417, 430, 485, 494, 497, 504, 505, 507, 521, 540, 550, 557, 560, 566, 569, 653
Maskin, E., **I:** 469, 471, 472, 481, 482, 485–487, 621, 632, 634, 647
Matsui, A., **I:** 470, 471, 473
Matthew (biblical), **I:** 475
Mayberry, J. P., **I:** 194, 209, 427, 436
Mayer, K. H., **I:** 164
Maynard Smith, J., **I:** 21, 43, 92, 93, 103, 415, 422, 477, 486
Mayr, E., **I:** 142
McGuire, C. B., **I:** 487
McKenzie, L. W., **II:** 164, 167, 171, 172, 177, 178, 180, 185, 187, 607, 617
McKinsey, J. C. C., **I:** 503–505, 507, 522
McShane, R. E., **I:** 237, 256
Megiddo, N., **I:** 45, 395, 417, 627, 628, 634, 631
Menger, K., **I:** 311, 312
Mertens, J.-F., **I:** 19, 25, 37, 41–43, 68, 84, 87, 98, 101, 103, 104, 119, 121, 124, 425, 426, 428, 436, 442, 470, 472, 473, 481, 482, 486, 597, 606, 614, 653, 663, 668. **II:** 219, 347, 357, 358, 383, 408, 430, 431, 444, 445, 549, 554, 558, 560, 563, 569, 653
Meyer, P. A., **II:** 667, 674
Michelson, A. A., **I:** 306, 310
Milgate, M., **I:** 47, 667, 668
Milgrom, P., **I:** 18, 22, 42, 43, 319, 444, 445, 471, 481, 486, 551, 552, 630, 633, 634, 650, 668
Milnor, J. W., **I:** 34, 43, 62, 66, 67, 87, 104, 140, 141, 250, 252, 256, 257, 264, 273. **II:** 86, 157, 161, 170, 277, 284, 287, 333, 555, 560
Mirman, L. J., **I:** 119
Mirrlees, J. A., **II:** 257, 284
Miyasawa, K., **II:** 40, 54
Moeschlin, O., **I:** 667. **II:** 521
Montgomery, D., **II:** 41, 55, 607, 617
Moore, E. H., **I:** 249
Morgenstern, O., **I:** ix, 3, 27 ff., 54 ff., 114, 116, 153, 185, 187, 189, 191, 231, 233, 247, 248, 250, 252, 253, 256, 257 ff., 295, 303, 312, 324, 353, 393, 395, 411, 414, 472, 491, 511, 522, 527, 532, 583, 589, 591, 609, 610, 614, 622. **II:** ix, 3, 5, 6, 9–11, 13, 17, 23, 26, 27, 29, 31 ff., 59, 60, 63, 79, 87, 89, 107, 110, 111, 113, 114, 122, 127, 132, 133, 157, 161, 167, 170, 242, 255, 256, 260, 282, 284, 289, 291, 292, 334, 337, 342, 360, 430, 448, 463, 487, 494, 509, 519, 521, 541, 544, 545, 551, 555, 560
Moriarity, S., **I:** 45, 46
Morley, E. W., **I:** 306, 310
Morse, P. M., **I:** 207, 208
Moschovakis, Y. N., **I:** 104
Moulin, H., **I:** 97, 119, 601, 614. **II:** 152
Muench, T., **II:** 335, 337
Muir, J., **I:** x, xi. **II:** x, xi
Müller, H., **I:** 126, 127. **II:** 384, 387, 390, 405, 429, 440, 445
Murnighan, J. K., **I:** 16, 44
Musgrave, R., **II:** 337
Mycielski, J., **I:** 51, 104
Myerson, R. B., **I:** 18, 43, 74, 81, 104, 428, 442, 473, 481, 482, 487, 597, 614. **II:** 219, 447, 449, 450, 456, 457, 460–463, 507, 521, 534, 535, 558–560

Nakayama, M., **I:** 36, 46
Nash, J. F., **I:** ix, x, 3, 18 ff., 50, 53, 55, 66, 67, 71, 72, 75, 76, 78, 79, 94, 96, 97, 99, 101–104, 139, 140, 153, 154, 260, 267, 274, 317, 318, 321, 324, 353, 356, 357, 359, 393, 395–397, 399, 412, 415, 416, 419, 431–434, 439, 442, 447, 448, 473, 477–479, 481, 485, 487, 491, 494, 495, 498, 501, 523, 524, 533, 555–557, 560, 562–565, 570, 571, 577, 585, 591, 597, 598, 600, 601, 609, 612, 614, 615, 620, 627, 632, 643, 647, 648, 649 ff. **II:** x, 10, 11, 13, 23, 26, 40, 41, 50, 52, 54, 55, 150, 152, 153, 230, 243, 256, 258, 263, 265, 266, 271, 272, 284, 291, 292, 332–334, 434, 445, 469, 473, 474, 483, 485, 488–490, 494, 507–509, 512, 516, 519, 521, 522, 533, 540, 543, 557, 558, 560
Nathan, Rabbi, **II:** 137
Negishi, T., **II:** 245, 256
Nelson, R., **I:** 422
Nering, E. D., **II:** 36, 55, 169, 170
Newman, P., **I:** 47, 667, 668
Newton, Sir Isaac, **I:** 7–9, 46, 66, 111
Neyman, A., **I:** 28, 35, 36, 39, 43, 68, 71, 75, 87, 95, 104, 119, 121, 319, 445, 467, 472, 473, 482–484, 486, 487. **II:** 218, 339–342, 344, 346, 348, 356, 358–360, 363, 375–377, 380–382, 408, 430, 432, 445, 485, 494, 507, 520, 521, 549, 550, 553–555, 558, 560, 566, 569, 653
Neyman, J., **I:** 295, 303
Nikodym, O. M., **I:** 606. **II:** 557
Novshek, W., **I:** 18, 43

O'Neill, B., **I:** 54, 104. **II:** 60, 143, 152, 153
Offenbacher, E., **I:** 46

Ordeshook, P. C., **II:** 507, 520
Osborne, M. J., **I:** 36, 43. **II:** 494, 507, 521
Ostroy, J., **I:** 24, 28, 43
Owen, G., **I:** 37, 43, 74, 97, 103, 104, 523, 530. **II:** 143, 153, 493, 494, 525, 527, 534, 555, 557, 558, 560

Pallaschke, D., **I:** 667
Papadimitriou, C. H., **I:** 483, 487
Papakyriakopoulos, C. D., **I:** 161, 181
Pareto, V., **I:** 20, 29, 36, 42, 59, 127, 419, 431, 439, 440, 448, 449, 461, 462, 466, 469, 470, 473, 498, 616. **II:** 13, 167, 168, 243, 280, 332, 333, 349, 350, 375, 379, 380, 383, 384, 394, 413, 439, 440, 458, 473, 480, 489, 495, 497–499, 505, 567, 568
Parker, T., **I:** 36, 44
Parker, W. N., **I:** 486
Pascal, A. H., **I:** 39
Pattanaik, P. K., **II:** 520
Pauly, M. V., **I:** 18, 44
Peacock, A., **II:** 337
Pearce, D. G., **I:** 613, 650, 668
Pecerskiy, S., **II:** 507, 521
Peleg, B., **I:** xi, 27, 43, 44, 76, 78, 82, 83, 97, 100, 103, 104, 253, 483, 485, 560. **II:** x, 3, 5, 13, 17, 26, 28, 29, 36, 37, 41, 47, 50, 54–55, 59, 60, 63, 87, 89, 91, 111, 113, 114, 122–124, 133, 148 ff., 158, 161, 167, 169, 170, 201, 244, 255, 401, 430, 469, 486, 494, 541, 543, 545, 604, 617
Pengry, D., **I:** 36, 43
Pennock, J. R., **I:** 44, 104
Perles, M. A., **I:** 72, 76, 103. **II:** 150, 153, 227, 240, 255, 315, 333, 412, 469, 574, 607, 617, 618, 623
Pettit, P., **I:** 636, 648
Phelps, E., **I:** 41, 415
Piniles, H. M., **II:** 151–153
Poisson, S. D., **I:** 7
Polak, B., **I:** 649
Polemarchakis, H., **I:** 658, 668
Pollock, Jackson, **I:** 17
Ponssard, J. P., **I:** 119, 427, 436, 437
Postlewaite, A., **I:** 119. **II:** 509, 520
Pratt, J. W., **I:** 245, 256, 300, 303. **II:** 258, 270, 284
Prescott, E. C., **I:** 18, 44
Ptolemy, **I:** 7

Quine, W. V. O., **I:** 307
Quinzii, M., **I:** 25, 44, 87, 101

Raanan, J., **I:** 37, 40, 100. **II:** 558, 559
Rabad. *See* Abraham ben David
Rabie, M., **I:** 32, 43, 60, 103
Rabin, M., **II:** 69, 581
Rabinovitch, N. L., **II:** 143, 153
Rabinowitz, P., **II:** 59, 91

Radner, R., **I:** 18, 22, 44, 421, 482, 483, 487, 492, 533, 536, 551, 552, 593, 596, 622, 630, 634
Radon, J., **I:** 606. **II:** 557
Raiffa, H., **I:** 15, 43, 60, 67, 79, 81, 103, 107, 112, 257, 259, 261, 274, 291, 294, 297, 298, 300, 301, 303, 321, 325, 328, 330, 338, 353, 498, 501, 530, 532. **II:** 5–8, 11, 17, 24, 26, 27, 32, 43, 52–54, 63, 65, 84, 86, 89, 94, 110
Rambam. *See* Maimonides
Ramsey, F. P., **I:** 54, 100, 104, 295, 300, 303
Ransmeier, J. S., **I:** 36, 44, 60
Rapoport, A., **I:** 3, 107 ff.
Rashbam. *See* Samuel ben Meier
Rashi, **I:** 422. **II:** 138, 143, 144
Rawls, John, **II:** 231, 256, 522
Regelmann, K., **I:** 92, 104
Reichardt, R., **I:** 501
Reidemeister, K., **I:** 165, 166, 175, 176, 181
Reiter, S., **II:** 535
Reny, P. J., **I:** 481, 487, 635, 636, 648, 650, 668
Ress, A., **I:** 39
Reynolds, B. A., **I:** 35
Richard, J.-F., **I:** 615
Richardson, M., **I:** 36, 55
Richter, H., **II:** 607, 608, 618, 620, 622
Richter, M. K., **I:** 208. **II:** 505
Riemann, G. F. B., **II:** 162, 314
Rigby, F. D., **I:** 259
Riker, W. H., **I:** 34, 35, 44, 87, 104. **II:** 560
Rob, R., **I:** 479, 486
Roberts, J., **I:** 18, 22, 42, 43, 119, 120, 319, 444, 445, 471, 481, 486, 630, 633, 634, 650, 668
Robinson, A., **I:** 24, 40, 84, 100, 104. **II:** 69, 417, 429
Robinson, S. M., **I:** 102, 105
Rockafellar, R. T., **II:** 395, 419, 430, 657, 663–665, 674
Roemer, J., **I:** 97
Rogers, C. A., **I:** 538, 552
Rosenmüller, J., **I:** 36, 44. **II:** 507, 521
Rosenthal, R. W., **I:** 22, 44, 67, 104, 105, 120, 421, 422, 492, 533, 536, 551, 552, 560, 627, 634, 645, 647. **II:** 218, 335, 337
Ross, S., **I:** 18, 44
Roth, A. E., **I:** 4, 16, 25, 37, 39, 43, 44, 74, 88, 98, 99, 105, 140, 157, 619, 620. **II:** 388, 430, 447, 462, 467, 469, 483, 485, 493, 494, 502, 505, 507 ff., 523, 524, 527, 529, 530, 532–535, 539, 540, 559, 560
Rothblum, U. G., **II:** 576, 645–648, 651
Rothenberg, J., **II:** 171
Rothschild, M., **I:** 18, 44
Rubinstein, A., **I:** 21, 22, 44, 55, 62, 67, 72, 95, 105, 318, 319, 417, 442, 467, 473, 483–485, 487, 650, 668
Russell, B., **I:** 113

Sage, R., **I:** 41, 415
Salles, M., **II:** 520
Salop, S. C., **I:** 18, 44
Samet, D., **I:** 19, 42, 442, 470, 471, 473, 481, 486, 597, 614, 635, 642, 648
Samuel ben Meier, Rabbi, **I:** 131
Samuel of Naharda'a, **II:** 146
Sanders, C. E., **I:** 35
Sanders, J. O., **I:** 35
Satterthwaite, M. A., **I:** 18, 44, 138
Sauermann, H., **II:** 89, 133
Savage, L. J., **I:** 63, 79, 94, 105, 189–191, 295, 300–303, 305, 307, 567, 568, 584, 588, 591, 598, 609, 614, 631, 634, 636, 662, 668. **II:** ix, 131, 133, 668, 670, 674
Scafuri, A. J., **II:** 467, 537, 539–541, 543–545
Scarf, H. E., **I:** 24–26, 40, 45, 62, 84, 88, 89, 98, 101, 105, 140, 141. **II:** 39, 47, 53, 54, 157, 160–162, 167–170, 205, 206, 213, 217, 245, 385, 407, 429
Schelling, T. C., **I:** 67, 105, 470, 473. **II:** 518, 522
Schlaifer, R., **I:** 300, 301, 303
Schmalensee, R., **I:** 668
Schmeidler, D., **I:** 18, 24, 27, 45, 53, 82, 83, 102, 105, 119. **II:** 60, 61, 113, 114, 118, 133, 148, 151, 153, 158, 619, 622
Schmittberger, R., **I:** 480, 486
Schoellkopf, W., **I:** 313
Schotter, A., **I:** 44
Schoumaker, F., **I:** 120
Schwartz, G., **I:** 421
Schwartz, J. T., **II:** 176, 183, 187, 613, 614, 618, 628, 638, 647, 651
Schwartz, N. L., **I:** 319, 475
Schwarz, K. H. A., **I:** 540, 546, 547
Schwarze, B., **I:** 480, 486
Schwodiauer, G., **I:** 44. **II:** 521
Scott, M., **I:** 36, 45
Seidenberg, A., **I:** 245, 256, 300, 303
Selten, R., **I:** 4, 19–21, 23, 45, 50, 55, 81, 92, 94, 102, 105, 153, 154, 400, 409, 415, 416, 422, 442, 471, 473, 481, 487, 563, 585, 591, 597, 614–616, 620, 635, 648. **II:** 40, 55, 114, 133, 344, 358, 453, 462, 463, 513, 522, 558, 560
Sen, A., **I:** 4, 145, 152
Shafer, W. J., **I:** 39, 99, 620. **II:** 467, 469, 483, 493–495, 507, 508, 517–519, 522, 527, 533, 535, 537–540
Shaked, A., **I:** 480, 486
Shapiro, N. Z., **I:** 34, 45, 194, 209. **II:** 413, 554, 560
Shapley, L. S., **I:** 3, 17, 18, 24–27, 33 ff., 53, 58, 60, 62, 66–68, 72, 74 ff., 116, 120, 121, 125 ff., 140, 141, 157, 188, 191, 259–261, 267, 272, 274, 279, 289, 311–313, 318, 357, 393, 395, 417, 419, 432, 433, 499, 501, 503, 552, 591, 597. **II:** 3, 5, 6, 10, 11, 27–29, 35, 36, 39–41, 47, 52, 53, 55, 60, 63, 107, 110, 111, 113, 116, 120, 122, 125, 132, 133, 143, 148, 149, 151, 153, 157, 158, 161, 167, 169–172, 186, 187, 198, 199, 205, 206, 212, 213, 217, 219, 221 ff., 258, 261, 263, 264, 266, 267, 272, 277, 280, 281, 283, 284, 285–288, 292, 302, 309, 311, 315, 316, 327, 330, 332–334, 335–337, 341, 344–347, 354, 357–360, 362–364, 373, 380, 382–385, 388, 389, 391, 394, 396, 406, 408, 409, 413, 420, 429–431, 433, 436, 437, 440, 441, 443, 447, 449–451, 457, 461–463, 467, 469, 470, 472 ff., 485, 489 ff., 495 ff., 507, 509, 520 ff., 534, 537–540, 542, 544, 545, 549, 550, 551 ff., 563 ff., 574, 624, 625, 629, 637, 639, 644, 645 ff., 653
Sharkey, W. W., **I:** 36, 45
Shavell, S., **I:** 18, 46
Shell, K., **II:** 350
Shelly, M. W., **I:** 232
Sheshinski, E., **II:** 231, 256, 257, 283, 284
Shitovitz, B., **I:** 25, 46, 87, 105, 119. **II:** 189, 191, 193, 198, 199, 227, 256
Shlein, A., **II:** 169
Shlomo Yitzhaki, Rabbi. *See* Rashi
Shubik, M., **I:** 18, 24–26, 34, 36, 40, 41, 44–46, 62, 67, 74, 75, 84, 85, 88, 89, 101, 104–106, 140, 312, 414. **II:** 5, 6, 11, 28, 29, 31, 53–55, 110, 111, 132, 167, 168, 170–172, 186, 187, 198, 199, 245, 255, 315, 334, 496, 505, 507, 522, 544, 545, 555, 557, 561, 624, 629, 637
Siegel, S., **I:** 16, 41
Sierpinski, M. W., **I:** 288, 289, 507, 522
Sigmund, K., **I:** 479, 486
Silberberg, M., **II:** 153
Simon, H., **I:** 139, 475, 487
Sims, M. Q., **I:** 35
Sloan, A. P., **II:** 527
Smale, S., **II:** 241, 256
Smith, H. L., **I:** 249
Smith, J. W., **I:** 185, 193, 194, 208, 209, 219, 229, 237, 256
Smorodinski, M., **I:** 72, 76, 102. **II:** 150, 153
Sobolev, A. I., **I:** 83, 97, 106. **II:** 60, 61, 148, 151, 153
Solomon, H., **I:** 237, 256
Sondermann, D., **I:** 24, 46, 119. **II:** 114
Sonnenschein, H., **I:** 18, 43, 208. **II:** 505
Sorin, S., **I:** 68, 70, 106, 119, 318, 427, 437, 439, 481, 482, 485, 632–634. **II:** 431, 445, 563
Sotomayor, M. A., **I:** 4, 157
Spence, M., **I:** 18, 46
Stacchetti, E., **I:** 650, 668
Stanford, W., **I:** 483, 486
Stearns, R. E., **I:** 32, 46, 79, 106, 318, 426, 427, 429, 436, 437, 483, 487. **II:** 36, 37, 54
Steinhaus, H., **I:** 51, 104
Stewart, F. M., **I:** 51, 101, 356, 357, 393
Stieltjes, T. J., **I:** 544

Stiglitz, J. E., **I:** 18, 44, 593, 596
Straffin, P. D., **I:** 36, 46, 100, 103
Strotz, R. H., **II:** 625
Sugden, R., **I:** 636, 648
Suppes, P., **I:** 300, 301, 303. **II:** 256
Sutton, J., **I:** 480, 486
Suzuki, G., **I:** 193, 194, 208, 209, 224, 227, 229, 237, 256
Suzuki, M., **I:** 36, 46
Szego, G. P., **I:** 41
Szidarovsky, F., **I:** 36, 40, 97, 100

Tan, T. C.-C., **I:** 613, 614, 643, 648, 650, 666, 668
Tauman, Y., **I:** 84, 106, 119. **II:** 356, 358, 408, 430, 549, 553, 555, 560, 561, 569
Thisse, J. F., **I:** 18, 40
Thompson, F. B., **I:** 329, 353
Thompson, G. L., **I:** 25, 46
Thomson, W., **I:** 97. **II:** 152, 539, 540
Thrall, R. M., **I:** 256, 257, 259, 265, 274. **II:** 63, 86, 89, 113, 129, 133
Tirole, J., **I:** 650, 668
Tolstoy, Leo, **I:** 17, 117
Tomioka, **I:** 431
Tucker, A. W., **I:** 39–41, 45, 46, 67, 69, 86, 101, 103–105, 239, 256, 322, 328, 479, 486, 501, 503, 552, 591, 614, 648. **II:** 3, 26, 27, 29, 55, 63, 110, 132, 133, 170, 217, 256, 284, 333, 334, 337, 358, 430, 445, 462, 463, 483, 494, 505, 522, 540, 560, 569, 624, 634, 638
Tulkens, H., **I:** 119
Turing, A. M., **I:** 471, 476, 482–484, 538
Tversky, A., **I:** 595, 596

Uzawa, H., **II:** 497, 505

Van Damme, E., **I:** 479, 481, 487
Verrecchia, R. E., **I:** 36, 43, 45, 46, 98, 105
Vial, J.-P., **I:** 119, 120, 601, 614
Vickrey, W., **I:** 18, 46. **II:** 63, 86, 89
Vietoris, L., **I:** 164
Ville, J., **I:** 506, 522
Ville, J. A., **I:** 66, 106
Vinacke, W. E., **I:** 64, 106, 110, 111
Vind, K., **I:** 24, 46, 84, 106, 119, 279, 289. **II:** 47, 54, 607, 617
Visscher, M., **I:** 18, 44
Vohra, R., **I:** 479, 486
Von Neumann, J., **I:** ix, 3, 18, 27 ff., 48, 52 ff., 114, 116, 153, 185, 187, 188, 191, 231, 233, 247, 248, 250, 252, 253, 256, 257 ff., 279, 281, 289, 295, 303, 324, 327, 353, 355 ff., 393, 395, 399, 491, 511, 522, 527, 532, 583, 589, 591, 609, 610, 614, 621, 622. **II:** ix, 3, 5, 6, 9–11, 13, 17, 22, 23, 26, 27, 29, 31 ff., 59, 60, 63, 79, 87, 89, 107, 111, 113, 114, 122, 127, 133, 161, 167, 170, 242, 255, 256, 260, 282, 284, 289, 291, 292, 326, 334, 337, 342, 360, 430, 448, 463, 487, 494, 509, 519, 521, 541, 545, 551, 555, 560, 608, 609, 618, 619, 622
Von Weiszäcker, C. C., **II:** 238
Vorobiev, N. N., **I:** 106, 497, 501. **II:** 61, 151, 153

Wagner, M., **I:** 46
Wald, A., **I:** 551, 552. **II:** 167, 170, 171, 187
Walras, L., **I:** 24, 42, 43, 45. **II:** 47, 167, 170, 206, 208, 210–213, 221, 245, 407, 431–433, 504
Washington, George, **II:** 152
Waternaux, C., **I:** 119
Watts, H. W., **II:** 283
Weber, R., **I:** 18, 42, 43, 551, 552, 597
Weber, S., **I:** 120
Weibull, J. W., **I:** 481, 485
Weierstrass, K., **I:** 545
Weiss, B., **I:** 492, 533. **II:** 420
Werlang, S. R. da Costa, **I:** 613, 614, 643, 648, 650, 666, 668
Wesberry, J. P., **I:** 35
Whinston, A. B., **I:** 36, 43
Whitehead, G. W., **I:** 163
Whitehead, J. H. C., **I:** 163, 174, 181
Whitney, H., **II:** 225, 241, 256
Wicksell, K., **II:** 336, 337
Wiener, N., **I:** 545–547
Wiles, P., **II:** 283, 284
Willig, R., **I:** 668
Wilson, R., **I:** 19, 22, 42, 46, 81, 106, 319, 442, 444, 445, 471, 473, 480, 486, 597, 614, 630, 633, 634, 650, 668. **II:** 494
Wolfe, P., **I:** 51, 101, 104, 106, 356, 393, 505, 522
Wolfowitz, J., **I:** 551, 552
Wolsey, L., **I:** 615
Wooders, M. H., **I:** 88, 90, 102, 106. **II:** 541, 545
Wright, F. B., **II:** 599

Yaari, M. E., **II:** 507, 522, 625, 638
Yadin, Y., **II:** 460
Yannelis, N. C., **II:** 467, 537, 539–541, 543–545
Younès, Y., **II:** 406, 430
Young, H. P., **I:** 73, 74, 97, 100, 106, 152, 479, 486, 487, 550. **II:** 152, 566, 569

Zame, W. R., **I:** 90, 106
Zamir, S., **I:** 119, 120, 124, 425–428, 436, 437, 606, 614, 653, 663, 668. **II:** 530, 531, 535
Zemel, E., **I:** 483, 487
Zermelo, E., **I:** 50–52, 76, 94, 106, 416, 621, 635. **II:** 52
Zeuthen, H., **I:** 101. **II:** 26, 54, 284
Zorn, M., **I:** 273
Zygmund, A., **I:** 545, 552

Subject Index

Many concepts are treated on several pages in the same chapter, often near each other. In these cases the index may mention only some of the pages, usually the first and/or that with the definition of the concept.

Occasionally, text items closely related to or substantially identical with an index entry are indexed also under that entry, or only under that entry. Thus "continuum of players" in the text appears in the index also under "continuum of agents"; and "axiomatic characterization" in the text appears in the index under "axiomatization" only.

Subentries may be understood as coming either before or after the main entry. Thus "additivity" has two subentries, "axiom" and "conditional"; the corresponding text items are "additivity axiom" and "conditional additivity." Also, some subentries—and entries—are conceptual in nature, and do not appear explicitly in the text. Thus the subentry "kernel" under "axiomatization" refers to an axiomatization of the kernel.

In the alphabetization, spaces come before "a," and hyphens are ignored. Thus "a posteriori" comes before "Abelian," which in turn comes before "a-core." Greek letters are treated as if spelled out. Thus "allocation" comes before "α-core."

A. *See* analytic set
a posteriori
 equilibrium, **I:** 578, 579
 preference, **I:** 579
a priori, **I:** 306
 plausibility, **I:** 5
Abel limit, **I:** 412
Abelian subgroup, **I:** 249
absolute risk aversion, **II:** 270
absolutely
 continuous function, **II:** 646
 purifying sequence, **I:** 538
acceptability, **I:** 359
acceptable
 payoff vector, **I:** 376
 point, **I:** 326, 349, 355, 356. **II:** 25
 c-acceptable, **I:** 336
accounting, **II:** 533
ACDA. *See* Arms Control and
  Disarmament Agency
acknowledgments, **I:** xi
a-core, **II:** 407
acreage restrictions, **I:** 114. **II:** 201
act, **I:** 301, 305
 rationality, **I:** 476, 480
acting together, **II:** 129
action, **I:** 447, 650, 652
 mixed, **I:** 447, 650
 recalling one's own, **I:** 469
 space, **I:** 509
additive structure, **I:** 236
additivity, **II:** 116, 223, 286
 axiom, **II:** 263
 conditional, **II:** 472, 473
adjusted normal form, **II:** 48
admissibility, **II:** 577, 579, 581–583, 600
admissible function, **II:** 209, 274
aerodynamics, **I:** 13
affine function, **II:** 660
aged and infirm, the, **I:** 422
agents
 continuum of. *See* continuum of agents
 symmetric, **II:** 542
aggregate
 preferred set, **II:** 175
 utility, **II:** 264
agreeing to disagree, **I:** 593
agreement, **I:** 352, 616
 binding, **I:** 21, 580
 enforceable, **I:** 21, 70, 317, 580
 enforcement of, **I:** 154, 618
 informal, **I:** 352
 Pareto-dominant, **I:** 616
 pre-play, **I:** 615
 self-enforcing, **I:** 23, 396, 434, 556, 615
airport landing, **I:** 37, 97. **II:** 558
airworthiness, **I:** 13
AKN, **II:** 359
algebraic topology, **I:** 161
algebra-of-combining, **I:** 263
allele, **I:** 93, 478
allocation, **I:** 185, 186, 193. **II:** 163, 173,
  227, 262, 294, 386, 408, 441
 commodity tax, **II:** 296, 308, 315
 competitive (equilibrium), **I:** 83. **II:** 206,
  222, 237, 243, 497, 557
 competitive tax, **II:** 297
 cost, **I:** 36, 37, 91, 97, 98
 decision, **I:** 195
 efficient, **II:** 298
 equal treatment, **II:** 442
 income tax, **II:** 320, 326, 329, 332
 procedure, **II:** 566
 symmetric, **II:** 566
 S-allocation, **II:** 294, 401
 of TVA projects, **I:** 36
 value, **II:** 221, 222, 227, 230, 236, 237,
  241, 266, 281, 388, 432, 437, 441, 496
 cardinal, **II:** 242, 243, 497, 538
 generalized, **II:** 394, 408
 ordinal, **II:** 497
 Walras, **II:** 210
 y-allocation, **II:** 399
all-player coalition, **I:** 25
almost
 all, **I:** 281
 strictly competitive game, **I:** 495, 496 (*see
  also* strictly competitive game)
α-core, **I:** 413, 414. **II:** 6, 24, 51
α-effectiveness, **II:** 6, 7, 22, 28

alternating knot, **I**: 161, 163, 181
alternative (or move)
  continuum of, **I**: 503, 533, 573, 574
  theorem of the, **I**: 239
altruism, **I**: 22, 68, 411, 415, 418, 422, 430
  true, **I**: 422
altruistic
  equilibrium, **I**: 415, 422, 430, 431
  perfect equilibrium, **I**: 430
amount or degree of irrationality, **I**: 556, 630, 631, 632
analytic
  aspects of game theory, **I**: 622
  set, **I**: 287. **II**: 609
Analytical Research Group, **I**: 185
animal conflict, **I**: 21
anti-trust law, **I**: 325. **II**: 128
approachability, **I**: 321, 334, 335
approximate
  purification, **I**: 533
  rationality, **I**: 630
approximating fixed points, **I**: 89
arbitrated outcome, **I**: 38
arbitration, **II**: 551
archeology, **I**: 13
archimedean
  assumption, **I**: 301
  axiom, **I**: 264
  principle, **I**: 263
Arms Control and Disarmament Agency, **I**: 318, 426
Arrow's impossibility theorem, **I**: 138
art, **I**: 16, 141
  forms, **I**: 16
as if, **I**: 12
a.s.c. *See* almost strictly competitive game
aspheric knot, **I**: 163
asphericity, **I**: 161, 163
assignment, **II**: 162, 173
  initial, **II**: 162, 173
  of interns to hospitals, **I**: 88, 140, 157
  problem, **I**: 186, 193
associated
  cardinal market, **II**: 243
  ordinal market, **II**: 243
  preference relation or scale, **II**: 242
asymmetric information, **I**: 79, 423, 593
asymptotic
  approach, **I**: 83. **II**: 205, 217, 385
  value, **II**: 224, 225, 228, 236, 288, 309, 311, 356, 362, 553
at least as desirable as another player, **II**: 101, 401
atoms, **I**: 25, 87. **II**: 164, 189 (*see also* monopoly; oligopoly; syndicate)
  continuum with. *See* continuum with atoms
attained (worth), **II**: 305
auction, **I**: 18, 139
  oil-lease, **I**: 140

Aumann–Peleg characteristic function, **II**: 87
A-unknotted knot, **I**: 181
automaton, **I**: 53, 421, 442, 461, 483, 641
  irrational, **I**: 441, 453, 454
Axiom of Choice, **I**: 51
axiomatization, **I**: 75. **II**: 483
  bankruptcy, **II**: 135
  cost allocation, **I**: 98
  kernel, **I**: 83
  Nash equilibrium, **I**: 97
  Nash solution to bargaining problem, **I**: 97
  NTU Shapley value, **II**: 469, 472
  nucleolus, **I**: 83
  strategic equilibrium, **I**: 97
  subjective probability, **I**: 295
  TU Shapley value, **II**: 563
  utility without completeness, **I**: 257, 261
  value, NTU Shapley, **II**: 469, 472
  value, TU Shapley, **II**: 563
axioms of set theory, **I**: 288

Babylonian Talmud, **I**: x, 37, 97. **II**: 60, 135
backward induction, **I**: 23, 626, 635. **II**: 453, 460
  paradox, **I**: 627
Baire class, **II**: 579
Baker *v.* Carr, **I**: 35
balancedness, **I**: 25, 26, 62, 88. **II**: 39, 92
  ε-balancedness, **II**: 97
Balinski's method, **II**: 94
ballet, **I**: 617
Banach class, **II**: 577, 578, 582, 586
  bounded, **II**: 578, 583, 586
bankruptcy, **II**: 135
  game, **I**: x. **II**: 148
  problem, **I**: 37. **II**: 135, 139
Banzhaf value, **II**: 555
bargaining, **I**: 11, 20, 498
  cooperative, **I**: 16
  curve, **I**: 33
  experiment, **I**: 16
  game, **I**: 96. **II**: 532
    with incomplete information, **I**: 22
  pre-play, **I**: 21
  problem, **I**: 71, 79. **II**: 243
  set, **I**: 27, 82, 83, 85, 97. **II**: 41, 63, 68, 84, 91, 114, 123, 161, 292, 541, 551
  solution of Nash, **I**: 36, 97. **II**: 292, 332, 473, 488
  strength, **II**: 228, 230
  theory, **I**: 90
Baroque, **I**: 16
basic alternatives, **I**: 257
battle of the sexes, **I**: 617
battleship, **I**: 47
Bayesian
  decision theory, **I**: 613
  rationality, **I**: 434, 602, 603

utility maximizer, **I:** 598
  rational, **I:** 598
  view of equilibrium, **I:** 433
  view of world, **I:** 434
b-core, **II:** 408
behavior
  off-path, **I:** 646
  strategy, **I:** 67, 328, 387, 503, 511, 520, 523
belief, **I:** 623
  and knowledge, **I:** 663
  with probability 1, **I:** 647
  second-order, **I:** 623
  system, **I:** 624, 653, 662
    infinite, **I:** 661
    interactive, **I:** 652 (*see also* information system)
believability, **I:** 401
Bell Telephone Laboratories, **I:** 185
β-core, **I:** 317, 413, 414. **II:** 6, 51, 52
β-effectiveness, **II:** 6, 7, 22, 28
BI. *See* backward induction
bi-affine function, **II:** 660
bibliographic references, **I:** ix
bi-convex
  function, **II:** 660
  hull, **II:** 653, 657
  set, **II:** 657
bi-convexity, **II:** 653, 657
Big Five, **I:** 34
bi-martingale, **II:** 653, 655
  binary, **II:** 657
binding agreement, **I:** 21, 580
biology, **I:** 21, 67, 91, 142, 415, 422
bi-relative interiority, **II:** 664
bi-separation, strong, **II:** 661
bi-simpliciality, local, **II:** 665
Blackwell strategy, **I:** 424, 427
blocking coalition, **I:** 11, 30, 32, 99
bluffing, **I:** 53
Board of Supervisors, **I:** 65
bold face, **I:** 639
boldness, **II:** 269
bordered Hessian, **II:** 225
Borel
  mapping, **II:** 581
  -measurable function, **II:** 178, 608, 633
  rectangle, **II:** 581
  set, **I:** 279, 286, 507, 550. **II:** 226, 335, 581, 602, 608, 625
  space, **II:** 581
  structure, **I:** 188. **II:** 581
bounded
  Banach class, **II:** 578, 583, 586
  complexity, **I:** 442
  differentiability, **II:** 226
  function, **II:** 619
  market, **II:** 294, 298
  memory, **I:** 421
  rationality, **I:** 95, 139, 319, 475
  recall, **I:** 318, 439, 442, 482

variation, **II:** 309, 353
boundedness, **II:** 185, 619
  away from zero, **II:** 226
  uniform, **II:** 294, 361
bread, **I:** 64, 86, 124
bribe, **I:** 291, 417. **II:** 32
bridge, **I:** 47, 50
Brouwer's theorem, **II:** 176
budget, **II:** 371
  line, **II:** 192
  set, **II:** 163, 173, 228, 387
burial, **I:** 422
"burning" of utility, **I:** 470
buses, **I:** 6
B-vector, **I:** 321

CA. *See* complementary analytic
c-acceptability, **I:** 326, 328, 336, 338, 386
c-acceptable
  payoff vector, **I:** 327, 382. **II:** 10
  points, **I:** 336
  strategy vector, **I:** 326–328
calculation, costly, **I:** 485
calculus
  of infinitesimals, **II:** 233
  propositional, **II:** 93
California, **I:** 34
Carathéodory's theorem, **I:** 549
cardinal
  market, **II:** 242
  transferable utility, **II:** 351
  utility, **II:** 241, 242, 416
  value allocation, **II:** 242, 243, 497, 538
  value equivalence theorem, **II:** 243
cardinality, **II:** 281
Carnegie Corporation, **I:** 495. **II:** 63, 171
cartel, **I:** 31, 32, 61. **II:** 71
catalogue, **II:** 49
  regular, **II:** 49
Cav, **I:** 423–425, 427
CEA. *See* constrained equal award
c.e.d. *See* correlated equilibrium distribution
Center for Operations Research and Econometrics, Catholic University of Louvain, **I:** 3, 119, 591. **II:** 218, 221, 383
center of symmetry, **I:** 84
centipede game, **I:** 625, 627
central limit theorem, **I:** 428
Cesaro limit, **I:** 412
CG. *See* contested garment
C-game, **I:** 81
chain, **II:** 353, 645
chance, **I:** 295
characteristic function, **I:** 99. **II:** 7, 16, 32, 63, 92, 351, 486
  form, **II:** 14, 16, 23, 33, 34, 114, 486
  value, **I:** 99
charity, **I:** 415, 430

checkers, **I:** 50
chess, **I:** 47, 48, 50, 616
Chief of Naval Operations, **I:** 211, 214
children, **I:** 13
chinese checkers, **I:** 50
choosing up, **I:** 49
civil law, **II:** 482
classical game, **II:** 13
classification, **I:** 14
CKR. *See* common knowledge of rationality
clear game, **II:** 27
closure, **I:** 293
  invariance, **II:** 472, 473, 568
CNO. *See* Chief of Naval Operations
coalition, **I:** 5, 56, 317, 340, 413, 414. **II:** 10, 92, 163, 287, 289, 342, 388, 408, 471
  all-player, **I:** 25
  blocking, **I:** 11, 30, 32, 99
  diagonal, **II:** 364, 554
  effective, **II:** 6
  even, **II:** 438
  flat, **II:** 108
  formation of, **I:** x
  formed, **II:** 454
  fuzzy, **I:** 74
  losing, **I:** 56. **II:** 100
  majority, **II:** 378
  near, **II:** 348, 365
  part-time, **I:** 26
  permissible, **II:** 64, 79, 80
  structure of, **I:** 119. **II:** 8, 86, 92, 113, 448
  superadditivity at, **I:** 88
  trivial, **II:** 78
  value of, **II:** 64
  winning, **I:** 57, 65. **II:** 100
    minimal, **I:** 30, 32, 65
coalitional
  form or game, **I:** ix, 54, 56, 57, 99, 317. **II:** 147, 286, 287, 518, 551 (*see also* NTU)
  game theory, **I:** x, 317
  rationality, **II:** 64, 65, 79
  worth, **I:** 25, 36, 99
Coase theorem, **II:** 378
coefficients of importance, **II:** 299
cold war, **I:** 112, 116
  strategy of, **I:** 67
college admissions, **I:** 25, 157
collusion, **I:** 612, 613
Colonel Blotto, **I:** 67.
commitment, **I:** 54, 530, 580, 582
  enforceable, **I:** 317
  irrevocable prior, **I:** 530
  self-, **I:** 529
commodity
  bundle, **II:** 162, 172
  endowments, monotonic in the, **II:** 405
  redistribution, **II:** 286, 294
  tax allocation, **II:** 296, 308, 315
  -wise saturation, **II:** 174

common
  interests, **I:** 439–442, 447, 453
  knowledge, **I:** 14–16, 20, 81, 423, 555, 593, 594, 598, 612, 625, 626, 650, 651, 653
    of conjectures, **I:** 652, 664
    of information partitions, **I:** 605
    of model, **I:** 82, 664
    of priors, **I:** 605
    of rationality, **I:** 476, 483, 556, 598, 608, 625, 635, 636, 642, 646, 650
  payoff, **I:** 440, 443
  prior, **I:** 594, 603, 608, 623, 628, 630–632, 651–653, 657 (*see also* Harsanyi doctrine)
communication, **I:** 580
  network, **II:** 447, 559
  pre-play, **I:** 615–617. **II:** 515
compact-valued theorem on selection, **II:** 620
comparative probability, **I:** 301
comparison
  endogenous utility, **II:** 539, 544
  function, **II:** 291, 308, 328, 329, 331, 344, 357, 408
  measure, **II:** 344, 363
  vector, **II:** 388, 441
compatible strategy, **I:** 451
competition, **I:** 27, 28, 131
  cut-throat, **I:** 65, 73
  free, **II:** 167
  long-term, **I:** 395, 397
  perfect, **I:** x, 18, 28, 129. **II:** 159, 380
competitive
  equilibrium, **I:** x, 18, 84, 135, 140. **II:** 39, 47, 159, 160, 163, 171, 173, 222, 335, 386
  (equilibrium) allocation, **I:** 83. **II:** 206, 222, 237, 243, 497, 557
  outcome, **I:** 24, 35
  price, **II:** 557
  of a vote, **I:** 35
  tax allocation, **II:** 297
  tax equilibrium, **II:** 297, 320, 333
competitiveness, **II:** 213
complement of a PCA set, **I:** 287
complementary analytic
  choice, **I:** 280
  choice hypothesis, **I:** 288
  graph, **I:** 288
  set, **I:** 287, 288
complete
  coalition, **II:** 454
  information, **I:** 15, 16, 19, 20, 423
completeness, **II:** 173
  axiom, **I:** 187, 258
  internal, **II:** 456
  utility without, **I:** 257, 275
complexity
  bounded, **I:** 442
  of strategy, **I:** 421
  theory, **I:** 143, 484

component, **I:** 161
connected, **II:** 450
composition, **I:** 60, 495, 497, 500. **II:** 9, 21, 23
compound lottery, **I:** 263, 296
comprehension, **I:** 6, 7
comprehensiveness, **II:** 471, 567
compromise, reasonable, **II:** 231
computation of minimax, **I:** 86
computational simplicity, **I:** 38, 482
computer, **I:** 483
  as player, **I:** 484
  science, **I:** 91
computing, **I:** 66
  distributed, **I:** 484
  of utilities, **I:** 243
concave
  function, **II:** 247, 626
  utility, **II:** 236, 386, 404
concavification, **I:** 423. **II:** 629
concavity, **I:** 423
cond, **I:** 513
conditional, **I:** 643
  additivity, **II:** 472, 473
  additivity axiom, **II:** 481
  counterfactual, **I:** 643
  material, **I:** 643
  payoff, **I:** 638
  strict, **I:** 643
  substantive, **I:** 643, 644
  sure-thing, **II:** 482
  utility, **I:** 306
conditionally atomless measure, **I:** 534, 535
  weakly, **I:** 536
cone, **I:** 268, 269
  dual, **I:** 269
  regular, **I:** 269, 281
conflict of interest, **I:** 442
conjecture, **I:** 650, 653, 664
  common knowledge of, **I:** 652, 664
  of i about j, **I:** 653
  independent, **I:** 665
connected
  component, **II:** 450
  drawing, **I:** 161
  set, **II:** 286, 293
  subgraph, **II:** 219
connections, **I:** 141
conscious randomization, **I:** 557, 611, 650
consequence, **I:** 301, 305
considerations, **I:** 291
consistency, **I:** x, 96, 251, 295. **II:** 141
  contested garment-, **II:** 141
  internal, **II:** 122
  principle, **I:** 77
  self-, **II:** 141
consistent
  case, **I:** 434, 584
  I-game, **I:** 81
constant partition, **I:** 323

constant-sum
  extreme game, **II:** 108
  game, **II:** 100
  majority game, **I:** x
constitution, **II:** 358
constrained
  equal award, **II:** 142
  game, **II:** 114
construction lines, **I:** 308
consumer preference theory, **I:** 187
contested garment, **II:** 138, 141
  -consistency, **II:** 141
continuity, **II:** 163, 173
  monotonic absolute, **II:** 646
continuous
  function, **II:** 178, 616, 664
  game, **II:** 224
  preference order, **I:** 285. **II:** 163, 173
  utility, **I:** 279
continuum, **I:** 25
  of agents, players, or traders, **I:** 37, 62, 66, 140, 188, 189. **II:** 574–576 (see also large market; oceanic game)
  with atoms ("mixed"), **I:** 25, 87, 140. **II:** 157, 161, 189, 555 (see also monopoly; duopoly; oligopoly)
  non-atomic, **I:** 24, 28, 31, 84, 86, 87, 140, 279. **II:** 157, 159, 171, 217, 221, 257, 285, 335, 339, 359, 385, 408, 431–439, 549, 552, 556, 558, 574–576, 625
  of alternatives or moves, **I:** 503, 533, 573, 574
  of goods, **I:** 24
  hypothesis, **I:** 288
  of pure strategies, **I:** 66
contract, **I:** 580, 582
  curve, **II:** 168, 209
contribution, **I:** 73
  marginal, **II:** 222, 235
  monotonic in the, **II:** 566
Contributions to the Theory of Games, **I:** 67
Convention, **I:** 555
conventions, **I:** 153
convergence
  indicator, **II:** 396
  rate of, **I:** 24, 427. **II:** 205
convex, **I:** 374. **II:** 7, 16, 33, 171, 250, 386, 471
  combination, **I:** 261
  hull, **II:** 42, 165, 233, 433, 608, 633, 653
  of the Nash equilibrium payoffs, **I:** 562
  structure, **I:** 261
convexification, **I:** 407, 414
convexity, **II:** 185
  assumption, **II:** 171
  even, **I:** 270
cooperation, **I:** 68, 411, 419, 439, 440, 443, 445
  as the consequence of selfishness, **I:** 137, 439
  evolution of, **I:** 444

cooperation *(cont.)*
  graph, **II:** 449
  rational, **I:** 439
  structure, **II:** 449
cooperative
  bargaining, **I:** 16
  game, **I:** 15, 21, 24, 54, 55, 123, 154, 317, 318, 321, 355, 581. **II:** 5, 13, 31
  game theory, **I:** x, 317
  n-person game theory, **I:** 53
  outcome, **I:** 440, 444
  strategy, **II:** 6
  supergame, **I:** 352
coordination, **II:** 514
CORE. *See* Center for Operations Research and Econometrics
core, **I:** x, 11, 21, 24, 27, 32, 53, 58, 59, 62, 66, 74, 75, 78, 83, 84, 87, 88, 90, 97, 119, 140, 317, 413. **II:** 6, 10, 13, 36, 39, 47, 65, 113, 114, 120, 159, 160, 186, 198, 205, 292, 335, 336, 384, 407, 444, 507, 509, 529, 541, 543, 551, 564
  a-core, **II:** 407
  α-core, **I:** 413, 414. **II:** 6, 24, 51
  b-core, **II:** 408
  β-core, **I:** 317, 413, 414. **II:** 6, 51, 52
  ε-core, **I:** 90. **II:** 186
  equivalence principle, **I:** 24, 27. **II:** 157, 159, 431, 444
  of a market, **I:** 62. **II:** 155, 157, 159, 189, 205
  non-emptiness of, **I:** 88, 89
  P-core, **II:** 8
  R-core, **II:** 17
  and superadditivity, **I:** 90
Cornell University, **I:** 37
corner, **I:** 374
corporation, **I:** 87
  with two large stockholders, **I:** 140
correlated
  equilibrium, **I:** x, 20, 53, 556, 559, 598, 600, 622, 627, 666
  equilibrium distribution, **I:** 600
  mixed strategy, **II:** 6, 14
  strategy, **I:** 397, 560, 580, 630. **II:** 51
  strategy B-vector, **I:** 330. **II:** 6
  strategy vector, **I:** 322, 325, 326, 356
correspondence *(see also* set-valued function)
  selection from a, **II:** 619
  Shapley, **II:** 472, 568
  value, **II:** 472, 568
cost
  allocation, **I:** 36, 37, 91, 97, 98
  to the punisher, **I:** 419
  sharing, **II:** 558
costly calculation, **I:** 485
counterfactual conditional, **I:** 643
counterintuitive example, **I:** 77. **II:** 508
counterobjection, **I:** 11, 82. **II:** 67, 123

coupon, **I:** 126 *(see also* dividend equilibrium)
endowment, **II:** 405
  monotonic in the commodity endowments, **II:** 405
  uniform, **II:** 405
equilibrium, **II:** 383, 384, 405, 406
covariance
  scale, **II:** 472, 568
  translation, **II:** 482
cover, superadditive, **II:** 125
Cowles Foundation for Research in Economics, Yale University, **I:** 289 *(see also* Yale University)
CPA. *See* common prior assumption
CPCA. *See* complement of a PCA set
cpo. *See* continuous preference order
crazy
  perturbation, **I:** 319, 476, 481, 632, 633
  player or type, **I:** 319, 488
critical line, **II:** 192
cross-fertilization, **I:** 120
cryptography, **I:** 96
c-strategy vector, **I:** 326, 329
cultural evolution, **I:** 93
current value, **I:** 412
cut-throat competition, **I:** 65, 73

Davis–Maschler "reduced game," **II:** 110, 148
de Finetti phenomenon, **II:** 239
debates, **I:** 107
decision
  -making, rational, **I:** 12
  theory, **I:** ix, 108
    Bayesian, **I:** 613
    statistical, **I:** 53
decisive
  voting game, **I:** 65, 99 *(see also* strong voting game)
  weighted voting game, **I:** 83
decomposable game, **II:** 120
decomposition, **I:** 495 *(see also* subgame)
decoupling, **II:** 385
decreasing marginal utility, **II:** 272
defection, **I:** 403, 405, 406, 408, 415, 417, 443
degenerate order, **I:** 269
degree or amount of irrationality, **I:** 556, 630–632
Delphi technique, **I:** 595
δ-kernel, **II:** 97
demand
  game, **I:** 498
  scheme, **I:** 498
democracy, **II:** 259, 283
  representative, **II:** 357
democratic society, **II:** 359
derived market, **I:** 126. **II:** 319
descriptive aspects of game theory, **I:** 11, 13, 14, 37, 108, 157, 662

desirability, **II:** 163, 173
  pattern, **II:** 101
  weak, **II:** 174
destroyer-submarine game, **I:** 14
determinacy, **II:** 49
determined, **I:** 50
  strictly, **I:** 49
deterrence, **I:** 67, 415
Deutsche Forschungsgemeinschaft, **II:** 653
deviation, **I:** 317, 405
  punishment of, **I:** 333, 405
diagonal
  coalition, **II:** 364, 554
  formula, **II:** 554
  neighborhood, **II:** 353, 398
  property, **II:** 552, 553
difference game, **I:** 495
differentiability, **II:** 225
  bounded, **II:** 226
differential
  equation, **I:** 428
  game, **I:** 52, 53
  information, **I:** 79, 602, 604, 605
diminishing returns, **I:** 242
diplomacy, **I:** 47
discount
  factor, **I:** 418
  varying, **I:** 464
discounted
  game, **I:** 482
  payoff, **I:** 403, 404
  sum, **I:** 397
discounting, **I:** 331, 396, 419, 429
discrete
  market, **I:** 25
  structure, **II:** 583
discreteness, **II:** 579
discrimination, **I:** 18, 32, 61, 129
  racial, **I:** 140
discriminatory (solution or stable set), **I:** 29, 116
distributed computing, **I:** 53, 96, 484
distribution, **I:** 317. **II:** 335, 656
  of a correlated equilibrium, **I:** 600
  normal, **I:** 428
  of payoff, **I:** 323
divide and rule, **II:** 68
dividend, **I:** 11. **II:** 384, 388, 441
  budget set, **II:** 388
  characterization, **I:** 74
  equilibrium, **II:** 388, 441 (*see also* coupon)
  monotonically concave, **II:** 411
division of payoff, **I:** 317
domain, **II:** 480
dominance, risk, **I:** 154
dominated convergence theorem, **II:** 609
domination, **II:** 6, 8, 13, 14, 17, 25, 122, 292, 519
  external, **II:** 122
  of outcome, **I:** 24, 28, 59
  weak, **I:** 421

successive, **I:** 421
Doppler effect, **I:** 10
double vote, **I:** 146
double-cross strategy, **I:** 420, 421, 435, 443
Drosophila, **I:** 62
dual, **II:** 287
  cone, **I:** 269
  f^*, **II:** 142
duality theorem of linear programming, **I:** 53
D-unknotted knot, **I:** 181
duopoly, **I:** 25 (*see also* continuum with atoms)
duration-independent strategy, **I:** 483
dynamic
  game, **I:** 67, 86
  programming, **I:** 23, 416
  theory of games, **I:** 479

ecology, **I:** 136
Econometric Research Program, Princeton University, **I:** 189, 231, 302. **II:** 171 (*see also* Princeton University)
Econometric Society, **I:** 4, 137, 145
Econometrica, **I:** 91
economics
  information, **I:** 136
  neoclassical, **I:** 411
  verbal, **I:** 16, 135
economy
  exchange or pure exchange, **II:** 83, 383, 496, 557 (*see also* market)
  finite, **II:** 537
  fixed price, **I:** 125. **II:** 383, 404
    value of, **I:** 3. **II:** 383, 405
  with many goods, **I:** 119
  productive, **I:** 24
  public goods, **I:** 3. **II:** 335, 342, 360, 436
    value of, **I:** 3. **II:** 335, 342, 360, 436
Edgeworth box, **II:** 209, 542
effective
  coalition, **II:** 6
  length, **I:** 447
effectively perfect information, **I:** 493, 494
effectiveness, **II:** 6, 7, 14, 28, 78, 163
  α-effectiveness, **II:** 6, 7, 22
  β-effectiveness, **II:** 6, 7, 22
  military, **I:** 197
efficiency, **I:** 59, 419. **II:** 126, 223, 287, 473, 552
  axiom, **II:** 263
  budget line, **II:** 193
  condition, **II:** 225
  pair, **II:** 298
    normalized, **II:** 298
  price, **II:** 193, 235
  price vector, **II:** 298
  relative, **II:** 116
  strong, **II:** 480
efficient
  allocation, **II:** 298

efficient *(cont.)*
  outcome, **I:** 73
egalitarian outcome, **II:** 273
egalitarianism, **II:** 231
elasticity of substitution, **II:** 238
elbow room assumption, **II:** 389, 406, 441
electing of fellows, **I:** 4, 138, 145
elections, **I:** x
  manipulation of, **I:** 20 (*see also* strategic voting)
Electoral College, **I:** 34
electric power, **I:** 36
electronic equipment, **I:** 193, 210
elementary questions, **I:** 186
empirics, **I:** 53
endogenous
  argument, **II:** 131
  utility comparison, **II:** 539, 544
enforceable
  agreement, **I:** 21, 70, 317, 580
  commitment, **I:** 317
enforcement
  of agreement, **I:** 154, 618
  mechanism, **I:** 69, 352
    external, **I:** 395, 413
engineering, **I:** 13
England, **I:** 91
entry and exit, **I:** 18
EP. *See* Nash equilibrium
epistemic
  concept, local, **I:** 659, 666
  condition, **I:** 649
  model, **I:** x
epistemology, **I:** 47
  interactive, **I:** 555
ε-balancedness, **II:** 97
ε-core, **I:** 90. **II:** 186
ε-equilibrium, **I:** 476, 482
ε-equilibrium point, **I:** 536
ε-equivalent, **I:** 535
ε-purification, **I:** 535, 536
ε-rationality, **I:** 630
equal
  opportunity, **I:** 129
  prior, **I:** 594, 603, 608, 623, 628, 630–632, 651–653, 657 (*see also* Harsanyi doctrine)
  treatment, **II:** 541, 543
    allocation, **II:** 442
    comparison vector, **II:** 396
    principle, **II:** 125, 206, 389
equilibrium, **I:** 18–20, 67, 317
  a posteriori, **I:** 578, 579
  allocation, **II:** 160, 163, 186
  altruistic, **I:** 415, 422, 430, 431
  argument, **I:** 527
  Bayesian view of, **I:** 433
  competitive, **I:** x, 18, 84, 135, 140. **II:** 39, 47, 159, 160, 163, 171, 173, 222, 335, 386
    existence of, **I:** 98. **II:** 171
  transferable utility, **II:** 246, 315
  competitive tax, **II:** 297, 320, 333
  correlated, **I:** x, 20, 53, 556, 559, 598, 600, 622, 627, 666
    subjective, **I:** 565 (2.9), 610, 613
  coupons, **II:** 383, 384, 405, 406
  c-point, strong, **I:** 333, 336, 338, 341, 362
  dividend, **II:** 388, 441 (*see also* coupon)
    uniform, **II:** 390
  ε-equilibrium, **I:** 476, 482
  finding of, **I:** 98
  intergenerational model, **I:** 422
  knowledge of, **I:** 664
  Lindahl, **II:** 335, 336
  mixed strategy, **I:** 19, 611
  m-point, strong, **I:** 392
  Nash, **I:** x, 3, 23, 97, 99, 140, 153, 260, 318, 357, 397, 435, 447, 477, 597, 598, 615, 627, 649. **II:** 507, 508, 512 (*see also* strategic equilibrium)
    pure strategy, **I:** 448
  payoff, **I:** 416
  population, **I:** 477
  perfect, **I:** 19, 21, 23, 400, 419, 429, 430, 442. **II:** 524
    altruistic, **I:** 430
    subgame, **I:** x, 50, 51, 318, 399. **II:** 453
  point, **I:** 266, 267, 333, 356, 413, 434, 435, 495. **II:** 50
    a posteriori, **I:** 579
    grim, **I:** 400, 404
    twisted, **I:** 495
  p-point, strong, **I:** 362
  p*-point, strong, **I:** 372, 378, 379
  price, **I:** 84, 85
  in pure strategies, **I:** 439
  refinement of, **I:** 90, 94, 476, 481
  selection, **I:** 4, 153
  sequential, **I:** 19
  stable, **I:** 477
    evolutionarily, **I:** 477
  strategic, **I:** ix, 3, 50, 53, 58, 80, 83, 92, 97–99, 119 (*see also* Nash equilibrium)
  strict, **I:** 19, 20
  strong, **I:** 20, 21, 23, 27, 317, 333, 338, 355, 357, 359, 413, 416, 419. **II:** 10, 24, 25, 51, 52
    pure strategy, **I:** 358
  subjective, **I:** 20, 21, 555
  unique, **I:** 20, 23, 153, 523, 524, 597. **II:** 508
  unstable, **I:** 479
equity, **II:** 228, 231
equivalence principle, **I:** 24, 28, 83, 140. **II:** 159, 221
equivalent representation, **II:** 101
errors, **I:** ix
ESS. *See* evolutionarily stable strategy
ether, **I:** 306, 310
ethics, **I:** 47, 98

Ethics of the Fathers, **I:** 68, 70
even
  coalition, **II:** 438
  convexity, **I:** 270
event, **I:** 594, 638, 652, 662
  objective, **I:** 568, 584
eventually probably
  close, **II:** 253
  small, **II:** 253
evolution, **I:** x, 13, 92, 415, 477
  of cooperation, **I:** 444
  cultural, **I:** 93
evolutionarily stable
  equilibrium, **I:** 477
  strategy, **I:** 21, 92, 422
evolutionary
  biology, **I:** 21, 142
  dynamics, **I:** 479
ex ante rationality, **I:** 646
ex post rationality, **I:** 646
excess, **I:** 82. **II:** 92, 149
exchange economy, **II:** 383, 496, 557 (*see also* market)
excludability, **I:** 321, 334, 336
exclusion, **II:** 368
existence, **I:** 143
  of competitive equilibrium, **I:** 98. **II:** 171
  of stable set, **I:** 86
expanding universe, **I:** 10
expectation, rational, **I:** 434
expected
  irrationality, **I:** 629–631
  payoffs, pitfalls of, **I:** 331
  utility, **I:** 63, 298
    hypothesis, **I:** 257
    maximization, **I:** 638
    property, **I:** 251, 257, 299
experience, **I:** 307
experimentation
  in game theory, **I:** 16, 70, 110–112, 157
  human, **I:** 92
extended game, **II:** 10, 17
extensive
  form, **I:** 48, 317, 415, 426, 489, 491, 493, 496, 523, 533, 621, 635. **II:** 32
  infinite, **I:** 503
  middle of, **I:** 532
  structure of a game, **I:** 329
external
  domination, **II:** 122
  enforcement mechanism, **I:** 395, 413
extreme game, **II:** 107, 108

falsifiability, **I:** 142
Fatou's lemma, **II:** 183
fear of ruin, **II:** 218, 258, 269, 273
  pure, **II:** 270, 282, 300, 435
feasible
  bundle of public goods, **II:** 343
  individually rational payoff, **I:** 412
  outcome, **I:** 21, 59, 69

payoff vector, **I:** 397, 412
fight, **I:** 617
fighter-fighter duel, **I:** 67
finding of equilibrium, **I:** 98
finite
  exchange economy, **II:** 537
  game, **II:** 552
  memory, **I:** 421
    model of, **I:** 431
  repetition, **I:** 445
    of prisoner's dilemma, **I:** 444, 626, 629
  state automaton, **I:** 95, 421, 467, 483, 484
  state space, **I:** 421
  stopping time, **II:** 667
fitness, **I:** 93, 477
fixed
  points, **II:** 213
    approximating, **I:** 89
    near to, **II:** 213
    nearly, **II:** 213
  -point theorem, **I:** 66, 89, 135
    Kakutani's, **I:** 414. **II:** 23
  price, **II:** 218, 383, 440
  -price economy, **I:** 125
  -price hyperplane, **II:** 405
  -price market, **I:** 121, 125. **II:** 404
  threats, **I:** 57, 64
flat coalition, **II:** 108
focal point, **II:** 518
Folk Theorem, **I:** 21, 69, 91, 95, 412, 413, 428, 440, 632
  perfect, **I:** 401, 442
forced labor, **II:** 260, 295
formed coalition, **II:** 454
forward-lookingness, **I:** 20, 51
foundations of mathematics, **I:** 51
Fourier coefficient, **I:** 544
France, **I:** 91
free
  competition, **II:** 167
  disposal
    of resources, **II:** 361
    of public goods, **II:** 361
freedom of choice, **I:** 604
frictionless motion, **I:** 139
friendly
  player, **I:** 70, 71, 95, 443
  strategy, **I:** 435
full set of traders, **II:** 165
function
  admissible, **II:** 209, 274
  affine, **II:** 660
  bi-affine, **II:** 660
  bi-convex, **II:** 660
  Borel-measurable, **II:** 178, 608, 633
  bounded, **II:** 619
  characteristic, **I:** 99. **II:** 7, 16, 32, 63, 92, 351, 486
    of Aumann–Peleg, **II:** 87
    of Harsanyi, **II:** 337
    of Thrall, **II:** 86

function *(cont.)*
  of von Neumann–Morgenstern, **II**: 337
  comparison, **II**: 291, 308, 328, 329, 331, 344, 357, 408
  generalized, **II**: 327, 328, 331, 408
  concave, **II**: 247, 626
  continuous, **II**: 178, 616, 664
  information, **I**: 330, 335, 509
  integrably bounded, **II**: 178, 608
  Lebesgue-measurable, **II**: 609
  lower-semicontinuous, **II**: 178, 616, 663
  measurable, **I**: 279
    simultaneously, **I**: 285
  measurable random, **II**: 606
  monotonic, **I**: 291
  objective, **I**: 186, 194
  payoff, **I**: 510, 652
  production, **I**: 56
  real, **II**: 624, 626
  representing, **I**: 285
  set-valued, **I**: 287. **II**: 607 (*see also* correspondence)
  social power, **II**: 283
  space, **II**: 599
  upper-semicontinuous, **II**: 178, 616, 619, 624, 664
fuzzy coalition, **I**: 74

galaxies, **I**: 10
game, **I**: 47, 509, 653. **II**: 7, 114, 223, 263, 563
  bankruptcy, **I**: x. **II**: 148
  bargaining, **I**: 96. **II**: 532
    n-person, **I**: 96
    two-person, **I**: 29, 33, 532
  centipede, **I**: 625, 627
  C-game, **I**: 81
  classical, **II**: 13
  clear, **II**: 27
  coalitional, **I**: ix, 54, 56, 57, 99, 317. **II**: 147, 286, 287, 518, 551 (*see also* NTU)
  in coalitional form, **II**: 351
  constant-sum, **II**: 100
  constrained, **II**: 114
  continuous, **II**: 224
  cooperative, **I**: 15, 21, 24, 54, 55, 123, 154, 317, 318, 321, 355, 581. **II**: 5, 13, 31
  decomposable, **II**: 120
  demand, **I**: 498
  destroyer-submarine, **I**: 14
  difference, **I**: 495
  differential, **I**: 52, 53
  discounted, **I**: 482
  dynamic, **I**: 67, 86
  dynamic theory of, **I**: 479
  extended, **II**: 10, 17
  extensive structure of, **I**: 329
  extreme, **II**: 107, 108
    constant-sum, **II**: 108
  finite, **II**: 552
  form, strategic, **I**: 652
  ideal, **II**: 310
  I-game, **I**: 80, 81, 663
    consistent, **I**: 81
  income redistribution, **II**: 261, 433
  with incomplete information, **I**: 318, 415, 524, 532, 621, 625
  infinitely long, **I**: 51, 503
  integral, **II**: 94
  linking, **II**: 452
  majority, **I**: x
    constant-sum, **I**: x
    three-person, **I**: 29, 56
    weighted, **I**: 15. **II**: 100, 104, 454, 554
  with many players, **I**: 90
  market, **I**: 37, 56, 89. **II**: 47
  monotonic, **II**: 100, 309, 353, 552
  non-atomic (*see* continuum, non-atomic)
    values of, **II**: 287
  non-cooperative, **I**: x, 15, 55, 317, 318, 321, 355, 384, 581, 618
  non-superadditive, **II**: 105, 128–130, 530
  non-voting, **II**: 362, 368, 374
  NTU, V corresponding to v, **II**: 567
  oceanic, **I**: 87, 140. **II**: 190, 197, 217, 555 (*see also* continuum of players)
  one-homogeneous, **I**: 28, 99
  one-shot, **I**: 62, 318
  ordinally equivalent, **I**: 617
  ordinary, **II**: 16
  with perfect information, **I**: 23, 317, 355, 356, 360, 386, 389, 616
  public goods, **II**: 340
  of pursuit, **I**: 52
  recursive, **I**: 67
  without a recursive structure, **I**: 428
  redistribution, **II**: 295, 305, 366
  reduced, **I**: 74, 77, 96. **II**: 102, 148
    of Davis–Maschler, **II**: 110, 148
  repeated, **I**: 21, 22, 53, 67, 68, 90, 95, 124, 315, 317, 321, 355, 395, 396, 411, 415, 419, 440, 476, 632 (*see also* supergame)
    with complete information, **I**: 98, 411
    with incomplete information, **I**: 3, 53, 81, 98, 411, 423
    with non-zero sum incomplete information, **I**: 428
  seven-person game with no solution, **II**: 37
  with side payments, **II**: 496, 537
  simple, **II**: 100
  stochastic, **I**: 22, 53, 66–68, 90, 358, 411, 425
    repeated, **I**: 119
  strategic, **I**: ix, 36, 48, 99, 317, 396, 397, 411, 621. **II**: 343, 362, 518 (*see also* normal form)
  strictly competitive, **I**: 48, 52, 64, 109, 495, 590
    almost, **I**: 495, 496

structure of perfect information, **I:** 355, · 384, 493
sum of games, **II:** 456
superadditive, **II:** 100, 104, 125
symmetric, **I:** 479
threat, **I:** 498
truncated, **II:** 312
T-unanimity game v_T, **II:** 117
twisted, **I:** 496
unanimity, **I:** 470. **II:** 471, 563
  NTU, **II:** 567
  on T, **II:** 471
utility of playing, **I:** 74
with varying opponents, **I:** 67
with vector payoffs, **I:** 53, 98
vocal, **II:** 515
voting, **I:** 25, 30, 34, 65. **II:** 340, 369
  decisive, **I:** 65, 99, 83
  mixed, **I:** 87
  strong, **I:** 30, 99
  three-person, **I:** 29, 56
  weighted, **I:** 56, 83
weighted, **II:** 490
with winning players, **II:** 106
zero-normalized, **II:** 115
zero-sum, **II:** 9
  of perfect information, **I:** 50, 375
  two-person, **I:** 48, 327, 523, 559, 560, 565
game theory
  aspects of analytic, **I:** 622
  descriptive, **I:** 11, 13, 14, 37, 108, 157, 662
  normative, **I:** 13, 14, 108, 109
  prescriptive, **I:** 157
  coalitional, **I:** x, 317
  cooperative, **I:** x, 317
  cooperative, n-person, **I:** 53
  experimentation in, **I:** 16, 70, 110–112, 157
  and g.e.t., **I:** 136
  informational, **I:** 621
Games and Decisions, **I:** 67
game-theoretic view of the world, **I:** 434
Gang of Four, **I:** 319, 444, 481
Gaussian curvature, **II:** 205, 211
  positive, **II:** 205, 238
gene, **I:** 478
general equilibrium theory, **I:** 65
generalized
  comparison function, **II:** 327, 328, 331, 408
  comparison vector, **II:** 394
  value allocation, **II:** 394, 408
genericity, **II:** 205
genetic endowment, **I:** 478
genus, **I:** 14
geology, **I:** 15
g.e.t., **I:** 136
globality, **I:** 659, 666
glove market, **I:** 26, 73, 74, 85. **II:** 565
gnim, **I:** 50

go, **I:** 50
good strategy, **I:** 496
goodness, **I:** 210, 211, 213
  order, **I:** 200
  rating, **I:** 203
  relative, **I:** 217
goods
  continuum of, **I:** 24
  economy with many, **I:** 119
government, **II:** 257, 339
graph, **I:** 74
  communication, stable, **II:** 453
  cooperation, **II:** 449
  of a knot, **I:** 166, 167
Gray v. Sanders, **I:** 35
great tit (Parus Major), **I:** 92
greedy player, **I:** 70, 71, 95, 443
grim
  equilibrium point, **I:** 400, 404
  strategy, **I:** 421
gross income, **II:** 258
group, **I:** 50
  rational payoff vector, **II:** 8, 17, 35
  rationality, **II:** 15
groups within groups, **I:** 61
growth, **II:** 229
Growth of Biological Thought, The, **I:** 142
guaranteed value argument, **I:** 527, 528
Gurk's method, **II:** 108

Harsanyi
  characteristic function, **II:** 337
  coalitional form, **II:** 291, 305, 331, 344
  doctrine, **I:** 587, 595, 603
  –Shapley–Nash NTU value. *See* NTU Shapley value
  –Shapley transferable utility value, **II:** 291
  stable bargaining solution, **II:** 485, 488
Hausdorff
  metric, **I:** 448
  topology, **I:** 418
health, **I:** 18
  insurance, **I:** 140
heavyweight, **II:** 390, 397, 442
Hebrew University of Jerusalem, **II:** 257, 285 (*see also* Institute for Advanced Studies)
heresy, **I:** 406
heretical imputation, **II:** 86
hex, **I:** 50
higher education, **I:** 18
high-order mutual knowledge, **I:** 625, 658
highway patrol, **I:** 417
Hilbert space, **I:** 520
history, **I:** 402, 403
  -independent strategy, **I:** 451
Holland, **I:** 91
Homo
  economicus, **I:** 64
  rationalis, **I:** 11, 12, 14
  sapiens, **I:** 11, 12, 14

homogeneous
  of degree one, **I**: 99. **II**: 311, 556
  weights, **I**: x, 30, 65
homosexuals, **I**: 88
homotheticity, **II**: 201–203
horse lottery, **I**: 296
hospitals, **I**: 88, 140, 157
hull
  bi-convex, **II**: 653, 657
  convex, **II**: 42, 165, 233, 433, 608, 633, 654
human experimentation, **I**: 92
hyperplane, fixed price, **II**: 405

i.C.m., **I**: 545
ideal game, **II**: 310
identification, **II**: 605
identity partition, **I**: 323
I-game, **I**: 80, 81, 662
ignorance, randomization as, **I**: 93
ignoring information, **I**: 424
IIA. *See* independence of irrelevant alternatives
i-independence, **I**: 568
impatience, **I**: 418
imperfect information, **I**: 430, 621
impossibility theorem of Arrow, **I**: 138
improvement, **I**: 59, 99, 201, 202, 219
imputation, **I**: 53, 59. **II**: 35, 86
  heretical, **II**: 86
IMSSS. *See* Institute for Mathematical Studies in the Social Sciences, Stanford University
incentive, **I**: 138
  compatibility, **I**: 429
income
  distribution, **II**: 229, 257
  redistribution, **II**: 280, 286, 296
  redistribution game, **II**: 261, 433
  tax allocation, **II**: 320, 326, 329, 332
incomplete information, **I**: 15, 16, 18, 79, 80, 120, 124, 430, 602
inconsistent case, **I**: 584, 611
increasing returns, **I**: 36
independence of irrelevant alternatives, **II**: 230, 472, 473, 490, 568
independent
  conjecture, **I**: 665
  set, **I**: 261
index
  numbers, **I**: 142
  of strength (or power)
    of a coalition, **I**: 57. **II**: 533
    of a player, **I**: 34, 38, 67, 72, 116
India, **I**: 91
indicator, **I**: 58. **II**: 470
  convergence, **II**: 396
indifference, **I**: 261. **II**: 173
indirect utility, **II**: 296
individual rationality, **I**: 21, 49, 53, 59, 432, 433. **II**: 15

individually rational payoff
  configuration, **II**: 92
  vector, **I**: 397, 412. **II**: 8, 17, 35
induced structure, **II**: 582
induction, backward, **I**: 23, 626, 635. **II**: 453, 460
inductive
  choice, **I**: 637
  outcome, **I**: 637, 638
industrial standardization, **I**: 153
inf sup, **I**: 426
infinite
  belief system, **I**: 661
  -dimensional partially ordered mixture space, **I**: 273
  expected utility, **I**: 312
  extensive form, **I**: 503
  regress, **I**: 80
  -stage supergame, **I**: 321, 355, 395, 411, 425, 446, 482
infinitely long game, **I**: 51, 503
infinitesimal probability, **I**: 416
infinitesimals, calculus of, **II**: 233
infirm and aged, the, **I**: 422
informal agreement, **I**: 352
information
  asymmetric, **I**: 79, 423, 593
  complete, **I**: 15, 16, 19, 20, 423
    repeated game with, **I**: 98, 411
  differential, **I**: 79, 602, 604, 605
  economics, **I**: 136
  function, **I**: 330, 335, 509
  ignoring, **I**: 424
  imperfect, **I**: 430, 621
  incomplete, **I**: 15, 16, 18, 79, 80, 120, 124, 430, 602
    bargaining with, **I**: 22
    game with, **I**: 318, 415, 524, 532, 621, 625
    repeated game with, **I**: 3, 53, 81, 98, 411
    repeated game with non-zero sum, **I**: 428
  leakage of, **I**: 470
  matrix, **I**: 426
  partition, **I**: 602, 625, 637
  perfect, **I**: 50, 51, 355, 414, 493, 621, 626–631, 635, 638, 642, 643
    effectively, **I**: 493, 494
    game structure of, **I**: 355, 384, 493
    game with, **I**: 23, 317, 355, 356, 360, 386, 389, 616
    two-person zero sum, **I**: 375
  revelation and concealment of, **I**: 318, 423, 429
  space, **I**: 509, 622
  state, **I**: 630
  structure, **I**: 433
  system, **I**: 556, 606, 622, 624, 625 (*see also* interactive belief system)
  vector, **I**: 329
informational game theory, **I**: 621
initial assignment, **II**: 162, 173
inner measure, **I**: 281

insult, **I:** 480
insurance, **I:** 18, 22
Institute for Advanced Studies, Hebrew University of Jerusalem, **I:** 533. **II:** 339, 381, 383, 520
Institute for Mathematical Studies in the Social Sciences (Economics), Stanford University, **I:** 5, 133, 311, 411, 439, 533, 591, 593, 597, 615, 634. **II:** 135, 189, 205, 221, 257, 285, 337, 339, 381, 383, 447, 469, 520, 535, 541, 559, 645, 653 (*see also* Stanford University)
Institute for Mathematics and Its Applications, University of Minnesota, **I:** 597. **II:** 135, 535
institutions, **I:** 65
integrability
  p-integrability, **II:** 625
  in t, **II:** 623
integrable sublinearity, **II:** 315
integrably
  bounded function, **II:** 178, 608
  sublinear market, **II:** 315
integral
  game, **II:** 94
  of a set-valued function, **II:** 607, 619
interactive
  belief system, **I:** 652 (*see also* information system)
  epistemology, **I:** 555
Interactive Decision Theory, **I:** 47
interchangeability, **I:** 267, 496
interest
  common, **I:** 439–442, 447, 453
  conflict of, **I:** 442
intergenerational equilibrium model, **I:** 422
interiority, bi-relative, **II:** 664
internal
  completeness, **II:** 456
  consistency, **II:** 122
International Game Theory Workshop, Jerusalem, **I:** 530
International Journal of Game Theory, **I:** 91, 124
interns, **I:** 88, 140, 157
interpersonal comparison of utility, **II:** 31, 66, 69
interpretation of the weights λ, **II:** 282
intra-company transfer price, **I:** 36
intransitivity, **I:** 259
invariance, closure, **II:** 472, 473, 568
invariant under π, **II:** 115
inventory policy, **I:** 266
invisible hand, **I:** 12, 131
irrational
  automaton, **I:** 441, 453, 454
  player, **I:** 319
irrationality, **I:** 94, 95, 471, 481, 621, 622
  amount or degree of, **I:** 556, 630, 631, 632
  expected, **I:** 629–631

irrelevant alternatives
  independence of, **II:** 230, 472, 473, 490
  principle of, **II:** 496, 502
irrevocable prior commitment, **I:** 530
i-secrecy, **I:** 568
isomorphic spaces, **II:** 578, 583, 602
isomorphism, **I:** 280. **II:** 585
Israel, **I:** 35, 134. **II:** 126
Israel National Council for Research and Development, **II:** 257, 285
iterated mutual knowledge, **I:** 625, 627

Japan, **I:** 91
Jewishness, **I:** 134
joint randomized strategy, **I:** 559, 597. **II:** 6
jointly producible bundle of public goods, **II:** 342
Journal of Economic Theory, **I:** 91
Journal of Mathematical Economics, **I:** 91
just
  price, **I:** 131
  taxation, **II:** 336
justified objection, **I:** 82. **II:** 78–80

Kakutani's fixed point theorem, **I:** 414. **II:** 23
kernel, **I:** x, 27, 81, 83, 97. **II:** 91–93, 113, 114, 124, 149, 150
  δ-kernel, **II:** 97
  pre-kernel, **II:** 149, 150
  pseudo-kernel, **II:** 102, 103, 106, 124
  tables, **II:** 103–106
khessed shel emmet, **I:** 422
kindness and truth, **I:** 422
knot, **I:** x, 161
  alternating, **I:** 161, 163, 181
  aspheric, **I:** 163
  A-unknotted, **I:** 181
  D-unknotted, **I:** 181
  graph of, **I:** 166, 167
  product of, **I:** 163
  projection of, **I:** 165
  separable, **I:** 181
  theory, **I:** ix, 161
  unknotted, **I:** 181
knowledge, **I:** x, 636
  and belief, **I:** 663
  common, **I:** 14–16, 20, 81, 423, 555, 593, 594, 598, 612, 625, 626, 650, 651, 653
  of equilibrium, **I:** 664
  mutual, **I:** 625, 649–651, 653
    high-order, **I:** 625, 658
    iterated, **I:** 625, 627
    to order m, **I:** 636, 658
    second-order, **I:** 625
  system, **I:** 637
  theory of, **I:** 553
  of one's own type, **I:** 663
König's lemma, **II:** 667
kriegsspiel, **I:** 50
Kuhn's theorem, **I:** 513. **II:** 580

Labor Party, **I:** 34
labor relations, **I:** 140
language of science, mathematics as, **I:** 136
large
  market, **I:** x, 31, 83, 120, 125. **II:** 157, 159, 171, 189, 205, 221, 383, 431, 557 (*see also* continuum of agents)
  numbers, law of, **I:** 331
  numbers, strong law of, **I:** 332, 340
  population, **I:** 189
law, **I:** 13. **II:** 135, 482
  civil, **II:** 482
  of large numbers, **I:** 331
leakage of information, **I:** 470
learning, **I:** 431
Lebesgue-measurable function, **II:** 609
lexicographic order, **I:** 253, 265
lightweight, **II:** 390, 397, 442
limit
  of value, **I:** 425–427
  value of, **I:** 425, 426
limiting average, **I:** 396, 397, 403, 419, 425, 482
Lindahl equilibrium, **II:** 335, 336
linear
  programming, **I:** 66, 185, 193, 209, 231
  utility, **II:** 271, 366
linearity, **I:** 291, 292
link, **II:** 219, 449, 645
linking game, **II:** 452
Lipschitz condition, **I:** 90
local
  bi-simpliciality, **II:** 665
  epistemic concept, **I:** 659, 666
  superadditivity, **I:** 25
locally transferable utility, **II:** 267
location, **I:** 18
logical probability, **I:** 295
logrolling, **I:** 123
long
  side of the market, **II:** 219, 413
  -term competition, **I:** 395, 397
losing coalition, **I:** 56. **II:** 100
lottery, **I:** 257, 262, 263, 566
  compound, **I:** 263, 296
  horse, **I:** 296
  roulette, **I:** 296
  simple, **I:** 263
  space, **I:** 259
  ticket, **I:** 311
loudest cry, **I:** 83
lower-semicontinuous function, **II:** 178, 616, 663
Lyapounov's theorem, **I:** 66. **II:** 620, 646

m-. *See* measurability
m-acceptability, **I:** 385, 386, 392
Mafia, **I:** 422
majority
  coalition, **II:** 378
  game, **I:** x
  rule, **II:** 259, 285
  vote, **I:** 110. **II:** 258
majorization, **II:** 8
malevolent player, **I:** 96
manipulation of elections, **I:** 20 (*see also* strategic voting)
many
  -commodity market, **II:** 119, 315
  players, **I:** 86, 90 (*see also* continuum of players; large market)
mapping
  Borel, **II:** 581
  structure-preserving, **II:** 595
marginal
  contribution, **II:** 222, 235
  monotonic in the, **II:** 566
  rate, **II:** 274
  tax rate, **II:** 282
market, **I:** 5. **II:** 39, 155, 159, 171, 174, 205, 226, 293 (*see also* exchange economy)
  bounded, **II:** 294, 298
  cardinal, **II:** 242
  associated, **II:** 243
  core of, **I:** 62. **II:** 155, 157, 159, 189, 205
  derived, **I:** 126. **II:** 319
  from M at prices p, **II:** 297
  discrete, **I:** 25
  failure, **I:** 137
  fixed price, **I:** 121, 125. **II:** 404
  game, **I:** 37, 56, 89. **II:** 47
  glove, **I:** 26, 73, 74, 85
  imperfections, **I:** 137
  integrably sublinear, **II:** 315
  large, **I:** x, 31, 83, 120, 125. **II:** 157, 159, 171, 189, 205, 221, 383, 431, 557 (*see also* continuum of agents)
  long side of, **II:** 219, 413
  many-commodity, **II:** 119, 315
  -by-market rationing, **II:** 406
  marriage, **I:** 87
  matching, **I:** 157
  mixed, **I:** 87, 120. **II:** 161, 190
  non-atomic, **I:** 28, 31, 84, 86, 140. **II:** 159, 171, 221, 385, 408, 431–433, 557, 574, 625
  one-commodity, **II:** 231, 317
  ordinal, **II:** 241, 242
  associated, **II:** 243
  price of a vote, **I:** 122
  with satiation, **II:** 386
  and fixed endowments, **II:** 410
  short side of, **II:** 413
  transferable utility, **II:** 230, 244, 625
  trivial, **II:** 298
marriage
  market, **I:** 87
  stability, **I:** 25, 157
Maschler–Owen example, **II:** 530
mass phenomena, **I:** 279
Massachusetts Institute of Technology, **I:** 163

Subject Index

matching, **I**: 4
  markets, **I**: 157
  two-sided, **I**: 157
material
  conditional, **I**: 643
  rationality, **I**: 645
Material Improvement Plan, **I**: 211–215
Mathematica, **I**: 495
mathematical programming, **I**: 53
Mathematical Sciences Research Institute, Berkeley, **I**: 138, 439, 597. **II**: 535, 541 (*see also* University of California, Berkeley)
mathematics
  foundations of, **I**: 51
  as the language of science, **I**: 136
Mathematics of Operations Research, **I**: 91
matrix form, **I**: 48
Matthew, **I**: 475
maximal
  consumption, **I**: 311
  partition, **II**: 81
maximization problems, **I**: 266
maxmin, **I**: 412, 414
  payoff, **I**: 403
mean, **I**: 3, 58, 74
  contribution, **I**: 73
  of a distribution, **II**: 533
measurability, **I**: 286, 433, 507. **II**: 163, 173, 620
  w.r.t. a partition, **I**: 603
measurable
  choice, **I**: 188
    principle, **I**: 280
    theorem, **I**: 188, 279, 281
  function, **I**: 279
  graph, **I**: 188
  random function, **II**: 606
  rectangle, **I**: 282
  space, **I**: 279, 281
  stochastic process, **II**: 606
  structure, **I**: 189
  transformation, **II**: 577
  utility, **I**: 188, 279, 285
    theorem, **I**: 286
measure
  polynomial, **II**: 310
  set-valued, **II**: 606
mechanisms, **I**: 120
median, **I**: 3, 58, 74, 75
medieval, **I**: 131
memory
  bounded, **I**: 421
  zero, **I**: 421
Mercury, perihelion of, **I**: 6
meshing, **I**: 251
meteorology, **I**: 13, 140
middle of extensive form, **I**: 532
military
  effectiveness, **I**: 197
  spare parts, **I**: 266

worth, **I**: 185, 186
mimicry, **I**: 472, 632
minimal winning coalition, **I**: 30, 32, 65
minmax, **I**: 14, 52, 317, 397, 399, 412, 414, 422, 423, 523
  computation of, **I**: 86
MIP. *See* Material Improvement Plan
mirror effect, **I**: 71
Mishna, **II**: 136
misprints, **I**: ix
mistrust, **I**: 318
MIT. *See* Massachusetts Institute of Technology
mixed
  action, **I**: 447, 650
  continuum. *See* continuum with atoms; monopoly; duopoly; oligopoly
  market, **I**: 87, 120. **II**: 161, 190
  voting game, **I**: 87
  strategy, **I**: 49, 93, 360, 504, 511
    equilibrium, **I**: 19, 611
    space, **I**: 327
    vector, **I**: 356
mixing value, **II**: 228
mixture space, **I**: 250, 251, 261, 267, 269
mode, **I**: 3
model
  common knowledge of, **I**: 82, 664
  politico-economic, **II**: 218
modulo the sets of measure 0, **II**: 640
monastery, **I**: 312
money, **I**: 291
monopoly, **I**: x, 31, 36, 47, 65. **II**: 189 (*see also* continuum with atoms; oligopoly; syndicate)
monopsony, **I**: 65
monotonic
  absolute continuity, **II**: 646
  in the commodity endowments, **II**: 405
  concave dividend, **II**: 411
  function, **I**: 291. **II**: 138
  game, **II**: 100, 309, 353, 552
  in the marginal contributions, **II**: 566
  in the net trade sets, **II**: 388
  production, **II**: 361
  rule, **II**: 145
  strongly, **II**: 566
  utility, **II**: 361
morality, **I**: 91, 98, 415
"more than half is like the whole," **II**: 144
motorist, **I**: 417
moves, **I**: 447
  continuum of, **I**: 503, 533, 573, 574
m-space, **II**: 599
MSRI. *See* Mathematical Sciences Research Institute
m-strategy vector, **I**: 385
multidimensional
  random variable, **II**: 253
  utility, **I**: 258
multiple allowances, **I**: 223

multiplicity of solution concepts, **I:** 3, 57–59
mutation, **I:** 478
mutual knowledge, **I:** 625, 649–651, 654
  to order m, **I:** 636, 658
μ-value, **II:** 288, 289, 308, 309, 356
Myerson value, **II:** 219, 449, 450

Nash
  bargaining solution, **I:** 36, 97. **II:** 292, 332, 473, 488
  equilibrium, **I:** x, 3, 23, 97, 99, 140, 153, 260, 318, 357, 397, 435, 447, 477, 597, 598, 615, 627, 649. **II:** 507, 508, 512 (*see also* strategic equilibrium)
  program, **I:** 55, 72, 154. **II:** 516
Nassau County, **I:** 65
National Science Foundation, **I:** 5, 311, 411, 439, 533, 591, 593, 597, 615, 634, 635. **II:** 135, 189, 205, 221, 257, 285, 337, 339, 381, 383, 447, 469, 520, 535, 541, 559, 645, 653
NATO, **II:** 219
natural
  admissible structure, **II:** 579, 583
  structure, **II:** 453
  topology, **II:** 590
naval electronics problem, **I:** 185, 186, 194, 209
Naval Research Logistics Quarterly, **I:** 237
near
  coalition, **II:** 348, 365
  to a fixed point, **II:** 213
nearly fixed point, **II:** 213
negotiation, **II:** 65, 128, 447
  set, **II:** 8, 17
neighborhood, diagonal, **II:** 353, 398
neoclassical economics, **I:** 411
net
  trade set, **II:** 384
  worth, **II:** 533
non-atomic continuum or non-atomic game. *See* continuum, non-atomic
non-atomic market, **I:** 28, 31, 84, 86, 140. **II:** 159, 171, 221, 385, 408, 431–433, 557, 574, 625
non-cooperative game, **I:** x, 15, 55, 317, 318, 321, 384, 581, 618
non-discrimination, **I:** 129
non-emptiness of the core, **I:** 88, 89
non-exclusive public goods, **I:** 35
non-existence of stable set, **I:** 79
non-levelness, **II:** 480
non-revealing strategy, **I:** 426
non-standard analysis, **I:** 24, 84
non-superadditive game, **II:** 105, 128–130, 530
non-transferable utility. *See* NTU
non-voting game, **II:** 362, 368, 374
non-zero sum incomplete information repeated game, **I:** 428

norm
  solution concept as, **I:** 76
  supremum, **II:** 309
  variation, **II:** 309, 554, 645
normal
  distribution, **I:** 428
  form, **I:** 36, 99, 411. **II:** 23, 32, 289 (*see also* strategic form or game)
normalized efficiency pair, **II:** 298
normative aspects of game theory, **I:** 13, 14, 108, 109
n-person bargaining game, **I:** 96
NSF. *See* National Science Foundation
NTU, **I:** 26, 32, 33, 37, 61, 62, 78, 82, 84, 86, 90, 621. **II:** 217, 219, 244, 352, 370, 471, 557, 567 (*see also* coalitional game; side payment)
  game V corresponding to v, **II:** 567
  unanimity game, **II:** 567
  Shapley value, **I:** 35–38, 76, 78, 79, 84, 619. **II:** 39–41, 215–333, 339–445, 465–545, 557, 567–569 (*see also* value in the no-side-payment case)
value. *See* NTU Shapley value
nucleolus, **I:** x, 27, 58, 74, 75, 81, 83, 90, 97. **II:** 113, 114, 117, 125, 135, 137, 148, 150
null
  player, **II:** 115, 223, 286
  -player condition, **II:** 116, 223, 287
  sequence, **II:** 205
numerical method, **I:** 227
v-value, **II:** 557

object, **I:** 11
objection, **I:** 82. **II:** 67, 123
  justified, **I:** 82. **II:** 78–80
objective
  event or strategy, **I:** 568, 584
  function, **I:** 186, 194
  probability, **I:** ix, 190, 568
  random device, **I:** 190
oceanic game, **I:** 87, 140. **II:** 190, 197, 217, 555 (*see also* continuum of players)
odds and evens, **I:** 49
Office of Naval Research, **I:** 67, 289, 302. **II:** 63, 91, 171, 535, 607, 623, 639
off-path behavior, **I:** 646
oil-lease auction, **I:** 140
oligopoly, **I:** x, 22, 24, 65, 120 (*see also* atoms; continuum with atoms; monopoly; syndicate)
one-commodity market, **II:** 231, 317
one-homogeneous game, **I:** 28, 99
one-man, one-vote, **I:** 35, 140
one-shot game, **I:** 62, 318
open vote, **I:** 123
operations research, **I:** 185
opponent, strategy of, **I:** 509
optimal
  quantity of a public good, **II:** 370

Subject Index

strategy, **I:** 50
threat strategy, **II:** 265, 437
order-preserving solution, **II:** 145
ordered absolutely continuous (set), **II:** 647
ordinal
  market, **II:** 241, 242
  transferable utility, **II:** 351
  utility, **II:** 242
  value allocation, **II:** 497
ordinally equivalent games, **I:** 617
ordinary game, **II:** 16
outer measure, **I:** 281
outside observer, **I:** 604
outweighing, **II:** 92

paradox, backward-induction, **I:** 627
Pareto
  optimality, **I:** 59, 419, 431, 448, 469
  -dominant agreement, **I:** 616
parliament, **I:** x, 56. **II:** 459, 517
partial (preference) order, **I:** 234
  pure, **I:** 262
  strong, **I:** 262
  weak, **I:** 262
partially ordered
  euclidean space, **I:** 270
  lottery space, **I:** 259
  mixture space, **I:** 262, 267, 268
partition
  constant, **I:** 323
  identity, **I:** 323
  information, **I:** 602, 625, 637
    common knowledge of, **I:** 605
    on the set of all terminals, **I:** 330
  maximal, **II:** 81
  and measurability, **I:** 603
partition function form, **II:** 113, 129
partition value, **II:** 356, 552, 553
partner, **II:** 67
part-time coalition, **I:** 26
past play, **I:** 645
pattern, **I:** 6, 7
  desirability, **II:** 101
  self-policing, **II:** 63, 86
payoff, **I:** 48. **II:** 6
  c-acceptable, **I:** 338
  common, **I:** 440, 443
  conditional, **I:** 638
  configuration, **II:** 41, 64
    individually rational, **II:** 92
  discounted, **I:** 403, 404
  distribution of, **I:** 323
  division of, **I:** 317
  equilibrium, **I:** 416
  feasible, **I:** 412
  function, **I:** 510, 650
  individually rational, **I:** 412
  maxmin, **I:** 403
  Nash equilibrium, convex hull of, **I:** 562
  profile, **I:** 48
  space, **I:** 510
  vector, **I:** 397. **II:** 6, 92, 114, 222, 471
PCA, **I:** 287
P-core, **II:** 8
penny-matching, **I:** 49, 360
  scheme for playing a long sequence, **I:** 384
PEP, **I:** 400
perfect
  competition, **I:** x, 18, 28, 129. **II:** 159, 380
  equilibrium, **I:** 19, 21, 23, 400, 419, 429, 430, 442. **II:** 524
  folk theorem, **I:** 401, 442
  gas, **I:** 139
  information, **I:** 50, 51, 355, 414, 493, 621, 626–631, 635, 638, 642, 643
  recall, **I:** 67, 503, 511, 523
perihelion of Mercury, **I:** 6
permissible coalition, **II:** 79, 80
permutation π, **II:** 115
persistency, **I:** 19
personal probability, **I:** 306, 307
perturbation, **I:** 441, 444, 445, 467, 471, 630
  crazy, **I:** 319, 476, 481, 632, 633
perturbed
  rationality, **I:** 318, 481
  supergame, **I:** 453
physics, **I:** 142
PI. *See* perfect information
piecewise algebraic set, **II:** 670
p-integrability, **II:** 625
pivoting, **I:** 75
plausibility, a priori, **I:** 5
play, **I:** 324
  past, **I:** 645
  single, **I:** 62, 68, 321
player, **I:** 47, 56
  at least as desirable as another player, **II:** 101, 401
  computer as, **I:** 484
  continuum of players. *See* continuum
  crazy, **I:** 319, 481
  friendly, **I:** 70, 71, 95, 443
  games with winning, **II:** 106
  greedy, **I:** 70, 71, 95, 443
  irrational, **I:** 319
  malevolent, **I:** 96
  many, **I:** 86, 90 (*see also* continuum of players, large market)
  null, **II:** 115, 223, 286
  rational, **I:** 623, 636, 638, 653
  risk-averse, **I:** 72
  symmetric, **II:** 101
  uninformed, **I:** 428
  veto, **II:** 101, 104
pluralism, **I:** 10
poker, **I:** 47, 50, 53
political science, **I:** x
politico-economic model, **II:** 218
pollution treatment, **I:** 36
population
  equilibrium, **I:** 477
  large, **I:** 189

population *(cont.)*
  measure, **II:** 293, 342
positive
  definite quadratic form, **II:** 238
  Gaussian curvature, **II:** 205, 238
  vector, **II:** 471
positivity, uniform, **II:** 226, 294
posterior probability, **I:** 428, 593
potential, **I:** 74. **II:** 566
power, **II:** 257, 339
  index of a player's, **I:** 34, 38, 67, 72, 116
  structure, **II:** 283
  struggle, **II:** 217, 281
precession, **I:** 6
preference, **I:** 261, 295. **II:** 173, 227
  a posteriori, **I:** 579
  -or-indifference, **I:** 261. **II:** 173
  order, **I:** 234, 252, 257, 260. **II:** 495
    continuous, **I:** 285. **II:** 163, 173
    partial, **I:** 234
    weakly additive, **I:** 242
  relation, **II:** 163
  scale, **II:** 242
  smooth, **I:** 84. **II:** 205, 206, 208, 221, 225
    uniformly, **II:** 221, 222, 225–227, 236, 240
  space, **I:** 249
  structure, **I:** 246
  subjective, **I:** 186
preferred set, **II:** 175
pre-kernel, **II:** 149, 150
pre-play
  agreement, **I:** 615
  bargaining, **I:** 21
  communication, **I:** 615–617. **II:** 515
preprint, **I:** ix
prescriptive aspects of game theory, **I:** 157
pressure group, **II:** 259
price
  competitive, **II:** 557
  efficiency, **II:** 193, 235
  equilibrium, **I:** 84, 85
    outcome, **I:** 24
  fixed, **II:** 218, 383, 440
  intra-company transfer, **I:** 36
  just, **I:** 131
  shadow, **I:** 53. **II:** 235
  theory, **I:** 131
  vector, **II:** 163, 173, 386
Princeton University, **I:** 54, 67, 185, 189
  (see also Econometric Research Program)
principal-agent problem, **I:** 18, 22
prior, **I:** 534, 602
  common or equal, **I:** 594, 603, 608, 623, 628, 630–632, 651–653, 657 *(see also* Harsanyi doctrine)
  common knowledge of, **I:** 605
priority, **I:** 210, 211, 213
  group, **I:** 196

rating, **I:** 203
prisoner's dilemma, **I:** 67, 69, 328, 397, 418, 419, 445
  repetition of, **I:** 22, 95, 476, 482, 556
  finite, **I:** 444, 626, 629
  hundredfold, **I:** 556
  with memory zero or one,* **I:** 421, 435, 443
probability, **I:** 295
  belief with probability 1, **I:** 647
  comparative, **I:** 301
  infinitesimal, **I:** 416
  logical, **I:** 295
  matching, **I:** 92
  objective, **I:** ix, 190, 568
  personal, **I:** 306, 307
  posterior, **I:** 428, 593
  subjective, **I:** ix, 185, 189, 295, 298, 306, 434, 560, 598
process of function space type, **II:** 606
procurement, **I:** 224
  -allocation problem, **I:** 224
product
  of knots, **I:** 163
  quality, **I:** 18
production
  correspondence, **II:** 342
  function, **I:** 56
  monotonic, **II:** 361
productive economy, **I:** 24
profile, **I:** 56, 99
programming
  dynamic, **I:** 23, 416
  linear, **I:** 66, 185, 193, 209, 231
    duality theorem of, **I:** 53
  mathematical, **I:** 53
  problems of, **I:** 254
  subjective, **I:** 186, 231
progressive tax, **II:** 273
projection
  of CA set, **I:** 287
  of a knot, **I:** 165
  principle, **I:** 280
  theorem, **I:** 284, 287. **II:** 620
properness, **I:** 19
property rights, **II:** 378
proportional division, **II:** 135, 142, 145
propositional calculus, **II:** 93
prospects, **I:** 312
pure, **I:** 257
pseudo-kernel, **II:** 102, 103, 106, 124
pseudo-value, **II:** 394
ψ-stable, **II:** 7, 13, 63, 84, 86
P-stable, **II:** 8
psychoanalysis, **I:** 13

* On p. 443, we used "memory one" for what we called "memory zero" on pp. 421 and 435.

psychology, **I**: 12, 613
public
 bads, **II**: 374
 goods, **I**: x, 25, 35, 36, 120, 122. **II**: 218, 257, 339, 350, 356, 368
 economy, **I**: 3. **II**: 335, 342, 360, 436
 feasible bundle of, **II**: 343
 free disposal of, **II**: 361
 game, **II**: 340
 jointly producible bundle of, **II**: 342
 non-exclusive, **I**: 35
 optimal quantity of, **II**: 370
 roulette, **I**: 570
 sector, **II**: 285
 utility pricing, **I**: 36
punishment, **I**: 68, 71, 396, 407, 408, 416, 417, 419, 430, 480
 and costs to the punisher, **I**: 419
 of deviation, **I**: 333, 405
 randomized, **I**: 417
 strategy, **I**: 422, 484
 threat of, **I**: 352, 412
 unrelenting, **I**: 402
pure
 exchange economy, **I**: 83
 fear of ruin, **II**: 270
 partial (preference) order, **I**: 262
 prospects, **I**: 257
 strategy, **I**: 48, 360, 534
 strategy Nash equilibrium, **I**: 448
 strategy strong equilibrium, **I**: 358
 supergame strategy, **I**: 355, 415
purifying sequence, **I**: 538
 absolutely, **I**: 538
pursuit, game of, **I**: 52

quadratic form, **II**: 238
quasi-concave, **II**: 225, 236
quasi-order, **II**: 168, 173
quota, **I**: 56
QWERTY, **I**: 478

racial discrimination, **I**: 140
radioactive dating, **I**: 9
Radon–Nikodym derivative, **I**: 606
Rand Corporation, **I**: 52, 67
random
 device, **I**: 360
 order, **II**: 312
 variable, **I**: 317, 322, 516
  multidimensional, **II**: 253
randomization
 conscious, **I**: 557, 611, 650
 as ignorance, **I**: 93
 subjective, **I**: 611
randomized
 punishment, **I**: 417
 strategy, **I**: 49, 91, 559, 560, 567, 611
 range, **II**: 601
 of a vector measure, **I**: 66

rate of convergence, **I**: 24, 427. **II**: 205
rational
 cooperation, **I**: 439
 decision-making, **I**: 12
 expectation, **I**: 434
 player, **I**: 623, 636, 638, 653
rationality, **I**: x, 395, 622
 act, **I**: 476, 480
 approximate, **I**: 630
 Bayesian, **I**: 434, 602, 603
 bounded, **I**: 95, 139, 319, 475
 coalitional, **II**: 64, 65, 79
 common knowledge of, **I**: 476, 483, 556, 598, 608, 625, 635, 636, 642, 646, 650
 ε-rationality, **I**: 630
 ex ante, **I**: 646
 ex post, **I**: 646
 group, **II**: 15
 individual, **I**: 21, 49, 53, 59, 397, 432, 433. **II**: 15
 material, **I**: 645
 perturbed, **I**: 318, 481
 postulates, relaxation of, **I**: 484
 rule, **I**: 476, 480
 substantive, **I**: 645
 at a vertex, **I**: 644
rationalizability, **I**: 613
rationing, **I**: 126. **II**: 385, 406
 market-by-market, **II**: 406
R-core, **II**: 17
reachability, **I**: 594
real world, **I**: 5, 157
real, **I**: 157
reasonable compromise, **II**: 231
recall
 bounded, **I**: 318, 439, 442, 481
 perfect, **I**: 67, 503, 511, 523
recalling one's own action, **I**: 469
rectangles, **I**: 283
recursive
 function theory, **I**: 484
 game, **I**: 67
redistribution, **II**: 218, 285, 339, 350, 356, 533
 game, **II**: 295, 305, 366
redistricting, **I**: 35
red shift, **I**: 10
reduced game, **I**: 74, 77, 96. **II**: 102, 148
refinement of equilibrium, **I**: 90, 94, 476, 481
reflexivity, **II**: 173
regular
 catalogue, **II**: 49
 cone, **I**: 269, 281
 space, **II**: 578, 583
relationships, **I**: ix, 5–7, 161
relative
 efficiency, **II**: 116
 goodness, **I**: 217
relativity, **I**: 6, 9

relaxation of rationality postulates, **I:** 484
relevant range, **II:** 366
religion, **I:** 134
repeated
  game, **I:** 21, 22, 53, 67, 68, 90, 95, 124, 315, 317, 321, 355, 395, 396, 411, 415, 419, 440, 476, 632 (*see also* supergame)
    of complete information, **I:** 98, 411
    of incomplete information, **I:** 3, 53, 81, 98, 411, 423
    prisoner's dilemma, **I:** 22, 95, 476, 482
      hundredfold, **I:** 556
      with memory zero or one,* **I:** 421, 435, 443
  stochastic game, **I:** 119
Repeated Games of Incomplete Information, **I:** 318
repetition, **I:** 318
  finite, **I:** 445
replication, **II:** 206, 389
representation, **I:** 260. **II:** 100
  equivalent, **II:** 101
representative democracy, **II:** 357
representing function, **I:** 285
reputation, **I:** 22
resistive medium, **I:** 17
resources, free disposal of, **II:** 361
restricted
  additivity assumption, **I:** 244
  bargaining set, **II:** 83, 85
retaliation, threat of, **I:** 352, 412
revelation and concealment of information, **I:** 318, 423, 429
revenge, **I:** 68, 411
Reynolds v. Sims, **I:** 35
rights
  property, **II:** 378
  voting, **II:** 340
risk
  aversion, absolute, **II:** 270
  -averse player, **I:** 72
  -dominance, **I:** 154, 616, 618
  -taking, **II:** 217
Roman camps around Metzada, **I:** 55
roulette, **I:** 570
  lottery, **I:** 296
  public, **I:** 570
  wheel, **I:** 295
ruin, **I:** 67
  fear of, **II:** 218, 258, 269, 273
  pure, **II:** 270, 282, 300, 435
rule, **II:** 141
  majority, **II:** 259, 285
  monotonic, **II:** 145

---

* On p. 443, we used "memory one" for what we called "memory zero" on pp. 421 and 435.

  of order, **II:** 452
  rationality, **I:** 476, 480
  self-consistent, **II:** 141
  of thumb, **I:** 480, 482

saddle point, **I:** 260
S-admissibility, **II:** 553
safety, **II:** 68
Sages of the Talmud, **II:** 60
S-allocation, **II:** 294, 401
II Samuel, **I:** 107
satiation, **II:** 383, 411
  market with, **II:** 386
satisficing, **I:** 11, 139, 476
saturation, **II:** 174
  commodity-wise, **II:** 174
  restriction, **II:** 175
scale covariance, **II:** 472, 568
scaling factor, **II:** 40
science, **I:** 3
search, **I:** 18
second-order
  belief, **I:** 623
  mutual knowledge, **I:** 625
secret vote, **I:** 123
section of Z at y, **II:** 116
security level, **I:** 49, 66
selection
  from a correspondence or set-valued function, **II:** 619
  von Neumann's theorem on, **I:** 66. **II:** 609
  compact-valued theorem on, **II:** 620
  equilibrium, **I:** 4, 153
self-commitment, **I:** 529
self-consistency, **II:** 141
self-consistent rule, **II:** 141
self-defeating theory, **I:** 153
self-duality, **II:** 144
self-enforcing agreement, **I:** 23, 396, 434, 556, 615
selfishness, **I:** 98, 115
  cooperation as the consequence of, **I:** 137, 439
self-policing patterns, **II:** 63, 86
sensations, **I:** 308
separability, **II:** 577, 582
separable
  knot, **I:** 181
  measurable space, **I:** 281
  topological space, **I:** 252
separating sequence, **II:** 288
separation, **II:** 224
sequential equilibrium, **I:** 19
S-equivalence, **II:** 94
set
  analytic, **I:** 287. **II:** 609
  complementary, **I:** 287, 288
    projection of, **I:** 287
    complement of, **I:** 287

bargaining, **I:** 27, 82, 83, 85, 97. **II:** 41, 63, 68, 84, 91, 114, 123, 161, 292, 541, 551
　restricted, **II:** 83
bi-convex, **II:** 657
Borel, **I:** 279, 286, 507, 550. **II:** 226, 335, 581, 602, 608, 625
budget, **II:** 163, 173, 228, 387
　dividend, **II:** 388
　connected, **II:** 286, 293
　independent, **I:** 261
　negotiation, **II:** 8, 17
　net trade, **II:** 384
　　monotonic in, **II:** 388
　ordered absolutely continuous, **II:** 647
　piecewise algebraic, **II:** 670
　preferred, **II:** 175
　　aggregate, **II:** 175
　stable. *See* stable set
theory, **I:** 289
　axioms of, **I:** 288
　-valued function, **I:** 287. **II:** 607 (*see also* correspondence)
　　selection from a, **II:** 619
　-valued measure, **II:** 607
seven-person game with no solution, **II:** 37
sewage disposal, **I:** 97
shadow price, **I:** 53. **II:** 235
Shapley
　correspondence, **II:** 472, 568
　index of power, **I:** 34, 38, 67, 72, 116
　NTU value. *See* NTU Shapley value
　value, **I:** 3, 33, 58, 98, 119, 121, 125. **II:** 40, 116, 125, 150, 161, 198, 215–569, 639
short side of the market, **II:** 413
shortsightedness, **I:** 115
side payment, **I:** 99, 291, 324. **II:** 5, 13, 31, 229, 379, 496, 537 (*see also* NTU)
　value in the no-side-payment case, **II:** 40 (*see also* NTU Shapley value)
signalling, **I:** 318, 429, 431
simple
　game, **II:** 100
　lottery, **I:** 263
simplex method, **I:** 186
simplicity, **I:** 7
　computational, **I:** 38, 482
simultaneously measurable function, **I:** 285
single play, **I:** 62, 68, 321
small worlds, **I:** 63, 96, 307, 588. **II:** 131
smiley, **I:** 448
smooth preference, **I:** 84. **II:** 205, 206, 208, 221, 225
smoothness, **II:** 471, 478, 567
social
　biology, **I:** 411
　choice, **I:** 4, 19, 20
　physics, **I:** 107–109
　power function, **II:** 283

productivity, **I:** 72, 73. **II:** 534
psychology, **I:** 69
science, **I:** 3
utility, **II:** 27
welfare, **II:** 217, 222, 229, 257, 283, 285, 339, 533
socialism, **I:** 132, 137
society, democratic, **II:** 359
socio-educational "gene," **I:** 431
sociology, **I:** 12, 415
solution, **II:** 6, 8, 36
　concept or notion, **I:** x, 5, 10, 11, 18, 57, 58, 76, 395. **II:** 148, 507
　Harsanyi's stable bargaining, **II:** 485, 488
　Nash bargaining, **I:** 36, 97. **II:** 292, 332, 473, 488
　n-person, **I:** 96
　Talmudic
　　constrained equal award, **II:** 142
　　contested garment-consistent, **II:** 139, 141
　　order-preserving, **II:** 145
　von Neumann–Morgenstern. *See* stable set
solvability, **II:** 9
span, **I:** 261
species, **I:** 14, 92
S-shaped production curve or returns, **I:** 25, 33
St. Petersburg paradox, **I:** 191, 311
stable
　bargaining solution of Harsanyi, **II:** 485, 488
　communication graph, **II:** 453
　evolutionarily, **I:** 21, 92, 422
　marriage, **I:** 25, 157
　payoff vector, **II:** 63, 91
　ψ-stable, **II:** 7, 13, 63, 84, 86
　P-stable, **II:** 8
　set (von Neumann–Morgenstern), **I:** 3, 27–29, 32, 33, 54, 58, 60, 62, 78, 81, 97, 116. **II:** 5, 35, 63, 87, 113, 114, 122, 161, 292, 448, 488, 541
　　discriminatory, **I:** 29, 116
　　existence of, **I:** 86
　　non-existence of, **I:** 79
　　symmetric, **I:** 31
stag hunt, **I:** 619
stage, **I:** 411, 412
standard
　deviation, **I:** 124
　measurable space, **I:** 280, 509. **II:** 226
standardization, industrial, **I:** 153
Stanford University, **I:** 625, 635 (*see also* Institute for Mathematical Studies in the Social Sciences)
state, **I:** 637
　finite state automaton, **I:** 95, 421, 467, 483–484
　group, **I:** 196, 211
　of the person, **I:** 307

state *(cont.)*
  steady, **I:** 357
  of the world, **I:** 301, 305, 307, 433, 566, 593, 602, 621, 623, 637, 652, 662
stationarity, **I:** 420, 425
stationary
  pure strategy, **I:** 359
  strategy, **I:** 358, 420
statistical decision theory, **I:** 53
steady state, **I:** 357
stochastic
  game, **I:** 22, 53, 66–68, 90, 358, 411, 425
  process, measurable, **II:** 606
stockholders, corporations with two large, **I:** 140
stop probability, **I:** 418
stopping time, **II:** 667
  finite, **II:** 667
strategic
  equilibrium, **I:** ix, 3, 50, 53, 58, 80, 83, 92, 97–99, 119 (*see also* Nash equilibrium)
  form or game, **I:** ix, 36, 48, 99, 317, 396, 397, 411, 621. **II:** 343, 362, 518 (*see also* normal form)
  game form, **I:** 652
  voting, **I:** 4, 138, 145 (*see also* manipulation of elections)
strategy, **I:** 48, 433, 534, 636, 644
  behavior, **I:** 67, 328, 387, 503, 511, 520, 523
  of Blackwell, **I:** 424, 427
  of cold war, **I:** 67
  compatible, **I:** 451
  complexity of, **I:** 421
  cooperative, **II:** 6
  correlated, **I:** 397, 560, 580, 630. **II:** 51
  double-cross, **I:** 420, 421, 435, 443
  duration-independent, **I:** 483
  friendly, **I:** 435
  good, **I:** 496
  grim, **I:** 421
  history-independent, **I:** 451
  mixed, **I:** 49, 93, 360, 504, 511
    correlated, **II:** 6, 14
    subjectively, **I:** 559, 560
  non-revealing, **I:** 426
  objective, **I:** 568, 584
  of opponent, **I:** 509
  optimal, **I:** 50
  punishment, **I:** 422, 484
  pure, **I:** 48, 360, 534
    continuum of, **I:** 66
    equilibrium in, **I:** 439
    stationary, **I:** 359
    supergame, **I:** 360, 368, 383, 392
  randomized, **I:** 49, 91, 559, 560, 567, 611
    joint, **I:** 559, 597. **II:** 6
  stable, **I:** 21, 422
    evolutionarily, **I:** 21, 92, 422

  stationary, **I:** 358, 420
  supergame, **I:** 334, 352, 399
    c-strategy, **I:** 329–331, 333
    non-cooperative, **I:** 390
    p-strategy, **I:** 361, 368
    pure, **I:** 355, 383, 414
  threat, **II:** 434
    optimal, **II:** 265, 437
  trigger, **I:** 484
  weakly dominated, **I:** 416
strength
  index of a coalition's, **I:** 57. **II:** 533
  index of a player's, **I:** 38, 67, 72, 116
strict
  conditional, **I:** 643
  equilibrium, **I:** 19, 20
strictly
  competitive game, **I:** 48, 52, 64, 109, 495, 590 (*see also* almost strictly competitive game)
  determined, **I:** 49, 51
strong
  bi-separation, **II:** 661
  efficiency, **II:** 480
  equilibrium, **I:** 20, 21, 23, 27, 317, 333, 338, 355, 357, 359, 413, 416, 419. **II:** 10, 24, 25, 51, 52
    c-point, **I:** 333, 336, 338, 341, 362
    m-point, **I:** 392
    p^*-point, **I:** 372, 378, 379
    p-point, **I:** 362
  law of large numbers, **I:** 332, 340
  monotonicity, **II:** 566
  partial (preference) order, **I:** 262
  voting game, **I:** 30, 99 (*see also* decisive voting game)
structure, **I:** ix. **II:** 577, 581
  additive, **I:** 236
  axiom, **I:** 301
  Borel, **I:** 188. **II:** 581
  coalition, **I:** 119. **II:** 8, 86, 92, 113, 448
  cooperation, **II:** 449
  discrete, **II:** 583
  induced, **II:** 582
  natural, **II:** 453
  natural admissible, **II:** 579, 583
  -preserving mapping, **II:** 595
structured situation, **I:** 116
subchain, **II:** 645
subgame, **I:** 19, 50, 89, 400, 405 (*see also* decomposition)
  perfect equilibrium, **I:** x, 50, 51, 318, 399. **II:** 453
  symmetry, **I:** 19
subgraph, connected, **II:** 219
subgroup, **I:** 50
subjective
  correlated equilibrium, **I:** 565 (2.9), 610, 613
  equilibrium, **I:** 20, 21, 555
  mixed strategy, **I:** 559, 560

preference, **I:** 186
probability, **I:** ix, 185, 189, 295, 298, 306, 434, 560, 598
programming, **I:** 186, 231
randomization, **I:** 611
value, **I:** 312
subjunctive mood, **I:** 642
sublinearity, **II:** 315
  integrable, **II:** 315
substantive
  conditional, **I:** 643, 644
  rationality, **I:** 645
substitute, **II:** 124, 223, 286, 541
substitution, elasticity of, **II:** 238
successive weak domination, **I:** 421
sum of games, **II:** 456
summable, **I:** 332, 367
  in the mean, **I:** 332, 367
sup inf, **I:** 426
sup norm. *See* supremum norm
superadditive
  cover, **II:** 125
  game, **II:** 100, 104, 125
superadditivity, **I:** 25, 26, 28
  at a coalition, **I:** 88
  and the core, **I:** 90
  local, **I:** 25
superdifferential, **II:** 395
supergame, **I:** 321, 326, 328, 338, 346, 357, 396, 413, 439, 446–448, 504, 505. **II:** 10, 15, 23, 24, 51, 52 (*see also* repeated game)
  cooperative, **I:** 352
  c-strategy, **I:** 329–331, 333
  infinite-stage, **I:** 321, 355, 395, 411, 425, 446, 482
  perturbed, **I:** 453
  p-strategy, **I:** 361, 368
  pure strategy, **I:** 360, 368, 383, 392
  strategy, **I:** 334, 352, 399
    non-cooperative, **I:** 390
  θ-discounted, **I:** 446
superplay, **I:** 325, 504
super-rationality, **I:** 481
supremum norm, **II:** 309
sure-thing principle, **I:** 301
surmise, **I:** 623
survival of the fittest, **I:** 12
S-vector, **II:** 15
symmetric
  agent, **II:** 542
  allocation procedure, **II:** 566
  game, **I:** 479
  player, **II:** 101
  stable set, **I:** 31
symmetry, **II:** 116, 223, 286, 541, 543, 552
  center of, **I:** 84
  subgame, **I:** 19
syndicate, **I:** 25, 119 (*see also* atoms; monopoly; duopoly)

tables, kernel, **II:** 103–106
tactical military, **I:** 67
Talmud, **I:** x, 37, 97, 131. **II:** 60, 135
TARK, **I:** 319
tax, **I:** 36. **II:** 229, 257, 274, 280, 285, 294, 339, 359, 431, 433
  just, **II:** 336
  progressive, **II:** 273
  rate, **II:** 282
    marginal, **II:** 282
telephone billing, **I:** 36, 97. **II:** 558
terminal, **I:** 324
  effects, **I:** 418
theorem of the alternative, **I:** 239
Theory of Games and Economic Behavior, **I:** 54
θ-discounted supergame, **I:** 446
Thrall characteristic function, **II:** 86
threat, **I:** 36, 411, 430. **II:** 289, 305
  fixed, **I:** 57, 64
  game, **I:** 498
  of punishment or retaliation, **I:** 352, 412
  strategy, **II:** 434
three-person majority or voting game, **I:** 29, 56
tic-tac-toe, **I:** 49
time
  horizon, **I:** 51
  phasing, **I:** 221
  stretching infinitely backwards, **I:** 420
tit for tat, **I:** 92, 443–445, 476, 481, 483, 633
top anchors, **I:** 220
topology, natural, **II:** 590
tracing procedure, **I:** 154
trade, **II:** 159
traders, **II:** 162, 172
  continuum of, **I:** 25, 31, 84, 87, 140, 188, 279. **II:** 157, 159, 171, 189, 221, 385, 408
  full set of, **II:** 165
trading of votes, **II:** 348
transferable utility, **I:** 25–27, 32–34, 37, 56, 57, 61, 62, 82, 99, 189, 291, 621. **II:** 31, 217, 229, 236, 245, 351, 370, 379, 439, 469, 533
  competitive equilibrium, **II:** 246, 315
  market, **II:** 230, 244, 625
transitivity, **I:** 255. **II:** 173
translation covariance, **II:** 482
tree form, **I:** 48
trembles, **I:** 476
trembling hand, **I:** 19, 94, 416, 442, 447, 471. **II:** 513
trigger strategy, **I:** 484
trivial
  coalition, **II:** 78
  market, **II:** 298
true altruism, **I:** 422
truncated game, **II:** 312
trust, **I:** 68, 137, 318

truth, **I**: ix, 8–11
  kindness and, **I**: 422
TU. *See* transferable utility
T-unanimity game v_T, **II**: 117
Turing machine, **I**: 471, 482, 483
twisted
  equilibrium point, **I**: 495
  game, **I**: 496
two-person
  bargaining game, **I**: 29, 33, 532
  zero-sum game, **I**: 48, 327, 523, 559, 560, 565
    of perfect information, **I**: 50, 375
  zero-sum theory, **I**: 53
two-sided
  matching, **I**: 157
  poker, **I**: 48
type, **I**: 31, 80, 90, 319, 625, 652. **II**: 206
  crazy, **I**: 319, 481
  knowledge of one's own, **I**: 663

umbrella, **I**: 305
U.N. Security Council, **I**: 34
unanimity
  axiom, **II**: 472, 473
  game, **I**: 470. **II**: 471, 563
understanding, **I**: ix
unemployment, **II**: 218, 383, 413, 440
unification, **I**: 6, 7
unified field, **I**: 47
uniform
  boundedness, **II**: 294, 361
  coupons endowments, **II**: 405
  dividend equilibrium, **II**: 390
  positivity, **II**: 226, 294
  smoothness, **II**: 225, 226, 240
  tax laws, **II**: 259
uniformly smooth preference, **II**: 221, 222, 225–227, 236, 240
uninformed player, **I**: 428
unique equilibrium, **I**: 20, 23, 153, 523, 524, 597. **II**: 508
universality, **I**: 608
universe, **I**: x
University of California, Berkeley, **II**: 541 (*see also* Mathematical Sciences Research Institute)
unknotted knot, **I**: 181
unrelenting
  ferocity, **I**: 338, 339
  punishment, **I**: 402
unrestrictedly transferable utility, **I**: 291. **II**: 5
unstable equilibrium, **I**: 479
upper-semicontinuous function, **II**: 178, 616, 619, 624, 664
U.S.
  Air Force Project RAND, **II**: 171
  Armed Forces, **I**: 185
  Supreme Court, **I**: 35

utility, **I**: ix, 11, 19, 63, 185, 238, 243, 251, 257, 262, 268, 279, 291, 311. **II**: 26
  aggregate, **II**: 264
  "burning" of, **I**: 470
  cardinal, **II**: 241, 242, 416
  without completeness, **I**: 257
  computing of, **I**: 243
  concave, **II**: 236, 386, 404
  conditional, **I**: 306
  continuous, **I**: 279
  decreasing marginal, **II**: 272
  expected, **I**: 63, 298
  infinite, **I**: 312
  horizon, **I**: 404
  indirect, **II**: 296
  interpersonal comparison of, **II**: 31, 66, 69
  linear, **I**: 291. **II**: 271, 366
  locally transferable, **II**: 267
  maximization, **I**: 11, 411
  maximizer, **I**: 598
    Bayesian, **I**: 598
    rational, **I**: 598
  measurable, **I**: 188, 279, 285
  monotonic, **II**: 361
  multidimensional, **I**: 258
  non-transferable. *See* NTU
  ordinal, **II**: 242
  of playing a game, **I**: 74
  social, **II**: 27
  theory, **I**: 187, 257
    expected, **I**: 54
    without completeness, **I**: 257, 275
  as tool, **I**: 247
  transferable, **I**: 25–27, 32–34, 37, 56, 57, 61, 62, 82, 99, 189, 291, 621. **II**: 31, 217, 229, 236, 245, 351, 370, 379, 439, 469, 533
    cardinal, **II**: 351
    ordinal, **II**: 351
  unrestrictedly, **I**: 291. **II**: 5
  of von Neumann–Morgenstern, **I**: 63, 247. **II**: 260

valuation, **II**: 41
value, **I**: x, 27, 50, 72, 74, 75, 83, 90, 97. **II**: 113, 263, 292, 333, 472, 529, 541, 552, 563
  allocation, **II**: 221, 222, 226, 230, 236, 237, 241, 266, 281, 388, 432, 437, 441, 496
  asymptotic, **II**: 224, 225, 228, 236, 288, 309, 311, 356, 362, 553
  Banzhaf, **II**: 555
  characteristic function, **I**: 99
  of a coalition, **II**: 64
  convergence theorem, **II**: 397
  correspondence, **II**: 472, 568
  current, **I**: 412
  equivalence principle, **I**: 84. **II**: 222, 431, 544
  of fixed price economy, **I**: 3

Harsanyi–Shapley–Nash NTU. *See* NTU Shapley value
Harsanyi–Shapley transferable utility, **II:** 291
  of limit, **I:** 425, 426
  limit of, **I:** 425–427
  mixing, **II:** 228
μ-value, **II:** 288, 289, 308, 309, 356
Myerson, **II:** 219, 449, 450
  of non-atomic game, **II:** 287
NTU. *See* NTU Shapley value
NTU Shapley. *See* NTU Shapley value
  in the no-side-payment case, **II:** 40 (*see also* NTU Shapley value)
v-value, **II:** 557
outcome, **II:** 341, 344, 352, 362, 364
partition, **II:** 356, 552, 553
pseudo-value, **II:** 394
of public goods economy, **I:** 3
Shapley, **I:** 3, 33, 58, 98, 119, 121, 125. **II:** 40, 116, 125, 150, 161, 198, 215–569, 639
Shapley NTU. *See* NTU Shapley value
subjective, **I:** 312
of a vote, **I:** 123
Values of Non-Atomic Games, **II:** 217
vanishing comparison weight, **II:** 480
variable threats bargaining model, **I:** 419
variation, **II:** 353
  bounded, **II:** 309, 353
  norm, **II:** 309, 554, 645
varying discount factor, **I:** 464
vector
  B-vector, **I:** 321, 330
  correlated strategy, **II:** 6
  c-acceptable, **I:** 326–328
  comparison, **II:** 388, 441
    equal treatment, **II:** 396
    generalized, **II:** 394
  correlated strategy, **I:** 322, 325, 326, 330, 356
  c-strategy, **I:** 326, 329
  efficiency price, **II:** 298
  information, **I:** 329
  measure, range of, **I:** 66
  mixed strategy, **I:** 356
  m-strategy, **I:** 385
  payoff, **I:** 397. **II:** 6, 92, 114, 223, 471
    acceptable, **I:** 376
    c-acceptable, **I:** 382. **II:** 10
    feasible, **I:** 397, 401, 412
    game with, **I:** 53, 98
    group rational, **II:** 8, 17, 35
    individually rational, **I:** 401, 412. **II:** 8, 17, 35
    stable, **II:** 63, 91
  positive, **II:** 471
  price, **II:** 163, 173, 386
  S-vector, **II:** 15
verbal economics, **I:** 16, 135

vertex, rationality at, **I:** 644
veto player, **I:** 25. **II:** 101, 104
vocal game, **II:** 515
von Neumann–Morgenstern
  characteristic function, **II:** 337
  solution. *See* stable set
  stable set. *See* stable set
  utility, **I:** 63, 247. **II:** 260
von Neumann's theorem on selection, **I:** 66. **II:** 609
vote, **I:** 5
  competitive price of, **I:** 35
  double, **I:** 146
  majority, **I:** 110. **II:** 258
  market price of, **I:** 122
  open, **I:** 123
  secret, **I:** 123
  trading of, **II:** 348
  value of, **I:** 123
voting, **I:** 64, 123. **II:** 374, 534
  game, **I:** 25, 30, 34, 65. **II:** 340, 369
  in a legislature, **I:** 123
  measure, **II:** 342
  rights, **II:** 340
  scheme, **II:** 357
  strategic, **I:** 4, 138, 145 (*see also* manipulation of elections)
  theory of, **I:** 4
  weight, **II:** 359
  weighted, **I:** 35

wage rigidities, **II:** 218
Walras allocation, **II:** 210
War and Peace, **I:** 108
water
  management, **I:** 36
  resource development, **I:** 36
  supply, **I:** 97
weak
  desirability, **II:** 174
  domination, **I:** 421
  partial (preference) order, **I:** 262
weakly
  additive preference order, **I:** 242
  conditionally atomless measure, **I:** 536
  dominated strategy, **I:** 416
Weierstrass approximation theorem, **I:** 545
weight, **I:** x, 30, 56. **II:** 100
  homogeneous, **I:** x, 30, 65
  -quota configuration, **I:** 64
  vanishing comparison, **II:** 480
  voting, **II:** 359
  zero, **II:** 529
weighted
  game, **II:** 490
  majority game, **I:** 15. **II:** 100, 104, 454, 554
  outcome, **I:** 78
  voting, **I:** 35
  voting game, **I:** 56

Wesberry v. Sanders, **I:** 35
winning, **I:** 56
  coalition, **I:** 57, 65. **II:** 100
world. *See* state of the world
worst case, **I:** 53
worth, **I:** 26, 56, 99
  attained, **II:** 305
  coalitional, **I:** 25, 36
  military, **I:** 185, 186
  net, **II:** 533
would, **I:** 642, 643

Yale University, **I:** 52, 289 (*see also* Cowles Foundation for Research in Economics)
y-allocation, **II:** 399

zero
  -normalized game, **II:** 115
  -sum game, **II:** 9
    of perfect information, **I:** 375
  weight, **II:** 529
Z-section, **II:** 587